# Geophysical Monograph Series

82  **Gravimetry and Space Techniques Applied to Geodynamics and Ocean Dynamics (IUGG Volume 17)** *Bob E. Schutz, Allen Anderson, Claude Froidevaux, and Michael Parke (Eds.)*

83  **Nonlinear Dynamics and Predictability of Geophysical Phenomena (IUGG Volume 18)** *William I. Newman, Andrei Gabrielov, and Donald L. Turcotte (Eds.)*

84  **Solar System Plasmas in Space and Time** *J. Burch, J. H. Waite, Jr. (Eds.)*

85  **The Polar Oceans and Their Role in Shaping the Global Environment** *O. M. Johannessen, R. D. Muench, and J. E. Overland (Eds.)*

86  **Space Plasmas: Coupling Between Small and Medium Scale Processes** *Maha Ashour-Abdalla, Tom Chang, and Paul Dusenbery (Eds.)*

87  **The Upper Mesosphere and Lower Thermosphere: A Review of Experiment and Theory** *R. M. Johnson and T. L. Killeen (Eds.)*

88  **Active Margins and Marginal Basins of the Western Pacific** *Brian Taylor and James Natland (Eds.)*

89  **Natural and Anthropogenic Influences in Fluvial Geomorphology** *John E. Costa, Andrew J. Miller, Kenneth W. Potter, and Peter R. Wilcock (Eds.)*

90  **Physics of the Magnetopause** *Paul Song, B.U.Ö. Sonnerup, and M.F. Thomsen (Eds.)*

91  **Seafloor Hydrothermal Systems: Physical, Chemical, Biological, and Geological Interactions** *Susan E. Humphris, Robert A. Zierenberg, Lauren S. Mullineaux, and Richard E. Thomson (Eds.)*

92  **Mauna Loa Revealed: Structure, Composition, History, and Hazards** *J. M. Rhodes and John P. Lockwood (Eds.)*

93  **Cross-Scale Coupling in Space Plasmas** *James L. Horwitz, Nagendra Singh, and James L. Burch (Eds.)*

94  **Double-Diffusive Convection** *Alan Brandt and H. J. S. Fernando (Eds.)*

95  **Earth Processes: Reading the Isotopic Code** *Asish Basu and Stan Hart (Eds.)*

96  **Subduction Top to Bottom** *Gray E. Bebout, David Scholl, Stephen Kirby, and John Platt (Eds.)*

97  **Radiation Belts: Models and Standards** *J. F. Lemaire, D. Heynderickx, and D. N. Baker (Eds.)*

98  **Magnetic Storms** *Bruce T. Tsurutani, Walter D. Gonzalez, Yohsuke Kamide, and John K. Arballo (Eds.)*

99  **Coronal Mass Ejections** *Nancy Crooker, Jo Ann Joselyn, and Joan Feynman (Eds.)*

100 **Large Igneous Provinces** *John J. Mahoney and Millard F. Coffin (Eds.)*

101 **Properties of Earth and Planetary Materials at High Pressure and Temperature** *Murli Manghnani and Takehiki Yagi (Eds.)*

102 **Measurement Techniques in Space Plasmas: Particles** *Robert F. Pfaff, Joseph E. Borovsky, and David T. Young (Eds.)*

103 **Measurement Techniques in Space Plasmas: Fields** *Robert F. Pfaff, Joseph E. Borovsky, and David T. Young (Eds.)*

104 **Geospace Mass and Energy Flow: Results From the International Solar-Terrestrial Physics Program** *James L. Horwitz, Dennis L. Gallagher, and William K. Peterson (Eds.)*

105 **New Perspectives on the Earth's Magnetotail** *A. Nishida, D. N. Baker, and S. W. H. Cowley (Eds.)*

106 **Faulting and Magmatism at Mid-Ocean Ridges** *W. Roger Buck, Paul T. Delaney, Jeffrey A. Karson, and Yves Lagabrielle (Eds.)*

107 **Rivers Over Rock: Fluvial Processes in Bedrock Channels** *Keith J. Tinkler and Ellen E. Wohl (Eds.)*

108 **Assessment of Non-Point Source Pollution in the Vadose Zone** *Dennis L. Corwin, Keith Loague, and Timothy R. Ellsworth (Eds.)*

109 **Sun-Earth Plasma Interactions** *J. L. Burch, R. L. Carovillano, and S. K. Antiochos (Eds.)*

110 **The Controlled Flood in Grand Canyon** *Robert H. Webb, John C. Schmidt, G. Richard Marzolf, and Richard A. Valdez (Eds.)*

111 **Magnetic Helicity in Space and Laboratory Plasmas** *Michael R. Brown, Richard C. Canfield, and Alexei A. Pevtsov (Eds.)*

112 **Mechanisms of Global Climate Change at Millennial Time Scales** *Peter U. Clark, Robert S. Webb, and Lloyd D. Keigwin (Eds.)*

113 **Faults and Subsurface Fluid Flow in the Shallow Crust** *William C. Haneberg, Peter S. Mozley, J. Casey Moore, and Laurel B. Goodwin (Eds.)*

114 **Inverse Methods in Global Biogeochemical Cycles** *Prasad Kasibhatla, Martin Heimann, Peter Rayner, Natalie Mahowald, Ronald G. Prinn, and Dana E. Hartley (Eds.)*

115 **Atlantic Rifts and Continental Margins** *Webster Mohriak and Manik Talwani (Eds.)*

116 **Remote Sensing of Active Volcanism** *Peter J. Mouginis-Mark, Joy A. Crisp, and Jonathan H. Fink (Eds.)*

117 **Earth's Deep Interior: Mineral Physics and Tomography From the Atomic to the Global Scale** *Shun-ichiro Karato, Alessandro Forte, Robert Liebermann, Guy Masters, Lars Stixrude (Eds)*

Geophysical Monograph 118

# Magnetospheric Current Systems

**Shin-ichi Ohtani**
**Ryoichi Fujii**
**Michael Hesse**
**Robert L. Lysak**
*Editors*

American Geophysical Union
Washington, DC

## Published under the aegis of the AGU Books Board

Roberta M. Johnson, Chair; John E. Costa, Jeffrey M. Forbes, W. Rockwell Geyer, Rebecca Lange, Douglas S. Luther, Walter H. F. Smith, Darrell Strobel, and R. Eugene Turner, members.

Library of Congress Cataloging-in-Publication Data

Magnetospheric current systems / Shin-ichi Ohtani ... [et al.], editors.
    p. cm. -- (Geophysical monograph ; 118)
  Papers derived from presentations given at the AGU Chapman Conference on Magnetospheric Current Systems held January 10-15, 1999 in Kona, Hawaii.
  ISBN 0-87590-976-0
  1. Magnetospheric currents--Congresses. I. Ohtani, Shin-ichi, 1962- II. AGU Chapman Conference on Magnetospheric Current Systems (1999 : Kona, Hawaii) III. Series.

QC809.M35 M322 2000
538'.76--dc21

                                          00-036207

Front and back cover photos courtesy Jack Finch.

ISBN 0-87590-976-0
ISSN 0065-8448

Copyright 2000 by the American Geophysical Union
2000 Florida Avenue, N.W.
Washington, DC 20009

    Figures, tables, and short excerpts may be reprinted in scientific books and journals if the source is properly cited.

    Authorization to photocopy items for internal or personal use, or the internal or personal use of specific clients, is granted by the American Geophysical Union for libraries and other users registered with the Copyright Clearance Center (CCC) Transactional Reporting Service, provided that the base fee of $1.50 per copy plus $0.35 per page is paid directly to CCC, 222 Rosewood Dr., Danvers, MA 01923. 0065-8448/00/$01.50+0.35.

    This consent does not extend to other kinds of copying, such as copying for creating new collective works or for resale. The reproduction of multiple copies and the use of full articles or the use of extracts, including figures and tables, for commercial purposes requires permission from the American Geophysical Union.

Printed in the United States of America.

# CONTENTS

**Dedication**
*Laurence J. Zanetti* .................................................................. ix

**Preface**
*Shin-ichi Ohtani, Ryoichi Fujii, Michael Hesse, and Robert L. Lysak* ........................ xi

## Frameworks for Describing Current Systems

**TUTORIAL: Newton, Maxwell, and Magnetospheric Physics**
*E. N. Parker* ....................................................................... 1

**Paradigm Transition in Cosmic Plasma Physics, Magnetic Reconnection and the Generation of Field-Aligned Current**
*Yan Song and Robert L. Lysak* ....................................................... 11

**FAST Observations of Electromagnetic Stresses Applied to the Polar Ionosphere**
*R. J. Strangeway, R. C. Elphic, W. J. Peria, and C. W. Carlson* .......................... 21

**Electric Current Approach to Magnetospheric Physics and the Distinction Between Current Disruption and Magnetic Reconnection**
*A. T. Y. Lui* ....................................................................... 31

## Modeling Magnetospheric Current Systems

**Global Geometry of Magnetospheric Currents Inferred From MHD Simulations**
*G. L. Siscoe, N. U. Crooker, G. M. Erickson, B. U. Ö. Sonnerup, K. D. Siebert, D. R. Weimer, W. W. White, and N. C. Maynard* ............................................................ 41

**Field-Aligned-Current Systems in the Numerically Simulated Magnetosphere**
*T. Tanaka* ......................................................................... 53

**Recent Progress in the Data-Based Modeling of Magnetospheric Currents**
*N. A. Tsyganenko* .................................................................. 61

**Field Line Mapping and Birkeland Currents**
*F. R. Toffoletto and T. W. Hill* ..................................................... 71

**Magnetotail Currents During the Growth Phase and Local Auroral Breakup**
*T. I. Pulkkinen, M. V. Kubyshkina, D. N. Baker, L. L. Cogger, S. Kokubun, T. Mukai, H. J. Singer, J. A. Slavin, and L. Zelenyi* ............................................................... 81

## Magnetosphere-Ionosphere Coupling

**TUTORIAL: Magnetosphere-Ionosphere Interactions: A Tutorial Review**
*S. W. H. Cowley* ................................................................... 91

# CONTENTS

**TUTORIAL: Field-Aligned Currents in Geospace: Substance and Significance**
*Takesi Iijima* . . . . . . . . . . . . . . . . . . . . . . . . . . . . . . . . . . . . . . . . . . . . . . . . . . . . . . . . . . . . . . . . . 107

**TUTORIAL: Ionospheric Electrodynamics: A Tutorial**
*A. D. Richmond and J. P. Thayer* . . . . . . . . . . . . . . . . . . . . . . . . . . . . . . . . . . . . . . . . . . . . . . . . 131

**The Role of Alfvén Waves in the Formation of Auroral Parallel Electric Fields**
*Robert L. Lysak and Yan Song* . . . . . . . . . . . . . . . . . . . . . . . . . . . . . . . . . . . . . . . . . . . . . . . . . 147

**A Three-Dimensional Simulation of the Kelvin-Helmholtz Instability**
*Kristi A. Keller, Robert L. Lysak, and Yan Song* . . . . . . . . . . . . . . . . . . . . . . . . . . . . . . . . . . . 157

**The Role of Space-Time Dependent Ionospheric Conductivity in the Evolution of Field Line Resonances: Relation to Auroral Arc**
*Manju Prakash and Robert Rankin* . . . . . . . . . . . . . . . . . . . . . . . . . . . . . . . . . . . . . . . . . . . . 165

**FAST Observations of Upward Accelerated Electron Beams and the Downward Field-Aligned Current Region**
*R. C. Elphic, J. Bonnell, R. J. Strangeway, C. W. Carlson, M. Temerin, J. P. McFadden, R. E. Ergun, and W. Peria* . . . . . . . . . . . . . . . . . . . . . . . . . . . . . . . . . . . . . . . . . . . . . . . . . . . . 173

**Characteristics of Field-Aligned Currents Near the Auroral Acceleration Region: FAST Observations**
*W. J. Peria, C. W. Carlson, R. E. Ergun, J. P. McFadden, J. Bonnell, R. C. Elphic, and R. J. Strangeway* . . . . . . . . . . . . . . . . . . . . . . . . . . . . . . . . . . . . . . . . . . . . . . . . . . . . . . . . . 181

**Auroral Surge Currents and Electrodynamics With FAST and VIS**
*S. A. Cummer, R. R. Vondrak, R. F. Pfaff, J. W. Gjerloev, C. W. Carlson, R. E. Ergun, W. J. Peria, R. C. Elphic, R. J. Strangeway, J. B. Sigwarth, and L. A. Frank* . . . . . . . . . . . . . . . . . . . . . . . 191

**A Synthetic View of the Magnetospheric-Ionospheric Current System Associated With Substorms**
*Gang Lu* . . . . . . . . . . . . . . . . . . . . . . . . . . . . . . . . . . . . . . . . . . . . . . . . . . . . . . . . . . . . . . . . . 199

**The Harang Discontinuity in Auroral Substorms**
*J. W. Gjerloev, E. Friis-Christensen, R. A. Hoffman, and S. A. Cummer* . . . . . . . . . . . . . . . . . 209

**The Effect of the January 10, 1997, Pressure Pulse on the Magnetosphere-Ionosphere Current System**
*E. Zesta, H. J. Singer, D. Lummerzheim, C. T. Russell, L. R. Lyons, and M. J. Brittnacher* . . . . . . . . . 217

**Ionospheric Shear Flow Situations Observed by the MIRACLE Network, and the Concept of Harang Discontinuity**
*O. Amm, P. Janhunen, H. J. Opgenoorth, T. I. Pulkkinen, and A. Viljanen* . . . . . . . . . . . . . . 227

**Statistical Characteristics of Field-Aligned Currents in the Earth's Inner Magnetosphere**
*Francis K. Chun and Christopher T. Russell* . . . . . . . . . . . . . . . . . . . . . . . . . . . . . . . . . . . . . 237

# CONTENTS

**Independency of the Dayside Field-Aligned Current System: A Restriction to Cusp Models**
M. Yamauchi, R. Lundin, L. Eliasson, S. Ohtani, P.-A. Lindqvist, and R. P. Lepping . . . . . . . . . . . . . 245

**Disappearance of Large-Scale Field-Aligned Current Systems: Implications for the Solar Wind-Magnetosphere Coupling**
S. Ohtani, T. Higuchi, T. Sotirelis, and P. T. Newell . . . . . . . . . . . . . . . . . . . . . . . . . . . . . . . 253

**Storm-Time Energetic Particle Penetration Into the Inner Magnetosphere as the Electromotive Force in the Subauroral Ion Drift Current Circuit**
J. De Keyser . . . . . . . . . . . . . . . . . . . . . . . . . . . . . . . . . . . . . . . . . . . . . . . . . . . . . . . 261

## Magnetospheric Electrodynamics

**Structured Currents Associated With Tail Bursty Flows During Turbulent Plasma Sheet Conditions**
L. R. Lyons, T. Nagai, J. C. Samson, E. Zesta, T. Yamamoto, T. Mukai, A. Nishida, and S. Kokubun . . . . . . . . . . . . . . . . . . . . . . . . . . . . . . . . . . . . . . . . . . . . . . . . . . . . 267

**Substorm Associated Tail Current Changes Inferred From Lobe Magnetic Field Observations**
Christian Jacquey . . . . . . . . . . . . . . . . . . . . . . . . . . . . . . . . . . . . . . . . . . . . . . . . . . . 275

**The Current Disruption Myth**
Joachim Birn and Michael Hesse . . . . . . . . . . . . . . . . . . . . . . . . . . . . . . . . . . . . . . . . . 285

**Near- and Mid-tail Current Flow During Substorms: Small- and Large-Scale Aspects of Current Disruption**
Michael Hesse and Joachim Birn . . . . . . . . . . . . . . . . . . . . . . . . . . . . . . . . . . . . . . . . . 295

**Intrinsic Variability in the Quiet-Time Magnetotail**
Vahé Peroomian, Maha Ashour-Abdalla, and Lev M. Zelenyi . . . . . . . . . . . . . . . . . . . . . . . 305

**Self-Consistent Model of 1D Current Sheet: The Role of Drift, Magnetization and Diamagnetic Currents**
Helmi V. Malova, Mikhail I. Sitnov, Lev M. Zelenyi, and Surja Sharma . . . . . . . . . . . . . . . . . 313

**Pressure Anisotropy and $B_y$ in the Magnetotail Current Sheet**
Richard L. Kaufmann, Bryan M. Ball, W. R. Paterson, and L. A. Frank . . . . . . . . . . . . . . . . . 323

**Formation of the Storm-Time Ring Current and the Dst Field: Some Recent Topics**
T. Iyemori . . . . . . . . . . . . . . . . . . . . . . . . . . . . . . . . . . . . . . . . . . . . . . . . . . . . . . . 331

## Current Systems in Other Magnetospheres

**TUTORIAL: Currents and Flows in Distant Magnetospheres**
Margaret Galland Kivelson . . . . . . . . . . . . . . . . . . . . . . . . . . . . . . . . . . . . . . . . . . . . 339

**Rotational Current Systems and the Offset Io Plasma Torus**
Duane H. Pontius, Jr. . . . . . . . . . . . . . . . . . . . . . . . . . . . . . . . . . . . . . . . . . . . . . . . . 353

# CONTENTS

**MHD Simulations of Current Systems in Planetary Magnetospheres: Mercury and Saturn**
*Tamas I. Gombosi, Darren L. DeZeeuw, Clinton P. T. Groth, Kenneth C. Hansen, Konstantin Kabin,
Kenneth G. Powell* .................................................. 363

**Currents in Mercury's Magnetosphere**
*Karl-Heinz Glassmeier* ................................................ 371

## New Analysis Techniques

**A New Technique for the Mapping of Ionospheric Field-Aligned Currents From
Satellite Magnetometer Data**
*Daniel R. Weimer* .................................................... 381

**Automatic Identification of Large-Scale Field-Aligned Current Structures and its Application
to Night-Side Current Systems**
*T. Higuchi and S. Ohtani* .............................................. 389

**Symmetry Breaking and Nonlinear Wave-Wave Interaction in Current Disruption: Possible
Evidence for a Phase Transition**
*Giuseppe Consolini and Anthony T. Y. Lui* ................................ 395

# DEDICATION

This volume is dedicated to Thomas A. Potemra, who over 15 years ago edited the previous Geophysical Monograph on magnetospheric currents. Tom passed away April 3, 1998. He was supervisor of the Space Physics Group at The Johns Hopkins University Applied Physics Laboratory (JHU/APL) in Laurel, Maryland, and had been a physicist there since 1965, specializing in ionospheric and magnetospheric physics, particularly field-aligned currents. He was fascinated with the history of Earth's magnetic field as well as the early arctic explorers and scientists. He led a large group of space scientists, published over 200 scientific papers, and directed many spacecraft experiments. He was deeply involved in defining and implementing major scientific missions for the space program.

Foremost among Tom's greatest scientific contributions was his exploration of how the vast, high-altitude current systems develop into the field-aligned (Birkeland) system and how they connect to the magnetosphere and beyond. Current signatures in U.S. Navy navigational satellite data were first identified by Tom's colleague at JHU/APL, Al Zmuda, and first interpreted as Birkeland currents by Alex Dessler. This was the discovery of the long disputed current system originally proposed by Birkeland in 1908 as the source of ground magnetic deviations. In partnership with Take Iijima of the University of Tokyo, the Birkeland current system was established via a series of papers in AGU's *Journal of Geophysical Research* from 1976 to 1978. These papers developed the concept of Region 1 and Region 2 statistical Birkeland current patterns and established them as the auroral current system paradigm. Tom edited two AGU monographs, both of which derived from Chapman Conferences on current systems, which he wholly or partially organized. He was also a member of the program committee for the Chapman Conference on Magnetospheric Current Systems, from which we derived the present monograph. We dedicated the conference to Tom and his life's work.

Tom Potemra participated in, directed and chaired advisory committees to the National Academy of Sciences, NASA, NSF, and the Department of Defense. For Tom's cumulative scientific contributions and leadership achievements he was elected into the International Academy of Astronautics. He served on the AGU Federal Budget Panel that assesses the geoscience research programs of NASA, NSF, NOAA, USGS, and DOE. In 1985 and 1986, he took special leave from JHU/APL to serve as Senior Policy Analyst in the Office of Science and Technology Policy, Executive Office of the President. In this post he aided the formulation of policy on the civil space program and provided scientific and technical support to President Reagan's Science Advisor, George A. Keyworth.

Tom Potemra possessed and practiced strong leadership, developed others' self-motivation and encouraged self-achievement. We (around the world) who have had the privilege and pleasure to work with Tom Potemra in space physics research see him as the ideal coworker: knowledgeable, open-minded, enthusiastic, prepared to share all his experience, interested in his fellow coworkers and genuinely a good friend. His pioneering work in describing current systems contributed to a major advance in humankind's understanding of how the Earth interacts with its space environment. In addition to Tom's scientific achievements, his personal qualities as colleague and group

leader were extraordinary and are deeply missed. Tom exuded an enthusiasm coupled with genuine warmth that made it a joyful and uplifting experience to be associated with him. His bright spirit was infectious and helped to remind us that science is about wonder and discovery. This was clearly how he felt, and whenever he was canvassing the group, checking up on how the various projects were progressing, his tenor voice could be easily heard up and down the hall bubbling with excitement. Perhaps the best memorial for Tom would be to remember his gift for wonder and enthusiasm and continue to research the mysteries of nature in this spirit even as he would have done.

Laurence J. Zanetti

# PREFACE

The magnetosphere is an open system that interacts with the solar wind. In this system, solar wind energy continuously permeates different regions of the magnetosphere through electromagnetic processes, which we can well describe in terms of current systems. In fact, our ability to use various methods to study magnetospheric current systems has recently prompted significant progress in our understanding of the phenomenon. Unprecedented coverage of satellite and ground-based observations has advanced global approaches to magnetospheric current systems, whereas advanced measurements of electromagnetic fields and particles have brought new insights about micro-processes. Increased computer capabilities have enabled us to simulate the dynamics not only of the terrestrial magnetosphere but also the magnetospheres of other planets. Based on such developments, the present volume revisits outstanding issues about magnetospheric current systems.

The present volume begins with a discussion of some importance and some controversy, "Frameworks for Describing Current Systems." We present various opinions about the issue in this section. The three subsequent sections focus on "Modeling Magnetospheric Current Systems," "Magnetosphere-Ionosphere Coupling," and "Magnetospheric Electrodynamics." We acknowledge that this classification is not appropriate for some papers, for which greater complexity is a byword, and which, taken singularly, reflect the fact that the study of magnetospheric current systems involves a variety of aspects. In response, we have intentionally mixed theoretical and observational studies. Papers in the section "Current Systems in Other Magnetospheres" compare current systems in other magnetospheres with those in the terrestrial magnetosphere. In the final section of the volume, "New Analysis Techniques," we present state-of-the-art papers that report new methods for studying current systems.

Papers included in this volume derive from presentations given at the AGU Chapman Conference on Magnetospheric Current Systems held January 10-15, 1999, in Kona, Hawaii. The National Science Foundation, the National Aeronautics and Space Administration, and the Office of Naval Research sponsored the conference. The success of the conference and this volume is due in part to the members of the program committee: S. W. H. Cowley, E. Friis-Christensen, R. A. Greenwald, G. Haerendel, Y. Kamide, J. R. Kan, J. G. Lyon, R. L. Lysak, G. T. Marklund, T. Mukai on behalf of T. Yamamoto (deceased), H. J. Opgenoorth, T. A. Potemra (deceased), G. Rostoker, C. T. Russell, J. Sauvaud, G. L. Siscoe, J. A. Slavin, N. A. Tsyganenko, R. J. Walker, R. A. Wolf. We thank Tina Powell of JHU/APL for her assistance with the conference. We are also grateful to the many individuals who served as reviewers, and whose comments were important to the papers now before you. We have listed reviewer names at the end of this preface. Finally, and most importantly, we thank all the authors for their diligence and expertise in helping to make this volume a reference for future research on magnetospheric current systems.

We dedicate this volume to the memory of a valued friend and colleague, Thomas A. Potemra.

Shin-ichi Ohtani
Ryoichi Fujii
Michael Hesse
Robert L. Lysak

### List of reviewers

B.-H. Ahn, P. C. Anderson, K. B. Baker, J. Birn, J. E. Borovsky, J. U. Brackbill, S. C. Buchert, J. Chen, S. P. Christon, C. R. Clauer, J. H. Clemmons, S. W. H. Cowley, S. A. Cummer, R. C. Elphic, R. E. Erlandson, G. M. Erickson, D. H. Fairfield, M. Fujimoto, H. A. Gallagher, J. W. Gjerloev, T. Gombosi, G. Haerendel, M. G. Henderson, T. Iyemori, A. M. Jorgensen, Y. Kamide, R. L. Kaufmann, K. A. Keller, A. J. Klimas, H. E. J. Koskinen, M. M. Kuznetsova, N. Lin, M. Lockwood, R. E. Lopez, W. Lotko, G. Lu, A. T. Y. Lui, L. R. Lyons, R. L. McPherron, T. Nagai, P. T. Newell, K. Niijima, K. Nishikawa, H. J. Opgenoorth, C. P. Paranicas, S. M. Petrinec, P. L. Pritchett,, T. I. Pulkkinen, J. Raeder, L. Rastaetter, A. D. Richmond, G. Rostoker, J. M. Ruohoniemi, C. T. Russell, E. R. Sanchez, P. E. Sandholt, J.-E. Sauvaud, K. Shiokawa, D. G. Sibeck, H. J. Singer, G. L. Siscoe, J. A. Slavin, T. Sotirelis, K. Stasiewicz, D. P. Stern, W. Sun, S. Taguchi, K. Takahashi, N. A. Tsyganenko, R. J. Walker, D. V. Vassiliadis, J.-E. Wahlund, D. R. Weimer, D. J. Williams, R. A. Wolf, M. Yamauchi, A. Yoshikawa.

# Newton, Maxwell, and Magnetospheric Physics

E. N. Parker

*Department of Physics and Enrico Fermi Institute, University of Chicago*

The goal of magnetospheric physics is to understand the active magnetosphere in terms of the principles of Newton and Maxwell. These principles include Ampere's law, the Faraday induction equation, and the dynamical interplay of the forces between the bulk plasma velocity **u**, the plasma pressures $p_\parallel$ and $p_\perp$, and the magnetic field **B**. It turns out, then, that these principles cannot be written in terms of the electric current density **j** and the electric field **E**, to the exclusion of **u** and **B**, in any generally useful form, indicating that the electric current and field are not the fundamental physical quantities sometimes claimed. Thus, the description of the active magnetosphere in electrical terms does not directly address the dynamics of the magnetosphere. The popular dynamical concepts of the electric circuit analog and the active role of the electric field $\mathbf{E} = -\mathbf{u} \times \mathbf{B}/c$ in the solar wind have not been established from Newton and Maxwell, and application to simple problems illustrates their restrictions and errors. In contrast, magnetospheric physics is making great strides forward using MHD modeling of the large-scale dynamical magnetosphere, and this tutorial emphasizes the generality of the MHD approach. We review the theoretical basis for the momentum equation and the induction equation for the large-scale nonrelativistic dynamics of the magnetosphere and ionosphere, showing the general MHD character of the contending stresses and momenta.

## INTRODUCTION

The subject of this volume is the electric current **j** in the terrestrial magnetosphere deformed by the varying external impact and drag of the solar wind and by the pressures of the internal plasmas and particles. The scientific question is why and how the magneto-sphere is obliged by the physical laws of Newton and Maxwell to develop the geotail and produce such phenomena as magnetospheric convection, the substorm, the aurora, etc. Extensive observations from both ground based and space borne instruments are available. The challenge is to infer the 3D activity from the observations and then to understand the physics of the activity from the principles of Newton and Maxwell.

The theoretical development begins, then, with the equations of Newton and Maxwell, deriving the form of the field equations appropriate for magnetospheric dynamics. Limiting the development to the large-scale dynamics of the magnetosphere, with a characteristic dimension $\Lambda$ that is large compared to the cyclotron radii of the ions and electrons, the derivation produces a momentum equation for the bulk velocity $\mathbf{u}(\mathbf{r},t)$ of the plasma and an induction equation for the magnetic field $\mathbf{B}(\mathbf{r},t)$. The equations are of the general form known as magnetohydrodynamics, or MHD, with additional terms arising from anisotropic plasma pressure and from the presence of neutral atoms. The form of the equations shows that magnetospheric dynamics represents the interaction of the Reynolds stress, or momentum flux, $\rho u_i u_j$ of the bulk motion, the particle pressures $p_\parallel$ and $p_\perp$ parallel and perpendicular to the magnetic field, and the Maxwell stress

$$M_{ij} = -\delta_{ij}\frac{B^2}{8\pi} + \frac{B_i B_j}{4\pi} \qquad (1)$$

in the magnetic field. Thus, **u** and **B**, along with $p_\parallel$ and $p_\perp$, are the primary physical variables describing the dynamics.

The compression of the geomagnetic dipole into the comet shaped magnetosphere – magnetotail is readily understood from these principles [*Chapman* and *Ferraro*, 1940; *Dungey*, 1958], as well as the inflation of the field by trapped particles [*Dessler* and *Parker*, 1959; *Dessler, Frances*, and *Parker* 1960; *Parker*, 1967; *Sckopke*, 1972]. More recently the principles have been employed in 2D and 3D numerical simulations of global magnetospheric dynamics as the increase in computing capability has risen to the task [cf, *Frank*, et al, 1995 and these Proceedings for further references] showing the relations between the several individual large-scale features of magnetospheric activity. The magnetospheric substorm is understood to follow from magnetic reconnection with the interplanetary magnetic field at the sunward magnetopause, stretching magnetic flux bundles back into the geotail and forcing a sunward convective flow in the outer magnetosphere [*Gold*, 1959; *Dungey*, 1961; *Axford* and *Hines*, 1961; *Siscoe*, 1988; *Zhu*, 1993, 1995]. It appears that the auroral sheets are a direct consequence of the singular properties of the Maxwell stress tensor and the irregular topology of the field in the magnetotail [*Parker*, 1994]. These effects all follow from the pushing and pulling of the field and fluid.

In contrast with the direct approach from Newton and Maxwell, there is the curious idea, widely held, that it is the electric current **j** and the electric field **E** that are the fundamental variables describing the dynamics of the magnetosphere. The idea begins with the declaration that **j** is the cause of **B**, and hence the more fundamental variable. In fact in the presence of interparticle collisions (resistivity), the energy flow is always from **B** to **j**, so that **B** is the prime mover. The electric current is present because the deformation of **B** from a potential form requires the presence of **j** in the amount $4\pi \mathbf{j} = c\nabla\times\mathbf{B}$ described by Ampere. In fact the mechanical gyrations of the individual electrons and ions in the inhomogeneous magnetic field automatically provide the current required by Ampere.

The electric current paradigm is hampered by the fact that there are no tractable general dynamical equations in terms of **j** and **E**. Of course the partial differential equations for **u** and **B** can be expressed in terms of **j** and **E** through the Biot-Savart integral

$$\mathbf{B}(\mathbf{r}) = \left(\frac{1}{c}\right)\iiint \frac{d^3r'\mathbf{j}(\mathbf{r}')\times(\mathbf{r}-\mathbf{r}')}{|\mathbf{r}-\mathbf{r}'|^3} \qquad (2)$$

and the electric drift velocity

$$\mathbf{u}_\perp = c\mathbf{E}\times\mathbf{B}/B^2. \qquad (3)$$

However, the result is a set of global integral equations, intractable except in special symmetric or linear systems where the dynamical equations reduce to simple forms [cf. *Song* and *Lysak*, 1994]. It must be understood that there is no objection to using **j** and **E** if the theoretical development originates properly with the equations of Newton and Maxwell. The objection is that in so many applications the proper equations are intractable and the theoretical development of **j** and **E** proceeds on principles that have no basis in Newton and Maxwell.

For instance, the idea that the electric current is the fundamental physical quantity has fostered the notion that magnetospheric dynamics can be described by simple electric circuit analogs [cf. *Alfven* and *Carlquist*, 1967]. Another arbitrary notion is that the electric field $\mathbf{E} = -\mathbf{u}\times\mathbf{B}/c$ in the solar wind actively penetrates into the magnetosphere and magnetotail, driving convection $\mathbf{u} = c\mathbf{E}\times\mathbf{B}/B^2$ within the magnetotail, etc.[cf. *Parker*, 1996].

The electric circuit equations are not derived. They are *declared*, by casual analogy between the time dependent net current in the magnetic field and the current in a fixed electric circuit in the laboratory. The mathematical solution of the ordinary differential equation for the analog circuit is relatively easy because the analog is based on a single function of time in each current loop. Hence it overlooks most of the crucial effects, e.g. the changing scale of the magnetic field, the varying plasma velocity, the viscous and wave drag of the wind at the magnetopause, the inertia and viscosity of the ionosphere, the transfer of magnetic flux from the sunward side and the forced return magnetoconvection, etc. all of which contend with the Maxwell stress in the magnetic field to determine the dynamics. One may declare that these effects are implicit in the chosen values for the inductance, capacitance and resistance, of course, but one is none the wiser for having proclaimed the analog.

*Spicer* [1982] gives a number of illustrative examples of the electric current patterns that arise in association with some of the more common quasi-static magnetic forms to be seen on the active Sun. He points out that in a dynamical system, the form of the electric circuit may change with time.

Specifically, the electric circuit analog overlooks the fact that the electric current in the plasma is carried in the frame of reference in which the electric field $\mathbf{E}'$ (= $\mathbf{E}$ + $\mathbf{u} \times \mathbf{B}/c$) vanishes (in the absence of resistivity and bulk acceleration), quite unlike the fixed electric circuit in the laboratory. Of course one can solve the problem fully and properly in terms of $\mathbf{u}$ and $\mathbf{B}$ and then go back and set up a circuit analog that mimics the main features of the solution. Examples are given in *Parker* [1996].

Applications of the analog sometimes overlook the fact that the isolated twisted flux bundle carries no net current. The bundle is declared to carry a current I and the author proceeds to formulate a current equation in terms of an I. The suggestive power of the current analog leads to a variety of scientific marvels, with the notion that magnetic energy from one region can be transmitted via the current for immediate released as heat at some distant point, forgetting that the current moves in the frame of reference in which there is no electric field ($\mathbf{E}'$ = 0 in the limit of small resistivity). The arbitrary circuit analog constructions have a way of discovering modes of oscillation that are unknown to the MHD equations deduced from Newton and Maxwell [cf. *Zaitsev*, et al 1998].

It is easy to illustrate the limitations of the electric circuit analog by applying it to simple cases where the result is readily worked out from the equations of Newton and Maxwell. Thus in a resistive fluid the evolution of the magnetic field is correctly described by the induction equation

$$\frac{\partial \mathbf{B}}{\partial t} = \nabla \times (\mathbf{u} \times \mathbf{B} - \eta \nabla \times \mathbf{B}) \quad (4)$$

where $\eta$ is the resistive diffusion coefficient $c^2/4\pi\sigma$ in terms of the conductivity $\sigma$. The scalar product with $\mathbf{B}/4\pi$ yields the energy equation

$$\left(\frac{\partial}{\partial t}\right)\frac{B^2}{8\pi} + \mathbf{u} \cdot \mathbf{F} + \nabla \cdot \left(\frac{\mathbf{u}_\perp B^2}{4\pi}\right) +$$

$$\nabla \cdot (\eta \mathbf{F}) + \frac{j^2}{\sigma} = 0 \quad (5)$$

where $\mathbf{u}_\perp$ is the plasma motion perpendicular to $\mathbf{B}$, $B^2/4\pi$ is the magnetic enthalpy, and $\mathbf{F}$ is the Lorentz force $(\nabla \times \mathbf{B}) \times \mathbf{B}/4\pi$. The electric circuit analog provides no means for determining the plasma motion $\mathbf{u}$, so to proceed farther put $\mathbf{u}$ = 0. The energy equation can then be converted easily into something that looks like an electric circuit equation in cases where the field and current topology are equivalent to single loops. Integrate the energy equation over the volume of the system, writing the total magnetic energy as $LI^2/2$, which defines the equivalent inductance L, and the total resistive dissipation as $RI^2$, which defines the equivalent resistance R. The term $\nabla \cdot (\eta \mathbf{F})$ vanishes if either $\eta$ or $\mathbf{F}$ vanishes on the surface of the system. The result is the "circuit" equation

$$\frac{1}{2}\frac{d}{dt}LI^2 + RI^2 = 0$$

for which the solution is

$$L(t)I(t)^2 = L(0)I(0)^2 \exp\left[-2\int_0^t ds \frac{R(s)}{L(s)}\right].$$

Unfortunately this approach fails to provide the time dependence of R and L as a consequence of the changing scale and topology of the field during the resistive decay. It should be obvious that the behavior of a 3D magnetic field, described by 3D partial differential equations, cannot be treated generally with an ordinary differential equation.

So we must go back to the induction equation (4) in terms of $\mathbf{B}$, for which the complete solution is often elementary. With $\mathbf{u}$ = 0 the evolution of the field is described by the simple diffusion equation

$$\partial \mathbf{B}/\partial t = \eta \nabla^2 \mathbf{B}. \quad (6)$$

Consider, then, the force-free magnetic field, satisfying $\nabla \times \mathbf{B} = \alpha \mathbf{B}$ with uniform $\alpha$. This field preserves its form and scale as it decays, so that R and L are constant in time. It follows that the field decays as $\exp(-\eta\alpha^2 t)$ and the energy decays as $\exp(-2\eta\alpha^2 t)$.

On the other hand, the scale of many field configurations varies as $(4\eta t)^{1/2}$, with the result that R and L are not constant in time. Thus, for instance, it follows from the diffusion equation for $\mathbf{B}$ that a plane slab of field asymptotically increases it thickness in proportion to $(4\eta t)^{1/2}$, and the total magnetic energy per unit area declines as $t^{-1/2}$. The energy of a rod of field declines asymptotically as $t^{-1}$. A column of azimuthal field, concentric about the z-axis for instance, exhibits an asymptotic energy decline in proportion to $t^{-3/2}$, all immediately evident from the diffusion equation, but beyond the grasp of the circuit analog.

Similarly, the electric circuit analogy falls into qualitative error when time dependent resistivity is introduced, because the changing resistivity alters the form and/or magnitude of R, as in the current interruption by the sudden appearance of a sheet of insulation blocking the current path. The creation of the high resistivity is a kinetic plasma dynamical phenomenon, of course, beyond the scope of MHD. However, the large-scale consequences of the electrical insulation are readily addressed by the MHD paradigm. The error of the circuit analogy arises because the topology of the actual current path simply changes so as to avoid flowing across the insulating sheet. The analog predicts an enormous potential drop across the insulating sheet because of the inductance in the circuit, delivering heat and particle acceleration in and around the insulating region which continues until the magnetic energy stored in the inductance is exhausted [cf. *Alfven* and *Carlquist*, 1967]. However, with the redirecting of the current, no such interesting and cataclysmic effect exists, because the current flows in a thin boundary layer around the insulator and continues on its way through the otherwise highly conducting medium. A simple example is worked out below to illustrate the process.

The essential point is that the behavior of the system following the current interruption is fully and correctly described by the resistive induction equation, and that equation is equivalent to the heat flow equation for each of the cartesian components of the magnetic field. Increasing the "thermal conductivity" to infinity (i.e. reducing the electrical conductivity to zero and increasing the resistive diffusion coefficient to infinity) in a thin sheet has no catastrophic global "thermal" effects throughout the region of otherwise poor "thermal conductivity".

Needless to say, the local rapid resistive diffusion of the magnetic field in the thin current layer flowing around the insulating obstacle generally upsets the local equilibrium of the magnetic field, with the transient emission of Alfven waves and a re-adjustment of the original field [see example in *Parker*, 1983]. However the result is quite unlike that of a large potential difference across the insulating region.

It is important that the theoretical studies of the magnetosphere be properly based on Newton and Maxwell. The theoretician is free to use any physical variables that best suit the problem, but it must be demonstrated how the principles employed in the analysis follow rigorously from the equations of Newton and Maxwell. The arbitrary invention of contrary principles and concepts does not contribute anything positive to the scientific enterprise.

The remainder of this lecture is devoted to a brief outline of the deduction of the momentum and induction equations from Newton and Maxwell in the collisionless plasma of the outer magnetosphere and in the partially ionized gases of the ionosphere. As already noted, the equations are in the general MHD form, describing the relatively simple physical principles of contending forces. In the penultimate section we work out an illustrative example of a current interruption.

It is important to note that the electric field **E** and current density **j** are readily computed once **u** and **B** are known, but generally not vice versa. For instance, **j** is needed to assess local resistive heating and to assess the role of plasma turbulence and anomalous resistivity [cf. *Spicer*, 1982; *Parker*, 1994]. The large-scale dynamics determines where intense thin current sheets lie, within which a more detailed plasma kinetic treatment or particle simulation may be required.

The theoretical development is in the nonrelativistic limit, neglecting terms second order in $u/c$ compared to one and treating the large-scale bulk dynamics of the magnetosphere on a spatial scale $\Lambda$ large compare to the cyclotron radii of the electrons and ions. Thus, for instance, we do not address the structure of the magnetopause [*Parker*, 1967a,b, 1969; *Lerche*, 1967; *Lerche* and *Parker*, 1967; *DeKeyser* and *Roth*, 1998] or the neutral sheet in the geotail or the auroral sheets. We also limit discussion to plasmas sufficiently dense that the electron conduction velocity in the large-scale magnetic field is small compared to the thermal velocities of the ions. Thus electron inertia is neglected and there is no plasma turbulence and anomalous resistivity. It should be noted that turning to **j** and **E** in no way circumvents the limitations of MHD. A full up kinetic plasma treatment is required to treat the small scales and/or high electron conduction velocities, etc.

## COLLISIONLESS PLASMA

Consider the simple case of a collisionless plasma, composed of N electrons and N singly charged ions per unit volume. Denote the velocity of the individual ion or electron by $v_i$, writing $v_i$ as the sum of the mean bulk velocity $u_i$ and the thermal velocity $w_i$ relative to the mean motion, so that $v_i = u_i + w_i$. Thus, within any fixed volume V whose dimensions are small compared to the large scale $\Lambda$ of the plasma and field distribution, but large enough to contain a large number of particles, the sum of the thermal momenta $mw_i$ vanishes. The thermal velocities $w_\perp$ perpendicular to **B** are statistically isotropic as a consequence of the cyclotron motion around **B**. The thermal momentum flux, or plasma pressure, perpendicular to **B** is given by $p_\perp = \Sigma mw_\perp^2/2V$, where the sum is over all particles in V. The thermal momentum flux, or plasma pressure, parallel to **B** is $p_\parallel = \Sigma mw_\parallel^2/V$, where $w_\parallel$ represents the thermal velocity parallel to B. The essential point is that the bulk motion $u_i$ of the plasma is made up of the sum of the motions of all the individual electrons and ions, and the electric current arises from the slight difference in the motions of the electrons and ions. The current must satisfy Ampere's law and the bulk motion is described by Newton's equation for the momentum flux $\rho u_i$.

To check this out we use the guiding center approximation, with which we assume the reader is familiar, to compute the motions of the individual particles, summing over the motions to obtain the bulk velocity **u** and the current **j**. The essential point is that the guiding centers of the particles move in the frame of reference in which the electric field **E**' vanishes, from which it follows that $\mathbf{E} = -\mathbf{u} \times \mathbf{B}/c$. The Faraday induction equation becomes

$$\partial \mathbf{B}/\partial t = \nabla \times (\mathbf{u} \times \mathbf{B}). \quad (7)$$

The Poynting vector is $\mathbf{u}_\perp B^2/4\pi$, indicating that the magnetic enthalpy, and hence the magnetic field, is carried bodily with the plasma velocity $\mathbf{u}_\perp$.

This is the appropriate place to point out that the large-scale electric field parallel to **B** in a quiescent plasma is generally weak because of the background of thermal ions and electrons with their free motion along **B**. Any significant thermal anisotropy, $p_\| - p_\perp \neq 0$, provides microscopic plasma instabilities that feed on the anisotropy in all but the most tenuous plasmas. Thus, for instance, the energetic particles of the Van Allen radiation belts are strongly anisotropic, as are the thermal velocities in the solar wind. But the quasistatic background thermal plasma in the magnetosphere is generally not significantly anisotropic.

The inhomogeneity of the large-scale magnetic field results in the slow drift of the guiding centers of the individual particles relative to the mean bulk motion **u**. For a particle with mass m and charge q there is the gradient drift

$$\mathbf{u}_g = \left(\frac{mw_\perp^2 c}{2qB^4}\right) \mathbf{B} \times \frac{\nabla B^2}{2},$$

the curvature drift

$$\mathbf{u}_c = \left(\frac{mw_\|^2 c}{qB^4}\right) \mathbf{B} \times [(\mathbf{B} \cdot \nabla)\mathbf{B}],$$

and the polarization drift

$$\mathbf{u}_p = \left(\frac{mc}{qB^2}\right) \mathbf{B} \times \frac{d\mathbf{u}}{dt}.$$

It is evident by inspection that the electrons and ions drift oppositely, and the net result is the electric current density

$$\mathbf{j}_\perp = \left(\frac{c}{B^2}\right) \mathbf{B} \times \left\{ \nabla p_\perp + [(\mathbf{B} \cdot \nabla)\mathbf{B}] \frac{(p_\| - p_\perp)}{B^2} + \rho \frac{d\mathbf{u}}{dt} \right\} \quad (8)$$

upon taking account of the geometrical factors, where $\rho$ is the plasma density NM. We note that this current flows in the frame of reference of the electrons and ions.

Noting that $(\nabla \times \mathbf{B})_\perp$ can be written as $\mathbf{B} \times [(\nabla \times \mathbf{B}) \times \mathbf{B}]/B^2$, the perpendicular component of Ampere's law becomes

$$\mathbf{B} \times \left\{ \rho \frac{d\mathbf{u}}{dt} + \nabla p_\perp + [(\mathbf{B} \cdot \nabla)\mathbf{B}] \frac{(p_\| - p_\perp)}{B^2} - \frac{1}{4\pi}(\nabla \times \mathbf{B}) \times \mathbf{B} \right\}$$
$$= 0$$

Thus the perpendicular component of $d\mathbf{u}/dt$ satisfies

$$\rho \frac{d\mathbf{u}_\perp}{dt} = -\nabla_\perp \left( p_\perp + \frac{B^2}{8\pi} \right) + \frac{[(\mathbf{B} \cdot \nabla)\mathbf{B}]_\perp}{4\pi} \left[ 1 - \frac{4\pi(p_\| - p_\perp)}{B^2} \right] \quad (9)$$

This is, of course, the perpendicular component of Newton's momentum equation. The term $4\pi(p_\| - p_\perp)/B^2$ represents the net centrifugal force of the excess field-aligned thermal motion along the curved field lines, vanishing in the case of isotropic thermal motions. Thus we have the familiar magnetohydrodynamic momentum equation with the possibility of an anisotropic plasma pressure [*Parker*, 1957]. These results tells us that in a plasma whose bulk velocity satisfies Newton's equation of motion, the gyrations of the electrons and ions around the magnetic field automatically provide the electric current required by Ampere. The electric currents are not driven by an applied electric field, whose only persisting consequence would be to change the bulk velocity $\mathbf{u}_\perp$. The electric current is an entirely local phenomenon, dictated by the local spatial variation of the magnetic field to satisfy automatically Ampere's relation between the current and the deformation of the magnetic field. Conditions elsewhere in the system do not determine the local current density.

The motion $\mathbf{u}_\|$ parallel to **B** is purely hydrodynamic, because the Lorentz force has no component parallel to **B**. The motions $\mathbf{u}_\|$ and $\mathbf{u}_\perp$ are related through the equation for conservation of matter,

$$\partial \rho/\partial t + \nabla \cdot [\rho(\mathbf{u}_\| + \mathbf{u}_\perp)] = 0.$$

In a stationary magnetic field the motion along **B** is described by the momentum equation

$$\rho\left(\frac{\partial}{\partial t} + u_\parallel \frac{\partial}{\partial s}\right) u_\parallel = -\frac{\partial p_\parallel}{\partial s} + \frac{p_\parallel - p_\perp}{B}\frac{\partial B}{\partial s} \quad (10)$$

and the continuity equation

$$\frac{\partial \rho}{\partial t} + \frac{\partial \rho u_\parallel}{\partial s} = \frac{\rho}{B} u_\parallel \frac{\partial B}{\partial s}. \quad (11)$$

In addition to these equations there is the equation of state, determining $p_\parallel$ and $p_\perp$ [cf. *Chew, Goldberger* and *Low*, 1956] in terms of the invariants of the motions of the individual particles in the inhomogeneous magnetic field. If anisotropy is not important, a simple polytropic relation between p and $\rho$ may be sufficient, or an energy equation involving heat flow may be required.

For a discussion of $j_\parallel$ and the possibilities and consequences of a significant $E_\parallel$ the reader is referred to *Schindler, Hesse,* and *Birn* [1991].

The bottom line is simply that the electric current in a dynamical plasma does not have the properties of the current in a fixed electrical circuit. The physics of the two are complementary, forming a duality. At any given instant in time the electric current at any position in a large-scale plasma is determined by the local conditions and not by conditions elsewhere around the current path; in contrast the current at any position in an electric circuit is controlled by the resistance, capacitance, and inductance distributed elsewhere around the circuit and not by the conditions at the given position.

## PARTIALLY IONIZED GAS

Consider next the dynamical behavior of a magnetic field in a partially ionized gas. e.g. the ionosphere. To illustrate the basic dynamical properties of the gas and field, consider the simple case that the gas is only slightly ionized, so that the number n of ions per unit volume is small compared to the number N of neutral atoms per unit volume. The mass of the neutral atom is denoted by M and the same mass is ascribed to the ion, with the small electron mass denoted by m. The electrons, ions, and neutral atoms are coupled by the frictional drag of collisions, and the development is a three fluid representation. The effective collision time for an ion with the neutral atoms is $\tau_i$, for an electron with the neutral atoms $\tau_e$, and for electrons with ions it is $\tau$. Note that $\tau_e \sim \tau_i(m/M)^{1/2}$ approximately. Thus $M/\tau_i \gg m/\tau_e$ and it is principally the drag of the ions moving through the neutral atoms that ties the ions and electrons to the neutral gas. The thermal pressures of the electrons, ions, and neutral atoms are denoted by $p_e$, $p_i$, and p, respectively and are taken to be isotropic as a consequence of frequent collisions. The electron, ion, and neutral gas bulk velocities are denoted by **u**, **w**, and **v**, respectively. So long as the relative velocities of the constituents are small compared to the thermal velocities, the effect of collisions is adequately represented by a linear drag force, and the equations of motion can be written

$$nm\frac{d\mathbf{u}}{dt} = -\nabla p_e - ne(\mathbf{E} + \mathbf{u} \times \mathbf{B}/c)$$
$$-\frac{nm}{\tau_e}(\mathbf{u} - \mathbf{v}) + \frac{nm}{\tau}(\mathbf{w} - \mathbf{u}) \quad (12)$$

$$nM\frac{d\mathbf{w}}{dt} = -\nabla p_i + ne(\mathbf{E} + \mathbf{w} \times \mathbf{B}/c)$$
$$-\frac{nM}{\tau_i}(\mathbf{w} - \mathbf{v}) - \frac{nm}{\tau}(\mathbf{w} - \mathbf{u}) \quad (13)$$

$$NM\frac{d\mathbf{v}}{dt} =$$
$$-\nabla p + \frac{nM}{\tau_i}(\mathbf{w} - \mathbf{v}) + \frac{nm}{\tau_e}(\mathbf{u} - \mathbf{v}) \quad (14)$$

The current density is $ne(\mathbf{w} - \mathbf{u})$, so that with Ampere's law it follows that

$$\mathbf{w} - \mathbf{u} = c\nabla\times\mathbf{B}/4\pi ne. \quad (15)$$

Assuming a low level of ionization (n << N) the inertia of the ions and electrons can be neglected. Thus there are three linear algebraic equations (12), (13), and (15) which can be solved for **u**, **w**, and **E** in terms of **v** and **B**. Adding equations (12) and (13) and using equation (15) to eliminate **w** − **u** from the result yields the momentum equation

$$NM\frac{d\mathbf{v}}{dt} = -\nabla(p + p_e + p_i) + \frac{(\nabla \times \mathbf{B}) \times \mathbf{B}}{4\pi} \quad (16)$$

for the bulk motion of the neutral gas. Alternatively, use equation (15) to eliminate only **u**, obtaining

$$\mathbf{w} = \mathbf{v} + \left(\frac{1}{nQ}\right)\mathbf{P} + \left(\frac{mc}{4\pi\tau_e neQ}\right)\nabla \times \mathbf{B}, \quad (17)$$

where **P** is the total force per unit volume exerted on the electrons and ions,

$$\mathbf{P} \equiv -\nabla(p_e + p_i) + (\nabla \times \mathbf{B}) \times \mathbf{B}/4\pi,$$

and $Q \equiv M/\tau_i + m/\tau_e \approx M/\tau_i$. It follows that

$$\mathbf{u} = \mathbf{v} + \left(\frac{1}{nQ}\right)\mathbf{P} - \left(\frac{Mc}{4\pi\tau_i neQ}\right)\nabla \times \mathbf{B}. \quad (18)$$

Then solve equation (12) or (13) for **E**, using equations (17) and (18) to eliminate **w** and **u**, and obtaining

$$\mathbf{E} = -\mathbf{v} \times \mathbf{B}/c$$
$$-\frac{1}{nQc}\mathbf{P} \times \mathbf{B} + \frac{\eta}{c}\nabla \times \mathbf{B} + \left(\frac{M}{\tau_i} - \frac{m}{\tau_e}\right)\left(\frac{1}{neQ}\right)\frac{(\nabla \times \mathbf{B}) \times \mathbf{B}}{4\pi}$$
$$+\left(\frac{m}{\tau_e enQ}\right)\nabla p_i - \left(\frac{M}{\tau_i enQ}\right)\nabla p_e, \quad (19)$$

where the resistive diffusion coefficient is

$$\eta \equiv \left(\frac{c^2}{4\pi ne^2}\right)\left(\frac{Mm}{\tau_i \tau_e Q} + \frac{m}{\tau}\right). \quad (20)$$

Substituting this expression for **E** into the Faraday induction equation yields

$$\frac{\partial \mathbf{B}}{\partial t} = \nabla \times (\mathbf{v} \times \mathbf{B})$$
$$-\nabla \times (\eta \mathbf{B}) - \nabla \times (h\mathbf{F}) + \nabla \times \left(\frac{\mathbf{P} \times \mathbf{B}}{nQ}\right)$$
$$-\frac{c}{e}\nabla \times \left\{\frac{1}{nQ}\left[\frac{m}{\tau_e}\nabla p_i - \frac{M}{\tau_i}\nabla p_e\right]\right\}, \quad (21)$$

where

$$h = \frac{c}{neQ}\left(\frac{M}{\tau_i} - \frac{m}{\tau_e}\right) \cong \frac{c}{ne} \quad (22)$$

is the Hall coefficient, and **F** is again the Lorentz force $(\nabla \times \mathbf{B}) \times \mathbf{B}/4\pi$. In the simple case that the spatial variations of n, $\tau$, $\tau_e$, and $\tau_i$ can be neglected, the induction equation reduces to

$$\frac{\partial \mathbf{B}}{\partial t} = \nabla \times \left[\left(\mathbf{v} - \frac{h}{4\pi}\nabla \times \mathbf{B} + \frac{\mathbf{P}}{nQ}\right) \times \mathbf{B}\right] + \eta\nabla^2\mathbf{B}. \quad (23)$$

The form of the induction equation shows that the field is transported with the velocity **V**, where

$$\mathbf{V} = \mathbf{v} - \left(\frac{h}{4\pi}\right)(\nabla \times \mathbf{B})_\perp + \left(\frac{1}{nQ}\right)\mathbf{P}_\perp,$$

so that

$$\partial \mathbf{B}/\partial t = \nabla \times (\mathbf{V} \times \mathbf{B}) + \eta\nabla^2\mathbf{B}. \quad (25)$$

The first term on the right hand side of equation (24) is the velocity of the neutral gas. The second term on the right hand side of equation (24) represents the Hall effect and the third term the Pedersen effect, both of which are a direct consequence of the force exerted on the ions and electrons by the Maxwell stress. First of all, the total force **P** pushes the ions and electrons through the neutral gas at a speed U such that the collisional drag opposes **P** in the steady state, with

$$n(M/\tau_i + m/\tau_e)\mathbf{U} = \mathbf{P} \quad (26)$$

so that $\mathbf{U} = \mathbf{P}/nQ$. This is the Pedersen effect, sometimes called ambipolar diffusion [cf. *Cowling*, 1957; *Parker*, 1979, p. 45]. It is a dissipative effect because it represents a slippage of the field through the neutral gas driven by **P** and opposed by friction, the energy coming from the relaxation of the magnetic field.

Consider, then, the Hall effect. Suppose that a force **f** perpendicular to **B** is applied to a point charge q. The initial response is to move in the direction of **f**, producing a transverse Lorentz force that accelerates the charge to a velocity **W** in the direction perpendicular to both **f** and **B** such that the Lorentz force q**W**x**B**/c exactly opposes the applied force **f**. Thus $\mathbf{W} = (c/qB^2)\mathbf{f} \times \mathbf{B}$. A familiar example is the electric drift $c\mathbf{E} \times \mathbf{B}/B^2$ as a consequence of the electric force $\mathbf{f} = q\mathbf{E}$. In the problem at hand the Lorentz force on the ions is $\mathbf{f} = \mathbf{F} \equiv (\nabla \times \mathbf{B}) \times \mathbf{B}/4\pi$ per unit volume and the charge is ne per unit volume, producing the drift velocity

$$\mathbf{W} = (c/4\pi neB^2)[(\nabla \times \mathbf{B}) \times \mathbf{B}] \times \mathbf{B}$$
$$= -(c/4\pi ne)(\nabla \times \mathbf{B})_\perp, \quad (27)$$

recognizable as the Hall drift across **B** and contributing the middle term on the right hand side of equation (24). Thus **V** = **v** + **U** + **W**.

The portion $U_F$ of the ambipolar drift **U** driven by the Lorentz force **F** ($U_F = F/nQ$) provides the dissipation and diffusion represented by the Pedersen resistive diffusion coefficient $B^2/4\pi nQ$. This may be seen by rewriting the induction equation (25) in the form

$$\frac{\partial \mathbf{B}}{\partial t} = \nabla \times \left[ (\mathbf{v} + \mathbf{W} + \mathbf{U}_P) \times \mathbf{B} \right] -$$
$$\nabla \times \left[ \left( \frac{B^2}{4\pi nQ} + \eta \right) \nabla \times \mathbf{B} \right], \quad (28)$$

where $nQ\mathbf{U}_P = -\nabla(p_e + p_i)$, and the Pedersen resistive diffusion coefficient is $B^2/4\pi nQ$.

It is instructive to form the scalar product of the induction equation with $\mathbf{B}/4\pi$ to obtain the energy equation

$$\frac{\partial}{\partial t}\left(\frac{B^2}{8\pi}\right) + \mathbf{F} \cdot \mathbf{V} + \nabla \cdot \left( \mathbf{V}_\perp \frac{B^2}{4\pi} \right)$$
$$+ \nabla \cdot \left[ \left( \eta + \frac{B^2}{4\pi nQ} \right) \mathbf{F} \right] + \frac{4\pi}{c^2}\left( \eta + \frac{B^2}{4\pi nQ} \right) j^2 = 0 \quad (29)$$

This equation is identical in form to the energy equation (5) for the ideal resistive fluid ($\mathbf{j} = \sigma \mathbf{E}$). In the present case of the partially ionized gas the field moves with the velocity **V**, transporting the magnetic enthalpy $B^2/4\pi$ and working against the Lorentz force **F**. The essential point is again the basic mechanical nature of the field and fluid interaction on large-scales. Electric currents are present, of course, but not part of the active stresses and momenta. Further discussion and application to illustrative examples can be found in *Parker* [1996].

## CURRENT INTERRUPTION

There has been considerable discussion of the consequences of a local current interruption, as may occur in the very tenuous plasma of the outer magnetosphere and geotail, or in the intense auroral current sheet when the electron conduction velocity exceeds the ion sound velocity and, particularly when the conduction velocity approaches the electron sound velocity, causing plasma turbulence and electric double layers. Dissipation of magnetic field may be enhanced locally. On the other hand, the electric circuit analog predicts a huge electrostatic potential drop across the resistive region as a consequence of the inductance in the electric circuit [*Alfven* and *Carlquist*, 1967]. The large potential drop and rapid dissipation continue until the entire magnetic field around the circuit is dissipated. On this basis a current interruption would have a catastrophic effect, dumping a major part of the total magnetic energy at the site of the current interruption.

In fact the example of the local current interruption provides a vivid illustration of the qualitative errors in the circuit analogy as it is commonly employed. For in the real world, the current can reroute itself at an instant's notice, and the system can be treated properly only by working from Maxwell's equations. The local anomalous resistivity that interrupts the current requires a full blown plasma kinetic treatment, of course, while the large-scale external region is properly treated with the MHD induction equation.

Consider the external consequences of abruptly locally blocking the flow of electric current along the axis of an azimuthal magnetic field $B_\varphi(\varpi)$ concentric about the z-axis, where $\varpi \equiv (x^2 + y^2)^{1/2}$ represents radial distance from the z-axis. The space is filled with a fixed static medium with very small resistive diffusion coefficient $\eta_0$. The total current out to a radius $\varpi$ is $I_z(\varpi) = c\varpi B_\varphi(\varpi)/2$ flowing from $z = -\infty$ to $z = +\infty$.

The characteristic scale of $B_\varphi$ is denoted by $\Lambda$, so that the magnetic field evolves very slowly, over the characteristic time $\Lambda^2/\eta_0$ (which does not concern us here) in the manner described by the diffusion equation (6). At time $t = 0$ the resistive diffusion coefficient in the thin disk, of radius a and small thickness $2\varepsilon$ ($\varpi < a$, $-\varepsilon < z < +\varepsilon$), is increased from the ambient $\eta_0$ to $\infty$ in the short time $\Delta t$ (~ $\varepsilon^2/4\eta_0 <<< a^2/4\eta_0$). The result is that $B_\varphi$ is entirely dissipated within the thin disk and $j_z$ can no longer penetrate across the disk. The evolution of the field outside the disk is described by equation (6) with the boundary condition $B_\varphi = 0$ as z declines to $+\varepsilon$ on the upper side and increases to $-\varepsilon$ on the lower side. Thus, for $t > \Delta t$ the field approaches the asymptotic form (for $\Delta t << t << a^2/4\eta_0$)

$$B_\varphi(\varpi, z, t) \sim B_\varphi(\varpi) \, \text{erf}\left[ (z-\varepsilon)/(4\eta_0 t)^{1/2} \right], \quad (30)$$

in $z > +\varepsilon$, with the corresponding solution in $z < -\varepsilon$, where $\text{erf}(x)$ is the error function. Needless to say, this simple solution is not valid within distances of about $2\varepsilon$ of the rim of the disk at $\varpi = a$, nor anywhere beyond the rim ($\varpi > a$). However, the solution is adequate across the face of the disk, where it shows that there is a radial current density

$$j_\varpi(\varpi, z, t) = -\frac{cB_\varphi(\varpi)}{4(\pi^3 \eta_0 t)^{1/2}} \exp\left( -\frac{(z-\varepsilon)^2}{4\eta_0 t} \right)$$

in the boundary layer at $z > \varepsilon$, and a similar but oppositely directed current in $z < -\varepsilon$. The net surface current density is $J_\varpi(\varpi) = (c/4\pi)B_\varphi(\varpi)$, so that the total current $2\pi\varpi J_\varpi(\varpi)$ flowing radially outward across any circle $\varpi = $ *constant* is equal to the current $I_z(\varpi)$ blocked within that circle.

There is a weak radial electric field

$$E_\varpi(\varpi, z, t) = j_\varpi(\varpi, z, t)/\sigma$$

where $\sigma \equiv c^2/4\pi\eta_0$ is large. The resistive dissipation rate per unit volume is

$$\frac{j_\varpi^2}{\sigma} = \frac{B_\varphi(\varpi)^2}{4\pi^2 t}\exp\left[-\frac{(z-\varepsilon)^2}{2\eta t}\right],$$

and the dissipation rate per unit area on the disk surface is

$$D(\varpi, t) = \int_\varepsilon^\infty dz \frac{j_z^2}{\sigma} = \frac{B_\varphi(\varpi)^2}{8\pi}\left(\frac{2\eta_0}{\pi t}\right)^{1/2}$$

for $t > \Delta t$.

The essential point is that the abrupt blocking of $j_z$ by the disk reroutes the current around the outside of the disk, with the associated rapid dissipation confined to the thin boundary layer with characteristic thickness $(4\eta_0 t)^{1/2}$. With small $\eta_0$ (i.e. large conductivity $\sigma$) this involves only a very small volume and only modest radial electric potential differences for $t > \Delta t$.

The radius a of the disk need not be small, of course. To take a specific example, suppose that the disk is a plane sheet of electrical insulation, extending to $\varpi = \infty$, while the column of azimuthal field is confined within some finite radius $\Lambda$, with $B_\varphi(\Lambda) = 0$. It follows from Ampere's law that $I_z(\Lambda) = 0$, with the current flowing one way in the vicinity of the z-axis and in the opposite direction at larger radius. The effect of the sheet of electrical insulation is simply to short circuit the currents flowing in the z-direction by introducing a radial current at $z = \varepsilon$.

It is obvious that the local rerouting of the current around a resistive barricade is not limited to the present case of an axially symmetric magnetic field and a disk or sheet of insulation. A resistive obstacle of any shape anywhere in a magnetic field with nonvanishing curl causes the electric current to be rerouted locally around the periphery of the obstacle, leaving the field elsewhere unaffected. No global inductive effect is achieved and the enormous potential difference expected from the electric circuit analog with its large inductance does not arise. The error in the circuit analog is that the current path is fixed, whereas in reality the current, representing no intrinsic energy or inertia, abruptly reroutes around the insulating obstacle. The induction equation that follows from Faraday's equation easily handles this detail, whereas the circuit analog is too simple to recognize it. The errors of the circuit analog only get worse when the motion of the plasma or neutral gas is included in the problem, because the motion is just one more essential feature that cannot be handled apriori by the circuit analog.

## CONCLUDING REMARKS

Faced with the challenge of working out the dynamics of the active magnetosphere, explored by spacecraft now to an impressive degree, the time has come to pay strict attention to the dynamical equations offered by Newton and Maxwell. It is possible, in some cases at least, to construct a circuit analog once the dynamical behavior of the system has been worked out from the partial differential equations of Newton and Maxwell, as already noted. However, the construction of an equivalent electric circuit after the fact serves in no way to illucidate those interactions. The idea that the electric current is somehow the fundamental physical variable is an unfortunate misunderstanding of the physics.

The second and third sections of this brief review outline the form of the dynamical equations that follow from the equations of Newton and Maxwell for the large-scale behavior of a collisionless plasma and also of a relatively dense gas with only a small degree of ionization. It is shown that the Pedersen and Hall effects represent the slippage of the magnetic field and the ionized component through the neutral gas, driven by the Lorentz force of the magnetic field on the ionized component and applicable to the ionosphere. The example of the collisionless plasma applies to the outer magnetosphere. The task ahead is to develop tractable dynamical equations that bridge the gap between, so that the present MHD models of the outer magnetosphere and magnetotail can be joined with MHD models of the dense ionosphere and with the net effects of the complex plasma physics within the magnetopause, geotail neutral sheet, etc.

This is perhaps the appropriate place to note again that the foregoing remarks are directed to establishing the theoretical connection of the diverse magnetospheric phenomena to the fundamental principles of Newton and Maxwell. However an intermediate stage of the scientific investigation of the magnetosphere should not go unnoticed, and that is the some-times arbitrary construction of idealized mathematical relations between various physical quantities as a guide to analysing observational data. Thus, for instance, *Akasofu* [1980] explored the energy input from the solar wind to the magnetosphere by comparing a variety of solar wind parameter products with the energy input inferred from observation. This empirical approach has developed over the years into systems of ordinary differential equations for four or more magnetospheric quantities [cf. *Horton* and *Doxas*, 1996, 1998] using observational behavior of the magnetosphere to evaluate the many free parameters employed in the construction. These efforts provide clues to guide the theoretical work from the equations of Newton and Maxwell, and are particularly useful for searching out predictive algorithms.

## REFERENCES

Akasofu, S. I.,The solar wind-magnetospheric energy coupling and magnetospheric disturbances, *Planet. Space Sci., 28,* 495, 1980.

Alfven, H. and Carlquist, P., Currents in the solar atmosphere and a theory of solar flares, *Solar Phys. 1,* 220, 1967.

Axford, W. I. and Hines, C. O., A unifying theory of high latitude phenomena and the geomagnetic storm, *Can. J. Phys., 39,* 1433, 1961.

Chapman, S. and Ferraro, V. C. A., The theory of the first phase of a geomagnetic storm, *Terrest. Magn. Atmos. Elec., 45,* 245, 1940.

Chew, G. F., Goldgerger, M. L., and Low, F. E., The Boltzmann equation and the one fluid hydrodynamic equation in the absence of collisions, *Proc. Roy. Soc. London, A, 236,* 112, 1956.

Cowling, T. G., Magnetohydrodynamics, Wiley (Interscience), New York, pp. 105-112. 1957.

DeKeyser, J. and Roth, M, Equilibrium conditions and magnetic field rotation at the tangential discontinuity magnetopause, *J. Geophys. Res., 103,* 6653, 1998.

Dessler, A. J., Frances, W. E., and Parker, E. N., Geomagnetic storm sudden- commencement rise times, *J. Geophys. Res., 65, 2715,* 1960.

Dessler, A. J. and Parker, E. N., Hydromagnetic theory of geomagnetic storms, *J. Geophys. Res., 64,* 2239, 1959.

Dungey, J. W., *Cosmic Electrodynamics*, Cambridge University Press, Cambridge, 1958, pp 137-146.

Dungey, J. W. Interplanetary magnetic field and the auroral zones, *Phys. Rev. Lett., 6,* 47, 1961.

Frank, L.A., Ashour-Abdalia, M., Berchem, J., Raeder, J., Paterson, W. R., Kokubun, S., Yamamoto, R. P., Leppino, R. P., Coroniti, F. V., Fairfield, D. H., and Ackerson, K.L., Observations of plasma and magnetic fields in Earth's distant magnetotail: Comparison with global MHD model, *J. Geophys. Res., 100,* 19177, 1995.

Gold, T., Motions in the magnetosphere of Earth, *J. Geophys. Res., 64,* 1219, 1959.

Horton, W. and Doxas, I., A low-dimensional energy conserving state space model of substorm dynamics, *J. Geophys. Res., 101,* 27,223, 1996.

Horton, W. and Doxas, I., A low dimensional dynamical model for the solar wind driven geotail-ionosphere system, *J. Geophys. Res., 103,* 4561, 1998.

Lerche, I. On the boundary layer between a warm streaming plasma and a confined magnetic field, *J. Geophys. Res., 72,* 5295, 1967.

Lerche, I and Parker, E. N., Nonequilibrium and enhanced mixing at a plasma-field interface, *Astrophys. J., 150,* 731, 1967.

Parker, E. N. Newtonian development of the dynamical properties of ionized gases of low density, *Phys. Rev., 107,* 924, 1957.

Parker, E. N., Confinement of a magnetic field by a beam of ions, *J. Geophys. Res., 72,* 2315, 1967a.

Parker, E. N., Small-scale nonequilibrium of the magnetopause and its consequences, *J. Geophys. Res., 72,* 4365, 1967b.

Parker, E. N., Solar wind interaction with the geomagnetic field, *Rev. Geophys., 7,* 3, 1969.

Parker, E. N., *Cosmical Magnetic Fields*, Clarendon Press, Oxford, 1979.

Parker, E. N., The propagation of torsion along flux tubes subject to dynamical nonequilibrium, *Geophys. Astrophys. Fluid Dyn., 24,* 245, 1983.

Parker, E.N., *Spontaneous Current Sheets in Magnetic Fields*, Oxford University Press, New York, 1994.

Parker, E. N., The alternative paradigm for magnetospheric physics, *J. Geophys. Res., 101,* 10,587, 1996.

Schindler, K. Hesse, M.and Birn, J. Magnetic field aligned electric potentials in nonideal plasma flows, *Astrophys. J., 380,* 293, 1991.

Sckopke, N., A study of self-consistent ring current models, *Cosmic Electrodyn. 3,* 330, 1972.

Siscoe, G. L., The magnetospheric boundary, in *Physics of Space Plasmas*, SPI Conf. Proc. and Reprint Seriues, No. 7, Scientific Publishers Inc., Cambridge, Massachusetts, ed. T. Chang, G. B. Crew, and J. R. Jasperse, 1988, pp70-73.

Spicer, D. S., Magnetic energy storage and conversion in the solar atmosphere, *Space Sci. Rev., 31,* 351, 1982.

Zaitsev, V. V., Stepanov, A. V., Urpo, S. and Pohjolainen, S., LRC-circuit analog of current carrying magnetic loop: diagnostics of electrical parameters, *Astron. Astrophys, 337,* 887, 1998.

Zhu, X., Magnetospheric convection pattern and its implications, *J. Geophys. Res., 98,* 21, 291, 1993.

Zhu, X., How the magnetosphere is driven into substorm, *J. Geophys. Res., 100,* 1847, 1995.

---

E. N. Parker, 1323 Evergreen Rd, Homewood, Illinois 60430.

# Paradigm Transition in Cosmic Plasma Physics, Magnetic Reconnection and the Generation of Field-Aligned Current

Yan Song and Robert L. Lysak

*School of Physics and Astronomy, University of Minnesota, Minneapolis, Minnesota*

The traditional theories of field-aligned current (FAC) generation, magnetic reconnection and mass, momentum and energy transfer within the magnetosphere were built on the basis of a convection picture and describe mainly large-scale, quasi-steady state phenomena occurring in a passive plasma region. However, these theories have not been able to account for many of the detailed observations that have been made in recent years. These incomplete, but widely accepted, theoretical models have led to numerous misleading or controversial issues in magnetospheric physics in the past few decades. In this paper, we point out that MHD wave propagation, in particular, wave packet dynamics, plays an important role for phenomena occurring in the active plasma region, such as FAC generation and reconnection. The topological approach, using the concept of helicity, is used to describe the structure of MHD wave packets. The theories of FAC generation as well as the concept of reconnection need to be revised under wave packet dynamics. The MHD wave packet paradigm gives a major change in our approach to cosmic plasma physics. The rich and varied implications of wave packet dynamics exceed those given by either the **u-B** paradigm or the **E-J** paradigm. This paradigm transition will bridge the gap between cosmic plasma physics and modern physics.

## INTRODUCTION

The traditional theories of FAC generation and magnetic reconnection are based on a convection picture and describe mainly large-scale, quasi-steady state phenomena occurring in a passive plasma region. Figure 1a shows the classical reconnection model proposed by Dungey (1961). Figure 1b schematically describes the FAC generation caused by the change of the field-aligned vorticity in association with the motion of convection cells which are generated by large scale viscous drag or magnetic drag (e.g., *Sato*, 1982). Both models shown in Figure 1 describe convection-related, large-scale and quasi-steady processes where the topological structure of the field lines do not change. However, these simple, smooth and topologically steady models require an anomalous viscosity $v^*$ or an anomalous resistivity $\eta^*$, neither of which has been adequately described theoretically (*Song and Lysak*, 1992). Therefore, the long-standing question how the kinetic energy of the plasma can be converted into electromagnetic energy in the form of the field-aligned current has not been resolved.

It has been noted (e.g., *Parker*, 1979, 1983; *Alfvén*, 1981, 1986) that the cosmic plasma environment is not just a passive place, but can be an active and rapidly changing environment where magnetic fields are born and grow, twisted magnetic flux ropes carrying current filaments penetrate space, and continuous, violent energy-release processes are occurring. The interaction of the solar wind and magnetosphere often occurs in a patchy and transient manner, as indicated by the presence of flux transfer events at the

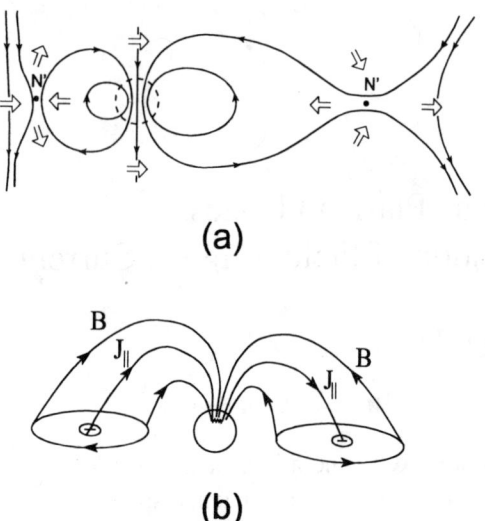

**Figure 1.** A schematic view showing (a) Dungey's steady state reconnection model (after Fig. 1, *Dungey*, 1961) and (b) a large scale discharge current of charges accumulated in the twin convection cells at the magnetospheric equator (after Fig. 1, *Sato*, 1982).

magnetopause (*Haerendel et al.*, 1978; *Russell and Elphic*, 1978). The traditional reconnection theory also cannot explain the nature of flux rope-like structures observed at the magnetopause and in the geomagnetic tail (e.g., *Scholer et al.*, 1985). At the magnetopause, magnetic shear is not the only factor that determines the reconnection-related high-speed plasma flow (*Sonnerup*, 1984; *Paschmann et al.*, 1986; *Phan et al.*, 1996). This implies that the interaction process is more delicate and complicated than the process described by the classical reconnection theory. It is now generally accepted that the initiation of the substorm current wedge (SCW) does not occur at the mid-tail reconnection region (~15-30 $R_E$), but rather it is linked to the thin intense current sheet in the near-Earth region (~6-10 $R_E$) (e.g., *Lui*, 1991; *Kennel*, 1992). Neither the classical reconnection theory (e.g., *Dungey*, 1961) nor the FAC generation theory (e.g., *Hasegawa and Sato*, 1979; *Vasyliunas*, 1984) are able to give a good explanation of the generation of FAC filaments and the shear Alfvén waves related to auroral particle acceleration or substorm onset.

Since observations conflict with and pose serious challenges to the existing theories of magnetic reconnection and FAC generation, the description of the physical processes in cosmic plasmas has drawn attention in the past few decades (e.g., *Alfvén*, 1977, 1981, 1983; *Fälthammar*, 1990; *Song and Lysak*, 1994, 1995; *Parker*, 1996; *Akasofu*, 1998). Alfvén was the first to distinguish between physical laws in passive and active plasma regions (e.g., *Alfvén*, 1981). Unlike passive plasmas, which carry little or no current and behave nearly classically, active plasmas carry a large current and involve strong localized and time-dependent plasma interactions (e.g., *Alfvén*, 1981; *Falthämmar*, 1997). It has been pointed out recently (*Falthämmar*, 1997) that active plasmas are characterized by inhomogeneity in space, variability in time, and by strong electric fields, including field-aligned electric fields. Alfvén pointed out (*Alfvén*, 1981) that some types of plasma phenomena occurring in the passive plasma regions, such as wave propagation, should be described by the magnetic field picture. Alfvén realized that the dynamics of active plasmas including FAC filaments, magnetic ropes, perturbed current sheets, inductive transfer of energy, electric double layers, etc., cannot be described by the classical theory and must be described by a current picture.

Alfvén emphasized that a revision of cosmic plasma physics and a revolution in its fundamental concepts are necessary (e.g., *Alfvén*, 1981, 1983). Alfvén warned us (*Alfvén*, 1981) that "Very few cosmic scientists seem to be interested in a critical analysis of the fundamentals of the theory of cosmic plasmas. Once a highly speculative model becomes 'generally accepted,' it easily becomes sacrosanct. The result of all of this is that the theoretical treatment of cosmic plasmas is not in a very healthy state. It seems that a rather drastic and inevitable painful revision of the theory is necessary." Fälthammar (1990) pointed out that "the real magnetosphere is controlled by complex plasma processes that were either neglected or not yet discovered."

Alfvén did not explicitly describe how to revise our classical theories. However, his criticism of the classical theories points out the direction to establish a new theory. Since the beginning of this century, modern physics concepts such as wave-particle dualism have been accepted by the physics society. However, it is not so well recognized how deeply this dualism penetrates into the field of cosmic plasmas (*Alfvén*, 1981). Alfvén emphasized that "The plasma dualism is somewhat analogous to the general particle-field dualism in physics" (*Alfvén*, 1981). A change of paradigm was predicted by Alfvén (e.g., *Alfvén*, 1983). Alfvén did not figure out the physical nature of "the particle aspect" and still discussed the particle-related phenomena in terms of a test particle (page 12, *Alfvén*, 1981). However, what he repeatedly emphasized about the particle aspect was not the description of a single charged particle, but rather an electric current description. He clearly pointed out (page 152, *Alfvén*, 1981) that "Plasma phenomena should be described not only by magnetic field models but also by explicitly accounting for electric currents and the circuits in which they flow." The phenomena occurring in the active plasma region "cannot be understood without explicitly accounting for the current."

Recently, Akasofu (*Akasofu*, 1998) strongly emphasized the necessity and importance of a paradigm transition. He

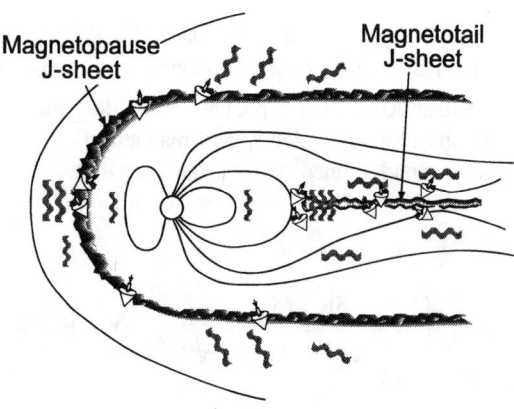

**Figure 2.** schematically describes an active magnetosphere. Overlapping on the convection picture, there is MHD waves propagation and the interaction between the wave packet and the current sheet (J-sheet). Small tops represent the current filaments carried by the shear Alfvén wave packets, which are generated by the interaction between the compressional fast mode wave packets and the current sheets. The arrows of small tops schematically represent the directions of the field aligned current filaments carried by the shear wave packets.

pointed out that "During the period of a powerful paradigm, the progress of its field is actually retarded and sometimes it regresses. ... we have lost about 30 years by pursuing the hypothesis of magnetic reconnection by believing that it is the only theory to explain substorms."

In a recent discussion about paradigm transitions for magnetospheric physics, Parker (1996) emphasized that the macrodynamics of the magnetospheric physics is more effectively treated in terms of the primary MHD variables **B** and **v** (the **B-v** paradigm) than in terms of the dynamics of the electric field **E** and the current **J** (the **E-J** paradigm). Parker directly derived the basic field equations for partially ionized gases from Maxwell's equations and Newton's law, and corrected some misleading but widely accepted concepts. His paper demonstrates how to do physics from first principles using fundamental physical laws. Despite its great contribution in dealing with the macrodynamics, however, Parker did not show an understanding of the urgent need for a paradigm transition in active plasmas. It seems that Parker only emphasized the description of physical process in the passive plasma region. Actually, Maxwell's equations and Newton's laws may not be enough for magnetospheric physics, as pointed out by Alfvén (1977), who noted that "it is just impossible to understand these phenomena" occurring in the active plasma region "by a magnetic formalism as it is to understand the photoelectric effect from Maxwell's equations alone." It seems that Parker did not show his understanding of Alfvén's intuitional picture of active plasmas and the urgent need to bring modern physics concepts into classical plasma physics. Parker's criticisms of the **E-J** paradigm, the current description and the circuit description may lead to more confusion and misunderstanding.

It has been noted (*Song and Lysak*, 1994, 1995, 1997; *Song*, 1998) that MHD wave packet dynamics plays an important role for phenomena occurring in the active plasma region, such as FAC generation and reconnection. To continue to explore and develop the new approach to deal with the physical processes in the active plasma region is an urgent task.

The large-scale evolution of plasmas and fields can be successfully treated by MHD theory, while the microphysics of plasma evolution is well described by plasma kinetic theory. The wave behavior of MHD mesoscale perturbations has been extensively studied, with primary emphasis being on the linear propagation properties and collective instabilities of the localized perturbation. These theories have been successful in explaining a number of phenomena occurring in passive plasma regions. In an active plasma, the nonlinear interaction between different wave packets and between the wave packets and the ambient magnetic field play a key role in the generation of field-aligned current, reconnection, energy conversion and transport processes. In this paper, we will briefly summarize MHD wave packet dynamics, and its application to the theory of reconnection, and the generation of FAC.

## MHD WAVE PACKET DESCRIPTION

Figure 2 shows a schematic picture of the magnetosphere including wave propagation and the interaction between wave packets and current sheets. The fast mode wave excited by an imbalance of forces across the magnetic field carries energy and momentum across magnetic field lines. Before the large scale plasma flows and the associated convection electric field physically arrive at the current sheet, fast mode wave packets will impinge on the current sheet. The interaction between the fast mode wave packet and the current sheet can generate the FAC filaments carried by shear Alfvén waves after localized reconnection (e.g., *Song and Lysak*, 1994; *Song*, 1998). This interaction is not directly related to the large-scale plasma flows and the associated convection electric field.

Hydromagnetic waves can be considered to be an example of electromagnetic waves in a dielectric medium (*Alfvén*, 1942; *Alfvén and Fälthammar*, 1963). The MHD equations can be rewritten in terms of an electromagnetic wave equation in a medium with a perpendicular dielectric constant given by $\varepsilon = 1 + c^2/V_A^2$ (*Alfvén*, 1942). If the photon is the particle corresponding to an electromagnetic wave, what is the particle corresponding to a MHD wave (see, Table 1)?

TABLE 1. Comparisons of the particle aspect
of electromagnetic wave and MHD wave

| Wave aspect | Particle aspect |
|---|---|
| Electromagnetic wave | Photon |
| MHD wave | MHD wave packet? |

The development of a fundamental theory of MHD mesoscale physics in active plasma regions is clearly needed to bridge this gap in our understanding of cosmic plasmas. The structure of MHD wave packets, which have been called "Alfvénons," has been derived from the linearized MHD equations in the cold plasma approximation (*Song and Lysak*, 1994). The fields in these wave packets can be expressed as oscillator variables, which are conserved in the purely linear propagation of the waves and illustrate the structure of the localized wave packet corresponding to that wave mode. The oscillator variables of the shear Alfvénon are the Elsässer variables $\delta \mathbf{Z}^\pm = \sqrt{\rho}\left(\delta\mathbf{u}_\perp \pm (\delta\mathbf{b}_\perp/\sqrt{4\pi\rho})\right)$ (*Elsässer*, 1950) and the curl of these variables $\delta\mathbf{M}^\pm_{sh} = \sqrt{\rho}(\delta\Omega \pm \delta\mathbf{j}')$, where the current is written as $\delta\mathbf{j}' = \delta\mathbf{j}(4\pi/c^2\rho)^{1/2}$ and the vorticity is $\delta\Omega = \nabla \times \delta\mathbf{u}$. These variables define quantities that are conserved in the linear propagation of the wave. This shear mode wave packet consists of loops of magnetic field and velocity field linked with filaments of FAC and vorticity. The FAC filaments are closed by polarization currents flowing perpendicular to the field. Note that the magnetic moment of the shear Alfvénon is zero. For the compressional mode, the radiation oscillators are expressed by $\delta\widetilde{\mathbf{M}}^\pm_{kcom} = (\sqrt{\rho}/k)(\delta\widetilde{j}_k \pm (\sin\theta\delta\widetilde{\gamma}_k + \cos\theta\delta\widetilde{\Omega}_k))\hat{\mathbf{e}}$, where θ is the angle between the wave vector **k** and $\mathbf{B}_0$, $\delta\gamma = \nabla \cdot \delta\mathbf{u}$ is the compression and $\hat{\mathbf{e}}$ is the polarization vector. Topologically, the compressional Alfvénon in real space consists of current and vorticity loops with a compressional flow linked with a dipolar magnetic field and velocity flow. Note that since the compressional Alfvénon is basically a current loop perpendicular to the background magnetic field, the magnetic dipole moment has a non-vanishing component in the $\mathbf{B}_0$ direction. The magnetic moment of the compressional Alfvénon provides the basis for the laws governing its interaction.

Since Alfvénons consist of linked δu, δB, δΩ, δγ and δJ filaments, a topological approach, in addition to an analytical approach, becomes necessary. For example, the magnetic helicity $K = \int \mathbf{A} \cdot \mathbf{B}\, dV$ describes the magnetic field self-linkage (e.g., *Moffatt*, 1969). Conservation of the magnetic helicity (e.g., *Berger*, 1984) implies that some topological structure will remain in a magnetic field after reconnection. It is known (e.g., *Moffatt*, 1969; *Berger and Field*, 1984; *Song*, 1988) that the degree of linkage of two solenoidal fields $\mathbf{P} = \nabla \times \mathbf{Q}$ and $\mathbf{Q}$ can be described by a helicity integral $K_{P,Q} = \int \mathbf{Q} \cdot \mathbf{Q}\, dV$. Since the shear Alfvén wave packet can be described as the linkage of the loops of velocity and magnetic field with the filaments of field-aligned current and vorticity, the total helicity of the shear Alfvén wave packet, $K_{sh}$ can be defined as

$$K_{sh} = \frac{\rho}{2}\int\left(\delta\mathbf{u} \pm \frac{\delta\mathbf{b}}{\sqrt{4\pi\rho}}\right)\cdot\left(\delta\mathbf{u} \pm \frac{\delta\mathbf{b}}{\sqrt{4\pi\rho}}\right) dV = W_T + K_C \quad (1)$$

where $W_T = \int(\rho\delta u^2/2 + \delta b^2/8\pi)dV$ is the total energy and $K_C = \pm\int\sqrt{\rho/4\pi}(\delta\mathbf{u}\cdot\delta\mathbf{b})dV$ is the cross helicity. Thus, $K_{sh}$ describes the linkage between the loops of velocity and magnetic field, described by $\delta\mathbf{Z}^\pm = \sqrt{\rho}\left(\delta\mathbf{u} \pm \left(\delta\mathbf{b}/\sqrt{4\pi\rho}\right)\right)$, and $\delta\mathbf{M}^\pm = \nabla \times \delta\mathbf{Z}^\pm = \sqrt{\rho}\left(\delta\Omega \pm \delta\mathbf{j}'\right)$, which describes the filaments of vorticity and current. The total helicity of the shear Alfvén wave packet $K_{sh}$ is a constant, since $W_T$ and $K_C$ are constants of motion for ideal MHD. The superposition of the background magnetic field and this magnetic loop gives a helical magnetic field, and so the shear Alfvénon carries magnetic helicity. The shear mode wave packet also carries angular momentum and rotational energy.

## WAVE PACKET DYNAMICS AND RECONNECTION

Reconnection is driven by the attractive force between the inertial current filaments produced during the interaction of a fast mode wave packet with the current sheet. Therefore, the threshold for reconnection onset depends on the magnetic shear and the kinetic energy carried by the wave packet (*Song and Lysak*, 1994, 1995).

Figure 3a schematically shows the conversion of the compressional wave packets into the shear wave packets during localized reconnection. Since the shear Alfvénons primarily carry $J_\|$ and have a near zero magnetic moment, they are nearly in a force-free ($\delta\mathbf{j}_\| \times \mathbf{B}_0 \approx 0$) and torque-free ($\delta\mathbf{m} \times \mathbf{B}_0 \approx 0$) state. Since a plasma attempts to relax to a lower energy state while conserving the helicity (e.g., *Berger*, 1984; *Taylor*, 1986), compressional Alfvénons tend to convert into circularly-polarized shear Alfvénons by twisting the magnetic field lines. The twisted magnetic structure radiates away as Alfvén waves carrying FAC filaments. This process can be described in terms of the interaction between magnetic flux tubes (Figure 3b), or the interaction between current loops (Figure 3c). When two magnetic flux tubes having a relative velocity and carrying different magnetic fields interact ($t = t_0$ in Figure 3b), the relative kinetic energy

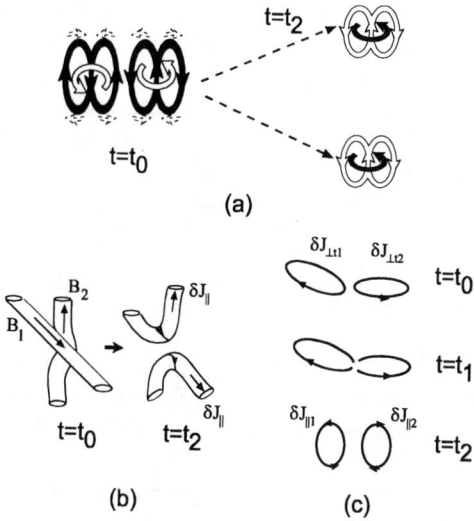

**Figure 3.** The generation of local reconnection-related FAC filaments are schematically described as the results of the interaction of (a) the compressional fast mode wave packets; (b) two magnetic flux tubes and (c) the two current loops. The dark arrows in (a) represent the velocity and magnetic perturbation, while the light arrows in (a) represent current and vorticity.

is converted into electromagnetic energy in the form of the current surrounding the two flux tubes. The current sheet becomes locally thin and intense. Localized and time-dependent driven reconnection can occur if the threshold conditions are satisfied. The flux tubes with one crossing at $t = t_0$ shown in Figure 3b have a relative mutual helicity $K_\ell = \pm \Phi^2$, where $\Phi$ is the reconnected magnetic flux and the sign is determined by the right hand rule (e.g., *Berger and Field*, 1984). By helicity conservation, each tube has a half twist after reconnection at $t = t_2$ (Figure 3a) (e.g., *Berger and Field*, 1984; *Song and Lysak*, 1989a; *Wright and Berger*, 1989). Figure 3c shows that FAC filaments can be irreversibly produced at $t = t_2$ due to the interaction of two current loops.

During reconnection, the scale size for the small diffusion region between closed neighboring loops of magnetic field $\delta \mathbf{B}$ is denoted as "$\delta_S$." The Ohm's law for the small diffusion region is then $\delta_S \mathbf{E} = \eta^* \delta_S \mathbf{J}$. The anomalous diffusivity $\eta^*$ can be caused by non-MHD effects where the current density is locally very high, for example, by the electron inertial effect. This fast diffusion serves only to release the topological constraints and to allow the interaction of the wave packets and the radiation of shear Alfvén waves to occur. The large scale, time-averaged electric field depends on the rate of entry of the time-averaged magnetic flux $\langle \Phi_B \rangle$ into the interaction region per unit length and not on the details of the anomalous diffusion.

During the conversion of a compressional wave packet into a shear wave packet, the input electromagnetic energy is carried away from the reconnection region by shear Alfvén waves. This result can be interpreted in terms of quanta of radiation. Poynting's theorem for harmonic fields (*Jackson*, 1975) can be used to discuss the "radiation resistance" related to the Alfvén wave radiation during reconnection. This radiation resistance is an impedance in nature and represents a current dynamo effect (*Song*, 1998). The voltage is balanced by the product of the current $I$ and an inductive impedance $Z$ given by the intermittent generation of the current filaments. In this case, there is a net electric field corresponding to the average entry of the intermittent magnetic flux (e.g., *Song and Lysak*, 1989b, 1995). The reconnection rate depends on the radiation resistivity as well as the global re-configuration.

In this model, magnetic reconnection is a wave-packet related phenomenon, and thus has a local and time-dependent nature. The onset of reconnection is associated with the travel time of the fast mode wave packet, not the convection time scale. Reconnection can occur in multiple regions, resulting in localized and transient magnetic field structures. Reconnection can occur in a closed field topology even without large-scale convection. Its occurrence is directly associated with a strong current sheet and fast mode wave packets carrying a large amount of kinetic energy, but not necessarily with an antiparallel magnetic field topology and convection electric field.

Reconnection causes the time-dependent local erosion of the current sheet. In general, the time-dependent erosion and collapse of the current sheet cannot be described by a neutral line. The magnetic and mechanical forces and torques acting on the eroded current sheet can cause a large-scale convection-related reconfiguration. Local reconnection, the erosion of the current sheet, and the reconfiguration of the magnetofluid are three distinct physical processes in space and time, although they are mutually correlated. Characteristics of the plasma flow and the magnetic structure after reconnection differ depending on the total force and torque acting on the tubes.

Magnetic reconnection is no more than a necessary condition that releases the topological constraints and allows a fast mode wave packet to convert into a shear Alfven wave packet. Reconnection is driven by the attractive force of inertial current filaments, which are produced by the interaction of compressional wave packets with the current sheet. The interaction of the MHD wave packets is more essential to the evolution of the plasma than reconnection.

In summary, reconnection is a current dynamo process that is related to the generation of FAC filaments and radiation of the shear Alfvén waves. The reconnection rate depends on the radiation resistivity, i.e., the energy flux of rotational energy carried by shear Alfvén waves.

## WAVE PACKET DYNAMICS AND FIELD-ALIGNED CURRENT GENERATION

A discussion of the generation of FAC must include a discussion of the dynamical processes that produce the FAC in time, thus an expression for $\partial J_\parallel/\partial t$ is necessary. However, previous theories of FAC generation (e.g., *Hasegawa and Sato*, 1979; *Sato and Iijima*, 1979; *Vasyliunas*, 1984), which are only based on force balance, describe a passive and quasi-steady process, where the field-aligned gradient of the FAC, $\partial(J_\parallel/B)/\partial s$, is evaluated. The field-aligned gradient of the FAC is usually evaluated using the current continuity condition, $B\partial(J_\parallel/B)/\partial s = -\nabla_\perp \cdot \mathbf{J}_\perp$, where $\mathbf{J}_\perp = (c\mathbf{B}\times\nabla P + c\mathbf{B}\times\rho\, d\mathbf{u}/dt + \rho\nu\nabla^2\mathbf{u}\times\mathbf{B})/B^2$. This quasi-steady picture implies that the rate of the energy input to drive the large-scale convection is nearly equal to the rate of the energy dissipation in the ionosphere. In this case, the electric field remains constant in the generator and the dissipation in the load controls the current generation. It must be noted that the force balance argument alone is not sufficient for a complete theory of FAC generation. This model does not address a number of key questions. First, the energy supply for the FAC generation is not clearly given in the previous theory. Although the diamagnetic current and the inertial current appear equally in the force balance equation, we will point out below that the inertial current, but not the diamagnetic current, can provide an energy source for the FAC generation. Secondly, a non-zero $\nabla_\perp \cdot \mathbf{J}_\perp$ is related to a vortex motion; however, the mechanisms to drag the large-scale vortex convection are not described by the previous model. We will point out later that the compression-related perpendicular current satisfies $\nabla_\perp \cdot \mathbf{J}_\perp = 0$. A dynamical process is needed to generate FAC from a divergence-free perpendicular current.

The dynamical equation for current generation is given by the curl of the magnetic induction equation

$$\frac{\partial \mathbf{J}'}{\partial t} = -\frac{c^2}{4\pi}\nabla\times(\nabla\times\mathbf{E}). \quad (2)$$

where $\mathbf{J}' = \mathbf{J} + (1/4\pi)(\partial\mathbf{E}/\partial t)$. If $c^2/V_A^2 \gg 1$, the perpendicular displacement current $(1/4\pi)(\partial\mathbf{E}_\perp/\partial t)$ can be ignored. Thus, $\mathbf{J}'_\perp \approx \mathbf{J}_\perp$ and the perpendicular diamagnetic constant $\varepsilon = 1 + c^2/V_A^2 \approx c^2/V_A^2$. If $\omega^2/\omega_{pe}^2 \ll 1$, the parallel displacement current $(1/4\pi)(\partial\mathbf{E}_\parallel/\partial t)$ can be ignored, thus, $J'_\parallel \approx J_\parallel$. The electric field is related to the perpendicular motion of the plasma through Ohm's law $\mathbf{E} \approx -1/c(\mathbf{u}\times\mathbf{B}) + (4\pi/\omega_{pe}^2)(\partial\mathbf{J}/\partial t)$. In the parallel direction, $\mathbf{E}_\parallel \approx (4\pi/\omega_{pe}^2)(\partial\mathbf{J}_\parallel/\partial t)$, which shows that changes in the FAC are always accompanied by a parallel electric field. In the perpendicular direction, if the spatial scale size is larger than electron inertial length, the electron inertial term can be ignored and the electric field is given by the frozen-in MHD condition $\mathbf{E}_\perp \approx -1/c(\mathbf{u}\times\mathbf{B})$. In this case, the relationship between the current generation and plasma motion can be re-written as

$$\frac{\partial \mathbf{J}'}{\partial t} = -\frac{c}{4\pi}(\nabla\times(\gamma\mathbf{B} - \mathbf{\Omega}\times\mathbf{B} + (\mathbf{u}\cdot\nabla)\mathbf{B}). \quad (3)$$

where $\mathbf{\Omega} = \nabla\times\mathbf{u}$ is the vorticity and $\gamma = \nabla\cdot\mathbf{u}$ is the compression. The change of the vorticity and compression are given by taking the curl and gradient of MHD momentum equation $\rho\, d\mathbf{u}/dt = -\nabla P + (1/c)\mathbf{J}\times\mathbf{B}$, respectively. The change of the pressure gradient can be derived from mass conservation $\partial\rho/\partial t + \nabla\cdot(\rho\mathbf{u}) = 0$ and the adiabatic assumption, e.g., $P = c\rho^\gamma$. The relationship between the perpendicular current and field-aligned current can be obtained from the nature of current continuity, i.e., $\partial\rho_e/\partial t + \nabla\cdot\mathbf{J} = \nabla\cdot\mathbf{J}' = 0$. These equations provide the basis for a further discussion of the FAC generation.

The three terms on the right hand side of equation (3) indicate that the current generation comes directly from the compression, twisting, or bending of magnetic field lines (*Song*, 1998). During the braking of a moving plasma ($\mathbf{u}\cdot(\rho d\mathbf{u}/dt) < 0$) by the Lorentz force and the pressure gradient force, current can be generated ($\mathbf{J}\cdot\mathbf{E} < 0$) and plasma can be compressed ($\mathbf{u}\cdot(-\nabla P) < 0$). The kinetic energy provided by slowing down the plasma is converted into electromagnetic energy in the form of current, as well as into thermal energy. If the FAC associated with the divergence of the diamagnetic current is connected to a load, the pressure gradient producing the diamagnetic current will relax, reducing the FAC, unless the pressure gradient is maintained by plasma flow. The main energy source in the magnetosphere is not internal energy but the kinetic energy and magnetic energy existing in the relatively moving plasmas. Hence it is the inertial current, it is not the diamagnetic current, that is responsible for FAC generation. To discuss FAC generation is to describe the dynamical processes by which the plasma interaction produces FAC. Current disruption by anomalous resistivity, instabilities or any other mechanism may be a necessary condition, but it does not give a mechanism for FAC generation.

During the plasma interaction, the conversion of the relative kinetic energy into electromagnetic energy is achieved mainly by the braking of a moving plasma. Thus, current generation is mainly caused by plasma compression. The interaction between a compressional fast mode wave packet with the current sheet thus provides an important source for the generation of FAC filaments (e.g., *Song and Lysak*, 1994, 1995; *Song*, 1998). The dynamic equations for current filaments carried by wave packets have been given by *Song and Lysak* (1994), where the nonlinear interaction has been discussed in terms of wave packet dynamics. We will use a simplified picture to explain the relationship between the wave packets and the current filaments.

Since the Lorentz force in the MHD momentum equation is only related to the perpendicular current $\mathbf{J}_\perp$, the MHD momentum equation gives only the relationship between the $\mathbf{J}_\perp$ and the inertial force, i.e.,

$$\mathbf{J}_{\perp in} = \frac{c^2}{4\pi V_A^2} \frac{d(-(1/c)\mathbf{u}\times\mathbf{B})}{dt} \approx \frac{c^2}{4\pi V_A^2} \frac{d\mathbf{E}_\perp}{dt} \quad (4)$$

where $\mathbf{E}_\perp \approx -1/c\,(\mathbf{u}\times\mathbf{B})$ has been assumed. Equation (2) and (4) describe the simultaneous changing of the inertial current and the electric field. Since the compressional mode has a divergence-free perpendicular electric field while the shear Alfvén mode is characterized by having an electric field whose curl has a zero component parallel to the background magnetic field (e.g., *Lysak*, 1997), we denote

$$q_E = \frac{\nabla_\perp \cdot \mathbf{E}_\perp}{4\pi} \approx -\frac{B}{4\pi c}\Omega_\parallel \quad (5)$$

$$q_B = \frac{(\nabla_\perp \times \mathbf{E}_\perp)\cdot \hat{\mathbf{z}}}{4\pi} \approx \frac{B}{4\pi c}\gamma \quad (6)$$

to describe the shear and compressional modes respectively, where $\Omega_\parallel = (\nabla \times \mathbf{u})_\parallel$ and $\gamma = \nabla \cdot \mathbf{u}$. For simplicity, $\mathbf{u}_\perp \cdot \mathbf{J}_\perp \approx 0$ and $(\mathbf{u}\cdot\nabla)\mathbf{B} \approx 0$ are assumed in the final forms in (5) and (6). Taking the perpendicular divergence of equations (2) and (4), the relationship between the FAC and vortex motion is

$$\frac{\partial \Omega_\parallel}{\partial t} \approx -\frac{4\pi c}{B}\frac{\partial q_E}{\partial t} = \frac{4\pi c}{\varepsilon B}\frac{\partial J'_\parallel}{\partial z} \quad (7)$$

$$\frac{\partial J'_\parallel}{\partial t} = -c^2 \frac{\partial q_E}{\partial z} \approx \frac{cB}{4\pi}\frac{\partial \Omega_\parallel}{\partial z} \quad (8)$$

where $\partial J'_\parallel/\partial z = -\nabla_\perp \cdot \mathbf{J}_\perp$ is used. Taking the perpendicular curl of equations (2) and (4), the relationship between the FAC and compression is

$$\frac{\partial \gamma}{\partial t} \approx \frac{4\pi c}{B}\frac{\partial q_B}{\partial t} = \frac{4\pi c}{\varepsilon B}(\nabla_\perp \times \mathbf{J}'_\perp)\cdot \hat{\mathbf{z}} \quad (9)$$

$$\frac{\partial (\nabla_\perp \times \mathbf{J}'_\perp)\cdot \hat{\mathbf{z}}}{\partial t} = c^2 \frac{\partial q_B}{\partial z} \approx \frac{cB}{4\pi}\frac{\partial \gamma}{\partial z} \quad (10)$$

Equations (7) and (8) indicate that the generation of FAC filaments is related to the field-aligned vorticity in the shear Alfvén mode. Equations (9) and (10) indicate that compression is related to the generation of a solenoidal current in the perpendicular direction in the compressional wave mode. The solenoidal perpendicular current generated by compression can not converted into FAC unless reconnection occurs. The generation of FAC filaments carried by the shear Alfvén wave caused by the mode conversion has been described in the last section. From a microphysics viewpoint, non-MHD effects, such as the electron inertial term and displacement current discussed earlier, break the frozen-in condition locally and time-dependently during the mode conversion in the interaction region. However, microphysics alone cannot describe the non-linear mode conversion. The twist of the flux tube during the interaction between the compressional fast mode wave packet and the current sheet is caused by the torque of Lorentz force and mechanical force; thus, the polarity of FAC filaments depends on, for example, the IMF By at the dayside magnetopause.

The generation of FAC filaments by the wave mode conversion provides a main source of the viscous and magnetic drag, which causes an enhanced large-scale convection producing the large-scale discharge FAC. Thus, the irreversibly generated FAC filaments overlap on the topologically steady FAC to form the total FAC. Since the FAC generation is related to the changes of the electric and magnetic fields, and the perturbations of the fields must propagate, the FAC generation is in general not a local process, the analysis of the whole flux tube or circuit is necessary. The shear Alfvén waves reflected between the generator, the auroral acceleration region and the load will change the generator condition and modify the resultant FAC filaments (e.g., *Lysak and Song*, 1998). The convection related FAC generation depends on the mechanical and electromagnetic drag from both the generator and the ionosphere.

## DISCUSSION

Alfvén pointed out (*Alfvén*, 1981) that "Since the beginning of this century, physics has been dualistic in the sense that some phenomena are described by a field formalism while others are treated in terms of particles." He emphasized that "it is not so well recognized how deeply this dualism penetrates into the field of cosmic plasmas." Unfortunately, facing the end of this century, our common theoretical treatment on some important dynamical processes in the magnetosphere, such as FAC generation, magnetic reconnection and auroral particle acceleration, are still based on the classical concepts. MHD mesoscale and time-dependent physics, in particular, the effect of the particle aspect of MHD waves, have not been properly included in discussing those processes.

A paradigm transition from a classical description to a description including the modern physics concept of wave-particle duality is necessary. A revision of the theories of the FAC generation and reconnection using wave packet dynamics is an important task. This paradigm transition may also benefit the research in other physical processes occurring in active plasma regions, such as the planetary dynamo problem and the auroral particle acceleration.

*Acknowledgments.* We have benefited from discussions with many of our colleagues, including D-H. Lee, N. Lin, T. Lui, E. N. Parker, T.-D. Phan, B. U. O. Sonnerup, W. Sun. This work has been supported in part by NSF Grants ATM-9301043, ATM-9502907 and by NASA grant NAG5-4466.

## REFERENCES

Akasofu, S.-I., The rise and fall of paradigms and some long-standing unsolved problems in solar-terrestrial physics, p. 21, *Substorm-4*, Edited by S. Kokubun and Y. Kamide, Terra Scientific Publishing Company / Kluwer Academic Publishers 1998.

Alfvén, H., Existence of electromagnetic-hydrodynamic waves, *Nature*, 3805, 405, 1942.

Alfvén, H., Electric current in cosmic plasmas, *Rev. Geophys. Space Phys.*, 15, 271, 1977.

Alfvén, H., *Cosmic plasma*, D. Reidel Publishing company, Dordrecht, 1981.

Alfvén, H., Paradigm transition in cosmic plasma physics, *Geophys. Res. Lett.*, 10, 487, 1983.

Alfvén, H., The plasma universe, *Physics Today*, 39, 22, 1986.

Alfvén, H. and Fälthammar, C.-G., *Cosmical Electrodynamics*, Int. Series of Monographs on Physics. Clarendon Press, Oxford, 1963.

Berger, M. A., Rigorous new limits on magnetic helicity dissipation in the solar corona, *Geophys. and Astrophys. Fluid Dynamics*, 30, 79, 1984.

Berger, M. A., and G. B. Field, The topological properties of magnetic helicity, *J. Fluid Mech.*, 147, 133, 1984.

Dungey, J. W., Interplanetary magnetic field and the auroral zones, *Phys. Rev. Lett*, 6, 47, 1961.

Elsässer, W. M., The hydromagnetic equations, *Phys. Rev.*, 79, 183, 1950.

Falthammar, Carl-Gunne, The plasma universe, *Basic plasma processes on the sun*, E. R. Priest and V. Krishan (eds.), 9-20, 1990.

Falthammar, Carl-Gunne, The heritage of Hannes Alfvén, *Phys. Chem. Earth*, 22, 599, 1997.

Haerendel, G., G. Paschmann, N. Sckopke, H. Rosenbauer and P. C. Hedgecock, The front side boundary layer of the magnetosphere and the problem of reconnection, *J. Geophys. Res.*, 83, 3195, 1978.

Hasegawa, A., and T. Sato., Generation of field-aligned current during substorm, *Dynamics of the magnetosphere*, S.-I. Akasofu (ed.), Reidel, Dordrecht-Holland, p. 529, 1979.

Jackson, J. D., *Classical electrodynamics*, John Wiley & Sons, Inc., 1975.

Kennel, C. F., The Kiruna conjecture: the strong version, *Proceeding of the ICS-1*, Kiruna, Sweden, ESP SP-335, p. 599, 23, 1992.

Lui, A. T. Y., A synthesis of magnetospheric substrom models, *J. Geophys. Res.*, 96, 1849, 1991.

Lysak, R. L., Propagation of Alfvén waves through the ionosphere, *Phys. Chem. Earth*, 22, 757, 1997.

Lysak, R. L. and Y. Song, Dynamics of auroral arc formation during substorms, Substroms-4, S. Kokubun and Y. Kamide (eds.), Terra Scientific Publishing Company, Tokyo, p.35, 1998.

Moffatt, H. K., The degree of knottedness of tangled vortex lines, *J. Fluid Mech.*, 35, 117, 1969.

Parker, E. N., *Cosmical magnetic fields*, Oxford University Press, 1979.

Parker, E.N., "Magnetic Fields in the Cosmos," *Scientific American*, 249 (August 1983) 44-54, p. 52, 1983.

Parker, E. N., The alternative paradigm for magnetospheric physics, *J. Geophys. Res.*, 101, 10587, 1996.

Paschmann, G., I. Papamastorakis, W. Baumjohann, N. Sckopke, C. W. Carlson, B. U. Ö. Sonnerup, and H. Lühr, The magnetopause for large magnetic shear: AMPTE/IRM observations, *J. Geophys. Res.*, 91, 11,099, 1986.

Phan, T. -D., G. Paschmann, and B. U. Ö. Sonnerup, Low-latitude dayside magnetopause and boundary layer for high magnetic shear, 2. Occurrence of magnetic reconnection, *J. Geophys. Res.*, 101, 7817, 1996.

Russell, C. T., and R. C. Elphic, Initial ISEE magnetometer results: magnetopause observations, *Space Sci. Rev.*, 22, 681, 1978.

Sato, T., and T. Iijima, Primary source of large-scale Birkeland currents, *Space Sci. Res.*, 24, 347, 1979.

Sato, T., Aurora physics, *Magnetospheric Plasma Physics* ed. by Atshkiro Nishida, Center for Academic Publications, Japan/Tokyo, D. Reidel Publishing Company, p. 197, 1982.

Scholer, M., B. Klecker, and D. Hovestadt, Energetic particle characteristics of magnetotail flux ropes, *Geophys. Res. Lett.*, 12, 191, 1985.

Song, Y., Self-Organization Process and Anomalous Transport in the Magnetosphere and Ionosphere, Ph. D. thesis, University of Minnesota, 1988.

Song, Y., Theoretical constraints on mechanisms for the substorm current wedge, Substroms-4, S. Kokubun and Y. Kamide (eds.), Terra Scientific Publishing Company, Tokyo, p.543, 1998.

Song, Y., and R. L. Lysak, Evaluation of twist helicity in FTE flux tubes, *J. Geophys. Res.*, 94, 5273, 1989a.

Song, Y. and R. L. Lysak, Dynamo effect of 3-d time-dependent reconnection in the dayside magnetopause, *Geophys. Res. Lett.*, 16, 911, 1989b.

Song, Y. and R. L. Lysak, Solar-wind/magnetospheric dynamos: MHD scale collective entry of the solar wind energy, momentum and mass into the magnetosphere, *Substorms 1*, European Space Agency, Paris, p. 149, Kiruna, Sweden, 1992.

Song, Y., and R. L. Lysak, Alfvénon, driven reconnection and the direct generation of the field-aligned current, *Geophys. Res. Lett.*, 21, 1755, 1994.

Song, Y., and R. L. Lysak, MHD mesoscale interactions at the magnetopause and driven reconnection, in *Physics of the Magnetopause*, P. Song, B. U. Ö. Sonnerup and M. F. Thomsen (eds.), AGU Monograph 90, AGU, Washington, p. 349, 1995.

Song, Y., and R. L. Lysak, Some theoretical aspects of the solar wind-magnetospheric interaction, *Phys. Chem. Earth*, 22, 715, 1997.

Sonnerup, B. U. Ö., Magnetic field reconnection at the magnetopause: an overview, *Magnetic Reconnection*, E. W. Hones, Jr. (ed.), AGU Monograph 30, American Geophysical Union, Washington, p. 92, 1984.

Taylor, J. B., Relaxation and magnetic reconnection in plasmas, *Rev. Mod. Phys., 58*, 745, 1986.

Vasyliunas, V. M., Fundamentals of current description, *Magnetospheric currents*, T. A. Potemra (ed.), AGU Monograph, Washington, p. 63, 1984.

Wright, A. N., and M. A. Berger, The effect of reconnection upon the linkage and interior structure of magnetic flux tubes, *J. Geophys. Res., 94,* 1295, 1989.

---

Yan Song, and Robert L. Lysak, School of Physics and Astronomy, University of Minnesota, 116 Church St. S.E., Minneapolis, MN 55455

# FAST Observations of Electromagnetic Stresses Applied to the Polar Ionosphere

R. J. Strangeway

*Institute of Geophysics and Planetary Physics, University of California at Los Angeles, Los Angeles, California*

R. C. Elphic

*Los Alamos National Laboratory, Los Alamos, New Mexico*

W. J. Peria and C. W. Carlson

*Space Sciences Laboratory, University of California at Berkeley, Berkeley, California*

The Fast Auroral Snapshot Explorer (FAST), with its 83° inclination orbit and 4000 km apogee, is ideally suited for investigation of the high latitude perturbations to the geomagnetic field. These data can be used to determine field-aligned currents, but here we emphasize the perturbations themselves, rather than their spatial gradient. This allows us to more readily visualize the forces applied to the ionosphere by the magnetosphere (and vice-versa). Our basic framework for interpreting the magnetic field perturbations is one in which flows in the magnetosphere and at the magnetopause apply stresses to the ionosphere where the imposed flows must overcome the collisional drag. Thus field-aligned currents flow in response to a requirement for an ionospheric $\mathbf{J} \times \mathbf{B}$ force to overcome the drag. We will interpret two intervals of polar data acquired by FAST in this framework, showing how the overall structure of the field perturbations can be understood in terms of applied stresses. We discuss briefly one implication of this approach, that the ionosphere may be important in braking substorm-related flow bursts.

## 1. INTRODUCTION

It has long been known that Field-Aligned Currents (FACs) flow into and out of the polar ionosphere. The earliest observations of low altitude field-aligned currents were by *Zmuda et al.* [1966], who reported transverse magnetic disturbances at 1100 km altitude, as measured by satellite 1963-38C. Although Zmuda et al. originally attributed the disturbances to hydromagnetic waves, *Cummings and Dessler* [1967] presented a convincing argument that the disturbances could best be attributed to field-aligned currents. Indeed in later studies using Triad magnetometer data, *Armstrong and Zmuda* [1973] and *Zmuda and Armstrong* [1974a,b] discussed the magnetic perturbations almost entirely in terms of field-aligned currents.

*Zmuda and Armstrong* [1974b] also organized their observations of FACs, showing a characteristic and now familiar distribution of currents, where the currents lie in two concentric circles roughly collocated with the auroral oval. *Iijima and Potemra* [1976] named these currents Region 1 and Region 2. Region 1 currents flow into the ionosphere on the dawnside of the high latitude auroral oval, and out on the dusk side. Region 2 currents flow in the opposite sense at lower latitudes. *Iijima and Potemra* [1976] and *Sugiura and Potemra* [1976] both pointed out that there need not be local closure of the field-aligned currents. *Iijima and Potemra* [1976] noted that Region 1 currents tended to be larger than region 2 currents, while *Sugiura and Potemra* [1976] noted a "steplike level shift."

The presence of a stepwise change in the transverse field on crossing the auroral oval can be interpreted in terms of electromagnetic stress applied to the polar ionosphere, presumably by some form of high altitude generator, either at the magnetopause or in the equatorial magnetosphere. Viewing transverse perturbations of the magnetic field in terms of stress, rather than in terms of field-aligned currents, is a theme also taken up by *Elphic et al.* [this issue], and reflects the arguments set forth by *Parker* [1996]. Note, however, that the use of the "B, v paradigm" does not preclude consideration of "E, j", but rather it lets us use a framework for determining the currents and electric fields that exist as a consequence of the applied flows and stresses.

In the next section we will present a simple cartoon relating high altitude magnetospheric and magnetopause drivers to the ionospheric response. We will then present examples of magnetometer data from the Fast Auroral Snapshot (FAST) explorer, emphasizing the interpretation of the observed signatures in terms of applied stress. In the concluding section we will summarize our analysis, and also address some comments to the role of ionospheric drag as a mechanism for flow braking within the inner magnetosphere.

## 2. IONOSPHERIC RESPONSE TO APPLIED STRESS

Figure 1 presents a simple cartoon of the ionospheric response to a driver at high altitudes. In this case we are assuming that the driver is a region of enhanced flux transport at the magnetopause. Such enhanced transport could arise from localized reconnection. Near the reconnection site the magnetopause flow is accelerated by the $\mathbf{J} \times \mathbf{B}$ "slingshot", but further downstream the reconnected flux tubes which thread the magnetopause will be transported by the magnetosheath flow, and it is this enhanced flux transport which ultimately drives field-aligned currents.

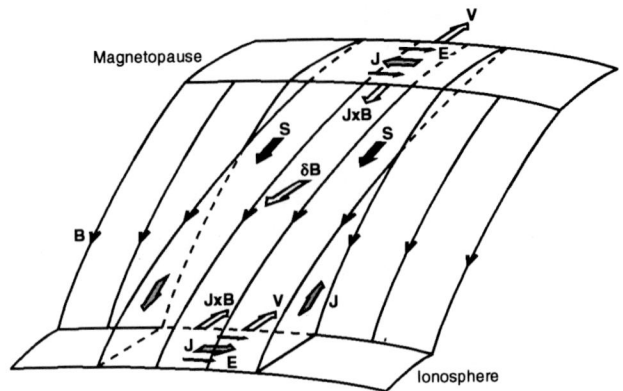

**Figure 1.** Cartoon showing the relationship between the applied flows at the magnetopause and the resultant stresses and flows at the ionosphere. Although drawn for the case of magnetopause flows, the sketch can be applied to magnetospheric flows. In the latter case the field lines should be traced to the equatorial region, and the high altitude flow would be earthward.

We note that this picture could as easily apply to processes such as substorms, where flow bursts, which are also regions of enhanced flux transport, appear to drive field-aligned currents into the ionosphere [*Shiokawa et al.* 1998]. In this case the high altitude region in Figure 1 should be mapped to the equator, and the flow would be directed earthwards.

At the high altitude end of the field lines in Figure 1 a region of enhanced tailward flow (**V**) carries field lines downstream (into the page in the figure perspective). The ambient magnetic field (**B**) has a normal component through the top surface (the magnetopause), which results in a convection electric field ($\mathbf{E} = -\mathbf{V} \times \mathbf{B}$). Furthermore, the downstream flow acts to stretch the field lines, giving a $\delta \mathbf{B}$ in the upstream direction. The current (**J**) associated with this $\delta \mathbf{B}$ opposes the convection **E**, and the magnetopause current layer is a generator, i.e., a source of electromagnetic energy. In addition, the $\mathbf{J} \times \mathbf{B}$ force in the layer opposes the flow. This is the drag on the flow due to the increased tension in the stretched field lines.

At this stage, we have not yet stated where the drag on the flow comes from. Increasing the tension along the field requires that at some location the convection of the field lines is retarded. In Figure 1 this is the ionosphere. For the ionosphere to act as a drag on the magnetopause flow there must be communication between the two regions. This communication can be thought to occur via either Poynting flux ($\mathbf{S} = \mathbf{E} \times \delta \mathbf{B}/\mu_0$) or the field-aligned currents that current continuity requires at the edges of the high flow region.

At the ionospheric end of the field lines the field-aligned currents close via a horizontal current. The resultant ionospheric **J**×**B** force will accelerate the ionospheric plasma, and we can consider the field lines to be pulling the plasma in the direction of the magnetopause flow. This motion will be retarded by the collisional drag of the ionosphere, as ionospheric ions are forced to move through the neutral atmosphere. For collisions to act as a drag, the flow must of course be in the same direction as the **J**×**B** force. Thus the motional electric field is in the same direction as **J** and the ionosphere is a load with **J**·**E** > 0.

Before considering the further ramifications of Figure 1, we should point out that this viewpoint is not particularly new. Indeed, *Coroniti and Kennel* [1973] used this methodology to discuss the role of ionospheric conductivity in controlling the rate of dayside magnetopause erosion. Similarly, *Cowley* [1981] discussed the effects of interplanetary magnetic field (IMF) $B_y$ on polar cap flows and field-aligned currents in terms of applied stresses. More recently, *Wright* [1996] used this framework to discuss the transfer of energy and momentum from the magnetosheath, but he also included the effects of finite Alfvén velocity. We have ignored this here, but is clearly important in establishing an equilibrium between the ionosphere and the magnetospheric and magnetopause drivers. Thus Figure 1 should be viewed as a framework for discussing the coupling between the ionosphere and the magnetosphere, and indeed we might expect many auroral and polar cap phenomena to be the signature of the negotiation that occurs as ionosphere and magnetosphere attempt to come to equilibrium.

Bearing in mind these limitations, we can nevertheless derive some useful scaling laws. First, mapping of the convection electric field requires

$$V_I B_0 L_I = f V_m B_n L_m \qquad (1)$$

where $V_I$ is the ionospheric flow velocity, $B_0$ is the vertical ionospheric magnetic field, $L_I$ is the ionospheric transverse scale-length (in the direction of the horizontal current), $V_m$ is the magnetopause flow velocity, $B_n$ is the normal component of the magnetic field, and $L_m$ is the magnetopause transverse scale-length. The factor $f$ takes into account the possibility of imperfect mapping of the electric field, with $0 \leq f \leq 1$. When $f = 1$ there is perfect mapping of the magnetopause convection electric field to the ionosphere. When $f \neq 1$ a parallel electric field is present which allows the ionosphere to decouple from the magnetosphere, and the ionosphere slips with respect to the magnetosphere.

The transverse scale-lengths in (1) are largely set by field line mapping.

Current continuity requires

$$J_I / B_0 L_I = J_m / B_n L_m \qquad (2)$$

where $J_I$ and $J_m$ are current intensities.

For a current sheet, the magnetic field perturbation is given by $\delta B = \mu_0 J$ and

$$\delta B_I / B_0 L_I = \delta B_m / B_n L_m \qquad (3)$$

Combining (1) and (3) we find

$$E_I \delta B_I / B_0 = f E_m \delta B_m / B_n \qquad (4)$$

where $E_I$ and $E_m$ are the convection electric fields. Equation (4) states that the Poynting flux into the ionosphere equals that fraction of the Poynting flux from the magnetosphere that is not dissipated by parallel electric fields, with the ratio $B_n / B_0$ taking into account the change in flux tube area.

Last,

$$J_I = \Sigma_p V_I B_0 \qquad (5)$$

where $\Sigma_p$ is the height-integrated Pedersen conductivity.

It should be noted that equations (1) and (5) both assume that $E_I = V_I B_0$. This implicitly assumes that although the ions do collide with neutrals, $v_{in} \ll \Omega_i$, where $v_{in}$ is the ion-neutral collision frequency and $\Omega_i$ is the ion gyrofrequency. In this case the ions and electrons move with nearly identical bulk velocities, with the difference in their velocities giving the current.

When $v_{in} \ll \Omega_i$ the strong equivalence between drag and conductivity is readily apparent. In steady state the force law shows that

$$n m_i V_I v_{in} = j_I B_0 \qquad (6)$$

where $n$ is the density, $m_i$ is the ion mass, and $j_I$ is the current density. Equation (6) is simply a statement that the momentum lost by the ions through collisions with neutrals is balanced by the **j**×**B** force. In making this statement we have implicitly assumed that $V_I$ is the velocity of the ions with respect to the neutral gas, and further that the neutrals are a drag on the ion flow. This need not always be the case. For example, *Kelley* [1989], in his Chapter 7, discusses acceleration of neutrals by the **j**×**B** force. He shows that on occasion the neutrals can flow at the same velocity as the ions, under conditions of extremely steady convection lasting for several hours. *Deng et al.* [1991] have also examined the flywheel effect, where the neutrals can drive convection in the polar cap, although the magnetosphere-imposed convection is necessarily weak for this to occur.

Notwithstanding the ability of the neutrals to sometimes drive ionospheric flows, equation (6) points out that it is the ionospheric drag which determines the size of the current: the larger the drag force the larger the perpendicular current. This then leads to the equivalence between drag and conductivity. On replacing $V_I$ with $E_I/B_0$, which assumes we can neglect the Hall term in the Ohm's law, we find

$$nm_i E_I v_{in}/B_0^2 = j_I \qquad (7)$$

Therefore

$$\sigma_p = ne^2 v_{in}/m_i\Omega_i^2 \qquad (8)$$

where $\sigma_p$ is the Pedersen conductivity. Equation (8) gives the standard form of $\sigma_p$ for $v_{in} \ll \Omega_i$.

We can combine equations (1), (3), and (5) to derive some useful scaling laws that relate ionospheric and magnetospheric parameters. From (1) and (5), since $\delta B_I = \mu_0 J_I$,

$$\delta B_I = f\left(\frac{L_m}{L_I}\right)\mu_0\Sigma_p V_m B_n \qquad (9)$$

while from (3) and (5)

$$V_I = \left(\frac{L_I}{L_M}\right)\frac{1}{\mu_0\Sigma_p}\frac{\delta B_m}{B_n} \qquad (10)$$

and from all three

$$\delta B_m = f\left(\frac{L_m}{L_I}\right)^2\left(\frac{B_n}{B_0}\right)\mu_0\Sigma_p V_m B_n \qquad (11)$$

Equation (9) states that the stress applied to the ionosphere increases for increasing ionospheric conductivity, but decreases if a parallel electric field is present ($f < 1$). Equation (10) states that for a given magnetic shear at the magnetopause, increasing the conductivity reduces the convection velocity in the ionosphere. Both of these statements are another way of saying that a highly conducting ionosphere acts as a drag on the higher altitude flows, and the forces required to move the ionosphere against this drag are larger.

Equation (11) relates the magnetopause stress to the magnetopause flow. Again, higher ionospheric conductivity results in higher stress. The stress is reduced if parallel electric fields are present.

Before discussing the consequences of these scaling laws it is worthwhile to determine if they provide reasonable estimates for the fields and flows. We shall assume that the flow velocity at the magnetopause ($V_m$) is 100 km/s, the normal component of the magnetic field ($B_n$) is 5 nT, and $\Sigma_p = 10$ S. For the purposes of estimating transverse scale lengths we will assume $L_m/L_I \approx (B_0/B_n)^{1/2}$, and $B_0 = 50,000$ nT, i.e., $L_m/L_I = 100$. From (9), assuming no slippage, $\delta B_I = 630$ nT. From (11), $\delta B_m = 6.3$ nT, while from (10), $V_I = 1$ km/s. That $V_m/V_I = \delta B_I/\delta B_m = L_m/L_I$ should not be surprising, this is simply a consequence of the mapping, but it is noteworthy that the inferred magnetic field perturbations are reasonable.

The scaling laws also show why the magnetosphere is usually thought to drive ionospheric convection, at least at higher latitudes. In their study *Deng et al.* [1991] found flywheel-driven currents of the order 0.04 μA/m². Although it is possible that the current densities were low because of the smearing inherent in the model they used, even assuming a current sheet of 10° width in latitude, we only obtain a current intensity of the order 40 mA/m. This gives $\delta B_I \approx 50$ nT, an order of magnitude less than the estimate given above.

The scaling laws may shed some light on the observation of *Newell et al.* [1996] that auroral electron acceleration events occur mainly when the ionosphere is in darkness. For the dark ionosphere $\Sigma_p$ is controlled largely by electron precipitation. Assuming a magnetospheric velocity shear, then the increased conductivity associated with the precipitating electrons carrying the upward current will require an increased current in the ionosphere. The current demand can be reduced by increasing the amount of slippage (decreasing $f$). At the same time, decreasing $f$, which requires a parallel electric field, will result in an increase in current density, and hence total current, as given by the Knight relation [*Knight*, 1973]. On the other hand, the sunlit ionosphere has a high conductivity and it might be expected that currents would preferentially flow to the sunlit hemisphere, rather than the dark ionosphere. For the sunlit ionosphere, however, the relatively uniform conductivity could allow an increase in field-aligned current intensity through a widening of the shear layer without increasing the field-aligned current density. Because the ionospheric currents in the dark ionosphere mainly flow in a high conductivity channel, perhaps the shear layer cannot increase in width in this case, and parallel electric fields are required. Thus, while the current may preferentially flow

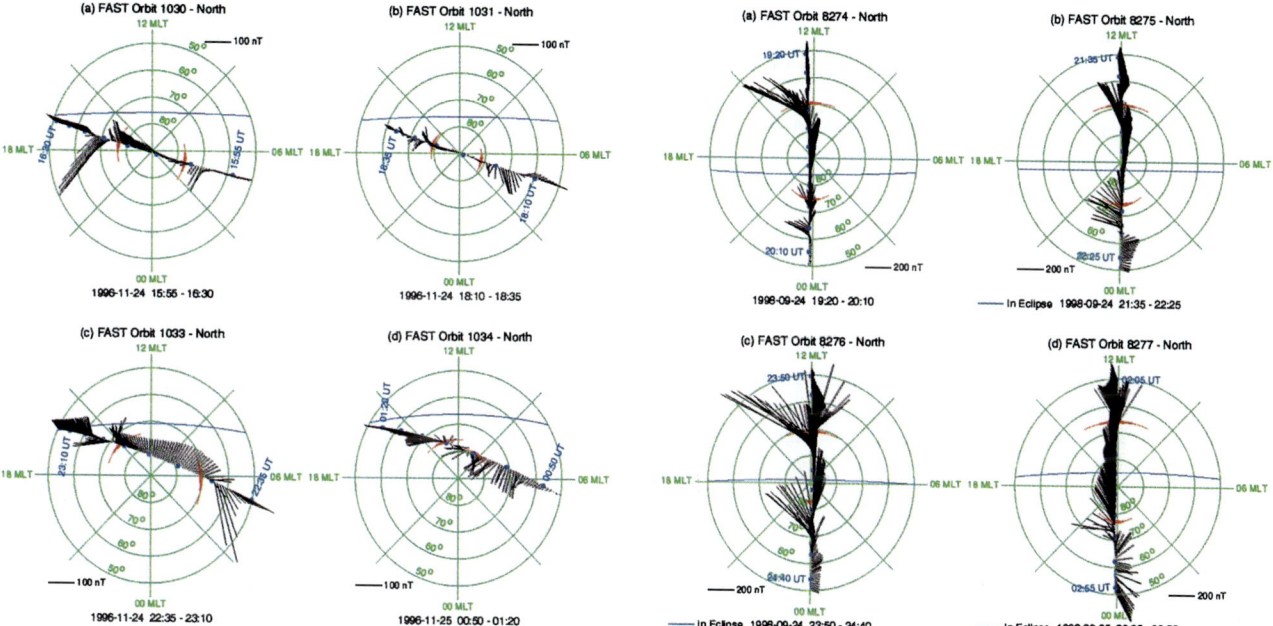

**Plate 1.** Polar projection of the magnetic field perturbations observed at FAST for the intervals indicated as solid bars in Figure 2. The data are plotted as a function of invariant latitude and magnetic local time along the spacecraft trajectory. The terminator is shown in blue, and the polar cap boundary, inferred from keV ion observations, is marked in red.

**Plate 2.** Polar projection of the FAST magnetic field perturbations. The data shown in panel c were acquired shortly after the Sudden Impulse (23:45 UT). For these data the polar cap boundary is inferred from keV electron observations.

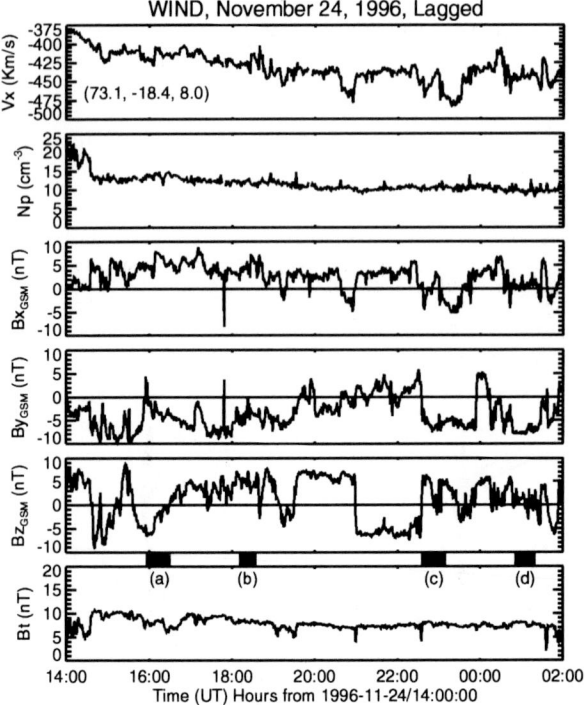

**Figure 2.** Solar wind data from the WIND spacecraft. The data have been lagged by 15 minutes, corresponding to nominal solar wind conditions. The top panel shows the x-component of the solar wind velocity. The position of the WIND spacecraft in GSE is given in parentheses in this panel. The next panel shows solar wind density. The bottom four panels show magnetic field data in GSM.

into the sunlit ionosphere, i.e., the field-aligned current *intensity* may be larger in the sunlit hemisphere, the field-aligned current *density* is likely to be larger in the dark hemisphere.

## 3. STRESSES IN THE POLAR CAP

As an example of the signatures observed in the polar cap we have analyzed FAST magnetometer data acquired on November 24th, 1996. This interval has been chosen by the Geospace Environment Modeling (GEM) community for an intense analysis and modeling effort. L. R. Lyons et al. ["Timing of Substorm Signatures During November 24, 1996 Geospace Environment Modeling Event", manuscript in preparation, 1999] describe the interval in greater detail, where two closely spaced substorm onsets occurred around 22:30 UT, following an extended interval of southward IMF. WIND data are shown in Figure 2. The data have been lagged by 15 minutes, which is an approximate lag-time for the prevailing solar wind conditions. The solid bars under the $B_z$ trace indicate those time intervals for which FAST data were acquired in the northern polar cap and auroral zone. The corresponding FAST data are shown in Plate 1.

In Plate 1 the data are plotted using a polar projection. In this projection we initially cast the spacecraft position and magnetic field perturbations into SM coordinates. The spacecraft radius vector then defines a magnetic meridian perpendicular to the SM equator, and the transverse deviations with respect to the model field (IGRF 95 plus secular variation) are used to derive the magnitude and angle of the projected field perturbation. In this projection a vector that points away from the origin lies within the magnetic meridian plane and points to lower latitude. At high latitudes, where the ambient magnetic field is nearly vertical, there is no ambiguity. For more equatorial latitudes the projected field can appear to point in the opposite direction to the observed field, but this is simply a matter of mapping the perturbation vector along the ambient magnetic field. We should emphasize that we have not mapped the magnitude of the field perturbation to the ionosphere. This mapping will typically increase the magnitude of the perturbation by a factor of two for data acquired near FAST apogee. Last, although the magnetometer data are acquired at a high rate by FAST, we have averaged the data to 20-s samples, for ease of visibility.

In each panel approximately 40 minutes of data are shown. For these passes the spacecraft altitude is increasing with time, starting around 1000 km at the beginning of each pass, and approaching apogee (~4000 km) at the end of each pass. The spacecraft orbit is near the dawn-dusk meridian. In addition to the magnetic field perturbations, shown in black, we also mark the terminator with a blue line. The red arc segments mark the polar cap boundary. For the data in Plate 1 we have taken a sharp decreases in the flux of keV ions observed on FAST to determine the polar cap boundary (electron data were unavailable for these intervals).

In Plate 1a we see large deflections of the magnetic field in the pre-dusk auroral oval. This corresponds to an interval of southward IMF, and we expect convection to be strong at this time. The deflections in the field are consistent with the standard Region-1, -2 current system, and there is some indication that the polar cap field lines are being pulled tailward. In interpreting northern hemisphere data, it should be remembered that a sunward perturbation of the field corresponds to the field line being pulled tailward, and the ionospheric flows should be in the opposite direction to the field perturbation. For Plate 1b the IMF has been northward for quite some time, and the polar cap is small and quiet, with very little forcing by the IMF. The auroral-zone currents are weak. This interval also demonstrates that the magnetometer calibration and spacecraft attitude are well determined, since the residual fields are very small. The next panel, Plate 1c, shows the field perturbations after an extended interval of southward IMF. As noted by Lyons et al., two substorm onsets occurred near 22:30 UT. Using the framework discussed in the previous section, we would say that there is strong forcing of the polar cap at this time. Last, Plate 1d shows the magnetic field perturbations after an extended interval of northward IMF. For this pass, FAST only briefly entered the polar cap, and there is some indication that the spacecraft is encountering cusp currents. Indeed the ion data show evidence of magnetosheath plasma entry, and these cusp ions make determination of the polar cap boundary somewhat difficult.

## 4. STRESSES IN THE CUSP

Figure 3 and Plate 2 show data for an interval studied by *Moore et al.* [1999], where intense ionospheric outflows are observed after the passage of an interplanetary shock and a coronal mass ejection (CME). Figure 3 shows WIND data. The interplanetary shock occurs at a lagged time of 23:45 UT, followed by the CME, marked by the roughly linear trend in IMF $B_y$ from 00:35 to 02:35 UT (lagged). The data have been lagged such that the shock occurs at the same time as the Sudden Impulse within the magnetosphere. For most of the interval prior to the shock passage the IMF is weakly southward, but immediately after the shock passage the IMF is predominantly in the +y direction. At the tail end of the CME the IMF is strongly southward, and remains so for the rest of the interval.

The FAST data in Plate 2 are similar in format to Plate 1, except that we have used keV electron data to determine the polar cap boundary. The FAST orbit has also evolved, lying in the noon-midnight meridian with the apogee (~4000 km) occurring near the middle of each pass. The data show many of the signatures we might expect on consideration of the forces applied at the magnetopause. For the first interval strong cusp currents are observed, with the delta-B pointing mainly in the +y direction, consistent with the IMF. The field lines in the polar cap are being pulled tailward, which is consistent with the weakly southward IMF at this time. In Plate 2b the cusp stresses are somewhat reduced, but after the shock passage (Plate 2c), the stresses are again large. The net change in the field is of the order 1000 nT on passing through the cusp current in Plate 2c. Taking into account the mapping along field lines, this gives an ionospheric signature about a factor of four larger than discussed in section 2. Given the increase in solar wind velocity and magnetic field strength after the shock passage, this increase is reasonable. Furthermore, the field perturbations are qualitatively similar in Plates 2b and 2c. We would therefore argue that the ion outflow reported by *Moore et al.* [1999] arises from enhanced dissipation associated with the increase in applied stress after the shock passage.

The last panel in Plate 2 shows data acquired at the end of the CME, where the IMF is strongly southward. The dominant signature is clearly a sunward field perturbation, consistent with field lines being dragged tailward, and the polar cap is much larger than on the previous orbit. There is an indication of cusp currents, but in this case the deflection of the field is towards dawn, which is consistent with the negative IMF $B_y$ observed at this time.

## 5. CONCLUSIONS

The major strength of FAST is the extensive suite of high resolution data, both particles and fields, which are being used to understand many of the processes occurring within the auroral acceleration region. The rapid precession in local time of the orbit (~ 3 hours per month) allows measurements to be acquired over the entire auroral oval, further enhancing the ability of FAST to investigate auroral processes. In this paper, however, we have concentrated on the FAST magnetometer observations to emphasize the usefulness of considering the stresses applied to the auroral ionosphere and polar cap. To do this we have used the deviations of the observed field from the model field, rather than converting the magnetometer data to an equivalent current density by taking the derivative of the transverse components along the spacecraft trajectory. This allows us to discuss the data in terms of the high altitude flows and the resultant stresses applied to the ionosphere, as shown schematically in Figure 1.

This "B, v" framework does allow us to place the overall magnetic field signatures in context. In doing so, however, we have neglected effects such as the finite time for information to travel from the magnetosphere and magnetopause to the ionosphere. We have acknowledged the effects of parallel electric fields, in terms of ionospheric slippage, but we have only discussed qualitatively how parallel electric fields could arise. Last, we have neglected neutral winds. Clearly, many auroral and polar cap phenomena are related to information travel times and parallel electric fields, as well as such effects as the neutral wind flywheel. The usefulness of cartoons such as Figure 1 is in specifying the idealized equilibrium. Much of the physics of the auroral zone and polar cap could then be understood in terms of the exchange of energy and momentum

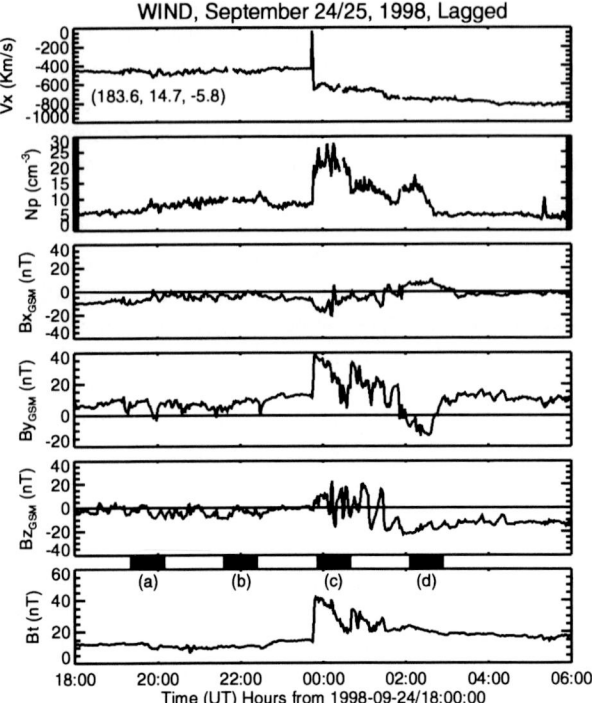

**Figure 3.** Solar wind data for an intense cusp region ion outflow event. The data have been lagged by 25 minutes. This lag-time is determined by the time of the Sudden Impulse observed within the magnetosphere. Similar in format to Figure 2.

between magnetosphere and ionosphere that occurs in trying to achieve this equilibrium.

The two intervals we have shown here demonstrate the clear correlation of the "delta-B's" observed on FAST with variations in the IMF. In many ways this complements the work of *Iijima and Potemra* [1982], who investigated the dependence of region-1 current strength on IMF parameters. As they noted, their results implied that reconnection was a major source of these currents. We would agree with that conclusion, but take it one step further. Reconnection may be a significant generator for region-1 currents, but the currents ultimately flow because of the need to accelerate the ionosphere against the drag caused by collisions with the neutral gas.

As a closing comment, taking Figure 1 and applying it the nightside auroral zone may have implications for the flow braking discussed by *Shiokawa et al.* [1998]. Figure 1 implies that the ionosphere will also act to brake any flows imposed by strong earthward streaming in the equatorial magnetosphere, and the equatorial current implied by Figure 1 is in the direction of the inertial current discussed by Shiokawa et al. Clearly this requires further analysis,

but we might speculate that the strong braking and enhanced field-aligned currents occur when the flow has extended sufficiently close to the earth that there is rapid communication between the magnetosphere and ionosphere. Consideration of the fundamental frequency for standing Alfvén waves indicates this may the case. *Cummings et al.* [1969] show that at L = 6.6 the fundamental toroidal mode frequency is ≈ 60 s for an equatorial density of 1 cm$^{-3}$. It is therefore possible that ionospheric drag acts as a brake on the flow in addition to field and plasma pressure gradients in the inner magnetosphere. As figure 1 shows, field-aligned currents are a natural consequence of the braking caused by ionospheric drag.

*Acknowledgments.* We thank the WIND experimenters for providing solar wind data. This work was supported by NASA grant NAG5-3596 to the University of California and NASA order number S-57795-F to the Los Alamos National Laboratory.

## REFERENCES

Armstrong, J. C., and A. J. Zmuda, Triaxial magnetic measurements of field-aligned currents at 800 km in the auroral region: Initial results, *J. Geophys. Res.*, 78, 6802–6807, 1973.

Coroniti, F. V., and C. F. Kennel, Can the ionosphere regulate magnetospheric convection?, *J. Geophys. Res.*, 78, 2837–2851, 1973.

Cowley, S. W. H., Magnetospheric asymmetries associated with the Y-component of the IMF, *Planet. Space Sci.*, 29, 79–96, 1981.

Cummings, W. D., and A. J. Dessler, Field-aligned currents in the magnetosphere, *J. Geophys. Res.*, 72, 1007–1013, 1967.

Cummings, W. D., R. J. O'Sullivan, and P. J. Coleman, Jr., Standing Alfvén waves in the magnetosphere, J. Geophys. Res., 74, 778–793, 1969.

Deng, W., T. L. Killeen, A. G. Burns, and R. G. Roble, The flywheel effect: ionospheric currents after a geomagnetic storm, *Geophys. Res. Lett.*, 18, 1845–1848, 1991.

Elphic, R. C., J. Bonnell, R. J. Strangeway, C. W. Carlson, M. Temerin, J. P. McFadden, R. E. Ergun, and W. Peria, FAST observations of upward accelerated electron beams and the downward field-aligned current region, this issue.

Iijima, T., and T. A. Potemra, The amplitude distribution of field-aligned currents at northern high latitudes observed by Triad, *J. Geophys. Res.*, 81, 2165–2174, 1976.

Iijima, T., and T. A. Potemra, The relationship between interplanetary quantities and Birkeland current densities, *Geophys. Res. Lett.*, 9, 442–445, 1982.

Kelley, M. C., *The earth's ionosphere: plasma physics and electrodynamics*, Academic Press, San Diego, 1989.

Knight, S., Parallel electric fields, *Planet. Space Sci.*, 21, 741–750, 1973.

Moore, T. E., W. K. Peterson, C. T. Russell, M. O. Chandler, M. R. Collier, H. L. Collin, P. D. Craven, R. Fitzenreiter, B. L.

Giles, and C. J. Pollock, Ionospheric mass ejection in response to a coronal mass ejection, *Geophys. Res. Lett.*, *26*, 2339–2342, 1999.

Newell, P. T., C.-I. Meng, and K. M. Lyons, Suppression of discrete aurorae by sunlight, *Nature*, *381*, 766–767, 1996.

Parker, E. N., The alternative paradigm for magnetospheric physics, *J. Geophys. Res.*, *101*, 10,587–10,625, 1996.

Shiokawa, K., W. Baumjohann, G. Haerendel, G. Paschmann, J. F. Fennell, E. Friis-Christensen, H. Lühr, G. D. Reeves, C. T. Russell, P. R. Sutcliffe, and K. Takahashi, High-speed ion flow, substorm current wedge, and multiple Pi 2 pulsations, *J. Geophys. Res.*, *103*, 4491–4507, 1998.

Sugiura, M., and T. A. Potemra, Net field-aligned currents observed by Triad, *J. Geophys. Res.*, *81*, 2155–2164, 1976.

Wright, A. N., Transfer of magnetosheath momentum and energy to the ionosphere along open field lines, *J. Geophys. Res.*, *101*, 13,169–13,178, 1996.

Zmuda, A. J., and J. C. Armstrong, The diurnal variation of the region with vector magnetic field changes associated with field-aligned currents, *J. Geophys. Res.*, *79*, 2501–2502, 1974a.

Zmuda, A. J., and J. C. Armstrong, The diurnal flow pattern of field-aligned currents, *J. Geophys. Res.*, *79*, 4611–4619, 1974b.

Zmuda, A. J., J. H. Martin, and F. T. Heuring, Transverse magnetic disturbances at 1100 kilometers in the auroral region, *J. Geophys. Res.*, *71*, 5033–5045, 1966.

---

C. W. Carlson, Space Sciences Laboratory, University of California at Berkeley, Berkeley, CA 94720

R. C. Elphic, NIS-1, Space and Atmospheric Sciences, Los Alamos National Laboratory, Los Alamos, NM 87545

W. J. Peria, Space Sciences Laboratory, University of California at Berkeley, Berkeley, CA 94720

R. J. Strangeway, Institute of Geophysics and Planetary Physics, University of California at Los Angeles, Los Angeles, CA 90095

# Electric Current Approach to Magnetospheric Physics and the Distinction Between Current Disruption and Magnetic Reconnection

A. T. Y. Lui

*The Johns Hopkins University Applied Physics Laboratory, Laurel, Maryland, U.S.A.*

The debate on whether the primary quantities in magnetospheric physics are the magnetic field combined with the plasma bulk flow or the electric field combined with the current density has been raging on and off for decades. The former approach is known as the Bu paradigm and the latter as the Ej paradigm. The first tutorial for the Conference on Magnetospheric Current Systems advocates that the correct approach to understanding magnetospheric problems is in terms of the Bu paradigm. In this paper, we address the limitations of the Bu paradigm and consider the merits of the Ej paradigm, with some emphasis on the magnetotail current. The concept of magnetic reconnection is formulated in the Bu paradigm and that of current disruption is formulated in the Ej paradigm. Since the usefulness of these two concepts and their distinction from each other is central to this debate, we discuss also the similarities and differences between them. We conclude by pointing out that each paradigm has its merits and limitations, and which one is the better approach should depend on the magnetospheric phenomenon to be investigated. Insisting on one paradigm being superior than the other for treating all magnetospheric problems would stifle innovative thinking in scientific pursuits.

## 1. INTRODUCTION

At the turn of the 20th Century, there were intense debates on whether electromagnetic emissions can only be treated as continuum waves or can also possess quantum characteristics like individual particles. In the latter viewpoint, it is considered that certain properties of the electromagnetic emissions (such as the photoelectric effect) can more readily be understood by such a treatment. As we know today, the latter viewpoint is the correct one in spite of many apparent successes of the former. Much like this aspect in classical physics, there is dualism in magnetospheric physics, as stressed by both *Alfvén* [1977] and *Song and Lysak* [1994], from which two paradigms emerge in treating magnetospheric dynamics. One is to adopt the magnetic field **B** and the plasma flow **u** as the primary quantities and derive electric field **E** and current density **j** from them based on a simplified form of Ohm's law and the Ampere's law. We shall refer to this paradigm hereafter as the 'Bu paradigm'. The other is to adopt the electric field and current density as the primary quantities and obtain the magnetic field and the plasma flow through the Biot-Savart law and calculation on the ensemble of single particles. Hereafter, this alternative paradigm is called the 'Ej paradigm'. Debates on which paradigm is the correct approach permeate throughout the four decades of magnetospheric research. Important to emphasize here is the assertion of the Bu paradigm that the Ej paradigm could lead to incorrect physics [*Parker*, 1996; denoted hereafter as P96]. On the other hand, the Ej paradigm in the practitioners' minds is considered to be an equivalent approach which provides invaluable physical insights on the underlying process in some cases [*Alfvén*, 1950, 1977; *Heikkila*, 1988, 1990, 1997; *Lui*, 1992, 1996; *Song and Lysak*, 1989, 1994, 1997; *Song*, 1998].

We examine in this paper the limitations of the Bu paradigm and the merits of the Ej paradigm in the hope that a more balanced perspective on the approach to treat magnetospheric physics may be gained. Much in the heart of the debate between these two paradigms is the concept of current disruption in the Ej paradigm and its distinction from magnetic reconnection in the Bu paradigm. For this reason, we discuss in some details the similarities and differences between these processes.

## 2. WHAT IS IN THE NAME?

It is important at the outset to elaborate in more details what each paradigm is meant in this paper to avoid misunderstanding. Formalism in the work of P96 is mainly based on the single-fluid theory, known as the magnetohydrodynamics (MHD) theory. In fact, it is stated (on p. 10587-10588 of P96) that "the macroscopic behavior ... of the magnetosphere is described by MHD which we refer to here as the Bu paradigm. The equations of MHD form a complete set of partial differential equations, providing a deductive approach to the theory of magnetospheric activity." In the above citation, the term "Bu" is substituted for "**B**, **v**" in the original statement for consistency with the notation in this paper. It is clear from these statements that the Bu paradigm in P96 is synonymous with the MHD approach. Certainly, inclusion of multiple fluid components and kinetic treatment for certain regions with violation of the frozen-in condition were mentioned as extension or exception to the general treatment. With these caveats, the Bu paradigm would of course be faultless. However, the Bu paradigm defines no specific criteria under its formalism to guide when and where multi-fluid or kinetic treatment should be applied. For example, it is stated that "if $j_\parallel$ proves to be so large as to require a strong $E_\parallel$, the consequences of $E_\parallel$ must be introduced into the calculation of **B**". What is meant by too large? In addition, one can equally say that the Ej paradigm, which considers plasma as made up of individual particles, is just as faultless as the Bu paradigm. The Ej approach could be extended to fluid treatment by taking ensemble averages in velocity space of individual particle properties to predict large-scale behavior of any system under Newton's and Maxwell's equations. With this line of argument, there would be no distinction between these approaches and there should never be any justification by those advocating the Bu paradigm to claim that the Ej paradigm leads to wrong physics.

If the Ej paradigm in P96 denotes the electrical circuit approach, then we would have no disagreement. However, it is incorrect to judge the Ej paradigm simply as the electrical circuit approach, as P96 did. First, researchers who focus on electric field and current as primary quantities in treating magnetospheric problems seldom rely on any electrical circuit. Second, even ones who use electrical circuit description in explaining the underlying physics are aware of its limitations [e.g., *Heikkila*, 1997]. In the following, we shall associate the Ej paradigm as the approach of single particle calculation or kinetic analysis (with full consideration of the collective effect) to distinguish the MHD approach of the Bu paradigm. The single particle approach is valid when the collective behavior of the plasma under consideration is relatively unimportant. We emphasize that the Ej paradigm, at least in here, does not state the Bu approach to be always invalid but its applicability has to be judged on a case-by-case basis.

## 3. LIMITATIONS OF THE BU PARADIGM

In the early years, the fluid approach has the clear advantage for a number of reasons. In spite of the collisionless nature of magnetospheric plasmas, the general configuration and large-scale dynamics are governed to a large extent by fluid-like behavior. The fluid approach provides the overall shape of the magnetospheric boundaries and a simple picture on the influence of the incident solar wind on the magnetosphere. These successes have bolstered the general acceptance of the MHD theory to treat magnetospheric dynamics. Unfortunately, the successes of MHD theory have brought forth also a general negligence on the inherent limitations on this approach.

There are several approximations/assumptions inherent in the MHD theory. Some are covered well in standard plasma physics textbooks [e.g., *Longmire*, 1963; *Krall and Trivelpiece*, 1973; *Parks*, 1991] while others, quite unique to the magnetospheric discipline, are not. Therefore, it is useful to elaborate the limitations here.

### 3.1 Fast Temporal and/or Small Spatial Scale Phenomena

The MHD equations are limited to problems involving slow time variations and long spatial scales. The MHD ordering is given by

$$(r_i/L)^2 \ll (T\Omega_i)^{-1} \ll 1, \tag{1}$$

where $r_i$ is the thermal ion gyroradius, $\Omega_i$ is the ion gyrofrequency, L and T are the spatial and time scales, respectively [*Krall and Trivelpiece*, 1973]. As a result, magnetospheric processes involving fast time variations and/or small spatial scales are outside the validity regime of the MHD theory. In particular, very interesting and important magnetospheric phenomena, such as substorm expansion onset, often involve such fast time variations (electric and magnetic fluctuations with frequencies comparable to the ion gyrofrequency) and/or small spatial scales (thin current sheets with thickness comparable to the thermal ion gyroradius). These processes are clearly beyond the scope of the MHD theory. Nevertheless, many MHD simulations have not been deterred by these limitations and proceed ahead with investigations of these phenomena. It is often argued that regions anomalous to the MHD theory occupy small volumes and would not expect to affect the macroscopic behavior. Unfortunately, the physical significance of any process cannot usually be judged by its size. For example, the diffusion region in magnetic reconnection is very small in volume (may be comparable to the electron inertial scale). However, without it, there can be no magnetic reconnection and the macroscopic behavior of the system would have been drastically different.

*3.2 Beam-Plasma Interaction*

Under the Bu paradigm, plasma is treated as a single fluid. This may be a deficient description of space plasmas because fast streaming particle beams (both ions and electrons) are often detected in the magnetosphere [e.g., *Williams*, 1981; *Yau et al.*, 1985]. These beams can excite waves which can heat local plasmas. Consider a simple example of two strong counter-streaming ion beams moving at the same speed but in opposite directions relative to a background of stationary electrons for charge-neutrality. In the Bu paradigm, the bulk flow is zero and no dynamic evolution of the system is anticipated. However, kinetic calculation shows that this is a classic example of the two-stream instability which involves wave-particle interaction. Nonlinear calculation invariably leads to a saturation state with wave excitation to a finite amplitude and a change in the second moment of the ion velocity distribution (temperature), resulting in an energy transfer from particles to waves [e.g., *Papadopoulos et al.*, 1971]. Another aspect of the beam interaction with ambient plasma is the production of electric field parallel to the magnetic field [e. g., *Kan*, 1975]. These dynamics of the system are not predicted by the Bu paradigm.

*3.3 Current Intensity Limit*

In the vocabulary of the Bu paradigm, there is no limit on the strength of current density. The current density is defined by the curl of B since in most space plasma situations the displacement current $\partial E/\partial t$ is negligible. However, there is no built-in check to ensure that a given current density implied by the curl of **B** could only be met by particles traveling at speeds higher than the light speed. Nor is there a current threshold specified to indicate the onset of kinetic processes which may lead to parallel electric field and/or wave excitation. One may argue that in some MHD simulations, the displacement current is included and with an artificial lowering of the light speed to impose a limit on the current intensity. There are also some MHD simulations which artificially increase the particle density to avoid running into this problem. However, these artificial fixes create unrealistic characteristic speeds of the medium with often undesirable consequences. For example, in the case of artificially increasing particle density, the Alfvén and the magnetosonic speeds are lowered, causing an unrealistic time delay in wave propagation from one magnetospheric region to another, which may then lead to misjudgment on the causal relationship between activity in different regions.

*3.4 Non-ideal Equation of State*

The set of equations in the Bu paradigm requires an equation of state for closure. The equation of state in some cases can be adopted with confidence based on some physical insights, e.g., isothermal or adiabatic process. However, observations indicate that plasmas in the real magnetosphere in general do not obey a simple equation of state, including the well-known CGL equation [*Chew et al.*, 1956] which P96 adopted without any reservation. There is at least one good reason for this departure. The magnetospheric plasma in an active region does not form a closed system. Particle loss through precipitation in the ionosphere and/or escape along the magnetic field to remote distances are commonplace. In fact, it has long been demonstrated by *Erickson and Wolf* [1980] that the near-Earth plasma pressure expected from the steady state magnetospheric convection is too high to be compatible with observations. Precipitation plays a significant role in relieving the near-Earth plasma pressure arising from convection [*Kivelson and Spence*, 1988]. As a result, the 'equation of state' associated with magnetospheric convection deviates considerably from that expected from adiabatic convection. It is prudent for any macroscopic

## 3.5 Arbitrary Artificial Coefficients

Artificial coefficients, such as diffusion or viscosity, are sometimes introduced in the equations without clear understanding of the underlying physical processes responsible for their generation. This approach precludes some essential physics which could provide predictions on the initiation of some dynamic magnetospheric processes. Surely, there are some MHD simulations which specify some of these coefficients with a physics-based algorithm. For example, anomalous resistivity dependent on the local current density can be introduced into MHD simulations. This is obviously an improvement over simulations adopting arbitrary values for the coefficients. This procedure, however, is not a complete solution since it oversimplifies the challenging problem of cross-scale coupling in the magnetosphere. The real situation may be more complex as the value and the onset threshold of anomalous resistivity may depend on not only the current density but also other parameters such as number density, temperature, and magnetic field all of which evolve in time. This limitation obviously does not apply to work which use kinetic theory to derive such coefficients [e.g., *Lysak and Dum*, 1983].

## 3.6 Frozen-in Condition

It is sometimes stated that MHD equations represent a set of conservation laws so that the MHD treatment cannot be wrong. However, this statement has two shortcomings: first, the conservation law argument is not valid in the open system where the transport of mass, momentum, and energy to outside the system is large; second, even though MHD is a single-fluid theory and no electron equation is explicitly given, the frozen-in condition ($\mathbf{E} + \mathbf{u} \times \mathbf{B} = 0$) is basically the momentum equation for electrons in the Bu paradigm. If there is any non-idealness in the MHD approximation, this electron momentum equation is invalid. The frozen-in condition also implies that all particles regardless of mass, charge, or energy move with the same velocity, a prediction which is clearly invalid in many magnetospheric regions. Furthermore, there is no straight-forward relationship between the validity of the MHD approximation and the plasma beta value [*Yamauchi and Blomberg*, 1997]. Since wave-particle interactions which can invalidate the MHD approximation can occur regardless of the plasma beta value, the frozen-in condition can be broken in both the low and high plasma beta regimes. In addition, as will be discussed later, current disruption in the Ej paradigm can break the frozen-in condition in regions where **B** is not zero, in contrast to the Bu paradigm which demands such a region for its breakdown.

The above points summarize some (and not an exhaustive list) of the limitations of the Bu paradigm which should be kept in mind when dealing with magnetospheric dynamics. Unfortunately, in the magnetospheric discipline, interesting physics almost invariably lies in the regime where the MHD approximation is violated, such as in auroral arc and substorm expansion onset regions. In recent years, the simple one-fluid MHD theory has been extended to treat both multiple fluid components and some of the non-idealness in the MHD approximation such as the inclusion of the Hall-effect and the finite Larmor radius correction [*Winglee*, 1998; *Ma and Bhattacharjee*, 1998; *Cheng and Johnson*, 1999]. These developments are certainly in the right direction to alleviate some of the limitations of the Bu approach discussed above.

## 4. MERITS OF THE EJ PARADIGM

Although the Ej paradigm in the early years is guided by laboratory experience and thus circuit analogy, the basic and recent approach calls for more fundamental use of physical laws rather than circuit elements. Formulation and solution of problems with this paradigm are admittedly more cumbersome and less tractable than the Bu paradigm. This is a major limitation of the Ej paradigm. However, scientists who prefer this paradigm are drawn in mainly because of the tremendous physical insights offered by this approach. Some of the merits are listed below.

## 4.1 Single Particle Behavior

In the consideration of charged particle energization in given electric and magnetic fields, each individual particle obeys the well-known equation of motion

$$m \, d\mathbf{v}/dt = q \, (\mathbf{E} + \mathbf{v} \times \mathbf{B}), \qquad (2)$$

where m is the particle mass, **v** is its velocity, and q is its charge. Note that we use **v** to denote single particle velocity and **u** to denote the bulk flow of the population. Forming the dot product of this equation with the particle velocity, one immediately recognizes that the magnetic field provides no particle energization. Only electric field, which includes the potential and the induced component $\partial \mathbf{B}/\partial t$, energizes particles. This concept is different from the notion based on the Bu paradigm which considers the magnetic stress causing acceleration of fluid elements. When the collective behavior is relatively unimportant, the single particle approach is appropriate. This simple approach leads to understanding of important phenomena such as velocity dispersion seen in substorm injections [*McIlwain*, 1974], plasma sheet boundary layer [*Williams*, 1981], and low-altitude particle precipitation region [*Zelenyi et al.*, 1990]. Moreover, the single particle approach is quite successful in some situations to determine the location relative to a spacecraft for the origin of energized particles [*Andrews et al.*, 1981] or the onset time of acceleration [*Kamide and McIlwain*, 1974] or the remote sounding of boundaries [*Williams*, 1979]. In contrast, the Bu approach gives no velocity dispersion effect at all since it is a fluid approach. The above important advances in our field would be absent if all scientists had adopted the Bu paradigm to address all magnetospheric problems.

*4.2 Differentiating Dynamo and Dissipation Regions*

The Ej paradigm provides a clear insight on where the driver of the dynamics is. Referring back to Eq. (2) and summing up the particles in a given volume of space, one finds that the energy exchange between particles and fields could be intuitively assessed simply by the dot product $\mathbf{j} \cdot \mathbf{E}$. The dissipation region (or load in electric circuit terminology) in which $\mathbf{j} \cdot \mathbf{E} > 0$ is where electromagnetic energy is converted to particles and, conversely, the dynamo region (or generator in electric circuit terminology) in which $\mathbf{j} \cdot \mathbf{E} < 0$ is where particle energy is converted to electromagnetic energy. This is not to say that the Bu paradigm cannot differentiate these two regions because the Poynting theorem which specifies the energy transfer is derivable from both paradigms. The point is that both $\mathbf{j}$ and $\mathbf{E}$ are primary quantities in the Ej paradigm and no further calculation is needed to differentiate these two regions whereas such is not the case in the Bu paradigm.

*4.3 Multi-component Plasmas*

Multiple components in a plasma are natural for the Ej paradigm since the very concept of electric current is the relative drift between positive-charged species (typically ions) and negative-charged species (typically electrons). The drift may be an 'apparent' drift due to a gradient in density and/or temperature as in the case of the diamagnetic current. Any need for a kinetic treatment of multi-component plasma processes (including analysis of instabilities not involving current such as drift-shear instability) or generation of waves in energy exchange between fields and particles will be immediately apparent from this approach. For example, one can check if the relative drift between species is sufficiently high to cause instability growth. A handy reference on thresholds for the general class of beam plasma instabilities is *Huba* [1994].

*4.4 Electric Current*

Electric currents considered under the Ej paradigm would never impose any particle traveling at speeds beyond the light speed. Careful consideration of particle motions contributing to the electric current also provides insight on the validity of fluid and guiding center approximation. For example, one can investigate whether the gyroradius of the particle is larger than or comparable to the gradient scale lengths involved. If it is, the guiding center approximation may not hold. Similarly, one can check whether the $\mathbf{E} \times \mathbf{B}$ drift is smaller than or comparable to gradient and curvature drifts. If it is, the fluid approximation may not be valid.

*4.5 No Equation of State or Artificial Coefficients Imposed*

The initial and boundary conditions provide the only input needed to examine the system evolution, requiring no equation of state to form closure of equations. No artificial coefficient such as resistivity, viscosity, and diffusion need be introduced to advance the equation of motion. Underlying physics for any anomalous transport will have to be worked out based on first principles from the Newton's and Maxwell's equations. In other words, the Ej paradigm forces us to consider the essential physics up front rather than to rely on black boxes as represented by artificial coefficients in the Bu paradigm.

*4.6 Cross-Scale Coupling*

The Ej paradigm allows one to treat the non-idealness of the MHD approximation in a general form, even for effects not representable by a resistive term $\eta \mathbf{j}$, where $\eta$ is the anomalous resistivity. For example, in the consideration of the global consequence of the cross-field current instability, the fractional change in the magnetic field normal to the current sheet $\Delta B_z/B_z$ at time t has been estimated by *Yoon and Lui* [1998] to be

$$\frac{t}{t_A} r_{ei} \frac{|\Omega_e|}{\omega_c} \left\langle \frac{\delta B^2}{B^2} \right\rangle \qquad (3)$$

where $t_A$ is the Alfvén transit time for the current sheet, $r_{ei}$ is the electron-to-ion temperature ratio, $\Omega_e$ is the electron gyrofrequency, $\omega_c$ is the characteristic wave frequency of the magnetic fluctuations $\delta B$, and <...> represents averaging over all phases of the fluctuations. A more detailed derivation of Eq. (3) is given in Appendix A, following the work of *Yoon and Lui* [1998]. Since, in the near-Earth tail, $r_{ei} \sim 0.1$, $\omega_c \sim \Omega_i$, and $\delta B \sim B$, one finds that $\Delta B_z/B_z \sim 10^2 \, t/t_A$, implying the time scale for the field change to be $\sim 10^2$ faster than the MHD time scale represented by $t_A$. This therefore can account for the fast temporal variations in the magnetic field seen during current disruptions by the non-idealness of MHD without introducing coefficients for resistivity, viscosity, or diffusion.

## 5. DISTINCTION BETWEEN CURRENT DISRUPTION AND MAGNETIC RECONNECTION

In the Bu paradigm, the breakdown of the frozen-in condition is often attributed to magnetic reconnection (MR) only. There is still no universal definition for MR. However, in general, it is a process whereby plasma flows across a surface separating regions with different magnetic field topology, and through which energy is typically exchanged between magnetic field and particles [*Vasyliunas*, 1975]. Sometimes, the definition of MR is broadened to encompass all processes violating the frozen-in condition [see, e.g., discussion in *Lundin*, 1997]. In this alternative definition, any process which violates the MHD approximation would be considered as MR. However, this alternative definition poses many undesirable implications. For instance, the essential element of MR envisaged by *Dungey* [1961] would be totally lost. The inner magnetosphere, which is a region where particles with different energy, mass, and charge undergo different trajectories due to curvature and gradient drifts, is a MR region at all times. Earlier work of interchange instability in the magnetosphere by *Gold* [1959], which predicts the breakdown of the frozen-in condition without involving plasma flow across a surface separating different magnetic field topology should then be credited as the first application of MR concept to the magnetospheric discipline. Therefore, the alternative, broadened definition of MR is not narrowly defined enough to be a useful concept. For the above reasons, the traditional definition will be adopted hereafter.

A standard schematic diagram for MR is shown in Figure 1a. An important feature to note here is that the magnetic field basically orders the plasma flow pattern. In the case of current disruption (CD), as shown in Figure 1b, the magnetic field geometry and its associated current pattern are complex and no constraint is imposed on the associated plasma flow. Even though the magnetic field geometry may be complex for CD, the process does not necessarily involve plasma flows across topologically different magnetic field regions. CD does not require the occurrence of a zero-**B** point/line either, in contrast to MR. Both processes are perceived to break the frozen-in condition and cause energy exchange between magnetic field and particles. Note that the Petschek model for MR with slow mode shocks is often considered to be necessary for fast explosive phenomena such as magnetospheric substorms and solar flares. Therefore, the presence of slow mode shocks becomes an essential feature for MR in explosive space phenomena. This can be contrasted with the CD concept in which the fast time variation is achieved by one or more local kinetic processes instead, as discussed in the previous section.

Figure 2 depicts the CD concept for the magnetotail. During quiet conditions, the current density varies rather monotonically throughout the current sheet. At the onset of CD, the current sheet becomes filamentary and the current density pattern is complex, even with embedded reversals in the current density direction. This is basically a turbulent medium and the complex current density pattern leads to highly fluctuating magnetic field. In the near-Earth region, where the ambient magnetic field normal to the neutral sheet (contributed by the Earth's dipole field, the ring current, and the magnetopause current) is relatively strong, the net magnetic field shows fluctuations typically above the zero level, i.e., the magnetic field is typically northward in spite of its high variability. However, in the mid-tail region beyond $\sim 20 \, R_E$ from the Earth, the ambient magnetic field normal to the neutral sheet is weak. The net magnetic field therefore frequently becomes negative.

In the near-Earth magnetotail, an estimate based on the quasilinear calculation of the cross field current instability with local approximation suggests that the development of this instability leads typically to an increase in the $\mathbf{j} \times \mathbf{B}$ force whereas the change in the perpendicular pressure remains small [*Lui et al.*, 1993; *Yoon and Lui*, 1993]. Note that we consider the $\mathbf{j} \times \mathbf{B}$ force to be basically the sum of Lorentz force acting on individual particles rather than the magnetic stress acting on a fluid element as in the Bu paradigm. This increase in force leads to the plasma typically accelerated earthward in the near-Earth magnetotail. This prediction of the force imbalance as a result of CD in the near-Earth region is later verified in 2-D and 3-D self-consistent magnetotail simulations by *Wiechen et al.* [1995, 1996]. They found that by putting an anomalous resistivity in the strong magnetic field region resembling the near-Earth tail environment, the $\mathbf{j} \times \mathbf{B}$ force indeed exceeds the pressure gradient force in the local anomalous resistivity region. This result is in agreement with the prediction made earlier by *Lui et al.* [1993]. In the mid-tail

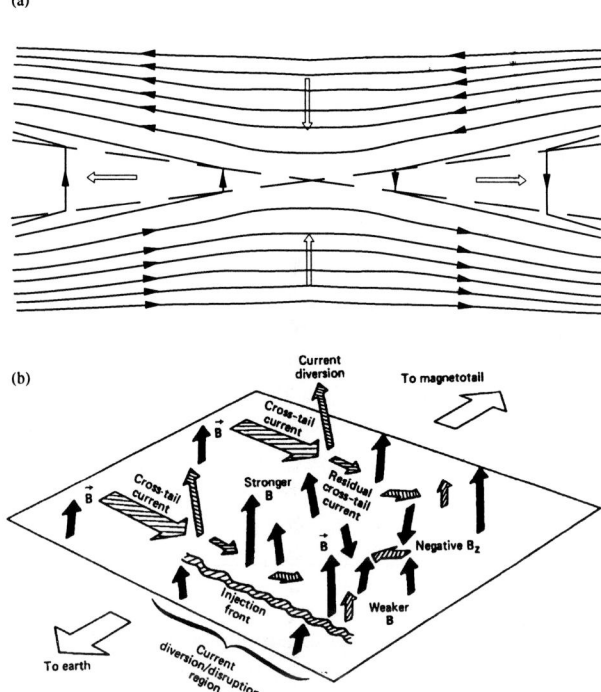

**Figure 1.** (a) A typical magnetic reconnection configuration. Plasma flows are indicated by the large arrows. (b) Illustration of the complex field configuration and particle signature (turbulence) for current disruption.

region where the current buildup prior to CD is weaker than that in the near-Earth region, the force imbalance leads typically to a tailward acceleration of plasma [*Lui*, 1999].

From the above discussion, it is expected that CD inside ~20 $R_E$ generally leads to earthward plasma acceleration and tailward beyond that distance. In a gross fashion, the direction of plasma acceleration predicted by CD in the tail resembles that of a MR region forming generally at ~20 $R_E$ in the magnetotail.

In the early theoretical study of the cross-field current instability as a plausible mechanism for CD [*Lui et al.*, 1991], local approximation is used in the linear analysis for the case of a thin current sheet with thickness comparable to the ion gyroradius. Two issues arise. Is the local approximation appropriate to the case of a thin current sheet and will the excited wave reside long enough within the current sheet to experience large growth? Appropriateness of local approximation implies $1 \ll k_z L_z$, where $k_z$ is the wavenumber of the excited wave in the direction normal to the current sheet and $L_z$ is the length scale of the thin current sheet. Enduring residence time implies that $2L_z/v_{gz} \gg \tau$, where $v_{gz}$ is the group velocity of the excited wave

normal to the current sheet and $\tau$ is the e-folding growth time of the excited wave. For excited waves at large propagation angles, the linear analysis indicates that $\tau \approx 38$ s, $2L_z/v_{gz} \approx 70 - 350$ s, and $k_z L_z \approx 2.0 - 2.5$. Therefore, the excited wave is found to reside long enough at the current sheet for a large growth and the local approximation is not violated. It should be noted that the wave propagation angle is sensitive to the ion-to-electron mass ratio [*McBride et al.*, 1972]. For a particle simulation with an artificially low ion-to-electron mass ratio (like 100), the unstable waves will propagate out of the current sheet before they grow to significant amplitudes.

MR includes a dissipation region where the frozen-in condition is violated. In the early years, anomalous resistivity is thought to be required for dissipation. However, recent numerical simulations indicate that the breakdown of the frozen-in condition is achieved with the electron inertial term or the off-diagonal terms in the electron pressure tensor [e.g., *Shay et al.*, 1998; *Lyons and Pridmore-Brown*, 1990], implying that no large magnetic fluctuation is expected for the dissipation region. These recent results in MR theories appear to be inconsistent with results from laboratory MR experiment which indicates the need for anomalous resistivity to account for the observed behavior [*Yamada et al.*, 1997; *Ji et al.*, 1998]. On the other hand, the major characteristics of current disruption is the large magnetic fluctuations.

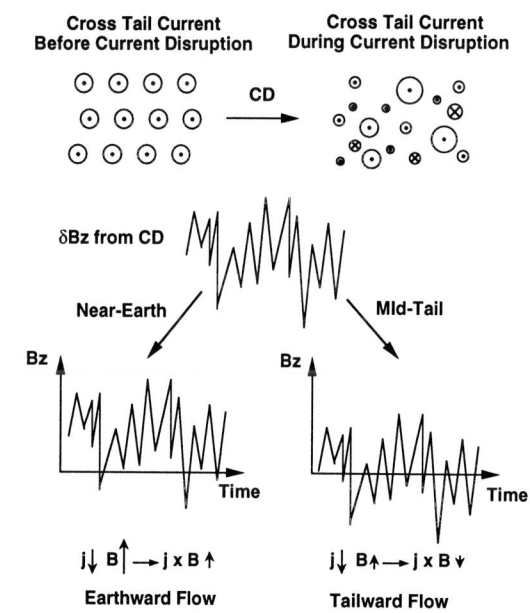

**Figure 2.** A schematic diagram to show expected signatures of current disruption in the near-Earth and the mid tail regions.

**Table 1.** Similarities and Differences between magnetic reconnection and current disruption

|  | Magnetic Reconnection | Current Disruption |
|---|---|---|
| Breakdown of frozen-in condition | yes | yes |
| Magnetic field is the primary energy source | yes | yes |
| Plasma acceleration | yes | yes |
| Plasma flow across separatrix essential | yes | no |
| A magnetic neutral line essential | yes | no |
| Plasma flow ordered by magnetic field | yes | no |
| Large magnetic fluctuations essential | no | yes |
| Slow mode shock required for fast energy conversion | yes | no |
| Potential mechanism for its onset | TI: $\omega \ll \Omega_i$; $k_x L_z \sim 1$ | KBI: $\omega \ll \Omega_i$; $k_y L_z \sim 1$ <br> CCI: $\omega \sim (0.1$ to $10)$ $\Omega_i$; $k_y L_z \sim 1$ to $10$ |

For MR, the tearing instability (TI) is the main mechanism under consideration [*Coppi et al.*, 1966; *Sitnov and Lui*, 1999]. The frequency of excited waves for tearing is low ($\ll \Omega_i$) and the wavenumber $k \approx k_x$, i.e., mainly in the sun-Earth direction with $k_x L_z < 1$. For current disruption, the present theories include the kinetic ballooning instability (KBI) [*Cheng and Lui*, 1998] and the cross-field current instability (CCI) [*Lui et al.*, 1993]. For the former instability, the wave frequency is expected to be low ($\ll \Omega_i$) and the wavenumber $k \approx k_y$, i.e. mainly in the dawn-dusk direction with $k_y r_i \sim 1$. For the latter instability, the wave frequency is expected to be high and in a broad range (0.1 to 10 $\Omega_i$) and the wavenumber $k \approx k_y$, i.e. mainly in the dawn-dusk direction with $k_y r_i \sim 1$ to $10$. Table 1 lists the main points discussed in this section.

## 6. SUMMARY

In this paper, we discuss briefly the limitations of the Bu paradigm and the merits of the Ej paradigm. We also elaborate on the similarities and differences between magnetic reconnection and current disruption since the former comes from the Bu paradigm while the latter comes from the Ej paradigm and so their distinction is closely tied to the debate between these two paradigms. We choose a physics approach in addressing these issues by not writing down a lot of equations. Otherwise, this paper will be unduly long. In addition, the distinction of physicists from mathematicians is that the former provide the physical intuition of the problem at hand with some basic equations while the latter often manipulate equations without questioning the relevance of assumptions being made in the process. A good historical case in point is the argument between Sydney Chapman and Hannes Alfvén on the existence and importance of field-aligned currents in the magnetosphere. The former used an elegant mathematical approach to describe the so-called equivalent current system in the ionosphere. The latter relied on ingenious physical insights to specify the current system in space. Which approach turns out to be a more accurate description of currents in space is now well known. Here the main point we wish to convey is that the two paradigms have their distinct advantages and limitations. Therefore, certain problems are better treated with one paradigm than with the other. Insisting on one paradigm being always better than the other is not well justified and is also undesirable because it would stifle innovation in scientific pursuit to understand nature.

## 7. APPENDIX A

In the study of CCI, the most important contribution to the non-idealness in MHD approximation comes from magnetic fluctuations and can be written for transverse fluctuations as [*Yoon and Lui*, 1998]

$$\mathbf{E} + \mathbf{u} \times \mathbf{B} = -\text{Re}\{\int d\omega \, d\mathbf{k} <\delta\mathbf{v}_e \times \delta\mathbf{B}^*>_{\mathbf{k}\omega}\}, \quad (A1)$$

where $\mathbf{u}$ is the fluid velocity, Re{...} denotes the real part, $<...>_{\mathbf{k}\omega}$ denotes the phase averaging over the fluctuations with the subscript $\mathbf{k}\omega$ indicating parameters inside the bracket to be functions of wavenumber $\mathbf{k}$ and frequency $\omega$

of the fluctuations, $\delta \mathbf{v}_e$ is the perturbed electron velocity, and $\delta \mathbf{B}^*$ is the complex conjugate of the perturbed magnetic field. Note that this is quite a standard procedure in kinetic quasilinear analysis [*Davidson*, 1972]. We consider the geometry in which $\mathbf{B} = B_z \mathbf{e}_z$, $\mathbf{v}_{i,e} = v_{i,e} \mathbf{e}_y$, and fluctuations varying as $\exp[i(\mathbf{k} \cdot \mathbf{r} - \omega t)]$. From previous linear analysis of CCI [*Lui et al.*, 1991], it is found that $\delta \mathbf{B} \approx \delta B \mathbf{e}_x$, $\delta \mathbf{v}_{i,e} \approx \delta v_{i,e} \mathbf{e}_z$. Since the important contribution is from the magnetic fluctuations, we consider for simplicity only the magnetic fluctuations in the electron momentum equation and obtain

$$\delta \mathbf{v}_e \approx -i \frac{e}{m_e} \frac{[\mathbf{b} \cdot (\mathbf{v}_e \times \delta \mathbf{B})]\mathbf{b}}{\omega - \mathbf{k} \cdot \mathbf{v}_e} \quad (A2)$$

where e is the elementary charge, $m_e$ is the electron mass, and **b** is the unit magnetic field vector. Taking the curl of Eq. (A1) and substituting $\delta \mathbf{v}_e$ from Eq. (A2), we get

$$\nabla \times (\mathbf{E} + \mathbf{u} \times \mathbf{B}) = \nabla \times \text{Re}\left\{\int d\omega \, d\mathbf{k} \left\langle i \frac{e}{m_e} \frac{[\mathbf{b} \cdot (\mathbf{v}_e \times \delta \mathbf{B})]\mathbf{b}}{\omega - \mathbf{k} \cdot \mathbf{v}_e} \times \delta \mathbf{B}^* \right\rangle_{\mathbf{k}\omega}\right\} \quad (A3)$$

With the use of $(\omega - \mathbf{k} \cdot \mathbf{v}_e)^{-1} = i\pi \, \delta(\omega - \mathbf{k} \cdot \mathbf{v}_e)$ and the Faraday's law, we can rearrange the expression as

$$\frac{\partial \mathbf{B}}{\partial t} = \nabla \times (\mathbf{u} \times \mathbf{B}) + \frac{\pi e}{m_e} \nabla \times \left[\int d\omega \, d\mathbf{k} \langle \mathbf{b} \cdot (\mathbf{v}_e \times \delta \mathbf{B} \delta \mathbf{B}^*) \times \mathbf{b} \, \delta(\omega - \mathbf{k} \cdot \mathbf{v}_e)\rangle_{\mathbf{k}\omega}\right] \quad (A4)$$

Noting that $\int d\omega \, d\mathbf{k} \langle \delta \mathbf{B} \, \delta \mathbf{B}^* \rangle_{\mathbf{k}\omega} \approx \langle \delta \mathbf{B} \, \delta \mathbf{B}^* \rangle / \omega_c$ for homogenous stationary turbulence, where $\omega_c$ is the characteristic wave frequency of the magnetic fluctuations, we find the temporal evolution of **B** to be

$$\frac{\partial \mathbf{B}}{\partial t} = \nabla \times (\mathbf{u} \times \mathbf{B}) + \frac{\pi e}{m_e} \nabla \times \left(\frac{\mathbf{b} \cdot \mathbf{v}_e \times \langle \delta \mathbf{B} \, \delta \mathbf{B}^* \rangle \times \mathbf{b}}{\omega_c}\right) \quad (A5)$$

Concentrating on the fluctuation component and with the characteristics of CCI as noted above, we can simplify the z-component of Eq. (A5) to

$$\left(\frac{\partial B_z}{\partial t}\right)_{\text{wave}} \approx -\frac{\pi e}{m_e} \frac{\partial}{\partial x} \left(\frac{v_{ey}}{\omega_c} \langle \delta B^2 \rangle\right) \quad (A6)$$

Let us denote the electron-to-ion temperature ratio by $r_{ei}$, the Alfvén speed by $v_A$, and the ion gyrofrequency by $\Omega_i$. For $\mathbf{v}_e = -r_{ei} \mathbf{v}_i$ as in the case of a Harris current sheet, $v_i \sim B/(\mu_0 n e L) = v_A^2/(L\Omega_i)$ for the equilibrium current sheet, and the spatial scale along the tail axis to be the ion inertial length, i.e. $\Delta x \sim c/\omega_{pi}$, we can then obtain Eq. (3) in the main text.

*Acknowledgments.* I am indebted to Peter Yoon, Yan Song, Bob Lysak, and Shin Ohtani for useful comments and discussion. This work was supported by the Atmospheric Sciences Division of the National Science Foundation (Grant ATM-9622080) and by National Aeronautics and Space Administration (Grant NAG5-7797) to The Johns Hopkins University Applied Physics Laboratory.

# REFERENCES

Alfvén, H., *Cosmical Electrodynamics*, Oxford University Press, 1950.
Alfvén, H., Electrical currents in cosmic plasmas, *Rev. Geophys.*, 15, 271, 1977.
Andrews, M. K., E. Keppler, and P. W. Daly, Plasma sheet motions inferred from medium-energy ion measurements, *J. Geophys. Res.*, 86, 7543, 1981.
Cheng, C. Z., and J. R. Johnson, A kinetic-fluid model, *J. Geophys. Res.*, 104, 413, 1999.
Cheng, C. Z., and A. T. Y. Lui, Kinetic ballooning instability for substorm onset and current disruption observed by AMPTE/CCE, *Geophys. Res. Lett.*, 25, 4091, 1998.
Chew, G. F., M. L. Goldberger, and F. E. Low, The Boltzmann equation and the one-fluid hydromagnetic equations in the absence of collisions, *Proc. R. Soc. London Ser. A*, 236, 112, 1956.
Coppi, B., G. Laval, and R. Pellat, Dynamics of the geomagnetic tail, *Phys. Rev. Lett.*, 16, 1207, 1966.
Davidson, R. C., *Methods in Nonlinear Plasma Theory*, Academic Press, New York, U.S.A., 1972.
Dungey, J. W., Interplanetary magnetic field and the auroral zones, *Phys. Res. Lett.*, 6, 47, 1961.
Erickson, G. M., and R. A. Wolf, Is steady convection possible in the Earth's magnetotail?, *Geophys. Res. Lett.*, 7, 897, 1980.
Gold, T., Motions in the magnetosphere of the earth, *J. Geophys. Res.*, 64, 1219, 1959.
Heikkila, W. J., Elementary ideas behind plasma physics, *Astrophys. Space Sci.*, 144, 85, 1988.
Heikkila, W. J., Magnetic reconnection, merging, and viscous interaction in the magnetosphere, *Space Sci. Rev.*, 53, 1, 1990.
Heikkila, W. J., Comment on "The alternative paradigm for magnetospheric physics" by E. N. Parker, *J. Geophys. Res.*, 102, 9651, 1997.
Huba, J. D., NRL Plasma Formulary, NRL/PU/6790-94-265, Department of the Navy, Washington, DC, 1994.
Ichimaru, S., Basic principles of plasma physics, W. A. Benjamin, Inc., Reading, MA, 1973.
Ji, H., M. Yamada, S. Hsu, and R. Kulsrud, Experimental test of the Sweet-Parker model of magnetic reconnection, *Phys. Rev. Lett.*, 80, 3256, 1998.
Kamide, Y., and C. E. McIlwain, The onset time of magnetospheric substorms determined from ground and

synchronous satellite records, *J. Geophys. Res., 79*, 4787, 1974.

Kan, J. R., Energization of auroral electrons by electrostatic shock waves, *J. Geophys. Res., 80*, 2089, 1975.

Kivelson, M. G., and H. E. Spence, On the possibility of quasi-static convection in the quiet magnetotail, *Geophys. Res. Lett., 15*, 1541, 1988.

Krall, N. A., and Trivelpiece, A. W., Principles of Plasma Physics, McGraw-Hill, New York, NY, U S A, 1973.

Longmire, C. L., Elementary Plasma Physics, Interscience Publ., New York, NY, U S A, 1963.

Lui, A. T. Y., Role of cross-field current instability in substorm onsets and intensifications, *Proceedings on the International Conference on Substorms (ICS-1)*, Kiruna, Sweden, 213, 1992.

Lui, A. T. Y., Current disruption in the Earth's magnetosphere: Observations and models, *J. Geophys. Res., 101*, 13067, 1996.

Lui, A. T. Y., Particle Acceleration in Disruption of the Tail Current Sheet, *Physics and Chemistry of the Earth*, 24, 259-267, 1999.

Lui, A. T. Y., C.-L. Chang, A. Mankofsky, H.-K. Wong, and D. Winske, A cross-field current instability for substorm expansions, *J. Geophys. Res., 96*, 11389, 1991.

Lui, A. T. Y., P. H. Yoon, and C.-L. Chang, Quasi-linear analysis of ion Weibel instability in the Earth's neutral sheet, *J. Geophys. Res., 98*, 153, 1993.

Lundin, R., Discussion 1: On merging and reconnection, *Phys. Chem. Earth*, 22, 605, 1997.

Lyons, L. R., and Pridmore-Brown, D. C., Force balance near an X line in a collisionless plasma, *J. Geophys. Res., 95*, 20903, 1990.

Lysak, R. L., and C. T. Dum, Dynamics of magnetosphere-ionosphere coupling including turbulent transport, *J. Geophys. Res., 88*, 365, 1983.

Ma, Z. W., and A. Bhattacharjee, Sudden enhancement and disruption of current sheets in the magnetotail, *Substorms-4*, ed. by S. Kokubun and Y. Kamide, Kluwer Academic Publ., Norwell, MA, p. 397, 1998.

Mauk, B. H., and C.-I. Meng, Dynamical injections as the source of near geostationary quiet time particle spatial boundaries, *J. Geophys. Res., 88*, 10011, 1983.

McBride, J. B., E. Ott, J. P. Boris, and J. H. Orens, Theory and simulation of turbulent heating by the modified two-stream instability, *Phys. Fluids, 15*, 2367, 1972.

McIlwain, C. E., Substorm injection boundaries, in *Magnetospheric Physics*, ed. by B. M. McCormac, p. 143, D. Reidel, Hingham, Mass., 1974.

Papadopoulos, K., R. C. Davidson, J. M. Dawson, I. Haber, D. A. Hammer, N. A. Krall, and R. Shanny, Heating of counterstreaming ion beams in an external magnetic field, *Phys. Fluids, 14*, 849, 1971.

Parker, E. N., The alternative paradigm for magnetospheric physics, *J. Geophys. Res., 101*, 10587, 1996.

Parks, G. K., Physics of Space Plasmas: An introduction, Addison-Wesley Publ. Co., Redwood City, CA, U S A, 1991.

Shay, M. A., J. F. Drake, R. E. Denton, and D. Biskamp, Structure of the dissipation region during collisionless magnetic reconnection, *J. Geophys. Res., 103*, 9165, 1998.

Sitnov, M. I., and A. T. Y. Lui, Cross-field current instability as a catalyst of the explosive reconnection in the geomagnetotail, *J. Geophys. Res., 104*, 6941, 1999.

Song, Y., and R. L. Lysak, Evaluation of twist helicity in FTE flux tube, *J. Geophys. Res., 94*, 5273, 1989.

Song, Y., and R. L. Lysak, Alfvénon, driven reconnection and the direction generation of field-aligned current, *Geophys. Res. Lett., 21*, 1755, 1994.

Song, Y., and R. L. Lysak, Some theoretical aspects of solar wind-magnetospheric interaction, *Phys. Chem. Earth*, 22, 715, 1997.

Song, Y., Theoretical constraints on mechanisms for the substorm current wedge, *Substorms-4*, ed. by S. Kokubun and Y. Kamide, Terra Scientific Publ. Co., Tokyo, p. 543, 1998.

Vasyliunas, V. M., Theoretical models of magnetic field line merging, 1, *Rev. Geophys. Space Phys.*, 13, 303, 1975.

Wiechen, H., J. Büchner, and A. Otto, Reconnection in a dipole-dominated magnetosphere: A two-dimensional model, *J. Geophys. Res., 100*, 19259, 1995.

Wiechen, H., J. Büchner, and A. Otto, Reconnection in the near-Earth plasma sheet: A three-dimensional model, *J. Geophys. Res., 101*, 24911, 1996.

Williams, D. J., Magnetopause characteristics inferred from three-dimensional energetic particle distributions, *J. Geophys. Res., 84*, 101, 1979.

Williams, D. J., Energetic ion beams at the edge of the plasma sheet: ISEE 1 observations plus a simple explanatory model, *J. Geophys. Res., 86*, 5507, 1981.

Winglee, R. M., Multifluid simulations of the magnetosphere: The identification of the geopause and its variation with IMF, *Geophys. Res. Lett., 25*, 4441, 1998.

Yamada, M., H. Ji, S. Hsu, T. Carter, R. Kulsrud, N. Bretz, F. Jobes, Y. Ono, and F. Perkins, Study of driven magnetic reconnection in a laboratory plasma, *Phys. Plasmas, 4 (5)*, 1936, 1997.

Yamauchi, M., and L. Blomberg, Problems on the concept and mapping of convection, *Phys. Chem. Earth*, 22, 709, 1997.

Yau, A. W., E. G. Shelley, W. K. Peterson, and L. Lenchyshyn, Energetic auroral and polar ion outflow at DE 1 altitudes: magnitude, composition, magnetic activity dependence, and long-term variations, *J. Geophys. Res., 90*, 8417, 1985.

Yoon, P. H., and A. T. Y. Lui, Nonlinear analysis of generalized cross-field current instability, *Phys. Fluids, B5 (3)*, 836, 1993.

Yoon, P. H. and A. T. Y. Lui, Cross-field current instability and substorm expansion onset, *Substorms-4*, ed. by S. Kokubun and Y. Kamide, Terra Scientific Publishing Co., Tokyo, Japan, p. 387-390, 1998.

Zelenyi, L. M., R. A. Kovrazhkin, and J.-M. Bosqued, Velocity-dispersed ion beams in the nightside auroral zone: AUREOL-3 observations, *J. Geophys. Res., 95*, 12119, 1990.

---

A. T. Y. Lui, The Johns Hopkins University Applied Physics Laboratory, 11100 Johns Hopkins Road, Laurel, MD, 20723-6099, U.S.A.

# Global Geometry of Magnetospheric Currents Inferred from MHD Simulations

G. L. Siscoe, N. U. Crooker, G. M. Erickson
*Center for Space Physics, Boston University, Boston, Massachusetts*

B. U. Ö. Sonnerup
*Thayer School of Engineering, Dartmouth College, Hanover, New Hampshire*

K. D. Siebert, D. R. Weimer, W. W. White, N. C. Maynard
*Mission Research Corporation, Nashua, New Hampshire*

This paper discusses the global geometry of magnetospheric current systems as determined by a global MHD code simulating the magnetosphere for four elemental settings of the interplanetary magnetic field: 1. zero field strength (giving the Chapman-Ferraro and tail current systems); 2. straight north (giving the NBZ current system); 3. straight south (giving the region 1 current system); and 4. straight dawn to dusk (giving the cusp-mantle current system). The Chapman-Ferraro and tail current systems have the topologies usually ascribed to them, but the MHD simulation reveals that they overlap across a significant area of the magnetopause. The NBZ current system closes in the distant magnetotail. The region 1 current system closes across the high-latitude magnetopause. The cusp-mantle current system threads most of the magnetosphere. It spirals earthward around the south magnetotail lobe until it reaches the magnetosphere proper where it catapults over the South Pole to descend into the ionosphere along the polar border of the dayside polar cusp where it becomes the mantle part of the cusp-mantle system. Once in the ionosphere, it jogs a little north then lifts off again along the equatorward border of the cusp as the cusp part of the cusp-mantle system. From there, it follows magnetic field lines in a dipolar arch to the equatorial magnetopause beyond which it mirrors its southern career in the northern hemisphere. The cusp-mantle current system joins smoothly and continuously onto the region 1 current system. They appear to form a single system.

## 1. BACKGROUND

Global electric current systems play a useful conceptual role in magnetospheric physics. They define the magnetosphere's "exoskeleton" and "endoskeleton" and they allow one to trace the course of momentum or energy from one structural unit of the magnetosphere to another, such as between the magnetopause and the ionosphere, which is especially useful when the units are not contiguous. For years, these uses motivated researchers to model the global 3D geometry of magnetospheric current systems. Such efforts include global MHD modeling, which is an approach that treats all current systems in the same way and not as separate cases. The results shown here are based on a global MHD code named ISM ("Integrated Space Weather Prediction Model"), which is being developed by Mission Research Corporation for the Defense Threat Reduction Agency (DTRA).

Other studies that have applied global MHD codes to study the geometry of magnetospheric current systems include Janhunen et al. [1996] and Tanaka [1995]. Janhunen et al., who simulated a southward IMF case, found that the region 1 current system closes on the high-latitude boundary layer. Tanaka, who simulated both a southward IMF case and a northward IMF case, presented figures of the global geometries of region1, region 2, and NBZ current systems. He, too, found region 1 currents closing over the high-latitude boundary layer. His region 2 currents closed primarily in the low-latitude, nightside magnetosphere as a partial ring current centered at midnight. His NBZ currents closed in the low-latitude magnetotail. Tanaka introduced the technique of color coding the current streamlines to reveal the local value of the electromechanical energy conversion term in Poynting's equation ($- J \cdot E$). This facilitated the use of the terms "dynamo" and "load" as a conceptual heuristic borrowed from electrical engineering. The 3D geometry of magnetospheric currents of the ISM code agrees with those of Janhunen et al. and Tanaka where comparisons are possible. New results presented here add the geometry of the Chapman-Ferraro, magnetotail, and cusp-mantle current systems.

## 2. ZERO IMF SHOWING CHAPMAN-FERRARO AND TAIL CURRENT SYSTEMS

The Chapman-Ferraro and magnetotail currents define the shape of the magnetosphere including its tail. Functionally, they shield solar wind plasma from magnetic field that has its origin within the Earth, and they transfer to the Earth the "ram force" that the solar wind exerts against the magnetopause [Siscoe, 1966]. To isolate these current systems in their pure forms, unmodified by the effects of magnetic merging that happens whenever the solar wind carries a magnetic field, the ISM code was run with the IMF set (essentially) to zero. A tail current results in this case even though the magnetotail is thought to be a product of magnetic merging with the interplanetary magnetic field, which is absent here. A tail nonetheless results because an initial plasma density exists on dipolar field lines before the solar wind is "turned on." Turning on the solar wind then compresses the initial plasma and squeezes it out into a tail. In all runs discussed here, the solar wind speed was set at 400 km/sec; solar wind density at 5 protons (and electrons)/$cm^3$; and solar wind temperature (protons and electrons) at 20 eV.

Plate 1 shows the resulting computed current streamlines, which have been placed to display the Chapman-Ferraro and tail current systems. The figure shows the northern dusk quadrant of a magnetosphere that is north-south symmetric and dawn-dusk symmetric. The tan contours that outline the magnetopause and paint a swath along the midplane in the magnetotail show where the current is strong. Chapman-Ferraro current streamlines, shown red in the figure, form a set of nested rings concentric with the cusp. They are topologically identical with current streamlines that had been inferred already before the 1970s for an ideal vacuum magnetosphere [Chapman and Ferraro, 1931, Midgley and Davis, 1963]. tail-current streamlines, shown blue in the figure, appear here as a row of current segments within the northern dusk quadrant of the magnetotail. When viewed from the Sun with all four quadrants present, each current streamline has the form of the Greek letter theta ($\theta$). The result is the same topologically as was described in the first papers on tail current topology [Axford et al., 1965; Dessler and Juday, 1965].

A feature seen here for the first time is the overlap within the current layer that defines the magnetopause of current streamlines from the two current systems (Chapman-Ferraro and tail). The overlap is most significant (in the sense that the currents are comparably strong) in the portion of the boundary where the magnetosphere transitions to the magnetotail. Note again for later reference that the Chapman-Ferraro currents circle around the cusp as a center. As we will see, merging with the IMF moves the center of the Chapman-Ferraro current rings away from the cusp, defined as a particle-entry feature. This is also a new finding.

## 3. SOUTHWARD IMF SHOWING REGION 1 CURRENT SYSTEM AND DISPLACED CHAPMAN-FERRARO CURRENT CENTER

Iijima and Potemra [1976] defined region 1 currents as a high-latitude ring of currents extending 24 hours in local time. The ring exhibits a dawn-dusk pair of field-aligned currents with a specific polarity: down on the dawn side and up and the dusk side. This holds for both northern and southern hemispheres. Plate 2 shows such a pair of field-aligned currents as contours (blue = up, tan = down) in the ionosphere as viewed looking down on the north pole. The contours are taken from an ISM run in which the IMF was straight south at 5 nT. (In this and all runs shown here, the ionospheric Pedersen conductance is a uniform 6 siemens and zero siemens for the Hall conductance. This is not meant to be a statement about the actual ionospheric situation, just a statement about the stage of development the ISM code had reached when the runs were made.) The contours have the same general shape (kidney bean) and distribution of intensity (strongest near noon) as inferred

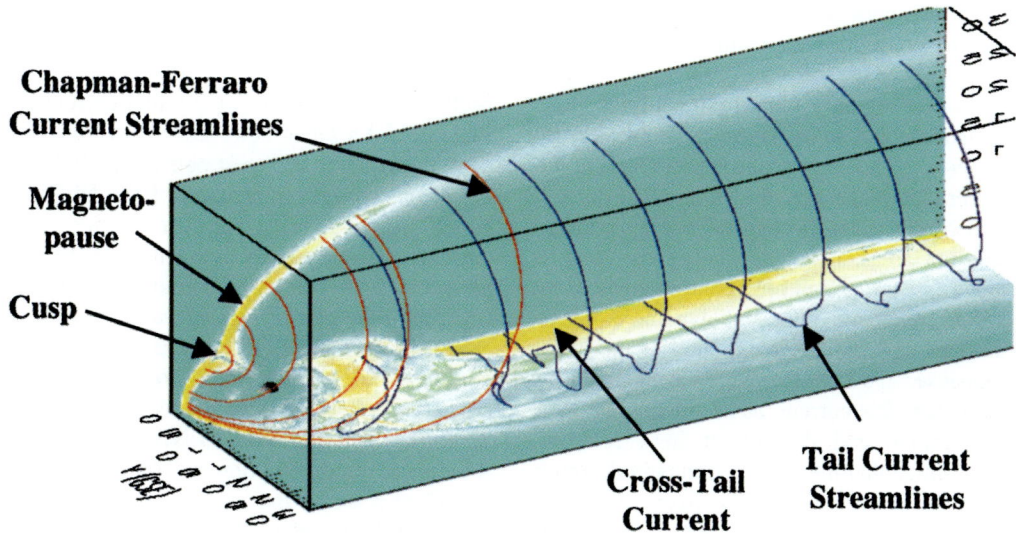

**Plate 1.** A 3D view of the northern-dusk quadrant of the magnetosphere. Chapman-Ferraro current streamlines (red) and tail current streamlines (blue) were generated with an MHD run with the IMF set equal to zero. Shading indicates the presence of current flowing perpendicular to the magnetic field. Axes are labeled GSE, but in all runs shown here, GSE and GSM are identical.

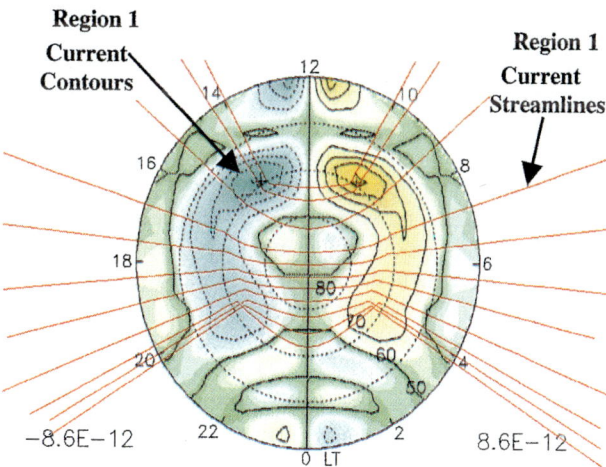

**Plate 2.** Contours of field-aligned currents drawn on the ionosphere as seen from above the north pole. The situation depicts the case of a purely southward IMF. Tan contours signify downward current and blue contours, upward current. Region 1 currents appear as a pair of kidney shaped contours. Current streamlines are in red. (The little pair of region-1 polarity currents straddling noon at 50° latitude is an artifact of the code as it was when this run was made. It reflects an imperfect match in numerical differencing at the interface between an inner and outer grid at 3.5 $R_e$.) Numbers at bottom give peak currents in amps/cm$^2$.

from observations. (Though the observed overlap in the noon and midnight sectors is absent.) To determine the 3D global geometry of this current system in particular where it closes, we traced a set of current streamlines, shown here in red.

Plate 3 shows that region 1 current streamlines close across the high-latitude boundary layer, as suggested by Siscoe et al. [1991]. The color-coded contours on the noon-midnight meridian plane give the value of "J dot E," which is the rate of electrical-mechanical energy conversion. In tan-colored areas, magnetic energy converts to flow energy ("load" areas), and in blue-colored areas, flow energy converts to magnetic energy ("dynamo" areas). Region 1 streamlines close in a dynamo area of the high-latitude magnetopause. To take the electrical engineering analogy at face value, the current picks up energy in a boundary layer dynamo and deposits it in a resistive ionosphere load. Notice that the J-dot-E contours defining the magnetopause are layered around the cusp. A "load" layer overlays a "dynamo" layer poleward of the cusp. This illustrates a peculiarity mentioned earlier: In the presence of magnetic merging, the geometrical focus of the Chapman-Ferraro current separates from the particle cusp.

Plate 4 is a close-up of the cusp region seen in the northern dawn quadrant. It shows how magnetic reconnection in this southward IMF case changes the geometry of the Chapman-Ferraro currents compared to the no IMF (pure Chapman-Ferraro) case. In the absence of an IMF, the Chapman-Ferraro currents close completely on the magnetopause as a nested set of closed curves concentric with the particle-entry cusp. They define the magnetopause. By contrast, here the closure geometry depends on depth in the current layer. Current streamlines lying in the inner part of the current layer still close on the magnetopause. But current streamlines lying farther out close progressively farther from the magnetopause, moving across the magnetosheath and, for a substantial part of the layer, to the bow shock. It is as if region 1 currents in their magnetopause closure have usurped in part at least the role of the Chapman-Ferraro currents leeward of the cusp. Chapman-Ferraro closure through the bow shock might come about in this IMF-southward case because the magnetosphere is now "stickier" because of magnetic coupling to the resistive ionosphere. Thus, from the shock's perspective, the magnetosphere is, in effect, a bigger object, which means that it must carry more current than in the IMF = 0 case. The extra current might be the Chapman-Ferraro current that closes through the shock.

Note that the center of the Chapman-Ferraro current rings that do close on the magnetopause is far from the particle-entry cusp. Seen in current geometry, the cusp actually takes a right-angle bend at the magnetopause and heads north to the center of the Chapman-Ferraro system. Associated with the current cusp is a new class of current streamlines, labeled "current vortex" in the figure.

## 4. NORTHWARD IMF SHOWING NBZ AND REGION 2 CURRENT SYSTEMS

The NBZ current system was generated by ISM using a 5 nT northward IMF as input. Plate 5, depicting the northern polar cap, shows parallel-current contours, as in Plate 2, and a set of color-coded current streamlines. The central-most and strongest pair of parallel-current contours is the ionospheric expression of the NBZ current system, which has a polarity opposite to the pair of region 1 currents that borders it along the $70°$ latitude line in the pre-noon and post-noon sectors [Iijima et al., 1984]. Encircling the pair of region 1 currents is a pair of weaker region 2 currents, with the same polarity as the NBZ currents. The streamlines are color coded to show connections between and among these two pairs of current systems. There is NBZ to NBZ (red), NBZ to region 1 (blue), and region 1 to region 1 (purple).

Plate 6 shows the current streamlines in the magnetosphere of the listed currents. The view is from a northern-hemisphere, dusk-side perspective from a position tailward of Earth. It shows the magnetosphere's northern dusk quadrant bounded by the noon-midnight meridian plane and the equatorial plane, both of which are "painted" in colors representing $J\bullet E$ as in Figure 3. The NBZ to NBZ current streamlines (red) extend into the deep magnetotail where, presumably, they close. The NBZ to region 1 current streamlines (blue) and the region 1 to region 1 streamlines (purple) close like region 1 streamlines in the southward IMF case (Plate 3). In contrast to the southward IMF case, however, in which region 1 current streamlines close through a high-latitude dynamo layer, in this northward IMF case, the streamlines close along the inside border of a high-latitude load layer. The high-latitude boundary layer is a load in this case because field lines merge tailward of the cusp forming a "slingshot" in which magnetic tension accelerates plasma tailward. In this case, the blue and purple current streamlines thread a dynamo region on the flanks of the magnetosphere, where numerical dissipation within the code provides a pseudo viscosity that allows magnetosheath flow along the magnetopause to stretch field lines in the low-latitude boundary layer.

## 5. DAWN-TO-DUSK IMF SHOWING SKEWED, SPIRAL CHAPMAN-FERRARO GEOMETRY, THE CUSP-MANTLE CURRENT SYSTEM, AND ITS RELATION TO THE REGION 1 CURRENT SYSTEM

Plate 7 shows the effect that a dawn-to-dusk IMF (i.e., IMF $B_y$) has on the geometry of Chapman-Ferraro current streamlines. The color shading in this meridian plane view shows plasma pressure, which defines the bow shock, the magnetopause, and the particle-entry cusps. The current streamlines in the figure, which were initialized on the x

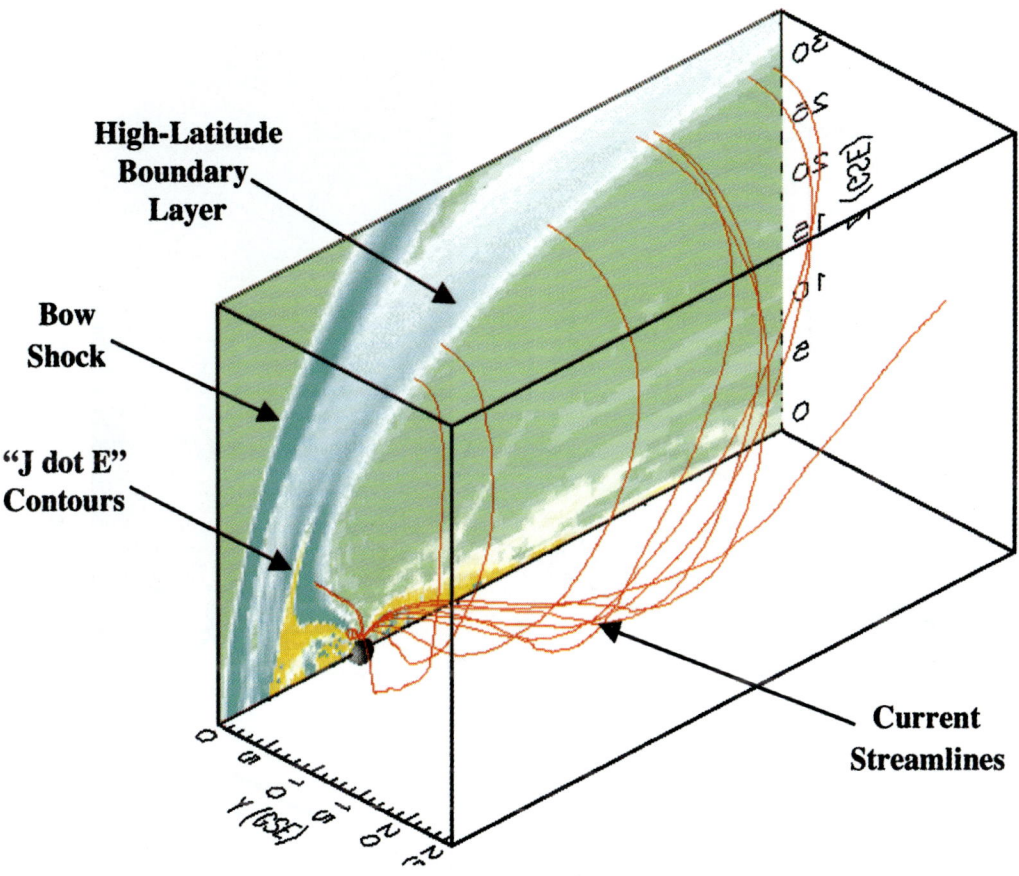

**Plate 3.** A 3D view of the northern-dusk quadrant of the magnetosphere showing contours in the noon-midnight meridian plane of J-dot-E. Region 1 current streamlines (red) are seen to close in a "dynamo" area of the high-latitude boundary layer of the magnetosphere.

46  GLOBAL GEOMETRY OF MAGNETOSPHERIC CURRENTS

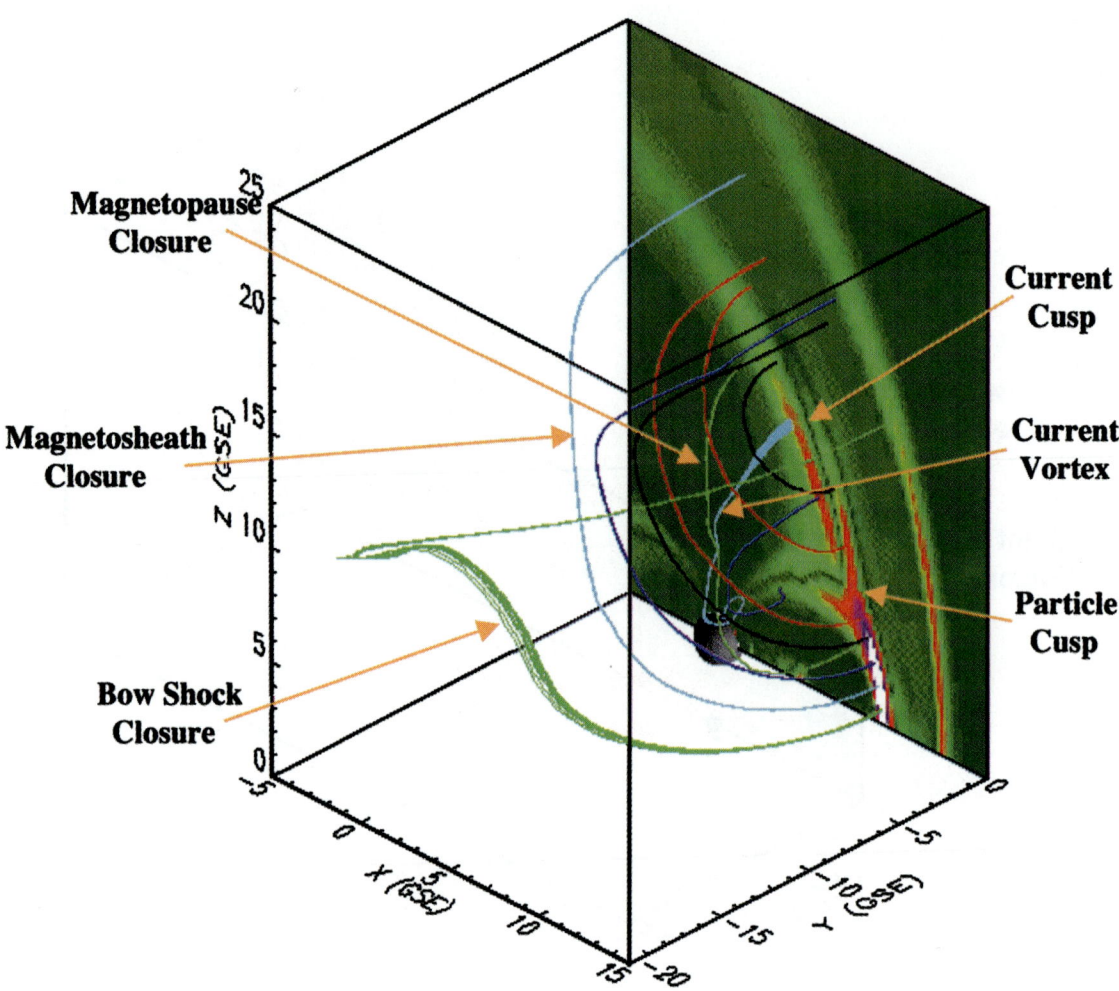

**Plate 4.** A 3D view of the northern-dawn quadrant of the magnetosphere showing current streamlines and contours associated with the Chapman-Ferraro current for the case of the IMF pointing straight south. The contours in the noon-midnight meridian plane show $J_y$, the y component of the current density, with tan contours positive current and blue contours negative current. Streamlines have been selected to show layering of closure paths and offset from the particle-entry cusp.

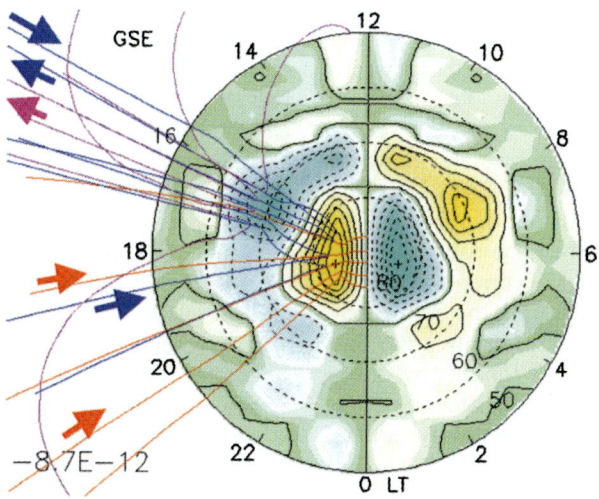

**Plate 5.** Similar to Plate 2 except for the case of a purely northward IMF. A variety of current streamlines are shown. Blue contours denote outward currents. Tan contours denote inward currents. Red streamlines connect NBZ to NBZ currents. Blue streamlines connect NBZ to region 1 currents. Purple streamlines connect region 1 currents to region 1 currents. Numbers at bottom give peak currents in amps/cm².

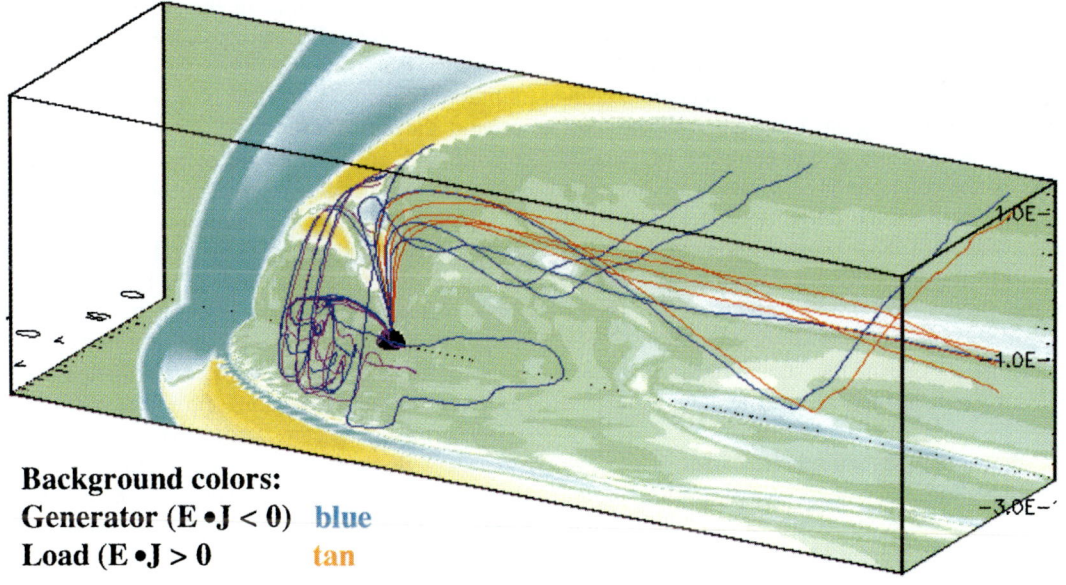

**Plate 6.** Similar to Plate 3 except for the northward IMF case and showing a variety of current streamline closure paths.

axis at different depths within the magnetopause current layer, generate a pair of nested spirals. These spirals correspond to the circles of the Chapman-Ferraro current system. The centers of the spirals are well separated from the particle-entry cusp both in the z direction seen here and in the y direction, being into the plane in the northern hemisphere and out of the plane in the southern hemisphere.

The field-aligned currents for this case are shown in Plate 8, which is the IMF-$B_y$ version of Plate 2 for the southward-IMF case and Plate 5 for the northward-IMF case. Unlike these previous cases, here the pattern of parallel currents does not consist of pairs of currents (region 1, region 2, and NBZ) symmetrically split by the noon-midnight meridian. Nonetheless, one recognizes the formal presence of region 1 and region 2 currents by imagining a traverse along the dawn-dusk terminator. Starting at the numeral 6, we encounter an upward (blue) current, which is here the polarity of region 2. Proceeding farther, we come to a downward (tan) current, which is here the polarity of region 1. Then in the middle of the polar cap, we find upward (blue) current again, this being the other half of the region 1 current. Our terminator traverse ends with a second encounter of downward (tan) current, this being the other half of the region 2 current. Note that the traditional classification that divides field aligned currents into distinct bipolar pairs--cusp-mantle, region 1, and region 2--is not what one would choose if one were starting with this pattern, which is more like a set of nearly concentric unipolar rings, the central one of which forms a filled-in circle. The dawnside region 1 current merges continuously with the duskside region 2 current. Where they merge around noon, there is a concentration of parallel current. This is the ionospheric expression of the cusp component of the cusp-mantle currents (first identified as such by Bythrow et al. [1988]). cusp-mantle currents arise whenever the IMF has a significant $B_y$ component, which is most of the time. We follow the two cusp-mantle current streamlines that are shown here in red. (The extraneous red line is one of the streamlines that passes over the pole of Earth at high altitude and so is seen in projection.)

Figure 1 shows the magnetospheric career of the two cusp-mantle streamlines. They run (in the sense of the arrows) continuously from the southern tail lobe (where they form part of the tail current system) into the southern-hemisphere ionosphere as the mantle current, then out again as the cusp current. They then follow dipolar field lines to the northern hemisphere where they enter the ionosphere as the cusp current. Then they leave as the southern mantle current to spiral into the northern tail lobe as part of the tail current system. Cusp-mantle current streamlines spend most of their career wrapping around the magnetotail as part of the tail current system.

Plate 9 shows the cusp-mantle currents from the Sun's view, which plainly displays their tail-wrapping geometry. The near-Earth parts of the current streamlines (shown here in gold) form two small loops near the center of the figure and two lines that "dive" into the southern polar cap and emerge symmetrically from the northern polar cap. The two small loops are the cusp segments of the streamlines, and the two diving lines are the mantle segments. At their closest approach in the center of the magnetotail, the streamlines are generally more than 2 $R_e$ apart. Here the plasma sheet separates them as shown by color contours of the value of the current density flowing perpendicular to B (i.e., J-perp.). The current-density contours, painted here on a yz plane located 30 $R_e$ tailward of Earth, show the plasma sheet and the tail boundary. The figure reveals where the cusp-mantle current streamlines intersect the 30 $R_e$ plane. Before the streamlines dis-appear behind the 30 $R_e$ plane, they wrap around the inner surface of the tail lobes and the plasma sheet boundary layer. They ensheathe the magnetotail lobes.

The plasma sheet exhibits a counterclockwise tilt with respect to the equatorial plane, which is associated with a positive y component of the IMF [Cowley, 1981]. The current contours that define the boundary also exhibit a marked $B_y$ signature in the diagonally symmetric gradation of strength that gives to the whole cross-sectional pattern an S shape (backwards in this case). The northern and southern mantles reside in the weak current parts of the pattern (the gaps that change a theta into a backwards S). Electromagnetically speaking, the mantle (a.k.a. the high-latitude boundary layer) is a dynamo. On each circuit around the tail, the current streamlines pass through the mantle and, so, presumably pick up the energy needed to maintain the current against ionospheric dissipation.

Consider again the continuity of cusp currents and region 1 currents seen in Plate 8. The two currents merged into each other with no sign of a break to separate them. Does this continuity extend to the global, 3D magnetospheric geometry of these currents? The question exposes a basic inconsistency in the picture presented thus far between generic region1 current geometry and generic cusp current geometry. According to the generic picture presented in Section 3, region 1 currents close over the high-latitude boundary layer, which means that they remain always within the same (northern or southern) hemisphere. Cusp currents, on the other hand, "close" by flowing into the opposite hemisphere. Therefore, it is impossible for generic region-1 geometry to merge continuously with generic cusp current geometry. They have different topologies. What, then, is the magnetospheric situation that corresponds to the continuity of the two currents at the ionospheric level in Plate 8? To answer this question, we trace three current streamlines from the dawn-side region 1 (tan) current segment. They were "launched"

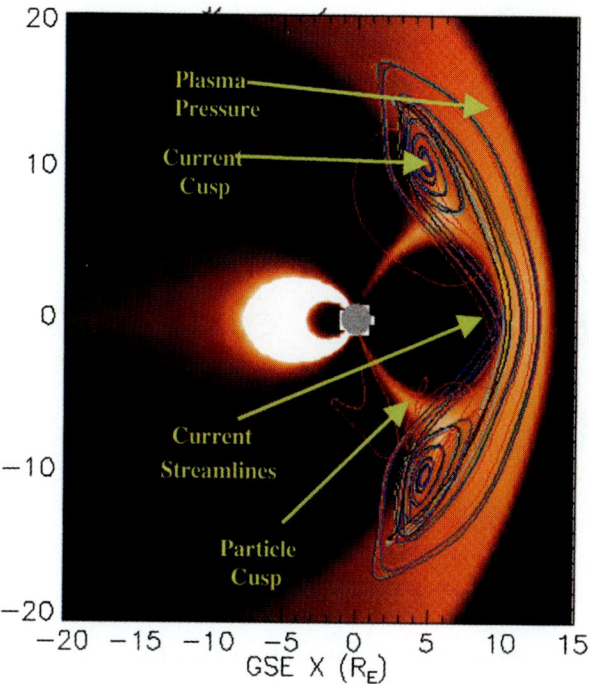

**Plate 7.** A dawnside view of the magnetosphere showing plasma pressure as color contours. Current streamlines, initiated from different depths within the Chapman-Ferraro layer, generate a set of nested spirals offset from the particle-entry cusp in both the y and z directions.

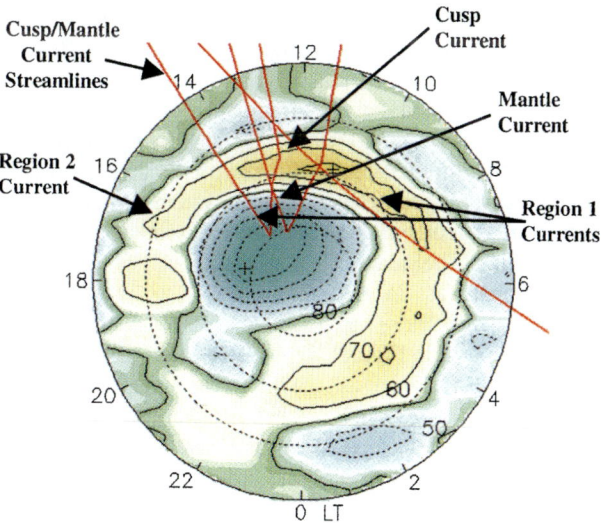

**Plate 8.** Similar to Plate 2 except for the case of a purely dawn-to-dusk (i.e., pure positive $B_y$) IMF. Two cusp-mantle current streamlines are shown.

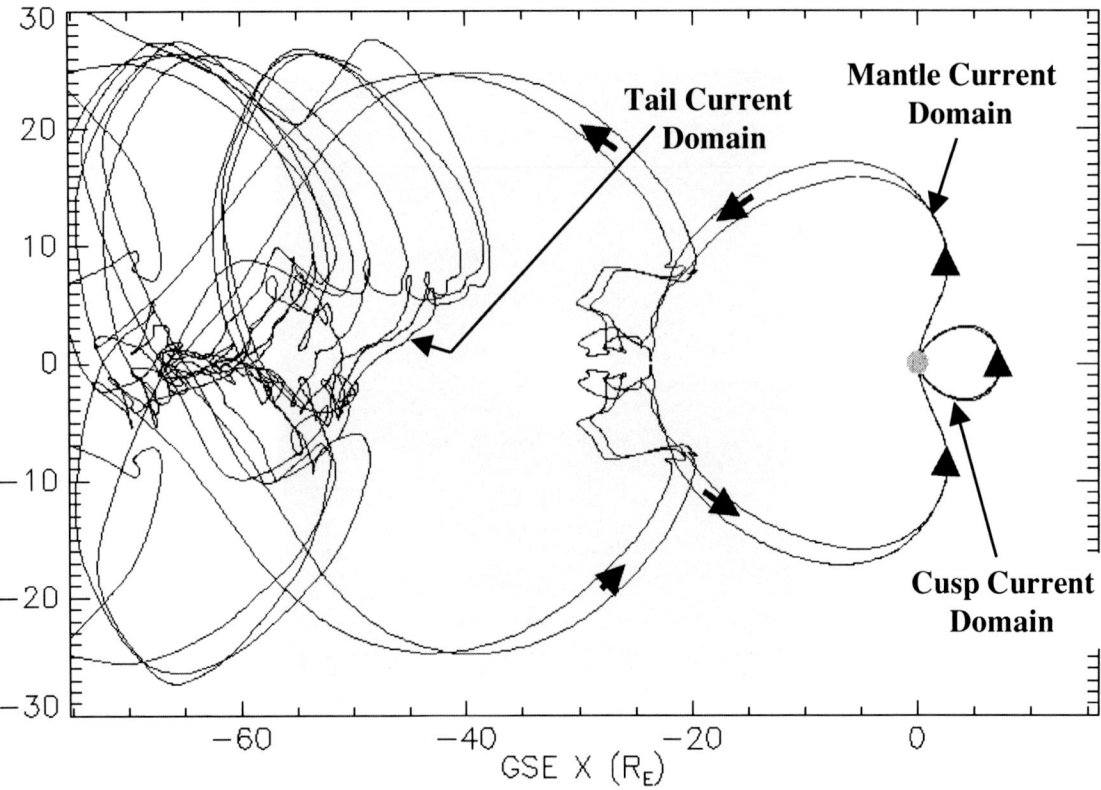

**Figure 1.** Dawnside view of the magnetospheric trajectories of the cusp-mantle current streamlines shown in Plate 8.

(in both directions to generate mantle and cusp segments) from the ridge of the region 1 contours at local times 6:00, 8:00, and 10:00. Together with the current streamlines already traced from around noon in Figure 1, they cover in equal, two-hour steps the local-time interval between dawn and noon.

Plate 10 shows the result. The currents from the ridge of the region 1 contour have cusp-mantle current geometry in the magnetosphere instead of region 1 geometry as determined for a pure negative IMF $B_z$. The mantle segments describe spirals around the lobes, as do the mantle segments in Plate 9. The cusp segments "close" to the opposite hemisphere, as do the cusp segments in Figure 1. They do not have same-hemisphere, region-1 closure as in Plate 3. An area poleward of the ridge of the region 1 contour but still within the contour does nonetheless have current streamlines that have generic region 1 geometry. They close within the same hemisphere across the mantle, which as Plate 10 shows for this IMF $B_y$ case is located to one side of the lobes. It remains for a future publication to describe how region 1 closure transitions to cusp-mantle closure as the IMF rotates from straight south to straight east. For the present, the message is that in the most common IMF situation ($|B_y| \gg |B_z|$), the dominant currents occupying the region 1 contour appear to be an extension of cusp-mantle currents. These currents emanate from the ridge of the contour. Poleward of the ridge generic region 1 streamlines emanate. This statement about the spatial dominance of cusp-mantle closure geometry should be further qualified to acknowledge that it ignores the part of the region 1 current system that couples to the region 2 current system, which here, as in all MHD codes to date, is under represented.

## 6. SUMMARY

This paper has explored five major current systems within the coupled ionosphere-magnetosphere system: Chapman-Ferraro, tail, region 1, NBZ, and cusp-mantle. For the case with IMF = 0, the classical Chapman-Ferraro and tail-current systems were found to overlap in the magnetosphere's midsection between the nose and the tail (i.e., its "waist") instead of abutting at a sharp interface on the magnetopause. The case with IMF purely southward was used to study the geometry of region 1 currents. They were found to close in the high-latitude boundary layer. The Chapman-Ferraro currents for this case were found to be layered, with the outer layer connecting to the bow

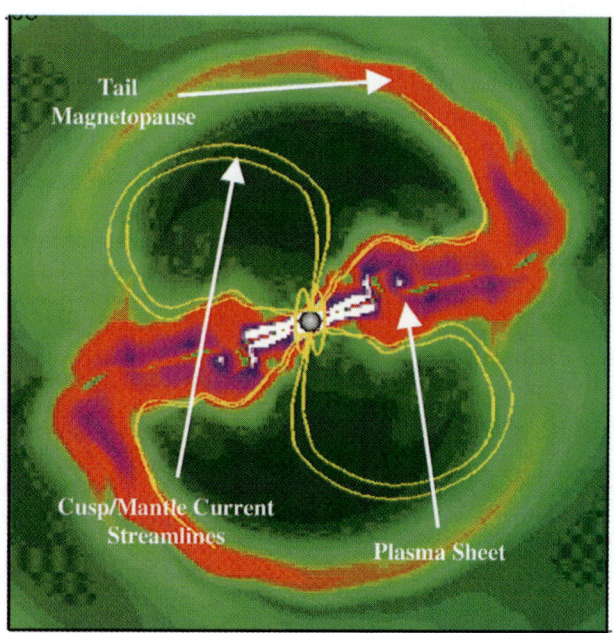

**Plate 9.** From-the-Sun view of the magnetospheric trajectories of the cusp-mantle current streamlines shown in Plate 8 with J-perp contours in color superimposed on a yz plane 30 $R_e$ tailward of Earth.

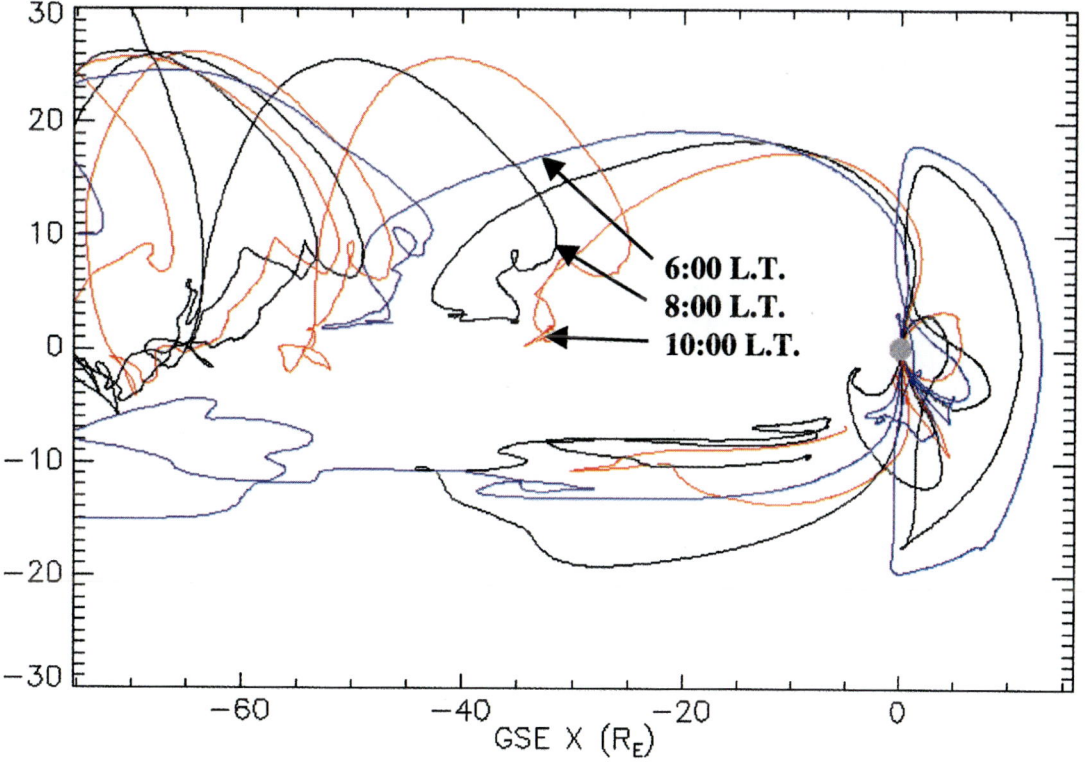

**Plate 10.** Similar to Plate 9 but with current streamlines initialized to pre-noon local times on the region 1 contour of Plate 8.

shock and the inner layer remaining within the magnetopause. A cusp-current vortex separated from the particle-entry cusp was identified. The case with IMF purely northward was used to study the geometry of NBZ currents. They were found to connect to the distant magnetotail. Classical region 2 current closure was also found. The case with IMF purely eastward was used to study the geometry of the cusp-mantle currents. They were found to integrate the entire magnetosphere—nose, waist, tail, ionosphere, and both hemispheres—with a single current structure. In the ionosphere, cusp currents and region 1 currents form a continuous band that has cusp-mantle closure geometry in the magnetosphere. Region 1-closure geometry was found poleward of the central ridge of the "region 1" current contour.

*Acknowledgments.* This work was supported in part by the NSF under grant ATM98-12678 and NASA under grant NAG5-8135. This work was also supported by the Defense Threat Reduction Agency, 45045 Aviation Drive, Dulles, VA 20166-7517.

## REFERENCES

Axford, W. I., H. E. Petschek and G. L. Siscoe: Tail of the magnetosphere, *J. Geophys. Res., 65*, 1231, 1965.

Bythrow, P. F., T. A. Potemra, R. E. Erlandson, L. J. Zanetti, and D. M. Klumpar, Birkeland currents and charged particles in the high-latitude prenoon region: A new interpretation, *J. Geophys. Res., 93*, 9791-9803, 1988.

Chapman, S., and V. C. A. Ferraro, A new theory of magnetic storms, *Terr. Mag. Atmos. Elec., 36*, 171-186, 1931.

Cowley, S. W. H., Magnetospheric asymmetries associated with the Y-component of the IMF, *Planet. Space Sci., 29*, 79-96, 1981.

Dessler, A. J., and R. D. Juday, Configuration of auroral radiation in space, *Planet. Space Sci., 13*, 63-72, 1965.

Iijima, T., and T. A. Potemra, The amplitude distribution of field-aligned currents at northern high latitudes observed by Triad, *J. Geophys. Res., 81*, 2165-2174, 1976.

Iijima, T., T. A. Potemra, L. J. Zanetti, and P. F. Bythrow, Large-scale Birkeland currents in the dayside polar region during strongly northward IMF: A new Birkeland current system, *J. Geophys. Res., 89*, 7441-7452, 1984.

Janhunen, P., H. E. J. Koskinen, T. I. Pulkkinen, Anew global ionosphere-magnetosphere coupling simulation utilizing locally varying time step, *Proc. Third International Conference on Substorms (ICS-3)*, pp. 205-210, EAS SP-389, October, 1996.

Midgley, J. E., and L. Davis, Calculation by a moment technique of the perturbation of the geomagnetic field by the solar wind, *J. Geophys. Res., 68*, 5111-5123, 1963.

Siscoe, G. L.: A unified treatment of magnetospheric dynamics, *Planet. Space Sci., 14*, 947-967, 1966.

Siscoe, G. L.: Energy coupling between regions 1 and 2 Birkeland current systems, *J. Geophys. Res., 87*, 5124-5130, 1982.

Siscoe, G.L., W. Lotko, and B.U.O. Sonnerup, A high-latitude, low- latitude boundary layer model of the convection current system, *J. Geophys. Res., 96*, 3487-3496, 1991.

Tanaka, T., Generation mechanisms for magnetosphere-ionosphere current systems deduced from a three-dimensional MHD simulation of the solar wind-magnetosphere-ionosphere coupling process, *J. Geophys. Res., 100*, 12,057-12,074, 1995.

---

G. L. Siscoe, N. U. Crooker, G. M. Erickson, Center for Space Physics, Boston University, 275 Commonwealth Ave., Boston, Massachusetts, 02215.

B. U. Ö. Sonnerup, Thayer School of Engineering, Dartmouth College, Hanover, New Hampshire, 03755.

K. D. Siebert, D. R. Weimer, W. W. White, N. C. Maynard, Mission Research Corporation, One Tara Blvd., Suite. 302, Nashua, New Hampshire, 03062.

# Field-Aligned-Current Systems in the Numerically Simulated Magnetosphere

T. Tanaka

*Communications Research Laboratory, Tokyo, Japan*

Configurations of the field-aligned-current (FAC) systems in the magnetosphere-ionosphere (M-I) coupling system are investigated through the analysis of the solutions obtained from a three-dimensional (3-D) magnetohydrodynamic (MHD) simulation. The results of calculation reproduce the region-1 and region-2 FAC systems for the southward interplanetary magnetic field (IMF) case, and additionally the NBZ FAC system for the northward IMF case. The energy budget associated with the generation of these FACs is discussed based on the numerical solution.

## 1. INTRODUCTION

The basic field-aligned-current (FAC) systems that connect the magnetosphere and the auroral ionosphere have been presented by *Iijima and Potemra* [1976], who referred to them as the region-1 and region-2 FAC systems. Later, *Iijima et al.* [1984] and *Iijima and Shibaji* [1987] showed an additional large-scale stable FAC system at higher latitude poleward of the region-1 system, during the period of northward interplanetary magnetic field (IMF). They called it as the NBZ FAC system.

In a steady current loop connecting the magnetosphere and the ionosphere, the ionosphere dissipates electromagnetic energy. In order to maintain a steady current system, therefore, the FAC must be connected to the magnetospheric dynamo region. This fact means that negative $\mathbf{J} \cdot \mathbf{E}$ must exists somewhere on the current loop in the magnetosphere. It is rather well known that the dynamo for the region-2 current loop distributes in the ring-current region [*Vasyliunas,* 1970; *Harel et al.,* 1981]. However, the problem is less clear for the region-1 current loop. A reason for this situation is that little is known as yet on where the FAC flows and closes in the magnetosphere despite the fact that a rather clear picture has been revealed for the spatial distribution of FAC at low altitudes. Recently, the magnetohydrodynamic (MHD) simulations of the solar wind-magnetosphere-ionosphere (S-M-I) system have been becoming more realistic. From these methods, *Tanaka* [1995] traced the FAC from the ionosphere to the magnetosphere and confirmed the previous expectation that the dynamo for the region-1 current loop distributes in the boundary layer [*Sonnerup,* 1980; *Stern* 1983].

In the magnetosphere-ionosphere (M-I) coupling system, large-scale plasma convection with convection electric field is essential to nearly all other phenomena occurring in this region, because it provides almost all free energy around there [*Caudal and Blanc,* 1988]. In this paper, the energy budget of the FAC generation is discussed in relation with the convection system based on the numerically reproduced FACs. This problem is considered under a complete self-consistent treatment of the three-dimensional (3-D) electordynamic structure in the S-M-I system obtained from an MHD simulation.

## 2. METHOD OF NUMERICAL SIMULATION

A numerical magnetohydrodynamic (MHD) simulation is adopted for the study of the present problem [*Tanaka,* 1995]. The FAC and plasma convection plays a central role in the M-I coupling, whereas the state of the energy source for these current systems depends on the solar wind-magnetosphere (S-M) interaction. Thus, a self-consistent treatment of the S-M-I coupling process is required for the investigation of FAC system. The calculation employs the finite volume (FV) total-variation diminishing (TVD) scheme with an unstructured grid system [*Tanaka,* 1995].

In this paper, the x-axis is pointing toward the sun, the y-axis is pointing toward the opposite direction of the earth's orbital motion, and the z-axis is pointing toward the north. The outer and inner boundaries for the simulation are at 90 $Re$ and 3 $Re$. A uniform solar wind with its speed of 350 km/sec and density of 5 cm$^{-3}$ is assumed at the upstream boundary. In the calculation of the M-I coupling process, dependent variables are projected along the field line from the inner boundary to the ionosphere. Under these circumstances, the number of effective grid points in the ionosphere becomes equal to the number of grid points on the inner boundary. Consequently, a sufficient number of grid points must be allocated on the inner boundary to resolve the structures generated in the ionosphere. Such a requirement is achieved through the use of an unstructured grid system.

In the ionosphere, Ohm's law is solved to match the divergence of the Pedersen and Hall currents with the FAC. The ionospheric conductivity is calculated from the solar EUV flux, diffuse precipitation modeled by the pressure and temperature, and discreet precipitation modeled by the upward FAC. Results of the simulation are obtained for two cases of northward and southward IMFs ($Bz = \pm 5$ nT). Plate 1 shows distributions of the Pedersen conductivity for the northward and southward IMF conditions obtained as a part of simulation results. In this paper, the Hall conductivity is set to be two times as larger as the Pedersen conductivity.

## 3. NORTHWARD IMF CASE

Plate 2 shows 3-D current lines in the M-I coupling system for the northward IMF case. Spheres in each panel show the size of the earth. In this plate, only current lines in the $+z+y$ region are drawn as seen from three directions, namely from dawn (upper left), from above the North Pole (lower) and from the sun (upper right). The color-codes on current lines show the plasma pressure normalized by the solar-wind pressure at where these currents are flowing.

In Plate 2, the region-1, region-2 and NBZ current systems are distinguishable from the difference in color tones. The region-1, region-2 and NBZ FAC systems are mainly mapped to the flank magnetopause, the equatorial region of the inner magnetosphere, and the nightside magnetosphere, respectively. It is interesting to note in Plate 2 that three current systems are distributing so as to occupy their position repelling with each other. Among them, the evening region-1 current flows outward from the earth to the magnetopause, and near the LLBL turns toward high latitudes. Then, evening region-1 current flows up the magnetopause toward the cusp. The region-1 current converges to a very narrow region in the noon-midnight meridian ($x$-$z$) plane. Passing through the bow shock, the pressure increases from the solar wind value to the magnetosheath value. A typical increase rate is from 1 to 15 in the color level shown in Plate 2. The pressure value on the region-1 current loop in Plate 2 (5-6 color level) is not so high as the magnetosheath value. This result indicates that the region-1 current loop is located inside the magnetopause.

While the region-1 current is closed through the cusp region, the region-2 current is closed in the inner edge of the plasma sheet. The evening region-2 current that starts from the inner plasma sheet flows longitudinally along the equatorial plane, and then turns earthward associated with a longitudinal pressure gradient. The color-codes on the region-2 current loop show that this current flows through the high-pressure regions in the inner magnetosphere.

The evening NBZ current flowing into the polar-cap ionosphere is mapped to the tail region. The NBZ current flows in the lobe toward the dusk magnetopause, then turns toward high latitudes near the magnetopause. Flowing up the magnetopause, it gradually turns the direction earthward and changes into the FAC. The color-codes on the NBZ current show that this current flows through the low-pressure region in the tail lobe.

## 4. SOUTHWARD IMF CASE

Plate 3 shows the 3-D configurations of the evening region-1 current and the Chapmann-Ferraro current for the southward IMF case. The views are from dusk sunward region (upper panel) and dusk anti-sunward region (lower panel) in the north of the equatorial plane. In this plate, red and black lines draw the region-1 and Chapmann-Ferraro currents, respectively. The region-2 current loop is not drawn here avoiding a mixedup. In addition to the current loops, Plate 3 also shows by color-codes distribution of $\mathbf{J} \cdot \mathbf{E}$ in the noon-midnight meridian ($x$-$z$) and equatorial ($x$-$y$) planes. In general, a stationary-state solution cannot be obtained for a steady southward IMF. Starting from a stationary solution for the northward IMF, the solution changes its configuration gradually after a southward turning of the IMF, until a sudden breakdown of monotonous tail structure. The results shown in Plate 3 is at the time just before the sudden breakdown, thus corresponds to the growth phase of the substorm. The most characteristic feature occurring in this interval is the thinning of the plasma sheet and the increase in the flaring angle of the magnetopause.

The configuration of the region-1 current loop under the southward IMF condition is topologically similar to that under the northward IMF. However, under the southward IMF condition, the region-1 current passes the noon-midnight ($x$-$z$) plane diverging widely in the $x$ direction. In Plate 3, the region-1 current loop is closed inside the Chapmann-Ferraro current. This feature does not coincides with the prediction by *Stern* [1983] that the driving force of region-1 FAC is the interplanetary electric field linked directly to the polar ionosphere along the open field lines.

In Plate 3, the color-codes for the $\mathbf{J} \cdot \mathbf{E}$ distribution are green to blue ($\mathbf{J} \cdot \mathbf{E} < 0$) at the bow shock, high-latitude side of the cusp, mantle and inner magnetosphere, whereas they are red to white

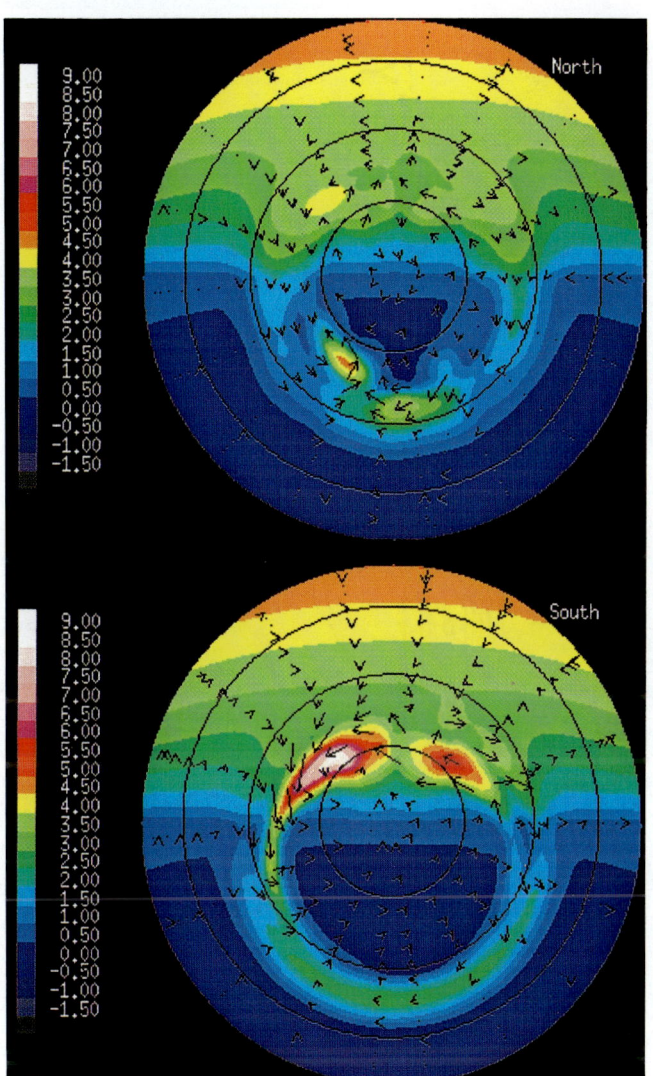

Plate 1. Distributions of the Pedersen conductivity (mho) obtained from the simulation for the northward (top) and southward (bottom) IMF cases. Noon is to the top. The three circles show the northern latitudes of 60°, 70°, and 80°.

## 56 FIELD-ALIGNED CURRENT SYSTEMS

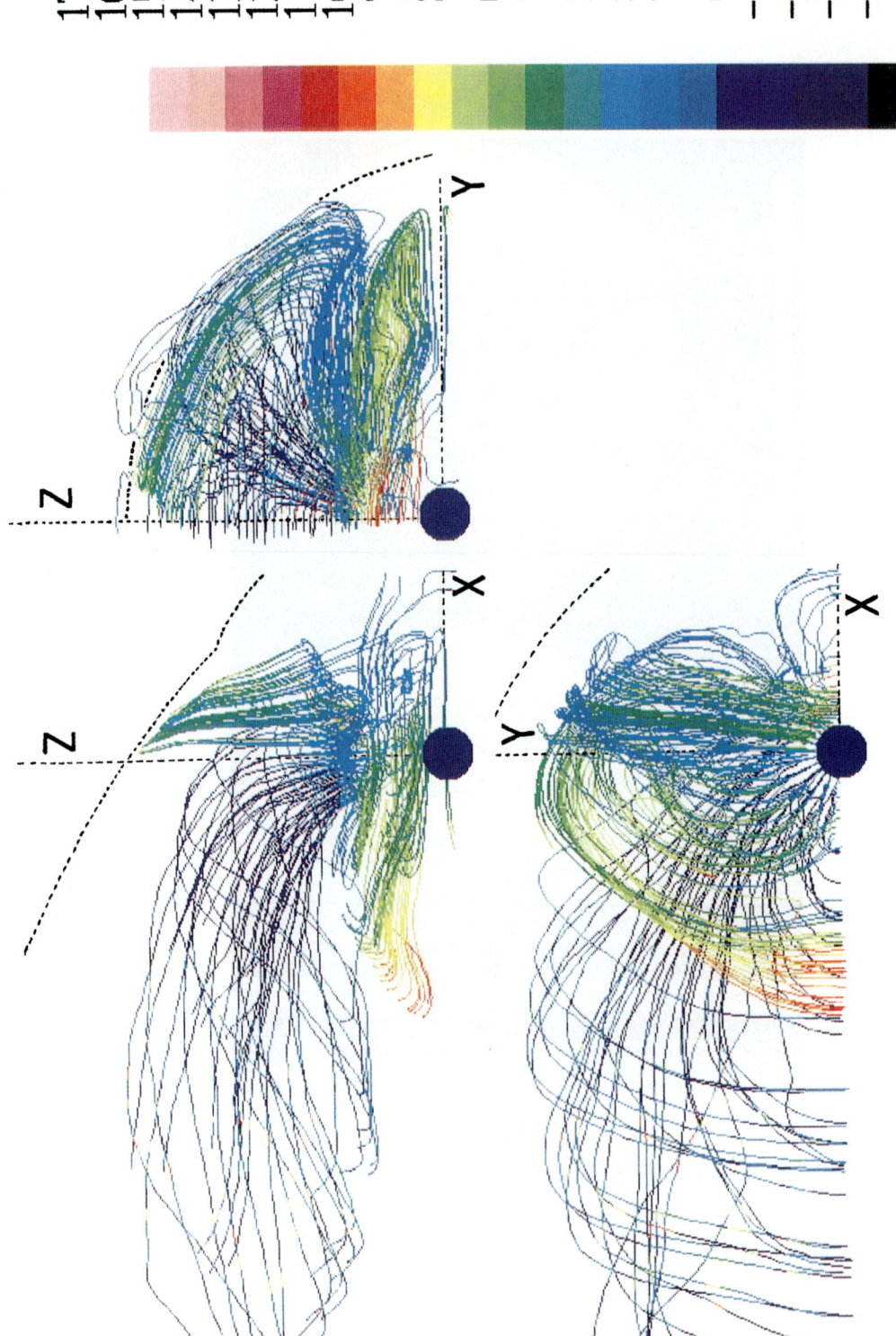

Plate 2. The 3-D current lines in the M-I coupling system for the northward IMF case. Spheres in each panel show the size of the earth. The color-codes on current lines show the plasma pressure normalized by the solar wind pressure at where these currents are flowing. Dotted curves show the magnetopause cross-positions in the planes including two axes drawn in each panel.

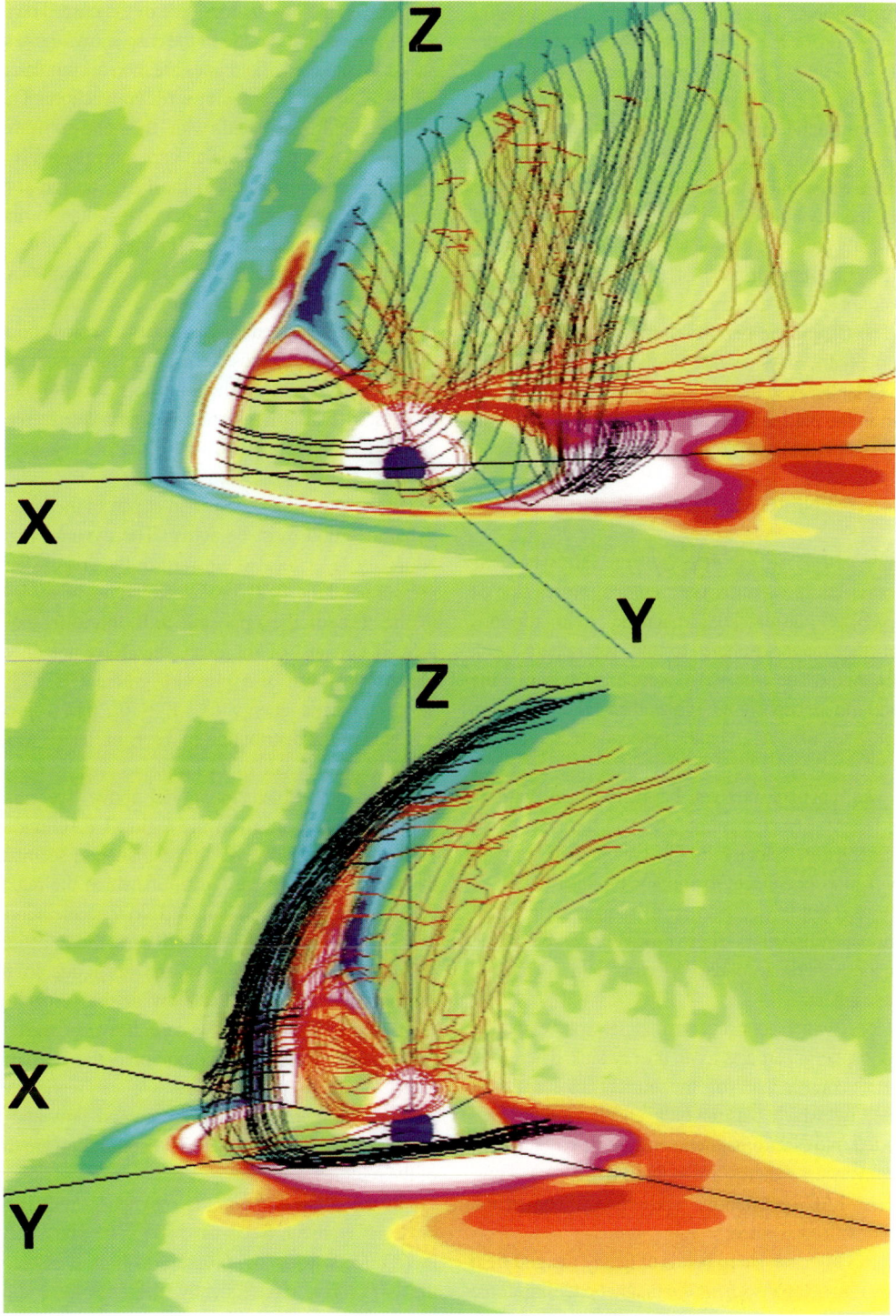

Plate 3. The 3-D configurations of the evening region-1 current and the Chapmann-Ferraro current for the southward IMF case with color-codes showing the distribution of **J · E** in the noon-midnight meridian (x-z) and equatorial (x-y) planes. Red and black lines draw the region-1 and Chapmann-Ferraro currents, respectively.

($\mathbf{J} \cdot \mathbf{E} > 0$) at the dayside magnetopause, low-latitude side of the cusp and the plasma sheet. The bow shock acts as a dynamo because the compression of the IMF occurs associated with the braking of the solar-wind flow. In the magnetosphere, the dynamos are in the mantle and in the ring-current region. A unique feature in the color distribution around the cusp region is that the cusp is clearly divided by two colors such as white color on the low-altitude side and blue color on the high-latitude side. This structure is intruding toward the low-altitude region. The dayside part of the region-1 FAC is mainly connected to the tailward edge of the blue-color region on the high-latitude side of the cusp, whereas the nightside part is connected to the green-color region in the lobe near the mantle. The Chapmann-Ferraro current is connecting the white-color region on the dayside magnetopause and the blue-color region in the mantle. Also the tail θ current is connecting the white-color region in the plasma sheet and the blue-color region in the mantle.

## 5. ENERGY BUDGET

The generation of FAC is closely related with the magnetospheric convection, which covers a global description of the way to organize plasma and magnetic field motions distributing from the magnetosphere to the ionosphere in a circulation configuration. In the M-I coupling system, the magnetosphere and the ionosphere recognize their relative motion through the exchange of FAC. Field perpendicular currents connected to this FAC adjust the flow configurations in such a way that in a steady state magnetic flux can be interchanged without flux pileup at all height. The dissipation effect in the ionosphere decelerates the convection velocity in the coupling system. In addition, the low-$\beta$ ionosphere acts to modify the distribution of the convection flow (electric field) so as to reduce a divergent component (rotational component). These processes can be said in other words that ionospheric secondary electric field is mapped back to the magnetosphere.

In the current loop, which connects the magnetosphere and the ionosphere, the ionospheric part acts as an energy sink. Therefore, the magnetospheric counter part of the current closure must include a dynamo, by which plasma bulk energy and internal energy are converted into electromagnetic energy. Apparently this kind of energy conversion requires a condition $\mathbf{J} \cdot \mathbf{E} < 0$. Thus, convection pattern, FAC generation, and energy conversion process are coupled with each other.

In the southward IMF case, sharp bends of newly merged field lines in the dayside magnetopause exert the magnetic tension force to accelerate the plasma (motor action), thus to convert electromagnetic energy into kinetic energy ($\mathbf{J} \cdot \mathbf{E} > 0$). The kinetic energy generated through the motor action is immediately converted to internal energy, because flow tends to stagnate toward the cusp. Consequently, the magnetic tension pumps up the cusp pressure on the low-latitude side of the cusp. The red color in Plate 3 on the low-latitude side of the cusp shows these situations.

Open magnetic field lines are then accumulated into the mantle in the course of anti-sunward convection. The dawn-to-dusk electric field associated with this anti-sunward convection is mapped down to the polar cap ionosphere. If the ionospheric current is not connected to the FAC, charge accumulation must appear on the polar cap boundary due to the dawn-to-dusk ionospheric Pedersen current [*Caudal and Blanc*, 1988]. Using the relation div$\mathbf{J}$=0, the present model automatically inhibits this charge accumulation and connects the ionospheric current to the FAC. Thus, a steady ionospheric convection is established following the magnetospheric convection. Seeing from a different angle, the $\mathbf{JxB}$ force in the ionospheric part of the region-1 current loop is such as to maintain the ionospheric plasma convection caused by the mapped down electric field against the atmospheric friction ($\mathbf{J} \cdot \mathbf{E} > 0$).

In the magnetospheric counter part of the region-1 current loop, dynamo action must be active. The dynamo region exists in the boundary layer located tailward of the cusp, where both dynamic pressure and the $\nabla P$ force keep the boundary-layer flow moving in the face of the opposing $\mathbf{JxB}$ force. In the boundary layer located tailward of the cusp, therefore, not only the braking of plasma flow but also pressure gradient is such as to generate electromagnetic energy ($\mathbf{J} \cdot \mathbf{E} < 0$). These processes are well reflected in the distribution of $\mathbf{J} \cdot \mathbf{E}$ in Plate 3 as the blue color on the high-latitude side of the cusp.

From the explanation given above, we can understand the generation mechanism of the region-1 FAC under the southward IMF condition. However this explanation seems not effective for the northward IMF case, even in which the region-1 FAC exists despite the lack of anti-sunward flow acceleration by reconnected field lines. In the northward IMF case, motor action is rather active on the high-latitude side of the cusp due to the high-latitude reconnection and sunward flow at the center of the polar cap [*Tanaka*, 1995]. Even in this case, however, the cusp reconnection loads anti-sunward momentum onto the open field lines in the LLBL [*Tanaka*, 1999]. Thus we can conclude that the anti-sunward momentum of the solar wind flow and its loading onto the open field lines through the reconnection are the primary cause of the region-1 FAC.

In the plasma sheet, the magnetic stress is balanced by the pressure gradient force as $\mathbf{JxB}=\nabla P$. The sunward convection there directs from low-pressure to high-pressure regions, against the $\nabla P$ force. Consequently, the $\mathbf{JxB}$ force directs toward the plasma convection and converts magnetic energy into internal energy ($\mathbf{J} \cdot \mathbf{E} > 0$). These configurations mean that the magnetic tension is acting as an electromagnetic pump that generates the high-pressure region in the inner edge of the plasma sheet.

The primary driver of the region-2 FAC is the high plasma pressure in the inner edge of the plasma sheet. The sunward convection from the inner plasma sheet diverts to both sides of the earth to the dayside. In this region, the $\nabla P$ force drives flow against the line-tying effects of the auroral oval. The region-2 FAC is generated from the azimuthal pressure gradient formed in the inner magnetosphere through plasma redistribution in the course of convection [*Harel et al.*, 1981]. When the MHD is applied to the inner magnetosphere, however, we must be aware that the MHD equivalently includes the $\nabla B$ drift but not the curvature drift. Due to this limitation, the generation rate of the region-2 current presented in this paper may not be complete.

It is interesting to note that the $\nabla P$ force is against the **JxB** force everywhere in the magnetosphere. Consequently, dynamo action takes place if the convection flow is anti-parallel to the **JxB** force ($\mathbf{J} \cdot \mathbf{E} < 0$), and pumping action takes place if the convection flow is parallel to the **JxB** force ($\mathbf{J} \cdot \mathbf{E} > 0$).

## REFERENCES

Caudal, G., and M. Blanc, Magnetospheric convection during quiet or moderately disturbed times, *Rev. Geophys.*, *26*, 809, 1988.

Harel, M., R. A. Wolf, P. H. Reiff, R. W. Spiro, W. J. Burke, F. J. Rich, and M. Smiddy, Quantitative simulation of a magnetospheric substorm, 1, Model logic and overview, *J. Geophys. Res.*, *86*, 2217, 1981.

Iijima, T., and T. A. Potemra, The amplitude distribution of field-aligned currents at northern high latitudes observed by Triad, *J. Geophys. Res.*, *81*, 2165, 1876.

Iijima, T., and T. Shibaji, Global characteristics of northward IMF-associated (NBZ) field-aligned currents, *J. Geophys. Res.*, *92*, 2408, 1987.

Iijima, T., T. A. Potemra, L. J. Zanetti, and P. F. Bythrow, Large-scale Birkeland currents in the dayside polar region during strong northward IMF: A new Birkeland current system, *J. Geophys. Res.*, *89*, 7441, 1984.

Sonnerup, B. U. O., Theory of the low-latitude boundary layer, *J. Geophys. Res.*, *85*, 2017, 1980.

Stern, D. P., The origin of Birkeland current, *Rev. Geophys.*, *21*, 125, 1983.

Tanaka, T., Generation mechanisms for magnetosphere-ionosphere current systems deduced from a three-dimensional MHD simulation of the solar wind-magnetosphere-ionosphere coupling processes, *J. Geophys. Res.*, *100*, 12,057, 1995.

Tanaka, T., Configuration of the magnetosphere-ionosphere convection system under northward IMF condition with non-zero IMF *By*, *J. Geophys. Res.*, *104*, 14,683, 1999.

Vasyliunas, V. M., Mathematical models of the magnetospheric convection and its coupling to the ionosphere, in Particle and Fields in the Magnetosphere, edited by B. M. McCormac, p. 60, D. Reidel, Norwell, Mass., 1970.

---

T. Tanaka, Communications Research Laboratory, Koganei-shi, Tokyo 184-8795, Japan

# Recent Progress in the Data-Based Modeling of Magnetospheric Currents

N. A. Tsyganenko

*Raytheon ITSS Corporation, NASA Goddard Space Flight Center, Greenbelt, Maryland*

Magnetospheric currents are very diverse and dynamical. The magnetopause shrinks, expands, and erodes, with varying degree of connection to the IMF. Intrusion of the plasma in the polar cusps results in diamagnetic currents which depress the field at high latitudes. The tail current sheet warps, bends, and twists, and its current density varies as the tail stretches and rebounds. Birkeland currents vary with the IMF and with the substorm cycle, and the ring current builds up and decays in the course of storms. During substorms, a dynamical current wedge develops on the nightside, resulting in a rapid reconfiguration of the near-Earth field. The data-based approach to magnetospheric modeling is to develop a flexible representation for the magnetic field vector and fit it to a large set of spacecraft data. All the above currents should be represented in the models in a meaningful way, and substantial advance was made recently in this direction. The new models feature a continuous parametric dependence on the solar wind conditions, include Birkeland currents, and have a solar-wind controlled magnetopause with a penetration of the IMF. New techniques are now available for modeling the twisting and warping of the cross-tail current, the collapse of the tail field during substorms, and the observed structure of the polar cusps. This article overviews some of those developments.

## 1. INTRODUCTION

The primary goal of data-based modeling of the magnetosphere is to extract full information on its structure from space magnetometer observations and to reconstruct its configuration as a function of external conditions. The electric current **j** and the magnetic field **B** in the magnetosphere are related by Maxwell's equations, so that after deriving **B** from data, one should in principle be able to reconstruct **j**. Using concurrent data from the upstream monitors of the interplanetary medium, one can relate the magnetospheric dynamics to the changing state of the incoming solar wind.

However, this seemingly simple plan turns out hard to implement because of the highly dynamical nature of the geomagnetic field and the lack of simultaneous measurements. So far, all existing data-based models have relied on observations made at different times and have implicitly assumed that similar solar wind conditions should result in similar geomagnetic field configurations, which is often not true because the "memory" of earlier conditions also plays a significant role.

In terms of the magnetospheric currents, the approach adopted in data-based modeling can be outlined as follows. If the current density **j** is known, the magnetic field at any location can in principle be derived as

$$\mathbf{B} = \frac{1}{\mu_0} \int \frac{\mathbf{j} \times (\mathbf{r} - \mathbf{r}')}{|\mathbf{r} - \mathbf{r}'|^3} d\mathbf{r}' \qquad (1)$$

Since what is observed is not **j** but **B**, this leads to the kind of problem usually referred to as an "inverse" problem. A standard approach is to choose for **j** a specific mathematical form involving adjustable parameters, based on physics and on existing *a priori* information, derive the corresponding

**B**, and fit the parameters to the data. The geometrical characteristics of the electric current systems and the way they respond to solar wind conditions are then given by the best fitting values of the model parameters.

Although the magnetic field at any point depends on the entire distribution of the electric current, the linearity of Maxwell's equations allows one to resolve the total **B** into "partial" fields and currents, each related to a different current system. This greatly facilitates the mathematical treatment of the problem. Thus, the field associated with the magnetopause current can be represented by an appropriate potential field, continuous over the entire space and ensuring the required distribution of the normal component $B_n$ on the boundary. Introducing an abrupt jump of **B** across the boundary automatically accounts for the magnetopause current required by Maxwell's equations. Likewise, the fields associated with the cross-tail and the ring current can be represented by separate "modules", providing the required geometry of **j**, full confinement of **B** inside the model magnetopause, and an appropriate parametric dependence of **j** on the state of the solar wind.

Such a modular approach was widely used in the past [e.g., *Voigt*, 1981; *Tsyganenko and Usmanov*, 1982; *Tsyganenko*, 1987, 1989, 1995, 1996] as an effective way of turning data into models. On the other hand, from the viewpoint of physics, different current systems in the magnetosphere are closely related. Such relationships should be treated consistently when modeling the total field. As an example, the observed magnetic features of the polar cusps represent the combined effect of at least three sources: the magnetopause currents, diamagnetic currents due to the injected magnetosheath plasma, and the field-aligned currents. Another example is the cross-tail current, whose global geometry cannot be separated from that of the magnetopause current.

During the last few years, significant progress was made in the development of realistic models of the magnetospheric field and electric currents, based on both new and old space magnetometer data. Because of space limitations, this paper focuses mainly on the most recent results and on problems still to be solved. Readers interested in a more detailed treatment of existing modeling methods are referred to reviews [*Tsyganenko*, 1990; *Stern*, 1994] and to the original papers, cited below.

## 2. MODELING OF THE MAGNETOPAUSE CURRENTS

Compared with other magnetospheric currents, that of the magnetopause has the largest spatial scale and the largest total magnitude. It dictates the configuration of the entire magnetosphere, confining the fields from all internal sources within a cavity in the solar wind.

In currently existing models, the magnetopause is assumed to be an axially symmetric surface with a prescribed shape, whose scale size is controlled by the dynamic pressure of the solar wind. Several data-based models of the magnetopause were recently proposed. *Sibeck et al.* [1991] and *Roelof and Sibeck* [1993] developed an ellipsoidal representation of the magnetopause, using a set of 1821 boundary crossings by spacecraft and parameterizing their model by the solar wind pressure and IMF $B_z$. *Petrinec and Russell* [1993, 1996] evaluated the magnetopause shape indirectly, using internal measurements of the tail field by ISEE spacecraft and a simplified form of the pressure balance condition. *Shue et al.* [1997, 1998] developed an alternative analytical magnetopause model, based on direct crossing data, and calibrated it by solar wind parameters. *Boardsen et al.* [1999] made a detailed statistical study of the high-latitude magnetopause shape, using the data of Hawkeye-1.

In this work, we focus on the modeling of the magnetic field of the magnetopause current, rather than on the shape of the boundary, and hence will not compare the merits of the above models. The basic approach is to represent the field of the magnetopause currents by a sum of simple and flexible potential fields, whose coefficients and scale factors are fitted by least squares to minimize the rms magnetic flux across the boundary (an idea first suggested by *Schulz and McNab* [1987]). Such a procedure can be applied to the "partial" contributions from all principal sources of the internal magnetospheric field, providing an accurate confinement of the total field for any combination of the amplitudes of individual sources [*Tsyganenko*, 1995, 1996].

Recent observations by the ISTP spacecraft Geotail and Polar revealed some interesting features, which allowed significant improvements. The first finding was that the average magnetopause was displaced northward and southward by the diurnal/seasonal tilt of the Earth's dipole. In fact, such an effect was predicted long ago from calculations of the pressure-balanced shape of the boundary for a tilted dipole [*Olson*, 1969; *Choe et al.*, 1973], confirmed by MHD simulations [*Walker et al.*, 1989] as well as in calculations of the equilibrium boundary by *Sotirelis and Meng* [1999]. However, it was not until recently that the tilt-related shift of the magnetopause was found in spacecraft observations, first reported in a statistical study of *Hammond et al.* [1994]. A clear indirect evidence of the tilt-induced magnetopause motion was demonstrated recently by *Tsyganenko et al.* [1998], based on the observed shape of the cross-tail current sheet as inferred from Geotail magnetometer data. In our later paper [*Tsyganenko*, 1998a], the effect of the dipole tilt upon the shape of the boundary was again confirmed by using direct magnetopause crossing data.

Figure 1 illustrates the dipole tilt effects upon the shape of the cross-section of the tail current sheet for two intervals of $X_{GSM}$. While at close distances the warping of the current sheet is quite pronounced, it gradually decreases tailward and becomes much weaker in the more remote bin with $-60 < X < -40 R_E$. At the same time, the amplitude of the

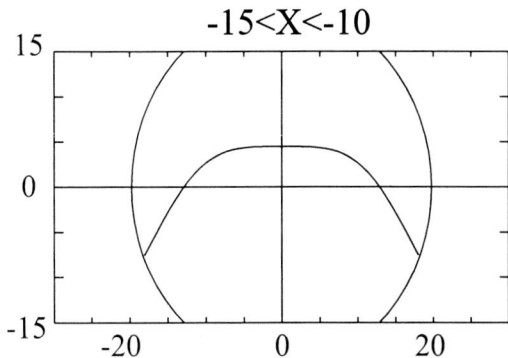 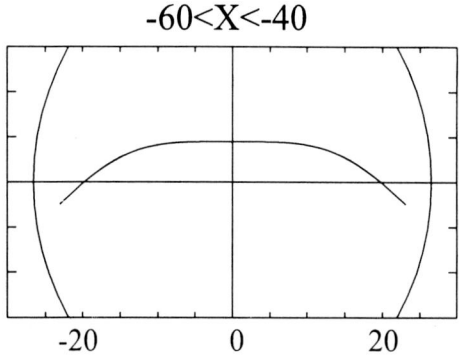

**Figure 1.** The shapes of the cross-tail current sheet for a tilted geodipole ($\Psi = 30°$), as derived from the Geotail and ISEE-1/2 data. Dashed lines show the unshifted position of a model magnetopause, which would imply an imbalance of the tail lobe magnetic fluxes [*Tsyganenko et al.*, 1998].

shift of the central current sheet remains virtually unchanged. Assuming no concurrent shift of the tail boundary (as shown in the plots) would imply a significant imbalance of the net magnetic flux between the northern and southern tail lobes. The only way to avoid that problem is to assume that the entire tail boundary is shifted in concert with the cross-tail current sheet, and observations fully corroborate that conjecture. Figure 2 shows three samples of the magnetopause crossing data in the interval of tailward distances between 20 and 40 $R_E$, for three intervals of the dipole tilt angle. Best fit ellipses are also shown, approximating the observed average cross-sections of the boundary. As can be seen, the centers of the ellipses (shown by crosses) shift upward, as the dipole tilt angle increases from negative to positive values. As already mentioned, this effect was found earlier in a data-based study of *Hammond et al.* [1994]. Although a direct comparison with their results is somewhat hampered by significant differences in the adopted binning by the tilt angle and by the tailward distance, the obtained amplitudes of the magnetopause shift are reasonably close to each other. Thus, for large tilt angles ($\sim 30°$) *Hammond et al.* [1994] found the overall shift $\Delta Z \sim 2\,R_E$, while our study gave $\Delta Z \sim 2.8\,R_E$, for the same tilt angle. Since our estimate was based on data taken, on the average, at larger distances, there is no major disagreement between the two results, given a relatively large statistical uncertainty. The same holds for the tail neutral sheet: although different data and mathematical forms were used by *Hammond et al.* [1994] and *Tsyganenko et al.* [1998], both studies provided nearly the same shapes of the sheet at $X \sim -25\,R_E$; in particular, the hinging distance values were found equal to $8.6\,R_E$ and $8.2\,R_E$, respectively.

The above described findings imply that the accurate modeling of the Earth's dipole tilt effects should include in a mutually consistent way deformations of both the magnetopause and the cross-tail current. A simple mathematical technique was recently developed [*Tsyganenko*, 1998a], allowing one

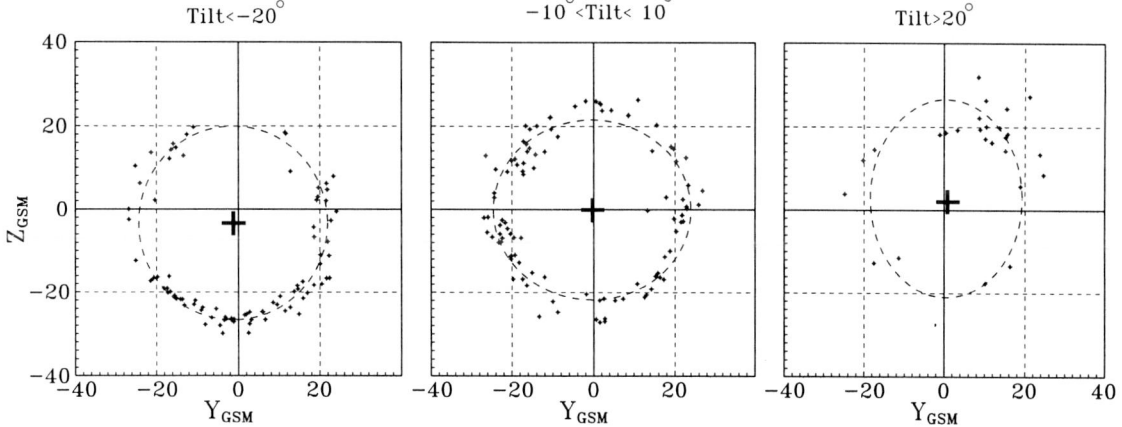

**Figure 2.** Observed crossings of the magnetopause at $-40 < X < -20 R_E$, binned into three intervals of the dipole tilt angle, with the elliptical best fit positions of the boundary. The centers of the ellipses (crosses), indicate the average position of the boundary, which clearly depends on the dipole tilt angle [*Tsyganenko*, 1998a].

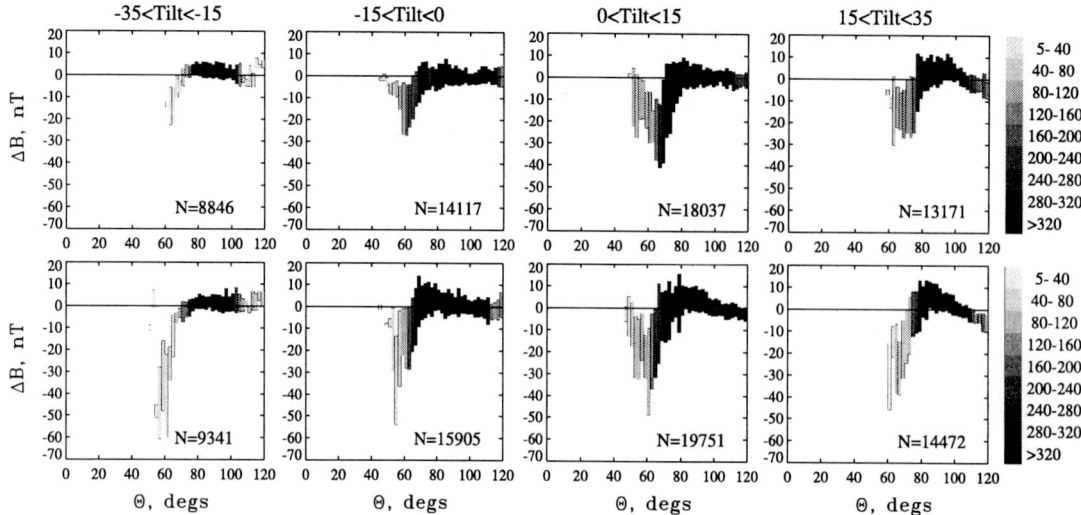

**Figure 3.** Difference $\Delta B$ between the magnitudes of the field measured by Polar and predicted by T96 model, plotted as a function of the angular distance from the sunward direction $\theta$, for positive (top) and negative (bottom) IMF $B_z$ and four intervals of the dipole tilt angle (indicated above the plots). Positions and heights of the shaded strips correspond to average values and rms deviations of $\Delta B$, respectively, and the intensity of the grey shading is proportional to the number of 1-min averages in each bin of $\theta$, as shown on the densitometer bar on the right. The plot corresponds to the farthest bin of the radial distance sampled by Polar. Note the depressed field inside the cusp, as well as a conspicuous positive excursion of $\Delta B$ polarward from the depression, due to enhanced ram pressure of the magnetosheath plasma flow in that region [*Tsyganenko and Russell*, 1999].

to represent the tilt effects by a compact transformation of the untilted configurations. The deformation method makes it possible to further improve the models by removing the currently adopted limitation of axial symmetry and allows one to take into account not only tilt-related effects, but also the distant tail flattening due to the IMF-related anisotropy of the external pressure upon the magnetosphere. More details of this method are given in the next section.

Another important feature of the magnetopause current, so far ignored by the existing models, is a pair of indentations on the dayside, associated with the polar cusps. Their existence was already predicted in early calculations of the magnetopause shape [e.g., *Mead and Beard*, 1964], based on assumption of the pressure balance between a shielded dipole field and an incoming flow of solar wind particles (the problem was revisited by *Sotirelis* [1996] and by *Sotirelis and Meng* [1999]). As discussed below, taking into account those indentations results in a significant change of the model field near the outer cusp; however, because of the limited spatial extent of the "dimples" and of the cusp-related diamagnetic currents, their low-latitude effects are small and cause no major modification in the global field pattern.

Figure 3 [*Tsyganenko and Russell*, 1999] shows distributions of the scalar difference $\Delta B$ between the total field observed by Polar and the one derived from a model [*Tsyganenko*, 1996], as a function of the angle $\theta$ between the position of the spacecraft and the solar-magnetic $X$ axis. The plots correspond to a relatively narrow interval $8 < R < 9$ of the radial distance near the Polar apogee and they display separately the profiles of $\Delta B$ for four intervals of the geodipole tilt angle. A well-pronounced depression, localized around $\theta \sim 60 - 75°$, is clearly seen in all panels, indicating a significantly weaker field in the cusps, than is predicted by the model, as well as some field compression poleward of the cusp. The source of that discrepancy is twofold: (i) the axially symmetric shape of the T96 boundary and, hence, an underestimate of the field compression near the cusps and (ii) the absence of the diamagnetic currents associated with the polar cusps in the T96 model.

Both effects can be taken into account in a relatively simple way. In the remaining part of this section we concentrate on modeling the effects of the magnetopause indentations, while the depression due to the diamagnetic currents is relegated to a separate section.

The indentations can be introduced as a perturbation of the initially axisymmetric model magnetopause, assumed to be an ellipsoid in the cislunar region, smoothly continued by a cylinder in the distant tail. The unperturbed boundary can be described as a surface of constant $\sigma(X, Y, Z) = \sigma_0$, where

$$\sigma = 0.5\left[\sqrt{(1+\xi)^2 + \zeta^2} + \sqrt{(1-\xi)^2 + \zeta^2}\right] \quad (1)$$

and

$$\xi = (X - X_0)/a + 1 \qquad \zeta = \sqrt{Y^2 + Z^2}/a$$

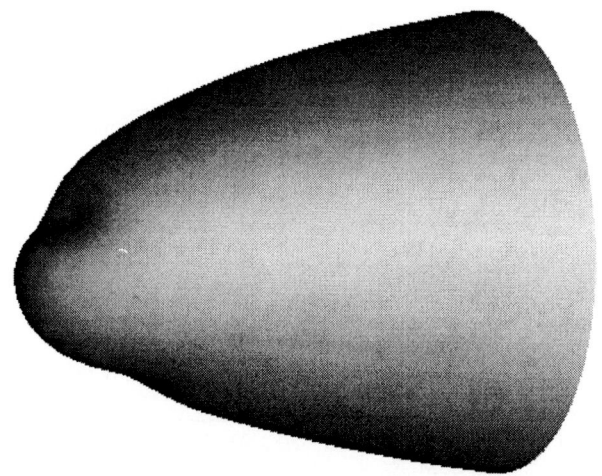

**Figure 4.** The surface of $\sigma' = \sigma_0$, illustrating the shape of the model magnetopause with indentations near the cusps. Due to the tilted dipole, the northern "dimple" is more exposed to the solar wind and hence deeper than the southern one.

Specifying $X_0 = 5.48 R_E$, $\sigma_0 = 1.08$, and $a = 70$ uniquely defines an average axisymmetric boundary, similar to that of *Sibeck et al.* [1991] for $P = 2$ nPa, and these values were used to approximate the boundary in the data-based model [*Tsyganenko*, 1995, 1996].

The cusp-related indentations on the boundary can be introduced by adding a perturbation to the righthand side of (1), so that the boundary is defined as a surface of constant

$$\sigma' = \sigma + \Delta\sigma(\tau, \phi) \quad (2)$$

where

$$\tau = 0.5\left[\sqrt{(1+\xi)^2 + \zeta^2} - \sqrt{(1-\xi)^2 + \zeta^2}\right] \quad (2a)$$

is the second ellipsoidal coordinate, varying from $+1$ at the subsolar point to 0 in the midtail, where the magnetopause smoothly joins the cylindrical tail boundary. The latitudinal position, angular width, and depth of the indentations should depend on the geodipole tilt angle and on the IMF conditions. A simple way to obtain the desired shift of the cusp position with the tilt angle is to deform the entire boundary by imposing a radially-dependent rotation of the coordinates around the $Y$ axis [*Tsyganenko*, 1998a]. Namely, we replace $(X, Y, Z)$ by $(X^*, Y, Z^*)$, where

$$\begin{aligned} X^* &= X\cos\Psi^* - Z\sin\Psi^* \\ Z^* &= X\sin\Psi^* + Z\cos\Psi^* \end{aligned} \quad (3)$$

In addition, the north-south asymmetry of the indentation depth and its latitudinal width for $\Psi \neq 0$ should be taken into account, as prompted by recent calculations of *Sotirelis and Meng* [1999]. This can be achieved by modulating the corresponding indentation parameters by the dipole tilt angle. Without entering into further details of the model, we reproduce in Figure 4 the resultant shape of the boundary as a 3-D view of the surface $\sigma' = \sigma_0$ for $\Psi = 30°$.

Once the model boundary is analytically described, we need to develop a mathematical representation of the magnetopause field, ensuring an accurate confinement of the total **B** within the surface, for any value of the dipole tilt angle. As found in our earlier works [*Tsyganenko*, 1995, 1996, 1998a], a simple and economical way to accurately shield the internal magnetospheric field is to employ a combination of the cartesian "box" potentials, having the form

$$\exp\left[\left(\frac{1}{p^2} + \frac{1}{q^2}\right)X\right]\cos\frac{Y}{p}\left\{\begin{array}{c}\sin(Z/q) \\ \cos(Z/q)\end{array}\right\} \quad (4)$$

The upper and lower factors in the braces correspond to two types of symmetry: the terms with $\sin(Z/q)$ are for the untilted magnetosphere, while the terms with $\cos(Z/q)$ represent the perturbation field, to the lowest order proportional to the tilt angle $\Psi$. In fact, as shown in detail by *Tsyganenko* [1998a], the extended form of the expansion with tilt-dependent perturbation terms is needed only for the geodipole shielding field, while contributions of external sources, including those of the tail current, can be treated in a much simpler way. Namely, it suffices to derive the confined tail field only for the case of zero tilt angle, $\Psi = 0$. The configurations for $\Psi \neq 0$ are then easily obtained by imposing a global deformation of coordinates, identical to (3) (so that the field from all magnetospheric sources be confined within the same tilt-dependent boundary). More details on the deformation method and examples of the resultant configurations are given in the next section.

Getting back to the effect of the polar cusp "dimples" on the magnetopause, the harmonic fields of the type (4) alone were found to be insufficient to accurately represent the local disturbance of the shielding field, even with many added terms. An effective remedy was found to add to the harmonic field (4) the field of four circular current loops, placed in pairs outside of the model magnetopause near the northern and southern polar cusps. The radii of the loops, their positions, attitudes, and the electric current magnitudes were parameterized as simple analytical functions of the Earth's dipole tilt angle, taking into account symmetry requirements.

The same form of the shielding field was used for the tail current terms, although, as explained above, in that case dipole tilt effects were treated by using the deformation method. Figure 5 illustrates the separate configurations of the shielded dipole and tail fields in the noon-midnight meridian plane; the shape of the new model magnetopause is shown by a broken line. The tail field displayed in the right panel of Figure 5 corresponded to one of the tail "modes" of the T96 model, with its current density peak at $X = -11 R_E$.

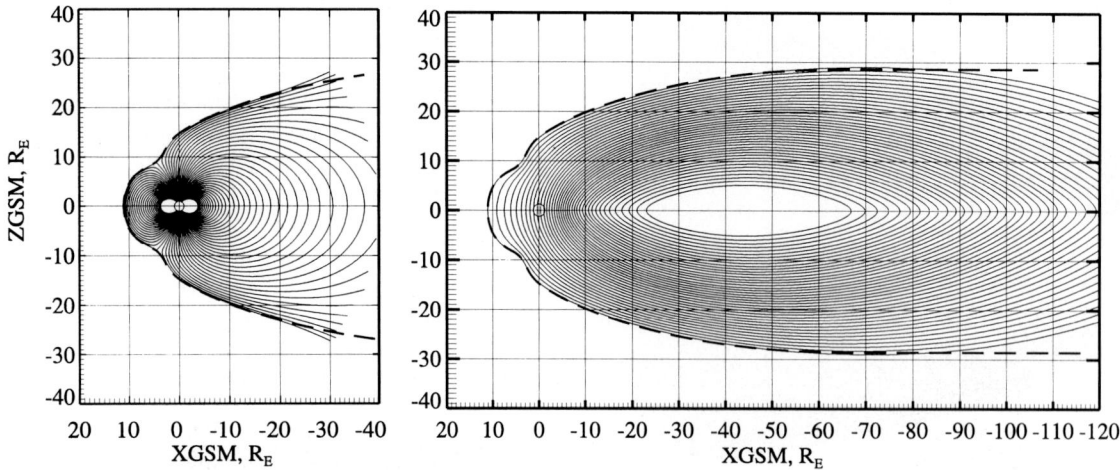

**Figure 5.** Configurations of the dipole (left) and of a tail field (right), confined within a model magnetopause.

The same procedure was used to obtain the shielding fields of other modes. Plate 1 shows the result of superposing the dipole and three tail modes to form a model field line configuration plotted against the color-coded background, displaying the distribution of the scalar difference $\Delta B$ between the strength of the total model field and that of pure dipole. Global isointensity maps of $\Delta B$ were first introduced by *Sugiura et al.* [1971] for visualizing the magnetospheric field as measured by OGO-3 spacecraft. They provide a simple and easy-to-grasp way to illustrate the contribution of external current systems to the total observed and/or model magnetic field. In particular, a near-Earth depression due to the ring current is clearly visible at low latitudes on the nightside, while on the dayside it is greatly reduced by the much larger positive contribution from the dayside magnetopause currents. Another region of a strongly depressed field is near the polar cusps. In this model calculation, the polar cusp depression was produced by another variant of the deformation method, described in more detail below. Note that the configuration in the figure does not correspond to any specific condition in the solar wind, nor was it fitted to any data set; the only purpose of the plot was to illustrate the great flexibility of the newly developed modeling methods.

The total field in Plate 1 includes an interplanetary magnetic field which interconnects with the magnetospheric field across the boundary. It is worthwhile to describe how this kind of model takes the IMF into account and how the transition from the magnetosphere to the interplanetary medium is modeled across the boundary, the more so because none of the previous publications [*Tsyganenko*, 1995, 1996] gave any details on that subject.

The main idea is to mathematically decouple the description of the relatively strong internal fields (like those of the cross-tail current or the geodipole) from the weak field penetrating from the outside and topologically connected with the IMF. This can be achieved in two steps. First, we need to fully shield the field of all internal sources within the model boundary. The harmonic field derived from the potentials (4) is continuous in all of space and, hence, in order to cancel the total model field outside the magnetosphere, one needs to multiply it by a factor $f$, equal to unity inside the magnetopause and to zero outside it:

$$f = \begin{cases} 1 & \sigma' < \sigma_0 \\ 0 & \sigma' \geq \sigma_0 \end{cases} \quad (5)$$

For the sake of computational feasibility (e.g., for tracing field lines across the magnetopause), it is more practical to define a boundary with a finite thickness, which can be easily achieved by introducing instead of the step function (5) a smooth variation of the factor f between 1 and 0 over a fixed distance $\delta\sigma$. In the T96 model, a relatively large value 0.01 was chosen for $\delta\sigma$, which corresponded to a magnetopause thickness of $\approx 0.7 R_E$ at the subsolar point and $\approx 1.8 R_E$ near the Moon's orbit.

Since the total field, by construction, is tangential to the boundary, multiplying **B** by $f$ does not violate Maxwell's equations [e.g., *Stern*, 1994]. Indeed, $\nabla \cdot (f\mathbf{B}) = f\nabla \cdot \mathbf{B} + \mathbf{B} \cdot \nabla f$, where not only the first, but also the last term in the righthand side is zero ($f$ is constant on the boundary and, hence, $\nabla f$ is normal to it). From the computational viewpoint, the factor $f$ also poses no problem, since the model magnetopause $\sigma' = \sigma_0$ is explicitly given by equations (1)–(4).

The second and final step is to "open up" the closed model magnetosphere by adding to it a relatively small current-free "interconnection field" [*Toffoletto and Hill*, 1989] with a specified distribution of $B_n$ on the boundary, controlled by

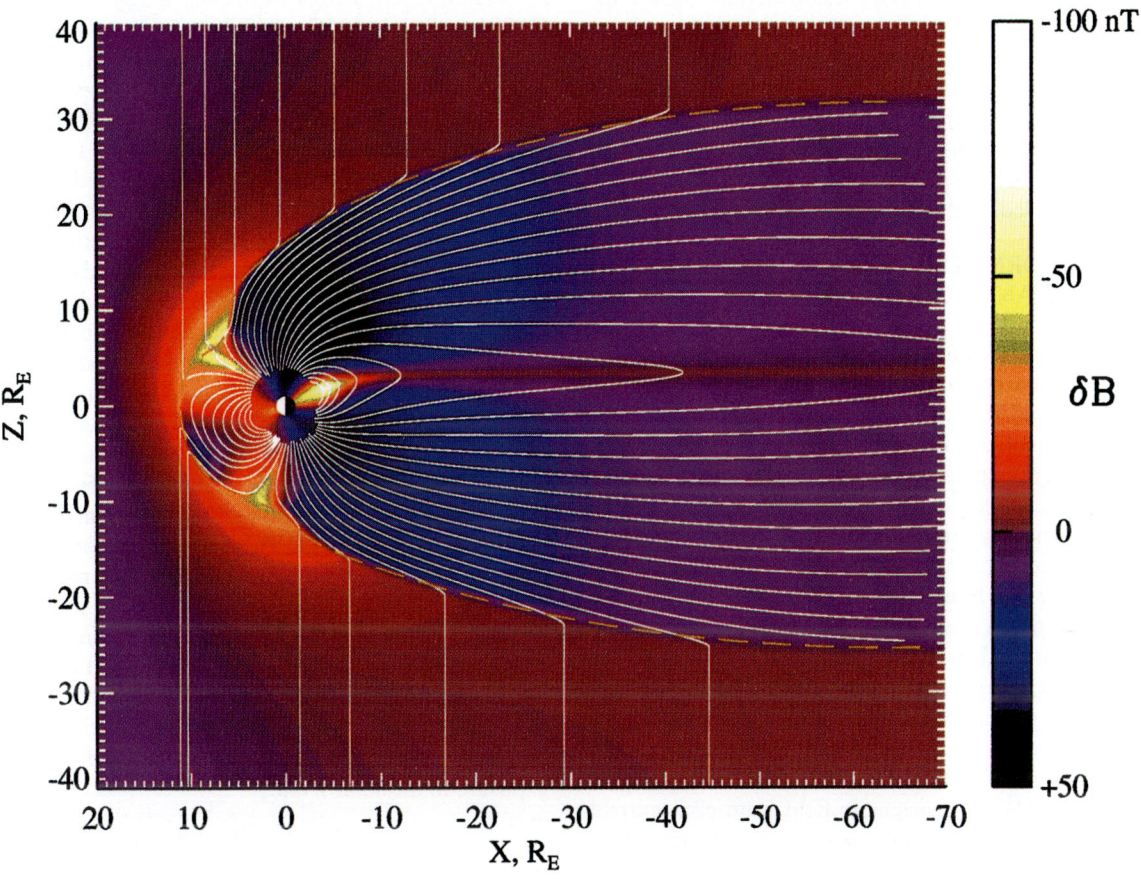

**Plate 1.** A model magnetic field configuration with a southward IMF, displayed against the color-coded distribution of $\Delta B = |\mathbf{B}_{mod}| - |\mathbf{B}_{dip}|$. Note a strongly depressed field in the polar cusps, modeled using the deformation method [*Tsyganenko and Russell,* 1999].

the magnitude and orientation of the IMF in the upstream solar wind flow. Inside the magnetosphere, the interconnection field can be represented by the same "box" harmonic potentials (4), whose parameters are fitted to provide a required distribution of $B_n$ on the magnetopause. In the T96 model [*Tsyganenko*, 1996], $B_n$ was specified by assuming a field $\mathbf{B}_{MS}$ in the magnetosheath as a vector, proportional to the observed transverse part of the IMF, $\mathbf{B}_y + \mathbf{B}_z$, whose magnitude peaked near the subsolar point and gradually decreased tailward. The interconnection rate was quantified by a factor $k$, giving the fraction of $\mathbf{B}_{MS}$ that penetrates inside the boundary, and derived from fitting the model field to a large set of data. Although the fitting provided generally reasonable values of the factor $k$, they varied over a wide interval (between 0.2 and 0.8) and were found [*Tsyganenko*, 1998b] to critically depend on the shape of the magnetopause, which is also controlled by the IMF orientation. Further studies in that area are needed, with more data from high latitudes.

## 3. THE DEFORMATION METHOD IN MODELING THE TAIL AND CUSP ELECTRIC CURRENTS

In contrast to the sources of Earth's main field, magnetospheric currents have a complex spatial structure, vary rapidly in time, and extend to large distances. As already noted, their modeling is complicated by the tilt of the geodipole axis, which results in asymmetrical boundaries and warped current sheets. Realistic models of the cross-tail current sheet should take into account its variable thickness, temporal changes in the tailward gradient of the electric current density, and the variable shape of the boundary confining the tail field inside the lobes. All these requirements should be met with a reasonable mathematical simplicity.

A convenient approach is to represent the tail current as a superposition of several current sheets of finite thickness, each with a different scale of the tailward variation of the electric current density [*Tsyganenko*, 1987, 1989; *Tsyganenko and Peredo*, 1994]. The magnetic field from each current sheet is matched by a shielding field, so that the total field of each of the partial "modules" is confined (as in Figure 5) within a common boundary. Combining these modules with different amplitude coefficients allows a great flexibility in fitting the total field to the data [*Tsyganenko*, 1995, 1996].

Modeling the effects of the dipole tilt and of the IMF upon the tail structure is greatly facilitated by applying deformations to the originally symmetric configurations. The essence of the method, as proposed by *Stern* [1987], is to replace the original coordinates $\{f, g, h\}$ in the Euler potential representation of the field

$$\mathbf{B} = \nabla \alpha(f, g, h) \times \nabla \beta(f, g, h) \qquad (6)$$

by modified ones, $\{\xi, \eta, \zeta\}$. The new magnetic field can be viewed as a deformation of the old one, and it can be shown that there is no need to know the Euler potentials themselves, since the components of the deformed field can be explicitly related to those of the undeformed field. A systematic description of the method and its specific applications to the modeling can be found elsewhere [*Tsyganenko*, 1998a]. Figure 6, reproduced from that paper, shows the effects of the tilt-related warping and IMF-induced twisting of the tail current sheet, simulated by applying a transformation of the angular coordinate $\phi = \tan Z/Y$. The twisting effect is observed in the tail magnetic field data, as was demonstrated by *Kaymaz et al.* [1995], using IMP-8 data, and by *Tsyganenko et al.* [1998] on the basis of Geotail observations in a wide range of tailward distances. MHD simulations of *White et al.* [1998] have demonstrated the formation of the S-shaped structure of the tail current, similar to that shown in Figure 6, under the influence of the azimuthal component of the IMF.

Another simple use of the deformation method (still to be implemented) allows one to model the flattening of the deep tail under the influence of the anisotropic external pressure due to the IMF draping around the tail boundary.

The deformation method is also ideally suited for modeling the polar cusp magnetic effects due to the diamagnetism of the injected magnetosheath plasma, already discussed in the preceding section. The idea behind the method [*Tsyganenko and Russell*, 1999] is to introduce a local rarefaction of the magnetic flux inside field line tubes containing the northern and southern polar cusps. This effect can be achieved by a local stretching in a direction approximately orthogonal to the magnetic field. The desired transformation can be conveniently performed in cylindrical coordinates $\{\rho, \phi, Y\}$, where the $Y$-axis coincides with the solar magnetic $Y_{SM}$, and the coordinate being deformed is the azimuthal angle $\phi$, corresponding to rotation around that axis. Leaving out the details, the final result of modeling the effect of the cusp currents appears in Plate 1 as a deep and narrow field depression, especially conspicuous in the northern cusp due to its greater exposure to the solar wind.

## 4. CONCLUSIONS

This brief review cannot provide a comprehensive coverage of all aspects of data-based modeling. Instead, attention was given to new methods of realistically representing the structure of the magnetospheric currents, and many important questions were left out, including data assimilation techniques and the response of the model electric currents to changing solar wind and magnetospheric conditions. Based on the recent advances and the continuing inflow of new ISTP data, one can expect significant further progress. Future work will focus on the development of dynamical models, including the substorm reconfiguration of the near-Earth field, on a variable solar-wind controlled magnetopause with a changing rate of interconnection with the IMF, on dynamical variations of the Birkeland currents and polar cusps, and on a realistic

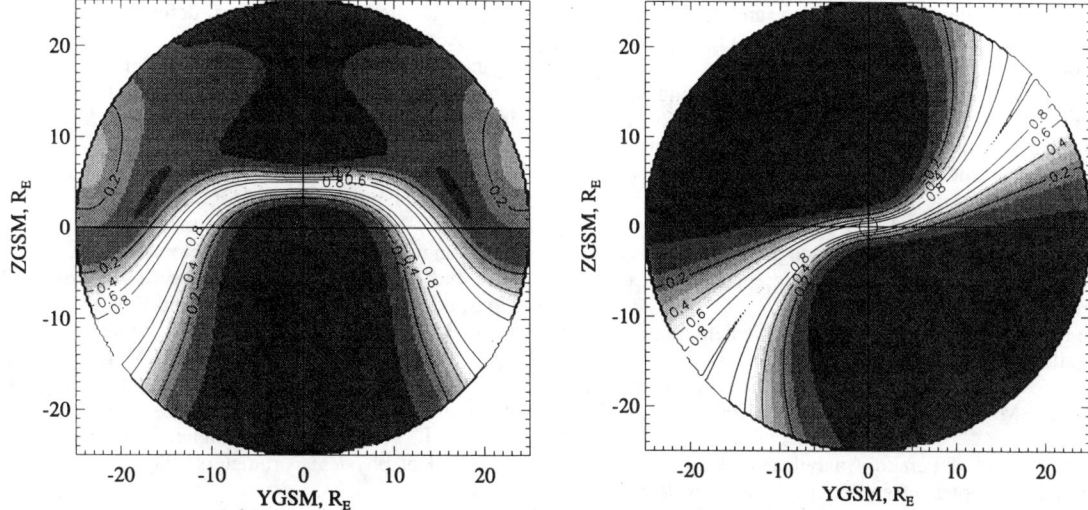

**Figure 6.** Plots of the electric current density in the cross sections of the magnetotail, demonstrating the effects of the tilt-related warping (left) and the IMF-induced twisting (right) of the current sheet, as replicated by using the deformation method [*Tsyganenko, 1998a*].

ring current model, taking into account storm-time changes of its amplitude, size, and local time asymmetry.

One can expect a major advance in the magnetospheric models following the launch of the "Magnetospheric Constellation" mission scheduled for the next decade. The existing data-based modeling techniques are perfectly suited for reconstructing the instantaneous configuration of magnetospheric currents from simultaneous data of tens (maybe hundreds) of flying space magnetometers, dispersed over a vast volume of space [*Tsyganenko, 1998c*]. Combined with simultaneous monitoring of the solar wind, this should allow one to trace and understand the actual impact of varying interplanetary conditions upon the magnetic environment of our planet.

*Acknowledgments.* The author thanks David Stern for his careful reading of the manuscript and many constructive comments. This work was supported by NASA grants NASW-97024 (ISTP GI Program), and NSF Magnetospheric Physics Program grant ATM-9501463.

## REFERENCES

Boardsen, S. A., T. E. Eastman, T. Sotirelis, and J. L. Green, An empirical model of the high latitude magnetopause, submitted to *J. Geophys. Res.*, 1999.

Choe, J. Y., D. B. Beard, and E. C. Sullivan, Precise calculation of the magnetosphere surface for a tilted dipole, *Planet. Space Sci., 21,* 485, 1973.

Hammond, C. M., M. G. Kivelson, and R. J. Walker, Imaging the effect of dipole tilt on magnetotail boundaries, *J. Geophys. Res., 99,* 6079, 1994.

Kaymaz, Z., G. L. Siscoe, J. G. Luhmann, J. A. Fedder, and J. G. Lyon, Interplanetary magnetic field control of magnetotail field: IMP 8 data and MHD model compared, *J. Geophys. Res., 100,* 17163, 1995.

Mead, G. D., and D. B. Beard, Shape of the geomagnetic field solar wind boundary, *J. Geophys. Res., 69,* 1169, 1964.

Olson, W. P., The shape of the tilted magnetopause, *J. Geophys. Res., 74,* 5642, 1969.

Petrinec, S. M., and C. T. Russell, An empirical model of the size and shape of the near-Earth magnetotail, *Geophys. Res. Lett., 20,* 2695, 1993.

Petrinec, S. M., and C. T. Russell, Near-Earth magnetotail shape and size as determined from the magnetopause flaring angle, *J. Geophys. Res., 101,* 137, 1996.

Roelof, E. C., and D. G. Sibeck, Magnetopause shape as a bivariate function of the interplanetary magnetic field $B_z$ and solar wind dynamic pressure, *J. Geophys. Res., 98,* 21421, 1993.

Schulz, M., and M. McNab, Source-surface model of the magnetosphere, *Geophys. Res. Lett., 14,* 182, 1987.

Shue, J.-H., J. K. Chao, H. C. Fu, C. T. Russell, P. Song, K. K. Khurana, and H. J. Singer, A new functional form to study the solar wind control of the magnetopause size and shape, *J. Geophys. Res., 102,* 9497, 1997.

Shue, J.-H., et al., Magnetopause location under extreme solar wind conditions, *J. Geophys. Res., 103,* 17,691, 1998.

Sibeck, D. G., R. E. Lopez, and E. C. Roelof, Solar wind control of the magnetopause shape, location, and motion, *J. Geophys. Res., 96,* 5489, 1991.

Sotirelis, T., The shape and field of the magnetopause as determined from pressure balance, *J. Geophys. Res., 101,* 15,255, 1996.

Sotirelis, T., and C.-I. Meng, Magnetopause from pressure balance, *J. Geophys. Res., 104,* 6889, 1999.

Stern, D. P., Tail modeling in a stretched magnetosphere 1. Methods and transformations, *J. Geophys. Res., 92,* 4437, 1987.

Stern, D. P., The art of mapping the magnetosphere, *J. Geophys. Res., 99,* 17,169, 1994.

Sugiura, M., B. G. Ledley, T. L. Skillman, and J. P. Heppner, Magnetospheric field distortions observed by Ogo 3 and 5, *J. Geophys. Res., 76,* 7552, 1971.

Toffoletto, F. R., and T. W. Hill, Mapping of the solar wind electric field to the Earth's polar caps, *J. Geophys. Res., 94,* 329, 1989.

Tsyganenko, N. A., and A. V. Usmanov, Determination of the magnetospheric current system parameters and development of experimental geomagnetic field models based on data from IMP and HEOS satellites, *Planet. Space Sci., 30,* 985, 1982.

Tsyganenko, N. A., Global quantitative models of the geomagnetic field in the cislunar magnetosphere for different disturbance levels, *Planet. Space Sci., 35,* 1347, 1987.

Tsyganenko, N. A., A magnetospheric magnetic field model with a warped tail current sheet, *Planet. Space Sci., 37,* 5, 1989.

Tsyganenko, N. A., Quantitative models of the magnetospheric magnetic field: Methods and results, *Space Sci. Rev., 54,* 75, 1990.

Tsyganenko, N. A., and M. Peredo, Analytical models of the magnetic field of disk-shaped current sheets, *J. Geophys. Res., , 99,* 199, 1994.

Tsyganenko, N. A., Modeling the Earth's magnetospheric magnetic field confined within a realistic magnetopause, *J. Geophys. Res., 100,* 5599, 1995.

Tsyganenko, N. A., Effects of the solar wind conditions on the global magnetospheric configuration as deduced from data-based field models, in *Eur. Space Agency Spec. Publication ESA SP-389,* 181, 1996.

Tsyganenko, N. A., S. B. P. Karlsson, S. Kokubun, T. Yamamoto, A. J. Lazarus, K. W. Ogilvie, C. T. Russell, Global configuration of the magnetotail current sheet as derived from Geotail, Wind, IMP 8 and ISEE 1/2 data, *J. Geophys. Res., 103,* 6827, 1998.

Tsyganenko, N. A., Modeling of twisted/warped magnetospheric configurations using the general deformation method, *J. Geophys. Res., 103,* 23551, 1998a.

Tsyganenko, N. A., Data-based models of the global geospace magnetic field: Challenges and prospects of the ISTP era, in *Geospace Mass and Energy Flow: Results From the International Solar-Terrestrial Physics Program, Geophys. Monogr. Ser.,* edited by J. L. Horwitz, D. L. Gallagher, and W. K. Peterson, vol. **104**, pp. 371-382, AGU, Washington, D. C., 1998b.

Tsyganenko, N. A., Toward real-time magnetospheric mapping based on multi-probe space magnetometer data, in *Science Closure and Enabling Technologies for Constellation Class Missions,* edited by V. Angelopoulos and P. V. Panetta, pp. 84-90, U. C. Berkeley, Berkeley, California, 1998c.

Tsyganenko, N. A. and C. T. Russell, Magnetic signatures of the distant polar cusps: Observations by Polar and quantitative modeling, to appear in *J. Geophys. Res.,* 1999.

Voigt, G. H., A mathematical magnetospheric field model with independent physical parameters, *Planet. Space Sci., 29,* 1, 1981.

Walker, R. J., T. Ogino, and M. Ashour-Abdalla, Simulating the magnetosphere: The structure of the magnetotail, in *Solar System Plasma Physics, Geophys. Monogr. Ser.,* vol.54, edited by J. H. Waite, Jr., J. L. Burch, and R. L. Moore, p.61, AGU, Washington, D.C., 1989.

White, W. W., G. L. Siscoe, G. M. Erickson, Z. Kaymaz, N. C. Maynard, K. D. Siebert, B. U. O. Sonnerup, and D. R. Weimer, The magntospheric sash and the cross-tail S, *Geophys. Res. Lett., 25,* 1605, 1998.

---

N. A. Tsyganenko, Raytheon ITSS Corporation, NASA Goddard SFC, Code 690.2, Greenbelt, MD 20771. (e-mail: kolya@nssdca.gsfc.nasa.gov)

# Field Line Mapping and Birkeland Currents

F. R. Toffoletto and T. W. Hill

*Department of Space Physics and Astronomy, Rice University, Houston, Texas*

We describe the preliminary development of an open magnetosphere model that includes the magnetic effects of Birkeland (magnetic-field-aligned) currents in a self-consistent way. This model is designed to satisfy two criteria: (1) when mapped to the ionosphere, the Birkeland current should be consistent with that computed by taking the divergence of the ionospheric Pedersen current that is driven by the mapped electric field, and (2) the magnetic field computed from the Birkeland current should be included in the magnetic mapping. Satisfying both of these criteria simultaneously requires an iterative procedure which is not automatically convergent. In particular, when we assume nominal values for the polar-cap size and potential drop (these are inputs to the model), a convergent solution occurs only if the nightside ionospheric Pedersen conductance is taken to be ~ 5 mho or less. This result supports the conjecture that the ionospheric conductance regulates the degree of openness of the magnetosphere: if a larger polar-cap potential were imposed (for a given conductance), it would drive Birkeland currents that disrupt the ordered magnetotail field geometry.

## INTRODUCTION

In this paper we address the deceptively simple question, "what are the effects of Birkeland (magnetic-field-aligned) currents on the mapping of the solar-wind electric field to the Earth's polar caps?". Our theoretical approach is to introduce Birkeland-current perturbations to the Toffoletto and Hill [1993] open magnetosphere model (TH93), which was developed to represent the mapping of the solar-wind electric field to the Earth's polar caps along open (interconnected) magnetic field lines, for a given dipole tilt and a given set of upstream solar-wind and interplanetary-magnetic-field (IMF) parameters. The most comprehensive observational test of this (or any other) theoretical model to date [Hill and Toffoletto, 1998] shows a satisfactory level of agreement on the global character of the mapping process for quasi-steady-state conditions.

The TH93 model contains, in addition to the geodipole, four of the global-scale current systems that shape the magnetosphere: the magnetopause or Chapman-Ferraro current which confines the geomagnetic field and excludes the IMF, the expansion-fan current of the tail lobes, the cross-tail current of the plasma sheet, and a small perturbation of the Chapman-Ferraro current that allows a physically realistic degree and pattern of interconnection between the IMF and the geomagnetic field. A geomagnetic ring current (among other model improvements) was added by Ding et al. [1996], but the present work takes TH93 as a starting point because of its greater computational efficiency (and because the ring current has little effect on the polar-cap mapping considered here). Apart from the ring current, the only global-scale magnetospheric currents that are not included in the TH93 model are Birkeland (magnetic-field-aligned) currents. This paper describes an exploratory attempt to incorporate such currents in the model. In particular, we address the "Region-1" driving currents that enter the ionosphere near the dawn side polar-cap boundary and exit the ionosphere near the dusk side polar cap boundary [Iijima and Potemra, 1976]. Because of their large magnitude and proximity to the polar cap, it is these Region-1 currents that should have the greatest effect on magnetic mapping within the polar cap.

The inclusion of Birkeland currents in magnetospheric field models has proven to be a difficult problem for several reasons. Perhaps the largest obstacle has been the lack of empirical information on the distribution of Birkeland currents in the distant magnetosphere. Most of what we know about the global distribution of Birkeland currents comes from observations at low altitudes just above the ionosphere [e.g., Iijima and Potemra, 1976]. The lower-latitude shielding currents ("Region 2" Birkeland currents) close primarily in the inner magnetospheric ring current [e.g., Wolf, 1983, and references therein], where they produce only a modest perturbation to the background (dipole) field. The higher latitude driving currents ("Region 1" Birkeland currents), on the other hand, are thought to close near the magnetopause and/or in the distant magnetotail [e.g., Stern, 1983], where they may produce field perturbations that are comparable in magnitude to the "background" field that would be expected in their absence.

Another fundamental difficulty is that, if a Birkeland current is postulated to flow parallel (or antiparallel) to the background field, it will not be parallel to the resultant field including its perturbation. To preserve the field-aligned character of the Birkeland current, it is necessary to employ an iterative approach which, as we shall show, is not convergent in the general case, particularly in the case of Region-1 currents where the perturbation field can become comparable in magnitude to that of the background field.

A less fundamental, but equally troublesome, difficulty lies in calculating the magnetic effects of a given Birkeland current distribution. The most straightforward approach, numerical integration of the Biot-Savart law, is numerically complex and very computer-intensive because it requires a six-dimensional grid (three dimensions for source points and three for field points). Other global magnetospheric current systems (e.g., ring current, magnetotail current sheet) can be modeled successfully by exploiting their relatively simple, quasi-two-dimensional geometry [e.g., Voigt, 1981; Toffoletto and Hill, 1993; Ding et al., 1996; Tsyganenko, 1993a], while the magnetic effects of the Chapman Ferraro (magnetopause) current can be obtained from the solution of a boundary-value problem for a magnetic scalar potential [e.g., Toffoletto et al., 1994, in addition to the references just cited]. These computational devices are not readily applicable to the inherently three-dimensional distribution of the global Birkeland current systems. Alternative approaches that have been employed include basis-function representations [e.g., Tsyganenko and Stern, 1996] and discrete wire approximations [e.g., Donovan, 1993; Peroomian et al., 1998]. These approaches suffer from the problem described in the previous paragraph, namely, that if the assumed current is aligned with the assumed background field, it will not be aligned with the resultant field.

A computational approach that avoids all these difficulties, in principle, is global numerical simulation based on the ideal mhd equations with an embedded, conducting ionosphere [e.g., Fedder et al., 1995; Raeder et al., 1998; White et al., 1998]. Such simulations are too computer-intensive for the type of exploratory work considered here. They also have difficulty simultaneously resolving both the ionosphere, where Birkeland currents must close through conduction currents, and the outer-magnetospheric boundaries where the distant closure occurs, at least in part, through non-mhd effects.

In this paper we introduce an efficient computational technique that enables us to study the magnetic effects of a particular subset of the global Birkeland current system, namely, the portion of the "Region-1" driving currents that close in the distant magnetotail, beyond a distance of 10 $R_E$ ($R_E$ = Earth's radius).

## APPROACH

We begin with the background magnetic and electric fields provided by the TH93 open-magnetosphere model. This model calculates the global magnetospheric magnetic-field configuration for a given set of (steady-state) upstream solar-wind/IMF parameters and dipole tilt, and the implied ionospheric polar-cap electric-potential distribution as mapped inward from the solar wind along interconnected (open) magnetic field lines. For the present study we restrict ourselves to the simple symmetric case of zero dipole tilt and purely southward (5 nT) IMF. We then calculate the divergence of the ionospheric Pedersen current resulting from the mapped electric potential distribution (assuming a uniform height-integrated ionospheric conductivity), to obtain a Birkeland-current distribution that is roughly consistent with those observed by low-altitude satellites. This low-altitude Birkeland current distribution $j_\parallel$ is given by

$$j_\parallel \sin I = -\nabla \cdot (\Sigma \nabla \Phi) = -\Sigma \nabla^2 \Phi \qquad (1)$$

where $\Phi$ is the potential, $\Sigma$ is the ionospheric height-integrated Pedersen conductivity, and $I$ is the dip angle ($\sin I \approx 1$ in the polar cap). We assume $\Sigma$ = constant, and we adopt a value $\Sigma$ = 5 mho except where noted otherwise.

In order to evaluate (1), we require the $\Phi$ distribution not only in the polar cap (as mapped from the solar wind by the TH93 model) but also in the adjacent lower-latitude region of return flow. To obtain an approximation of $\Phi$ for the return-flow region, we solve the Laplace equation between the polar-cap boundary and a zero-potential boundary placed a few degrees equatorward of the polar-cap boundary. This "Region-2 ring" [Siscoe, 1982] represents the inner edge of the ring current in steady-state shielding theory [e.g., Wolf, 1983, and references therein], where the bulk of the Region-1 current is closed through the Region-2 system. This zero-potential boundary is taken to be a displaced circle of latitude that crosses the noon meridian at 65° latitude and the midnight meridian at 60°.

In the unmodified TH93 model, most of the Birkeland current from (1) would be confined to an infinitesimally thin sheet at the polar-cap boundary. In the present application, however, this sheet has a finite thickness determined by the numerical grid spacing, which is adjusted to give it a realistic thickness ~1°.

This low-altitude Birkeland-current distribution is then mapped along model field lines under the assumption $j_\parallel/B$ = constant until the field line crosses either the "dayside" magnetopause (at $x > -10\ R_E$) or the near-Earth boundary of the magnetotail, defined as the tail cross-section at $x = -10\ R_E$. (The coordinate "x" refers to the geocentric solar-magnetospheric system in which, for the symmetric case considered here, $\hat{x}$ is sunward, $\hat{y}$ is duskward and $\hat{z}$ is northward; see Figure 1.) The dayside portion of the Region-1 current (that which crosses the magnetopause sunward of $x = -10\ R_E$) is ignored in the present analysis. The nightside portion is used to construct a two-dimensional model of the resulting field perturbation $\delta \mathbf{B}$ in the tail, as follows.

The Birkeland current in the tail is approximately in the x direction and the resultant field perturbation is approximately in the yz plane. (The tail boundary in TH93 is a circular cylinder given by $y^2 + z^2 = R^2$.) We assume that $\delta \mathbf{B}$ is precisely in the yz plane and can therefore be derived from a vector potential **A** having only an x component:

$$A = A(x,y,z)\hat{x} \qquad (2)$$

$$\delta \mathbf{B} = \nabla \times \mathbf{A} = \hat{y}\frac{\partial A}{\partial z} - \hat{z}\frac{\partial A}{\partial y} \qquad (3)$$

and the electric current

$$\mu_0 \mathbf{j} = \nabla \times \delta \mathbf{B} = -\hat{x}\left(\frac{\partial^2 A}{\partial y^2} + \frac{\partial^2 A}{\partial z^2}\right) + \hat{y}\frac{\partial^2 A}{\partial x \partial y} + \hat{z}\frac{\partial^2 A}{\partial x \partial z} \qquad (4)$$

We further assume that

$$A(x,y,z) = A_1(y,z)e^{-(x_1-x)/D} \qquad (5)$$

where D is chosen to approximate the empirical result that the Birkeland current decreases significantly between $x = -10\ R_E$ and $x = -35\ R_E$ [Tsyganenko, et al., 1993]. (A value of $D = 40\ R_E$ was used in the results presented here.). The x component of (4) is a two-dimensional Poisson equation which we solve for $A_1(y,z)$ at $x = x_1 = -10\ R_E$ (Figure 1), using the ionospheric Birkeland current mapped to $x = -10\ R_E$ as the source term $j_x$ on the left-hand side. We apply

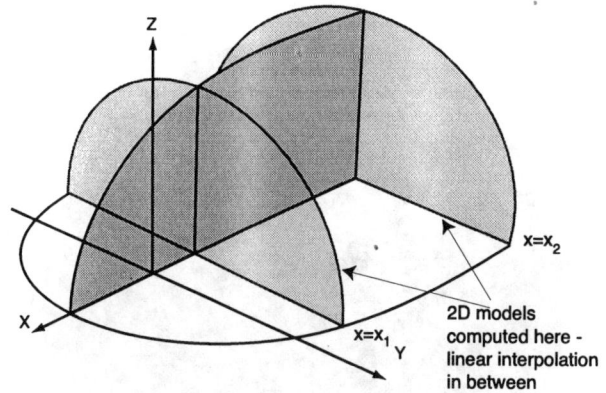

**Figure 1.** The coordinate system and geometry (not to scale) of the magnetopause in the TH93 magnetosphere model. The magnetopause is a hemisphere for $x > x_1 = -10\ R_E$, and a cylinder for $x < x_1$. The radius of the hemisphere and of the cylinder is 20 $R_E$. The Birkeland currents are assumed to close within the tail, largely in the interval $x_1 > x > -50\ R_E$.

the boundary condition A = 0 at the magnetopause in order to preserve the (carefully prescribed) distribution of the normal magnetic-field component there. The Poisson equation is solved numerically using a finite difference approximation on a polar grid. The implied closure currents are given by the y and z components of (4). Note that $\nabla \cdot \mathbf{j} = 0$ although $\nabla \cdot \mathbf{A} \neq 0$. The desired perturbation field is then given by (3).

This perturbation field is then added to the background field and the process is repeated from the beginning, starting with the mapping of the solar-wind electric field to the polar cap using the perturbed field. Within two or three iterations it becomes obvious that the procedure either is or is not converging on a stable solution, depending on the size of the perturbation (i.e., depending on the assumed ionospheric conductivity).

## RESULTS FOR Σ = 5 MHO

Figure 2 shows the ionospheric convection pattern for the unperturbed TH93 model with a southward (5 nT) IMF. The equipotentials are drawn at 4 kV intervals; the total cross-polar-cap potential is about 60 kV. The total current into (or out of) the ionosphere is $9 \times 10^5$ amps. The gray scale shows the Birkeland current density computed from (1), for an ionospheric conductivity Σ = 5 mho, with lighter (darker) shading indicating current into (out of) the ionosphere. The gray scale key at the bottom is marked in units of $\mu A/m^2$. (To remove extraneous numerical noise from the grayscale plot, the current density magnitude has been set to zero whenever it is less than 10% of the maximum magnitude.) The jagged line crossing the polar cap near noon is the boundary between

# 74  FIELD LINE MAPPING AND BIRKELAND CURRENTS

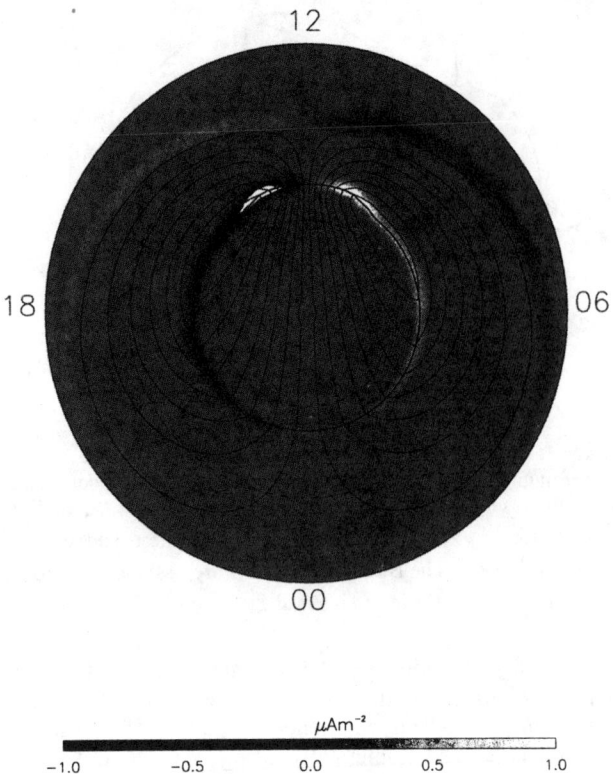

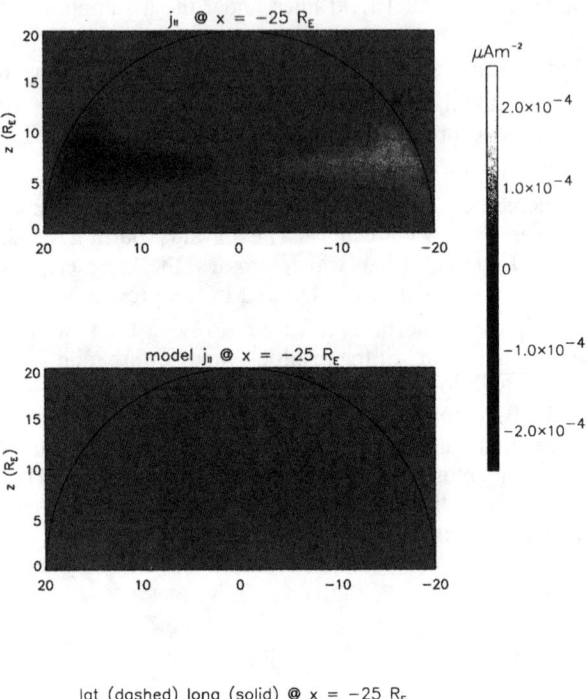

**Figure 2.** The polar-cap potential pattern as mapped from the solar wind in the first iteration step (unperturbed TH93 model), for a 5 nT southward IMF. The equipotential spacing is 4 kV. The gray scale shows the (Region-1) Birkeland current computed from the divergence of the ionospheric Pedersen current, with a uniform height-integrated Pedersen conductivity $\Sigma = 5$ mho. In this and all subsequent grayscale plots, we have eliminated noise by setting $j_\parallel$ equal to zero wherever it is less than 10% of the maximum magnitude. The total upward (or downward) current is about $9 \times 10^5$ amperes.

magnitude threshold used in constructing the grayscale plot. It becomes visible in cross sections taken closer to Earth (not shown here). To give a visual impression of the mapping geometry, the bottom panel of Figure 3 shows lines of constant geomagnetic latitude (dashed lines, at 2° intervals) and longitude (solid lines, at 20° intervals) mapped to the same ($x = -25\ R_E$) tail cross section. As expected, the largest Birkeland current density in the tail occurs near the outer boundary of the plasma sheet, the tailward projection of the polar-cap boundary.

dayside and nightside currents as defined above, i.e., currents Sunward of this line cross the magnetopause Sunward of $x = -10\ R_E$ and are ignored in the present analysis. Note that the "dayside" Region-1 currents, by this definition, intersect the ionosphere within about ± 2.5 hr of noon LT.

The nightside Birkeland currents of Figure 2 are mapped onto the tail cross section at $x = -25\ R_E$ in the top panel of Figure 3. The view is toward the Sun, and positive $j_\parallel$ (lighter shading) is Earthward while negative $j_\parallel$ (darker shading) is tailward. The second panel shows, for comparison, the accidental $j_\parallel$ that was already present in the TH93 model by virtue of the cross-tail current having a small field-aligned component. The fact that nothing is visible in this panel does not indicate that this accidental component is zero, only that it is below the 10%

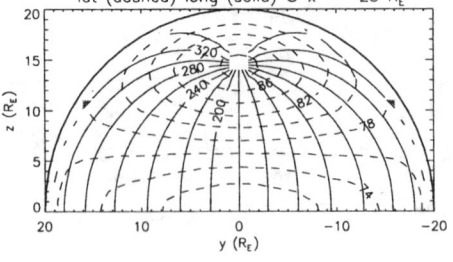

**Figure 3.** The mapping of the Birkeland current density of Figure 2 onto the tail cross section at $x = -25\ R_E$ (top panel), and the accidental Birkeland current density that was already present at that location in TH93 (second panel). The latter is blank because the pre-existing $j_\parallel$, while not zero, was less than the 10% threshold used in the grayscale plot. The bottom panel shows the mapping to the same tail cross section of lines of geomagnetic latitude (at 2° intervals) and longitude (at 20° intervals). The view is toward the Sun.

of the closure current is in the dawn-dusk direction in the lobes. However, this current makes a small (<10%) perturbation to the existing dusk-dawn current that is present in the model due to the expansion fan [Toffoletto and Hill, 1993].

The twisting of magnetotail field lines produces noticeable effects in the mapping of the solar-wind electric field to the polar cap, as shown in Figure 5 (a, second iteration step, and b, third iteration step). Figure 5 is in the same format as Figure 2 above. There is a pronounced equatorward bulge of the dawn and dusk edges of the polar cap, with a compensating poleward migration of the midnight edge. (The total polar-cap area, or more precisely its enclosed magnetic flux, is constrained by the magnetopause boundary condition, which is unchanged from one iteration to the next.) The polar-cap shape seems to be converging on a stable solution in the sense that the change from the second to the third iteration step (from Figure 5a to 5b) is generally smaller in magnitude and opposite in sense to the change between the first and second steps (from Figure 2 to 5a). The potential contours, on the other hand, exhibit an increasing dawn-dusk deflection between the second and third steps; this may be a numerical artifact, or it may be an early indication of an inherent instability in the system that manifests itself clearly for larger ionospheric conductivities, as described further in the next section.

Figure 6 shows the $j_{\parallel}$ distributions of Figure 5 as mapped to the tail cross section at $x = -25\ R_E$, in the same format as Figure 3. The top panels show the ionospheric $j_{\parallel}$ distributions as mapped to the tail cross section in the second (Figure 5a) and third (Figure 5b) iteration steps. The second panels show the actual $j_{\parallel}$ distributions present in the same tail cross section (obtained from $\nabla \times \mathbf{B}$) at the same iteration steps. The fact that the first and second panels are starting to resemble each other at least qualitatively by the third iteration step (Figure 5b) is an encouraging sign that the configuration is beginning to converge toward a stable, self-consistent solution.

## RESULTS FOR $\Sigma = 25$ MHO

The results of the previous section provide some hope that a convergent solution can be found by this method when the nightside ionospheric conductivity has a nominal value ~ 5 mho. For larger conductivities, however, all hope is lost. This is illustrated in Figure 7, showing the corresponding results on the polar cap when the conductivity (and hence the magnetic perturbation field) is increased by a factor of 5. Figure 7a shows the second iteration step and Figure 7b the third iteration step, in the same format as Figure 5; the gray scale for $j_{\parallel}$ has been eliminated in (b) in order to reveal the gross distortion of the polar-cap boundary itself. The spiraling of the tail field

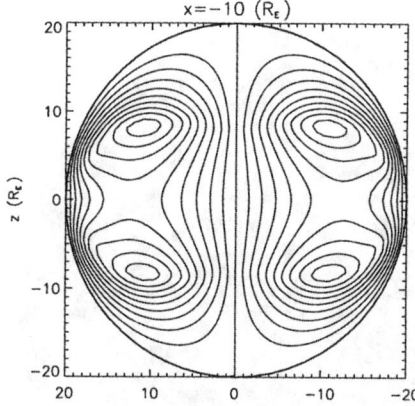
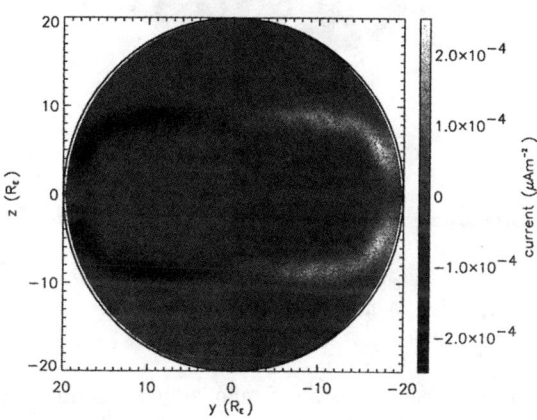

**Figure 4.** The top panel shows the perturbation magnetic field (contours of constant A) in the tail cross section at $x = -10\ R_E$, as calculated from equations (3) and (4). The field direction is northward on the tail axis ($y = z = 0$). The bottom panel shows the x component of the current density that was used as the source term in (4) to compute the field perturbation.

Figure 4 (top panel) shows the perturbation magnetic field lines (contours of constant A) in the tail cross section at $x = -10\ R_E$, as computed from (3) and (4) above. The bottom panel shows, in the same cross section, the current density component $j_X$ used as the source in (4). (This is not exactly the same as $j_{\parallel}$ because the background field is not exactly in the $\pm$ x direction.) The field direction is generally northward (+z direction) near the tail axis and southward around the edges, which introduces a helical twist when added to the background tail field. The magnitude of the field perturbation is ~ 3 nT at this distance, decreasing exponentially to zero in the far tail. The closure current, as implied by (4), is proportional to $\nabla A$. From Figure 4 (top panel), this suggests that while much of the current closes though the magnetopause, some

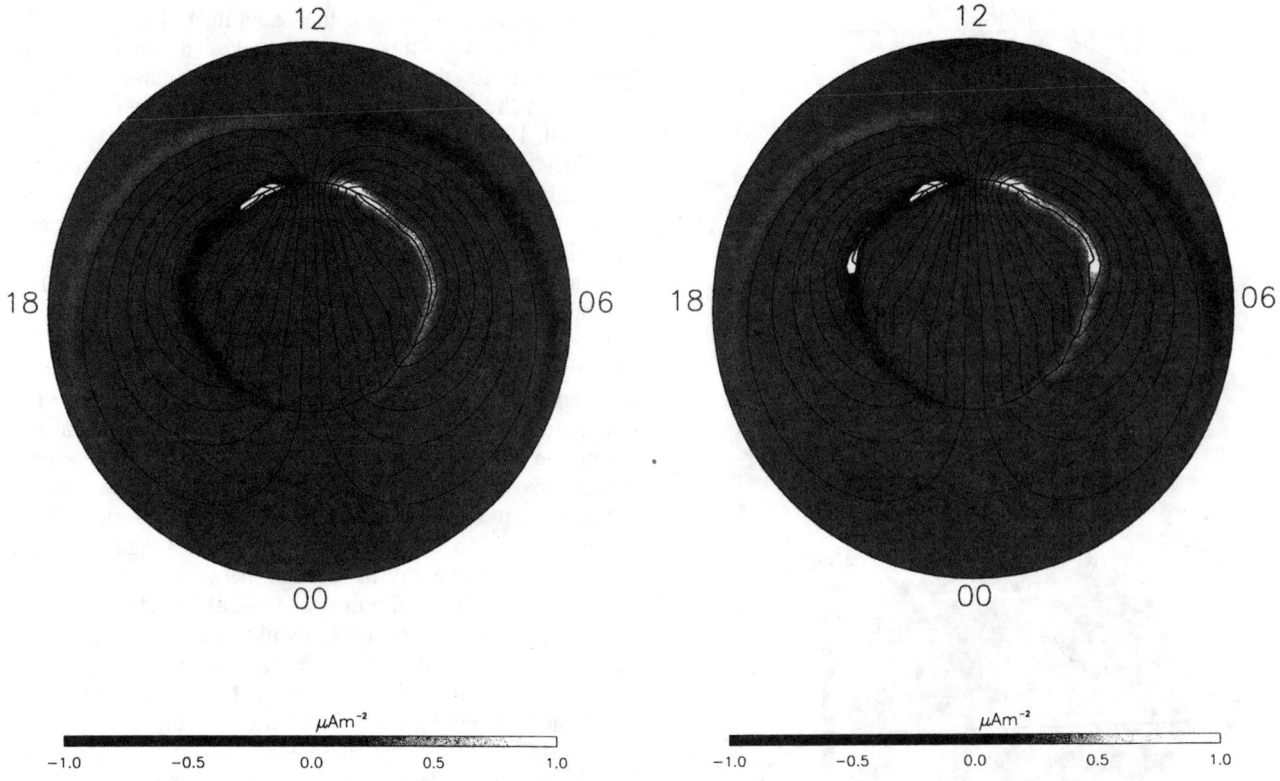

**Figure 5.** Polar-cap potential and Birkeland current patterns, as in Figure 2, for (a) the second iteration step and (b) the third step.

in this case causes a reduction of the polar-cap area (in spite of holding the magnetopause boundary condition fixed), and a significant distortion of its shape. This is clearly a non-physical result: it violates not only our intuition but also our empirical knowledge of possible polar-cap shapes. It is a result worth showing, however, because it draws attention to a fundamental instability of the system that may affect the real magnetosphere, not just our simplified representation thereof.

Consider the effect of increasing ionospheric conductivity on the stability of our model. If Σ were effectively zero (as, for example, in Mercury's magnetosphere), the original TH93 model would provide a physically reasonable configuration that is self-consistent in the sense that the Birkeland current in the magnetotail (essentially zero) is consistent with that supplied by the ionosphere in response to the electric field mapped down from the solar wind. (The model lacks self-consistency in other respects, for example, it does not contain an explicit plasma population that is in equilibrium with the imposed $\mathbf{j} \times \mathbf{B}$ forces, but at least it would be self-consistent with respect to Birkeland currents, if they had vanishingly small magnitude.) The results of the previous section suggest that this level of self-consistency can be maintained for conductivities as large as Σ ~ 5 mho, although further work is needed to demonstrate this conclusively. The results of Figure 7, however, clearly show that such self-consistency is not attainable for conductivities ~ 25 mho. For sufficiently large Σ, the perturbation magnetic field in the tail necessarily becomes comparable in magnitude to the background tail field. The resultant distortion of the polar-cap boundary produces extra layers of Birkeland current that exacerbate the mapping distortions; there is a positive feedback that precludes any hope of convergence toward a stable configuration. We suggest, though we cannot prove, that this positive-feedback problem also exists for the real magnetosphere, differing in detail but not in essence from the problem that arises in our model.

## DISCUSSION

This paper represents a progress report on our ongoing effort to develop a theoretical open-magnetosphere model that incorporates Region-1 Birkeland currents in a self-consistent way. The specific criterion of self-consistency that we aspire to is that the Birkeland current implied by the model field configuration $[\mathbf{B} \cdot (\nabla \times \mathbf{B})/\mu_0 B]$ agrees with the Birkeland current implied by the divergence of the

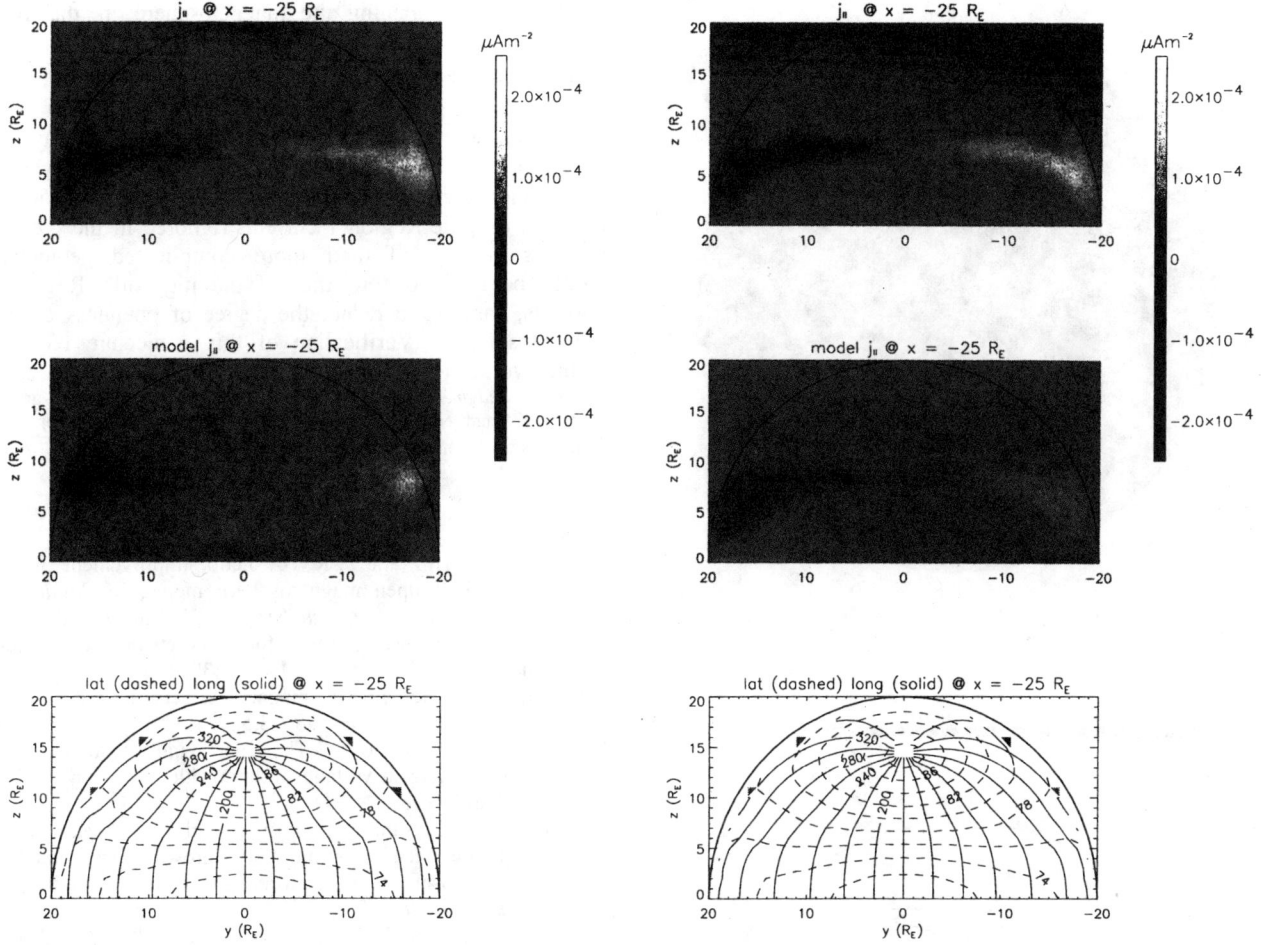

**Figure 6.** Birkeland current density in the tail cross section at x = -25 $R_E$, as in Figure 3, for (a) the second and (b) the third iteration steps.

ionospheric conduction current that flows in response to the electric field mapped from the solar wind to the ionosphere within the same model field configuration.

We have utilized a simplified but quantitative model of the perturbation $\delta\mathbf{B}$ associated with a given distribution of $j_\parallel$ in the magnetotail as mapped outward from the ionosphere. Application of this model for a nominal polar-cap potential drop ~ 60 kV, appropriate to a 5 nT southward IMF, suggests that the self-consistency criterion may be achievable for a nominal nightside ionospheric conductivity ($\Sigma \sim 5$ mho), but is clearly not achievable for much larger conductivities ($\Sigma \sim 25$ mho).

It has been suggested [Hill et al., 1976] that the ionospheric conductivity may regulate, through the mechanism of Birkeland currents, the degree of openness of the magnetosphere, and hence the magnitude of the potential drop that the solar wind can impose across the polar cap. A dedicated series of global mhd simulations with an embedded ionosphere [Fedder and Lyon, 1987] has supported this suggestion. The present results lend further support to this suggestion, and also point to a potential mechanism whereby the conductivity may exert its control over the polar-cap potential drop. If, for a given conductivity, the solar wind imposes too large a potential drop on the polar cap, the resulting nightside Region-1 Birkeland currents disrupt the orderly magnetotail field configuration whereby the solar-wind potential is mapped to the polar cap.

If we were to relax the assumption A = 0 ($\delta\mathbf{B}\cdot\hat{\mathbf{n}} = 0$) on the magnetotail boundary (for example, by applying the same boundary condition on a larger cylindrical boundary well outside the magnetopause surface), we would find that the field perturbation $\delta\mathbf{B}$ associated with $j_\parallel$ has a northward component over most of the tail magnetopause surface (imagine Figure 4 without the magnetopause confinement). This would tend to reduce the normal component that is

# 78 FIELD LINE MAPPING AND BIRKELAND CURRENTS

responsible for opening the magnetosphere in the first place. In the present analysis, this would introduce an inconsistency because the distribution of $\mathbf{B} \cdot \hat{\mathbf{n}}$ on the tail magnetopause is specified, in the TH93 model, to be consistent with that already specified on the dayside magnetopause, according to magnetic flux conservation along magnetosheath streamlines. It seems likely that the effect of dayside Region-1 currents (ignored in the present analysis because of their more complicated geometry) would be analogous to that of the nightside Region-1 currents, namely, to reduce the degree of openness of the magnetosphere. Verification of this conjecture requires further work.

*Acknowledgments.* This work was supported in part by NASA grants NAG5-4726 and NAG5-6432. We wish to thank both referees for their comments on the manuscript.

## REFERENCES

Ding, C., T. W. Hill, and F. R. Toffoletto, Improvement of the Toffoletto-Hill open magnetosphere model, in *Physics of Space Plasmas (1995), Number 14*, (T. Chang and J. R. Jasperse, eds.), MIT Center for Theoretical Geo/Cosmo Plasma Physics, Cambridge, MA, p. 639, 1996.

Donovan, E. F., Modeling the magnetic effects of field-aligned currents, *J. Geophys. Res., 98*, 13,529, 1993.

Fedder, J. A., and J. G. Lyon, The solar wind-magnetosphere-ionosphere current-voltage relationship, *Geophys. Res. Lett., 14*, 880, 1987.

Fedder, J. A., S. P. Slinker, J. G. Lyon, and R. D. Elphinstone, Global numerical simulation of the growth phase and the expansion onset for a substorm observed by Viking, *J. Geophys. Res., 100*, 19,083, 1995.

Hill, T. W., and F. R. Toffoletto, Comparison of empirical and theoretical polar-cap convection patterns for the January 1992 GEM interval, *J. Geophys. Res., 103*, 14,811, 1998.

Hill, T. W., A. J. Dessler, and R. A. Wolf, Mercury and Mars: the role of ionospheric conductivity in the acceleration of magnetospheric particles, *Geophys. Res. Lett., 3*, 429, 1976.

Iijima, T., and T. A. Potemra, The amplitude distribution of field-aligned currents at northern high latitudes, *J. Geophys. Res., 81*, 2165, 1976.

Peroomian, V., L.R. Lyons, and M. Schulz, Inclusion of shielded Birkeland currents in a model magnetosphere, *J. Geophys. Res., 103*, 151, 1998.

Raeder, J., J. Berchem, and M. Ashour-Abdalla, The Geospace Environment Modeling grand challenge: results from a global geospace circulation model, *J. Geophys. Res., 103*, 14787, 1998.

Siscoe, G. L., Energy coupling between regions 1 and 2 Birkeland current systems, *J. Geophys. Res., 87*, 5124, 1982.

Stern, D. P., The origins of Birkeland currents, *Rev. Geophys. Space Phys., 21*, 125, 1983.

Toffoletto, F. R., and T. W. Hill, A nonsingular model of the open magnetosphere, *J. Geophys. Res., 98*, 1339, 1993.

Toffoletto, F. R., R. V. Hilmer, T. W. Hill, and G. -H. Voigt, Solution of the Chapman-Ferraro problem with an arbitrary magnetopause, *Geophys. Res. Lett., 21*, 621, 1994.

Tsyganenko, N. A. A global analytical representation of the magnetic field produced by the region 2 Birkeland currents

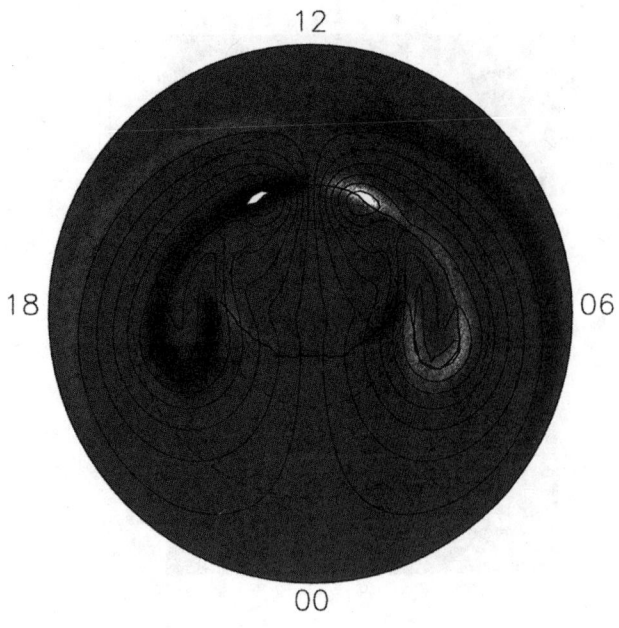

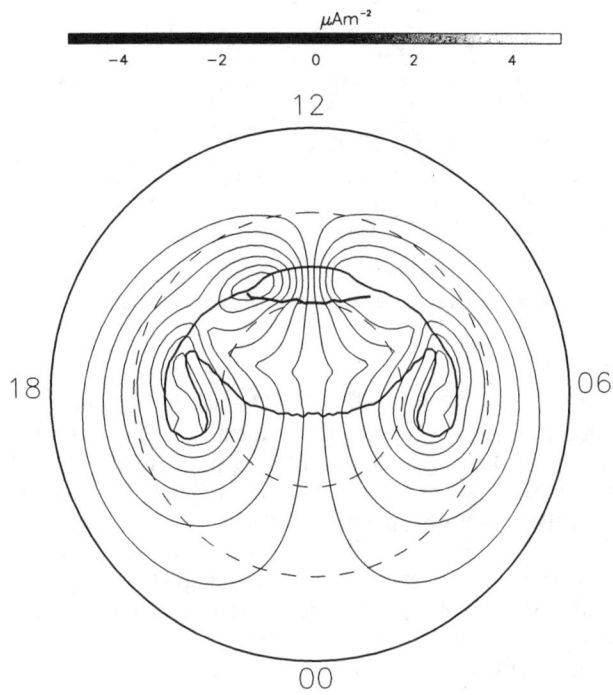

**Figure 7.** Polar-cap potential patterns, as in Figures 2 and 5, for the (a) second and (b) third iteration steps, but with a larger conductivity $\Sigma = 25$ mho. The Birkeland current density is also shown in (a) but deleted from (b) in order to reveal the detailed distortion of the polar cap boundary. This result, though clearly unrealistic, is useful in that it illustrates a basic instability of the system for large $\Sigma$, when the Birkeland current field perturbation becomes comparable in magnitude to the background tail field.

and the partial ring current. *J. Geophys. Res.*, *98*, 5677, 1993a.

Tsyganenko, N., D.P. Stern, and Z. Kaymaz, Birkeland currents in the plasma sheet, *J. Geophys. Res.*, *98*, 19,455-19,464, 1993b.

Tsyganenko, N. A., and D. P. Stern, Modeling the global magnetic field of the large-scale Birkeland current systems, *J. Geophys. Res.*, *101*, 27,187, 1996.

Voigt, G.-H., A mathematical magnetospheric field model with independent physical parameters, *Planet. Space Sci.*, *29*, 1, 1981.

White, W. W., G. L. Siscoe, G. M. Erickson, Z. Kaymaz, N. C. Maynard, K. D. Siebert, B. U. Ö. Sonnerup, and D. R. Weimer, The magnetospheric sash and the cross-tail S, *Geophys. Res. Lett.*, *25*, 1605, 1998.

Wolf, R. A., The quasi-static (slow-flow) region of the magnetosphere, in *Solar Terrestrial Physics*, ed. R. L. Carovillano and J. M. Forbes, Series, D. Reidel, Hingham, MA, 303-368, 1983.

---

T. W. Hill and F. R. Toffoletto, Department of Space Physics and Astronomy, MS 108, 6100 Main Street, Rice University, Houston, TX 77005-1892

# Magnetotail Currents During the Growth Phase and Local Auroral Breakup

T. I. Pulkkinen,[1] M. V. Kubyshkina,[2] D. N. Baker,[3] L. L. Cogger,[4]
S. Kokubun,[5] T. Mukai,[6] H. J. Singer,[7] J. A. Slavin,[8] L. Zelenyi[9]

The current distribution in the nightside magnetosphere during a substorm event on Dec 10, 1996, is modeled using a local magnetic field model and two empirical large-scale magnetic field models. The empirical models utilizing both high-altitude spacecraft data and low-altitude energetic particle measurements suggest that a thin current sheet, which formed during the substorm growth phase extended from the inner magnetotail at least out to 20-30 $R_E$. It is shown that the results from the three different modeling approaches are consistent with each other, which is an important test for the validity of the empirical modeling methods. The models are compared and contrasted and their validity and use in magnetospheric studies are addressed.

## 1. INTRODUCTION

Large-scale current systems are of key importance for the magnetospheric dynamics as well as for their influence on the magnetospheric morphology. Electric currents in space are difficult to measure directly, hence they are in most cases inferred from magnetic field measurements using Ampere's law. However, it is impossible to monitor the large-scale current systems with the few available measurement at any given time in the vast region of near-Earth space. Therefore, models for the current systems and their temporal evolution are key to understanding magnetotail dynamics.

Magnetotail currents include slow, large-scale variations that can be described by magnetohydrodynamics as well as smaller scale and faster variations that require kinetic treatment of the plasmas. Because of the vast size of the system and the small temporal and spatial scales of many of the dynamic processes, no one method is sufficient for describing the entire system. Due to the difficulties in solving the self-consistent equations of the magnetotail dynamics, empirical models for the magnetospheric currents and magnetic fields are widely used in the analysis of magnetospheric problems. The data-based models by [*Tsyganenko*, 1987, 1989, 1995] describe the magnetospheric currents as a function of the level of magnetospheric activity, interplanetary conditions, and dipole orientation relative to the Sun-Earth line. Based on this statistical representation of the magnetospheric configuration, *Pulkkinen et al.* [1992] and *Kubyshkina et al.* [1999] modified the magnetospheric configuration to best represent the observed features

---

[1] Finnish Meteorological Institute, Helsinki, Finland
[2] Department of Physics, University of St. Petersburg, St. Petersburg, Russia
[3] Laboratory for Atmospheric and Space Physics, University of Colorado, Boulder, CO
[4] Department of Physics, University of Calgary, Alberta, Canada
[5] Solar Terrestrial Environment Laboratory, Nagoya University, Japan
[6] Institute of Space and Astronautical Science, Sagamihara, Japan
[7] NOAA Space Environment Center, Boulder, CO
[8] NASA Goddard Space Flight Center, Greenbelt, MD
[9] Space Research Institute, Moscow, Russia

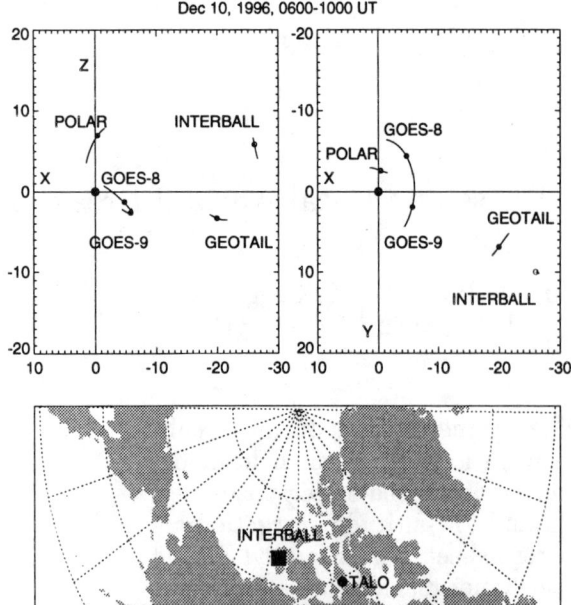

**Figure 1.** (Top) Spacecraft locations in GSM $X-Z$ (left) and $X-Y$ (right) projections during Dec 10, 1996, 0600-1000 UT. The dots indicate the spacecraft locations at 0730 UT. (Bottom) Locations of the CANOPUS magnetometer stations (black dots) and the projections of the spacecraft locations using the T89 model. The map is oriented such that magnetic midnight is roughly downward at 0730 UT. The grid shown is in geographic coordinates. (From Pulkkinen et al. [1999]).

during individual events. These event-oriented models allow for more accurate mapping between the ionosphere and the magnetosphere as well as examination of the detailed structure of the tail current system during dynamically interesting periods.

The study of the substorm onset process in many ways focusses around examination of the properties of the near-Earth tail current (and plasma) sheet [*Baker and Pulkkinen*, 1991; *Lui*, 1996; *Nagai et al.*, 1998]. It is widely accepted that the current sheet intensifies during the substorm growth phase [*McPherron et al.*, 1973]. Recent hybrid modeling results show that the current intensification in the model occurs mainly within a thin layer at the center of the pre-existing current sheet [*Hesse et al.*, 1996]. This implies that toward the end of the growth phase the plasma sheet hosts two current sheets, a thin, strongly peaked current embedded within a thicker and weaker current sheet.

One of the key problems in developing event-oriented models is addressing their accuracy. The only quantitative test for these models is a direct comparison with the data that was used as model input, but this does not necessarily guarantee good accuracy further away from the region where the observations were made. Each model provides a unique solution to the error minimizing procedure, but it is not self-evident that models using different inputs and/or different time periods would yield consistent results. Therefore, it is critically important to compare various methods to model the current systems to examine how robust the results are. Proving that the empirical models are reliable methods to analyze the magnetotail dynamics requires that the models give comparable results under a variety of conditions and for different sets of input data. In this paper we show results from three different modeling approaches and demonstrate that the results are in good agreement with each other.

## 2. EVENT DESCRIPTION

The substorm on December 10, 1996, was well-observed by two geostationary spacecraft in the inner magnetotail (GOES 8 and GOES 9) and two spacecraft in the midtail (INTERBALL Tail Probe in the northern tail lobe and GEOTAIL in the plasma sheet). The substorm was also monitored by ground-based instrumentation in Canada. WIND was in the upstream solar wind. Figure 1 shows the locations of the magnetospheric spacecraft and ground stations (from *Pulkkinen et al.* [1999]).

The CANOPUS magnetometer chain recorded initial signs of intensification of the westward electrojet currents at 0644 UT; this time is taken as the beginning of the growth phase. The CANOPUS magnetometers show two clear onsets: The onset at 0731 UT initiated to the west of Gillam, but the expansion did not reach the highest-latitude stations (data not shown). A second onset at 0800 UT initiated to the east of the first one, and expanded quickly over the entire latitude range of CANOPUS. The first expansion was about 600 nT in magnitude, whereas the maximum disturbance during the second activation was close to 900 nT. The top panels in Figures 2, 3, and 4 show the magnetogram from Gillam (GILL) for reference.

In the midtail, the growth phase was seen as a gradual rise of the lobe field as monitored by INTERBALL Tail Probe and by gradual thinning of the plasma sheet and intensification of the cross-tail current as observed by GEOTAIL. Figure 2 shows the magnetic field measurements as well as the plasma $\beta$ and flow velocity in the $X$ direction from GEOTAIL (the model results also shown in the figure will be discussed in later). Magnetic

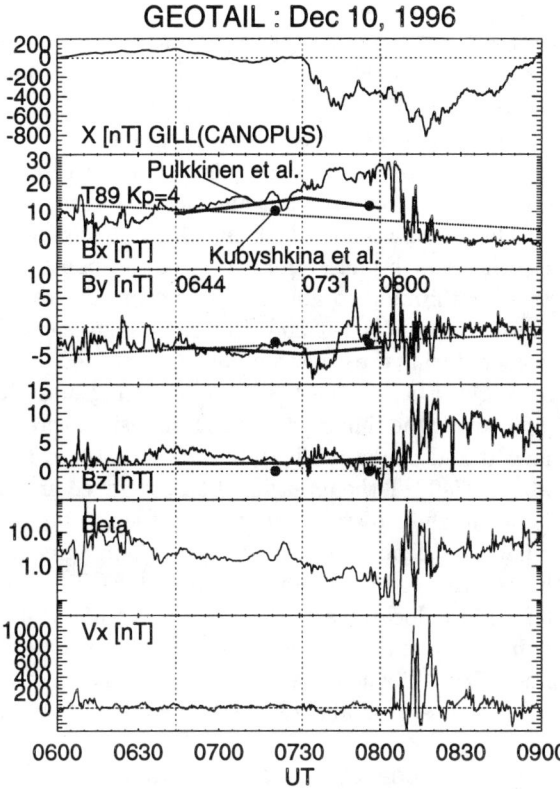

**Figure 2.** (top) $X$ component from the GILLAM station of the CANOPUS magnetometer chain. (middle) Magnetic field data from GEOTAIL at 12-s resolution. The solid dots show the time-evolving magnetic field model for the growth phase. The filled circles show the model results using the method developed by Kubyshkina et al. [1999]. The dotted line shows the Tsyganenko [1989] model for $Kp = 4$. (bottom) Plasma $\beta$ and flow velocity $V_X$ from GEOTAIL LEP measurements. All measurements are presented in GSM coordinates.

field data from INTERBALL Tail Probe are displayed in Figure 3. After 0731 UT, the growth phase signatures seemed to continue, and it was only after the 0800 UT intensification that the lobe field decreased, the plasma sheet expanded ($\beta$ increased), and strong Earthward flows were detected. These data suggest that a global reconfiguration of the magnetotail occurred only at the time of the second onset after 0800 UT.

In the inner magnetotail, GOES 8 and GOES 9 were close to the midnight sector, and hence were well-positioned to monitor the substorm evolution (Figure 4). After 0731 UT, both spacecraft recorded a change in the $Y$ component, which indicates the formation of a field-aligned current pattern. However, the field dipolarization was relatively weak and followed about 10 min later than the onset time. The second onset at 0800 UT similarly caused a $Y$ component disturbance beginning around the onset time, but again the field dipolarization (that now was observed only by GOES 9) was delayed by about 10 min.

## 3. LOCAL MODEL RESULTS

The local model results have been reported in detail by *Pulkkinen et al.* [1999], and are only summarized here. Having INTERBALL Tail Probe in the lobe and GEOTAIL in the plasma sheet allows for local examination of the current sheet evolution during the substorm growth phase. Assuming that the plasma sheet thins and that a new thin current sheet forms at the center of the plasma sheet during the growth phase, a two-sheet model was adopted. The magnetic field was described as a linear combination of the magnetic field from two Harris-type current sheets as

$$B = B_1 \tanh\left(\frac{Z - Z_0}{\lambda_1}\right) + B_2 \tanh\left(\frac{Z - Z_0}{\lambda_2}\right)$$

where $B_1$ and $B_2$ give the plasma sheet and thin current sheet intensities, $Z_0$ defines the location of the current sheet center, and $\lambda_1$ and $\lambda_2$ give the plasma sheet and thin current sheet thicknesses. Figure 5 shows the

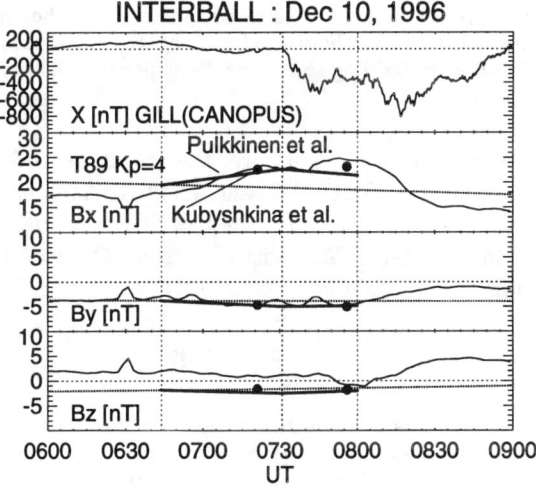

**Figure 3.** (top) $X$ component from the GILLAM station of the CANOPUS magnetometer chain. (bottom) Magnetic field data from INTERBALL Tail Probe. The measurements shown are two-minute averages. The solid dots show the time-evolving magnetic field model for the growth phase. The filled circles show the model results using the method developed by Kubyshkina et al. [1999]. The dotted line shows the Tsyganenko [1989] model for $Kp = 4$.

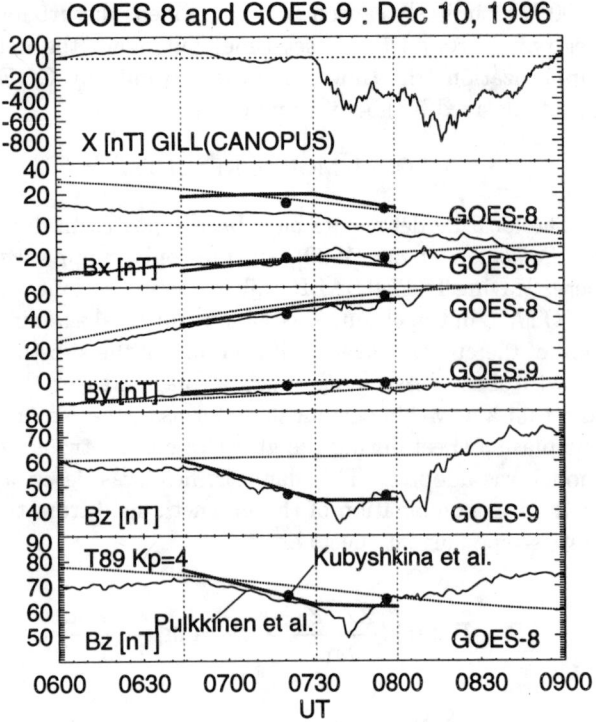

Figure 4. (top) $X$ component from the GILLAM station of the CANOPUS magnetometer chain. (bottom) Magnetic field measurements from GOES 8 and GOES 9 at geosynchronous orbit. The data are shown in GSM coordinates. The solid dots show the time-evolving magnetic field model for the growth phase. The filled circles show the model results using the method developed by Kubyshkina et al. [1999]. The dotted line shows the Tsyganenko [1989] model for $Kp = 4$.

model geometry. Furthermore, the radial distance difference between GEOTAIL and INTERBALL was accounted for by transforming the INTERBALL magnetic field data to GEOTAIL distance using the statistical formula by *Fairfield and Jones* [1996]

$$B = \frac{C}{R^{1.46}} + 7.47,$$

which gives the field in nanoteslas when the radial distance is inserted in Earth radii. Using the observed field value and radial distance at the INTERBALL location the constant $C$ can be determined under these particular conditions. Then, the field value at the radial distance of the GEOTAIL location can be computed.

Two free parameters can be resolved from the two-spacecraft observations; these were chosen to be the plasma sheet thickness ($\lambda_1$) and thin current sheet intensity ($B_2$). Of the remaining parameters, the thin current sheet thickness was set to $\lambda_2 = 0.1 R_E$ and the plasma sheet current intensity was set to the lobe field value before the beginning of the growth phase, $B_1 = B_{IB}(t = 0630 UT)$. The current sheet center was determined from the *Tsyganenko* [1989] model.

The bottom panels of Figure 5 shows the resulting plasma sheet thickness and thin current sheet intensity. The plasma sheet thickness did not vary much during the growth phase, but remained at about $2 R_E$. A stronger thinning to about $1 R_E$ occurred after the first onset during 0731-0800 UT. The thin current sheet intensified during the growth phase, and reached a maximum slightly before the first onset. During and after the first onset, the thin current sheet intensity decreased somewhat and the intensification stopped. However, after about 0740 UT the intensification continued and the thin current sheet disrupted only after the second onset after 0800 UT. These values gave an excellent agreement with the observed field values both at GEOTAIL and INTERBALL locations (see *Pulkkinen et al.* [1999]). The model is considered unreliable during the period around 0725 UT when the current sheet thickness was over $5 R_E$, most probably caused by flapping of the tail.

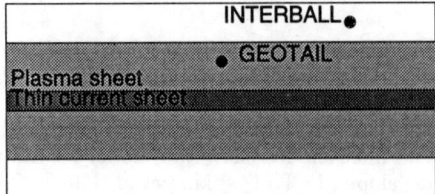

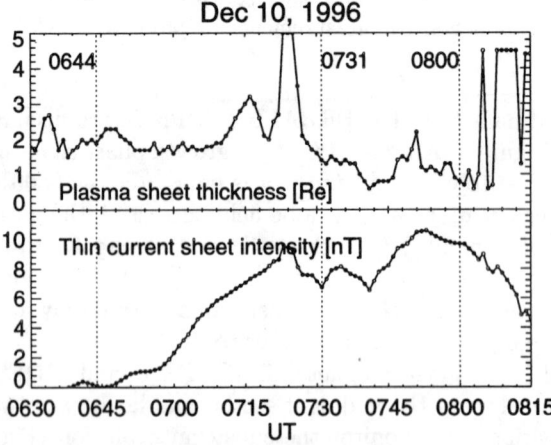

Figure 5. (Top) Schematic of the two-sheet model and relative spacecraft locations. (Bottom) Local model results. Plasma sheet thickness ($\Lambda_1$) in units of $R_E$. Thin current sheet intensity ($B_2$) in units of nT.

## 4. TIME-EVOLVING MAGNETIC FIELD MODEL RESULTS

The large-scale magnetic field configuration and magnetic connectivity in the inner and midtail was modeled using the technique described in *Pulkkinen et al.* [1992]. This model assumes the *Tsyganenko* [1989] model as the initial state, after which the current systems are modified with linear time-dependence to account for the growth-phase associated changes in the field configuration. Two periods were modeled, the growth phase (0644–0731 UT) and the period between the two onsets (0731–0800 UT). By assuming linear time dependence, one set of model parameters is sufficient to describe the field evolution during each of the modeled time periods.

In developing the growth phase model, the field configuration at the beginning of the growth phase (0644 UT) was assumed to be that given by the *Tsyganenko* [1989] field model for $Kp = 4$, which was in best agreement with the observations from the four spacecraft in the tail. The dotted lines in Figures 2, 3, and 4 show the T89 model prediction for the entire period. During the growth phase, the model tail current sheet was gradually thinned

$$D(X,Y) \to D(X,Y) \left(1 - \frac{A(t)}{\cosh^2 \frac{X-X_0}{\Delta X}} \frac{1}{1 + (\frac{Y-Y_0}{\Delta Y})^2}\right)$$

$$A(t) = A_{max} \frac{t - t_{min}}{t_{max} - t_{min}}$$

where $D(X,Y)$ gives the current sheet thickness in the T89 model, $A(t)$ gives the minimum thickness at $X_0$ and $Y_0$, and $\Delta X$ and $\Delta Y$ define the scale size of the thinned region. The tail current was intensified

$$B_T = B_T(1 + f_T(t))$$

$$f_T(t) = f_{T,max} \frac{t - t_{min}}{t_{max} - t_{min}}$$

where $B_T$ gives the tail current intensity in the T89 model and $f_T$ the current intensification. The ring current (which also gives the currents near the inner edge of the plasma sheet) was similarly intensified

$$B_{RC} = B_{RC}(1 + f_{RC}(t))$$

$$f_{RC}(t) = f_{RC,max} \frac{t - t_{min}}{t_{max} - t_{min}}$$

where $B_{RC}$ gives the ring current intensity in the T89 model and $f_{RC}$ the current intensification. In addition, the current sheet was allowed to rotate around the $Y$ axis by letting the tilt angle of the dipole to be a free parameter in the model. The parameter values were chosen by minimizing the error function

$$ERR = \sum_{k=1}^{k_{max}} \sum_{j=1}^{n_{sc}} \sum_{i=1}^{3} (B_{ij}^{obs}(t_k) - B_{ij}^{mod}(t_k))^2$$

between the model values and the measured values at each of the four spacecraft, where the summations are over all timesteps ($k$), over all spacecraft ($j$), and over all field components ($i$). Hence, one set of model parameters describes the entire time period. The thick dotted lines give the model field at the spacecraft locations in Figures 2, 3, and 4. Note how the intensification and thinning of the current sheet reproduces the typical growth phase signatures, increase of lobe flux ($B_X$ in the tail spacecraft) and Earthward motion and intensification of the inner edge current (depression of $B_Z$ at the geostationary spacecraft). The parameter values used for the modeling are given in Table 1.

The second period between the onsets was modeled assuming the initial values from the growth phase model so that the field changes remain continuous. This means that only the time-evolving parameters were free parameters in determining the fit; others were determined by the model for the growth phase. For example, the location of the thinned region did not change, only the degree of current sheet thinning was changed. During this period, the dynamics associated with the first breakup as well as the strong intensification of the GEOTAIL field possibly associated with a rotation of the current sheet lead to a somewhat poorer fit. However, the model is able to capture the lobe flux leveling, some of the plasma sheet dynamics, and the average geostationary orbit behavior. After the second onset, the dynamics of the expansion phase dominated the field behavior, and no attempt was made to model the field evolution using this technique.

## 5. HYBRID INPUT ALGORITHM MODEL RESULTS

The hybrid input algorithm (HIA) is an extension of the modeling presented above [*Kubyshkina et al.*, 1999]. HIA uses similar modifications of the T89 model, but, in addition to the high-altitude magnetic field measurements, the HIA takes advantage of energetic particle measurements from low-altitude polar-orbiting spacecraft. As the low-altitude measurements are available only at one time instant, this modeling method does not give the time evolution of the field, but rather describes the field configuration at one given time.

Table 1. Comparison of model parameters for Dec 10, 1996 substorm. (Models: (1) Maximum values for the time-evolving model 0644-0731 UT, (2) Maximum values for the time-evolving model 0731-0800 UT, (3) Time-evolving model at 0724 UT, (4) Time-evolving model at 0756 UT, (5) Hybrid input algorithm at 0724 UT, (6) Hybrid input algorithm at 0756 UT).

| Model | $f_{TC}$ | $f_{RC}$ | $A$ | $X_0$ | $\Delta X$ | $Y_0$ | $\Delta Y$ | $\Delta \Psi$ |
|---|---|---|---|---|---|---|---|---|
| 1) TEM 0731 UT | 0.22 | 0.15 | 0.88 | -13.8 | 9.0 | 3.0 | 6.5 | 1.5 |
| 2) TEM 0800 UT | 0.19 | 0.08 | 0.98 | -13.8 | 9.0 | 3.0 | 6.5 | 2.0 |
| 3) TEM 0724 UT | 0.21 | 0.14 | 0.84 | -13.8 | 9.0 | 3.0 | 6.5 | 1.4 |
| 4) TEM 0756 UT | 0.19 | 0.09 | 0.97 | -13.8 | 9.0 | 3.0 | 6.5 | 1.9 |
| 5) HIA 0724 UT | 0.3 | 0.1 | 0.87 | -7.0 | 3.0 | 3.0 | 7.0 | 7.0 |
| 6) HIA 0756 UT | 0.4 | 0.0 | 0.96 | -7.0 | 6.3 | 2.3 | 6.0 | 7.0 |

Low-altitude measurements frequently show a sudden isotropization of the previously trapped particle population, and an ordering where the highest energy populations isotropize at lower latitudes than the lower-energy ones. This has been interpreted as the onset of pitch-angle scattering in the equatorial plane [*Sergeev et al.*, 1993], which can be related to the particle gyroradius and field-line curvature by

$$\frac{R_C}{\rho} = \frac{1}{G} \frac{B_n^2}{\partial B_t / \partial n} = 8$$

where $G = mv/q$ is the particle rigidity, and $B_t$ and $B_n$ are the magnetic field components parallel and perpendicular to the current sheet, respectively [*Kubyshkina et al.*, 1999]. Hence, the low-altitude particle measurements made over a range of energies define a magnetic field profile along the current sheet at the longitude of the low-altitude spacecraft, thus giving more information about the global field configuration than the high-altitude field measurements alone. If no low-altitude measurements are available, the method reduces to that described in *Pulkkinen et al.* [1992], but only for one time instant, and therefore these models do not necessarily give exactly similar results. The free parameter values are evaluated by minimizing the error function

$$ERR = \frac{W_B}{3n_{sc}} \left( \sum_{j=1}^{n_{sc}} \sum_{i=1}^{3} (B_{ij}^{obs} - B_{ij}^{mod})^2 \right)^{1/2}$$
$$+ \frac{W_{IB}}{n_{IB}} \left( \sum_{j=1}^{n_{IB}} (CGLAT_j^{obs} - CGLAT_j^{mod})^2 \right)^{1/2}$$

[*Kubyshkina et al.*, 1999], where $W_B$ and $W_{IB}$ are the weight functions which allow different weighting between the observations, $CGLAT$ are the observed and modeled corrected geomagnetic latitudes of the isotropic boundaries, and the summations are over the number of spacecraft ($j$ in the first term), over the three field components ($i$ in the first term), and over the number of isotropic boundaries ($j$ in the second term).

Two specific time instants, one before each onset, were modeled using this technique. At 0724 UT the NOAA spacecraft crossed the southern auroral region at about 22 MLT. The observed isotropic boundaries are shown in Figure 6 together with comparisons with the various models. Obviously, the HIA model, which uses these observations as input data, gives the best fit to the observed isotropic boundaries, but the time-evolving empirical model also produces significant improvement over the statistical picture. The field components are shown in Figures 2, 3, and 4 as large dots. The HIA model gives a consistent representation at the INTERBALL Tail Probe and at geosynchronous orbit. The model parameter values are given in Table 1.

During the second period modeled using this approach at 0756 UT, no isotropic boundary observations were available, and hence the model is based on field observations only. Here the differences between the time-evolving model arise from the fact that the time-evolving model includes the entire period 0731-0800 UT in the modeling, whereas the HIA-method was applied to only single time step at 0756 UT. However, the results are qualitatively consistent with each other and with the local model results.

## 6. COMPARISON OF MODEL RESULTS

During the growth phase, both empirical models indicate a thin current sheet extended from the geostationary orbit out to at least 30 $R_E$. This is consistent

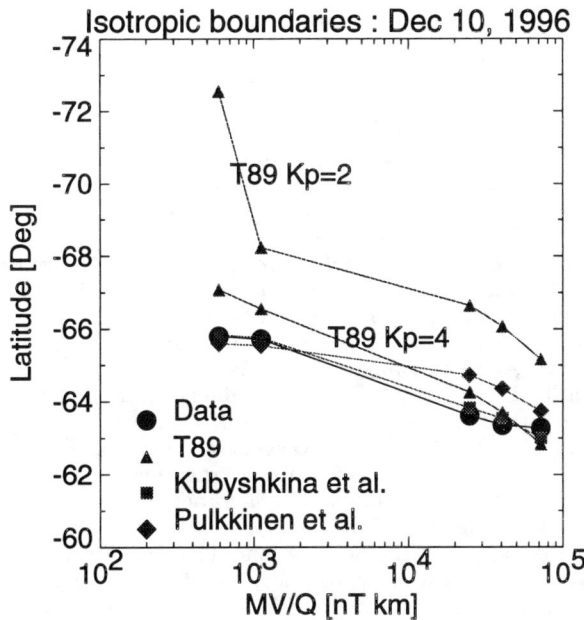

**Figure 6.** Isotropic boundaries as measured by the NOAA spacecraft (black circles). Isotropic boundaries as predicted by the various field models:HIA model (grey large squares), time-evolving model (grey diamonds), T89 $Kp = 2$ (triangles), and T89 $Kp = 4$ (triangles).

with the local model result and its extension to the near-Earth tail [*Pulkkinen et al.*, 1999]. A difference with the time-evolving model and the HIA model is the location of the minimum thickness region: while the time-evolving model suggests that the minimum thickness appears near $X = -14R_E$, the HIA-model suggests a more near-Earth location at $X = -7R_E$. However, both models suggest a minimum thickness of about 0.4 $R_E$. The HIA-model also suggests somewhat stronger current intensification than the time-evolving model.

During the period between the two onsets, all models suggest that the thin current sheet persisted and continued to thin. The local model follows closely the field measurements, indicating first a small reduction and then further slight intensification of the current. Both the HIA model and the time-evolving model again are very consistent with the thickness estimate, reduced to only about 0.1 $R_E$, which was also the assumed input for the local model.

The largest difference between the time-evolving and the HIA model is the tilt-angle change. Whereas the time-evolving model only supports a less than 2° decrease of the tilt angle, the HIA-model shows that the tail was tilted 7° from its nominal orientation. This difference in model results is due to different input data:

The HIA-model uses as input only data from one time instant, whereas the time-evolving model uses measurements from the entire growth phase time period. Comparing the unmodified T89 model and observations at the beginning of the growth phase, one can see quite a good agreement between the model and the data (with the exception of $B_X$ at GOES 8). Therefore, the time-evolving model does not produce a strong twist in the tail, as this would significantly decrease the fit quality at this time instant. The situation is different during the late growth phase when the unmodified T89 model and the observations deviate considerably from each other. Therefore, the two models arrive at a different end configuration and the difference in tail rotation also affects the values of the other parameters.

In the magnetosphere, it is important that models for the current systems address the question of closure of the current systems properly. This is especially critical in regions where more than one current system contributes to the measured total field. In this study, all the spacecraft were either close to or within the tail plasma/current sheet. Therefore, the majority of the field variations are caused by variations in the cross-tail current. The local field model results were obtained from a single current system, where the current has an infinite extent in the $Y$ direction, and hence the current system is not closed within the magnetosphere. On the other hand, both event-oriented large-scale models close the enhanced tail currents along the tail magnetopause and enhanced ring currents around the dayside plasma sheet. INTERBALL in the tail lobe would be the only spacecraft at a location where it could see effects of the magnetopause currents. The good correspondence between the local and large-scale empirical models indicates that INTERBALL is sufficiently close to the plasma sheet that a majority of the field disturbances actually come from currents flowing below the satellite within the plasma sheet.

## 7. DISCUSSION

As empirical models are widely used in magnetospheric studies, addressing their accuracy is critically important. The models give a global picture of the magnetospheric topology and smaller-scale information about the current densities and the radial distribution of the current systems. Event-oriented models can use many different data sources, the examples here utilized high-altitude magnetic field measurements and low-altitude precipitation boundaries. It is important that the empirical models give consistent results for a variety of inputs, as the available data sources and spacecraft constellations are different for each event.

The three models shown in this paper demonstrate that the empirical models can capture the main features of the growth phase evolution quite similarly even if the model details are different. The similarity of the results would indicate that the models are quite robust and represent the field configuration also outside the region where the measurements were made. These results quite significantly increase the confidence to the results obtained using the large-scale event-oriented magnetic field models (see also *Kubyshkina et al.* [1999]), and indicate that they can reliably be used in interpreting multi-spacecraft and ground-based observations. Furthermore, the good agreement in the current sheet thickness estimates given by the two large-scale empirical models shows that these models can also be used to examine the current sheet parameters.

It is a fairly well-established fact that a thin current sheet forms during every substorm, but from single-spacecraft observations it is difficult to determine the scale size of the thin current sheet. The results here further support the idea that current sheet intensification and thinning is a large-scale process, which covers the entire inner and midtail regions from near-geostationary distance out to at least 30 $R_E$.

*Sergeev et al.* [1994] noted that thin current sheets can remain stable for several hours during steady magnetospheric convection periods, when the solar wind is very steady and no large-scale substorm activity occurs. In this paper we have shown that a thin current sheet can also remain stable throughout a relatively strong local auroral breakup, which does not develop to a large-scale reconfiguration. This indicates that the thin current sheets are quite large scale and robust in nature, not something that can remain stable only within a very narrow parameter regime.

Based on the model results shown here, the two onsets differ only in terms of the minimum current sheet thickness (or maximum current density) in the magnetotail. It is possible that the maximum current density could be a control parameter for the global expansion onset, as has been suggested by some substorm models [*Lui*, 1996]. If the thin current sheet indeed extends from the geosynchronous region to the midtail, it would tie together processes discussed in the current disruption model with those proposed in the NENL model.

*Acknowledgments.* We thank the NSSDC personnel for maintaining the online ISTP key parameter facility. The work at CU/LASP was supported by funding from the NASA ISTP program.

## REFERENCES

Baker, D. N., T. I. Pulkkinen, V. Angelopoulos, W. Baumjohann, and R. L. McPherron, The neutral line model of substorms: Past results and present view, *J. Geophys. Res.*, *101*, 12,975, 1996.

Baker, D. N., and T. I. Pulkkinen, The Earthward edge of the plasma sheet in magnetospheric substorms, in *Magnetospheric Substorms, Geophys. Monogr. Ser., 64*, edited by J. R. Kan, T. A. Potemra, S. Kokubun, and T. Iijima, p. 147, AGU, Washington, D.C., 1991.

Fairfield, D. H., and J. Jones, Variability of the tail lobe field strength, *J. Geophys. Res.*, *101*, 7785, 1996.

Hesse, M., D. Winske, M. Kuznetsova, J. Birn, and K. Schindler, Hybrid modeling of the formation of thin current shetes in magnetotail configurations, *J. Geomagn. Geoelectr.*, *48*, 749, 1996.

Kubyshkina, M., V. A. Sergeev, and T. I. Pulkkinen, Hybrid input algorithm – Possibility to get a realistic event-oriented magnetospheric model, *J. Geophys. Res.*, paper number 1999JA900222, 1999.

Lui, A. T. Y., Current disruption in the Earth's magnetosphere: Observations and models, *J. Geophys. Res.*, *101*, 13,067, 1996.

McPherron, R. L., C. T. Russell, and M. P. Aubry, Satellite studies of magnetospheric substorms on August 15, 1968, *J. Geophys. Res.*, *78*, 3131, 1973.

Nagai, T., M. Fujimoto, Y. Saito, S. Machida, T. Terasawa, R. Nakamura, T. Yamamoto, T. Mukai, A. Nishida, and S. Kokubun, Structure and dynamics of magnetic reconnection for substorm onsets with Geotail observations, *J. Geophys. Res.*, *103*, 4419, 1998.

Pulkkinen, T. I., D. N. Baker, R. J. Pellinen, J. Büchner, H. E. J. Koskinen, R. E. Lopez, R. L. Dyson, and L. A. Frank, Particle scattering and current sheet stability in the geomagnetic tail during the substorm growth phase, *J. Geophys. Res.*, *97*, 19,283, 1992.

Pulkkinen, T. I., D. N. Baker, L. L. Cogger, L. A. Frank, J. B. Sigwarth, S. Kokubun, T. Mukai, H. J. Singer, K. Ogilvie, J. Slavin, and L. Zelenyi, Spatial extent and dynamics of a thin current sheet during the substorm growth phase on Dec 10, 1996, *J. Geophys. Res.*, paper number 1999JA900240, 1999.

Sergeev, V. A., M. Malkov, and K. Mursula, Testing the isotropic boundary algorithm method to evaluate the magnetic field configuration in the tail, *J. Geophys. Res.*, *98*, 7609, 1993.

Sergeev, V. A., T. I. Pulkkinen, R. J. Pellinen, and N. A. Tsyganenko, Hybrid state of the tail magnetic configuration during steady convection events, *J. Geophys. Res.*, *99*, 23571, 1994.

Tsyganenko, N. A., Global quantitative models of the geomagnetic field in the cislunar magnetosphere for different disturbance levels, *Planet. Space Sci.*, *35*, 1347, 1987.

Tsyganenko, N. A., Magnetospheric magnetic field model with a warped tail current sheet, *Planet. Space Sci.*, *37*, 5, 1989.

Tsyganenko, N. A., Modeling the Earth's magnetospheric magnetic field confined within a realistic magnetopause, *J. Geophys. Res.*, *100*, 5599, 1995.

---

D. N. Baker, Laboratory for Atmospheric and Space Physics, University of Colorado, 1234 Innovation Drive, Boulder, CO 80303

L. L. Cogger, Department of Physics, 2500 University Dr. N.W., Calgary, Alberta, T2N 1N4, Canada

S. Kokubun, Solar Terrestrial Environment Laboratory, Nagoya University, Honohara 3-13 Toyokawa, Aich 442, Japan

M. V. Kubyshkina, University of St. Petersburg, Institute of Physics, 198904 St. Petersburg, Russia

T. Mukai, Institute of Space and Astronautical Science, 3-1-1 Yoshinodai, Sagamihara, Kanagawa 229, Japan

T. I. Pulkkinen, Finnish Meteorological Institute, POBox 503, FIN-00101 Helsinki, Finland. e-mail tuija.pulkkinen@fmi.fi

H. J. Singer, NOAA Space Environment Center, 325 Broadway, Boulder, CO 80303

J. A. Slavin, NASA Goddard Space Flight Center, Code 696, Greenbelt, MD 20771

L. Zelenyi, Space Research Institute of Russian Academy of Sciences, 84/32 Profsoyuznaya Str, Moscow, 117810, Russia

# Magnetosphere-Ionosphere Interactions: A Tutorial Review

## S. W. H. Cowley

*Department of Physics and Astronomy, University of Leicester, Leicester, UK*

We review the basic physics of the field-aligned current (FAC) systems which transmit energy and stress between the magnetosheath-magnetosphere system, and the ionosphere-thermosphere system. The specific topics covered include (a) ionospheric flow and currents, (b) the large-scale Region 1/2 current system associated with Dungey-cycle flow, (c) cusp currents and their relation to the interplanetary magnetic field; (d) travelling convection vortices, and (e) substorm-related current systems.

## 1. INTRODUCTION

The large-scale system of field-aligned currents (FACs) which transmit stress between the magnetosheath-magnetosphere, and ionosphere-thermosphere were first detected as "transverse magnetic disturbances" by the low-altitude polar-orbiting satellite 1963 38C (Zmuda et al., 1966). They were not immediately recognised as the effect of FACs, however, and it was Cummings and Dessler (1967) who first suggested a link with the current system which had been proposed by Birkeland sixty years earlier (Birkeland, 1908; see also Dessler, 1984). A further ten years had to pass before the overall morphology of the FAC system became clear, as presented in a number of papers by Iijima and Potemra (1976 a, b; 1978), using triaxial magnetic data from the Triad satellite. The overall pattern consists of two contiguous rings of current, "Region 1" at higher latitudes and "Region 2" at lower latitudes, with opposite polarities at dawn and dusk and some overlap in the pre-midnight Harang region. A third system at higher latitudes than Region 1 on the dayside is associated with the dayside cusp (Iijima and Potemra, 1976b; Wilhjelm, et al., 1978; Iijima et al., 1978; McDiarmid et al., 1978). This current pattern has been found to be an almost permanent feature of the magnetosphere-ionosphere system, though it is modulated in size, strength, and form both by the orientation and magnitude of the interplanetary magnetic field (IMF) and by geomagnetic disturbance. During relatively quiet times the total current flowing in the Region 1 system is ~1.6 MA, while that flowing in Region 2 is ~1.1 MA. During more disturbed times ($|AL|>100$ nT) the pattern expands typically by a few degrees, and the Region 1 and 2 currents increase to ~2.7 and ~2.5 MA respectively (Iijima and Potemra, 1978). Typical current intensities are ~0.5 to ~2 $\mu A\,m^{-2}$, again somewhat lower for Region 2 than for Region 1.

It is the principal purpose of this paper to discuss the physical origins of these currents, and some of their consequences. The central framework for our discussion will be Dungey's (1961) open model of the magnetosphere, in which plasma flow is generated principally by reconnection at the magnetopause between the terrestrial field and the IMF, and consequent related phenomena in the geomagnetic tail.

## 2. IONOSPHERIC FLOW, CONDUCTIVITY, AND CURRENTS

The flow imposed on the ionosphere by Dungey-cycle convection is shown schematically in Fig. 1, where the dashed line indicates the boundary between open and closed field lines. The flow consists of twin vortices, with antisunward flow over the polar cap, which maps to the magnetospheric tail lobes, and return sunward flow in the auroral zone, which maps mainly to the hot plasma sheet and ring current regions. When the ionosphere participates in such flow the plasma particles are subject to collisions with neutral atmospheric particles at lower altitudes in the E region, which causes a drag on the flow and heats the neutral gas. Assuming that the gas is stationary in the Earth's frame, an assumption which is usually valid as a first approximation, the force-balance equation for ions which determines the drift velocity $V_i$ is

$$e(E + V_i \times B) = m_i \nu_{in} V_i , \qquad (1)$$

where $E$ and $B$ are the electric and magnetic fields, $m_i$ the ion mass, and $\nu_{in}$ the ion-neutral collision frequency. The solution for the field-perpendicular flow is

$$V_{i\perp} = \frac{1}{\left(1 + \left(\frac{\nu_{in}}{\Omega_i}\right)^2\right)} \left[ \frac{E \times B}{B^2} + \left(\frac{\nu_{in}}{\Omega_i}\right) \frac{E}{B} \right] , \qquad (2)$$

where $\Omega_i = eB/m_i$ is the gyrofrequency. The first term is the $E \times B$ drift slowed by collisions, while the second describes mobility in the direction of $E$ produced by them. It can be seen that the drift magnitude and direction depend on the ratio of the collision frequency to the gyrofrequency, though it is the former parameter by far which varies the most rapidly through the ionosphere, since the magnetic field strength is almost constant in the appropriate range of altitudes (~100-200 km). Because the neutral density increases rapidly with decreasing altitude, so does the ion-neutral collision frequency, with the condition $(\nu_{in}/\Omega_i) \approx 1$ being reached at an altitude of ~125 km (see the paper by Richmond, this volume, for further details). In the region somewhat above ~125 km, therefore, $(\nu_{in}/\Omega_i)$ is small, such that the ion drift in the direction of $E \times B$ is not substantially diminished, while the ion mobility in the direction of $E$ increases with decreasing height proportional to $(\nu_{in}/\Omega_i)$. Similarly, in the region somewhat below ~125 km, $(\nu_{in}/\Omega_i)$ becomes increasingly large compared with unity, such that the drift in the direction of $E \times B$ becomes negligible, and the ion drifts approximately in the direction of $E$ with diminishing speed, inversely proportional to $(\nu_{in}/\Omega_i)$. The ion mobility in the direction of $E$ peaks at the speed $\frac{1}{2}(E/B)$ at the height where $(\nu_{in}/\Omega_i) = 1$ (i.e. at ~125 km), at which altitude the drift in the direction of $E \times B$ is also reduced to the same value, so that the ions drift at 45° to $E \times B$, towards the direction of $E$.

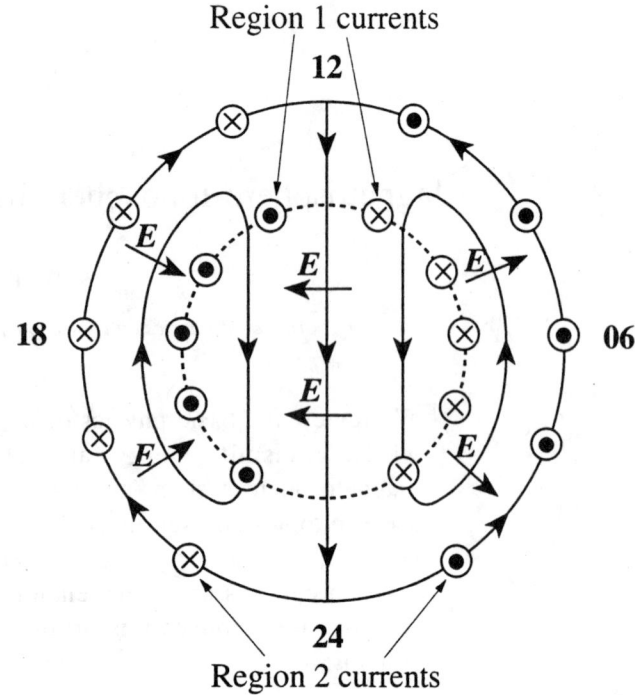

**Figure 1.** Schematic of Dungey-cycle flow mapped into the ionosphere, where the arrowed solid lines are the plasma streamlines, the short arrows give the direction of the electric field, and the dashed line is the open-closed field line boundary. The Hall current flows round the plasma streamlines opposite to the flow, while the Pedersen current flows in the direction of $E$. The direction of FAC flow associated with the horizontal divergence of the latter currents is indicated by the circular symbols, where circled dots indicate upward currents out of the ionosphere, while circled crosses indicate downward currents into the ionosphere.

In principle a similar discussion also applies to ionospheric electrons, in terms of the ratio of the electron-neutral collision frequency to the electron gyrofrequency. However, this ratio remains small throughout the whole region of the ionosphere where appreciable plasma densities are present (above ~90 km). Thus the electrons $E \times B$ drift at all ionospheric heights. The immediate consequence is that a field-perpendicular electric current must flow in the lower ionosphere, whose density is

$$j_\perp = \frac{ne\left(\frac{\nu_{in}}{\Omega_i}\right)}{\left(1 + \left(\frac{\nu_{in}}{\Omega_i}\right)^2\right)} \left[ \frac{E}{B} - \left(\frac{\nu_{in}}{\Omega_i}\right) \frac{E \times B}{B^2} \right] \qquad (3)$$

$$= \sigma_P E + \sigma_H \hat{B} \times E ,$$

where $n$ is the ion and electron number density (for simplicity we assume the dominance of one singly-charged ion species only), and $\hat{B}$ is the unit vector along $B$. The first term is the Pedersen current in the direction of $E$, which is dominant above ~125 km where both species approximately $E \times B$ drift, but where the ions have some mobility in the direction of $E$. The second term is the Hall current in the direction $-E \times B$ which is dominant below ~125 km where the electrons $E \times B$ drift but the ions become increasingly immobile. The two current densities are equal at ~125 km where $(v_{in}/\Omega_i) = 1$, which is also approximately where the Pedersen current peaks (depending a little on the height profile of $n$). Upon integrating with altitude, the total height integrated field-perpendicular current intensity is thus

$$i_\perp = \Sigma_P E + \Sigma_H \hat{B} \times E \quad , \quad (4)$$

where $\Sigma_P = \int dz \, \sigma_P$ is the height-integrated Pedersen conductivity, and $\Sigma_H = \int dz \, \sigma_H$ is the height-integrated Hall conductivity. In the sunlit ionosphere these conductivities are of order ~10 mho. On the nightside they depend on the intensity and energy of precipitating plasma particles from the magnetosphere, and may vary by at least an order of magnitude in either direction, with Hall conductivities exceeding Pedersen conductivities typically by factors of 2 to 4.

The implication of this discussion is that when the magnetosphere drives an ionospheric plasma through the neutral atmosphere, currents must flow in the lower ionosphere due to ion-neutral collisions. The $j \times B$ force of the currents just balances the neutral drag force of the atmosphere (the height integral of minus the RHS of Eq. 1), and consists of two components, one associated with the Pedersen current, the other with the Hall current. The $j \times B$ force associated with the Pedersen current just balances the drag force in the direction opposite to the $E \times B$ drift, while the force associated with the Hall current just balances the drag force in the direction opposite to $E$ associated with the Pedersen mobility. Equal and opposite drag forces also act, of course, on the neutral gas, which are thus just equal to $j \times B$, and which tend to excite winds in the thermosphere. As just indicated, these forces have both a "Hall" component in the direction of $E$, and a "Pedersen" component in the direction of $E \times B$, and despite the fact that they are of comparable magnitude, the "Pedersen" component is much more effective in exciting winds than the "Hall" component because the Pedersen currents flow at a somewhat higher altitude where the neutral densities are significantly less.

In addition to requiring a mechanical force to maintain the flow against neutral air drag, electromagnetic energy is also dissipated and heats the neutral gas. The height-integrated Joule heating rate per unit area of the ionosphere is $i_\perp . E = \Sigma_P E^2$ W m$^{-2}$, where we note that the Hall current is non-dissipative ($j.E = 0$) and does not enter. These considerations inescapably imply that the ionospheric "load" must be coupled to a "generator" in the magnetosphere/magnetosheath via a large-scale current system, and that energy and momentum must flow from the latter to the former via Poynting flux and Maxwell field stress respectively.

We now turn specifically to consider the flow system associated with the Dungey cycle, shown in Fig. 1. In principle, if the ionosphere were uniformly conducting, the Hall current would close wholly within the ionosphere, flowing around the plasma streamlines ($E \times B$ drift paths) opposite to the direction of plasma flow. However, the Pedersen currents flowing in the direction of $E$ cannot close within the ionosphere, but instead their divergence must be accommodated by a system of currents flowing into and out of the ionosphere along the field. The sense of those currents is indicated by the circled dot and cross symbols in Fig. 1, where circled dots indicate current flow out of the ionosphere, and circled crosses current flow into the ionosphere. These clearly provide a basic explanation of the Region 1-Region 2 currents described in the introduction. The Region 1 system flowing in the vicinity of the open-closed field line boundary is fed by Pedersen currents flowing from dawn to dusk across the polar cap, as well as by Pedersen currents flowing north-south in the auroral zone. The Region 2 currents ensure continuity of the auroral zone currents alone in the lower latitude regions of the flow cells, and consequently carry a lower total current than do the Region 1 currents, as previously noted.

Figure 1 may be used directly to estimate the total Joule heat production rate in the ionosphere. Using the fact that the ionospheric electric field is essentially curl-free (due to the strength and incompressibility of the ionospheric magnetic field) and hence describable as the gradient of a scalar potential $\varphi$, together with the divergence-free condition for the total current $j$, we find

$$R_J = \int_V j.E \, d\tau = -\int_V j.\nabla\varphi \, d\tau = -\int_S \varphi \, j.dS \quad , \quad (5)$$

where the final integral is over the upper surface of the ionosphere such that the total current $j$ which appears within it is effectively the FAC flowing into and out of the ionosphere. If we then take the outer streamline in Fig. 1 to be at zero volts, such that the focus of the dawn flow cell is

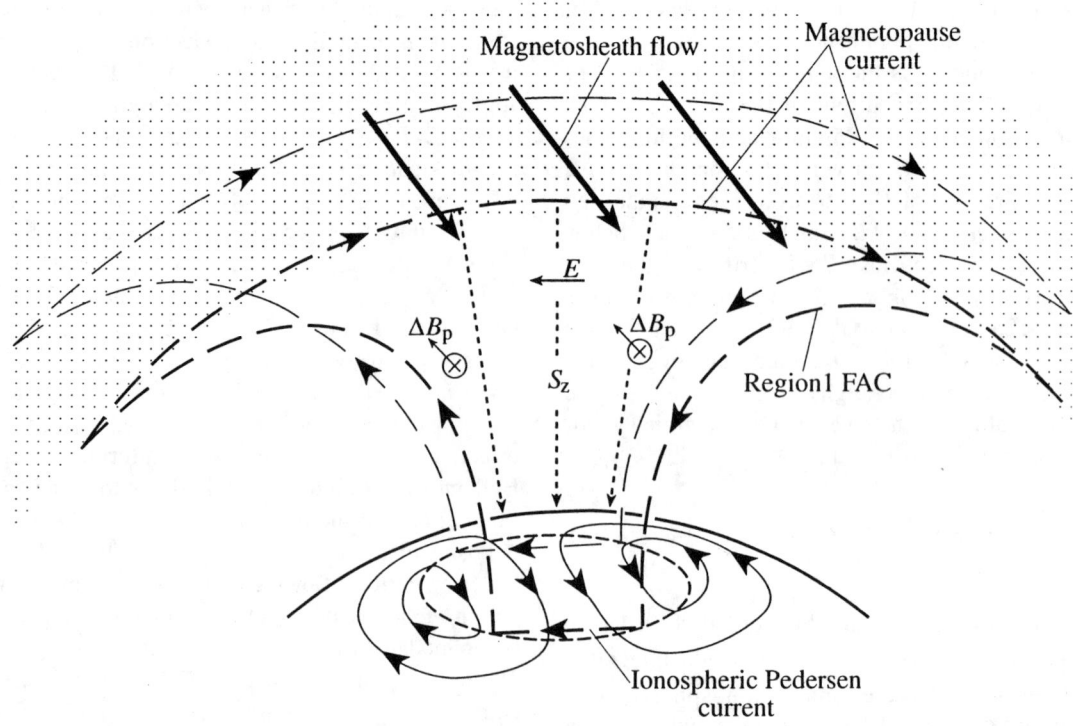

**Figure 2.** Sketch of the polar cap current circuit (long dashed lines), in which the dawn-to-dusk Pedersen current in the ionosphere closes in the magnetopause current via Region 1 FAC flowing on the outer surface of the plasma sheet. The current circuit produces sunward-directed perturbation magnetic fields $\Delta B_p$ in the polar region (in the northern hemisphere), which, combined with the dawn-to-dusk electric field $E$ associated with the flow, produces a net downward Poynting flux $S_z$ of electromagnetic energy into the ionosphere (short-dashed lines).

at potential $\varphi = \Phi/2$ volts, while that of the dusk cell is at $\varphi = -\Phi/2$ volts, where $\Phi$ is the total transpolar voltage associated with the flow, then it is easy to see that $R_J \approx I\Phi$ W, where $I$ is the total Region 1 current. For typical values $\Phi \approx 50$ kV and $I \approx 2$ MA we thus have $R_J \approx 10^{11}$ W. This represents ~10% of the energy consumed by the magnetosphere in cislunar space, and ~1% of the total kinetic energy of the solar wind which is incident on the magnetospheric cross-section. With regard to the total force exerted on the ionosphere by the magnetosphere, it is easy to show that if the conductivities are uniform, the total $j \times B$ force integrated around each ionospheric streamline is zero. The net force on the ionosphere will thus depend upon the distribution of conductivity, and will in general be directed sunward, due to the larger conductivity, and hence drag, in the auroral zone. The total antisunward force acting on the polar cap ionosphere is typically ~$10^8$ N, comparable to the total ram pressure of the solar wind acting over the magnetospheric cross-section, while the total sunward force acting on the auroral zone ionosphere is typically about double this.

## 3. MAGNETOSPHERE-IONOSPHERE CURRENT CIRCUITS

As indicated above, the currents flowing in the ionospheric "load" must close in a magnetosphere-magnetosheath "generator", involving a large-scale system of FACs flowing between these regions. Figure 2 shows the large-scale circuit associated with the polar cap current, where the ionospheric Pedersen currents close in the tail lobe magnetopause via Region 1 FACs flowing on the outer surface of the plasma sheet. The magnetopause currents are the "generator" currents where $j.E < 0$, the ionospheric Pedersen currents are the "load" where $j.E > 0$, and there is a net downward Poynting flux from one region to the other via the perturbation magnetic field produced by the current circuit. In the northern hemisphere the perturbation fields are directed opposite to the flow, while in the southern hemisphere they are directed parallel to the flow. These fields constitute the "transverse magnetic disturbances" originally observed by Zmuda *et al.* (1966). Just above the conducting layer of the ionosphere the field perturbation due to the Pedersen current is $\Delta B_P = \mu_0 \Sigma_P E$, so that the vertical component of the Poynting vector is $S_z = (E \Delta B_P)/\mu_0 = \Sigma_P E^2$, i.e. $S_z$ is just equal to the ionospheric Joule heating rate per unit area of the ionosphere, as required by energy conservation (Poynting's theorem). In mechanical terms, the magnetosheath is slowed by the sunward $j \times B$ force of the magnetopause current and provides energy to the electromagnetic field. The stress is fed by the tilted field to the ionosphere, where the $j \times B$ force balances the frictional drag on the ions and in turn accelerates the neutral atmosphere in the direction of the plasma flow.

In general, because currents in space plasmas are always essentially divergence-free (otherwise the build-up of space charge implied by the continuity equation would be enormous), we can consider current tubes (like flux tubes of the magnetic field) around which the total current $dI$ is constant. In some regions of the tube $j.E > 0$ and energy flows from the field to the plasma, while in others $j.E < 0$ and the energy flows from the plasma to the field. If we integrate $j.E$ over the whole tube, it is easy to show that the integral is equal to $dI$ times the emf around the tube, where the latter is equal to the rate of change of magnetic flux through the tube by Faraday's law (Cowley, 1991). In the steady state, therefore, the integral of $j.E$ over the tube is zero, and the "generators" in the tube exactly balance the "loads". Then the Poynting flux output from the generator regions is equal to the Poynting flux input into the loads, as implied by Fig. 2, though in general there is no guarantee that the Poynting flow will be direct. In the time-dependent case, however, the loads and generators need not balance, in which case energy is either stored or extracted from the changing field configuration. If the loads predominate, such that the volume integral of $j.E$ is positive, then the magnetic flux threading the current circuit decreases with time, while if the generators predominate, such that the integral of $j.E$ is negative, then the flux threading the current circuit increases with time.

One word of caution should be introduced, however, before concluding the above general discussion of energy flow, concerning frames of reference. It is obvious that the kinetic energy of an element of the plasma depends upon the frame of reference, and that an element which is gaining kinetic energy in one frame may be losing it in another. Similarly, the electric field in the plasma, given approximately by $E = -V \times B$ when collisions are absent, is frame-dependent, such that the Poynting flux is also frame-dependent, and a $j.E<0$ "generator" region in one frame may transform into a $j.E>0$ "load" in another. While the laws of physics, including Maxwell's equations, Poynting's Theorem, and conservation of energy, are of course valid in any frame, such that the above discussion of energy flow can be applied equally to any frame, and will make equal physical sense in any frame, it should therefore be understood that the physical terms of that discussion may well change from one frame to another. In discussing overall energy flow, therefore, we need to choose, and stick to, a particular frame of reference. Throughout this paper we choose the (non-rotating) rest frame of the Earth. While perhaps parochial, this choice nevertheless has virtues for terrestrial observers.

Having discussed above the flow of energy in the polar cap current circuit (in the Earth's frame), we now turn to the current circuit associated with the auroral zone. Geometrically it is clear that the Region 1 current must flow in the outer part of the plasma sheet, while the Region 2 current must flow in the inner part of the plasma sheet and ring current region. To examine the closure of the latter current, therefore, we must consider the current flow in the hot plasma of the quasi-dipolar magnetosphere. The essential physical principle to be applied is that the FAC flowing into or out of the ionosphere must just balance the flux-tube integrated divergence of the field-perpendicular current carried by the hot magnetospheric plasma, such that the divergence of the total current is zero. From the continuity equation, this is exactly equivalent to considering the flow of charge which must take place along the field lines in order to maintain the charge-neutrality of the hot plasma, which is an equivalent and sometimes simpler way to think about the problem. Let us therefore consider the contributions to the field-perpendicular current in the magnetosphere, that is to say plasma magnetisation and particle drifts, and also equivalently consider the particle motions which may produce charge-separation in the plasma. First, magnetisation currents are exactly divergence-free (given by the curl of the magnetisation), and therefore make no contribution to the discussion. Since these currents are associated with particle gyration around the field lines at a microscopic level, they also clearly cannot relate to charge separation in the plasma. Second, turning to the drifts, the $E \times B$ drift at any point is the same for all particles such that it produces no current at all in a charge-neutral plasma. Equally clearly this drift cannot produce charge-separation either. Third, inertia currents and drifts associated with the changing bulk velocity in the plasma are generally (though not invariably) small in the inner magnetosphere. It therefore becomes clear that the principal origins of current divergence and hot plasma charge-separation in the inner magnetosphere must be associated with the gradient and curvature drifts of the magnetospheric particles.

We will now outline some basic physical ideas following the discussion given by Wolf (1983), which is in turn based on the earlier ideas presented by Schield et al. (1969). For

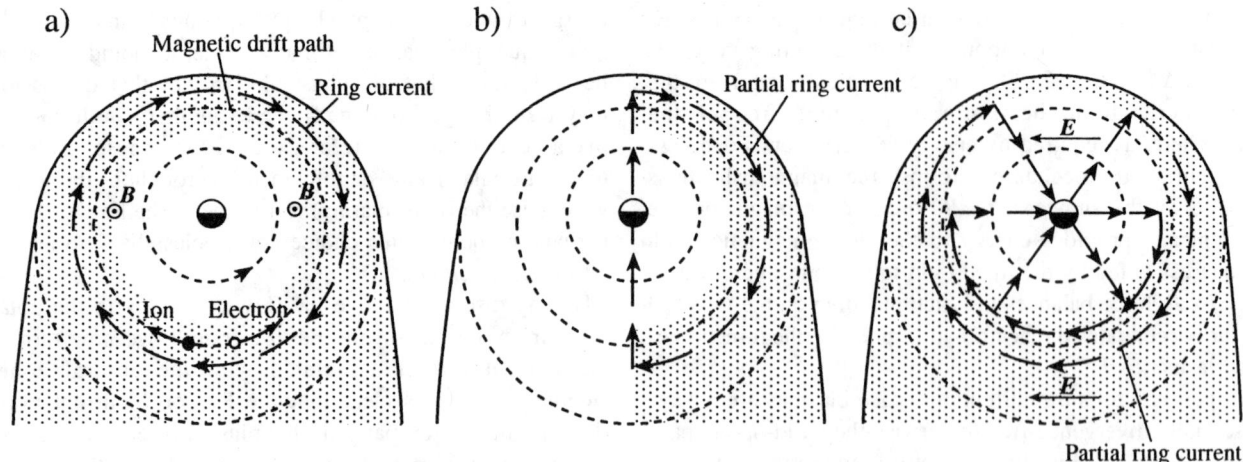

**Figure 3.** Sketches of the equatorial magnetosphere showing the FAC flow which connects the magnetospheric and ionospheric current systems (long-dashed lines) required by current continuity, for various spatial distributions of hot ring current plasma (dotted regions). The short-dashed lines represent the magnetic drift paths of ions and electrons, with ions drifting to the west and electrons to the east. In sketch (a) the plasma is distributed uniformly around the drift paths, such that the drift current is divergence-free in the magnetosphere and no FAC flows. In sketch (b) the initial plasma distribution has higher densities at dawn than at dusk, such that the partial ring current at dawn must close in the ionosphere via downward FACs at midnight and upwards FACs at noon. Sketch (c) shows the situation produced from an initial equilibrium by an interval of sunward flow imposed by a dawn-to-dusk electric field $E$. A partial ring current is formed centred on midnight, which closes via downward FAC at dusk and upward FAC at dawn. After Wolf (1983).

simplicity, this discussion neglects the time-varying magnetic field perturbations due to the hot plasma currents, which is correct only for a low-beta plasma. Nevertheless, the essential physical ideas remain valid in the more general case. In Fig. 3 we thus view the equatorial plane of the inner magnetosphere and its hot plasma population (dotted areas), where the plasma is assumed initially charge-neutral. For simplicity we first assume that there is no $E \times B$ drift of the plasma, so that the particles simply move along gradient and curvature drift paths, ions to the west and electrons to the east. These paths are shown by the short-dashed lines. For particles with 90° pitch angle these paths are contours of constant field strength. For particles with 0° pitch angle they are contours of constant field line length. For a population maintained isotropic by strong pitch-angle scattering, as generally assumed in modelling, they are lines of constant flux tube volume per unit magnetic flux, $V = \int ds/B$ (the integral extends over the length of the flux tube from the southern to the northern ionosphere). Figure 3a illustrates the situation in which the hot plasma flux tube content per unit magnetic flux is constant around each drift path. In this case the macroscopic plasma configuration does not change at all as the individual particles drift. Consequently, no charge-separation of the hot plasma occurs, the hot plasma current (a westward ring current) is divergence-free around the drift-paths in the magnetosphere, and there is no requirement for current flow to or from the ionosphere. Suppose instead, however, that the initial hot plasma density is higher at dawn than at dusk, as shown in Fig. 3b. Now the drift of ions to the west and electrons to the east would result in the development of a positive space charge in the plasma near midnight, and a negative space charge near noon. We therefore require a flux of cold electrons out of the ionosphere to neutralise the positive space charge at midnight (or hot ions in), and a flux of cold ions out (or hot electrons in) at noon. In current circuit terms, then, a net partial ring current flows westward in the hot plasma in the dawn magnetosphere, which is fed by an upward FAC at noon, and is closed by a downward FAC at midnight, as shown by the long-dashed lines in the figure. These FAC directions would be reversed if the hot plasma was more dense at dusk than at dawn.

These are hypothetical situations. The question we have to ask concerns the nature of the plasma distributions which would be set up by Dungey-cycle flow. Suppose we start

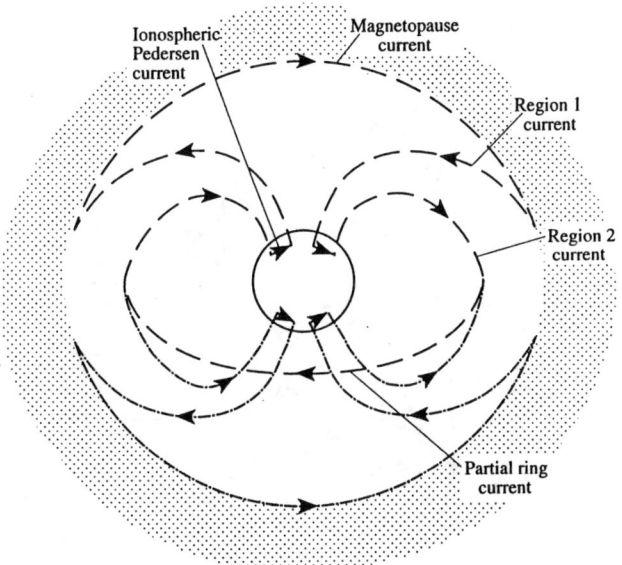

**Figure 4.** Sketch of the overall auroral zone current circuit looking at the Earth from the tail, showing both the northern (dashed lines) and southern (dot-dashed lines) branches of the circuit.

with an equilibrium distribution with no FAC such as that shown in Fig. 3a, set up by some earlier episode of hot plasma inflow from the tail, in which the hot plasma content per unit magnetic flux on each drift path decreases as we move towards the Earth. If we then apply a dawn-to-dusk electric field across the system, the $E \times B$ drift will displace the plasma sunward everywhere, with the result shown in Fig. 3c. On each drift path the flux tube content is now maximum at midnight and minimum at noon. The maintenance of charge neutrality, or equivalently current continuity, therefore requires current flow into the ionosphere at dusk, and out of the ionosphere at dawn. That is, we require a FAC flow in the same sense as the Region 2 current. We therefore infer that the latter currents are closed in the inner magnetosphere by a westward partial ring current flowing in the sunward-propagating inner plasma sheet population. We note that this inference is in accord with the equatorial current distribution determined from magnetic measurements made by the AMPTE-CCE spacecraft (Iijima et al., 1990). The overall auroral zone current circuit is therefore as shown in Fig. 4. The magnetospheric partial ring current flowing in the nightside inner plasma sheet region closes in the ionosphere by Region 2 FACs, the current then flows across the auroral zone ionosphere as north-south Pedersen currents, then out as Region 1 currents flowing in the outer layers of the plasma sheet to the magnetopause, where it then closes in

the magnetosheath plasma. In the steady state the magnetosheath "generator" feeds Poynting flux into both the dissipative ionospheric Pedersen currents, and into the energy stored in the compressed and heated hot magnetospheric plasma. In the absence of the magnetosheath "generator", the circuit could also be powered by the decay of the tail magnetic flux which threads through it.

The above discussion is qualitative. In reality (and in modelling) the flow in the system must adjust in order to ensure that the divergence of the hot plasma current in the magnetosphere is matched by the divergence of the horizontal current in the ionosphere. From Eq. 4, the FAC density into the ionosphere required by the continuity of the field-perpendicular ionospheric current is

$$j_{\|_i} = \nabla_h \cdot i_\perp = \nabla_h \cdot (\Sigma_P E + \Sigma_H \hat{B} \times E) , \quad (6)$$

where $\nabla_h$ is the two-dimensional horizontal gradient operator, and for simplicity we have assumed a vertical polar magnetic field. Current continuity in the magnetosphere requires

$$j_{\|_i} = -\frac{1}{2}\int_V d\tau \; \text{div}\, j_\perp = -\frac{1}{2}B_i \int \frac{ds}{B} \text{div}\, j_\perp , \quad (7)$$

where the integrals extend over the whole magnetospheric flux tube from the southern to the northern ionosphere, $j_\perp$ is the field-perpendicular magnetospheric plasma current density, $B_i$ is the ionospheric field strength, $s$ is distance along a field line, and we have assumed equal parallel current density into the ionosphere in both hemispheres. With the neglect of the inertia current we have

$$j_\perp = \frac{B \times \nabla p}{B^2} ,$$

where $p$ is the plasma pressure (assumed isotropic and hence constant along a field line). Substitution into Eq. 7 can then be shown to yield

$$j_{\|_i} = -\frac{B_i}{2B}\hat{B}\cdot\nabla p \times \nabla V = -\frac{B_i}{2BV^{5/3}}\hat{B}\cdot\nabla(pV^{5/3})\times\nabla V , \quad (8)$$

where the $V$ is the flux tube volume per unit magnetic flux as before, and the gradients can be evaluated at any point on the field line in the magnetosphere. Equating $j_{\|_i}$ between Eqs. (6) and (8) then yields the condition for continuity of the magnetosphere-ionosphere current, which can be solved for the self-consistent electric field and flow. This equation was first derived by Vasyliunas (1970), and

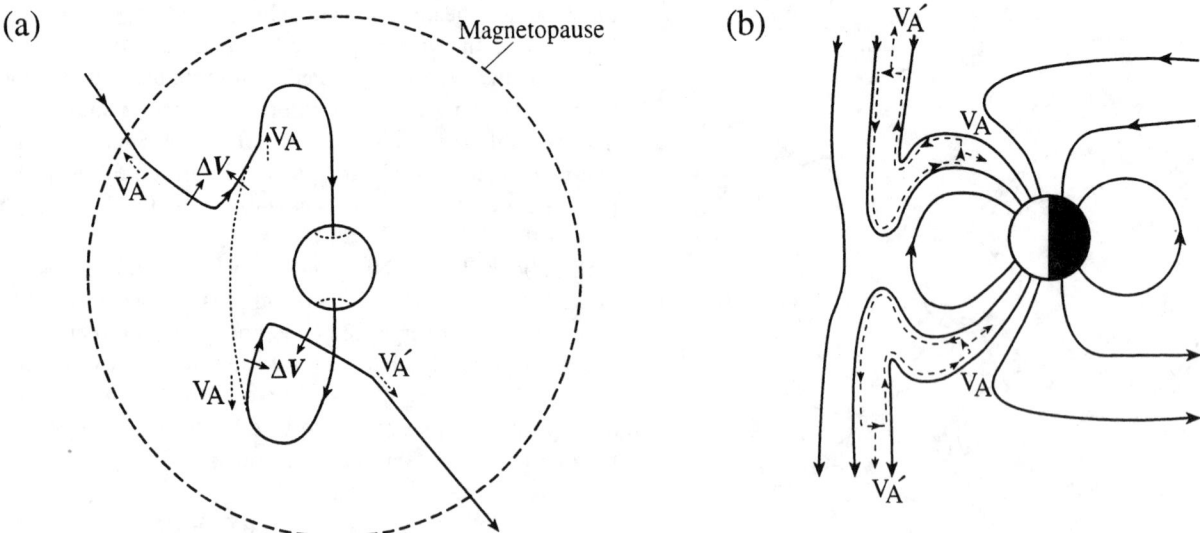

**Figure 5.** (a) Sketch of newly-opened field lines following subsolar reconnection with an IMF having negative Z and positive Y components, showing the field tilting effects in the magnetospheric and magnetosheath boundary layers due to the tension in the magnetic field. The view is looking at the Earth from the direction of the Sun. The short solid arrows marked $\Delta V$ show the associated velocity perturbations transverse to the magnetic field, while the short dashed arrows marked $V_A$ indicate the propagation of the disturbance along the open field lines at the Alfvén speed. (b) View projected onto the noon-midnight meridian, showing the associated FAC and cross-field closure current systems (arrowed dashed lines) propagating along the open field lines at the Alfvén speed.

is the condition on which self-consistent models such as the Rice convection model are based (Wolf, 1983). The physical content of the equation is equivalent to the discussion which we made in relation to Fig. 3. The final form of Eq. 8 is interesting because it shows that in any region where $pV^{5/3}$ is a constant (as will result from lossless adiabatic convection from a uniform source), there will be no FAC flow between the magnetosphere and ionosphere.

## 4. CUSP CURRENTS

Having discussed the Region 1/2 current system associated with large-scale twin-cell convection, we now turn to look at the origins of the third FAC component mentioned in the introduction, namely the cusp currents, which flow on open field lines poleward of the Region 1 system on the dayside. These currents relate to the stresses exerted on newly-opened field lines following reconnection at the magnetopause, and the consequent motion of the open flux tubes. Two factors influence this motion, namely the tension in the reconnected magnetic field lines, and the flow of the magnetosheath plasma around the magnetopause away from noon. For near-subsolar reconnection with a southward-pointing IMF, the field tension effect will be the most important initially, while the effect of the flow will exert itself as the magnetosheath plasma becomes super-Alfvénic in the downstream region. An important consequence of the initial dominance of the field tension force is that the motion of the newly-opened flux tubes responds strongly to the Y component of the IMF, as first discussed by Jørgensen et al. (1972). Figure 5a shows open field lines shortly after subsolar reconnection has taken place with a magnetosheath field which has positive Y and negative Z components, in a view looking towards the Earth from the direction of the Sun. In the magnetosphere, the field tension force pulls the open lines towards dawn in the northern hemisphere, and simultaneously towards dusk in the southern hemisphere, such that the field tilts over in the boundary layer towards the direction of the magnetosheath field outside. This disturbance propagates down the open field lines as an Alfvén wave, which we note is the MHD mode specifically associated with the propagation of field-aligned current. A similar disturbance also propagates out into the magnetosheath, which results in the sheath field being

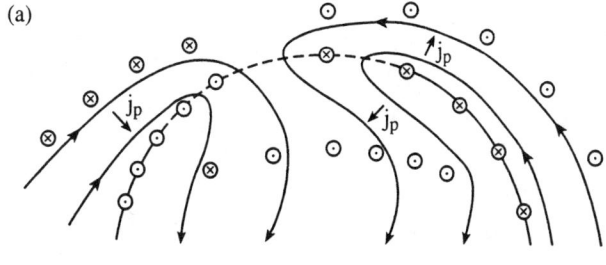

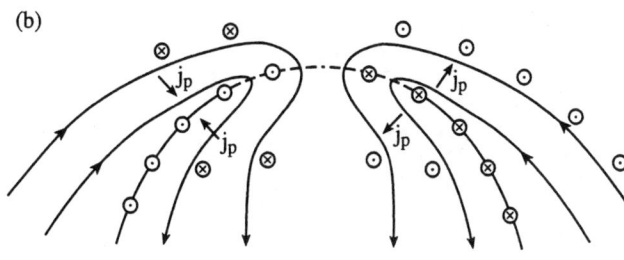

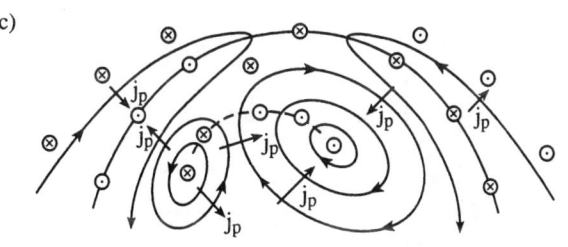

**Figure 6.** Sketches looking down on the northern hemisphere ionosphere showing the plasma streamlines (arrowed solid lines) for various IMF orientations, together with the sense of FAC flow. Circled dots indicate upward flow, and circled crosses downward flow. The short arrows marked $j_p$ indicate the closure Pedersen currents in the ionosphere. The solid lines without arrows indicate the open-closed field line boundary, while the dashed lines map along the field to the magnetopause reconnection sites. Sketch (a) is for an IMF with negative Z and positive Y components, (b) for negative Z and near-zero Y, and (c) for positive Z and positive Y (in the presence of continued tail reconnection).

pulled towards the magnetospheric field direction. The form of the current system is shown schematically in the side view in Fig. 5b, and consists of a system of oppositely-directed sheets of FAC in both hemispheres, bounding the region of tilted field, and terminated by propagating field-transverse inertia currents in the "head" of the wave. After ~2 min the magnetospheric "head" arrives at the ionosphere, and (after a bounce or two due to the impedance mis-match between the wave and the ionosphere) establishes a westward flow of open field lines in the northern cusp, and a similar eastward flow in the southern cusp. At ionospheric heights the flow is associated with paired sheets of FAC as indicated in Fig. 5b, which in the northern hemisphere are directed downward, into the ionosphere, on the equatorward boundary of the cusp (essentially the open-closed field line boundary), and upward, out of the ionosphere, on the poleward boundary, and vice versa in the southern hemisphere.

The dayside pattern of FAC in the northern hemisphere for steady reconnection with IMFs of various orientations is shown in Fig. 6. The situation for positive Y and negative Z is shown in Fig. 6a, where, as just discussed, the cusp currents are predominantly downward on the equatorward border and upward on the poleward border, and are closed by poleward-directed Pedersen currents in between. The flow between the sheets is predominantly westward, and thus associated with an eastward Hall current which provides most of the magnetic effect seen on the ground. We note that the poleward cusp FAC sheet will be co-located with the region where the plasma flow rotates from westward to antisunward. It thus represents the point on the flow streamlines where the field tension effect ceases to be dominant and gives way to the effect of the anitsunward flow of the magnetosheath plasma (Saunders, 1989).

For an IMF with negative Y and negative Z, the sense of the east-west flow asymmetry is reversed from that shown in Fig. 6a, together with the predominant sense of the cusp FAC, and is not shown here. Rather, in Fig. 6b we show the symmetrical situation for negative Z and near-zero Y. Here the newly-opened tubes are swept symmetrically away from noon towards dawn and dusk by the magnetosheath flow before turning antisunward. The cusp currents are correspondingly symmetrical, with the third FAC sheet at highest latitude having opposite polarity to the Region 1 current.

Figure 6c illustrates the fact that "cusp" currents also flow when IMF Z is positive. Here we show, as an example, the simultaneous presence of a "reversed" twin vortex flow on open field lines driven by lobe reconnection in the presence of positive IMF Y, together with the continued presence of "normal" twin-vortex flow at lower latitudes driven by open flux closure in the tail, such that the open-closed field line boundary contracts. The "reversed" twin vortex, first inferred from ground magnetic measurements by Maezawa (1976), is associated with a paired FAC system, termed the "NBZ" currents, in which the FAC flows downwards in the dusk vortex, and upwards in the dawn vortex (McDiarmid et al., 1980; Saflekos and Potemra, 1980). The origins and

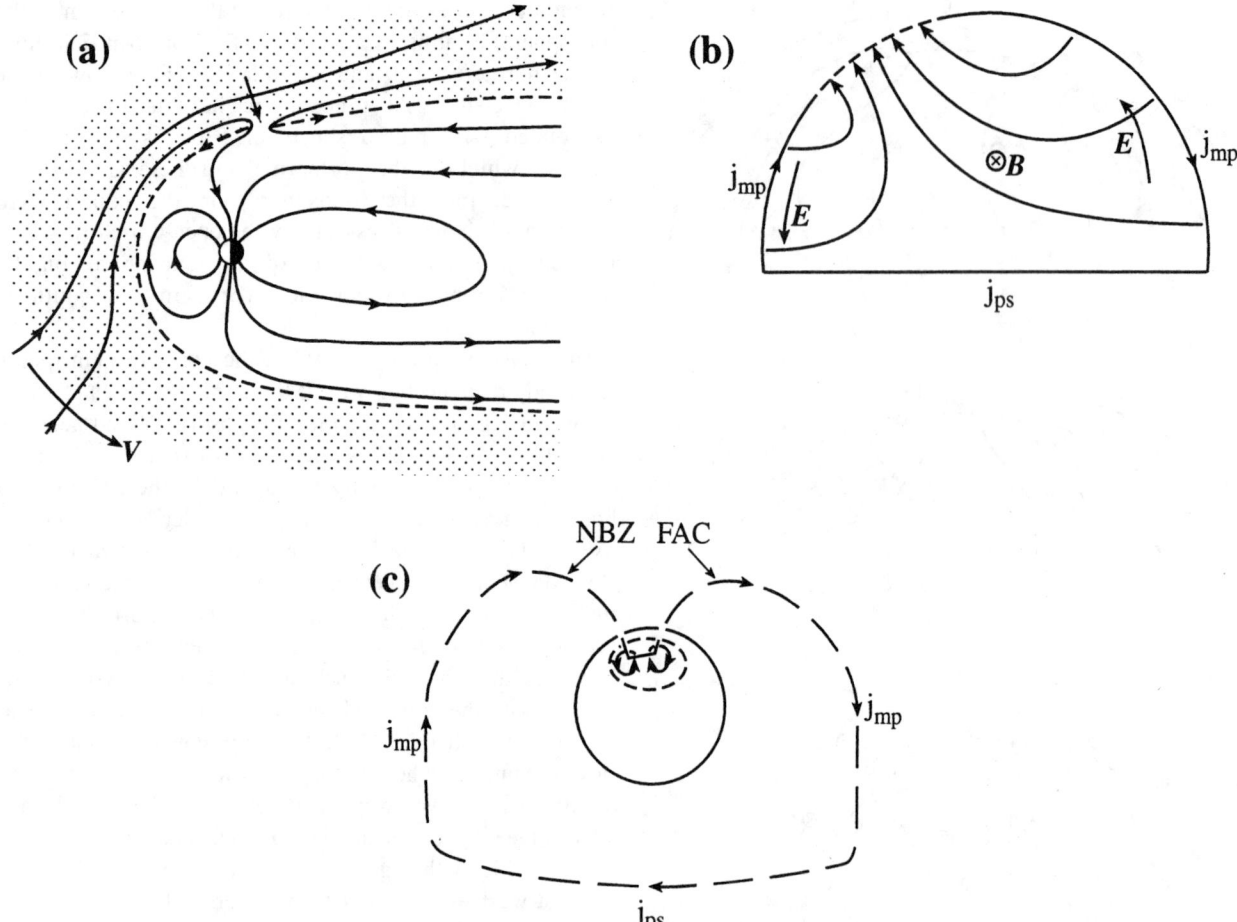

**Figure 7.** Sketches showing the fields and flows associated with single-lobe reconnection for an IMF with positive Z, positive Y, and negative X components. In sketch (a) reconnection in the northern lobe produces "new" open field lines draped over the dayside (without changing the amount of open flux), which are subsequently swept into the tail by the magnetosheath flow (preferentially on the dawn side in this case). Sketch (b) shows a cross-section through the northern tail looking towards the Earth, showing the flow of open flux from the sides of the tail (preferentially the dawn side) to the duskside lobe magnetopause reconnection site. The current flows clockwise around the northern lobe. Sketch (c) shows the "NBZ" FACs which flow into and out of the central regions of the "reversed" polar cap vortices, and which close through the flank magnetopause and plasma sheet.

closure of this system are illustrated in Fig. 7, where for simplicity we have neglected the effects of simultaneous tail reconnection. In Fig. 7a single-lobe reconnection in the northern hemisphere produces "new" open flux tubes draped over the dayside magnetopause, which initially contract sunward due to the field tension (also moving to dawn or dusk in the presence of an IMF Y component), and are then swept into the tail by the magnetosheath flow. Figure 7b shows the flow in a cross-section through the northern hemisphere tail lobe looking towards the Earth, such that the magnetopause current flow is clockwise from dusk to dawn, closing from dawn to dusk in the plasma sheet. For the case with a positive IMF Y component as shown (as in Fig. 6c), the lobe reconnection site will be located preferentially on the dusk side of the tail in the northern hemisphere, while the "new" open field lines will be swept preferentially towards dawn. The open tubes then flow from the flank magnetopause, where $j \cdot E$ is negative and $j \times B$ slows the magnetosheath plasma flow, into the reconnection site at higher latitudes, where $j \cdot E$ is positive.

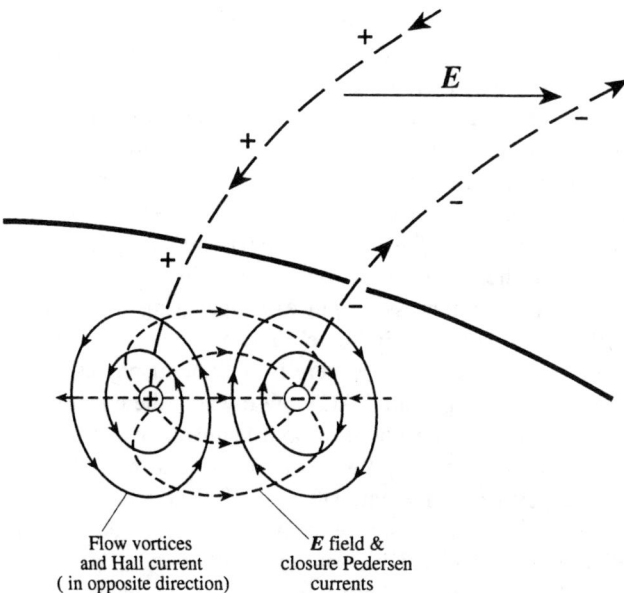

**Figure 8.** Sketch showing the electric field (arrowed short-dashed lines) and flow patterns (arrowed solid lines) in the northern hemisphere associated with a matched pair of oppositely-directed FACs (long-dashed lines). The FACs are closed in the ionosphere by the Pedersen currents flowing in the direction of the electric field. The plus and minus symbols indicate the senses of the slight space charge distributions associated with the electric field.

This flow corresponds to the antisunward part of the "reversed" twin vortices which appear in the ionosphere. The "NBZ" currents then tap part of the tail lobe current system, as shown in Fig. 7c, and thus close through the magnetopause "generator" currents on the tail flanks, and then through the essentially "inactive" (in this case) plasma sheet. Poynting flux flows from the tail flank magnetopause into the polar ionosphere.

## 5. TRAVELLING CONVECTION VORTICES

Reconnection between the IMF and the terrestrial field is not the only mechanism by which the solar wind may perturb and transfer momentum into the magnetosphere, though it is usually the most important. A second class of phenomena, termed "travelling convection vortices" (TCVs) are also observed (e.g. Friis-Christensen *et al.*, 1988), in which one or more east-west aligned pairs of oppositely-directed flow vortices propagate through the dayside ionosphere east or west away from noon at high latitudes. Each vortex has a spatial scale of ~1000 km, such that at any instant the twin vortices encompass several hours of local time, and they propagate over a few tens of minutes at phase speeds of 5 km s$^{-1}$. From our previous discussion it is evident that an ionospheric flow vortex must be associated with FAC flow at its centre. In the northern hemisphere, the FAC flows upward from the centre of a clockwise vortex, and downward into the centre of an anticlockwise vortex (and vice versa in the southern hemisphere). The basic system of ionospheric electric field and flow for such a system of paired currents is shown in Fig. 8. The FAC is closed in the ionosphere by the Pedersen current driven by an electric field which is dipolar in form, such that the region of downward current is associated with a (slight) net positive space charge, while the region of upward current is associated with a (slight) net negative space charge. The flow then consists of a pair of oppositely-directed vortices, around which the Hall current flows in the direction opposite to $E \times B$. For a vertical field, the magnetic effects of the FAC and the Pedersen currents exactly cancel under the ionosphere, such that the magnetic disturbance on the ground is dominated by the Hall current vortices. Typical FACs associated with each vortex in observed events are a few hundred kA.

While the basic form of TCVs at ionospheric heights is thus reasonably well understood, their physical origin as manifestations of solar wind-magnetosphere coupling at large distances remains to be clarified. Most theoretical discussion has centred on the effect of sudden changes in compressive plasma pressure in the magnetosheath, but while some of these events are associated with precursory changes of the dynamic pressure in the solar wind, this is by no means always the case. Indeed, Sibeck *et al.* (1999) have recently discussed one event which was associated with the interaction between the magnetosphere and a tangential discontinuity propagating in an otherwise undisturbed solar wind. This interaction produced a "hot flow anomaly" event in the dawn magnetosheath, and a sudden localised expansion of the equatorial magnetopause by ~5 $R_E$, which propagated tailward. It therefore appears that the TCV phenomenon can have more than one precursory signature in the solar wind. Whatever the origin of the pressure change and boundary motion, however, we may still enquire how the system of paired FACs come to be generated. The first simple thing we can say is that they are not generated in a direct way by compressions or rarefactions of the magnetosphere. A uniform contraction or expansion of the magnetosphere would produce almost no field or flow effect at ionospheric heights, because the field there is strong and almost incompressible. Equivalently, we may say that compressive (fast mode) MHD waves propagating in the magnetosphere are almost perfectly reflected by the ionosphere. We therefore must consider the effect of pressure fronts propagating over the

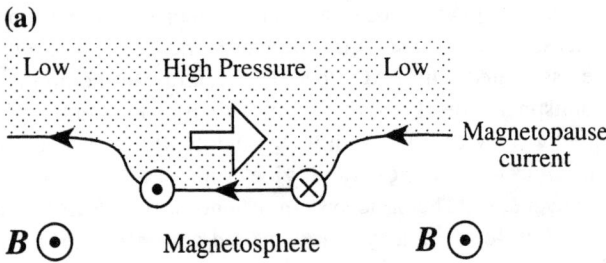

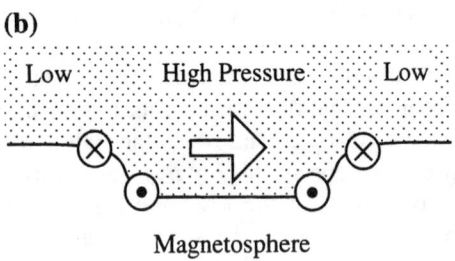

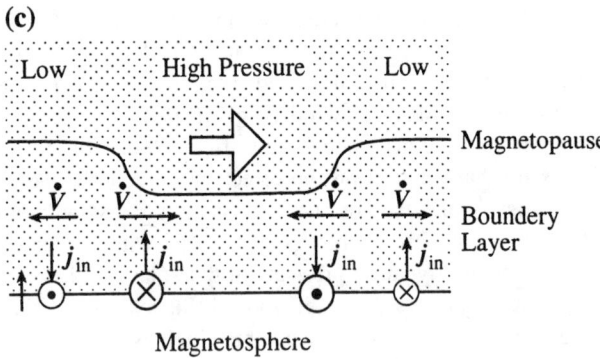

Figure 9. Sketches showing proposed FAC patterns associated with a single antisunward-propagating (left to right) compressive in-out motion of the magnetopause. The plane of the diagrams is the equatorial plane, such that the magnetospheric field points outwards. The dotted regions correspond to the high-density magnetosheath (upper two diagrams) and boundary layer (lower diagram) regions. Circled dots indicate current flow away from the equator towards the ionosphere in both hemispheres, while circled crosses indicate current flow towards the equator away from the ionosphere in both hemispheres. The former FACs are associated with a clockwise flow vortex in the plane of the sketch, while the latter are associated with an anticlockwise flow vortex. Sketch (a) follows the discussion of Glassmeier (1992), sketch (b) is after Kivelson and Southwood (1991), while sketch (c) is after Lühr et al. (1996). In the latter sketch we also show the directions of the inertia currents in the boundary layer, and the associated accelerations ($\dot{V}$) of the plasma.

magnetopause which may generate vortical flows associated with FAC, which can propagate to the ionosphere as Alfvén waves.

Various suggested mechanisms are compared in Fig. 9, where we show the effect of a single compressive pulse propagating antisunward on the magnetopause. Each figure shows an equatorial cut through the dawn-side boundary region (for definiteness) perpendicular to the magnetospheric magnetic field, with the magnetosheath plasma (and compressive pulse) propagating from left to right. Circled dots indicate FAC flow away from the equator towards the ionosphere in both hemispheres, while circled crosses indicate FAC flow towards the equator and away from the ionosphere in both hemispheres. Figure 9a is due to Glassmeier (1992), who considers the continuity of the perturbed magnetopause current, and suggests one FAC (vortex) at each end of the perturbed region. Figure 9b is due to Kivelson and Southwood (1991), who consider the flow vorticity introduced at the magnetopause by the in-out boundary motions, and predict paired currents at each end. Both pictures therefore locate the FACs at the magnetopause, which will map to the open-closed field line boundary in the ionosphere. Recent work by Moretto and Yahnin (1998), however, shows that these currents are centred well inside the region of closed field lines, which then seems more in line with the suggestion of Lühr et al. (1996) shown in Fig. 9c. These authors suggest that the FACs are associated with the divergence of the inertia current at the density gradient at the inner edge of the magnetopause boundary layer. The inertia current is given by $j_\perp = \rho(B/B^2) \times (dV/dt)$, where $V$ is the bulk velocity produced in the magnetospheric plasma by the propagating boundary perturbation. This produces a central pair of FACs which are opposite in sense to those proposed by Glassmeier (1992), plus two "outliers" of smaller amplitude. Overall, there is as yet no consensus on which of these proposed patterns, if any, matches the observed pattern for an impulsive compression, but it is clear that there exists sufficient diversity in the predicted outcome that some could be eliminated as the dominant effect.

## 6. SUBSTORM CURRENTS

The magnetopause reconnection processes which generate new open flux, whose effects were described in Sect. 4 above, initiate the growth phase of the reconnection cycle by causing the transfer of open flux from the dayside magnetopause to the tail lobes. Eventually, reconnection of the lobe field in the tail centre plane must also occur,

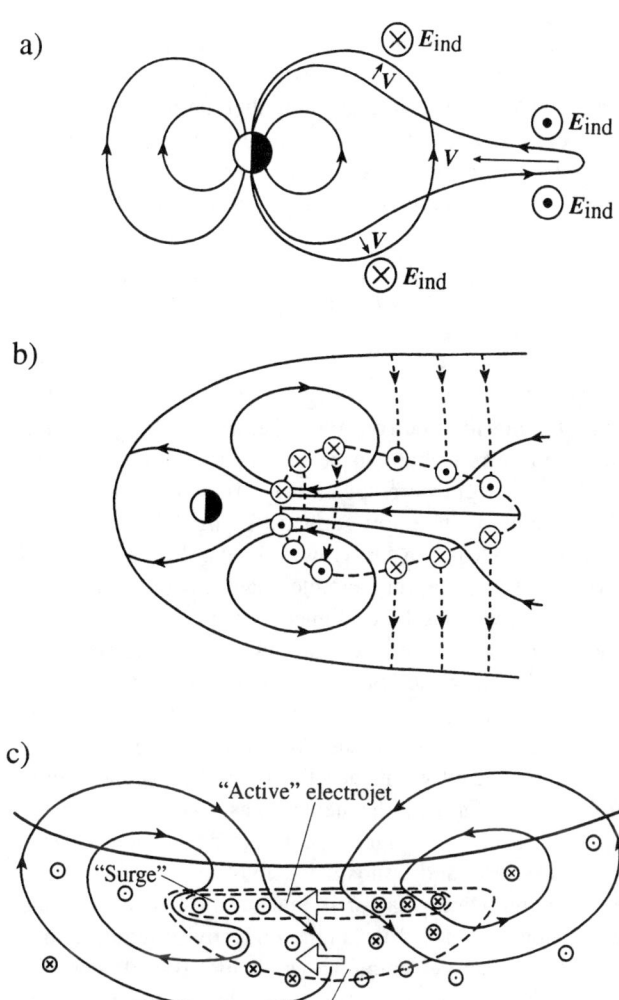

**Figure 10.** (a) Sketch showing the effect of substorm dipolarization on field lines in the near-Earth tail. Field lines move rapidly inwards near the equatorial plane and outwards at higher latitudes, associated with large cross-system inductive electric fields as shown. (b) Pattern of FACs in the near-Earth tail associated with substorm dipolarization within the dashed-line region. Circled dots indicate FAC flow away from the equator and into the ionosphere in both hemispheres, while circled crosses indicate FAC flow into the equator and away from the ionosphere in both hemispheres. The arrowed solid lines indicate plasma streamlines. (c) Sketch of the flow and currents in the conjugate ionosphere, where the outer dashed line indicates the dipolarized region, corresponding to the substorm expansion phase auroral bulge. The inner dashed line indicates the region of the "active" electrojet in the poleward part of the bulge. Arrowed solid lines are streamlines, and the pattern of FACs is again indicated by the circled dot and cross symbols.

forming new closed field lines in the plasma sheet which return towards Earth and close the Dungey cycle. The substorm expansion phase is believed to play a central role in this latter process, during which the key feature is an inward collapse or "dipolarization" of the growth phase-enhanced tail field, associated with bursts of rapid earthward flow in the plasma sheet (Baumjohann et al., 1990; Angelopoulos et al., 1992). It remains controversial whether the collapse is initiated by tail reconnection directly, or by some other process (e.g. an instability in the plasma sheet) which excites reconnection as a subsequent effect, though recent results from the Geotail spacecraft have shown that reconnection typically begins in the pre-midnight plasma sheet at down-tail distances between 20 and 30 $R_E$ at times close to substorm expansion phase onset (Nagai et al., 1998). In either case, the principal features of the expansion phase field and current effects are illustrated in Fig. 10. Figure 10a illustrates the expansion phase dipolarization of a field line. The growth phase field line is highly distorted away from a dipolar form by the presence of a thin but intense current sheet located in the plasma sheet in the near-Earth tail. After expansion phase onset these field lines collapse inwards at the equator, and outwards at high latitudes, to assume a more dipolar form associated with a much reduced tail current. These inductive effects do not, however, produce correspondingly large motions in the ionosphere, though to the extent that the process contributes to inward flux transport in the tail, it will excite twin-vortex Dungey-cycle flow.

The effect of this process on the tail current system is shown in Fig. 10b, which is a view of the equatorial plane of the magnetosphere. Here the azimuthally-limited dipolarized region is bounded by the heavy dashed line, and the circular symbols near its periphery indicate the direction of FAC flow, circled dots representing current flow away from the equator towards the ionosphere (in both hemispheres), and circled crosses current flow towards the equator away from the ionosphere. Two effects are illustrated (following the results of Lu et al., 1997). Within the near-Earth tail (typically at distances from ~8 to ~30 $R_E$), the cross-tail current is reduced within the azimuthally-restricted dipolarized region, such that the tail current on either side is diverted along the field, towards the Earth on the dawn side of the region, and away from the Earth on the dusk side. This current flow just accommodates the shear in the field direction across the boundary between the dipolarized field inside the region and the remaining tail-like field outside. These FACs close through the ionosphere at one end (as further described below), forming the "substorm current wedge" first described by McPherron *et al.* (1973), and over the tail magnetopause at the other. They thus form a special type of auroral zone "Region 1" current, though in this case the ionospheric dissipation in the circuit is powered more by the emf provided by the reducing flux of the tail lobes, as outlined in Sect. 3, than by the magnetopause "generator". In this region the tail plasma flows rapidly inwards towards the Earth, where it is eventually "braked" by the increasing pressure of the compressed field and plasma at the outward-moving boundary between quasi-dipolar and tail-like fields, the "braking" being associated with a dawnward-directed inertia current in the plasma which contributes to the formation of the substorm current wedge in the near-Earth system (Shiokawa *et al.*, 1997). The hot injected plasma in the new quasi-dipolar region then forms a partial ring current centred near midnight which closes via "Region 2" currents, as described previously in Sect. 3.

In Fig. 10c illustrates conditions in the conjugate ionosphere, where, in addition to the above, we have also drawn on the discussion by Fujii *et al.* (1994) and Weimer *et al.* (1994). The outer dashed line corresponds to the dipolarized region in the tail, within which the accelerated plasma precipitates to form the substorm auroral bulge. This precipitation strongly enhances the conductivity of the bulge ionosphere to values (typically many tens of mhos) much higher than that of the surrounding region. The quadripolar pattern of FAC associated with the bulge follows that shown in Fig. 10b. In the poleward region the ("substorm wedge"-Region 1 type) current flows upwards on the dusk side of the bulge (corresponding to the "surge"), and downwards on the dawn side. These currents close principally via the "active" westward substorm electrojet flowing within the poleward part of the bulge (interior dashed line region), which carries typically ~1 MA of current. In the equatorward region the enhanced "Region 2" currents flow in the opposite sense, and presumably close as before principally via north-south Pedersen currents in the oppositely-directed FAC on its poleward side. The flow streamlines are shown by the solid lines, where we depict a distorted twin-cell flow centred around the "substorm wedge"-Region 1 currents, which is excited by the transport in the tail (Cowley *et al.*, 1998; Opgenoorth *et al.*, 1998).

The nature of the flow observed at ionospheric heights is, however, influenced by several complicating effects. The first is that magnetospheric and ionospheric flows can be partially decoupled by field-aligned voltages at intermediate (~1 $R_E$) heights, which may be required to drive the FACs themselves. Since FACs are mainly carried by highly mobile plasma electrons, upward FACs in particular are carried by hot magnetospheric electrons moving downwards into the mirror field geometry near the Earth. Field-aligned voltages may then be needed to draw sufficient current from the magnetospheric population. According to the formula first derived by Knight (1973), the upward current provided by magnetospheric electrons is

$$j_\| \approx j_o \left( 1 + \left( \frac{e\Phi_\|}{kT_e} \right) \right), \qquad (9)$$

where $\Phi_\|$ is the field-aligned voltage, $T_e$ is the magnetospheric electron temperature, and $j_o$ is the maximum current that can be obtained without a voltage. The latter current corresponds to a full downward-going loss cone and an empty upward-going loss cone, and amounts to ~1 µA m$^{-2}$. However, the FACs observed flowing in the surge, for example, may be several times this value, thus requiring field-aligned voltages of order several times $(kT_e/e)$, i.e. of order a few kV. Such voltages produce characteristic features in the electron distributions, variously known as "inverted-V" or "BPS"-type precipitation. The accelerated precipitating electrons may in turn significantly enhance the ionospheric conductivity through ionization of the neutral gas, and thus alter the pattern of FACs. A complex non-linear feedback between magnetosphere and ionosphere may then occur. The second complicating factor is that the flow at ionospheric heights tends to avoid the high-conductivity regions, such as that in the surge region shown in Fig. 10c. Radar data consistently show that the flow in such regions is suppressed relative to the surrounding regions, at least over few-minute intervals (e.g. Morelli *et al.*, 1995). One factor which could be involved is the field-aligned voltages just discussed, since these will close off magnetospheric equipotentials above the ionosphere. Another could be the effect of enhanced ion-neutral drag. The third complicating factor is the apparent "polarization" effects which lead to an overall eastward flow (i.e. equatorward electric field) and westward electrojet current within the high-conductivity bulge, as we have shown in Fig.10c. The usual story here is that the westward electric field in the bulge associated with the equatorward flow drives a poleward Hall current $\Sigma_H E_W$ across the high-conductivity bulge which cannot close by FAC along its poleward and equatorward borders. Instead, an equatorward polarization electric field is developed whose Pedersen current cancels the northward Hall current, such that the equatorward electric field is given by $E_S = (\Sigma_H / \Sigma_P) E_W$. The Pedersen current of the westward electric field and the Hall current of the southward electric field then add to produce a westward electrojet ("Cowling") current given by

$$i_C = \left(1 + \left(\frac{\Sigma_H}{\Sigma_P}\right)^2\right) \Sigma_P E_W . \qquad (10)$$

However, no obvious reason is given in this argument as to why the initial Hall current cannot close by FACs at the equatorward and poleward borders of the bulge, while at the same time the intense electrojet current which is formed thereby can close by FAC at its dusk and dawn "ends". As in Sect. 4, the answer to this question must surely lie in the conditions required for closure of the entire magnetosphere-ionosphere current circuit, and we note from Fig. 10b that the natural form of the magnetospheric FAC tends to favour the latter closure rather than the former. Whereas there has been much overall progress in understanding magnetosphere-ionosphere interactions in recent years, as evidence by the contents of former sections of this paper, there is clearly much left to be understood of the complexities of substorm electrodynamics.

## REFERENCES

Angelopoulos, V., W. Baumjohann, C.F. Kennel, F.V. Coroniti, M.G. Kivelson, R. Pellat, R.J. Walker, H. Lühr, and G. Paschmann, Bursty bulk flows in the inner central plasma sheet, *J. Geophys. Res.*, 97, 4027, 1992.

Baumjohann, W., G. Paschmann, and H. Lühr, Characteristics of high-speed ion flows in the plasma sheet, *J. Geophys. Res.*, 95, 3801, 1990.

Birkeland, K., *The Norwegian Aurora Polaris Expedition, 1902-1903, Vol. 1*, H. Aschehoug & Co., Christiania, 1908.

Cowley, S.W.H., Acceleration and heating of space plasmas: basic concepts, *Ann. Geophysicae*, 9, 176, 1991.

Cowley, S.W.H., H. Khan, and A. Stockton-Chalk, Plasma flow in the coupled magnetosphere-ionosphere system and its relationship to the substorm cycle, in *Substorms-4* edited by S. Kokobun and Y. Kamide, p. 623, Terra Sci. Publ. Co., Tokyo, 1998.

Cummings, W.D., and A.J. Dessler, Field-aligned currents in the magnetosphere, *J. Geophys. Res.*, 72, 1007, 1967.

Dessler, A.J., The evolution of arguments regarding the existence of field-aligned currents, in *Magnetospheric Currents*, edited by T.A. Potemra, *Geophys. Monograph 28*, p. 22, AGU, Washington, , 1984.

Dungey, J.W., Interplanetary field and the auroral zones, *Phys. Rev. Lett.*, 6, 47, 1961.

Friis-Christensen, E., M.A. McHenry, C.R. Clauer, and S. Vennerstrøm, Ionospheric traveling convection vortices observed near the polar cleft: A triggered response to sudden changes in the solar wind, *Geophys. Res. Lett.*, 15, 253, 1988.

Fujii, R., R.A. Hoffman, P.C. Anderson, J.D. Craven, M. Sugiura, L.A. Frank, and N.C. Maynard, Electrodynamic parameters in the nighttime sector during aurural substorms, *J. Geophys. Res.*, 99, 6093, 1994.

Glassmeier, K.-H., Travelling magnetospheric convection twin vortices: observations and theory, *Ann. Geophysicae*, 10, 547, 1992.

Iijima, T., and T.A. Potemra, The amplitude distribution of field-aligned currents at northern high latitudes observed by Triad, *J. Geophys. Res.*, 81, 2165, 1976a.

Iijima, T., and T.A. Potemra, Field-aligned currents in the dayside cusp observed by Triad, *J. Geophys. Res.*, 81, 5971, 1976b.

Iijima, T., and T.A. Potemra, Large-scale characteristics of field-aligned currents associated with substorms, *J. Geophys. Res.*, 83, 599, 1978.

Iijima, T., R. Fujii, T.A. Potemra and N.A. Saflekos, Field-aligned currents in the south polar cusp and their relationship to the interplanetary magnetic field, *J. Geophys. Res.*, 83, 5595, 1978.

Iijima, T., T.A. Potemra, and L.J. Zanetti, Large-scale characteristics of magnetospheric equatorial currents, *J. Geophys. Res.*, 95, 991, 1990.

Jørgensen, T.S., E. Friis-Christensen, and J. Wilhjelm, Interplanetary magnetic field direction and high-latitude ionospheric currents, *J. Geophys. Res.*, 77, 1976, 1972.

Kivelson, M.G., and D.J. Southwood, Ionospheric traveling vortex generation by solar wind buffeting of the magnetosphere, *J. Geophys. Res.*, 96, 1661, 1991.

Knight, S., Parallel electric fields, *Planet. Space Sci.*, 21, 741, 1973.

Lu, G., G.L. Siscoe, A.D. Richmond, T.I. Pulkkinen, N.A. Tsyganenko, H.J. Singer, and B.A. Emery, Mapping of the ionospheric field-aligned currents to the equatorial magnetosphere, *J. Geophys. Res.*, 102, 14467, 1997.

Lühr, H., M. Lockwood, P.E. Sandholt, T.L. Hansen, and T. Moretto, Multi-instrument ground-based observations of a travelling convection vortices event, *Ann. Geophysicae*, 14, 162, 1996.

Maezawa, K., Magnetospheric convection induced by the positive and negative Z components of the interplanetary magnetic field: Quantitative analysis using polar cap magnetic records, *J. Geophys. Res.*, 81, 2289, 1976.

McDiarmid, I.B., J.R. Burrows, and M.D. Wilson, Magnetic field perturbations in the dayside cleft and their relationship to the IMF, *J. Geophys. Res.*, 83, 5753, 1978.

McDiarmid, I.B., J.R. Burrows, and M.D. Wilson, Comparison of magnetic field perturbations and solar electron profiles in the polar cap, *J. Geophys. Res.*, 85, 1163, 1980.

McPherron, R.L., C.T. Russell, and M.P. Aubry, Satellite studies of magnetospheric substorms on August 5, 1968, 9, Phenomenological model for substorms, *J. Geophys. Res.*, 78, 3131, 1973.

Morelli, J.P., R.J. Bunting, S.W.H. Cowley, C.J. Farrugia, M.P. Freeman, E. Friis-Christensen, G.O.L. Jones, M. Lester, R.V. Lewis, H. Lühr, D. Orr, M. Pinnock, G.D. Reeves, P.J.S. Williams, and T.K. Yeoman, Radar observations of auroral zone flows during a multiple-onset substorm, *Ann. Geophysicae*, 13, 1144-1163, 1995.

Moretto, T., and A. Yahnin, Mapping travelling convection vortex events with respect to energetic particle boundaries, *Ann. Geophysicae, 16*, 891, 1998.

Nagai, T., M. Fujimoto, Y. Saito, S. Machida, T. Terasawa, R. Nakamura, T. Yamamoto, T. Mukai, A. Nishida, and S. Kokobun, Structure and dynamics of magnetic reconnection for substorm onsets with Geotail observations, *J. Geophys. Res., 103*, 4419, 1998.

Opgenoorth, H.J., and R.J. Pellinen, The reaction of the global convection electrojets to the onset and expansion of the substorm current wedge, in *Substorms-4*, edited by S. Kokobun and Y. Kamide, , p. 663, Terra Sci. Publ. Co., Tokyo, 1998.

Saflekos, N.A., and T.A. Potemra, The orientation of Birkeland current sheets in the dayside polar region and its relationship to the IMF, *J. Geophys. Res., 85*, 1987, 1980.

Saunders, M.A., Origin of the cusp Birkeland currents, *Geophys. Res. Lett., 16*, 151, 1989.

Schield, M.A., J.W. Freeman, and A.J. Dessler, A source for field-aligned currents at auroral latitudes, *J. Geophys. Res., 74*, 247, 1969.

Shiokawa, K., W. Baumjohann, and G. Haerendel, Braking of high-speed flows in the near-Earth tail, *Geophys. Res. Lett., 24*, 1179,1997.

Sibeck, D.G., N.L. Borodkova, S.J. Schwartz, C.J. Owen, R. Kessel, S. Kokobun, R.P. Lepping, R. Lin, K. Liou, H. Lühr, R.W. McEntire, C.-I. Meng, T. Mukai, Z. Nemecek, G. Parks, T.D. Phan, S.A. Romanov, J. Safrankova, J.-A. Sauvaud, H.J. Singer, S.I. Solovyev, A. Szabo, K. Takahashi, D.J. Williams, K. Yumoto, and G.N. Zastenkar, Comprehensive study of the magnetospheric response to a hot flow anomaly, *J. Geophys. Res., 104*, 4577, 1999.

Vasyliunas, V.M., Mathematical models of magnetospheric convection and its coupling to the ionosphere, in *Particles and Fields in the Magnetosphere*, edited by B.M. McCormac, D. Reidel Publ. Co., Dordrecht, p. *60*, 1970.

Weimer, D.R., J.D. Craven, L.A. Frank, W.B. Hanson, N.C. Maynard, R.A. Hoffman, and J.A. Slavin, Satellite measurements through the center of a substorm surge, *J. Geophys. Res., 99*, 23639, 1994.

Wilhjelm, J., E. Friis-Christensen, and T.A. Potemra, The relationship between ionospheric and filed-aligned currents in the dayside cusp, *J. Geophys. Res., 83*, 5586, 1978.

Wolf, R.A., The quasi-static (slow-flow) region of the magnetosphere, in *Solar-Terrestrial Physics*, edited by R.L. Carovillano and J.M. Forbes, p. 303, D.Reidel Publ.Co., Dordrecht, 1983.

Zmuda, A.J., J.H. Martin, and F.T. Heuring, Transverse magnetic disturbances at 1100 km in the auroral region, *J. Geophys. Res., 71*, 5033, 1966.

---

S.W.H. Cowley, Department of Physics and Astronomy, University of Leicester, Leicester LE1 7RH, United Kingdom

# Field-Aligned Currents in Geospace: Substance and Significance

Takesi Iijima

*Department of Earth and Planetary Sciences, Graduate School of Sciences, Kyushu University, Japan*

More than a quarter century has passed since in situ observation in geospace was carried out by spacecraft and the magnetospheric age began. The elucidation of electric current systems by magnetic field measurements characterizes the magnetospheric age, and the study of field-aligned currents (FACs) has become the distinctive methodology. In fact, FACs play a crucial role in the transfer, conversion, and circulation of momentum and energy between the solar wind, the magnetosphere, and the ionosphere, and they link this whole system. Global plasma convection in the entire region of the magnetosphere and ionosphere is absolutely indebted to FACs. In this article, I will describe the fundamentals of observation, momentum and energy transmission, generation mechanism, and recent developments related to FACs.

## INTRODUCTION

Global plasma convection controls the magnetic field configuration and plasma environment in the Earth's magnetosphere. Various kinds of plasma population regimes are known: magnetospheric boundary layers, the polar cusp, the plasma mantle, the polar wind, the plasma sheet, trapped radiation belts (ring current), and the plasmasphere. These plasma regimes are formed by plasmas of the solar wind and the ionosphere that are subject to convection and associated acceleration processes in the magnetosphere. Various types of electric currents are suggested to exist, for example, magnetospheric boundary currents, the cross-tail/plasma sheet current, and a ring current. These currents are generated so that the current-associated $J \times B$ force balances the pressure gradient force and inertia force occurring in the plasma regimes. The direct momentum transfer from the solar wind medium at magnetopause boundary layers, along with the pressure gradient force occurring internally in the magnetosphere, are suggested to be the principal causes for the convection

[*Southwood and Kivelson*, 1991]. In either case, transverse momentum (that is, perpendicular to the magnetic field) should be transmitted from the causal region to the ionospheric level so that plasma population in a flux tube convects as a whole. The field-aligned currents (FACs) play a crucial role in this process: They transmit the transverse momentum (i.e., magnetic tangential stress) and the electromagnetic energy (i.e., Poynting flux) from the source region to the other end along magnetic field lines in a flux tube [*Southwood and Hughes*, 1983; *Haerendel*, 1990]. By virtue of the transmission of FACs, plasma convection occurs in the collisional ionosphere and ionospheric Pedersen and Hall currents flow. In fact, plasma convection is ascribable to the $J \times B$ force that is associated with the Pedersen current and caused by the transverse momentum carried by FACs. The Poynting flux carried by FACs also causes the energy consumption occurring in the ionosphere that is associated with Joule heating and the modulation of the neutral wind velocity field. Thus, FACs are closely involved with the transfer, conversion, and circulation process of the momentum, energy, and matter in the whole magnetospheric system, including the ionosphere and that part bounded by the loose boundary of the magnetopause. In this article I will review the large-scale characteristics of FACs, along with the physics of momentum and energy transmission, and the

Magnetospheric Current Systems
Geophysical Monograph 118
Copyright 2000 by the American Geophysical Union

generation mechanism of FACs in the magnetohydrodynamic (MHD) limit with the assumption of an isotropic plasma pressure and no significant voltage (field-aligned) maintaining the FACs in the circuit.

## FUNDAMENTALS OF FIELD-ALIGNED CURRENTS

### Observations

The existence of FACs has been elucidated by magnetic field measurements using spacecraft above the ionosphere. After the first in situ observation by the 1963 38C satellite [*Zmuda et al.*, 1970; *Armstrong and Zmuda*, 1970], the experimental study of FACs in geospace really started in the 1970s. The studies included observations of magnetic disturbances at high latitudes (mapped to the ionosphere) by various magnetometer systems using the Azur satellite [*Theile and Praetourius*, 1973] at an altitude between 400 and 3100 km, by the Imp4 and 5 satellites in the magnetotail [*Fairfield*, 1973], and by the Ogo5 satellite [*Sugiura*, 1975] at altitudes of >4 $R_E$. Using the only continuous vector magnetic field measurements on board the Triad satellite at relatively low altitudes (~800 km) in the high-latitude regions at that time, *Armstrong and Zmuda* [1973] and *Zmuda and Armstrong* [1974a and b] identified the spatial location of transverse magnetic disturbances and suggested for the first time the flow direction pattern of FACs in the northern polar region. The FACs are determined by the transverse magnetic disturbance they produce in space as

$$J_{//} = \frac{1}{\mu_0}(\nabla \times \mathbf{B}) \cdot \mathbf{e}_B = \frac{1}{\mu_0}(\nabla \times (\mathbf{B}_0 + \Delta \mathbf{B})) \cdot \mathbf{e}_B \quad (1)$$

$$= \frac{1}{\mu_0}(\nabla \times (\Delta \mathbf{B}_\perp)) \cdot \mathbf{e}_B$$

where $\mathbf{B}_0$ is the background magnetic field and $\Delta \mathbf{B}$ is the disturbance field. $\mathbf{B}_0$ is represented by the IGRF reference field at low altitudes. Examples of transverse magnetic disturbance profile that are observed along the dawn-dusk track at low altitudes (~300–400 km) are shown in Figure 1. Spatial location and flow direction of FACs are determined by these magnetic disturbances by identifying the principal shear region in the $\Delta \mathbf{B}_\perp$ vectors along an individual orbit and/or on multiple orbits. A large number of studies fully used the Triad magnetometer data and determined various characteristics of extraterrestrial FAC systems, which have provided basic knowledge to the field of magnetosphere-ionosphere coupling physics. The studies include the basic pattern of spatial distribution of flow direction [*Iijima and Potemra*, 1976a and b] and its relationships with polar geomagnetic substorm activity [*Iijima and Potemra*, 1978], with ionospheric auroral electrojets [*Kamide and Rostoker*, 1977], with ionospheric conductivity conditions [*Fujii et al.*, 1981; *Fujii and Iijima*, 1987], and with the interplanetary magnetic field (IMF) $B_y$ and $B_z$ [*Wilhjelm et al.*, 1978; *Iijima et al.*, 1978; *Saflekos and Potemra*, 1980; *Iijima and Potemra*, 1982]. Figure 2 (upper) depicts a basic model of the FAC system, which is shown by the global distribution of flow direction pattern and (lower) magnetic local time (MLT) variation of current density; both were determined from data obtained from 439 passes of Triad during geomagnetically less active conditions ($|AL| < 100$ nT). The distribution of FACs as observed at low altitudes encircles the magnetic pole as an auroral oval belt. The poleward currents, denoted as region 1 currents, flow into the ionosphere in the dawn sector and flow away from the ionosphere in the dusk sector. The equatorward currents, denoted as region 2 currents, flow into the duskside ionosphere and flow away from the dawnside ionosphere. The intensity and density of region 1 currents are statistically larger than the currents in region 2 in all substorm phases and in all MLT sectors, except in the midnight sector of ~2100–0100. The total current (summed over all regions) flowing into the ionosphere is equal to the total current flowing away from the ionosphere (within 10% of the total amount). The total current values in one hemisphere are ~3 MA (= $10^6$ A) for less active conditions ($|AL| < 100$ nT) and more than ~5 MA for active conditions ($|AL| > 100$ nT) of geomagnetic disturbances. In the midnight sector, the pattern of region 1 and region 2 current is highly modulated and FACs often exhibit multiple flow direction patterns. In the midday sector, distinctive FAC systems appear poleward of and adjacent to the region 1 system. The nature of FACs in both the midday and midnight sectors will be discussed later.

### Momentum and Energy Transfer

Figure 3 shows the role of FACs, namely, they transmit transverse momentum along $\mathbf{B}_0$ (background magnetic field). Figure 3 illustrates an idealized case of a three-dimensional current system that is established after magnetosphere-ionosphere (M-I) coupling processes have finished through passive and active response of the ionosphere to the magnetosphere. This current system comprises magnetospheric perpendicular current (flowing transverse to background magnetic field), a pair of FACs, and an ionospheric Pedersen current (closure current). By taking up one flux tube from this current system, forces exerted on some certain volume elements numbered as ①②③④⑤ will be discussed. First is the volume element ③. It starts from the stress balance relation between the magnetic field and plasma in a certain magnetic flux-tube volume containing perpendicular currents and FACs. Using the orthogonal coordinate system ($x_1$, $x_2$, $x_3$; $x_1$, $x_2$ perpendicular to $\mathbf{B}_0$, $x_3$ parallel to $\mathbf{B}_0$ and positive toward the northern ionosphere) and approximating a flux-tube volume to be a column, an overall stress balance in the $x_1$ direction is described by the following equations in different form:

$$\int \left(\rho \frac{\partial \mathbf{v}}{\partial t}\right) dV + \int (\rho (\mathbf{v} \cdot \nabla)\mathbf{v}) dV = \int (-\nabla p) dV + \int (\mathbf{J} \times \mathbf{B}) dV \quad (2)$$

$$\int \left(\frac{1}{\mu_0} B_3 \frac{\partial \mathbf{B}_\perp}{\partial x_3}\right) dV = \int \left(\rho \frac{\partial \mathbf{v}}{\partial t}\right) dV + \int (\rho(\mathbf{v} \cdot \nabla)\mathbf{v}) dV + \int (\nabla p) dV + \int \left(\nabla \left(\frac{B_3^2}{2\mu_0}\right)\right) dV \quad (3)$$

supplemented by

$$\int (\mathbf{J} \times \mathbf{B}) dV = \int \left(-\nabla\left(\frac{B^2}{2\mu_0}\right) + \frac{1}{\mu_0}(\mathbf{B} \cdot \nabla)\mathbf{B}\right) dV$$
$$= \int T_{ij} n_j dS \quad (3')$$
$$= \int \left(-\frac{B^2}{2\mu_0} \delta_{ij} + \frac{1}{\mu_0} B_i B_j\right) n_j dS$$

Here, $\mathbf{B} = \mathbf{B}_1 + \mathbf{B}_2 + \mathbf{B}_3$, $\mathbf{B}_1 = \Delta \mathbf{B}_\perp$, $\mathbf{B}_3 = \mathbf{B}_0 + \Delta \mathbf{B}_\parallel$, and $\mathbf{B}_2$ is assumed for simplicity to be zero. The $\Delta \mathbf{B}_\perp$ and $\Delta \mathbf{B}_\parallel$ denote the transverse and parallel magnetic disturbance, respectively, associated with the current systems. The intermediate altitude region between the near-equatorial generator plasma and the collisional ionosphere is described by the low β plasma, because plasma pressure is thought to be much less than magnetic pressure. The production of $\Delta \mathbf{B}_\parallel$ due to perpendicular currents and its contribution to FACs are presumably quite small, and the contribution of $\Delta \mathbf{B}_\perp$ to the total magnitude $B$ is also very small; in turn, $\mathbf{B} \approx \Delta \mathbf{B}_\perp + \mathbf{B}_0$ and the total magnitude $B \sim B_3 \sim B_0$ in the intermediate altitude region. (The contribution of $\Delta \mathbf{B}_\parallel$ is not negligible in the generator plasma.) The magnetic force on plasma per volume is written as the divergence of the Maxwell stress tensor $T_{ij}$ and, by applying the divergence theorem to it, Eq. (3′) comes up. The term $(-B^2/2\mu_0)\delta_{ij} n_j$ is the normal stress per area, and $(B_i B_j /\mu_0) n_j$ is the tangential stress per area, where $n_j$ is the outward normal direction to the surface $dS$ and $B_i$ is the field component along the surface. The left-hand term of Eq. (3) comes from an off-diagonal component of the Maxwell stress tensor (the tangential stress component). This magnetic force component remains exclusively in the vicinity of the magnetospheric perpendicular current if FACs are not derived from it. This force component can, however, be transmitted in a whole magnetic flux-tube if FACs flow between the magnetosphere and ionosphere. The integral of this force component along a flux tube from the equatorial plane (at $x_3 = 0$, strictly the symmetric plane of FACs toward the northern and southern hemisphere) up to the cross section at a certain altitude equals $B_0 \Delta B_\perp/\mu_0$ per unit area. This is exactly the net magnetic tangential stress exerted on this volume element ③ through both the northern and southern cross sections, strictly speaking, from the ionospheric side toward the magnetosphere. (Note: $\Delta B_\perp$ is zero and the tangential stress is necessarily zero on the equatorial plane. The integral in the southern hemisphere is pursued toward the negative $x_3$ direction.) Since the force per unit area is the same as the momentum per area per time, then $B_0 \Delta B_\perp/\mu_0$ is a flux of transverse momentum that is derived from magnetospheric perpendicular current and is transmitted toward both the northern and southern hemispheres along the magnetic field over the whole flux tube by FACs. The integral of $\mathbf{J} \times \mathbf{B}$ force over any volume is exactly equal to the surface integral of $T_{ij}$ over the surface of the volume, which is given by the sum of normal and tangential stresses over the surface. Then, the sum of (tangential stress) × (area) on the northern and southern cross sections equals ($\mathbf{J} \times \mathbf{B}$ force) × (volume element ③) if, for simplicity, the background magnetic field is uniform in the $x_1$ direction and the net sum of normal stress in the $x_1$ direction is negligible. The stress balance relation demands that these magnetic stresses (i.e., total $\mathbf{J} \times \mathbf{B}$ force) should balance (pressure gradient force plus inertia force) × (volume element ③). As for a very small volume element ② (and ④) in the intermediate region, the $\mathbf{J} \times \mathbf{B}$ force is equivalent to the magnetic tangential stresses exerted on two cross sections. Tangential stresses have the same magnitude but directions opposite each other. Net $\mathbf{J} \times \mathbf{B}$ is almost zero, which holds in the intermediate region. (Note: The direction of tangential stress is determined by the sign of a unit vector component $n$ defined to be outward normal to the cross section, along with the sign of $B_0$ and $\Delta B_\perp$ in the expression $(B_0 \Delta B_\perp/\mu_0)n$.) Volume element ① (and ⑤) corresponds to the ionospheric part. Pedersen current $\mathbf{J}_P$ flows so as to complete a three-dimensional current system through the ionosphere. The $\mathbf{J}_P \times \mathbf{B}_0$ force occurs there, and its forcing sustains plasma convection in the collisional ionosphere by accelerating the plasma medium and being impeded by momentum transfer to the neutral wind so as to keep the force-balance relation $\mathbf{J}_P \times \mathbf{B}_0 = \rho(d\mathbf{v}/dt) + a(\mathbf{v} - \mathbf{v}_n)$, where $a$ is a correlation factor and $\mathbf{v}_n$ is neutral wind velocity. $\mathbf{B}_0$ is the geomagnetic reference field at the ionosphere. Using a boundary condition that magnetic perturbations produced by FACs and by ionospheric currents are the same at the magnetosphere-ionosphere interface, ($\mathbf{J}_P \times \mathbf{B}_0$ force) × (volume element ① with a unit cross-sectional area) equals $\Sigma_P(\mathbf{E} + \mathbf{v}_n \times \mathbf{B}_0) \times \mathbf{B}_0$ and turns out to be exactly the stress $B_0 \Delta B_\perp/\mu_0$ that is transmitted via FACs and applied to

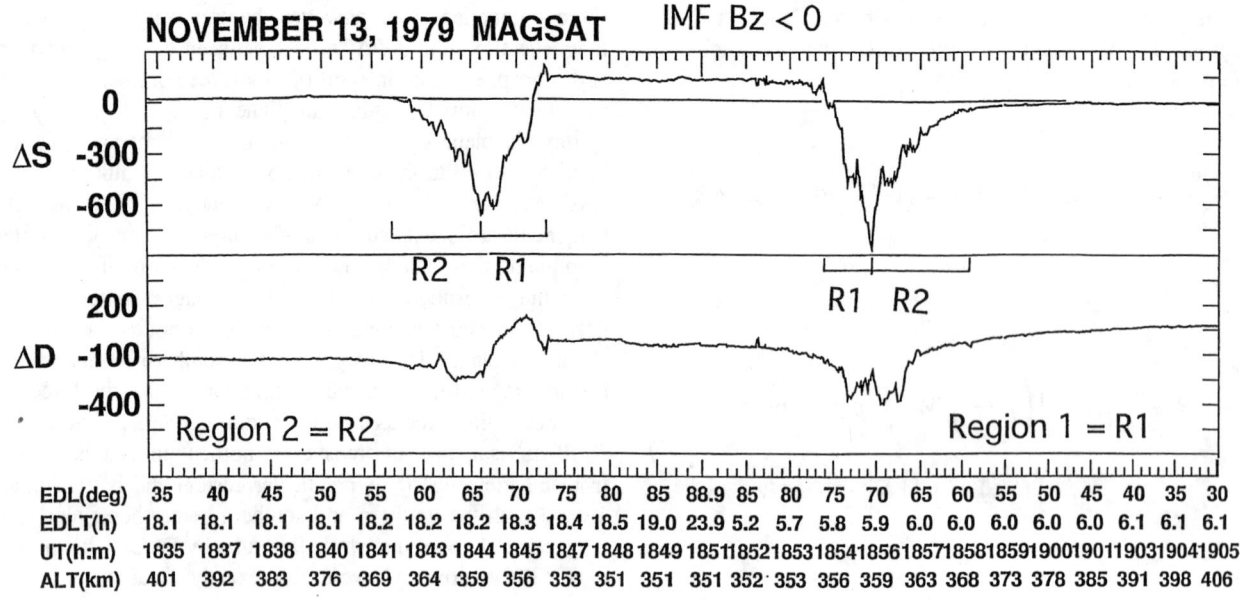

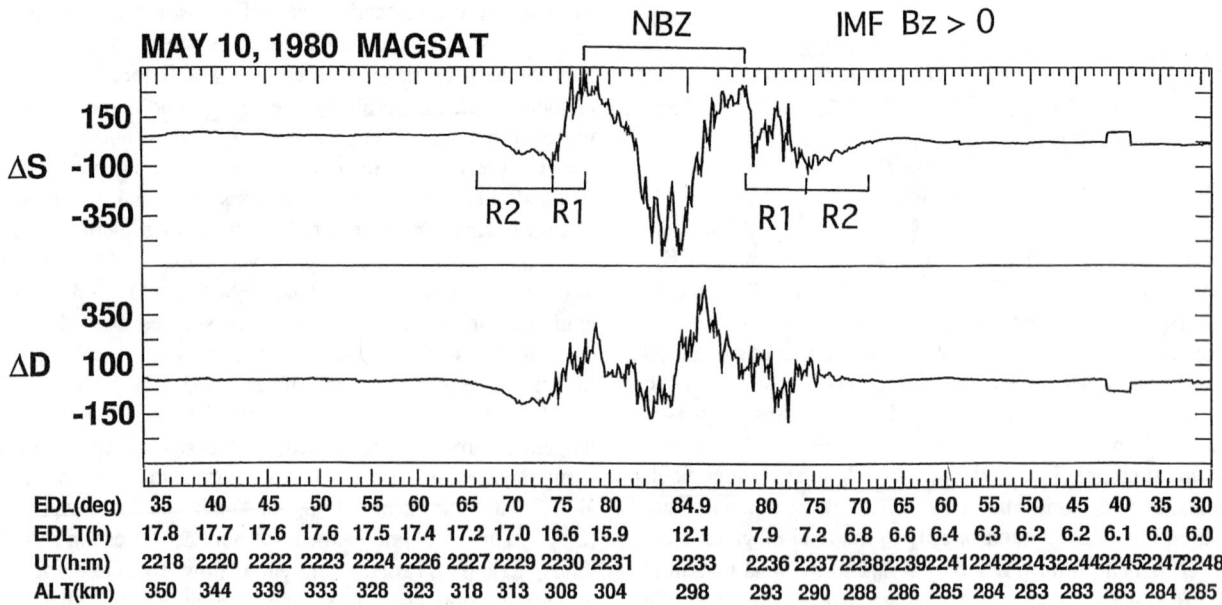

**Figure 1.** Examples of transverse magnetic disturbances observed by the MAGSAT satellite along dawn-dusk tracks at low altitudes in the north pole. $\Delta S$ and $\Delta D$ denote sunward and dusk-to-dawn component, respectively. (Upper) For the southward IMF condition; (Lower) for the northward IMF condition.

the magnetosphere-ionosphere interface, by virtue of completion of a three-dimensional current system via the ionospheric Pedersen current. Magnetic field disturbance is shielded from penetrating into the neutral atmosphere by the Pedersen current, and magnetic stress is zero at the ionosphere-atmosphere interface.

Figure 4 (upper) shows an example of the spatial distribution of the flow direction pattern of FACs that were obtained from data of nine consecutive passes, parallel to the dawn-dusk meridian at an altitude of ~400 km during the prolonged period of southward IMF and active polar magnetic disturbance [*Iijima et al.*, 1982]. The pattern shows ordinary region 1 and region 2 currents. The lower part displays the spatial distribution of transverse magnetic disturbance $\Delta \mathbf{B}_\perp$ for some selected passes. As seen in Figure 3, $\Delta \mathbf{B}_\perp$ observed in the southern hemisphere manifests

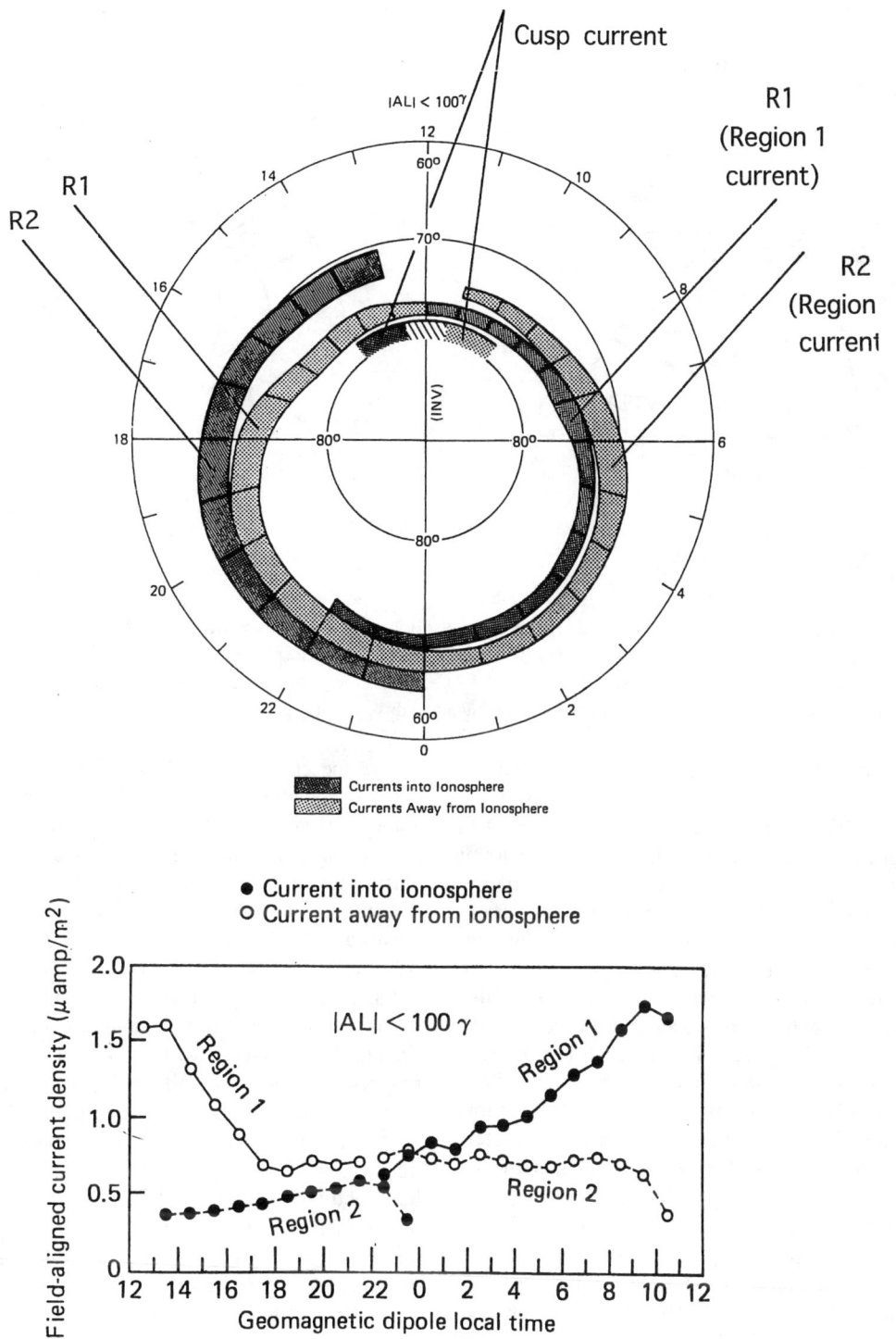

**Figure 2.** Observation model of large-scale FACs. (Upper) Spatial configuration of the flow direction pattern (after *Iijima and Potemra*, 1976b). (Lower) Density distribution in MLT (after *Iijima and Potemra*, 1978). Both are from the same data set for geomagnetically less active conditions.

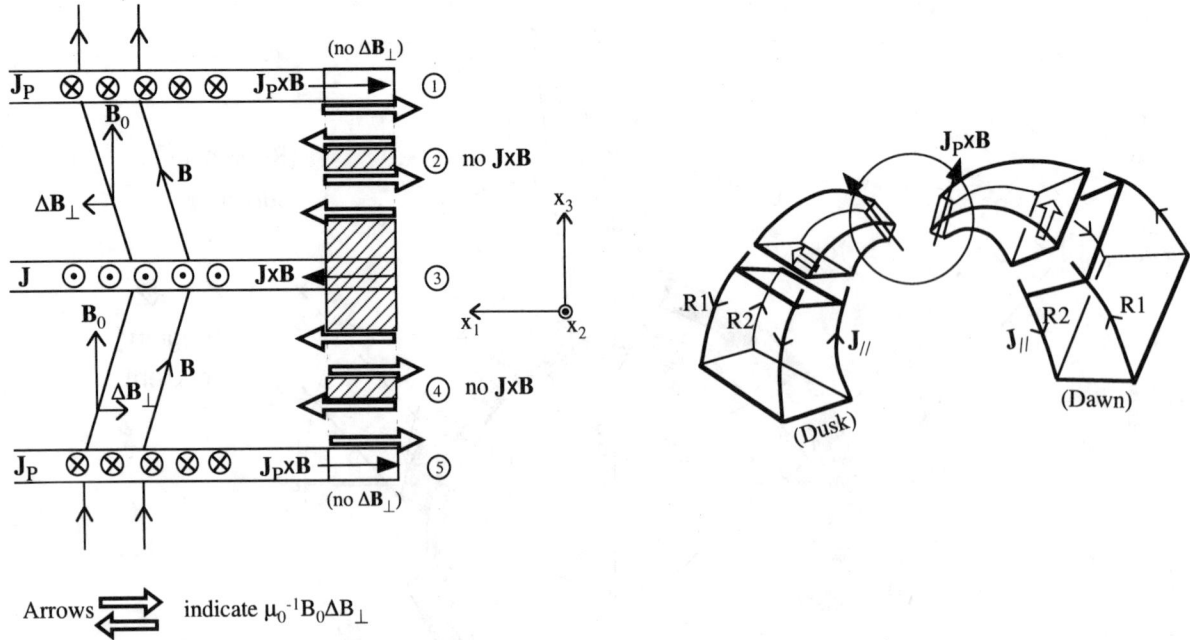

**Figure 3.** Illustration depicting the relation between transverse magnetic disturbance $\Delta \mathbf{B}_\perp$ and magnetic tangential stress $B_0 \Delta B_\perp / \mu_0$ (i.e., transverse momentum) transmitted by FACs from the magnetospheric generator plasma to the collisional ionosphere. An idealized case associated with a pair of FACs symmetric between northern and southern hemispheres is given, which resembles region 1 and region 2 currents in dawn and dusk sectors.

magnetic tangential stress $B_0 \Delta B_\perp / \mu_0$ that is exerted on the cross section of the ionospheric part of a flux tube, corresponding to the volume element ⑤ in Figure 3. Data show that magnetic stress is exerted sunward on the auroral zone ionosphere in both the dusk and dawn sectors and exerted antisunward over the polar cap ionosphere. These stresses sustain plasma convection in the collisional ionosphere that conforms to a well-known two-cell pattern consisting of antisunward flow over the polar cap bounded by sunward flows at both the dawn and dusk auroral zones [e.g., *Heppner and Maynard*, 1987]. Ionospheric Hall current flows in the opposite direction to the plasma flow. Ionospheric Pedersen current flows to keep the relation that magnetic stresses exerted on the ionosphere are consistent with the distribution of $\int (\mathbf{J}_P \times \mathbf{B}_0) dV = \mathbf{I}_p \times \mathbf{B}_0$, the Ampere force occurring in the ionosphere. Ionospheric Pedersen currents flow poleward in the duskside auroral zone, dawn-to-dusk in the polar cap, and equatorward in the dawnside auroral zone; all are perpendicular to both the magnetic stresses and the background magnetic field. These patterns of ionospheric Hall and Pedersen currents are quite similar to those of ionospheric currents that were inferred from ground-based geomagnetic data by combining the mathematical inversion method and the ionospheric electrodynamics by a number of authors [e.g., *Kamide and Akasofu*, 1981].

In the collisionless magnetosphere, magnetic flux and plasma contained in it move together with the same velocity of approximately $\mathbf{v} = (\mathbf{E} \times \mathbf{B})/B^2$. The $\mathbf{J} \times \mathbf{B}$ force in the magnetospheric generator plasma acts to oppose plasma convection velocity $\mathbf{v}$ so that the generator serves as a source region of electromagnetic energy. Then, conditions of $\mathbf{J} \cdot \mathbf{E} = (\mathbf{J} \times \mathbf{B}) \cdot \mathbf{v} < 0$ and $-\mathbf{J} \cdot \mathbf{E} = \nabla \cdot (\mathbf{E} \times \Delta \mathbf{B}_\perp)/\mu_0 > 0$ come up and electromagnetic energy can radiate, where temporal variation of magnetic field energy density is neglected in the stationary state. The amount of Poynting flux divergence emanating from a unit volume is reduced to be $(\mathbf{E} \times \Delta \mathbf{B}_\perp) \cdot \mathbf{e}_B / \mu_0$ per unit area, that is, Poynting flux along $\mathbf{B}_0$ passing through a cross section perpendicular to $\mathbf{B}_0$. Using the boundary condition at the magnetosphere-ionosphere interface, this flux turns out to be exactly as $\mathbf{I} \cdot (\mathbf{E} + \mathbf{v}_n \times \mathbf{B}_0) + \mathbf{v}_n \cdot (\mathbf{I} \times \mathbf{B}_0)$, where $\mathbf{I}$ is height-integrated ionospheric currents. The first and second terms designate, respectively, the Joule heating and the mechanical energy transfer rate, both per unit area of the ionosphere. Thus, Poynting flux transmitted by FACs brings about energy consumption due to the ionospheric currents. As seen in Figure 1 (upper), the profile of transverse magnetic disturbances observed along the dawn-dusk track at low altitudes often exhibits a triangular disturbance (directed antisunward) at both the dusk and dawn auroral latitudes and a hill-shaped sunward

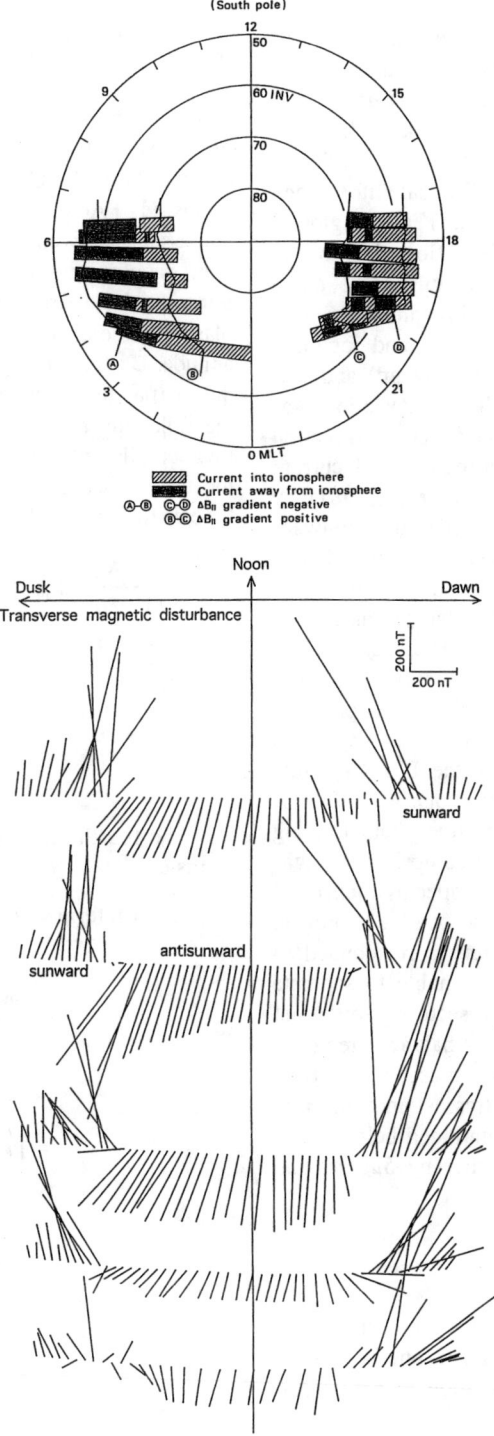

**Figure 4.** Examples of region 1 and region 2 currents observed just above the ionosphere in the south pole during southward IMF and geomagnetically active periods. (Upper) Spatial configuration of the flow direction pattern (after *Iijima et al.*, 1982). (Lower) Corresponding transverse magnetic disturbances $\Delta\mathbf{B}_\perp$ for selected passes. Both results are seen from above the north pole through "transparent" Earth. $\Delta\mathbf{B}_\perp$ in the south pole manifests tangential stress $B_0\Delta B_\perp/\mu_0$ that is exerted on the southern ionospheric cross section of the volume element ⑤ in Figure 3.

disturbance over the polar cap in the northern hemisphere. The former indicates the intensity-balanced region 2 and region 1 current regimes, and the latter associates the net (or excess) region 1 current regime. *Gary et al.* [1994] determined Poynting flux along the dawn-dusk track at low altitudes by magnetic field and ion drift velocity measurements acquired with the DE2 satellite. They demonstrated that when ordinary region 1 and region 2 currents were observed, Poynting flux flowed into the ionosphere over the entire regions of dusk and dawn auroral zones and polar cap. The flux intensity was strongest in the intensity-balanced region 2 and region 1 domains (where plasma convection was sunward) and was much weaker over the polar cap (where convection was antisunward). The fact that reversal of the flow direction pattern of plasma convection occurs on the region 1 current domain was observed earlier by *Bythrow et al.* [1981] by the AE satellite data. Their data suggested that sunward convection occurs on the intensity-balanced region 2 and region 1 current regimes both in the dusk and dawn sectors and antisunward convection spans the polar cap and dawn and dusk net region 1 current regimes. The fact that the intensity-balanced region 2 and region 1 current regimes almost reside on the precipitation regimes of particles characterizing the plasma sheet/ring current was suggested earlier by *McDiarmid et al.* [1978a] by the Isis 2 satellite data. Using the Viking satellite data, *Woch et al.* [1993] confirmed that the ion precipitation regime characterizing the low-latitude boundary layer (LLBL) comprises its high-latitude portion (denoted as the LLBL proper by them) and the low-latitude portion that overlaps the plasma sheet/ring current regime. The net region 1 current corresponded to their LLBL proper. It seems to be reasonable to separate the FAC systems in the dawn and dusk sectors into the net region 1 current regime and the intensity-balanced region 2 and region 1 current regimes by reasons of their combination of different magnetic stresses and different ionospheric conductivity control as mentioned below. The net region 1 current was observed earlier by *Sugiura and Potemra* [1976] from the Triad data.

*Generation*

Using Eq. (3), we can determine the conditions required to generate FACs in the magnetospheric plasma. By applying mathematically vector rotation to both sides, Eq. (3) turns to be

$$\int \left( \frac{1}{\mu_0} \frac{\nabla B}{B} \times B \frac{\partial (\Delta \mathbf{B}_\perp)}{\partial s} \right) \left( \frac{ds}{B} \right) + \frac{1}{\mu_0} \nabla \times (\Delta \mathbf{B}_\perp)$$

$$= \int \left[ \frac{\nabla \rho}{\rho} \times \left( \rho \frac{\partial \mathbf{v}}{\partial t} + \rho (\mathbf{v} \cdot \nabla) \mathbf{v} \right) \right] \quad (4)$$

$$+ \rho \frac{\partial}{\partial t} (\nabla \times \mathbf{v}) + \rho (\mathbf{v} \cdot \nabla)(\nabla \times \mathbf{v}) \bigg] \left( \frac{ds}{B} \right)$$

where $\int dV = \int (dx_3/B) = \int (ds/B)$, with $B_3 = B$, and $x_3 = s$ as used often. On the left-hand side, the second term $= \nabla \times (\Delta \mathbf{B}_\perp)/\mu_0$ exactly shows FACs. Eqs. (3) and (4) indicate at least two things: (1) FACs observed at a certain altitude on a flux tube are associated with the rotation (and/or shear) of magnetic stress appearing on its cross section; in turn, FACs have transmitted the stress up to that altitude. (2) FACs combine plasma motions contained in the flux tube volume up to that altitude. By further applying $(1/B)\mathbf{e}_B\cdot$ to Eq. (4) and denoting the inertia current as, $\mathbf{J}_{ine} = (B/B^2) \times \rho(d\mathbf{v}/dt)$, we can derive the following equation that relates the FAC density per unit magnetic flux observed at a certain altitude on a flux tube and the causal factors integrated along the magnetospheric flux-tube up to that altitude:

$$\frac{J_{//}}{B} = \int \left( -\frac{\nabla \rho}{\rho} \cdot \mathbf{J}_{ine} \right) \left( \frac{ds}{B} \right) + \int \left[ \rho \frac{\partial}{\partial t} \left( \frac{\mathbf{e}_B}{B} \cdot (\nabla \times \mathbf{v}) \right) \right] \left( \frac{ds}{B} \right)$$

$$+ \int \left[ \rho (\mathbf{v} \cdot \nabla) \left( \frac{\mathbf{e}_B}{B} \cdot (\nabla \times \mathbf{v}) \right) \right] \left( \frac{ds}{B} \right) \quad (5)$$

$$- \int \left[ \frac{\mathbf{e}_B}{B^2} \cdot \left( \frac{\nabla B}{\mu_0} \times B \frac{\partial (\Delta \mathbf{B}_\perp)}{\partial s} \right) \right] \left( \frac{ds}{B} \right)$$

The fourth term of Eq. (5) can be replaced by various forms as

$$\text{Fourth term} = \mu_0 \int \left( \frac{\mathbf{e}_B}{B^3} \cdot (\mathbf{M} \times \mathbf{T}) \right) \left( \frac{ds}{B} \right)$$

$$= \mu_0 \int \left( \frac{\mathbf{e}_B}{B^3} \cdot (\mathbf{T} \times \mathbf{F}) \right) \left( \frac{ds}{B} \right)$$

$$= \mu_0 \int \left( \frac{\mathbf{e}_B}{B^3} \cdot (\mathbf{F} \times \mathbf{M}) \right) \left( \frac{ds}{B} \right) \quad (6)$$

$$= \int \left( \frac{\nabla B}{B} \cdot \mathbf{J}_\perp \right) \left( \frac{ds}{B} \right)$$

where,

$$\mathbf{F} = -\nabla p - \rho \frac{d\mathbf{v}}{dt},$$

$$\mathbf{M} = -\nabla \left( \frac{B^2}{2\mu_0} \right)$$

and

$$\mathbf{T} = \frac{1}{\mu_0} B \frac{\partial (\Delta \mathbf{B}_\perp)}{\partial s}$$

The force-balance relation $\mathbf{J} \times \mathbf{B} + \mathbf{F} = (\mathbf{M} + \mathbf{T}) + \mathbf{F} = 0$ is used, which should hold everywhere in the magnetospheric plasma. In Eq. (5), the second term on the right-hand side displays the transmission of Alfven waves via FACs, which is retained here for general usage. In the stationary state, the terms related with $\partial/\partial t$ are omitted. As $\mathbf{E} = -\mathbf{v} \times \mathbf{B}$,

$$\frac{\mathbf{e}_B}{B} \cdot (\nabla \times \mathbf{v}) = -\left(\frac{1}{B^2}\right) \nabla \cdot \mathbf{E}$$

holds. From an inspection of Eqs. (5) and 6, we notice the following noteworthy characteristics: (1) In Eq. (5) three terms, from the first to the third, can also be summarized as

$$\int \left(-\frac{\mathbf{e}_B}{B} \cdot \nabla \times \left(-\rho \frac{d\mathbf{v}}{dt}\right)\right) \left(\frac{ds}{B}\right)$$

and FACs are generated by inertia forces if the force field exhibits rotation and/or shear in the plane perpendicular to the magnetic field in the generator plasma. In the stationary state FACs are mainly associated with the third term, which requires that the plasma convection pattern should include the velocity shear in the plane perpendicular to $\mathbf{B}$ with the intensity of the shear changing along the streamline [e.g., *Sonnerup*, 1980]. (2) The fourth term of Eq. (5), which is equivalent to all of Eq. (6), manifests another mechanism of FACs. As shown by the last term of Eq. (6), if the gradient of the magnetic field intensity occurs in the flow direction of the magnetospheric perpendicular current, $\mathbf{J}_\perp$ diverges (or converges) there and FACs occur to compensate for it [e.g., *Vasyliunas*, 1970, 1972; *Southwood*, 1977; *Sato and Iijima*, 1979; *Harel et al.*, 1981; *Peymirat and Fontaine*, 1994]. Using the first three terms, we use a different interpretation, that FACs occur if three forces of $\mathbf{F}$, $\mathbf{M}$, and $\mathbf{T}$ balance so as to make a triangle and so that the vector product of any combination of two forces is not zero, as first demonstrated by *Haerendel* [1990]. (3) Once FACs are generated by any mechanism, they derive plasma motions from the generator plasma and transmit the field-aligned component of vorticity of plasma convection along the field lines by virtue of magnetic stress transmission.

Figure 5 (upper) gives an example of spatial distribution of the flow direction pattern of FACs that were observed at low altitudes in the conjugate hemisphere during nearly the same period as an example in Figure 4. Figure 5 (lower) shows the spatial distribution of the transverse magnetic disturbance $\Delta \mathbf{B}_\perp$ for some selected passes. As seen from Figure 3, $\Delta \mathbf{B}_\perp$ observed in the northern hemisphere manifests magnetic tangential stress $B_0 \Delta B_\perp / \mu_0$ that is exerted on the northern cross section of the magnetospheric part of a flux tube, corresponding to the volume element ③ in Figure 3. Data show that antisunward stresses are exerted on the flux tube at auroral latitudes in both the dusk and dawn sectors, and sunward stresses are exerted on the flux tube at polar cap latitudes which span net region 1 currents. It sounds reasonable to identify the generation of FACs in the dawn and dusk sectors with the causes of antisunward magnetic stresses at auroral latitudes (carried by intensity-balanced region 2 and region 1 currents) and sunward magnetic stresses over the polar cap (carried by net region 1 currents).

Figure 6 illustrates a generation mechanism of the intensity-balanced region 2 and region 1 currents in the dawn and dusk sectors, which are suggested to reside on magnetic field lines emanating from the ionosphere at auroral latitudes and containing the hot and energetic particles characterizing the plasma sheet and trapped radiation belts (ring current) domain. Using an example of Figure 5, antisunward magnetic stress $B_0 \Delta B_\perp / \mu_0$ is shown on the cross section of a flux tube emanating from just above the auroral ionosphere, corresponding to the northern cross section of the volume element ③ in Figure 3. This stress is derived from force $\mathbf{T}$ of $\mathbf{J} \times \mathbf{B}$ forces occurring in the generator plasma (i.e., plasma sheet/ring current). Using the implication of Eqs. (5) and (6), the sunward component of $\mathbf{P}$ ($= -\nabla p$) is required to drive approximately dusk-to-dawn perpendicular currents (closure currents in the causal region). In order to generate region 1 and region 2 currents in a pair, three forces $\mathbf{M}$, $\mathbf{T}$, and $\mathbf{P}$ should balance so as to make a triangle; in turn, any combination of two forces produces a nonzero vector product. Force $\mathbf{P}$ should have not only a sunward component (causing sunward convection) but also a transverse component to it [*Iijima et al.*, 1997]. Once region 1 and region 2 currents are generated, they transmit the field-aligned component of the vorticity (equivalent to $-\nabla \cdot \mathbf{E}$) of sunward plasma motions along the field lines up to the ionosphere by virtue of magnetic stress transmission. Thus, global sunward convection system is completed on the whole magnetic flux tube at auroral latitudes in the magnetosphere-ionosphere system. In the generator plasma, as $\mathbf{J} \cdot \mathbf{E} = \mathbf{v} \cdot (\nabla p) < 0$, in turn, $\mathbf{v} \cdot (-\nabla p) > 0$, $\mathbf{P}$ acts to accelerate the background plasma medium to convect with its assigned velocity of $(\mathbf{E} \times \mathbf{B})/B^2$ (sunward in this case). At the same time, $\mathbf{P}$ sets up that plasma thermal energy is converted to electromagnetic energy. Electromagnetic energy is released there and transmitted to the auroral-zone ionosphere by region 1 and region 2 currents.

Net region 1 current is suggested to reside on magnetic field lines that comprise closed field lines threading the magnetospheric LLBL proper region and open field lines connecting the so-called high-latitude boundary layer (HLBL), both in the vicinity of the magnetopause as illustrated in Figure 7. Referring to data in Figure 5, sunward magnetic stress $B_0 \Delta B_\perp / \mu_0$ is exerted on the cross

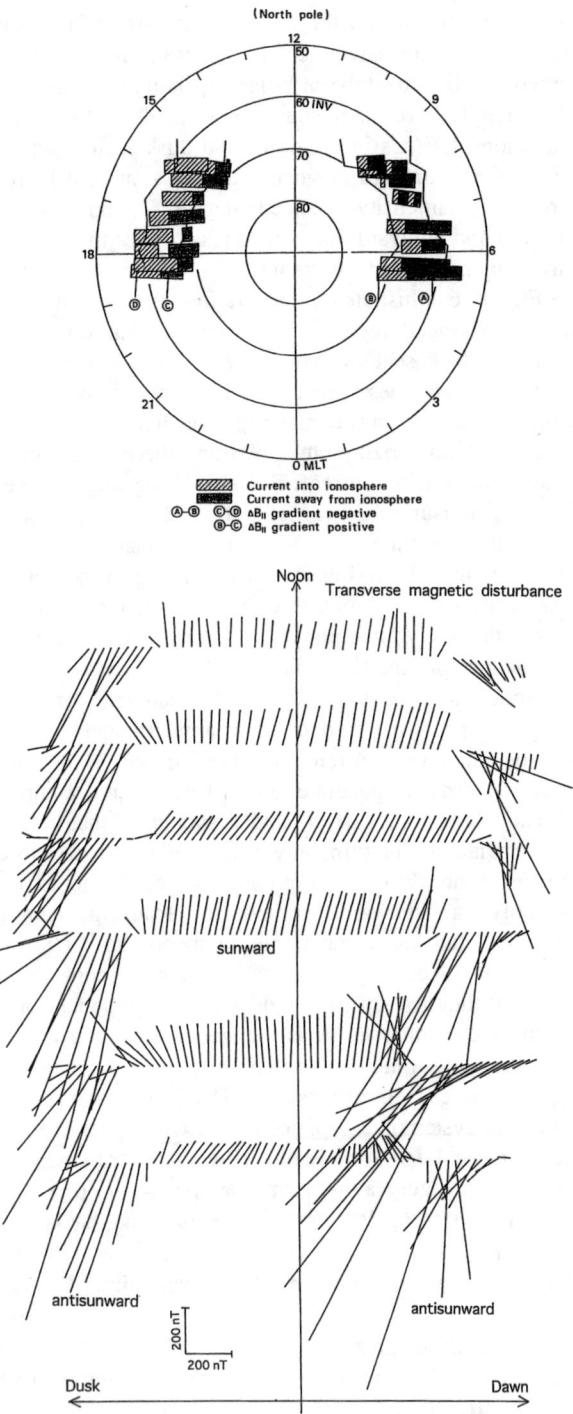

**Figure 5.** Examples of region 1 and region 2 currents observed in the conjugate hemisphere (in the north pole) for the same period as Figure 4. (Upper) Spatial configuration of the flow direction pattern (after *Iijima et al.*, 1982). (Lower) Corresponding transverse magnetic disturbances $\Delta B_\perp$ for selected passes. In the north pole, $\Delta B_\perp$ manifests tangential stress $B_0 \Delta B_\perp / \mu_0$ that is exerted on the northern cross section of the volume element ③ in Figure 3.

section of a flux tube emanating from just above the polar cap ionosphere, corresponding to the northern cross section of the volume element ③ in Figure 3. This stress really manifests force **T** of **J** × **B** forces occurring in both the generator regions of LLBL and HLBL. By Eq. (5), it is suggested that net region 1 currents are generated if antisunward streaming plasma associates the velocity shear in the plane perpendicular to background magnetic field and its magnitude of shear varies along the streamline in both the LLBL and HLBL generator plasmas. This is a result of the momentum acquisition from the magnetopause plasma flow, triggered by either the viscous interaction in the LLBL or the momentum coupling at the HLBL ultimately due to the magnetic reconnection [e.g., *Siscoe and Maynard*, 1991]. Net region 1 currents transmit the field-aligned component of vorticity of antisunward plasma motions along the field lines up to the ionosphere by virtue of magnetic stress transmission. The global antisunward convection system is completed on the whole flux tube at polar cap latitudes in the magnetosphere-ionosphere system. In the generator plasma, as

$$\mathbf{J} \cdot \mathbf{E} = \frac{d}{dt}\left(\frac{1}{2}\rho v^2\right) \approx \rho(\mathbf{v} \cdot \nabla)v^2 < 0$$

in turn, kinetic energy of plasma convection is lost and is converted to the source of Poynting flux divergence.

### R1 and R2 FACs and sunward convection

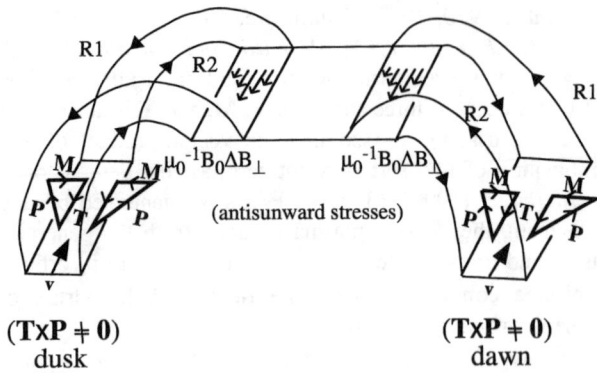

**Figure 6.** Illustration depicting the generation of intensity-balanced region 1 and region 2 currents. Flux tubes carrying FACs are cut just above the auroral ionosphere, corresponding to the volume element ③ in Figure 3. Antisunward magnetic stresses are exerted on their cross sections (refer to the observation in Figure 5). These stresses should be derived from force **T** of **J** × **B** forces produced in the generator plasma. It is suggested that pressure gradient force **P** determines the perpendicular currents (closure currents) and that three forces **P**, **M**, and **T** are balancing to make a triangle in the generator regions.

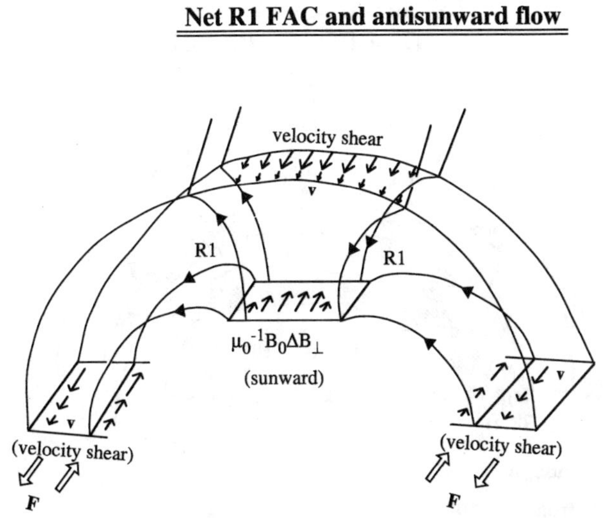

### Net R1 FAC and antisunward flow

**Figure 7.** Illustration depicting the generation of net region 1 currents. Flux tubes at polar cap latitudes are cut just above the ionosphere, corresponding to the volume element ③ in Figure 3. Sunward magnetic stresses are exerted on their cross sections (see the observation in Figure 5). These stresses are derived from $\mathbf{J} \times \mathbf{B}$ forces in the generator plasma. It is suggested that inertia force determines the perpendicular currents and that velocity shears are occurring in the generator regions.

Electromagnetic energy is released there and transmitted to the polar cap ionosphere by net region 1 currents. In this context, the following observations just above the ionosphere are noteworthy. Under winter polar cap conditions the intensity-balanced region 2 and region 1 currents occur on auroral latitudes without an association of net region 1 current, and net region 1 current appears mostly under sunlit polar cap conditions [*Smidday et al.*, 1980, from S3-2 satellite data]. The intensity of region 1 current is far beyond that of the region 2 current, especially in the daytime sector of ~0600 to 1600 MLT. Furthermore, much more transverse momentum and energy can be derived from perpendicular currents in the LLBL and HLBL generator plasmas under sunlit polar cap conditions of conductivities, whereas the derivation of transverse momentum and energy from perpendicular currents in the plasma sheet/ring current does not depend much on the ionospheric conditions of conductivities [*Fujii and Iijima*, 1987, from Triad data].

In Figures 6 and 7, flux tubes are illustrated very schematically. Meridional segments on the ionospheric level are presumably traced back into the deep magnetosphere, not in the radial direction but rather in the $X_{GSM}$ direction. Directions of magnetic stresses would never fade by particular choice of field-line tracing.

## ADVANCES IN OBSERVATION

### IMF $B_y$-Dependent Midday FACs

In the midday sector of ~0930–1430 MLT, distinctive FACs, denoted as cusp currents, appear at ~78°–80° ML poleward of and adjacent to region 1 currents and show an flow direction opposite to that of region 1 current in the same MLT meridian [*Iijima and Potemra*, 1976b]. They suggested that these FACs are located within the region associated with the dayside magnetospheric cusp. Using simultaneous measurements of magnetic field, plasma particle flux, electric field, and ion drift velocity with various spacecraft, later studies have analyzed the relationship of spatial distributions between FACs and particle precipitation/plasma population regimes and plasma convection. The morphology is represented by Figure 8 (upper) and includes the following characteristics: (1) FACs in the midday sector are identified unambiguously with two regimes; one is ordinary region 1 and region 2 currents residing in the precipitation domain of LLBL/plasma sheet and the other is midday region 1 and "traditional" cusp currents corresponding to cusp/mantle (cusp plume) precipitation domain. The traditional cusp currents from *Iijima and Potemra* [1976b] really correspond to the plasma mantle regime and were referred to as region 0 currents by *Bythrow et al.* [1988] from DMSP and HILAT data. (2) Cusp/mantle currents develop exclusively depending on IMF $B_y$ polarity and, in the north pole, the flow-direction pattern on the prenoonside overwhelms the pattern of the postnoonside during the period of $B_y > 0$ and the pattern of the postnoonside overwhelms that of the prenoonside for $B_y < 0$ [*Erlandson et al.*, 1988, from the Viking data; *Taguchi et al.*, 1993, from DE2 data]. This kind of $B_y$ dependence of midday FACs in the north pole was suggested earlier by *Wilhjelm et al.* [1978] from Triad data and by *McDiarmid et al.* [1978b] from Isis2 data. In the south pole, midday FACs exhibit the same flow-direction pattern as that in the north pole during the period of reversed IMF $B_y$ polarity [*Iijima et al.*, 1978; *Saflekos and Potemra*, 1980, both from Triad data]. (3) The overall pattern of flow direction in the midday sector comprises the ordinary region 1 and region 2 current pair and the $B_y$-dependent cusp/mantle current pair. $B_y$-dependent cusp/mantle currents also exhibit poleward dislocation, as suggested by *Yamauchi et al.* [1993] from Viking data. Thus, the overall configuration tends to exhibit three current patterns in the prenoon sector and four current patterns in the afternoon sector in the north pole when IMF $B_y$ is positive, while in the same hemisphere it exhibits four current patterns in the prenoon sector and three current patterns in the postnoon sector when IMF $B_y$ is reversed to

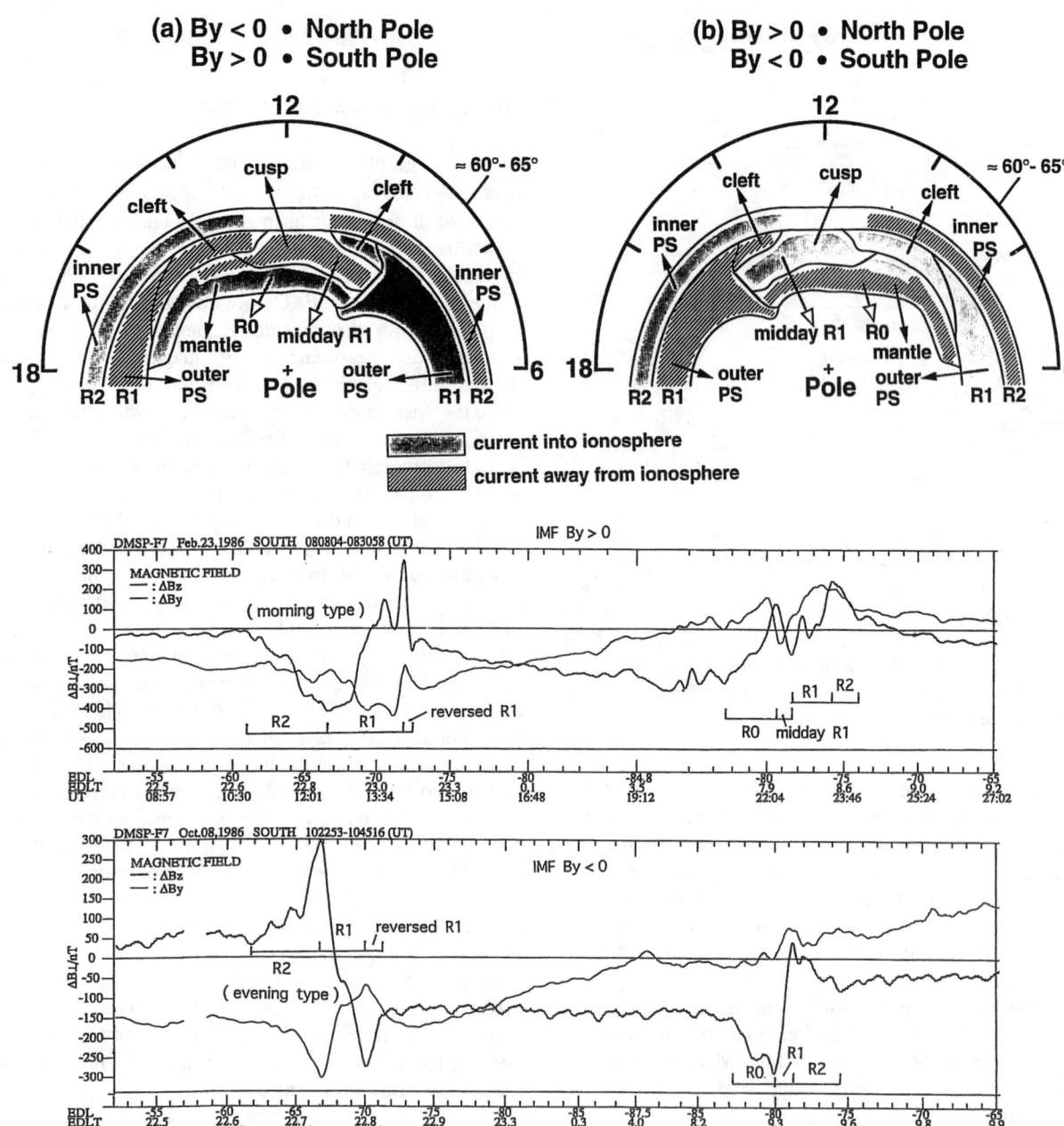

**Figure 8.** Synthesis model of $B_y$-dependent midday FACs. (Upper) Spatial configuration of the flow direction pattern is shown superimposed on the concurrent particle precipitation pattern. Associated IMF $B_z$ was southward (after *Watanabe et al.*, 1996). (Lower) Examples of transverse magnetic disturbances used for the upper diagram. Disturbances were observed along approximately noon-midnight tracks at low altitudes in the south pole. $\Delta B_z$ and $\Delta B_y$ denotes the dusk-to-dawn and noon-to-midnight component, respectively.

be negative [*Taguchi et al.*, 1993; *Ohtani et al.*, 1995a and b, from Viking and DMSP F7 data]. These three current patterns and four current patterns are exactly seen in the south pole for reversed polarity of IMF $B_y$ to that of the north pole, which was confirmed by *Watanabe et al.* [1996] using data from northern 265 passes and southern 245 passes of DMSP F7. In a synthesis model of Figure 8 (upper), the so-called cusp current (occurring on the cusp precipitation regime) is denoted as midday region 1, and the so-called mantle current (corresponding to "traditional

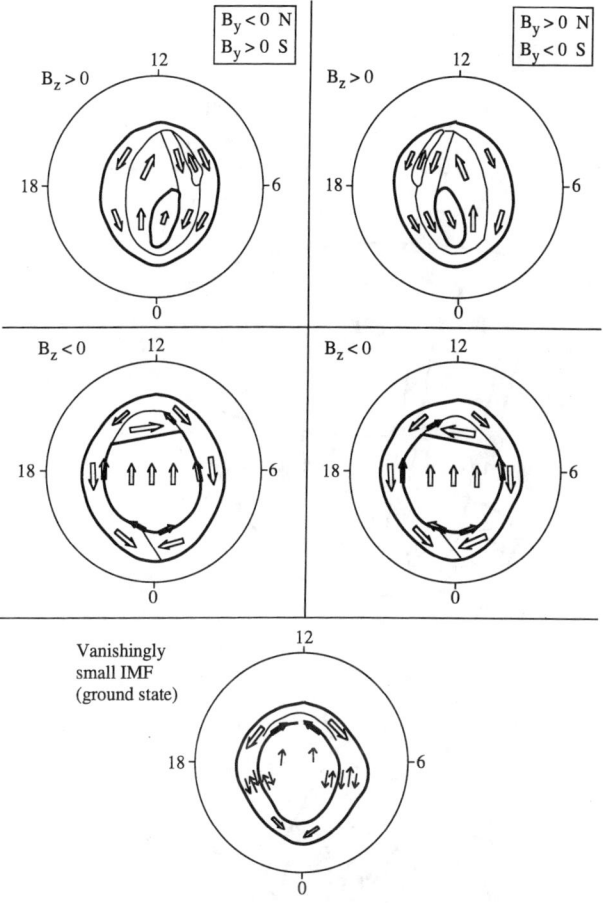

Figure 9. A summary of large-scale characteristics of FACs. Characteristics are represented in terms of magnetic tangential stresses carried by FACs. Diagrams are organized systematically by polarity of IMF $B_z$ and $B_y$ components and hemisphere exposed by IMF. The format displays the stresses that are exerted on the magnetospheric flux-tube element that emanates from just above the ionosphere, corresponding to the volume element ③ in Figure 3. These stresses justly manifest magnetic tangential stresses that are derived from $\mathbf{J} \times \mathbf{B}$ forces occurring in the magnetospheric generator plasma. Magnetic stresses exerted on the ionosphere are reversed to these stresses because of the nature of the stress itself.

cusp" current and occurring on the mantle precipitation regime) is denoted as region 0 current. Figure 8 (lower) gives two examples of transverse magnetic disturbances that were used for their analysis. (4) The amplitude of the midday region 1/region 0 currents is more strongly correlated with $B_y$ than with $B_z$ of IMF [*McDiarmid et al.*, 1978b] and is best correlated with solar wind quantities of $nvv_A$, where $nv$ is related to solar wind momentum and $v_A$ is the Alfven velocity in the solar wind and related to momentum transfer rate via magnetic reconnection process [*Iijima and Potemra*, 1982].

Here, Figure 9 is introduced which is a summary of the spatial configuration of magnetic tangential stresses associated with FACs, organized systematically by polarity (positive or negative) of IMF $B_z$ and $B_y$ components and the hemisphere (northern or southern) exposed by the IMF. The format displays the stress that is exerted on the northern (and southern) cross section of the magnetospheric flux tube emanating from just above the ionosphere, corresponding to the cross section of the volume element ③ in Figure 3. $B_y$-dependent midday FACs correspond to the middle of Figure 9. For example, when the magnetosphere is exposed by positive $B_y$, magnetic stress is exerted eastward in the north pole and westward in the south pole on the cross sections of magnetospheric flux tubes carrying cusp/mantle currents. These stresses should manifest just magnetic tangential stresses that are derived from $\mathbf{J} \times \mathbf{B}$ forces occurring in the perpendicular current in the generator plasma near the magnetopause, as illustrated in Figure 10 (left). It is suggested that midday region 1 and region 0 currents (i.e., cusp/mantle currents) are generated if three forces $\mathbf{F}$, $\mathbf{M}$, and $\mathbf{T}$ balance to make a triangle in the generator plasma. These magnetic stresses are exerted in a reverse fashion on the ionosphere and cause ionospheric plasma flows with flow directions opposite to stresses in Figure 9 (middle). This is consistent with $B_y$-dependent characteristics of plasma flows that were observed at midday high latitudes in both the northern and southern hemispheres by *Doyle et al.* [1981] with the S3-2 satellite.

## FACs Associated with Northward IMF

Using the continuous measurements of vector magnetic field with the MAGSAT satellite that provided data over the full latitudes at altitudes of ~300–600 km primarily in the dawn and dusk sectors, characteristics of FACs occurring in the central polar cap were determined, which were not done thoroughly with the other satellites. Characteristics include a new FAC system reported by *Iijima* [1984] and denoted as NBZ (northward IMF $B_z$) current by *Iijima et al.* [1984], which occurs during the period of definitely northward IMF. In Figure 1, two examples of transverse magnetic disturbances compare FACs observed along the dawn-dusk track in the north pole for IMF $B_z < 0$ and $B_z > 0$ periods. For $B_z < 0$ (upper), the $\Delta S$ disturbance (sunward component) is antisunward at dusk auroral latitudes (indicating intensity-balanced region 2 and region 1 currents), sunward in the polar cap (including net region 1 currents both at dusk and dawn polar cap-auroral zone boundaries), and again antisunward at dawn auroral latitudes (intensity-balanced region 1 and region 2 currents). Whereas, for $B_z > 0$ (bottom), the $\Delta S$

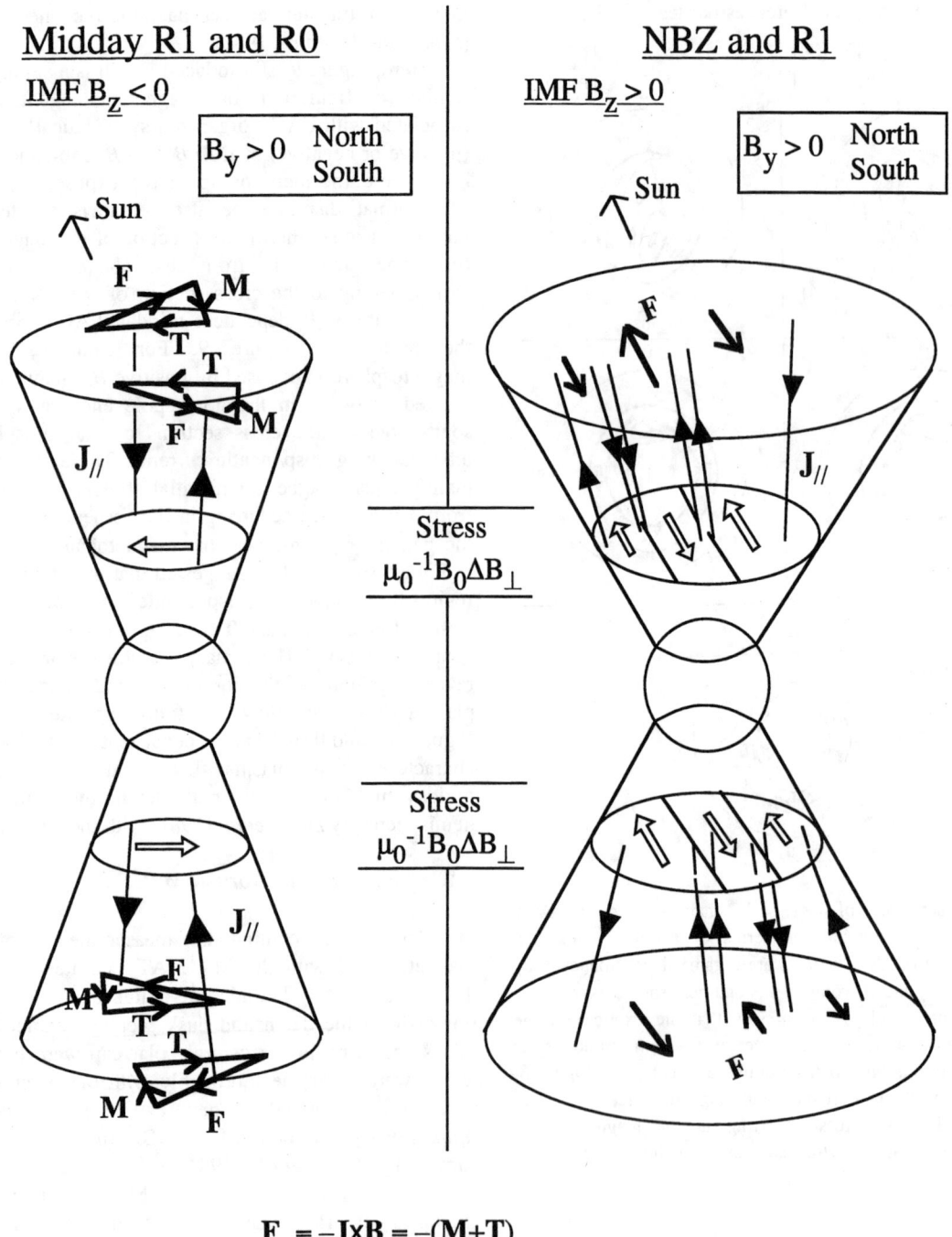

**Figure 10.** Illustration depicting the generation of FACs. (Left) For midday region 1 and region 0 currents (i.e., cusp/mantle currents); (Right) for NBZ currents and region 1 currents (coupled with NBZ currents). For both cases, the magnetosphere is exposed by positive IMF $B_y$. Magnetic tangential stresses are shown as exerted on the cross sections of the volume element ③ in Figure 3 (see observation in Figures 8 and 9). These stresses should be derived from force **T** of **J** × **B** forces produced by perpendicular currents in the generator plasma in the vicinity of the magnetopause.

disturbance exhibits a sharp wedge toward antisunward in the central polar cap that is indicative of a new FAC system distinctive to the region 1 current. This is the NBZ current. The occurrence of sun-aligned transverse magnetic disturbances over the polar cap under northward $B_z$ conditions was suggested by earlier studies [*Saflekos and Potemra*, 1980, from Triad data; *McDiarmid et al.*, 1980, from Isis2 data]. Figure 11 shows a model of the NBZ current system along with region 1 and region 2 currents. The upper left displays spatial configuration of flow direction pattern, and its right section corresponds to the distribution of current density [*Iijima and Shibaji*, 1987]. These were determined from data of 30 passes of MAGSAT in the southern summer hemisphere. All events occurred under the condition of IMF $B_z > B_y > 0$, with individual $B_z$ being $8 > B_z > 2$ nT and individual $B_y$ being $5 > B_y > 0$ nT. The determined area is bounded by the dawn-dusk track crossing the noon-midnight meridian at 82.7° ML on the dayside and 70.1° ML on the night side. Noteworthy points are as follows. (1) NBZ currents occur poleward of region 1 currents with reversed flow directions. NBZ currents flow into the ionosphere on the duskside and flow away from the ionosphere on the dawnside of their system. (2) NBZ currents change their configurations depending systematically on IMF $B_y$ polarity and show almost mirror images with respect to the noon-midnight meridian between the images for $B_y > 0$ and for $B_y < 0$. For example, in the south pole and for $B_y > 0$, morning NBZ currents are confined in a smaller region and evening NBZ currents occupy a large area of the polar cap, as seen in Figure 11. (3) The intensities of NBZ currents flowing into and away from the ionosphere are an order of ~0.3–0.8 A/m. The total NBZ current (dominant current on the dayside of sunlit hemisphere, either flowing into or away from the ionosphere) is generally 0.3–0.4 MA, and for region 1 the total current is 0.2–0.3 MA. The total NBZ current on the dayside amounts to more than 1 MA for strongly northward IMF. The peak magnitude of the NBZ current on an individual dayside pass correlated well with solar wind parameters in the form of

$$\left(B_y^2 + B_z^2\right)^{\frac{1}{2}} \cos\frac{\theta}{2},$$

where $\theta$ is an angle of the IMF vector measured from the positive $z$ axis in the $y$-$z$ plane.

Figure 11 (bottom) corresponds to the spatial configuration of transverse magnetic disturbances for some selected passes that were seen from just above the north. As seen in Figure 3, $\Delta \mathbf{B}_\perp$ observed in the southern hemisphere manifests magnetic tangential stress that is exerted on the ionospheric cross section of the volume element ⑤ in Figure 3. This stress sustains ionospheric plasma flow. Profiles show sunward stress in the polar cap and antisunward stresses along the dawn and dusk flanks of the polar cap, which are bounded by sunward stresses at dawn and dusk auroral latitudes. This configuration of magnetic stress resembles well the plasma convection pattern suggested by electric field measurement by the S3-2 [*Burke et al.*, 1979] and by ion drift vector measurement by the DE2 [*Heelis et al.*, 1986] under northward IMF conditions. Their convection patterns on the dayside sunlit hemisphere conforms to four cells, two of them confined to the polar cap with sunward flow in the central polar cap and antisunward flow along the flanks of polar cap, and the other cells bounded by sunward flow at auroral latitudes. We have noticed that convection (in turn, stress) cells degenerate into a smaller number of cells (only two or three) in accordance with $B_y$ dependence of the NBZ and region 1 currents. The convection pattern includes structured flows on the nightside. Examples of transverse magnetic disturbances were also reported by *Araki et al.* [1984].

Later studies [*Rich et al.*, 1990; *Nakagawa*, 1995] used simultaneous measurements of magnetic field and particle flux by the DMSP F6 and F7 satellites and provided new insights into the NBZ currents. These include the following: The NBZ current and associated plasma convection system are entirely located within the boundary plasma region (BPR) precipitation, which is the region of cooler, magnetosheathlike plasma mapping to the various boundary layers. The BPR is surrounded by the central plasma sheet (CPS) precipitation (or inner plasma sheet by Nakagawa), which is characterized by hot/energetic plasma. There was no indication of the intrusion of CPS toward high latitudes so as to bifurcate the polar cap. Adding particle flux data at low altitudes by the NOAA and TIROS satellites, Nakagawa further showed that energetic ions (with energy of 30–80 keV and higher) exhibit nearly isotropic precipitation, which coincides approximately with the BPR on the dayside and the CPS on the nightside. He suggested that dayside NBZ currents reside on open field lines where energization and acceleration of ions and isotropization of their pitch angle distribution occur via the magnetopause current sheet as suggested by *Lyons et al.* [1987]. For northward $B_z$, the spatial configuration of magnetic tangential stresses is represented by Figure 9 (top) as exerted on the northern cross section of the volume element ③ in Figure 3. These stresses manifest magnetic tangential stresses that are derived from $\mathbf{J} \times \mathbf{B}$ forces in the perpendicular current in the generator plasma presumably in the vicinity of the magnetopause. Figure 10 (right) depicts the flux tubes in the northern and southern

122 FIELD-ALIGNED CURRENTS IN GEOSPACE: SUBSTANCE AND SIGNIFICANCE

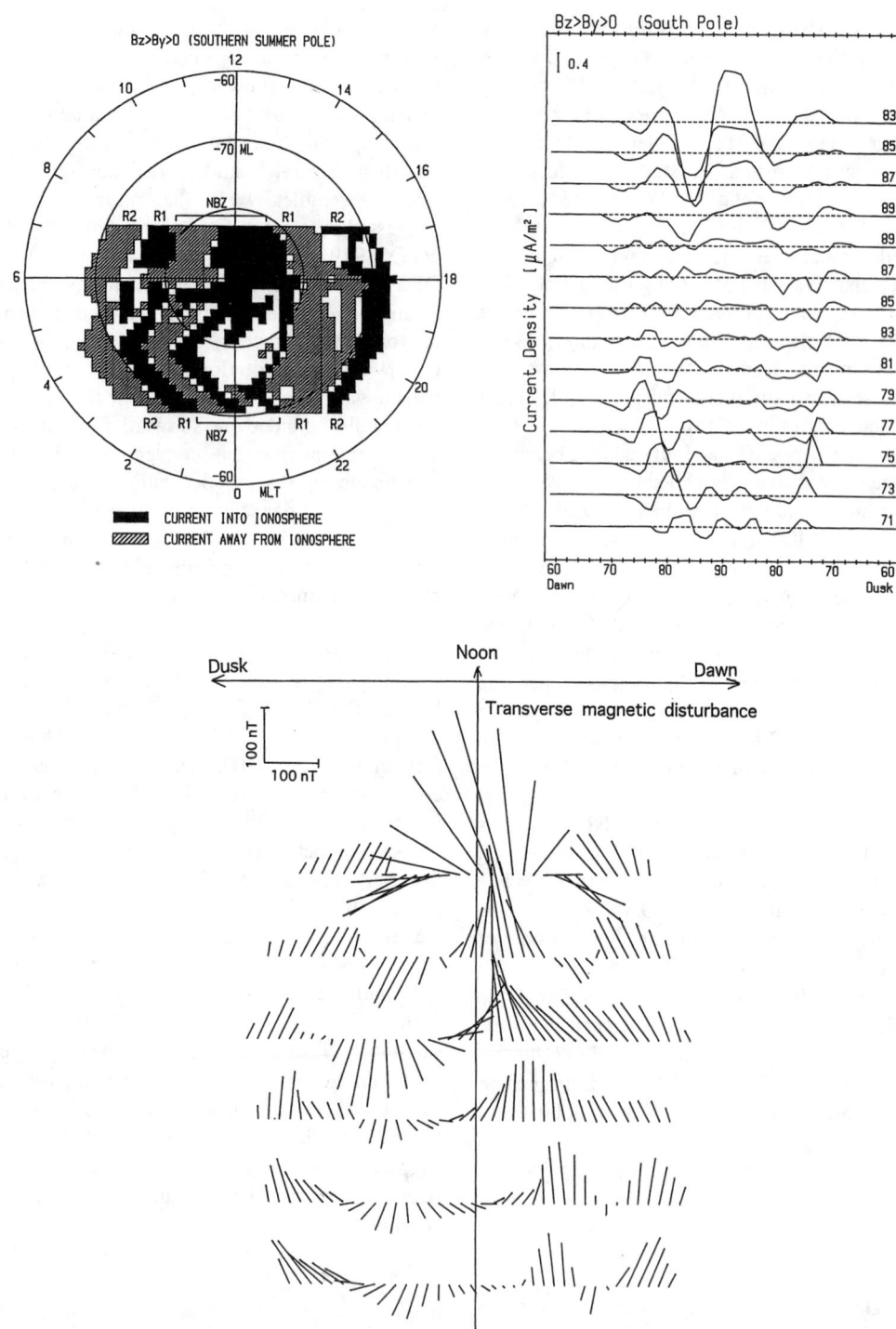

**Figure 11.** Observation model of NBZ FACs determined at low altitudes in the south pole for definitely northward IMF conditions. (Upper left) Spatial configuration of flow direction pattern. (Upper right) Corresponding current density distribution. Both are seen from above the south pole (after *Iijima and Shibaji*, 1987). (Lower) Corresponding transverse magnetic disturbances $\Delta \mathbf{B}_\perp$ (seen from above the north pole). $\Delta \mathbf{B}_\perp$ in the south pole manifests tangential stress that is exerted on the ionospheric cross section of the volume element ⑤ in Figure 3.

hemispheres (corresponding to the volume element ③) carrying NBZ and region 1 currents under $B_y > 0$ conditions. It is suggested that plasma forces **F** inevitably associate sunward/antisunward components to produce the perpendicular current (closure current) and to balance $\mathbf{J} \times \mathbf{B}$ forces in the generator plasma.

*Nightside FACs Associated with Substorms*

Earlier studies by *Iijima and Potemra* [1976a and 1978] on FACs associated with geomagnetic disturbances and substorms were followed by detailed analysis of magnetic field, electric field, particle precipitation, and plasma flow measurements, which include studies by *Fukunishi et al.* [1993] from Akebono data, *Fujii et al.* [1994] from DE 1 and 2 data, *Senior et al.* [1982] from Triad and radar data, and *Robinson et al.* [1985] from S3-2 and radar data. The studies of both *Fukunishi et al.* and *Fujii et al.* suggested that FACs generally correspond to almost the entire region of hot electron precipitation (> ~1 keV) and that distinctive FACs appear poleward of and adjacent to ordinary region 1 and region 2 currents, in the vicinity of the precipitation regime of the plasma sheet boundary layer (PSBL). Noteworthy results by *Fujii et al.* are summarized in Figure 12. They conjoined the configurations of the auroral bulge (using the measurements of auroral images with the 39 passes of DE1) and FACs and plasma flows (using magnetic field and electric field measurements by DE2 on the same conjugate passes with DE1); all were observed during the substorm expansion phase. Figure 12 (upper) illustrates the spatial configuration of the flow direction pattern of FACs along with the latitudinal profile of the plasma flow velocities. Figure 12 (lower) gives statistical distributions of the flow direction of FACs and relative amplitudes of FACs in six sectors on the nightside; sectors 1 ~ 4 correspond to the west side of the so-called Harang discontinuity line before midnight, and sectors 5 ~ 6 correspond to the east side in the early morning hours. Characteristics determined here include the following: (1) Plasma flows are generally sunward in nearly the whole auroral bulge, bounded by a flow of reversed direction at the highest latitudes (streaming toward the midnight meridian both in the premidnight and postmidnight sectors). (2) In the midnight sector characterized by the surge structure in global auroral images, the pattern of FACs along a latitudinal slice cut comprises, from the equator toward the pole, current flowing into the ionosphere, current flowing away from the ionosphere, and current flowing into the ionosphere again. The flow direction of the first two currents corresponds to that of ordinary region 2 and region 1 current, respectively, while the last one (appearing at the highest latitudes) shows a reversed flow direction to region 1 current and was denoted as region 0 by *Fujii et al.* They suggested that the region-1-sense current can be divided into two parts: one is its equatorward part coupled with region 2 current and residing on the sunward convection region, and the other is its poleward part coupled with region 0 current and residing on the convection streaming toward the midnight sector (eastward). In the early morning sector characterized by auroral images exhibiting simple bulge structure and not associating surge, the latitudinal pattern of the FACs consists of ordinary region 2 current (flowing away from the ionosphere), ordinary region 1 current (flowing into the ionosphere), and region 0 current at the highest latitudes (flowing away from the ionosphere in an opposite sense to region 1 current). The region-1-sense current is divided into its equatorward part coupled with region 2 current and its poleward part coupled with region 0 current.

From the results of *Fujii et al.*, we can suggest the spatial configuration of magnetic tangential stresses associated with FACs. Figure 13 illustrates the distribution of magnetic stresses that are exerted on the northern cross section of the near-equatorial flux-tube element (corresponding to the volume element ③ in Figure 3). On the west side of the Harang discontinuity line, magnetic stress is exerted eastward in the domain closer to the Earth and westward in the more distant domain away from the Earth. However, on the east side of the Harang discontinuity, line magnetic stress is exerted westward in the domain closer to the Earth and eastward in the distant domain. FACs on the nightside conform quasi-persistently to these configurations of magnetic stresses even if they exhibit a multiple and/or complex flow direction pattern. Region 1 currents span both regions of magnetic stresses that are exerted eastward and westward. Using Eq. (6), the configuration of plasma forces **F** is suggested and shown on the right side of Figure 13, which is required to produce FACs on the nightside.

*FACs for Vanishingly Small IMF*

During periods of geomagnetically and aurorally quiet, magnetic field measurements did not prove the definite existence of large-scale FACs but only manifest small-scale structures throughout the auroral oval over the polar cap [*Hoffman et al.*, 1988, from DE2 data] except for the occurrence of midday cusp/mantle currents in the summer hemisphere [*Rich et al.*, 1987, from DMSP F7 data]. Later, by defining the quietness as a period of vanishingly small IMF, for example, when the equinoctial northern and southern hemispheres were exposed by $|B_y| < 1.5$ nT and $-0.5$ nT $< B_z < 1.5$ nT of IMF, *Watanabe et al.* [1998] confirmed from MAGSAT data that ordinary region 1, region 2, and traditional cusp currents occur in dawn and dusk sectors of both the conjugate hemispheres. These

124 FIELD-ALIGNED CURRENTS IN GEOSPACE: SUBSTANCE AND SIGNIFICANCE

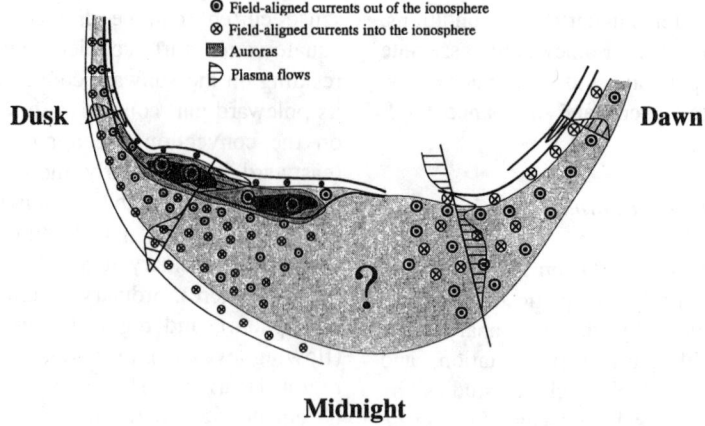

**Figure 12.** Observation model of nightside FACs associated with the expansion phase of a substorm. (Upper) Spatial distribution of the FAC flow direction pattern and concurrent plasma flow pattern superimposed on auroral bulge structure. (Lower) Flow direction pattern of FACs and relative intensities of FACs determined along latitudinal slice cuts that were grouped into six characteristic sectors of auroral images (after *Fujii et al.*, 1994).

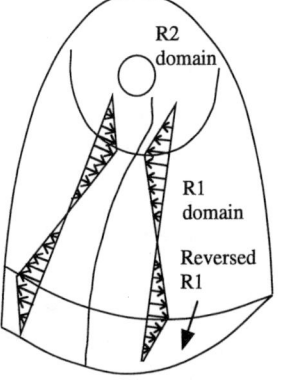

**Figure 13.** Near-equatorial plane schematic of spatial distributions of magnetic tangential stresses and plasma forces associated with nightside FACs. (Left) Stresses inferred from low-altitude observation (in Figure 12) are projected on the cross sections of the near-equatorial magnetospheric region, corresponding to the volume element ③ in Figure 3. These stresses exactly manifest force **T** of **J** × **B** forces produced in the generator region. (Right) Plasma forces **F** inferred from the requirement of production of perpendicular currents and generation of FACs (see Eq. (6) in the text).

FACs manifest the physics of solar wind-magnetosphere-ionosphere coupling processes under conditions of the minimization of the momentum and energy transfer from the solar wind into the magnetosphere, namely, a ground state of the magnetosphere. The spatial configuration of magnetic tangential stresses shown at the bottom of Figure 9 corresponds to these conditions, which exhibit the stresses exerted on the northern cross section of the volume element ③ as in Figure 3.

*FACs at Middle and High Altitudes*

Studies using the magnetic field measurements at high altitudes by the ISEE 1 and 2 satellites manifested various aspects of FACs, which include the following identification: the flow direction pattern of region-1-sense and region-2-sense currents at a radial distance of ~2.4 to 7 $R_E$ [*Kelly et al.*, 1986], the domain of region-1-sense current extending from the outermost toward the near-Earth (<15 $R_E$) in the magnetotail plasma sheet [*Ohtani et al.*, 1988], and distinctive currents to region-1-sense currents in the vicinity of the PSBL [*Elphic et al.*, 1985]. The identification of FACs at middle and high altitudes in the magnetosphere was preceded by earlier studies by *Fairfield* [1973] and *Sugiura* [1975]: the existence of region-1-sense current at the plasma sheet/tail lobe boundary by Fairfield from Imp 4 and 5 data and of region-1-sense current at the polar cap/auroral zone domain and region-2-sense current inside the auroral zone by *Sugiura* from Ogo5 data. Recent magnetic field measurements by the AMPTE CCE and Geotail satellites made it possible for us to determine the spatial configuration of the flow direction pattern and the amplitude distribution of FACs on a statistical basis. Figure 14 shows distributions of FACs that were determined from the Geotail data available for the geomagnetically active period (Kp < 3+) from March 1995 to April 1996, in the near-equatorial magnetosphere beyond ~10 $R_E$ in radial distance (*Yoshimura and Iijima*, 1999, manuscript in preparation). FACs were projected on the planes parallel to the neutral sheet plane for three different conditions of dipole tilt angle. The arrows pointing toward and away from the center of the Earth denote, respectively, the current flowing into and away from the ionosphere; current density is indicated by their length. Characteristics determined from the CCE data by *Iijima et al.* [1990] and the Geotail data are summarized next. (1) Region-2-sense currents occur in the relatively inner magnetosphere within the radial distance of ~10 $R_E$ at all local time sectors. The dayside magnetosphere (with a subsolar distance of ~11 $R_E$) is almost entirely dominated by region 2 currents; (2) region-1-sense currents occur beyond the radial distance of ~10 $R_E$ and occupy mostly the nightside domain of the magnetosphere.

## CONCLUDING REMARKS

Since the substantial success of in situ observation of FACs in geospace, vast efforts have been devoted to studying various problems relevant to FACs. However, it seems to me unreasonable to review all of them. I have tried to focus my article on the most important role that FACs play in the physics of the solar wind-magnetosphere-ionosphere coupling processes: that is, FACs transmit magnetic tangential stress (transverse momentum) and energy (Poynting flux) along field lines from the causal region to the other end of a flux tube. I think that this nature is not well recognized and not used in the studies of FACs. Most of the studies have preferentially addressed the generator region of each FAC system rather arbitrarily and separately from various regions in the magnetosphere, for example, region 1 currents to the magnetospheric boundary domain and region 2 currents separately to the plasma sheet/ring current domain. We should, however, be careful about the following: for example, the intensity-balanced region 1 and region 2 currents transmit together the common magnetic stresses and Poynting flux on the

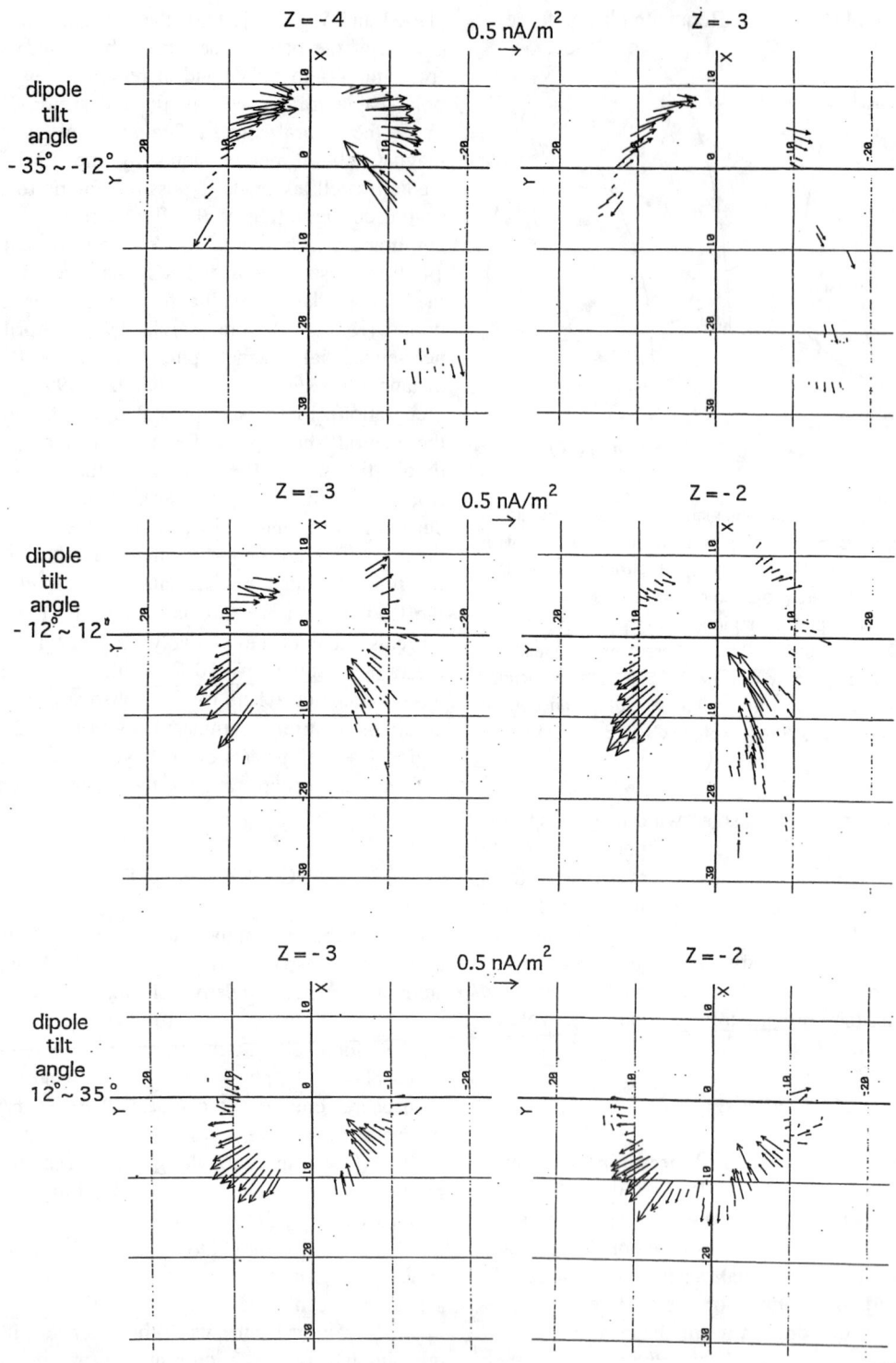

**Figure 14.** Examples of FACs determined in the magnetospheric region of ~10–30 $R_E$ in radial distance and within 5 $R_E$ from neutral sheets for three different conditions of dipole tilt angles from Geotail data. Region 2 currents occur in the inner magnetosphere within ~10 $R_E$, while region 1 currents occur in the distant magnetosphere beyond ~10 $R_E$ in radial distance (after Yoshimura and Iijima, manuscript in preparation).

collisional ionosphere, which are inevitably traced back to the common region producing $\mathbf{J} \times \mathbf{B}$ forces and $\mathbf{J} \cdot \mathbf{E}$ (<0) in the magnetospheric generator plasma. The same situation holds for cusp/mantle currents in pairs, NBZ currents in pairs (flowing into and away from the polar cap ionosphere), and intensity-balanced NBZ and region 1 currents in pairs and net region 1 currents in pairs. It would not be excellent physics if we try to connect region 2, region 1, cusp, mantle, NBZ current, etc., only by the ionospheric paths as we like, without thinking of the roots of magnetic stresses exerted on the ionosphere (i.e., $\mathbf{J} \times \mathbf{B}$ force associated Pedersen current) and energy consumption occurring in the ionosphere. Finally, I will summarize this article by representing large-scale characteristics of FACs by associated magnetic tangential stresses as given in Figure 9. These stresses manifest magnetic tangential stresses that are derived from $\mathbf{J} \times \mathbf{B}$ forces occurring in the generator plasma. Generation mechanisms of FACs are addressed absolutely to the configuration of plasma forces $\mathbf{F}$ (plasma pressure gradient force and inertia force) that balance $\mathbf{J} \times \mathbf{B}$ (= $\mathbf{M}$ + $\mathbf{T}$) forces and satisfy the requisite condition to generate FACs, for example, forces $\mathbf{F}$, $\mathbf{M}$, and $\mathbf{T}$ balance to make a triangle in the generator plasma. Furthermore, mechanisms of FACs should justly conform to the actual observation in the deep magnetosphere.

*Acknowledgments.* I am grateful to two conveners of this Chapman Conference, R. Fujii and S. Ohtani, for their efforts that produced the conference with great success. I am also indebted to them for their kind suggestions and patience for my submittal of this article.

## REFERENCES

Araki, T., et al., Polar cap vertical currents associated with northward interplanetary magnetic field, *Geophys. Res. Lett.*, 11, 23–26, 1984.

Armstrong, J. C., and A. J. Zmuda, Field aligned current at 1100 km in the auroral region measured by satellite, *J. Geophys. Res.*, 75, 7122–7127, 1970.

Armstrong, J.C., and A. J. Zmuda, Triaxial magnetic measurements of field-aligned currents at 800 kilometers in the auroral region: Initial results, *J. Geophys. Res.*, 78, 6802–6807, 1973.

Burke, W. J., et al., Polar cap electric field structures with a northward interplanetary magnetic field, *Geophy. Res. Lett.*, 6, 21–24, 1979.

Bythrow, P. F., et al., Observational evidence for a boundary layer source of dayside region 1 field-aligned currents, *J. Geophys. Res.*, 86, 5577–5589, 1981.

Bythrow, P. F., et al., Birkeland currents and charged particles in the high-latitude prenoon region: A new interpretation, *J. Geophys. Res.*, 93, 9791–9803, 1988.

Doyle, M. A., et al., Field-aligned currents and electric fields observed in the region of the dayside cusp, *J. Geophys. Res.*, 86, 5656–5664, 1981.

Elphic, R. C., et al., Observations of field-aligned currents at the plasma sheet boundary: A ISEE-1 and 2 survey, *Geophys. Res. Lett.*, 12, 631–634, 1985.

Erlandson, R. E., et al., IMF By dependence of region 1 Birkeland currents near noon, *J. Geophys. Res.*, 93, 9804–9814, 1988.

Fairfield, D. H., Magnetic field signatures of substorms on high-latitude field lines in the nighttime magnetosphere, *J. Geophys. Res.*, 78, 1553–1562, 1973.

Fujii, R. et al., Seasonal dependence of large-scale Birkeland currents, *Geophys. Res. Lett.*, 8, 1103–1106, 1981.

Fujii, R., and T. Iijima, Control of the ionospheric conductivities on large-scale Birkeland current intensities under geomagnetic quiet conditions, *J. Geophys. Res.*, 92, 4505–4513, 1987.

Fujii, R., et al., Electrodynamic parameters in the nighttime sector during auroral substorms, *J. Geophys. Res.*, 99, 6093–6112, 1994.

Fukunishi, H., et al., Latitudinal structures of nightside field-aligned currents and their relationships to the plasma sheet regions, *J. Geophys. Res.*, 98, 11,235–11,255, 1993.

Gary, J. B., et al., Field-aligned Poynting flux observations in the high-latitude ionosphere, *J. Geophys. Res.*, 99, 11,417–11,427, 1994.

Haerendel, G., Field-aligned currents in the Earth's magnetosphere, in *Physics of Magnetic Flux Rope, Geophys. Monogr. Ser.*, vol. 58, pp. 539–553, AGU, Washington, DC, 1990.

Harel, M., et al., Quantitative simulation of a magnetospheric substorm. I. Model logic and overview, *J. Geophys. Res.*, 86, 2217–2241, 1981.

Heelis, R. A., et al., Ionospheric convection signatures observed by DE 2 during northward interplanetary magnetic field, *J. Geophys. Res.*, 91, 5817–5830, 1986.

Heppner, J. P., and N. C. Maynard, Empirical high-latitude electric field models, *J. Geophys. Res.*, 92, 4467–4489, 1987.

Hoffman, R. A., et al., Electrodynamic patterns in the polar region during periods of extreme magnetic quiescence, *J. Geophys. Res.*, 93, 14,515–14,541, 1988.

Iijima, T., and T. A. Potemra, The amplitude distribution of field-aligned currents at northern high latitudes observed by Triad, *J. Geophys. Res.*, 81, 2165–2174, 1976a.

Iijima, T., and T. A. Potemra, Field-aligned currents in the dayside cusp observed by Triad, *J. Geophys. Res.*, 81, 5971–5979, 1976b.

Iijima, T., and T. A. Potemra, Large-scale characteristics of field-aligned currents associated with substorms, *J. Geophys. Res.*, 83, 599–615, 1978.

Iijima, T., et al., Field-aligned currents in the south polar cusp and their relationship to the interplanetary magnetic field, *J. Geophys. Res.*, 83, 5595–5603, 1978.

Iijima, T., et al., Transverse and parallel geomagnetic perturbations over the polar regions observed by Magsat, *Geophys. Res. Lett.*, 9, 369–372, 1982.

Iijima, T., and T. A. Potemra, The relationship between interplanetary quantities and Birkeland current densities, *Geophys. Res. Lett.*, 9, 442–445, 1982.

Iijima, T., Field-aligned currents during northward IMF, in

*Magnetospheric Currents, Geophys. Monogr. Ser.*, vol. 28, pp. 115–122, AGU, Washington, DC, 1984.

Iijima, T., et al., Large-scale Birkeland currents in the dayside polar region during strongly northward IMF: A new Birkeland current system, *J. Geophys. Res., 89*, 7441–7452, 1984.

Iijima, T., and T. Shibaji, Global characteristics of northward IMF-associated (NBZ) field-aligned currents, *J. Geophys. Res., 92*, 2408–2424, 1987.

Iijima, T., et al., Large-scale characteristics of magnetospheric equatorial currents, *J. Geophys. Res., 95*, 991–999, 1990.

Iijima, T., et al., Contribution of pressure gradients to the generation of dawnside region1 and region2 currents, *J. Geophys. Res., 102*, 27,069–27,081, 1997.

Kamide, Y., and G. Rostoker, The spatial relationship of field-aligned currents and auroral electrojets to the distribution of nightside auroras, *J. Geophys. Res., 82*, 5589–5608, 1977.

Kamide, Y., and S.-I. Akasofu, Global distribution of the Pedersen and Hall currents and the electric potential pattern during a moderately disturbed period, *J. Geophys. Res., 86*, 3665–3668, 1981.

Kelly, T. J., et. al., ISEE1 and 2 observations of Birkeland currents in the Earth's inner magnetosphere, *J. Geophys. Res., 91*, 6945–6958, 1986.

Lyons, L. R., et al., Ion precipitation from the magnetopause current sheet, *J. Geophys. Res., 92*, 6147–6151, 1987.

McDiarmid, I. B., et al., Comparison of magnetic field perturbations at high latitudes with charged particle and IMF measurements, *J. Geophys. Res., 83*, 681–688, 1978a.

McDiarmid, I. B., et al., Magnetic field perturbations in the dayside cleft and their relationship to the IMF, *J. Geophys. Res., 83*, 5753–5756, 1978b.

McDiarmid, I. B., et al., Comparison of magnetic field perturbations and solar electron profiles in the polar cap, *J. Geophys. Res., 85*, 1163–1170, 1980.

Nakagawa, M., Characteristic of field-aligned currents associated with northward IMF, Master's Thesis, University of Tokyo, March 1995.

Ohtani, S., et al., Field-aligned current signatures in the near-tail region. 1. ISEE observations in the plasma sheet boundary layer, *J. Geophys. Res., 93*, 9709–9720, 1988.

Ohtani, S., et al., Simultaneous prenoon and postnoon observations of three field-aligned current systems from Viking and DMSP-F7, *J. Geophys. Res., 100*, 119–136, 1995a.

Ohtani, S., et al., Four large-scale field-aligned currents systems in the dayside high-latitude region, *J. Geophys. Res., 100*, 137–153, 1995b.

Peymirat, C., and D. Fontaine, Numerical simulation of magnetospheric convection including the effect of field-aligned currents and electron precipitation, *J. Geophys. Res., 99*, 11,155–11,176, 1994.

Rich, F. J., and M. S. Gussenhoven, The absence of Region 1/Region 2 field-aligned currents during prolonged quiet times, *Geophys. Res. Lett., 14*, 689–692, 1987.

Rich, F. J., et al., Northward IMF and patterns of high-latitude precipitation and field-aligned currents: The February 1986 storm, *J. Geophys. Res., 85*, 7893–7913, 1990.

Robinson, R. M., et al., Chatanika radar and S3-2 measurements of auroral zone electrodynamics in the midnight sector, *J. Geophys. Res., 90*, 8487–8499, 1985.

Saflekos, N. A., and T. A. Potemra, The orientation of Birkeland current sheets in the dayside polar region and its relationship to the IMF, *J. Geophys. Res., 85*, 1987–1994, 1980.

Sato, T., and T. Iijima, Primary sources of large-scale Birkeland currents, *Space Sci. Rev., 24*, 347–366, 1979.

Senior, C., et al., Relationship between field-aligned currents, diffuse auroral precipitation and the westward electrojet in the early morning sector, *J. Geophys. Res., 87*, 10,469–10,477, 1982.

Siscoe, G., and N. Maynard, Distributed two-dimensional region 1 and region 2 currents: Model results and data comparisons, *J. Geophys. Res., 96*, 21,071–21,085, 1991.

Smiddy, M., et al., Effects of high-latitude conductivity on observed convection electric fields and Birkeland currents, *J. Geophys. Res., 85*, 6811–6818, 1980.

Sonnerup, B. U. O., Theory of the low-latitude boundary layer, *J. Geophys. Res., 85*, 2017–2026, 1980.

Southwood, D. J., The role of hot plasma in magnetospheric convection, *J. Geophys. Res., 82*, 5512–5520, 1977.

Southwood, D. J., and W. J. Hughes, Theory of hydromagnetic waves in the magnetosphere, *Space Sci. Rev., 35*, 301–366, 1983.

Southwood, D. J., and M. G. Kivelson, An approximate description of field-aligned currents in a planetary magnetic field, *J. Geophys. Res., 96*, 67–75, 1991.

Sugiura, M., Identifications of the polar cap boundary and the auroral belt in the high- altitude magnetosphere: A model for field-aligned currents, *J. Geophys. Res., 80*, 2057–2068, 1975.

Sugiura, M., and T. A. Potemra, Net field-aligned currents observed by Triad, *J. Geophys. Res., 81*, 2155–2164, 1976.

Taguchi, S., et al., Characterization of the IMF By-dependent field-aligned currents in the cleft region based on DE 2 observations, *J. Geophys. Res., 98*, 1393–1407, 1993.

Theile, B., and H. M. Praetorius, Field-aligned currents between 400 and 3000 km in auroral and polar latitudes, *Planet. Space Sci., 21*, 179–187, 1973.

Vasyliunas, V. M., Mathematical models of magnetospheric convection and its coupling to the ionosphere, in *Particles and Fields in the Magnetosphere*, pp. 60–71, D. Reidel, Norwell, MA, 1970.

Vasyliunas, V. M., The interrelationship of magnetospheric processes, in *Earth's Magnetospheric Processes*, pp. 29–38, D. Reidel, Norwell, MA, 1972.

Watanabe, M., et al., Synthesis models of dayside field-aligned currents for strong interplanetary magnetic field By, *J. Geophys. Res., 101*, 13,303–13, 319, 1996.

Watanabe, M., et al., Field-aligned current systems in the magnetospheric ground state, *J. Geophys. Res., 103*, 6853–6869, 1998.

Wilhjelm, J., et al., The relationship between ionospheric and field-aligned currents in the dayside cusp, *J. Geophys. Res., 83*, 5586–5594, 1978.

Woch, J., et al., The low-latitude boundary layer at mid-altitudes: Relation to large-scale Birkeland currents, *Geophys. Res. Lett., 20*, 2251–2254, 1993.

Yamauchi, M., et al., The interplanetary magnetic field By effects on large-scale field-aligned currents near local noon: Contribution from cusp part and non-cusp part, *J. Geophys. Res.*, 98, 5761–5767, 1993.

Zmuda, A. J., et al., Characteristics of transverse magnetic disturbances observed at 1100 km in the auroral oval, *J. Geophys. Res.*, 75, 4757–4762, 1970.

Zmuda, A. J., and J. C. Armstrong, The diurnal variation of the region with vector magnetic field changes associated with field-aligned currents, *J. Geophys. Res.*, 79, 2501–2502, 1974a.

Zmuda, A. J., and J. C. Armstrong, The diurnal flow pattern of field-aligned currents, *J. Geophys. Res.*, 79, 4611–4619, 1974b.

Takesi Iijima, Graduate School of Sciences, Kyushu University, Hakozaki 6-10-1, Higashiku, Fukuoka 812-8581, Japan

# Ionospheric Electrodynamics: A Tutorial

### A. D. Richmond

*High Altitude Observatory, National Center for Atmospheric Research, Boulder, Colorado*

### J. P. Thayer

*Geoscience and Engineering Center, SRI International, Menlo Park, California*

This paper gives a tutorial overview of ionospheric electrodynamics, including the observed behavior of ionospheric electric fields and currents, the physics of ionospheric electrical conductivity and Ohm's law, the operation of the ionospheric wind dynamo, and the transfer of energy between the magnetosphere and the ionosphere. The ionosphere forms an important part of the magnetospheric electrodynamic system. It is a region where ion-neutral collisions cause ions and electrons to move at different velocities across magnetic field lines, thereby violating the frozen-in flux condition and resulting in significant flow of ohmic current. Ionospheric conductivity is a function of the geomagnetic field, the plasma density, and the collision rate. Neutral winds cause generation of electric current through a dynamo effect. The winds result from diurnally varying solar heating, from upward-propagating global atmospheric waves, and from the Ampère force and Joule heating resulting from the electric current flow. Electromagnetic energy flow is normally directed from the magnetosphere into the ionosphere, as can be evaluated with the aid of Poynting's theorem, but strong thermospheric winds can sometimes reverse the direction of this energy flow.

## INTRODUCTION

The ionosphere is an electrically conducting medium, and carries a substantial portion of the electrical current flowing in the Earth's space environment. It forms a critical part of the global magnetospheric current system, providing a closure path for geomagnetic-field-aligned currents that extend to the outer magnetosphere. The magnetic stresses associated with these currents result in significant momentum transfer between the ionosphere and the magnetosphere. Important amounts of electromagnetic energy are also transferred between the magnetosphere and the ionosphere by the electric fields and currents, leading to dissipation of magnetospheric energy and to heating of the upper atmosphere. The ionosphere is a region where the frozen-in flux approximation of magnetohydrodynamics breaks down, owing to collisions between charged and neutral particles. It is also the seat of current generation produced by the dynamo effect of winds in the thermosphere, at altitudes above about 90 km. Those winds are produced not only by solar heating and by upward-propagating global atmospheric waves like tides, but they are also produced through the Ampère force and

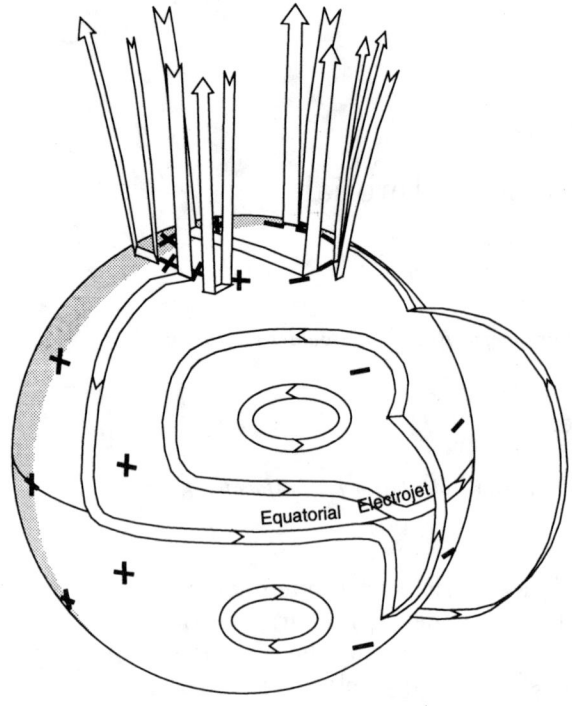

**Figure 1.** Schematic of global ionospheric electric currents (ribbons with arrows) and electric potentials (+ and −), viewed from the day side of the Earth.

the Joule heating exerted on the medium by the current itself. There is therefore an interesting mutual coupling between the currents and the winds. This paper presents a tutorial overview of the phenomenology and physical processes associated with ionospheric electric fields and currents.

## DESCRIPTION OF GLOBAL ELECTRIC FIELDS AND CURRENTS

Figure 1 sketches the global ionospheric currents and electric potential, with the currents illustrated by ribbons and the potential with + and −. (On global scales, the ionospheric electric field is essentially a potential field for phenomena that vary on time scales longer than a minute or so, which are the only phenomena considered in this paper.) The Earth's main magnetic field has a dominant influence on the ionospheric conductivity and on the flow of current between the magnetosphere and the ionosphere. Electric fields and currents are therefore strongly organized with respect to the geomagnetic field. For convenience, currents at high magnetic latitudes and those at middle and low latitudes are often considered separately, although in reality the currents and electric fields at all latitudes are coupled. Because the Sun is ultimately responsible not only for the ionospheric conductivity but also for the drivers of ionospheric currents, the patterns of electric potential and current tend to be organized in a coordinate system of magnetic local time (MLT) and magnetic latitude. As the Earth rotates, there is therefore a daily variation in the direction and strength of currents and electric fields over any given location on the ground.

At high latitudes the ionospheric currents are joined with currents flowing along geomagnetic field lines into the magnetosphere, and the electrodynamics is dominated by the influences of magnetospheric processes. The total current flow is on the order of $10^7$ A. At midlatitudes much of the ionospheric current is generated by the ionospheric wind dynamo, which on the average produces global current vortices on the dayside of the Earth, counterclockwise in the northern hemisphere and clockwise in the southern hemisphere. The total current flow in each vortex is on the order of $10^5$ A. The two hemispheres are electrically coupled by currents flowing along geomagnetic field lines whenever there is an imbalance in the dynamo forcing between the hemispheres. Near the magnetic equator there is a substantial intensification of the eastward current known as the equatorial electrojet, which is associated with the highly anisotropic conductivity of the ionosphere and the presence of a nearly horizontal geomagnetic field.

There are strong electric fields at high latitudes, on the order of several tens of millivolts per meter or more, associated with the magnetospherically produced currents. On the average, these electric fields are represented by an electric potential having a high on the morning side of the polar region and a low on the evening side, with a total potential drop that ranges from 20 kV to 200 kV. Figure 2 shows a specific example of the high-latitude potential pattern in the northern hemisphere. Both the pattern and the strength of the high-latitude potential pattern have been found to depend strongly on the direction and strength of the interplanetary magnetic field (IMF).

At middle and low latitudes electric fields are considerably smaller, typically a few millivolts per meter during magnetically quiet periods. Along the magnetic equator there is a potential high around dawn and a low around dusk, with a total potential drop on the order of 7 kV. At midlatitudes the MLT of the potential high and low tend to shift more toward evening and midday, respectively.

The quiet-day geomagnetic variations associated with the overhead ionospheric currents have traditionally

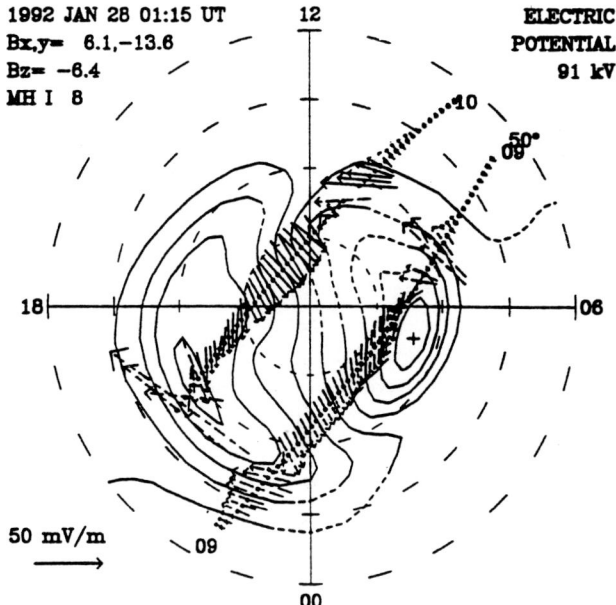

**Figure 2.** Electric potential in the northern hemispheric polar region estimated from ion velocities measured by two Defense Meteorological Satellite Program (DMSP) spacecraft, whose data are labeled 09 and 10, as well as from numerous ground magnetometers, on 1992 January 28, at 0115 UT. The coordinates are magnetic latitude, from 50° to the pole, and magnetic local time, with local noon at the top. The contour interval is 10 kV, with dashed lines indicate regions where the uncertainty in the large-scale electric field exceeds 50%. The potential high and low are marked by + and −, respectively, and combine for a total potential drop of 91 kV. The measured ion velocities are multiplied here by the magnetic field strength in order to give magnitudes in units of electric field. The electric-field directions are rotated 90° clockwise from the displayed vectors. From *Lu et al.* [1994].

been called $S_q$, $S$ for solar (as opposed to a much smaller lunar magnetic variation), and $q$ for quiet-day. At middle latitudes they have magnitudes on the order of 30 nT. The associated ionospheric currents are called the $S_q$ currents. The centers of the current vortices seen in Figure 1 lie around 30° north or south magnetic latitude. The $S_q$ magnetic variations at solar maximum are nearly twice as large as at solar minimum, due primarily to increased ionospheric conductivity but also to stronger upper atmospheric winds at solar maximum. The variations under the equatorial electrojet are more than a factor of two larger than those at other stations, and are larger at equinox than at either of the solstices. The magnitudes of the variations at midlatitudes are larger in the summer than in the winter. Both the $S_q$

currents and their associated electric fields exhibit considerable day-to-day variability, even on magnetically quiet days, that is believed to be caused by day-to-day variability in the thermospheric winds that drive the ionospheric dynamo.

During magnetic storms the global ionospheric electric fields and currents and their associated magnetic variations increase in magnitude, and exhibit rapid fluctuations. The disturbed magnetic perturbations are associated only partly with overhead ionospheric currents, since a substantial portion comes from more distant magnetospheric currents like the ring current and field-aligned currents. Figure 3 shows an example assembled by *Fejer* [1990] of a two-day period with some disturbances. Negative excursions of the $B_z$ component of the IMF correspond well with the periods of auroral disturbances in the AU/AL indices. The largest

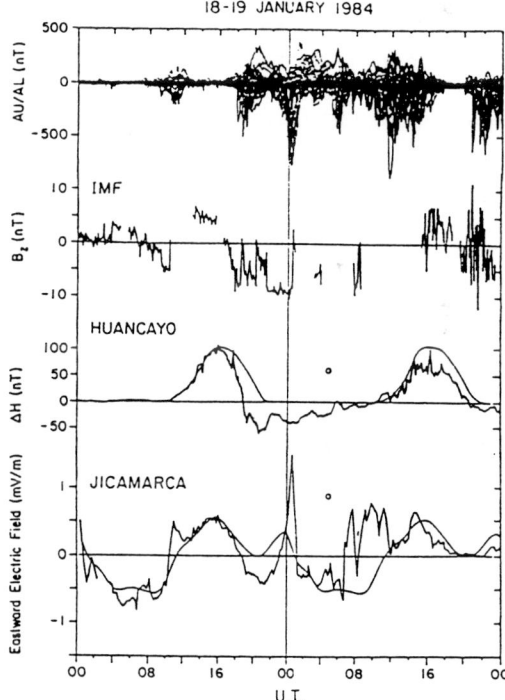

**Figure 3.** Interplanetary, auroral, and equatorial electrodynamic features on 1984 January 18-19. MLT is approximately UT − 5 hours. The AU/AL indices are obtained by superposing auroral-zone magnetograms. The IMF $B_z$ component was measured by the IMP-8 spacecraft in the solar wind. The magnetogram from Huancayo, under the equatorial electrojet, is referenced to the average quiet-day variation (the smooth line). The eastward electric field over the Jicamarca incoherent-scatter radar is also referenced to the average quiet-day variation (smooth curve). The small circles are at local midnight. From *Fejer et al.* [1990].

equatorial electric-field disturbances measured at Jicamarca occurred at night, between 23 UT on January 18 and 11 UT on January 19. Because the overhead ionospheric conductivity is very small at night, the nighttime electric-field disturbances do not show up on the Huancayo magnetogram, which senses rather only the effects of distant magnetospheric currents.

The ionospheric electric field $\mathbf{E}$ is essentially perpendicular to the geomagnetic field $\mathbf{B}$, and at middle and high latitudes it is approximately constant in altitude. The electric current density $\mathbf{J}$, however, varies strongly with altitude through the ionosphere. Plate 1 illustrates the height and time variations of the component of $\mathbf{J}$ perpendicular to $\mathbf{B}$, between 100 km and 130 km altitude, for a four-hour period on 1997 March 25, measured by the Sondrestrom radar (74.2° magnetic latitude). Details of how the electrodynamic parameters are derived from the monostatic radar are provided by *Thayer* [1998a]. The electric-field vector is also shown in the inset for one particular time, along with idealized current vectors calculated at three altitudes as outlined in the next section (the idealized current vectors are different from the measured values of $\mathbf{J}$ in that the former are computed neglecting the effects of winds and height variations of electron density). When $\mathbf{E}$ is strong, $\mathbf{J}$ tends to be approximately parallel to $\mathbf{E}$ at high altitudes, but the two vectors tend to become more nearly perpendicular at 100 km.

Also shown in Plate 1 is the Joule heating rate per unit volume. Joule heating can contribute significantly to the energy budget of the thermosphere at high latitudes, at times exceeding the heating produced by solar insolation [*Banks*, 1977]. The Joule heating rate is structured in altitude owing to the influence of height-varying winds and conductivities. It peaks around 120 km in this example, and is closely tied to the peak in current density. The rate of local thermospheric temperature change due to Joule heating is more directly related to the heating rate per unit mass, that is, to the heating rate per unit volume divided by the mass density. For example, at 125 km an estimate of the neutral temperature increase after 20 minutes of Joule heating (using a value of $1.0 \times 10^{-6}$ W m$^{-3}$) is less than 100 K. At higher altitudes, say above 200 km, the temperature increase over the same time period can be many hundreds of kelvins. This is due to the fact that although the volumetric Joule heating rate is decreasing with increasing height the neutral density is decreasing more rapidly, and so less energy is needed to significantly heat the more tenuous neutral gas.

## IONOSPHERIC ELECTRICAL CONDUCTIVITY AND OHM'S LAW

In order to understand the physics of ionospheric electrodynamics, it is essential to understand the nature of ionospheric conductivity. Three elements are critical for determining the conductivity: the plasma density, the geomagnetic field, and the rate at which the charged particles collide with neutral atmospheric molecules. Figure 4a shows typical midlatitude density profiles of atmospheric neutral molecules and of charged particles; for the latter, the densities of electrons and singly-charged positive ions are essentially the same. Note that the neutral density exceeds the plasma density by several orders of magnitude at all heights. The neutral constituents and their dynamics are obviously going to have a major impact on the plasma dynamics.

Let us examine the mean motion of the charged particles in the frame of reference of the neutral gas. The main forces acting on a charged particle are the Lorentz force and the frictional forces due to collisions with other particle species. For ions and electrons the force-balance conditions, averaged over the particle distribution functions, are respectively:

$$N_e e(\mathbf{E}' + \mathbf{v}_i \times \mathbf{B}) - N_e m_i \nu_{in} \mathbf{v}_i - N_e m_i \nu_{ie}(\mathbf{v}_i - \mathbf{v}_e) = 0 \quad (1)$$

$$-N_e e(\mathbf{E}' + \mathbf{v}_e \times \mathbf{B}) - N_e m_e \nu_{en} \mathbf{v}_e + N_e m_e \nu_{ei}(\mathbf{v}_i - \mathbf{v}_e) = 0 \quad (2)$$

where $N_e$ is the electron density, $e$ is the magnitude of the electron charge, $m_i$ and $\mathbf{v}_i$ are the ion mass and bulk velocity, $m_e$ and $\mathbf{v}_e$ are the electron mass and velocity, $\nu_{in}$, $\nu_{ie}$, $\nu_{en}$, and $\nu_{ei}$ are the ion-neutral, ion-electron, electron-neutral, and electron-ion collision frequencies, which are assumed to be independent of the bulk velocities, and $\mathbf{E}'$ is the electric field in the frame of reference of the neutral gas. In reality, the effective electron-neutral collision frequency is somewhat different for electron motions parallel or perpendicular to $\mathbf{B}$.

Analysis of (1) and (2) [e.g., *Richmond*, 1995b] shows that parallel to $\mathbf{B}$ the electron velocity dominates over the ion velocity and is given to a good approximation by

$$v_{e\parallel} = -\frac{eE'_\parallel}{m_e(\nu_{en\parallel} + \nu_{ei\parallel})} \quad (3)$$

while perpendicular to $\mathbf{B}$ the ion and electron velocities are given to a good approximation by

$$\mathbf{v}_{i\perp} = \frac{\nu_{in}\Omega_i \mathbf{E}'_\perp - \Omega_i^2 \mathbf{b} \times \mathbf{E}'_\perp}{B(\nu_{in}^2 + \Omega_i^2)} \quad (4)$$

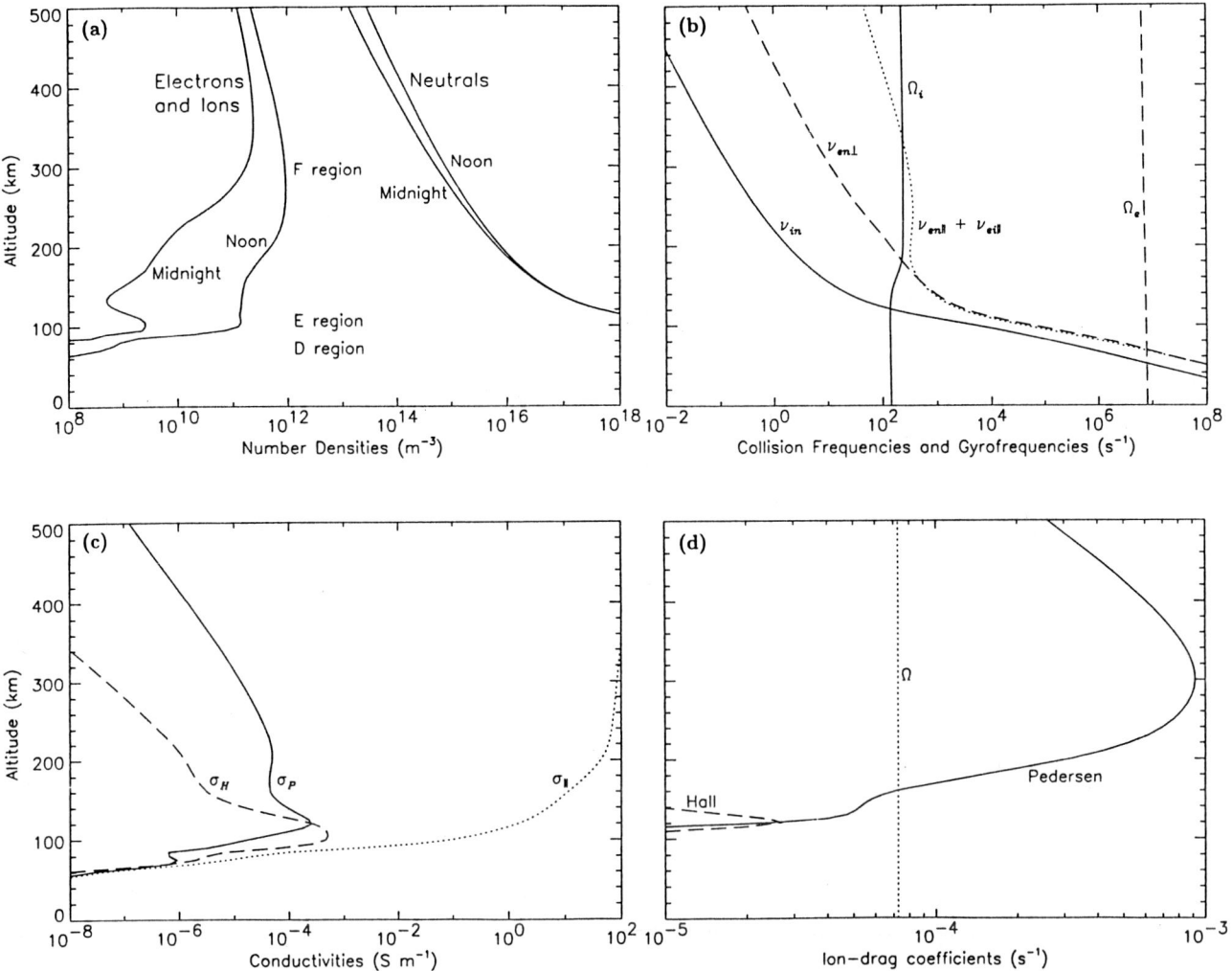

**Figure 4.** Typical ionospheric parameters at 44.6°N, 2.2°E on March 21 for medium solar activity (10.7 cm radio flux of $120 \times 10^{-22}$ W cm$^{-2}$ Hz$^{-1}$; sunspot number 67) and low magnetic activity ($A_p = 4$), obtained from the 1990 International Reference Ionosphere [*Bilitza*, 1990] and the MSISE-90 neutral-density model [*Hedin*, 1991]. (a) Number densities of electrons, ions, and neutrals at noon and midnight. The altitudes of the ionospheric regions conventionally called $D, E,$ and $F$ are also indicated. (b) Collision frequencies $\nu_{in}, \nu_{en\perp}$ (for motion perpendicular to **B**), and $\nu_{en\parallel} + \nu_{ei\parallel}$ (for motion along **B**), and gyrofrequencies $\Omega_i$ and $\Omega_e$. Collision frequencies are from *Richmond* [1995b]. (c) Noontime parallel ($\sigma_\parallel$), Pedersen ($\sigma_P$), and Hall ($\sigma_H$) conductivities. (d) Ion drag coefficients $\sigma_P B^2/\rho$ (Pedersen) and $\sigma_H B^2/\rho$ (Hall), and the angular rotation rate of the Earth, $\Omega$.

$$\mathbf{v}_{e\perp} = \frac{-\nu_{en\perp}\Omega_e \mathbf{E}'_\perp - \Omega_e^2 \mathbf{b} \times \mathbf{E}'_\perp}{B(\nu_{en\perp}^2 + \Omega_e^2)} \quad (5)$$

where **b** is a unit vector in the direction of **B**, and where

$$\Omega_i = eB/m_i \quad (6)$$

$$\Omega_e = eB/m_e \quad (7)$$

are the angular gyrofrequencies of the ions and electrons, describing their gyration in the geomagnetic field.

Figure 4b shows typical noontime midlatitude profiles of the collision and gyro-frequencies for electrons and positive ions. The collision frequency of electrons with neutrals determines the mobility of electrons along **B** below about 200 km, while above that height collisions with ions become more important. In any case,

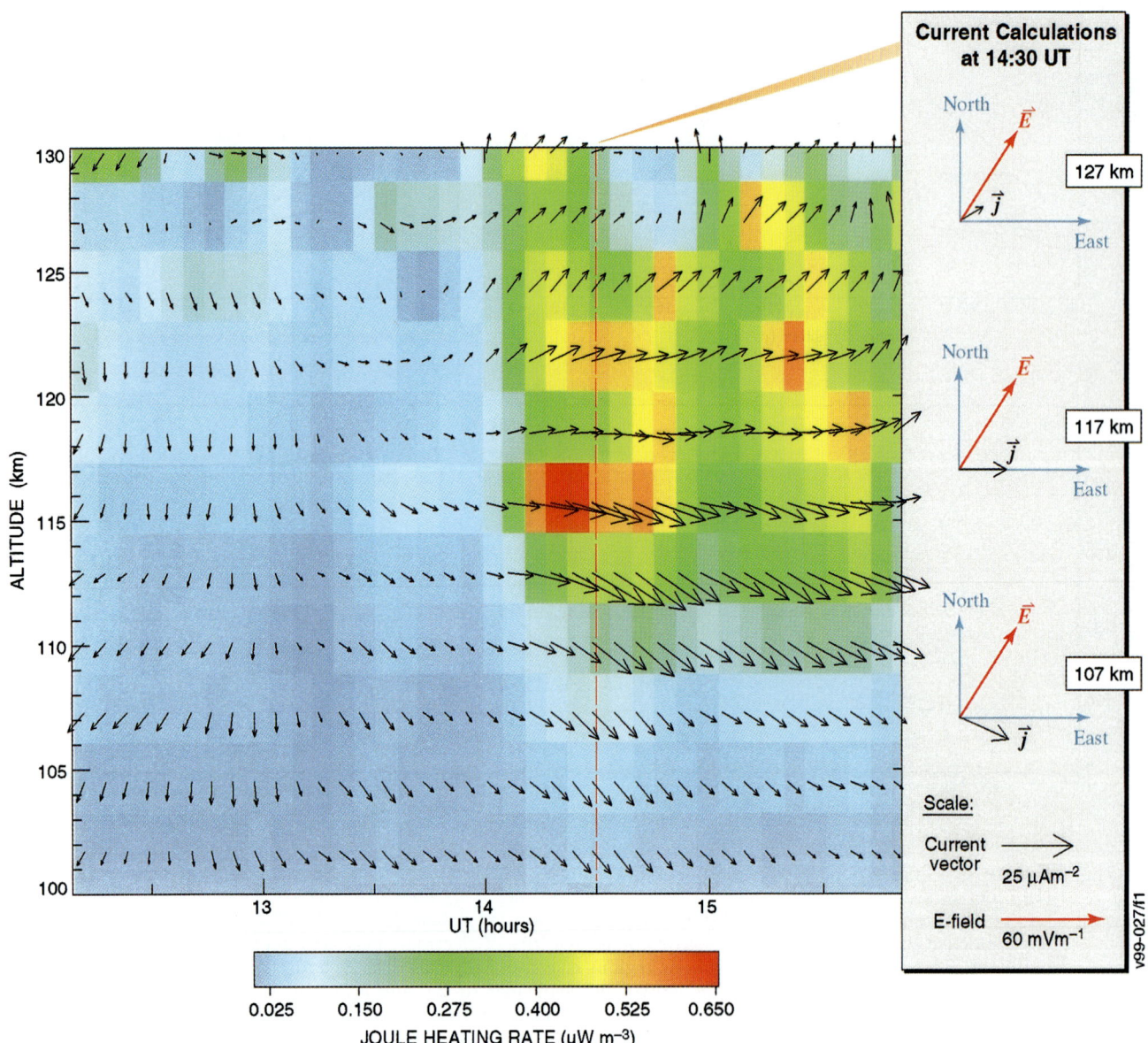

**Plate 1.** An example of the variations of electrodynamic parameters with height and time in the high-latitude $E$ region, measured by the Sondrestrom incoherent-scatter radar at a resolution of 3 km in height and 5 minutes in time. The vectors are the measured horizontal current density, in $\mu A/m^2$. Northward is up and eastward is toward the right. The scalar image behind the current vectors is the volumetric Joule heating rate in $\mu W/m^3$. The inset shows numerical calculations of the horizontal current density **J** at three different altitudes, corresponding to the radar electric-field measurement **E** at 1430 UT, accounting for changes in the ion-neutral collision frequency and the ion gyrofrequency with height, while neglecting neutral winds and using a constant electron density at all altitudes.

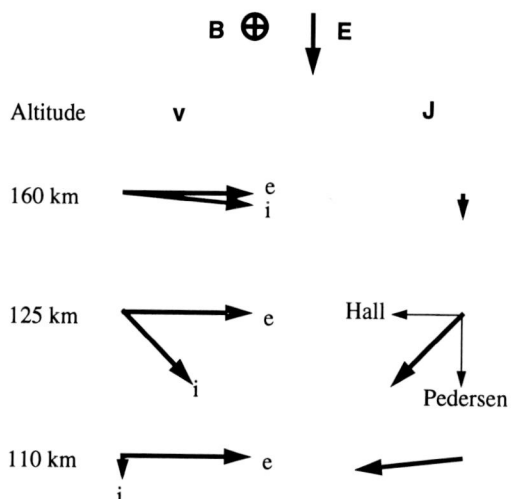

**Figure 5.** Schematic of height variations of electron (e) and ion (i) velocities, **v**, and of the electric current **J**. The magnetic field **B** is into the page, and the electric field **E**' is downward. The Pedersen component of the current is downward, and the Hall component is toward the left.

the electron mobility parallel to **B** is sufficiently large to produce a very large electrical conductivity in that direction. This large conductivity tends largely to short out any parallel electric field in the ionosphere, i.e.,

$$E'_\parallel \approx 0, \qquad (8)$$

for phenomena with scale sizes perpendicular to **B** of about 1 km or more. In the plane perpendicular to **B**, the situation is very different. The geomagnetic field has a very strong influence on the charged particle motion perpendicular to the magnetic field, and therefore on the perpendicular conductivity at all altitudes above 70 km. It tends to constrain charged particles to spiral around field lines. Notice that at high altitudes, where $\nu_{in} \ll \Omega_i$ and $\nu_{en\perp} \ll \Omega_e$, (4) and (5) reduce to

$$\mathbf{v}_{i\perp} = \mathbf{v}_{e\perp} = \frac{\mathbf{E}' \times \mathbf{B}}{B^2} \qquad (9)$$

which represents the **E** × **B** (E-cross-B) drift velocity of charged particles in crossed electric and magnetic fields. Thus at high altitudes in the ionosphere the ions and electrons essentially move together in the direction perpendicular to the magnetic field. Together with the approximation (8), this leads to the condition of "frozen-in magnetic flux" at these high altitudes, whereby all charged particles on a common magnetic field line at one time remain on a common field line at all future times. However, collisions of charged particles with neutrals break the condition of frozen-in magnetic flux, so that the charged particles along a given field line no longer move to neighboring field lines in unison. What is important in determining the degree to which charged particles are tied to magnetic field lines is the ratio of the collision frequency with neutrals to the angular gyrofrequency in the magnetic field. The electron gyrofrequency is nearly $10^7$ rad/s, while the ion gyrofrequency is only a little over 100 rad/s, with some height dependence due to the varying mean molecular mass of the ions. The electron collision frequency equals the gyrofrequency around 70 km, well below heights where there is enough plasma density to carry significant current. As a consequence, the electrons are essentially tied to the magnetic field throughout the entire conducting ionosphere. For ions, collisions are relatively much more important. Even though the ion collision frequency with neutrals is about two orders of magnitude smaller than the electron-neutral collision frequency, it does not decrease to the level of the ion gyrofrequency until an altitude of about 125 km. It is only above about 150 km that the ions become strongly tied to the magnetic field, and that the frozen-in flux condition is approximately valid. Below 110 km the motion of the ions is strongly coupled with that of the neutral air through collisions. The intermediate altitude range, 110 - 150 km, is where the ions gain the ability to move at a velocity substantially different from either the **E** × **B** velocity or the velocity of the neutrals.

Figure 5 illustrates the variations of ion motion and electric current with altitude, in the neutral frame of reference. At all three altitudes shown, the electrons essentially move at the **E** × **B** velocity, toward the right. At 160 km the ions move nearly in that direction, but have a small component of velocity in the direction of **E**'. At 125 km the ion-velocity component parallel to **E**' becomes approximately equal to the component in the **E** × **B** direction. At 110 km the ions are nearly immobilized, but do still have a small velocity component nearly parallel to **E**'.

The electric current density **J** is given by

$$\mathbf{J} = N_e e (\mathbf{v}_i - \mathbf{v}_e). \qquad (10)$$

By convention, the component of **J** in the direction of **E**', in the plane perpendicular to **B**, is called "Pedersen" current, while the component perpendicular to both **E**' and **B** is called "Hall" current. At 150 km the velocity difference between ions and electrons is small but lies approximately in the direction of $\mathbf{E}'$, and so the resulting electric current is mainly Pedersen. At 110 km, on the other hand, the current is carried mainly by negatively charged electrons moving at the $\mathbf{E}' \times \mathbf{B}$ velocity, and the current is mainly Hall, flowing opposite to the electron velocity. Around 125 km the Pedersen and Hall current components are comparable.

By combining (3)-(5) with (10), we obtain an expression for Ohm's law:

$$\mathbf{J} = \sigma_P \mathbf{E}'_\perp + \sigma_H \mathbf{b} \times \mathbf{E}'_\perp + \sigma_\| E'_\| \mathbf{b} \qquad (11)$$

$$\sigma_\| = \frac{N_e e^2}{m_e (\nu_{en\|} + \nu_{ei\|})} \qquad (12)$$

$$\sigma_P = \frac{N_e e}{B}\left(\frac{\nu_{in}\Omega_i}{\nu_{in}^2 + \Omega_i^2} + \frac{\nu_{en\perp}\Omega_e}{\nu_{en\perp}^2 + \Omega_e^2}\right) \qquad (13)$$

$$\sigma_H = \frac{N_e e}{B}\left(\frac{\Omega_e^2}{\nu_{en\perp}^2 + \Omega_e^2} - \frac{\Omega_i^2}{\nu_{in}^2 + \Omega_i^2}\right) \qquad (14)$$

where $\sigma_\|$, $\sigma_P$, and $\sigma_H$ are respectively the parallel, Pedersen, and Hall conductivities.

Ohm's law seems to work very well in the ionosphere for time scales considerably longer than the inverse collision and gyro-frequencies, that is, longer than a minute or so, although the assumed linearity between $\mathbf{J}$ and $\mathbf{E}'$ may fail if the electric field becomes so large that it affects the values of the collision frequencies. Ohm's law does not say anything about cause and effect; that is, it does not say that the electric field is the source of the current or that the current is the source of the electric field. It merely states that the electric field and current are linearly related. If one exists then the other must also exist. Any mechanism that drives current through the medium must be accompanied by an electric field, and any mechanism that creates an electric field in the medium must be accompanied by current flow.

By appropriately defining a conductivity tensor $\tilde{\sigma}$, dimensioned $3 \times 3$, we can write Ohm's law in a more compact form:

$$\mathbf{J} = \tilde{\sigma} \mathbf{E}' \qquad (15)$$

The components of the conductivity tensor are shown in Figure 4c for typical noontime midlatitude conditions. The principal anisotropy of the conductivity is the very large difference between the conductivity along $\mathbf{B}$ and the conductivity perpendicular to $\mathbf{B}$ at all heights above 80 km. Although one might be tempted to conclude that this large difference would result in currents along $\mathbf{B}$ that were much larger than those perpendicular to $\mathbf{B}$, that is not the case in the lower ionosphere. The reason is that parallel currents cannot continue flowing into the poorly conducting lower atmosphere, and must find a continuation path that traverses magnetic field lines. The parallel and perpendicular current densities are therefore linked together, and the parallel current density is severely limited in its magnitude. As a consequence, the parallel electric field must be very small, as represented by (8). The ratio of parallel to perpendicular electric field strengths is roughly of the order of the ratio of perpendicular to parallel conductivity, typically $10^{-5}$. Another important feature of the anisotropy of the conductivity is the changing ratio of Hall to Pedersen conductivity with height: $\sigma_H$ is larger below about 125 km, while $\sigma_P$ is larger above that height. Around 100 km $\sigma_H$ is about 30 times larger than $\sigma_P$. The conductivities have a great deal of variability as the ionospheric plasma density changes, and, to a lesser extent, as the neutral density of the upper atmosphere changes. There is a large day-night difference, and also an important change with the solar cycle. There is great variability in the auroral zone, due to the irregular nature of auroral ionization by precipitating energetic particles.

At magnetic high latitudes, where geomagnetic field lines are approximately vertical, the electric field is approximately horizontal, and an electric field mapped from the magnetosphere is approximately constant with height over the few-hundred-kilometer thickness of the ionosphere. Under these conditions we can often treat the ionosphere as a thin conducting shell, with shell conductances given by the height integrals of the Pedersen and Hall conductivities. Figure 6 shows an example of the Hall conductance over the northern polar region, estimated by combining a variety of data for the auroral conductivity component, plus a model of the conductance produced by solar extreme ultraviolet radiation on the dayside of the Earth. The magnitudes of the solar and auroral contributions are roughly comparable, although the auroral component is highly variable. The Pedersen conductance tends to have a magnitude comparable with that of the Hall conductance. The ratio of Hall to Pedersen conductance in the auroral region increases with the mean energy of the ionizing auroral particles [e.g., *Robinson et al.*, 1987].

## IONOSPHERIC WIND DYNAMO

It is important to remember that Ohm's law applies to the reference frame of the material medium, in this case essentially the neutral gas that provides the ion collisions. The electric field $\mathbf{E}'$ has to be that measured in this frame. If there is a wind of velocity $\mathbf{U}$, then $\mathbf{E}'$ is

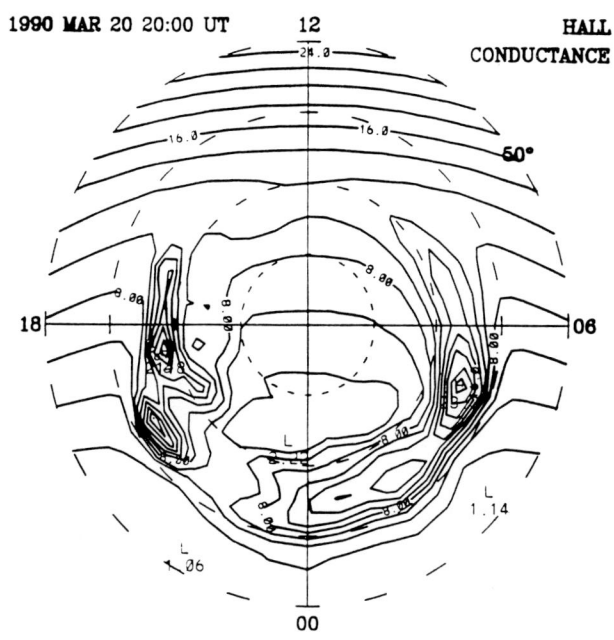

**Figure 6.** Example of Hall conductance, in units of siemens, poleward of 50° magnetic latitude on 1990 March 20 at 2000 UT. Local noon is at the top. The contour interval is 2 S.

related to the electric field in the frame of reference of the Earth, **E**, by the (nonrelativistic) transformation:

$$\mathbf{E}' = \mathbf{E} + \mathbf{U} \times \mathbf{B}. \quad (16)$$

On time scales longer than a minute or so **E** is electrostatic:

$$\mathbf{E} = -\nabla \Phi \quad (17)$$

Unlike **E**, the so-called "dynamo electric field" $\mathbf{U} \times \mathbf{B}$ is not constrained to be either a potential field or constant along **B**. When it is not, which is normally the case, then it cannot be canceled by **E**, so that **E**' must be non-zero and current must flow. This is the essence of the ionospheric wind dynamo effect: the motion of the conducting medium through the geomagnetic field by winds usually leads to current generation. On time scales longer than a fraction of a second, the current must be divergence-free:

$$\nabla \cdot \mathbf{J} = 0 \quad (18)$$

Combining (15)-(18) results in a partial differential equation for $\Phi$:

$$\nabla \cdot [\tilde{\sigma} \nabla \Phi] = \nabla \cdot [\tilde{\sigma} \mathbf{U} \times \mathbf{B}] \quad (19)$$

With suitable boundary conditions, (19) can be solved for $\Phi$, and the distributions of electric fields and currents can be determined everywhere in the ionosphere. In practice, (19) can be simplified considerably by taking advantage of the fact that geomagnetic field lines are essentially equipotentials. Since $\Phi$ varies only in the two spatial dimensions transverse to **B**, (19) can be reduced to a partial differential equation in only two dimensions by integrating it along field lines, all the way from one hemisphere to the other for closed field lines. When that is done, it is the field-line integrals of the transverse components of $\tilde{\sigma}$ and of $\tilde{\sigma}\mathbf{U} \times \mathbf{B}$ that become important, rather than the height-varying values of of $\tilde{\sigma}$ and $\tilde{\sigma}\mathbf{U} \times \mathbf{B}$ themselves. Those altitudes that give the dominant contribution to the field-line integrals, mainly 90-200 km at day, define what is called the "dynamo region."

The appropriate boundary condition for (19) at the base of the ionosphere is that the vertical current density goes to zero, since estimates of current flow between the lower atmosphere and the ionosphere indicate that it is negligible on a global scale, if perhaps not always so locally above thunderstorms. The upper boundary condition is much more complicated, as it essentially requires knowledge of how the magnetosphere behaves and how it reacts to changes in the ionospheric conditions. Typically, modelers of the ionospheric dynamo either ignore the magnetosphere altogether, or else treat it as a simple voltage generator or current generator. However, it is possible to take into account the electrodynamic interaction between the ionosphere and the inner part of the magnetosphere in a relatively simple way by using the concept of "shielding" [e.g., *Southwood*, 1977]. As magnetospheric electric fields convect plasma toward or away from the Earth, gradient-curvature drifts of the modified distributions of hot particles generate electric currents that produce field-aligned currents into and out of the ionosphere at high latitudes, the so-called "region-2" field-aligned currents. These alter the electric potential in the ionosphere and, because of the tight electrical connection with the magnetosphere, the magnetospheric electric field is also altered. The alteration is such that further inward or outward motion of the plasma is strongly diminished. This corresponds to a weakening of the east-west electric field, so that the low-latitude boundary of the region-2 current tends to become more nearly equipotential, and the penetration of electric field to middle and low latitudes is suppressed. The magnetospheric plasma takes several tens of minutes to redistribute after any change in the potential at the high-latitude boundary,

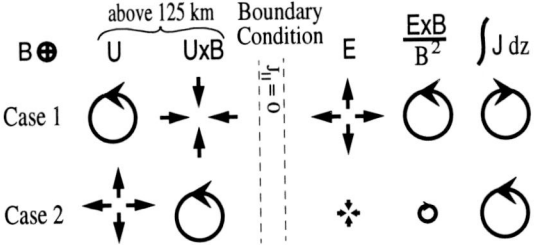

**Figure 7.** Two cases of dynamo action driven by winds above 125 km. The magnetic field **B** is into the page. In case 1 the wind velocity **U** is a counterclockwise vortex, while in case 2 the horizontal wind is divergent. The assumed boundary conditions are that the vertical current density $J_\parallel$ at the top and bottom of the ionosphere are zero, and that the electric potential vanishes at the horizontal boundaries. The polarization electric field is **E**, and the height-integrated horizontal current density is $\int \mathbf{J} dz$.

so shielding is not effective for rapidly varying fields. In the steady state, the electric fields in the auroral zone do not penetrate much into midlatitudes, and an approximate boundary condition that is sometimes used in dynamo modeling is to set the potential around the equatorward edge of the shielding region to zero. A slightly more sophisticated way to treat the shielding in a steady state was shown by *Vasyliunas* [1972] to be simply to replace the hot magnetospheric plasma by an effective Hall conductor, with a conductance many times larger than that of the ionosphere at the foot of the respective fields lines.

From inspection of (19), one might expect that in regions of the ionosphere where magnetospheric influences are relatively weak, the solution for $\Phi$ will tend to yield an electric field $-\nabla\Phi$ that is of the same order of magnitude as $\mathbf{U} \times \mathbf{B}$. Another way of saying this is that the resultant $\mathbf{E} \times \mathbf{B}$ drift velocities might be expected to be on the order of magnitude of **U**. In fact, that expectation is not too far from reality, with a few caveats. First, the high conductivity along geomagnetic field lines tends to average the dynamo effects of winds along field lines, so that winds with vertically oscillating structure tend to be ineffective in generating electric fields, and winds in regions of relatively low conductivity along the field line are also ineffective, like the daytime $F$ region or the nighttime valley between the $E$ and $F$ regions. Second, the peculiar conditions in the lower ionosphere near the magnetic equator, where the ratio of field-line-integrated Hall to Pedersen conductivities is very large, give rise to a significant Cowling effect and a strong vertical polarization electric field that is much larger than typical magnitudes of $\mathbf{U} \times \mathbf{B}$. This vertical polarization field drives the strong horizontal Hall current of the equatorial electrojet. The peak current density in the equatorial electrojet is on the order of $10^{-5}$ A/m$^2$, comparable to that in the auroral electrojets. However, the equatorial electrojet current is confined to a relatively small altitude region, so that its height-integrated value tends to be less than that of typical auroral electrojets.

Figure 7 illustrates the way that electric fields and currents respond to two different idealized forms of the wind, for a laterally bounded region with zero electric potential around the boundary, and for upper and lower boundary conditions requiring zero vertical current. The geometry is plane, with a vertically downward magnetic field. For both cases the wind exists mainly above 125 km, where the Pedersen conductivity dominates. However, the electric field that is generated extends down to lower heights, where it can drive Hall current. The first case is that of a counterclockwise wind vortex. The dynamo electric field $\mathbf{U} \times \mathbf{B}$ drives Pedersen current toward the center of the vortex, which must be offset by an equal amount of outward Pedersen current driven by an outward polarization electric field which is immediately established. Although the Pedersen current is effectively canceled, the electric field causes electrons to circulate counterclockwise at all heights, which in the lower part of the dynamo region (where ion motion is impeded by collisions) gives rise to a clockwise Hall current. The resultant net height-integrated current is therefore a clockwise vortex. (Note that if the wind vortex had existed only at lower altitudes, in the Hall conductivity region, instead of at the higher altitudes, the winds would convect the ions in a counterclockwise vortex of Hall current, and the height-integrated current would be reversed.)

The second case is that of a divergent wind, which drives a non-divergent counterclockwise Pedersen current. In reality, since the Hall conductivity does not entirely vanish above 125 km, a small outward-directed Hall current will also exist, which must be offset by a small inward-directed polarization electric field and Pedersen current. However, the dominant current is just the counterclockwise wind-driven Pedersen current. In this case the generation of the electric field is weak: it is theoretically possible to have current flow without any significant electric field (in the Earth frame of reference).

The winds that drive the dynamo effects have a variety of sources. Figure 8 shows a model simulation of winds in the upper atmosphere driven by a combination of solar heating and upward-propagating atmospheric

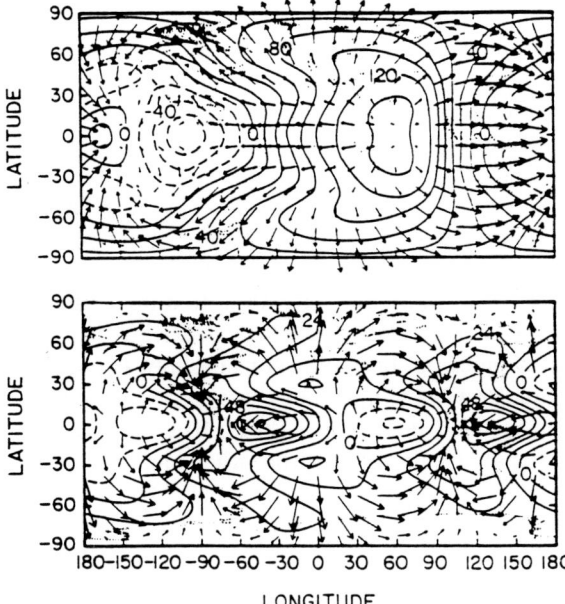

**Figure 8.** Temperatures (contours) and winds (arrows) at 12 UT for equinox, solar-minimum conditions, at atmospheric pressure levels of 6.8 $\mu$Pa (approximately 300 km, top) and 2.7 mPa (approximately 125 km, bottom). Temperatures are expressed as departures from the global mean. Contour intervals are 20 K (top) and 12 K (bottom). The maximum wind arrows are 166 m/s (top) and 71 m/s (bottom). (Adapted from *Fesen et al.* [1986].)

tides from the lower atmosphere. The daily heating of the atmosphere above 100 km causes dayside expansion and a pressure bulge that drives winds toward the night side. These winds generally increase in strength with height through the dynamo region, so they are largest where the Pedersen conductivity dominates over the Hall conductivity. In Figure 8 the character of this wind component is clearly manifested in the upper thermosphere, at 300 km, but it is also present at lower altitudes. On the day side, the midlatitude poleward wind in the dynamo region drives westward current. A westeast polarization electric field develops in the low latitude dayside ionosphere that helps to close this midlatitude westward current and form the global $S_q$ vortices. The current driven by this wind component is supplemented by current driven by semidiurnal tides (that is, global waves with 12-hour period) that are generated at lower atmospheric levels and propagate upward into the thermosphere. These waves have oscillatory characteristics in all three spatial dimensions as well as in time, and dominate the simulated winds at 125 km shown in Figure 8. Their dynamo effects are more complicated

to characterize than those of the wind driven by thermospheric solar heating, but they provide a significant contribution to the $S_q$ current. Observations of tides in the lower thermosphere have revealed a large amount of variability, not only on a seasonal basis, but also on a day-to-day basis. This wind variability can be expected to contribute to the day-to-day variability observed in the $S_q$ magnetic variations.

Another driver of thermospheric winds, especially important at high latitudes, is the electric current. There are two ways that the current affects the winds: through the Ampère force (or $\mathbf{J} \times \mathbf{B}$ force) and through Joule heating. The Ampère force represents the sum of Lorentz forces on the ions and electrons, forces that are transferred to the neutrals through collisions. It therefore balances the collisional force exerted by the neutrals on the ions and electrons as represented by the second terms in (1) and (2). The force per unit mass on the neutrals, or Ampère acceleration, is

$$\frac{1}{\rho}(N_e m_i \nu_{in} \mathbf{v}_i + N_e m_e \nu_{en} \mathbf{v}_e),$$

where $\rho$ is air mass density and where, as in (1) and (2), the velocities are those with respect to the frame of reference of the neutral air. This acceleration is mainly important above 100 km, where the ion term dominates. For this reason it is often called "ion drag." Use of (4), (5), (13), (14), (11), and (16), with some manipulation, shows that this acceleration can also be written as

$$\begin{aligned}\frac{\mathbf{J} \times \mathbf{B}}{\rho} &= \frac{\sigma_P B^2}{\rho}\left(\frac{\mathbf{E} \times \mathbf{B}}{B^2} - \mathbf{U}_\perp\right) \\ &+ \frac{\sigma_H B^2}{\rho}\mathbf{b} \times \left(\frac{\mathbf{E} \times \mathbf{B}}{B^2} - \mathbf{U}_\perp\right)\end{aligned} \quad (20)$$

where $\mathbf{U}_\perp$ is the component of $\mathbf{U}$ perpendicular to $\mathbf{B}$. The first term on the right-hand side simultaneously accelerates the wind towards the $\mathbf{E} \times \mathbf{B}$ drift velocity and acts as a frictional drag; or viewed in the reference frame of the neutral air, it accelerates the wind towards the $\mathbf{E}' \times \mathbf{B}$ drift velocity. It is effective on a time scale determined by the inverse of the coefficient $\sigma_P B^2/\rho$. The second term on the right-hand side of (20) simultaneously accelerates the wind in the direction of $\mathbf{E}$ and causes an acceleration perpendicular to the wind velocity that generally opposes the Coriolis acceleration associated with the Earth's rotation.

Figure 4d shows typical midlatitude profiles of the coefficients $\sigma_P B^2/\rho$ and $\sigma_H B^2/\rho$. For reference, the angular rotation rate of the Earth is also shown, which is representative of the Coriolis acceleration or of the time-rate-of change of a diurnally varying wind. The Pedersen coefficient $\sigma_P B^2/\rho$ is much more important than the Hall coefficient $\sigma_H B^2/\rho$. Above 130 km it varies with height as the electron density. These daytime midlatitude profiles are also roughly representative of values in the nighttime auroral zone, though of course the actual values vary with the electron density. A value of $10^{-4}$ s$^{-1}$ represents about a 3-hour acceleration time for the wind.

At high latitudes, where the magnitude of $\mathbf{E} \times \mathbf{B}/B^2$ typically exceeds that of $\mathbf{U}_\perp$, the Ampère force above 125 km tends to accelerate the neutral wind towards the $\mathbf{E} \times \mathbf{B}/B^2$ velocity. Figure 9 shows an example of the influence of ion convection on the wind, taken from a numerical simulation. Figure 9a is the electric potential over the northern high latitudes, showing the assumed pattern of two-cell ion convection. Figure 9b is the steady-state wind response at 145 km altitude, with a vector scale 4 times as sensitive as that on the left. There is a large-scale day-to-night flow, in addition to an imprint of the two-cell ion-convection pattern. The convection-driven component of the wind velocity is considerably smaller than the $\mathbf{E} \times \mathbf{B}/B^2$ velocity, and the pattern is turned counter-clockwise about 2 hours. This rotation is in the direction of the Earth's rotation and corresponds roughly to the time scale for ion drag to be effective. To a large extent, that rotation also prevents the wind from having a chance to accelerate up to the ion velocity, because the direction and strength of the ion convection is continually changing.

What would now happen if we cut off the current flow between the ionosphere and the magnetosphere? These winds would continue to have a dynamo effect. With the model we can do such experiments. In Figure 9c, the neutral wind pattern of Figure 9b has been retained, while all current flow with the magnetosphere has been canceled. As in Case 1 of Figure 7, the wind vortices generate an electric potential that tends to cause the ions to drift with the neutrals, or more specifically, with the height-averaged neutral velocity, weighted by the Pedersen conductivity. To zeroth order, the potential has the appearance of the potential that had been imposed from the magnetosphere earlier, although the electric potential is about 25% as large, 7.5 kV vs. 30 kV, and again the pattern is rotated in the sense of Earth's rotation. This effect has been called the "flywheel" effect [Banks, 1972], since the spinning mass of the neutral atmospheric vortices tends to maintain the ion convection that originally spun it up.

The Joule heating has a different form of influence on thermospheric dynamics than does the Ampère force. Joule heating causes the air to rise and flow equatorward above 125 km. At middle and low latitudes the air subsides, and there is a return poleward circulation below 125 km to preserve mass continuity, though the air velocity there is much smaller, because the air is much more dense. The Coriolis force acting on the equatorward wind causes westward zonal winds to develop encircling the poles and extending to midlatitudes. For a storm lasting many hours, the westward winds continue to build up, becoming considerably stronger than the equatorward winds, and these westward winds can continue many hours after the heating subsides. The meridional circulation, on the other hand, shuts down fairly quickly after the heating stops. These westward winds can have an important influence on the ionospheric wind dynamo, an effect that has been called the "disturbance dynamo" [Blanc and Richmond, 1980].

## ELECTROMAGNETIC ENERGY TRANSFER BETWEEN THE MAGNETOSPHERE AND THE IONOSPHERE

In examining the role of the ionosphere-thermosphere system in the energy exchange with the magnetosphere at high latitudes, we need to recognize that the net current and electric field are the result of all processes occurring along the field line, including both ionospheric and magnetospheric influences. Separating these two contributions in terms of currents and electric field is not feasible - measuring the winds, conductivities, and electric fields simultaneously over a range of spatial scales to evaluate the ionospheric influences at high latitudes is extremely challenging. This has led to the concept of applying Poynting's theorem to the high-latitude ionosphere. The theorem, derived from Maxwell's equations, reads

$$\frac{\partial}{\partial t}\left(\frac{B^2 + E^2/c^2}{2\mu_0}\right) + \nabla \cdot \left(\frac{\mathbf{E} \times \mathbf{B}}{\mu_0}\right) + \mathbf{J} \cdot \mathbf{E} = 0 \quad (21)$$

where $\mu_0$ is the permeability of free space and $c$ is the speed of light. A clear treatment of how Poynting's theorem can be applied to describing the electromagnetic energy flow within the magnetosphere has been given by Hill [1983]. Cowley [1991] presented a general view of electromagnetic energy exchange between the magnetosphere and ionosphere, applying a source-sink concept.

Poynting's theorem applies to all types of electromagnetic interactions, ranging from electromagnetic waves

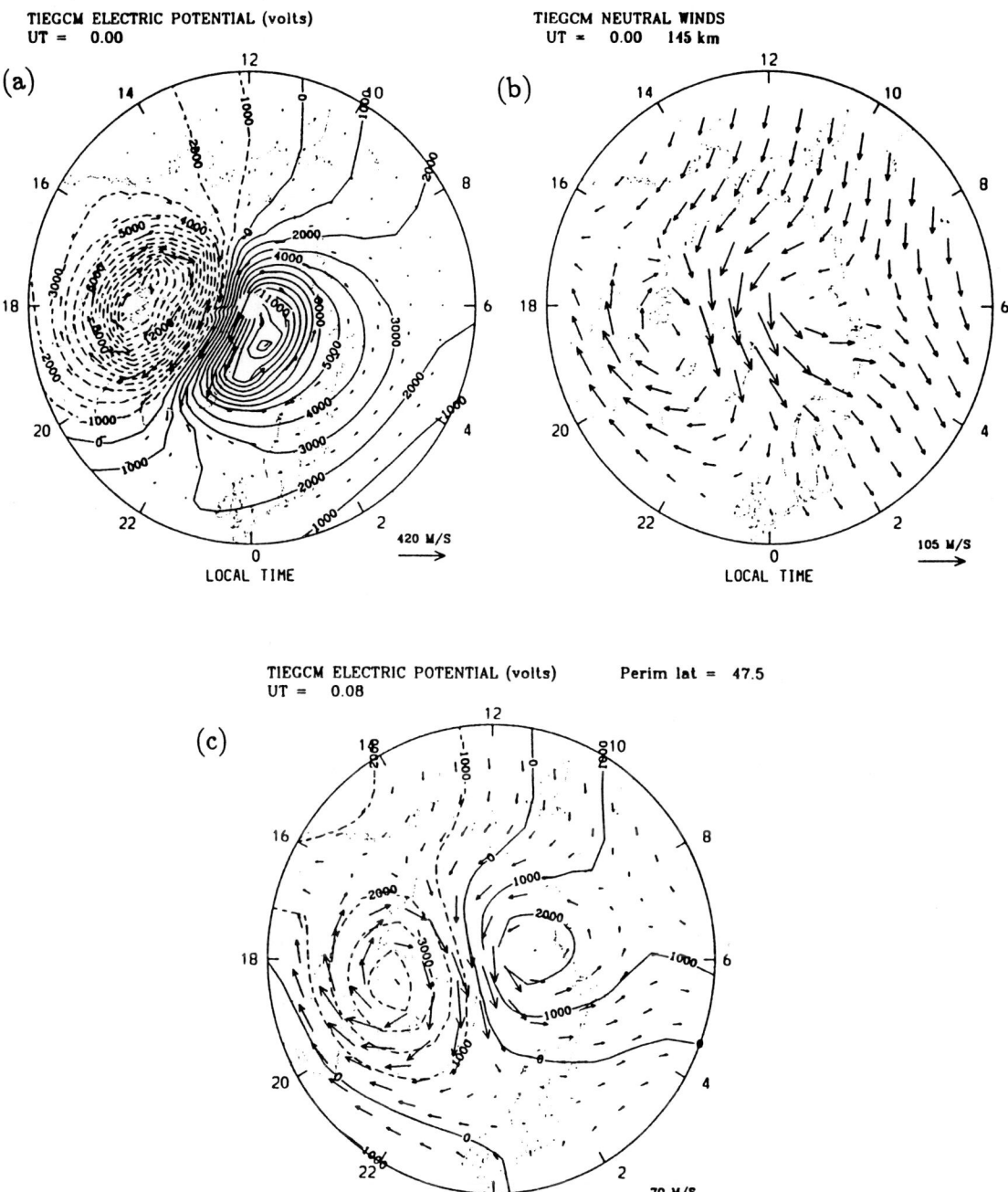

**Figure 9.** Results of a simulation illustrating the "flywheel" effect for solar-minimum equinox conditions. Contour intervals are 1000 V; vector velocity scales vary, as shown at the lower right of each plot. (a) Electric-potential contours and $\mathbf{E} \times \mathbf{B}$ velocity vectors in geographic coordinates between 47.5° and the North Pole, at 0 UT, for a diurnally reproducible simulation with an imposed cross-polar-cap potential of 30 kV. (b) Corresponding neutral wind vectors at 145 km altitude. (c) Electric potential contours and $\mathbf{E} \times \mathbf{B}$ velocity vectors one time step later, after field-aligned current between the ionosphere and the outer magnetosphere has been cut off. From *Richmond* [1995a].

to steady-state fields. Here we consider only the energy flow associated with quasi-steady-state ionospheric fields, varying on time scales longer than about 10 minutes. For such fields the time-rate-of-change of electromagnetic energy density $\partial[(B^2 + E^2/c^2)/2\mu_0]/\partial t$ is generally negligible in comparison with the other terms in (21) between 100 km and 200 km altitude, and we can consider a balance between the convergence of $\mathbf{E} \times \mathbf{B}/\mu_0$ and the rate of electromagnetic energy transfer to the medium, $\mathbf{J} \cdot \mathbf{E}$. For such fields $\mathbf{E}$ is also essentially a potential field, as expressed by (17). $\mathbf{E} \times \mathbf{B}/\mu_0$ is often called the Poynting vector, but the Poynting vector can also be defined in other ways, as long as it has a divergence equal to the divergence of $\mathbf{E} \times \mathbf{B}/\mu_0$. For ionospheric purposes it is convenient to represent $\mathbf{B}$ as the sum of the main geomagnetic field, which is given by the negative gradient of a scalar magnetic potential $V_0$, and a perturbation $\delta\mathbf{B}$:

$$\mathbf{B} = -\nabla V_0 + \delta\mathbf{B} \qquad (22)$$

Then the Poynting vector $\mathbf{S}$ can be defined by

$$\mathbf{S} = \frac{\mathbf{E} \times \delta\mathbf{B}}{\mu_0}, \qquad (23)$$

since $\nabla \cdot (\nabla\Phi \times \nabla V_0)$ vanishes identically.

Plate 2 illustrates the geometry of the electromagnetic fields and Poynting vector at high latitudes, where geomagnetic field lines are approximately vertical. At an altitude of 600 km, a typical height for low-Earth-orbiting (LEO) scientific spacecraft, $\delta\mathbf{B}$ is predominantly horizontal where significant field-aligned currents flow. $\mathbf{E}$ is also predominantly horizontal by virtue of its orthogonality with the main geomagnetic field. $\mathbf{S}$ is therefore predominantly vertical. At the bottom boundary of the ionosphere, where field-aligned current is absent, $\mathbf{E}$ and the horizontal component of $\delta\mathbf{B}$ tend to lie in roughly the same direction, and so the component of $\mathbf{S}$ parallel to $\mathbf{B}$, $S_\parallel$, tends to be considerably smaller than it is on the top side of the ionosphere. Thus there tends to be a divergence of $\mathbf{S}$ through the ionosphere, which is usually negative and therefore corresponds to a dissipation of electromagnetic energy. There can also be horizontal contributions to the divergence of $\mathbf{S}$, but they are generally much smaller than the vertical contributions. If we integrate (21) over a cylindrical volume aligned with a geomagnetic field line, as illustrated in Plate 2, and if we neglect both $S_\parallel$ at the bottom and the horizontal contributions to the divergence of $\mathbf{S}$, then we find that

$$S_\parallel^{600\text{km}} \approx \pm \int_{90\text{km}}^{200\text{km}} \mathbf{J} \cdot \mathbf{E}\, dz, \qquad (24)$$

where $z$ is altitude, and where the $+$ ($-$) sign applies to the northern (southern) hemisphere. Thus spacecraft measurements of $S_\parallel$ can give us an estimate of the height-integrated electromagnetic energy transfer to the ionosphere [*Kelley et al.*, 1991].

The electromagnetic energy transfer to the ionosphere/thermosphere can be divided into two components: Joule heating and acceleration of the medium. That is, by making use of (16), we find that

$$\mathbf{J} \cdot \mathbf{E} = \mathbf{J} \cdot \mathbf{E}' + \mathbf{U} \cdot \mathbf{J} \times \mathbf{B}. \qquad (25)$$

The first term on the right-hand side of (25) is the electromagnetic energy transfer rate in the frame of reference of the medium, and corresponds to Joule heating. Ohm's law gives it a value of $\sigma_P E_\perp'^2 + \sigma_\parallel E_\parallel^2$, which is always positive. The second term on the right of (25) is the scalar product of the velocity and the Ampère force, which corresponds to the feeding of kinetic energy of the medium by this force, or, if negative, to the extraction of kinetic energy from the medium. When kinetic energy is extracted, it may be converted locally to Joule heat, or it may contribute negatively to $\mathbf{J} \cdot \mathbf{E}$, in which case it would be transferred through the Poynting vector to another ionospheric altitude or even back to the magnetosphere. *Thayer* [1998b] showed that the two terms on the right-hand side of (25) can be comparable in magnitude at a given height in the polar region. However, when integrated in height through the thickness of the conducting ionosphere, the second term is usually considerably smaller than the first [*Lu et al.*, 1995; *Thayer et al.*, 1995]. LEO spacecraft observations have shown that the Poynting vector is primarily downward throughout the high latitudes [e.g., *Gary et al.*, 1994, 1995], which is compatible with a dominance of the Joule heating term in (25). Nonetheless, that study also found upward values of $S_\parallel$ over an extended region on numerous polar passes of the Dynamics Explorer spacecraft, suggesting that the second term on the right of (25) can, at times and in certain regions, dominate over the first. Modeling studies [e.g., *Thayer and Vickery*, 1992; *Thayer et al.*, 1995] indicate that the winds spun up by the Ampère acceleration have a tendency to contribute to an upward $S_\parallel$ in cases where the large-scale electric field has rapidly decreased.

## CONCLUDING REMARKS

The coupling between ionospheric electrodynamics and thermospheric dynamics, and their coupling with magnetospheric electrodynamics and with lower-atmospheric dynamics, create a rich variety of phenomena which pose an abundance of research questions to the

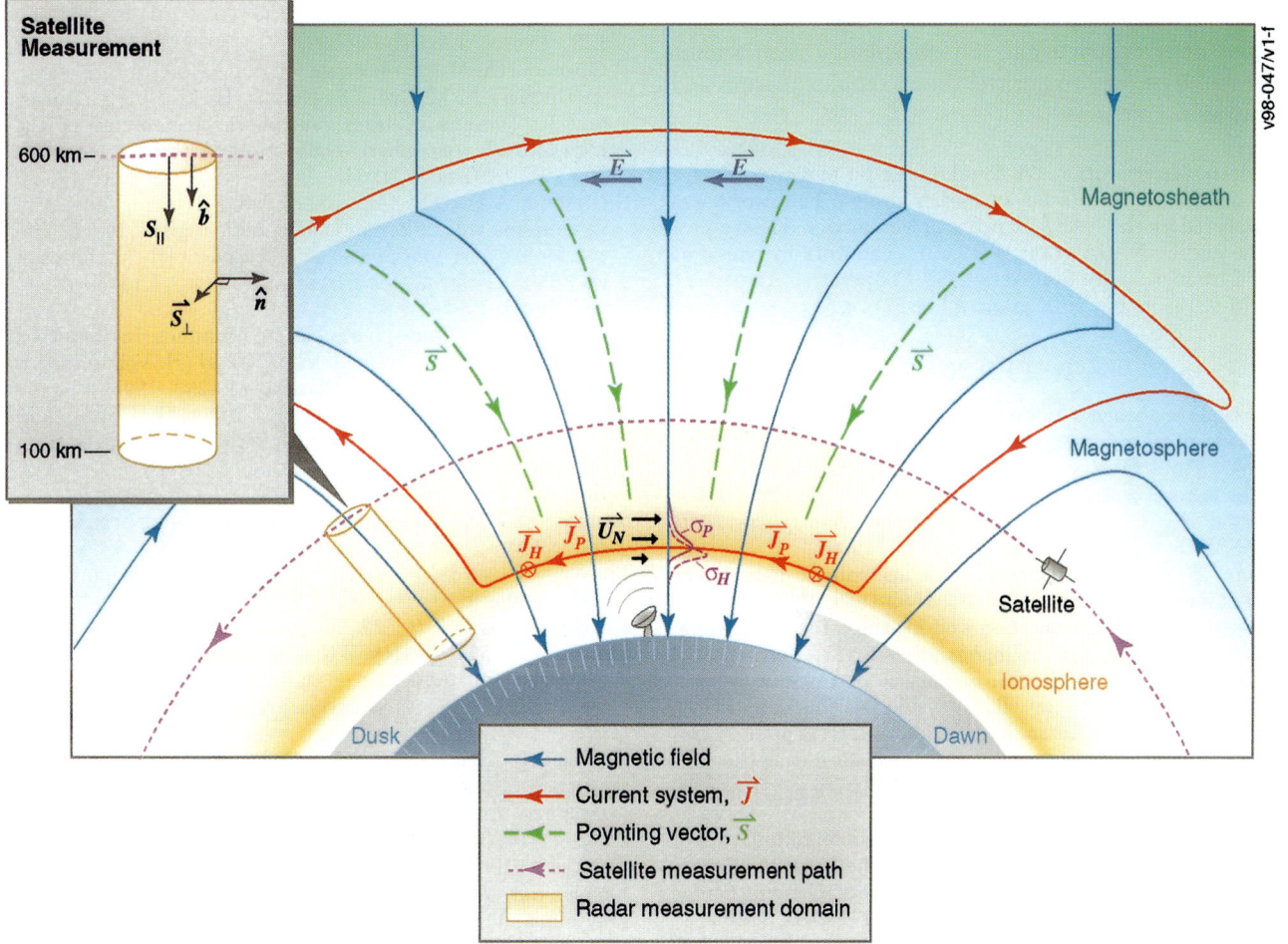

Plate 2. A schematic of the "region 1" current system associated with the transfer of energy and momentum into the ionosphere. The view is looking towards the Sun over the northern hemisphere, with spacecraft and radar measurement domains depicted accordingly. Over the polar cap the magnetic field is downward, **E** and the Pedersen current $\mathbf{J}_P$ are directed from dawn to dusk, while the Hall current $\mathbf{J}_H$ flows into the page in this plane. The Poynting vector **S** is primarily downward. Typical height profiles of $\sigma_P$ and $\sigma_H$ are indicated over the pole. The neutral wind $\mathbf{U}_N$ is a response to multiple forces, and has components both in the plane and out of the plane of the figure. In the inset, $\hat{\mathbf{b}}$ is a unit vector parallel to **B**, $\hat{\mathbf{n}}$ is a unit vector normal to the side of the cylindrical volume, and $\mathbf{S}_\parallel$ and $\mathbf{S}_\perp$ are the components of **S** parallel and perpendicular to **B**, respectively. This figure is a modification of Figure 7 of *Cowley* [1991].

scientific community. It is because of these interactions that observations of ionospheric electrodynamics can tell us a great deal about magnetospheric processes and about global atmospheric dynamics. Furthermore, the ionosphere plays an active role in the electrodynamics of the magnetosphere, and the impact of the interactive electrodynamics on the upper atmosphere can be profound. It is a challenge to the ionosphere-thermosphere community to try to unravel these complex plasma and neutral interactions.

*Acknowledgments.* We thank Gang Lu and two referees for helpful comments on an earlier draft. This work was supported by the NASA Sun-Earth Connection Theory program and by NASA grant W-19,201. The work by one of us (JPT) was partially funded by NSF Cooperative Agreement ATM-9813556 and NSF grant ATM-9714705.

## REFERENCES

Banks, P.M., Magnetospheric processes and the behaviour of the neutral atmosphere, *Space Res., 12*, 1051-1067, 1972.

Banks, P.M., Observations of Joule and particle heating in the auroral zone, *J. Atmos. Terr. Phys., 39*, 179, 1977.

Bilitza, D., ed., *International Reference Ionosphere 1990*, NSSDC 90-22, Greenbelt, Maryland, 1990.

Blanc, M., and A.D. Richmond, The ionospheric disturbance dynamo, *J. Geophys. Res., 85*, 1669-1686, 1980.

Cowley, S.W.H., Acceleration and heating of space plasma: Basic concepts, *Ann. Geophys., 9*, 176-187, 1991.

Fejer, B.G., M.C. Kelley, C. Senior, O. de la Beaujardière, J.A. Holt, C.A. Tepley, R. Burnside, M.A. Abdu, J.H.A. Sobral, R.F. Woodman, Y. Kamide, and R. Lepping, Low- and mid-latitude ionospheric electric fields during the January 1984 GISMOS campaign, *J. Geophys. Res., 95*, 2367-2377, 1990.

Fesen, C.G., R.E. Dickinson, and R.G. Roble, Simulation of thermospheric tides at equinox with the National Center for Atmospheric Research Thermospheric General Circulation Model, *J. Geophys. Res., 91*, 4471-4489, 1986.

Gary, J.B., R.A. Heelis, W.B. Hanson, and J.A. Slavin, Field-aligned Poynting flux observations in the high-latitude ionosphere, *J. Geophys. Res., 99*, 11,417-11,427, 1994.

Gary, J.B., R.A. Heelis, and J.P. Thayer, Summary of field-aligned Poynting flux observations from DE 2, *Geophys. Res. Lett., 22*, 1861-1864, 1995.

Hedin, A.E., Extension of the MSIS thermosphere model into the middle and lower atmosphere, *J. Geophys. Res., 96*, 1159, 1991.

Hill, T.W., Solar-wind magnetosphere coupling, in *Solar-Terrestrial Physics*, edited by R.L. Carovillano and J.M. Forbes, pp. 261-302, D. Reidel, Norwell, Mass., 1983.

Kelley, M.C., D.J. Knudsen, and J.F. Vickrey, Poynting flux measurements on a satellite: A diagnostic tool for space research, *J. Geophys. Res., 96*, 201-207, 1991.

Lu, G., A.D. Richmond, B.A. Emery, P.H. Reiff, O. de la Beaujardière, F.J. Rich, W.F. Denig, H.W. Kroehl, L.R. Lyons, J.M. Ruohoniemi, E. Friis-Christensen, H. Opgenoorth, M.A.L. Persson, R.P. Lepping, A.S. Rodger, T. Hughes, A. McEwin, S. Dennis, R. Morris, G. Burns, and L. Tomlinson, Interhemispheric asymmetry of the high-latitude ionospheric convection pattern, *J. Geophys. Res., 99*, 6491-6510, 1994.

Richmond, A.D., The ionospheric wind dynamo: effects of its coupling with different atmospheric regions, in *The Upper Mesosphere and Lower Thermosphere* (R. M. Johnson and T. L. Killeen, eds.), Am. Geophys. Union, Wash. DC, 49-65, 1995a.

Richmond, A.D., Ionospheric electrodynamics, in *Handbook of Atmospheric Electrodynamics*, Vol. II (H. Volland, ed.), CRC Press, Boca Raton, Florida, 249-290, 1995b.

Robinson, R.M., R.R. Vondrak, K. Miller, T. Dabbs, and D. Hardy, On calculating ionospheric conductances from the flux and energy of precipitating electrons, *J. Geophys. Res., 92*, 2565-2569, 1987.

Southwood, D.J., The role of hot plasma in magnetospheric convection, *J. Geophys. Res., 82*, 5512-5520, 1977.

Thayer, J.P., and J.F. Vickrey, On the contribution of the thermospheric neutral wind to high latitude energetics, *Geophys. Res. Lett., 19*, 265-268, 1992.

Thayer, J.P., J.F. Vickrey, R.A. Heelis, and J.B. Gary, Interpretation and modeling of the high-latitude electromagnetic energy flux, *J. Geophys. Res., 100*, 19,715-19,728, 1995.

Thayer, J.P., Height-resolved Joule heating rates in the high-latitude E region and the influence of neutral winds, *J. Geophys. Res., 103*, 471-487, 1998a.

Thayer, J.P., Radar measurements of the energy rates associated with the dynamic ionospheric load/generator, *Geophys. Res. Lett., 25*, 469-472, 1998b.

Vasyliunas, V.M., The interrelationship of magnetospheric processes, in *Earth's Magnetosphere Processes*, edited by B. M. McCormac, pp. 29-38, D. Reidel, Norwell, Mass., 1972.

---

A. D. Richmond, High Altitude Observatory, National Center for Atmospheric Research, 3450 Mitchell Lane, Boulder, CO 80301. (e-mail: richmond@ucar.edu)

J. P. Thayer, Geoscience and Engineering Center, SRI International, 333 Ravenswood Avenue, Menlo Park, CA 94025. (e-mail: thayer@sri.com)

# The Role of Alfvén Waves in the Formation of Auroral Parallel Electric Fields

Robert L. Lysak and Yan Song

*School of Physics and Astronomy, University of Minnesota, Minneapolis, MN*

The visual aurora is a manifestation of a system of field-aligned currents that couple the magnetosphere and ionosphere. Changes in these currents are mediated by the propagation of shear mode Alfvén waves along auroral field lines. The development of parallel electric fields on these field lines is considered in terms of the propagation characteristics of Alfvén waves. Auroral parallel electric fields occur in low-density regions where the ratio $B/n$ maximizes. On the scale of individual auroral arcs, the electron inertial effect becomes critically important, and inhomogeneities in the Alfvén speed both parallel and perpendicular to the field must be considered to develop a complete theory of the formation of auroral arcs. Mechanisms such as nonlinear effects, ionospheric feedback or phase mixing at horizontal gradients in the Alfvén speed can act to produce the narrow scales observed in auroral arcs.

## INTRODUCTION

The dynamics of the aurora is intimately entwined with the evolution of the field-aligned current system. This current system transmits electromagnetic energy generated in the outer magnetosphere to the low-altitude auroral zone where it is dissipated by Joule heating in the ionosphere or by the acceleration of auroral particles. The mechanism for the parallel electric fields that accelerate auroral particles is not yet firmly established; however, most if not all of the plausible mechanisms for this acceleration depend on the existence of the field-aligned current system. The interplay between the dynamics of the current system and the formation of parallel electric fields remains one of the important unsolved problems of magnetospheric physics.

In the MHD regime, the evolution of the field-aligned currents is mediated by the passage of shear mode Alfvén waves along auroral field lines. These waves are the normal mode of an effective "auroral transmission line" that carries electromagnetic energy into the low-altitude auroral zone. This transmission line is terminated by the conducting ionosphere and, since the terminating resistance is rarely perfectly matched to the impedance of the transmission line, Alfvén waves are reflected back up the field line and interfere with the incident waves. This process is complicated by the fact that the Alfvén speed is strongly inhomogeneous, trapping Alfvén waves in a resonant cavity that has been termed the "ionospheric Alfvén resonator." The resulting multiple reflections of the Alfvén waves structure the field-aligned current and give rise to the multiplicity of forms found in auroral arcs.

The remainder of this paper will investigate the important physical processes associated with the evolution of auroral currents, the formation of parallel electric fields, and the formation of auroral arcs. We will concentrate on aspects of these processes that have not been adequately addressed in previous work. First, the fundamental theory of magnetosphere-ionosphere coupling by Alfvén waves will be reviewed, including the effects of inhomogeneities in the Alfvén speed. Second, the causes and consequences of parallel electric fields will be considered, and the resulting critical scale lengths will be defined. Finally, results from simulations of small perpendicular scale Alfvén waves will be considered, and some mechanisms that can lead to the evolution of narrow auroral arcs will be suggested.

## MAGNETOSPHERE-IONOSPHERE COUPLING BY ALFVÉN WAVES

The theory of Alfvén wave propagation in the magnetosphere has been studied by various authors, and is summarized in a review by *Lysak* [1990]. One aspect that is not generally recognized is the fact that the Alfvén speed reaches very high values above the auroral zone, so that the displacement current must be taken into account. Writing the incompressible MHD momentum equation, and using Ampere's Law, including the displacement current term, to replace the current in the **j**×**B** force, we find

$$\left(1+\frac{c^2}{V_A^2}\right)\frac{\partial \mathbf{v}}{\partial t} = \frac{c^2}{B_0^2}\mathbf{B}_0 \cdot \nabla \mathbf{b} \quad (1)$$

where $V_A^2 = B_0^2/\mu_0\rho$ is the usual Alfvén speed. Including this effect modifies the Alfvén speed, now denoted $c_A$ to distinguish it from the usual definition, to become

$$c_A^2 \equiv \frac{c^2}{1+c^2/V_A^2} = \frac{V_A^2}{1+V_A^2/c^2} \quad (2)$$

Note that this speed reduces to $V_A$ is the limit that $V_A \ll c$, and approaches the speed of light in the opposite limit.

Assuming a plane wave state, where $\omega/k_\parallel = \pm c_A$ for propagation parallel or anti-parallel to the background magnetic field $\mathbf{B}_0 = B_0\hat{\mathbf{z}}$, gives the relation $\mathbf{v}^\pm/c_A = \mp\mathbf{b}^\pm/B_0$, or, in terms of the electric field

$$\mathbf{E}_\perp^\pm = \pm c_A \mathbf{b}^\pm \times \hat{\mathbf{z}} \quad (3)$$

where the ± superscripts refer to propagation parallel or anti-parallel to $\hat{\mathbf{z}}$, respectively. Taking the divergence of (3) leads to a relationship for the field-aligned current:

$$j_z^\pm = \pm\frac{1}{\mu_0 c_A}\nabla\cdot\mathbf{E}_\perp^\pm \equiv \pm\Sigma_A\nabla\cdot\mathbf{E}_\perp^\pm \quad (4)$$

where the last step defines the Alfvén conductance $\Sigma_A$. Note that the magnitude of the total field-aligned current integrated over the perpendicular wavelength is just $I_\parallel = \Sigma_A E_\perp$.

The ionosphere provides a terminating resistance for waves propagating along the auroral transmission line that in general will reflect incident Alfvén waves. In the simple case when the ionosphere is considered an incompressible, two-dimensional slab and when neutral wind effects can be neglected, current continuity at the ionosphere leads to a relation between the field-aligned current and the electric field [e.g., *Lysak*, 1990]

$$j_z = \Sigma_P \nabla\cdot\mathbf{E}_\perp + \mathbf{E}_\perp\cdot\nabla\Sigma_P - (\nabla\Sigma_H \times \mathbf{E}_\perp)\cdot\hat{\mathbf{z}} \quad (5)$$

where $\Sigma_P$ and $\Sigma_H$ are the height-integrated Pedersen and Hall conductivities. Note that the sign here is for the northern hemisphere where the magnetic field is downward. Combining equations (4) and (5) for the case when the conductivity gradients are negligible gives the expression for the reflection coefficient, defined as $R \equiv E_\perp^-/E_\perp^+$ [*Scholer*, 1970; *Mallinckrodt and Carlson*, 1978]

$$R = \frac{\Sigma_A - \Sigma_P}{\Sigma_A + \Sigma_P} \quad (6)$$

Note that for the southern hemisphere, the sign in equation (5) is switched since the magnetic field points out of the ionosphere, but the roles of $E_\perp^+$ and $E_\perp^-$ are also switched, so that equation (6) remains valid.

Although this expression is very useful, there are a number of limitations that should be noted. First of all, conductivity gradients are often important, especially on the night side, and so the full expression (5) should be used in general. Secondly, equation (5) is written in the neutral wind rest frame. Since the neutral wind does not change much over the Alfvén wave time scales, this may not be too restrictive; however, work by *Thayer* [1998] indicates that the neutral wind speed may in fact have vertical gradients, which would make the definition of a neutral rest frame problematical. At higher frequencies, such as the 1 Hz fluctuations discussed below, the vertical structure of the ionosphere must be taken into account since the electromagnetic skin depth becomes smaller than the thickness of the ionosphere [*Lysak*, 1991; 1997; 1999]. In addition at these frequencies, the effect of compressional perturbations must be taken into account since the Hall conductivity couples the shear and compressional MHD modes, and the compressional mode can carry energy away in the ionospheric waveguide [e.g., *Greifinger and Greifinger*, 1968]. These complications can be included in the formalism of equation (6) by defining an effective Pedersen conductance that can be frequency and wavelength dependent [*Lysak*, 1991]. However, in what follows we shall adopt the simplifying model given by (6) while keeping the limitations of the expression in mind.

One difficulty that does have important consequences is the fact that the Alfvén speed varies strongly above the ionosphere, and so it is not clear what Alfvén conductance to use in the above expressions. This arises due to the fact that the mass density decreases exponentially with increasing altitude above the ionosphere, dropping by 5 or 6 orders of magnitude between the ionosphere and 5000 km altitude, while the magnetic field strength, varying as $1/r^3$, changes by less than a factor of ten over the same range. This gives rise to an effective resonant cavity, known as the ionospheric Alfvén resonator [*Polyakov and Rapaport*, 1981; *Trakhtengertz and Feldstein*, 1991; *Lysak*, 1991, 1993] and is also responsible for the waveguide for compressional waves noted in the previous paragraph. This resonant cavity has eigenfrequencies that are given by multiples of $V_{AI}/4\pi h$, where $V_{AI}$ is the Alfvén speed at the ionosphere and $h$ is the scale height of the mass density. These frequencies are typically in the range of 0.1-1.0 Hz for auroral parameters. Waves in this frequency range are often observed on the ground [e.g., *Arnoldy et al.*, 1988; *Koskinen et al.*, 1993] as Pc1 or Pi1B pulsations, and have also been observed by Viking [*Block and Fälthammar*, 1990; *Marklund et al.*, 1990], Freja [*Louarn et al.*, 1994; *Stasiewicz et al.*, 1997], FAST [*Sigsbee et al.*, 1998; *Chaston et al.*, 1999], as well as sounding rocket missions [*Boehm et al.*, 1990; *Knudsen et al.*, 1990]. The prevalence of these 1 Hz waves is an important feature of magnetosphere-ionosphere coupling that has a number of consequences discussed below.

## CAUSES AND EFFECTS OF PARALLEL ELECTRIC FIELDS

The formation of parallel electric fields in the auroral zone is one of the fundamental unsolved problems of space plasma physics. While a number of suggestions have been made as to the cause of such fields, there is as yet no consensus as to the dominant mechanism. Plasma double layers are self-consistent nonlinear structures that can support a parallel electric field on scales of the electron Debye length. Observations of "strong" double layers, with potentials greater than the electron temperature, have been recently made by Polar [*Mozer and Kletzing*, 1998]; however, these remain controversial due to the difficulty of making these measurements. The so-called "weak double layers," with potentials the order of the electron temperature, have been observed by S3-3 and Viking [*Temerin et al.*, 1982; *Koskinen et al.*, 1990; *Mälkki et al.*, 1993]; however, they do not appear to have enough total potential drop to account for the auroral acceleration [*Mälkki et al.*, 1993]. Related observations of electron and ion phase space holes have been recently made by FAST [*Ergun et al.*, 1998], and it is quite possible that further studies of these detailed measurements may lead to a better understanding of how parallel electric fields form.

Parallel electric fields can also be described by including the electron inertia term in the generalized Ohm's Law, as was first introduced by *Goertz and Boswell* [1979]. This model has the advantage that it is theoretically very simple, with no nonlinear plasma physics involved, and that it is easy to incorporate into fluid models describing the evolution of auroral currents and fields. The idea here is that as an Alfvén wave propagates along auroral field lines, the shear in the magnetic field requires a field-aligned current to flow. Since the electrons are the prime current carriers, their finite mass implies that an electric field is required to accelerate them to the velocity needed to carry the current. Including this effect modifies the dispersion relation for the Alfvén wave to become

$$\omega^2 = \frac{k_\parallel^2 c_A^2}{1+k_\perp^2 \lambda^2} \quad (7)$$

It should be noted that electron pressure effects can also be included in (7), but it has been found that this term is not important below about 4 $R_E$ in the auroral acceleration region [*Lysak and Carlson*, 1981]. A full description of the Alfvén wave dispersion including all kinetic effects supports this conclusion [*Lysak and Lotko*, 1996] and indicates that equation (7) is an adequate description of the dispersion relation in the auroral acceleration region.

This effect gives a relation between $E_z$ and $\mathbf{E}_\perp$

$$\left(1-\lambda^2 \nabla_\perp^2\right)E_z = -\lambda^2 \frac{\partial}{\partial z}\nabla \cdot \mathbf{E}_\perp \quad (8)$$

which, for a plane wave, becomes

$$E_z = \frac{k_z}{k_\perp}\frac{k_\perp^2 \lambda^2}{1+k_\perp^2 \lambda^2}E_\perp \quad (9)$$

This expression emphasizes that the importance of the inertial effect depends on the quantity $k_\perp^2 \lambda^2$. When this parameter is large, the relation $k_\perp E_z = k_z E_\perp$ holds and the wave becomes quasi-electrostatic. The electron inertial length given above depends only on the electron density, and is given by $\lambda = 5$ km $/ n^{1/2}$, where the density is given in cm$^{-3}$. The perpendicular wave number will scale inversely with the distance between field lines, and so is proportional to $B^{1/2}$. Thus, $k_\perp \lambda$ will scale as $(B/n)^{1/2}$ along the auroral field line. Figure 1 gives the parameters for a model auroral flux tube in which the density is modeled by an exponential plus a power law in density, together with a dipole magnetic field. Figure 1c gives the profile of the electron inertial length for this profile. Figure 1d,

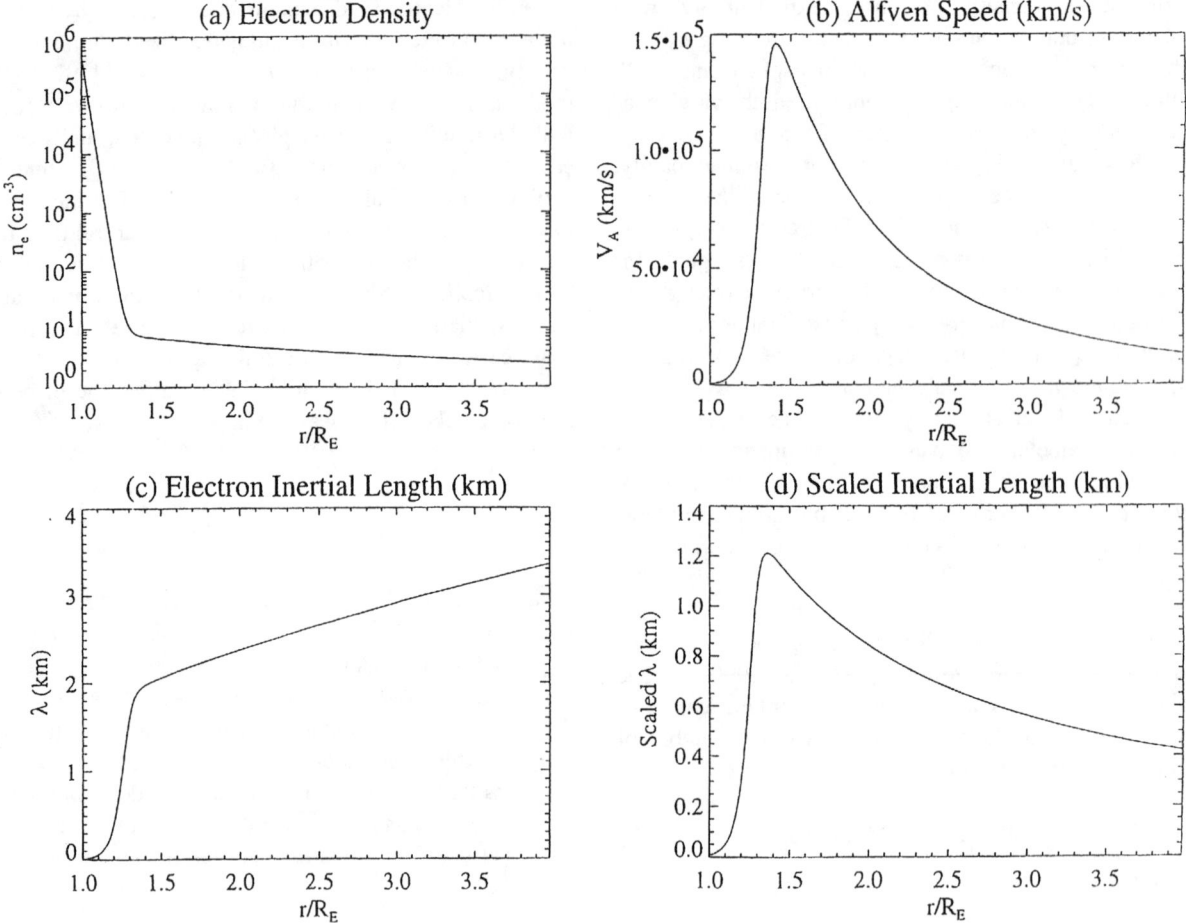

**Figure 1.** Characteristic profiles of relevant parameters along a model auroral field line. (a) Electron density, modeled as an exponential plus a power law; (b) Alfvén speed profile, which scales as $B/n^{1/2}$; (c) Profile of the electron inertial length, which scales as $n^{-1/2}$; (d) Electron inertial length mapped to the ionosphere, which scales as $(B/n)^{1/2}$.

however, gives this inertial length scaled to ionospheric altitudes by mapping along the dipole field, which is the parameter that scales like $(B/n)^{1/2}$. It can be seen that the maximum value of this parameter for this model is about 1.2 km. This quantity reaches its maximum at altitudes of 3000-6000 km, in the observed region of auroral acceleration. This correspondence suggests that the inertial effect may play an important role in the acceleration process.

More generally, the quantity $B/n$ plays a significant role for the flow of current along field lines. Because of the magnetic field convergence, a time-independent field-aligned current must have a current density that scales like $B$. This implies that the drift velocity between electrons and ions is proportional to $B/n$. Thus, the observed location of the auroral acceleration region suggests that this drift velocity is important for the development of parallel electric fields. It should be noted that this holds for double layers and for anomalous resistivity models since these models require a threshold drift velocity to be surpassed. It might also be noted that *Temerin and Carlson* [1998] have developed a quasi-neutral model for the downward current region that also incorporates such a scaling.

It is worth noting that no matter what the mechanism for the parallel potential drop, its existence will have an effect of the motions of auroral particles. *Knight* [1973] has shown that in the presence of a parallel potential drop, the field-aligned current due to precipitating hot electrons follows an S-shaped curve reminiscent of the current-voltage relationship of a diode. It is useful to note that this Knight relation contains a linear regime for potential drops between about 100 eV and about 10 keV, in which the current density is simply given by $j_\parallel = K\Phi_\parallel$, where $K$ is a constant. It has been noted [e.g., *Lysak*, 1985] that this relation implies that if a potential structure with characteristic wave number $k_\perp$ is imposed on the ionosphere, the electrostatic condition $\nabla \times \mathbf{E} = 0$ and the ionospheric Ohm's Law (5) imply that

$$\Phi_I = \frac{\Phi_0}{1 + k_\perp^2 L_{MI}^2} \quad (10)$$

where $\Phi_0$ is the imposed potential, $\Phi_I$ is the ionospheric potential and $L_{MI} = (\Sigma_P/K)^{1/2}$ is a characteristic scale length for magnetosphere-ionosphere coupling. For typical values of $\Sigma_P$ of 10 mho and $K = 10^{-9}$ mho/m$^2$, this scale length is 100 km. On larger scales, $\Phi_I \approx \Phi_0$, and there is no parallel potential drop. Thus, parallel electrostatic fields can only occur on scales smaller than $L_{MI}$.

Applying this model to Alfvén wave reflection, *Vogt and Haerendel* [1998; see also *Lysak and Song*, 1998] showed that the reflection coefficient for Alfvén waves could be written in the same form as equation (6) but with a reduced effective Pedersen conductance

$$\Sigma_{eff} = \frac{\Sigma_P}{1 + k_\perp^2 L_{MI}^2} \quad (11)$$

Note that even though the Pedersen conductance is generally higher than the Alfvén conductance, this reduction implied by (11) indicates that the reflection coefficient can become zero and then positive, indicating that the electric field is enhanced upon reflection, rather than being reduced as in the previous case. When the reflection coefficient becomes zero in this case, it implies that all of the Alfvén wave energy is absorbed by auroral acceleration in the parallel electric field. This occurs when $\Sigma_A = \Sigma_{eff}$. Setting these two quantities equal and defining $L_A$ as the inverse of the wave number at which this takes place gives

$$L_A^2 = \frac{\Sigma_A}{K} \frac{\Sigma_P}{\Sigma_P - \Sigma_A} \quad (12)$$

When $\Sigma_P \gg \Sigma_A$, this scale length becomes $L_A = (\Sigma_A/K)^{1/2}$. For a typical value of $\Sigma_A$ of 0.1 mho, this gives a scale of 10 km.

Thus, the coupling of the magnetosphere and ionosphere including parallel electric fields introduces a hierarchy of scale lengths: $L_{MI}$, the order of 100 km below which parallel potential drops can be formed; $L_A$, the order of 10 km, below which Alfvén wave reflection changes sign and the structures become more electrostatic; and $\lambda$, the order of 1 km, where electron inertial effects become important.

The interaction between electric fields and currents on these various scales no doubt plays a role in the multiplicity of scales observed in auroral arcs.

## FORMATION OF SMALL-SCALE STRUCTURE IN AURORAL CURRENTS AND FIELDS

Finally, we will consider some new results on the evolution of auroral currents and fields. In this section, we will concentrate on the formation of small-scale structure, i.e., on the electron inertial scale relevant to individual auroral arcs. As noted above, the Alfvén wave develops an electrostatic character on small spatial scales. The wave propagation characteristics also develop a resonance cone structure [*Bellan*, 1996; *Stasiewicz et al.*, 1997] in which the group velocity is a function of frequency. Differentiation of the dispersion relation (7) implies that the group velocity is given by

$$\mathbf{v}_g = \hat{\mathbf{z}} \frac{c_A}{\sqrt{1 + k_\perp^2 \lambda^2}} - \hat{\mathbf{x}} \frac{k_\| k_\perp \lambda^2 c_A}{\left(1 + k_\perp^2 \lambda^2\right)^{3/2}} \quad (13)$$

where the wave vector perpendicular to the field is assumed to be in the $x$ direction. Taking the ratio of the group velocity components and using the dispersion relation we find

$$\tan \theta_{res} = \frac{v_{g\perp}}{v_{g\|}} = -\frac{\omega \lambda}{c_A} \frac{k_\perp \lambda}{\sqrt{1 + k_\perp^2 \lambda^2}} \quad (14)$$

It can be seen that this angle becomes a function only of frequency in the short wavelength limit, $k_\perp \lambda \gg 1$.

The difficulty with the previous discussions of Alfvén resonance cones is that the inhomogeneity of the plasma was not taken into account. First of all, it should be noted that the resonance cone model is only valid in the cold plasma limit, when the electron thermal speed is much less than the Alfvén speed. This restricts the region of validity to below about 4 $R_E$ (geocentric distance) along auroral field lines [*Lysak and Carlson*, 1981]. A second point is that the reflection of Alfvén waves from the gradients in the Alfvén speed and from the ionosphere has not been included in previous resonance cone models.

Figure 2 shows the results from a numerical simulation that models linear Alfvén wave propagation along auroral field lines using the density profile shown in Figure 1. In these runs, a source that is a Gaussian in $x$ with a scale length of 0.25 km mapped to the ionosphere is introduced at 4 $R_E$ geocentric distance. In Figures 2a, 2b, and 2c, this

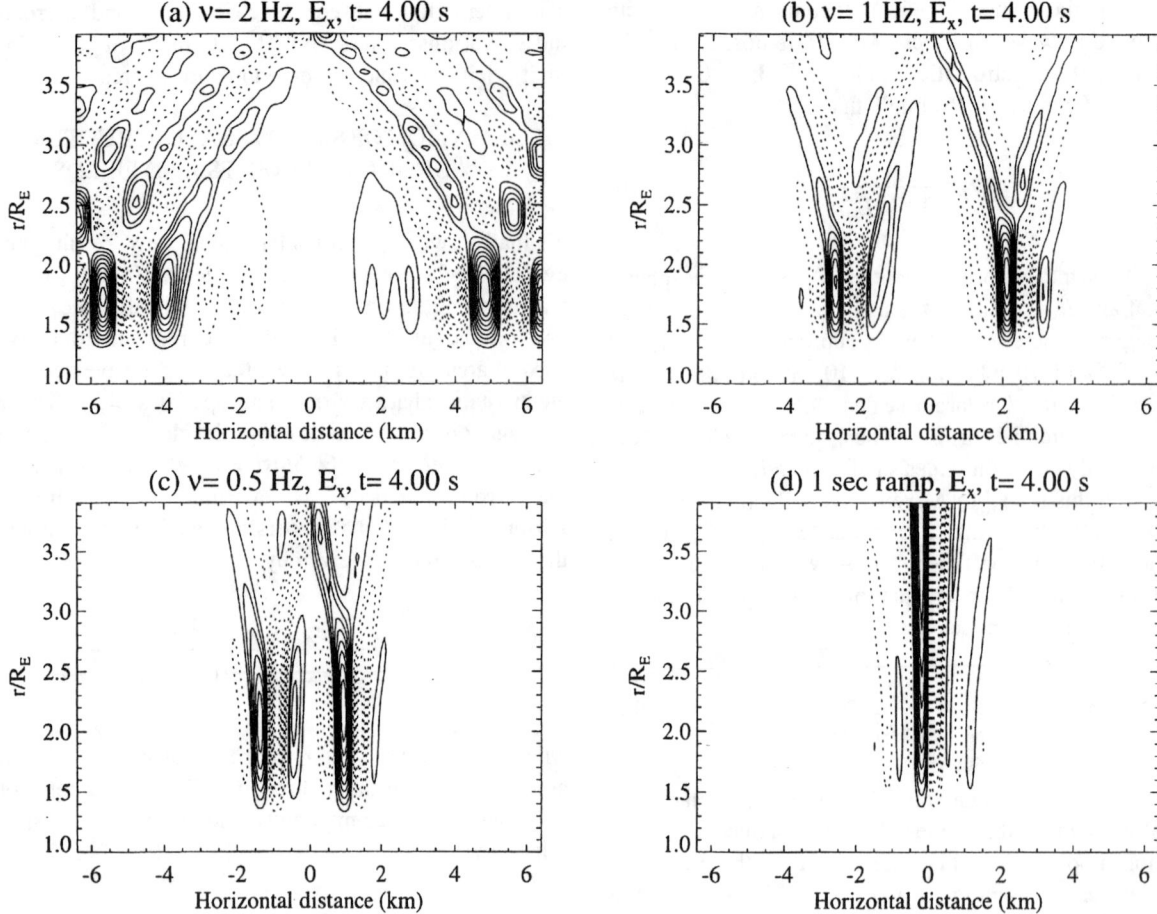

**Figure 2.** Perpendicular electric field profiles for 4 runs using the background profiles shown in Figure 1. Solid contours indicate positive values, dotted contours negative values. These runs used a 0.25 km source of field-aligned current oscillating at a frequency of (a) 2.0 Hz; (b) 1.0 Hz; and (c) 0.5 Hz; and (d) a run in which the current was ramped from 0 to its maximum value over a time of 1 second.

source oscillates in time at frequencies of 2.0, 1.0 and 0.5 Hz, respectively, while in Figure 2d, the source ramps up to its maximum value over a period of 1 second. The characteristic resonance cone feature can be seen in which the angle of propagation is largest for the higher frequency waves, in agreement with equation (14). It can also be seen that there is a large amount of interference between up and downgoing waves, especially below about 2.5 $R_E$. The resonance cone pattern is not too distinct in the case of a ramped-up current structure.

While these resonance cone structures are theoretically possible, a very small source, in this case 0.25 km, must be assumed in order to produce distinct resonance cone structures. A convincing model for such a small-scale source has not yet been clearly determined. Thus, the resonance cone models as they stand presently do not in themselves explain the existence of small-scale structures in the aurora. One must invoke other phenomena in order to understand this small structure. One possibility is the presence of nonlinear effects in the Alfvén waves. *Seyler* [1988] has shown that the electron inertial effect can give rise to a tearing mode that gives current structure on the scale of the electron inertial length. Another possibility lies in the so-called ionospheric feedback instability [*Atkinson*, 1970; *Miura and Sato*, 1980]. In this model, electron precipitation associated with the field-aligned current can locally increase the ionospheric conductivity, and the resulting conductivity gradients lead to additional field-aligned currents. This instability can be coupled with the ionospheric Alfvén resonator model [*Lysak*, 1991], which reduces the time scale from the order of minutes to the order of seconds, and reduces the spatial scale from the

10-km scale to scales less than a kilometer. Thus, the ionospheric feedback instability combined with the Alfvén resonator could provide an explanation for narrow auroral structures.

A third explanation lies in the interaction of Alfvén waves with spatial gradients perpendicular to the magnetic field. In such a case, the Alfvén wave phase fronts travel at different speeds on either side of the gradient. This leads to a shearing of the phase fronts that reduces the perpendicular scale length of the wave. When coupled to the Alfvén resonator, the phase mixing effect can occur at a rapid rate. Figure 3 shows results from a simulation similar to the one shown above, but with a horizontal gradient in the density such that the density is lower, and thus the Alfvén speed higher, on the right-hand sides of the figures (i.e., positive values of the horizontal distance). This run is initialized with a 5 km input pulse (mapped to the ionosphere), much larger than in the previous run. However, it can be seen that the electric field structures itself on a much smaller spatial scale on the low-density side. The resulting waves have a scale even smaller than the electron inertial length, which has a maximum of about 2.5 km scaled to the ionosphere on the low-density side compared to a 1 km maximum inertial length on the high-density side. Note however, that the magnetic field, and thus the current, is not affected by this structuring, reflecting the fact that these very narrow spatial structures are effectively electrostatic. As a final point, it can be seen that the parallel electric field in these structures is quite small, less than 0.015 mV/m. This emphasizes the fact that while the electron inertial effect does give a parallel electric field, it is quite small, and most likely the plasma nonlinear effects must be invoked to provide a sufficient acceleration for auroral particles.

## SUMMARY

The preceding discussion indicates the importance of Alfvén wave dynamics on the formation and evolution of auroral arcs. While it does not appear that Alfvén waves combined with the electron inertial effect is sufficient to explain stable arcs, these waves are clearly important for an understanding of the initial evolution and internal dynamics of auroral structures. These effects suggest a scenario for the initial stages of auroral arc formation that starts with the generation of field-aligned currents in the outer magnetosphere, which launches shear Alfvén waves that propagate along field lines toward the ionosphere. These Alfvén waves can give rise to a parallel electric field in the region where the density is small, or more precisely, in the region where $B/n$ maximizes. At this point, the ionospheric feedback instability can lead to a narrow

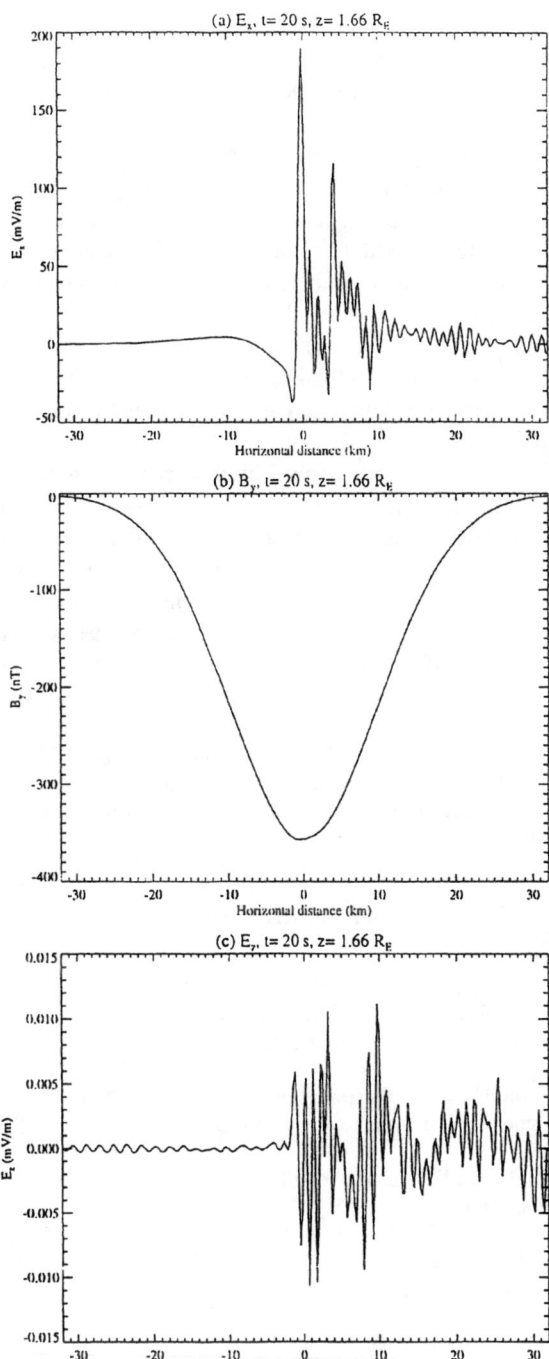

**Figure 3.** Electric and magnetic field profiles that would be observed near FAST apogee (4000 km altitude, or 1.66 $R_E$ radial distance) in a model run in which the density for positive values of the horizontal distance was depleted compared to negative distances. The original pulse was 5 km in scale and was ramped up over 1 second. The perpendicular (a) and parallel (c) electric fields show structuring on scales less than the inertial length in the low-density region, while the magnetic perturbation (b) remains smooth, indicating an electrostatic structure.

structuring of the auroral current system. Another feedback mechanism is initiated by the fact that even a transient parallel electric field can evacuate a region above the aurora [*Thompson and Lysak*, 1996]. This can be described in terms of the ponderomotive force of the Alfvén wave coupling to the ion acoustic mode [*Rankin and Tikhonchuk*, 1998]. In any case, the reduced density in the evacuated region increases $B/n$ and so enhances the parallel electric field. In addition, the reduced density will increase the electron inertial length and thus the parallel electric field. Phase mixing at the boundaries of the density cavity can rapidly decrease the scale of the Alfvén waves, again enhancing the inertial effect and producing a positive feedback.

Certainly, the details implied by this scenario are sketchy at present and much further work needs to be done. Nevertheless, these considerations yield a plausible scenario that is most likely of critical importance in the evolution of auroral arcs. It seems unavoidable to conclude that Alfvén wave propagation and the resulting non-ideal MHD effects are critically important to the formation of parallel electric fields in auroral arcs.

*Acknowledgments.* This work was supported by NASA grant NAG5-4466 and NSF grant ATM-9502907. Supercomputer resources were provided by the Minnesota Supercomputing Institute.

## REFERENCES

Arnoldy, R. L., M. J. Engebretson, and L. J. Cahill, Bursts of Pc1-2 near the ionospheric footprint of the cusp and their relationship to flux transfer events, *J. Geophys. Res., 93,* 1007, 1988.

Atkinson, G., Auroral arcs: result of the interaction of a dynamic magnetosphere with the ionosphere, *J. Geophys. Res. 75,* 4746, 1970.

Bellan, P. M., New model for ULF Pc5 pulsations: Alfvén cones, *Geophys. Res. Lett., 23,* 1717, 1996.

Boehm, M. H., C. W. Carlson, J. P. McFadden, J. H. Clemmons, and F. S. Mozer, High-resolution sounding rocket observations of large-amplitude Alfvén waves, *J. Geophys. Res., 95,* 12,157, 1990.

Block, L. P., and C.-G. Fälthammar, The role of magnetic-field-aligned electric fields in auroral acceleration, *J. Geophys. Res., 95,* 5877, 1990.

Chaston, C. C., C. W. Carlson, W. J. Peria, R. E. Ergun, and J. P. McFadden, FAST observations of inertial Alfvén waves in the dayside aurora, *Geophys. Res. Lett., 26,* 647, 1999.

Ergun, R. E., C. W. Carlson, J. P. McFadden, F. S. Mozer, G. T. Delory, W. Peria, C. C. Chaston, M. Temerin, I. Roth, L. Muschietti, R. Elphic, R. Strangeway, R. Pfaff, C. A. Cattell, D. Klumpar, E. Shelley, W. Peterson, E. Moebius, and L. Kistler, FAST satellite observations of large-amplitude solitary structures, *Geophys. Res. Lett., 25,* 2041, 1998.

Goertz, C. K., and R. W. Boswell, Magnetosphere-ionosphere coupling, *J. Geophys. Res., 84,* 7239, 1979.

Greifinger, C., and P. Greifinger, Theory of hydromagnetic propagation in the ionospheric waveguide, *J. Geophys. Res., 73,* 7473, 1968.

Knight, S., Parallel electric fields, *Planet. Space Sci., 21,* 741, 1973.

Knudsen, D. J., M. C. Kelley, G. D. Earle, J. F. Vickrey, and M. Boehm, Distinguishing Alfvén waves from quasi-static field structures associated with discrete aurora: sounding rocket and HILAT measurements, *Geophys. Res. Lett., 17,* 921, 1990.

Koskinen, H., R. Lundin, and B. Holback, On the plasma environment of solitary waves and weak double layers, *J. Geophys. Res., 95,* 5921, 1990.

Koskinen, H. E. J., R. E. Lopez, R. J. Pellinen, T. I. Pulkkinen, D. N. Baker, and T. Bösinger, Pseudobreakup and substorm growth phase in the ionosphere and magnetosphere, *J. Geophys. Res., 98,* 5801, 1993.

Louarn, P., J.-E. Wahlund, T. Chust, H. deFeraudy, A. Roux, B. Holback, P. O. Dovner, A. I. Eriksson, and G. Holmgren, Observation of kinetic Alfvén waves by the Freja spacecraft, *Geophys. Res. Lett., 21,* 1847, 1994.

Lysak, R. L., Auroral electrodynamics with current and voltage generators, *J. Geophys. Res., 90,* 4178, 1985.

Lysak, R. L., Electrodynamic coupling of the magnetosphere and ionosphere, *Space Sci. Rev., 52,* 33, 1990.

Lysak, R. L., Feedback instability of the ionospheric resonant cavity, *J. Geophys. Res., 96,* 1553, 1991.

Lysak, R. L., Generalized model of the ionospheric Alfvén resonator, in *Auroral Plasma Dynamics*, R. L. Lysak (ed.), AGU Monograph 80, p. 121, 1993.

Lysak, R. L., Propagation of Alfvén waves through the ionosphere, *Phys. Chem. Earth, 22,* 757, 1997.

Lysak, R. L., Propagation of Alfvén waves through the ionosphere: Dependence on ionospheric parameters, *J. Geophys. Res., 104,* 10,017, 1999.

Lysak, R. L., and C. W. Carlson, Effect of microscopic turbulence on magnetosphere-ionosphere coupling, *Geophys. Res. Lett., 8,* 269, 1981.

Lysak, R. L., and W. Lotko, On the kinetic dispersion relation for shear Alfvén waves, *J. Geophys. Res., 101,* 5085, 1996.

Lysak, R. L., and Y. Song, Dynamics of auroral arc formation during substorms, in *Substorms-4*, S. Kokubun and Y. Kamide (eds.), Terra Scientific, Tokyo, p. 35, 1998.

Mälkki, A., A. I. Eriksson, P.-O. Dovner, R. Boström, B. Holback, G. Holmgren, and H. E. J. Koskinen, A statistical survey of auroral solitary waves and weak double layers, 1., occurrence and net voltage, *J. Geophys. Res., 98,* 15,521, 1993.

Mallinckrodt, A. J., and C. W. Carlson, Relations between transverse electric fields and field-aligned currents, *J. Geophys. Res., 83,* 1426, 1978.

Marklund, G. T., L. G. Blomberg, C.-G. Fälthammar, R. E. Erlandson, and T. A. Potemra, Signatures of the high-altitude polar cusp and dayside aurora regions as seen by the Viking electric field experiment, *J. Geophys. Res., 95,* 5767, 1990.

Miura, A., and T. Sato, Numerical simulation of the global formation of auroral arcs, *J. Geophys. Res., 85,* 73, 1980.

Mozer, F. S., and C. A. Kletzing, Direct observation of large, quasi-static, parallel electric fields in the auroral acceleration region, *Geophys. Res. Lett., 25,* 1629, 1998.

Polyakov, S. V., and V. O. Rapoport, Ionospheric Alfvén resonator, *Geomag. Aeronomy, 21,* 816, 1981.

Rankin, R., and V. T. Tikhonchuk, Numerical simulations and simplified models of nonlinear electron inertial Alfvén waves, *J. Geophys. Res., 103,* 20,419, 1998.

Scholer, M., On the motion of artificial ion clouds in the magnetosphere, *Planet. Space Sci., 18,* 977, 1970.

Seyler, C. E., Nonlinear 3-d evolution of bounded kinetic Alfvén waves due to shear flow and collisionless tearing instability, *Geophys. Res. Lett., 15,* 756, 1988.

Sigsbee, K., C. A. Cattell, R. L. Lysak, C. W. Carlson, R. E. Ergun, J. P. McFadden, F. Mozer, R. C. Elphic, R. J. Strangeway, K. Tsuruda, T. Yamamoto, S. Kokubun, D. Fairfield, R. Pfaff, G. Parks, and M. Brittnacher, FAST-Geotail correlative studies of magnetosphere-ionosphere coupling in the nightside magnetosphere, *Geophys. Res. Lett., 25,* 2077, 1998.

Stasiewicz, K., G. Gustafsson, G. Marklund, P.-A. Lindqvist, J. H. Clemmons, and L. Zanetti, Cavity resonators and Alfvén resonance cones observed on Freja, *J. Geophys. Res., 102,* 2565, 1997.

Temerin, M., K. Cerny, W. Lotko, and F. S. Mozer, Observations of double layers and solitary waves on auroral zone field lines, *Phys. Rev. Lett., 48,* 1175, 1982.

Temerin, M., and C. W. Carlson, Current-voltage relationship in the downward auroral current region, *Geophys. Res. Lett., 25,* 2365, 1998.

Thayer, J. P., Height-resolved Joule heating rates in the high-latitude $E$ region and the influence of neutral winds, *J. Geophys. Res., 103,* 471, 1998.

Thompson, B. J., and R. L. Lysak, Electron acceleration by inertial Alfvén waves, *J. Geophys. Res., 101,* 5359, 1996.

Trakhtengertz, V. Yu., and A. Ya. Feldstein, Turbulent Alfvén boundary layer in the polar ionosphere, I, Excitation conditions and energetics, *J. Geophys. Res., 96,* 19,363, 1991.

Vogt, J., and G. Haerendel, Reflection and transmission of Alfvén waves at the auroral acceleration region, *Geophys. Res. Lett., 25,* 277, 1998.

---

Robert L. Lysak and Yan Song, School of Physics and Astronomy, University of Minnesota, 116 Church Street SE, Minneapolis, MN 55455

# A Three-Dimensional Simulation of the Kelvin-Helmholtz Instability

Kristi A. Keller, Robert L. Lysak and Yan Song

*School of Physics and Astronomy, University of Minnesota, Minneapolis, MN*

The Kelvin-Helmholtz instability has been invoked to describe the formation of auroral curls and spirals. In this paper we present results about the generation of field-aligned currents (FACs) by the Kelvin-Helmholtz instability using a three-dimensional compressible simulation at the dayside low-latitude boundary layer. We explore the relationship of the height-integrated Pedersen conductivity with the formation of the Kelvin-Helmholtz instability and the generation of FACs. A higher conductivity stabilizes the system but generates a larger integrated FAC. When the two ionospheres have different conductivities, an interhemispherical current can be generated.

## INTRODUCTION

The Kelvin-Helmholtz instability is a possible mechanism for viscous interaction and field-aligned current (FAC) generation at the magnetopause. Recently, the effect of ionospheric conductivity on the generation of FACs and the effect of ionospheric conductivity on the formation of auroral arcs has been considered. *Newell et al.* [1996, 1998] used the ionospheric conductivity feedback mechanism to explain the suppression of discrete aurora under sunlit conditions. *Stenbaek-Nielsen and Otto* [1997] suggested that an interhemispherical current in the region of the aurora might cause the observed hemispherical differences in discrete aurora. *Sato et al.* [1998] discussed nonconjugate auroral breakups and suggested that differences in ionospheric conductivity might explain the time lag of auroral breakup in the two hemispheres.

There has been a lot of work done on the Kelvin-Helmholtz instability in two dimensions, but due to computational requirements less work has been done in three dimensions. *Miura* [1996] did a stability analysis for of a sheared flow equilibrium against the Kelvin-Helmholtz instability for a magnetosphere-ionosphere coupled system using a box-shaped magnetosphere. He found that the necessary condition for the instability in his model was

$$\mu_0 \Sigma_P V_A < 1 + M_A^{-1} \left| \frac{k_z}{k_y} \right| \quad (1)$$

where $\mathbf{B}_0$ was in the $z$ direction, the initial velocity shear was in the $y$ direction, $k_y$ and $k_z$ were the wave numbers in the direction of the velocity shear and magnetic field respectively, $V_A$ was the Alfvén speed, $M_A$ was the Alfvén Mach number, and $\Sigma_P$ was the height-integrated Pedersen conductivity. If $M_A^{-1} |k_z/k_y|$ is small then larger values of $\mu_0 \Sigma_P V_A$ will stabilize the system.

Previous simulations studied the Kelvin-Helmholtz instability with magnetospheric-ionospheric coupling. *Keskinen et al.* [1988] studied the effect of Pedersen conductivity with a height-integrated magnetosphere. *Lotko and Shen* [1991], *Wei and Lee* [1993], and *Lysak and Song* [1996] used a current-voltage relationship to couple the magnetosphere and ionosphere. *Rankin et al.* [1993] studied the effect of the Kelvin-Helmholtz instability on field line resonances. They modeled the ionosphere as a perfect conductor with boundary conditions appropriate to a standing Alfvén wave. In our model we model the magnetosphere in three dimensions and couple the magnetosphere to the ionosphere through ionospheric

Magnetospheric Current Systems
Geophysical Monograph 118
Copyright 2000 by the American Geophysical Union

reflection coefficients that depend on the height-integrated Pedersen conductivity.

After briefly discussing the computational method of our model, we investigate the generation of FAC by the Kelvin-Helmholtz instability. We will present results showing the effect of different Pedersen conductivity on FACs in the ionosphere and perpendicular currents in the equatorial region. In particular we show how having different conductivities in the northern and southern ionospheres increases the FAC in the higher conductivity hemisphere and can produce an interhemispherical current.

## COMPUTATIONAL METHOD

We use an explicit second-order-accurate finite-difference MHD code developed by *Ryu and Jones* [1995] that is an extension of the Roe-type upwind scheme. The code uses a Total Variation Diminishing (TVD) scheme [*Harten*, 1983] extended to MHD to update the velocity **v**, magnetic field **B**, density ρ, and energy $E$ at the center of each grid using the following ideal MHD equations:

$$\partial \rho / \partial t + \nabla \cdot (\rho \mathbf{v}) = 0 \qquad (2)$$

$$\frac{\partial (\rho \mathbf{v})}{\partial t} + \nabla \cdot (\rho \mathbf{v}\mathbf{v} - \mathbf{B}\mathbf{B}) + \nabla \left( p + \frac{B^2}{2} \right) = 0 \qquad (3)$$

$$\partial \mathbf{B} / \partial t - \nabla \times (\mathbf{v} \times \mathbf{B}) = 0 \qquad (4)$$

$$\frac{\partial E}{\partial t} + \nabla \cdot \left[ \left( E + p + \frac{B^2}{2} \right) \mathbf{v} - \mathbf{B}(\mathbf{B} \cdot \mathbf{v}) \right] = 0 \qquad (5)$$

where $E = \rho u^2/2 + p/(\gamma-1) + B^2/2$ and $p$ is the pressure and the adiabatic constant $\gamma=5/3$. The magnetic field is in rationalized units so that the magnetic pressure $P_b = B^2/2$ and the Alfvén speed $c_A = B/\rho^{1/2}$. *Ryu and Jones* [1995] and *Ryu et al.* [1995] presented details and tests of the TVD code for MHD. The dissipation used by the code is only numerical. The code is dimensionally split so that the variables are updated in time for each dimension with a different pass [*Ryu et al.*, 1995]. For one time step the order of integration will be $x$, $y$, and $z$. In the next five time steps the remaining five permutations of $x$, $y$, and $z$ will be used. Since the code uses dimensional splitting, the divergence of the magnetic field is not equal to zero at the end of this step so the magnetic field at the grid interfaces is updated by using the fluxes that were calculated in the TVD step to calculate the electric field at the grid edges. This electric field is then used to update the magnetic field at the grid interfaces [*Ryu et al., 1999*]. The magnetic field at the grid interfaces is then averaged to get a value of the magnetic field at the grid center for use in the next TVD step.

In these runs we use a box model for our computational grid. The boundary conditions are chosen to be periodic in the $x$ direction and reflecting in the $y$ direction. The boundaries in the $z$ direction are the northern and southern ionosphere. It is assumed that when Alfvén waves are incident on the ionosphere that they are reflected by

$$\frac{B_x^{ref}}{B_x^{inc}} = \frac{\mu_0 \Sigma_P V_A - 1}{\mu_0 \Sigma_P V_A + 1} \qquad (6)$$

$$\frac{V_x^{ref}}{V_x^{inc}} = \frac{1 - \mu_0 \Sigma_P V_A}{1 + \mu_0 \Sigma_P V_A} \qquad (7)$$

$B_y$ and $V_y$ are reflected similarly [*Scholer*, 1970; *Maltsev et al.*, 1974; *Mallinckrodt and Carlson*, 1978]. The reflection coefficients for the Alfvén wave are derived by *Scholer*, [1970] by assuming a finite thickness for the ionosphere with conductivity $\sigma_P$ perpendicular to the magnetic field and parallel conductivity that is infinite. Ohm's law **j** = $\sigma_P(\mathbf{E}+\mathbf{v}_n \times \mathbf{B})$ is used in the ionosphere where the electric field is given by the motion of the magnetosphere **E** = -**v**x**B**$_0$ and **v**$_n$x**B** in the ionosphere is dropped where **v**$_n$ is the neutral wind speed. At the boundary $V_z = 0$ and $B_z = B_0$.

For our initial conditions we have a constant density, pressure, and $B_z$. The initial velocity profile is

$$V_x(y,z) = -\frac{V_{0x}}{2} \tanh\left(\frac{y - l_y/2}{a}\right) \exp\left[-\frac{(z - l_z/2)^2}{w_z}\right] \qquad (8)$$

$$V_y(x,y,z) = -\frac{V_{0y}}{2} \cos(kx) \exp\left[-\frac{(y - l_y/2)^2}{w_y}\right] \times \exp\left[-\frac{(z - l_z/2)^2}{w_z}\right] \qquad (9)$$

where $w_z = l_z^2$, $a = 0.1$, $w_y = 0.0075$, and $k = 2\pi/l_x$ and $l_x$, $l_y$, and $l_z$ are the lengths of the box in their respective directions. For these runs $l_x = 1.4$, $l_y = 1.25$, and $l_z = 5$. The variation in velocity along the $z$ direction causes an

Alfvén wave to propagate from the equatorial region to the ionosphere. The variation of $V_y$ in the $x$ direction is the seed to start the Kelvin-Helmholtz instability. The variation was chosen so that $2ka \approx 0.9$. This corresponds to a maximum growth rate for the two-dimensional transverse case from *Miura and Pritchett* [1982]. The Alfvén transit time from the equatorial plane to the ionosphere $\tau_A$ is 12.5 times larger than the eddy turnover time $\tau_{eddy} = 2a/V_{0x}$. For $2a = 4000$ km and $V_{0x} = 400$ km/s, $\tau_A = 2.1$ min, $l_x = 4.4$ $R_E$, $l_y = 3.9$ $R_E$, and $l_z = 15.6$ $R_E$. In these runs $M_A = 1.0$ and $M_S = 1.0$ where $M_S$ is the sound Mach number.

## RESULTS

In this section we will compare cases with different conductivities at the ionosphere. Two cases have symmetric boundary conditions with $\mu_0 \Sigma_P V_A < 1$. Low values of $\mu_0 \Sigma_P V_A$ correspond to night and winter. Two cases have asymmetric boundary conditions that could correspond to hemispheric differences. In particular we compare FAC generation at the ionosphere for the different cases. We also compare perpendicular currents in the equatorial region to see how these currents are related to FAC generation at the ionosphere.

We ran the simulation with symmetric boundary conditions with $\mu_0 \Sigma_P V_A = 0.11$ and $0.33$. For both cases the Kelvin-Helmholtz instability occurs. In both cases $j_z$ is small in the equatorial region and larger at the ionosphere. A comparison of the integrated $j_z$ at $z = 0$ for $t = 2.0, 3.0, 4.0$ $\tau_A$ shows that the current was larger when $\mu_0 \Sigma_P V_A = 0.33$. The higher conductivity allows larger currents but also has a larger dissipation since the reflection coefficient is smaller. Because the $\mu_0 \Sigma_P V_A = 0.11$ case has a smaller dissipation, the vorticity in the $z$ direction at the equatorial plane is larger than for the $\mu_0 \Sigma_P V_A = 0.33$ case. The impact of the higher dissipation for the $\mu_0 \Sigma_P V_A = 0.33$ case can also be seen in the integrated $j_z$ at $z = 0$. The $\mu_0 \Sigma_P V_A = 0.33$ case has a larger decrease in the FAC from $t = 3.0$ to $4.0$ $\tau_A$ than the $\mu_0 \Sigma_P V_A = 0.11$ case. Figure 1 shows the time evolution of $j_z$ at the $z = 0$ plane for $\mu_0 \Sigma_P V_A = 0.33$.

For the $\mu_0 \Sigma_P V_A = 0.11$ case at $t = 2.0$ $\tau_A$, the perpendicular currents in the equatorial plane forms a vortex in the same region that corresponds to the area of largest $j_z$ in the ionosphere (Figure 2). In this region the diamagnetic current also has a vortex but it is directed in the opposite direction to the total $\mathbf{j}_\perp$. At $t = 3.0$ $\tau_A$ the total $\mathbf{j}_\perp$ forms a vortex but the diamagnetic current is very small (Figure 3). At these two times the inertial currents dominate the vortex region. At $t = 4.0$ $\tau_A$ there is a vortex structure in the diamagnetic current that is similar to the vortex at $t = 2.0$ $\tau_A$. The large vortex structure is not present in the total $\mathbf{j}_\perp$ at the center of box, but smaller vortices are present near the $y$ boundary (Figure 4).

For the $\mu_0 \Sigma_P V_A = 0.33$ case at $t = 2.0$ $\tau_A$, the total $\mathbf{j}_\perp$ and diamagnetic current are very similar to the $\mu_0 \Sigma_P V_A = 0.11$ case. At $t = 3.0$ $\tau_A$ the total $\mathbf{j}_\perp$ is similar to the previous case but the diamagnetic current is larger in the region where the vortex forms in the total current. The larger diamagnetic currents are in the opposite direction to the total $\mathbf{j}_\perp$. At $t = 4.0$ $\tau_A$ the vortex structures are present in the diamagnetic current. The total $\mathbf{j}_\perp$ in this case is similar in structure to the previous case.

We also did two runs with different values of $\mu_0 \Sigma_P V_A$ for each boundary. In the first case $\mu_0 \Sigma_P V_A = 0.11$ at the top boundary ($z = 5$) and $\mu_0 \Sigma_P V_A = 0.33$ at the bottom boundary ($z = 0$). The Kelvin-Helmholtz instability forms in the whole simulation box. At both boundaries the direction of $j_z$ is away from the ionosphere. The magnitude of $j_z$ is larger for the bottom boundary. At $t = 3.0$ $\tau_A$ the integrated $j_z$ in the higher conductivity hemisphere is 3.9 times larger than in the other hemisphere and is also larger than the integrated $j_z$ in the symmetric cases. In the equatorial region the perpendicular currents are similar to the symmetric case with $\mu_0 \Sigma_P V_A = 0.33$. The magnitude of the perpendicular currents falls between the two symmetric cases.

In the second run with asymmetric boundaries, $\mu_0 \Sigma_P V_A = 0.33$ at the top boundary and $\mu_0 \Sigma_P V_A = 3.0$ at the bottom boundary. We did runs with $\mu_0 \Sigma_P V_A = 3.0$ at both boundaries and these were stable to the Kelvin-Helmholtz instability. This is consistent with results from *Miura* [1996] for a more specific case. In this run the vortex only forms in the upper region of the box. There is a larger $j_z$ at the bottom boundary but very little vortex structure. In this case $j_z > 0$ for the whole box at $t = 3.0$ $\tau_A$. This is not true at all times but $j_z > 0$ for most of the simulation box at other times. Unlike the previous case there is no vortex formation in the perpendicular currents that correspond to the area of largest $j_z$ at the ionosphere. There is a vortex signature in the diamagnetic currents at $t = 2.0$ and $3.0$ $\tau_A$, but this corresponds to a minimum in $j_z$. The other difference is that the total $\mathbf{j}_\perp$ and the diamagnetic currents are in the same direction in this region at $t = 3.0$ $\tau_A$ (Figure 5). There is a rough correspondence between the $\nabla \cdot \mathbf{j}_\perp$ in the equatorial plane and $j_z$ at the ionosphere (Figure 6).

## SUMMARY

The lower conductivity runs have a larger vorticity in the $z$ direction at the equatorial plane but smaller FAC. Runs done with different conductivities in each hemisphere

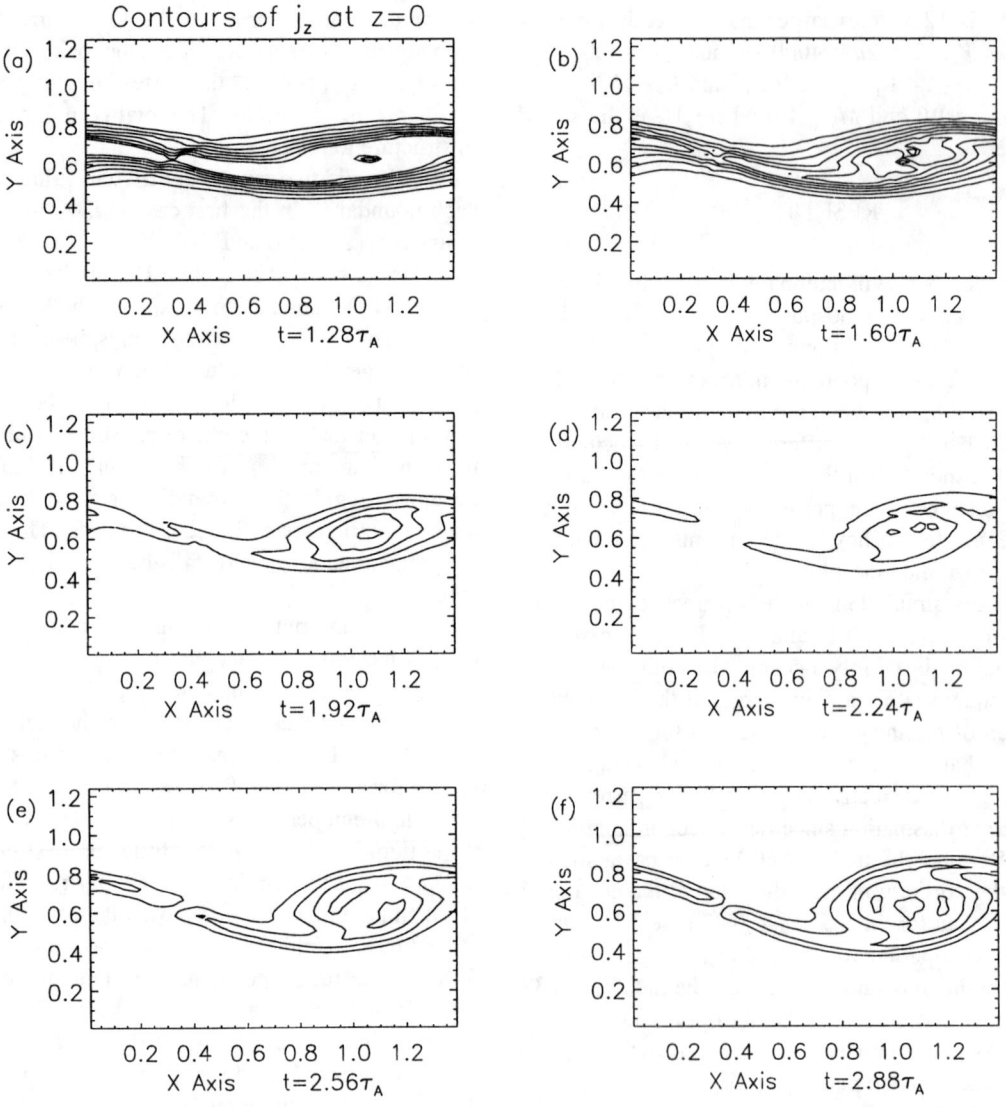

**Figure 1.** Contour plots of $j_z$ in the x-y plane at $z = 0$ at (a) $t = 1.28\ \tau_A$; (b) $1.60\ \tau_A$; (c) $1.92\ \tau_A$; (d) $2.24\ \tau_A$; (e) $2.56\ \tau_A$; and (f) $2.88\ \tau_A$.

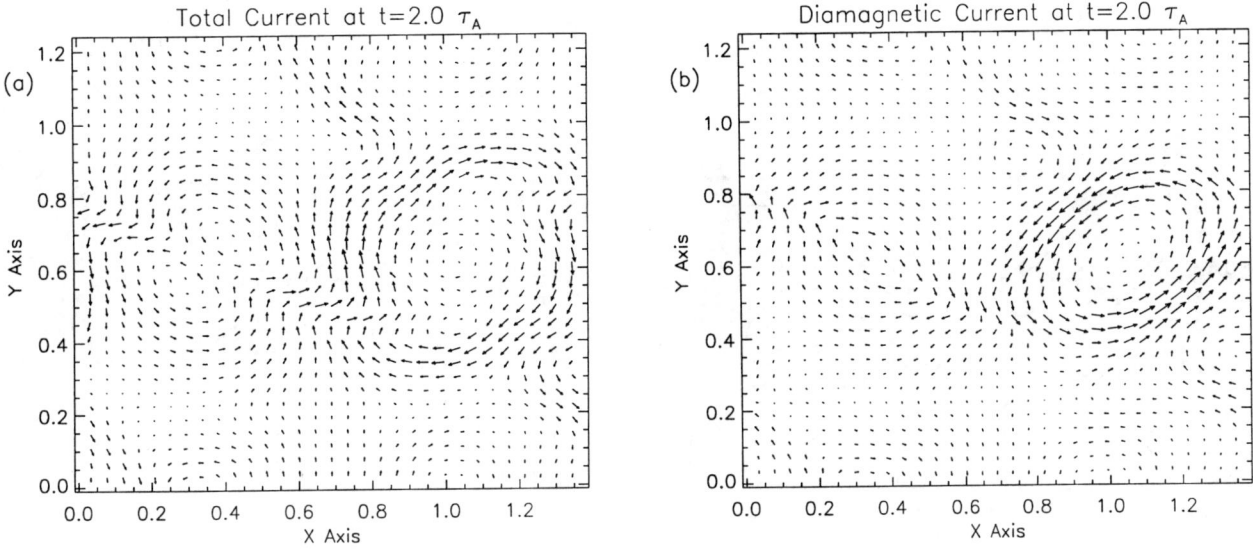

**Figure 2.** Vector plots of (a) $j_\perp$; and (b) the diamagnetic current in the $x$-$y$ plane at $z = 2.5$ for 3.0 $\tau_A$ for the symmetric run with $\mu_0 \Sigma_P V_A = 0.11$.

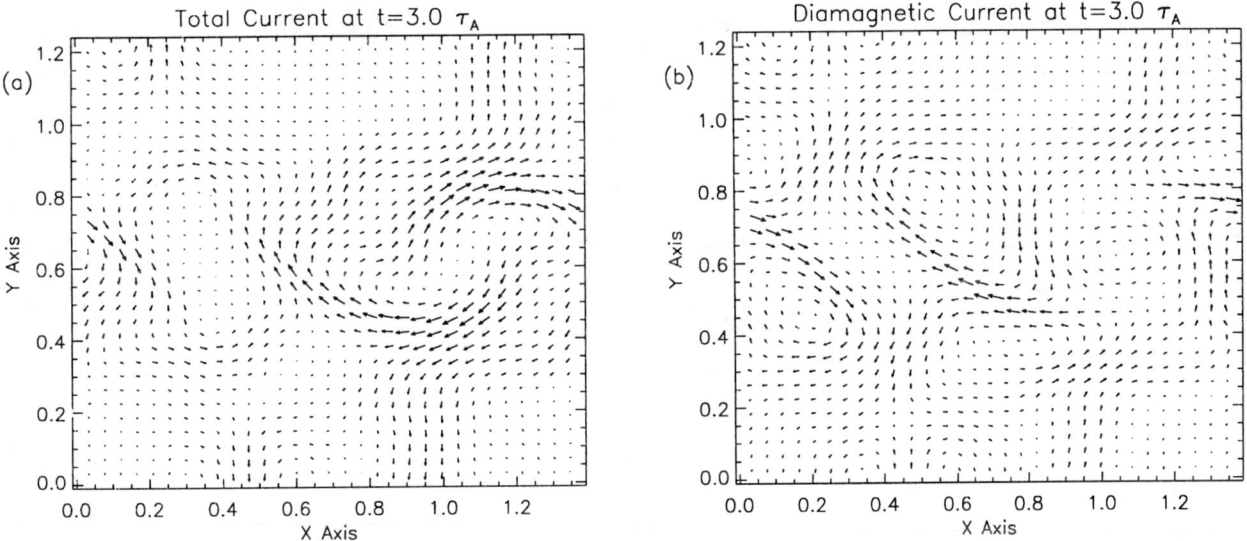

**Figure 3.** Vector plots of (a) $j_\perp$; and (b) the diamagnetic current in the $x$-$y$ plane at $z = 2.5$ for 3.0 $\tau_A$ for the symmetric run with $\mu_0 \Sigma_P V_A = 0.11$.

## 162 KELVIN-HELMHOLTZ INSTABILITY

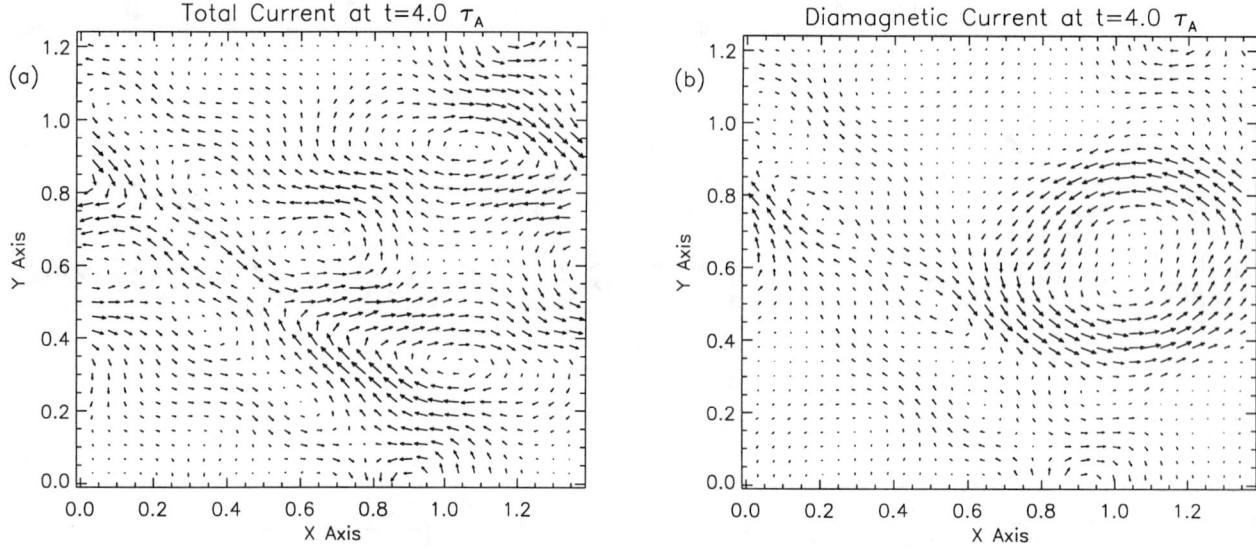

**Figure 4.** Vector plots of (a) $\mathbf{j}_\perp$; and (b) the diamagnetic current in the $x$-$y$ plane at $z = 2.5$ for 4.0 $\tau_A$ for the symmetric run with $\mu_0 \Sigma_P V_A = 0.11$.

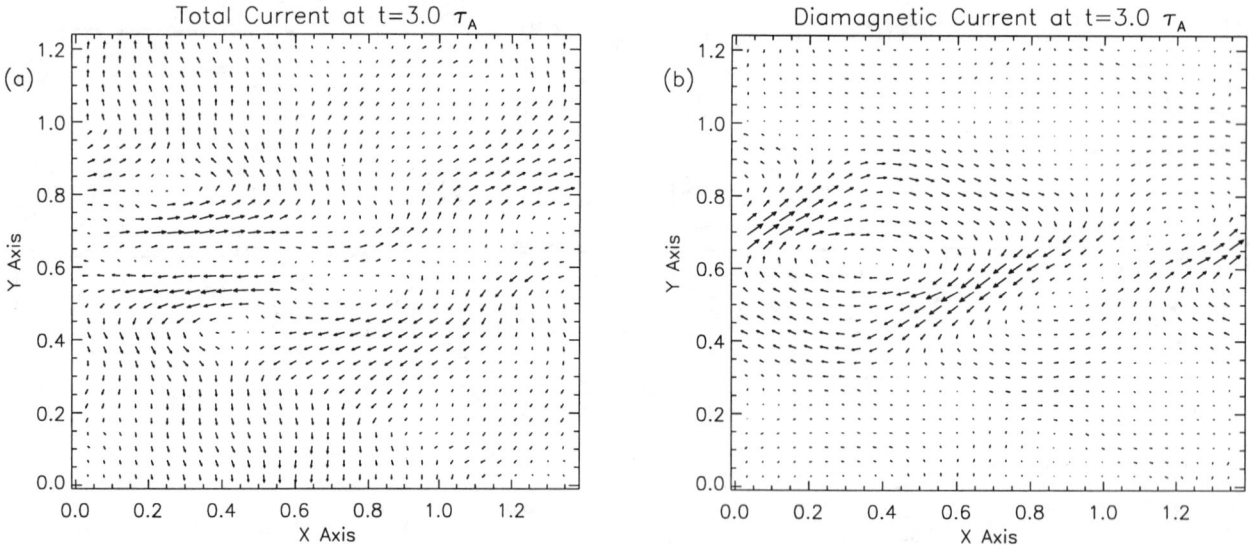

**Figure 5.** Vector plots of (a) $\mathbf{j}_\perp$; and (b) the diamagnetic current in the $x$-$y$ plane at $z = 2.5$ for 3.0 $\tau_A$ for the asymmetric run with $\mu_0 \Sigma_P V_A = 3.0$ at $z = 0$ and $\mu_0 \Sigma_P V_A = 0.33$ at $z = 5$.

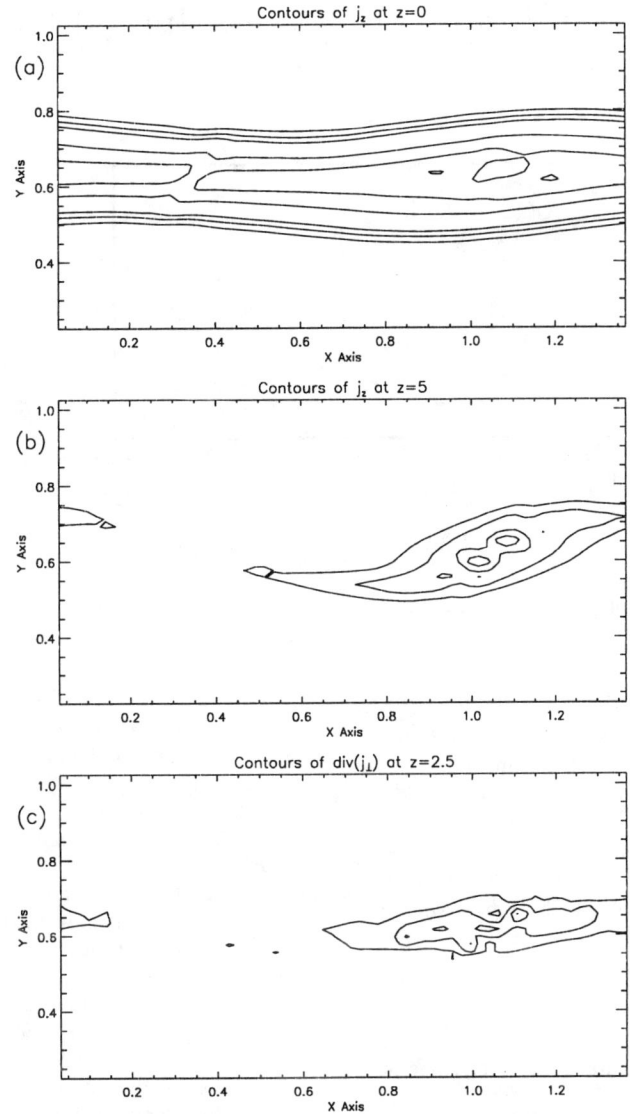

**Figure 6.** Contour plots of (a) $j_z$ at $z = 0$; (b) $j_z$ at $z = 5$; and (c) $\text{div}(\mathbf{j}_\perp)$ at $z = 2.5$ at $t = 3.0\ \tau_A$ for the asymmetric run with $\mu_0 \Sigma_P V_A = 3.0$ at $z = 0$ and $\mu_0 \Sigma_P V_A = 0.33$ at $z = 5$.

develop a larger FAC in one hemisphere since the lower conductivity hemisphere has the larger growth rate for the instability while the higher conductivity produces the FAC. Figure 7 shows the time evolution of the FAC integrated over the $x$-$y$ plane at $z = 0$, demonstrating that the asymmetric runs have a larger current than the symmetric run at the high conductivity hemisphere. The run with $\mu_0 \Sigma_P V_A = 3.0$ at $z = 0$ has a larger dissipation at the top boundary than the run with $\mu_0 \Sigma_P V_A = 0.33$ at $z = 0$ so that the FAC drops more at the end. An interhemispherical FAC is generated in the asymmetric case with $\mu_0 \Sigma_P V_A = 0.33$ and 3.0. The conductivity in this run could correspond to a case at solstice where it is summer in one hemisphere and winter in the other hemisphere.

Song [1998] has pointed out that the main energy source in the magnetosphere is not internal energy, but the kinetic energy and magnetic energy existing in the relatively moving plasmas. The kinetic energy provided by slowing down the plasma is converted into electromagnetic energy in the form of current, as well as into thermal energy. Therefore, it is the inertial current, not the diamagnetic current, which corresponds to FAC generation. Our study of the perpendicular current in the equatorial regions shows that the inertial current is very significant in the equatorial region that corresponds to the FAC in the ionosphere. The case with $\mu_0 \Sigma_P V_A = 3.0$ and $\mu_0 \Sigma_P V_A = 0.33$ has a large region where the diamagnetic current and the inertial current are in the same direction but this region does not

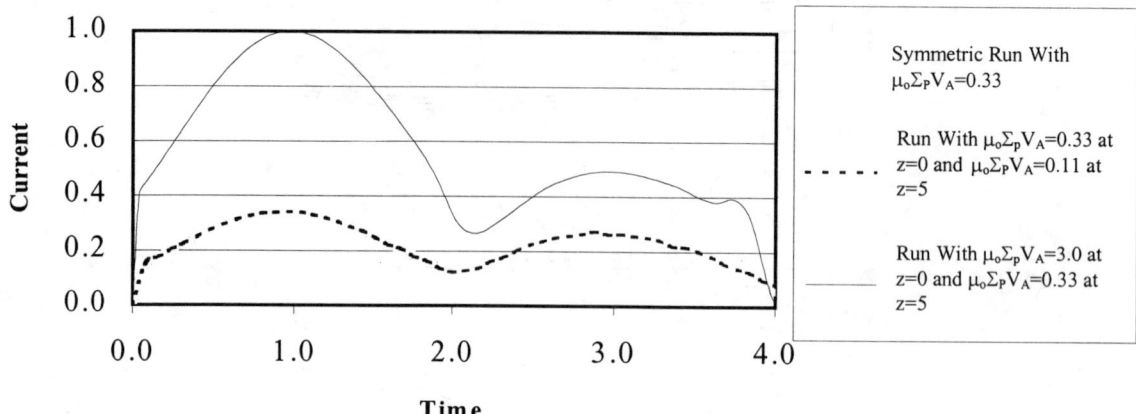

**Figure 7.** Time evolution of the field-aligned current integrated over the $x$-$y$ plane at $z = 0$. Time is in units of $\tau_A$. Current is normalized to the maximum value of current for the case with $\mu_0 \Sigma_P V_A = 3.0$ at $z = 0$ and $\mu_0 \Sigma_P V_A = 0.33$ at $z = 5.0$.

correspond to the region of largest FAC at the ionosphere. While there is a general correspondence of the $\nabla \cdot \mathbf{j}_\perp$ in the equatorial region and the $j_z$ at the ionosphere, this correspondence is not exact.

*Acknowledgments.* This work was supported by NASA Grants NGT5-50186 and NAG5-4466 and NSF Grant ATM-9502907. Supercomputing time was provided by the Minnesota Supercomputer Institute.

## REFERENCES

Harten, A., High resolution schemes for hyperbolic conservation laws, *J. Comput. Phys., 49,* 357, 1983.

Keskinen, M. J., H .G. Mitchell, J. A. Fedder, P. Satyanarayana, S. T. Zalesak, and J. D. Huba, Nonlinear evolution of the Kelvin-Helmholtz instability in the high-latitude ionosphere, *J. Geophys. Res., 93,* 137, 1988.

Lotko, W., and M.-M Shen, On large-scale rotational motions and energetics of auroral shear layers, *J. Geophys. Res., 96,* 9549, 1991.

Lysak, Robert L. and Yan Song, Coupling of Kelvin-Helmholtz and current sheet instabilities to the ionosphere: A dynamic theory of auroral spirals, *J. Geophys. Res., 101,* 15,411, 1996.

Mallinckrodt, A. J., and C. W. Carlson, Relations between transverse electric fields and field-aligned currents, *J. Geophys. Res., 83,* 1426, 1978.

Maltsev, Yu. P., S.V. Leontyev, and W. B. Lyatsky, Pi2 pulsations as a result of evolution of an Alfvén impulse originating in the ionosphere during a brightening of aurora, *Planet. Space Sci., 22,* 1519-1533, 1974.

Miura, A., Stabilization of the Kelvin-Helmholtz instability by the transverse magnetic field in the magnetosphere-ionosphere coupling system, *Geophys. Res. Lett., 23,* 761, 1996.

Miura, A. and P. L. Pritchett, Nonlocal stability analysis of the MHD Kelvin-Helmholtz instability in a compressible plasma, *J. Geophys. Res., 87,* 7431, 1982.

Newell, P. T., C.-I. Meng, and K. M. Lyons, Suppression of discrete aurora by sunlight, *Nature, 381,* 766, 1996.

Newell, P. T., C.-I. Meng, and S. Wing, Relation of solar activity of intense aurorae in sunlight and darkness, *Nature, 393,* 342, 1998.

Rankin, R., B. G. Harrold, J. C. Samson, and P. Frycz, The nonlinear evolution of field line resonances in the Earth's magnetosphere, *J. Geophys. Res., 98,* 5839, 1993.

Ryu, D. and T. W. Jones, Numerical magnetohydrodynamics in astrophysics: Algorithm and tests for one-dimensional flow, *Astrophys. J., 442,* 228, 1995.

Ryu, D., T. W. Jones and A. Frank, Numerical magnetohydrodynamics in astrophysics: Algorithm and tests for multidimensional Flow, *Astrophys. J., 452,* 785, 1995.

Ryu, D., F. Miniati, T. W. Jones and A. Frank, A divergence-free upwind code for multidimensional magnetohydrodynamic flows, *Astrophys. J., 509,* 244, 1998.

Sato, N., T. Nagaoka, K. Hashimoto, and T. Saemundsson, Conjugacy of auroral arcs and nonconjugate auroral breakups, *J. Geophys. Res., 103,* 11,641, 1998.

Scholer, M., On the motion of artificial ion clouds in the magnetosphere, *Planet. Space Sci., 18,* 977, 1970.

Song, Y., Theoretical considerations on mechanisms for the substorm current wedge, *Substorm-4,* S. Kokubun and Y. Kamide (eds.), Terra Scientific, Tokyo, p. 543, 1998.

Stenbaek-Nielsen, H. C., and A. Otto, Conjugate auroras and the interplanetary magnetic field, *J. Geophys. Res., 102,* 2223, 1997.

Wei, C. Q., and L. C. Lee, Coupling of the magnetopause-boundary layer to the polar ionosphere, *J. Geophys. Res., 98,* 5707, 1993.

Kristi A. Keller, Robert L. Lysak and Yan Song, School of Physics and Astronomy, University of Minnesota, 116 Church Street SE, Minneapolis, MN 55455

# The Role of Space-Time Dependent Ionospheric Conductivity in the Evolution of Field Line Resonances: Relation to Auroral Arc

Manju Prakash

*Department of Physics and Astronomy, SUNY at Stony Brook, New York*

Robert Rankin

*Department of Physics, University of Alberta, Edmonton*

The present work examines the nonlinear evolution of field line resonances (FLRs) in the presence of electron inertia and the space-time dependent ionospheric conductivity. The nonlinear space-time evolution results from the ponderomotive forces that are exerted by the large amplitude FLRs that are excited during magnetospheric substorms. The space-time evolution is studied using density and field amplitude equations. These nonlinear equations are derived using the MHD equations and the generalized Ohm's law. The effects arising from the non-uniform and time-dependent Pedersen conductivity are incorporated through the boundary conditions on field-aligned currents (FAC) that close the ionospheric currents. The space-time dependent ionospheric conductivity can increase the FAC associated with FLRs. This increase leads to intensification of the pre-existing auroral arc observed during substorms. The increase in the FAC can also enhance the nonlinear effects arising from the ponderomotive forces. These effects can result in density perturbations (in the auroral plasma) that can structure the FLRs. The structuring of the FLRs can lead to the restructuring of the auroral arc observed during substorms. These studies indicate that the ionospheric manifestation of the substorms is modified by the space-time dependent ionospheric conductivity.

## 1. INTRODUCTION

The effect of the ionosphere on the detectability, polarization structure, and decay of ultra low frequency (ULF) magnetospheric waves (frequency range 1mHz–1 Hz) has been a topic of interest for more than thirty years [*Allan and Knox*, 1979; *Southwood and Hughes*, 1983]. The degree of ionospheric conductivity controls the amplitude of the ULF signal, its reflection, and its subsequent decay on the ground. The increasing use of the ground based networks such as CANOPUS (to study auroral substorms) provides further impetus to understand the modification of the ULF signals between their point of generation in space and their detection on the ground. The finite conductivity of the ionosphere leads to partial reflections of the shear Alfvén waves (SAWs). The SAW is a transverse ULF wave which is guided along the geomagnetic field lines. It carries field-aligned current, momentum, and electromagnetic energy into and out of the Earth's magnetosphere. This energy is converted into the kinetic energy of the auroral particles and is subsequently dissipated in the form of Joule heating of the particles at the polar ends of the auroral flux tubes [*Louarn et al.*, 1994; *Lysak*, 1990; *Wahlund et al.*, 1994].

During substorms, the dynamic changes in the plasma sheet are communicated to the ionosphere by the field-aligned currents (FAC) associated with the SAWs. On reflection from the ionosphere the SAWs transfer ionospheric drag back to the magnetosphere. Thus Alfvén waves play an important role in coupling the dynamics of the inter-regions of the Earth's magnetosphere [*Louarn et al.*, 1994; *Lysak and Dum*, 1983; *Lysak*, 1990; *Prakash and Lysak*, 1992; *Wahlund et al.*, 1994]. The degree of space-time dependent ionospheric conductivity can alter the FACs and restructure the auroral arc [*Ellis and Southwood*, 1983; *Glassmeier*, 1983]. The non-uniform conductivity can also modify the transport of energy and momentum within the magnetosphere [*Ellis and Southwood*, 1983]. Therefore, it is important to examine the role of space-time dependent ionospheric conductivity in magnetosphere-ionosphere coupling.

The SAWs play an important role in the coupling of the magnetosphere and ionosphere through the excitation of FLRs. The FLRs are standing wave structures (frequency range 1-10mHz) which are formed along the closed geomagnetic field lines when the SAWs are reflected at the conjugate ends of the ionosphere. The FACs associated with FLRs can intensify the pre-existing auroral arc. This intensification is an important signature of the onset of substorms [*Samson et al.*, 1991; 1992a;b]. Near the ionosphere, the FLRs have a latitudinal scale size of 50 km. This size is comparable to the latitudinal size of the auroral arc [*Rankin et al.*, 1994]. These observations indicate that the FLRs can play an important role in the dynamics of the auroral region during substorms. The observations of the vortices in the auroral zone that are caused by the coupling of the shear flows (resulting from the $\mathbf{E} \times \mathbf{B}$ drift in FLRs) with the pressure gradients [*Rankin et al.*, 1997; *Samson et al.*,1996; *Voronkov et al.*, 1997] further support the role of FLRs in the auroral dynamics during substorms.

During substorms, a significant amount of energy is pumped into the excitation of the FLRs. This leads to excitation of large amplitude FLRs that can exert ponderomotive forces on the ambient plasma. The ponderomotive forces can narrow the meridional scale size of the FLRs by various physical effects listed in section 2.1 of this paper [*Rankin et al.*, 1994; 1995]. When the scale size is comparable to the electron inertial length ($\sim$ 1km) in the auroral zone, the parallel electric fields $E_z$ develop [*Hasegawa*, 1976; *Wei et al.*,1994]. The parallel electric fields can accelerate electrons to several hundred eV and can lead to spatial modulation of the auroral arc [*Wei et al.*, 1994]. The nonlinear effects resulting from the ponderomotive forces can narrow the auroral arc to a scale size of less than 5 km [*Bellan and Stasiewicz*, 1998; *Rankin et al.*, 1995; *Trondsen et al.*, 1997; *Wei et al.*, 1994]. The present work will focus on the structuring of the auroral arc related to substorms only. It will not address various mechanisms which lead to formation of arc sizes down to 100m [*Stasiewicz et al.*, 1997].

It has been reported that some discrete auroral arcs may be spatially modulated with FLR frequencies in the range 1-4mHz [*Samson et al.*, 1996]. The characteristic features of the auroral arc as well as the dynamics of the other observable physical effects can be understood theoretically by examining the nonlinear evolution of the FLRs. The current studies on the nonlinear evolution of FLRs, however, assume that the ionospheric conductivity is constant and uniform [*Rankin et al.*, 1993]. On the dayside, ionosphere conductivity is uniform and constant, because it is primarily produced by the impact of the solar UV radiation. On the nightside, the ionospheric conductivity is caused by the precipitating particles and energy flux from the magnetosphere. This results in a non-uniform and time dependent conductivity. The FAC resulting from the conductivity can lead to feedback interactions which can intensify the auroral arc observed during substorms [*Lysak*, 1991; *Trakhtengertz and Feldstein*, 1984]. Therefore, the current theoretical studies on the nonlinear evolution of FLRs should be extended to include effects arising from the inhomogeneous and time-dependent conductivity [*Ellis and Southwood*, 1983; *Glassmeier*, 1983]. The present work is an effort in this direction.

In section 2 of the paper, we discuss the mathematical and the physical aspects of the FLR wave model. In section 3, the ionospheric model and the FAC observed during substorms will be studied. In section 4, the equations describing the nonlinear evolution of FLRs in the presence of electron inertia and space-time dependent conductivity will be derived. The significant results and directions for the future research will be summarized in section 5 of the paper.

## 2. THE FIELD LINE RESONANCE WAVE MODEL

### 2.1. Physical Model

The formation of FLRs along the closed field lines of the terrestrial magnetosphere can be understood using magnetohydrodynamic (MHD) equations [*Southwood and Hughes*, 1983]. The MHD equations, when coupled with the boundary of the magnetopause, yield compressional mode solutions with discrete frequencies. The discrete modes are formed when the compressional Alfvén modes (normal modes of the magnetosphere) are reflected at the magnetopause boundary (at one end) and at the wave turning points (on the other end). These modes constitute a discrete spectrum of the compressional waves in the Earth's magnetospheric cavity [*Kivelson and Southwood*, 1986]. The energy to excite these cavity modes may be derived from pressure pulses in the solar wind, transient dayside reconnection, Kelvin Helmholtz instability, or other substorm-related instabilities.

These cavity modes can couple their energy with the SAWs at the resonance points. The coupling results in the formation of standing wave structures (along the closed geomagnetic field lines) when the SAWs are reflected at the northern and southern ionospheres. These standing wave structures are known as FLRs. The observations based on the ground based magnetometers, radar, and optical data support that FLRs are standing SAWs that are formed along the closed field lines [*Greenwald and Walker*, 1980; *Samson et al.*, 1991; *Walker et al.*, 1992].

We remark that the present theoretical formalism underlying the formation of FLRs is based on ideal MHD equations which neglect the parallel electric field. The parallel electric field effects are incorporated through the Hall term in the generalized Ohm's law. This approach has been criticized by *Bellan* [1994; 1996] who claims that when the parallel electric field is incorporated into the fluid equations the FLRs do not exist. Therefore, FLR is an artifact of ideal MHD plasma and does not exist in reality. In view of this, all observations pertaining to FLRs should be re-evaluated. We believe that MHD formalism with Hall term is a suitable approach at wavelengths of hundreds of kilometers of FLRs. It is meaningless to apply fluid models at such large scales.

During substorms, large amount of energy is pumped into the excitation of the FLRs. The resulting large amplitude FLRs can exert ponderomotive force defined as:

$$F = -\nabla(\frac{<B^2>}{2\mu_0}). \quad (1)$$

Here, the average is taken over the SAW period. The force moves the ambient magnetospheric plasma along the field lines, away from the ionosphere and towards the equatorial region. This redistribution and the spatial movement of the plasma can lead to following effects:

*Changing the frequency of SAWs:* The above change in plasma density can result in frequency shifts of FLRs. These shifts are time-dependent. The shifts are different for different field lines. This leads to a spatial variation in the phases of the FLRs. The temporal dephasing on adjacent field lines can narrow the FLRs [*Rankin et al.*, 1995]. When the width of the resonances is comparable to electron inertial length $\lambda_e$, the mode conversion of SAW into inertial Alfvén wave can take place. The sharp density gradient due to ponderomotive forces can facilitate the mode conversion process. The electron inertial wave propagates the energy away from the resonance and saturates the amplitudes of the FLRs. The dispersion relation of the inertial wave is given as:

$$\omega^2 = \frac{k_z^2 V_A^2}{1 + k_\perp^2 \lambda_e^2}. \quad (2)$$

Here, $\lambda_e = \frac{c}{\omega_{pe}}$ is the electron inertial length and $V_A$ is the Alfvén wave velocity. One of the 3D computer simulations including nonlinear effects and electron inertia was carried out by *Wei et al.* [1994].

*Generation of higher harmonics:* The density perturbations due to the ponderomotive forces result in the excitation of slow magnetosonic wave (SMW) along the ambient magnetic field. The nonlinear interaction between the SMW and FLRs results in the generation of higher harmonics. These higher harmonics can restructure the pre-existing auroral arc and can lead to the filamentation of the FAC that intensifies the auroral arc. [*Rankin et al.*, 1995]

Auroral ion acceleration and the formation of density cavities (in the ionosphere) by the ponderomotive forces has been studied in the past [*Bellan and Stasiewicz*, 1998; *Li and Temerin*, 1993; *Rankin et al.*, 1999]. The goal of the present work is to examine the nonlinear structuring of the auroral arc (by ponderomotive forces) in the presence of electron inertia and the space-time dependent conductivity. In contrast to linear theory the nonlinear effects can lead to rapid structuring and mode conversion of the FLRs [*Hasegawa*, 1976]. These nonlinear effects due to the ponderomotive forces occur on the time scale $t_{nl}$ [*Rankin et al.*, 1994]. The observed SAW fields are generally of the order of 300-400 nT near the ionosphere. The FLR amplitudes (in the frequency range 1-1.3 mHz) can result in the density perturbation of order one in a time interval of of 5-10 SAW period [*Rankin et al.*, 1994; 1999]. The density perturbations can lead to structuring of the auroral arc due to the physical effects listed above.

## 2.2. Mathematical Model

To examine mathematically the nonlinear evolution of FLRs (based on the above physical model), Earth's magnetosphere is approximated to a rectangular "box" with straight magnetic lines [*Rankin et al.*, 1994]. The density and the field gradients are assumed in the $x$ direction (radial direction). The $y$ axis is along the azimuthal direction and the $z$ axis is along the ambient magnetic field. The MHD equations with the generalized Ohm's law are:

$$\frac{\partial \rho}{\partial t} + \nabla \cdot (\rho \mathbf{V}) = 0 \tag{3}$$

$$\rho \frac{d\mathbf{V}}{dt} = -\nabla P + \mathbf{J} \times \mathbf{B} \tag{4}$$

$$\frac{\partial \mathbf{B}}{\partial t} = -\nabla \times \mathbf{E} \tag{5}$$

$$\nabla \times \mathbf{B} = \mu_0 \mathbf{J} \tag{6}$$

$$\mathbf{E} + \mathbf{V} \times \mathbf{B} = \eta \mathbf{J} + \lambda_e^2 \mu_0 \frac{\partial \mathbf{J}}{\partial t}, \tag{7}$$

where $\rho(x)$ is the plasma density, $\mathbf{V}$ is the flow velocity, P is the plasma pressure, and $\mathbf{J}$ is the current density. The quantity $\mathbf{B(x)}$ is the magnetic field, $\mathbf{E}$ is the electric field, $\eta$ is the resistivity in the magnetosphere due to the electron conductivity, and $m_e$ is the electron mass. The last term in Eq.(7) describes the electron inertial effects [*Lysak*, 1990]. These effects play an important role when the characteristic length scale of the system is comparable to the electron inertial length $\lambda_e$. The electron inertial effects lead to the development of a parallel electric field $E_z$. The $E_z$ field can accelerate auroral particles up to hundreds of eV in energies during substorms.

In the next two sections we will study the nonlinear evolution of FLRs in the presence of space-time dependent ionospheric conductivity.

## 3. THE ROLE OF SPACE-TIME DEPENDENT CONDUCTIVITY ON THE EVOLUTION OF FLRs

### 3.1. Plane Polar Ionosphere

We model the Earth's ionosphere as a slab with height integrated Pedersen and Hall conductivities $\Sigma_P$ and $\Sigma_H$, respectively. The sheet ionosphere approach is justified because the extension of the ionosphere is a few hundred kilometers while the wavelength of the typical ULF pulsations is of the order of tens of thousands of kilometers. We define the $z$ axis as the positive upward direction. It should be noted that the conclusions of our studies are not significantly changed by assuming that the geomagnetic field lines are straight. The effects of the dipolar geometry can be incorporated by including an appropriate cosine factor.

### 3.2. Field-aligned Current During Substorms:

During substorms, an enhanced magnetospheric convection sets up an electric field E in the ionosphere. The electric field drives a large ionospheric current. In the presence of non-uniform Pedersen and Hall conductivities, the ionospheric current is closed through the FAC in the magnetosphere [*Lysak*, 1990]. The upward FAC is given as:

$$J = \Sigma_p \nabla_\perp^2 \phi + \nabla_\perp \Sigma_p \cdot \nabla_\perp \phi - \nabla_\perp \Sigma_H \times \nabla_\perp \phi \cdot \mathbf{z}. \tag{8}$$

Here, the conductivities $\Sigma_p$ and $\Sigma_H$ are space-time dependent. The FAC ($J_{alf}$) associated with FLRs is given as:

$$J_{alf} = \frac{1}{\mu_0} \frac{\partial B_y}{\partial x}. \tag{9}$$

Using dispersion relation for Alfvén waves:

$$E_x = V_A B_y. \tag{10}$$

$$J_{alf} = \frac{1}{\mu_0 V_A} \frac{\partial E_x}{\partial x}. \tag{11}$$

The Alfvén travel time between the two ends of the ionosphere is 30s to few minutes. Therefore, in the steady state the current continuity equation for the auroral plasma implies that the FAC ($J_{alf}$) of the FLRs is equal to the ionospheric current given in Eq.(8). During substorms, the FAC is enhanced due to space-time dependent ionospheric conductivities. The enhanced FAC can intensify the pre-existing auroral arc. The ponderomotive forces associated with the field amplitudes (associated with enhanced current) can structure the auroral arc. In a steady state, the Hall current closes within the ionospheric E-region and does not contribute to the FAC. However, when the steady state assumption is relaxed, the Hall current driven by the inductive electric field becomes divergent and closes through the magnetospheric FAC. These effects can become important at high frequencies or at scale lengths of hundreds of kilometers (the scale size of FLRs). The magnetospheric component of the Hall current can significantly modify the dynamic evolution of the auroral arc during substorms [*Buchert and Budnik*, 1997; *Yoshikawa and Itonaga*, 1996]. The role of the divergent Hall current in the evolution of the auroral arc is a complex problem by itself. The present work will not address these studies further.

We will further assume that either Hall conductivity is uniform or the gradient in the Hall conductivity is parallel to the applied electric field. Therefore, the contribution of the Hall term in Eq.(8) is negligible. For uniform ionospheric conductivity, each reflection of Alfvén waves at the ionospheric end can enhance the FAC by a factor of two [*Kan*, 1991]. The enhanced FAC can intensify the pre-existing auroral arc. The ponderomotive forces due to the enhanced magnetic field (associated with the current) increase and lead to restructuring of the auroral arc due to the effects listed in section 2.1 of the paper. The typical height integrated conductivity of the ionosphere is 3-10 mho. The conductivity gradient occurs on the length scale of 20 km. This length scale is much larger than the electron inertial length (a few km) in the auroral zone. Hence, the conductivity gradient does not introduce a new length scale which will lead to fine scale structuring of the auroral arc.

## 4. NONLINEAR EVOLUTION OF FLRs IN THE PRESENCE OF INHOMOGENEOUS AND TIME-DEPENDENT CONDUCTIVITY

Following the approach developed in the literature, we derive amplitude equations describing the nonlinear evolution of FLRs in the presence of electron inertia and the space-time dependent ionospheric conductivity [*Frycz et al.*, 1998; *Rankin et al.*, 1995]. Since the magnetospheric plasma is coupled to the auroral region by FAC, the effects due to the space-time dependent conductivity are incorporated through the boundary conditions on the FAC.

We make following assumptions:
(1) The variation in the field quantities and ionospheric conductivity is along the $x$ direction only, and (2) the nonlinear effects enter through the azimuthal component $B_y$ of the magnetic field [*Rankin et al.*, 1995]. We have assumed $B_z \sim B_0$ and $B_y$ is much less than $B_0$.

### 4.1. Nonlinear evolution of the FLRs in the presence of non-uniform Pedersen Conductivity:

The FAC associated with FLRs is given as:

$$J_{alf} = \Sigma_p \nabla_\perp^2 \phi + \nabla \Sigma_p \cdot \nabla_\perp \phi \qquad (12)$$

Here, $E_x = -\nabla_\perp \phi$.

$$J_{alf} = -\Sigma_p \frac{\partial E_x}{\partial x} - \frac{\partial \Sigma_p}{\partial x} E_x. \qquad (13)$$

Using Faraday's law:

$$\frac{\partial B_y}{\partial t} = -(\nabla \times \mathbf{E})_\mathbf{y}. \qquad (14)$$

$$\frac{\partial B_y}{\partial t} = \frac{\partial E_z}{\partial x} - \frac{\partial E_x}{\partial z}. \qquad (15)$$

$$\frac{\partial^2 B_y}{\partial t^2} = \frac{\partial^2 E_z}{\partial x \partial t} - \frac{\partial^2 E_x}{\partial z \partial t}, \qquad (16)$$

where the parallel electric field is given as:

$$E_z = \lambda_e^2 \mu_0 \frac{\partial J_z}{\partial t}. \qquad (17)$$

From Ampere's law:

$$J_z = \frac{1}{\mu_0} \frac{\partial B_y}{\partial x}. \qquad (18)$$

This leads to :

$$E_z = \lambda_e^2 \frac{\partial^2 B_y}{\partial x \partial t}. \qquad (19)$$

$$\frac{\partial^2 E_z}{\partial x \partial t} = \lambda_e^2 \frac{\partial^4 B_y}{\partial x^2 \partial t^2}. \qquad (20)$$

Next, we proceed to determine $\frac{\partial^2 E_x}{\partial z \partial t}$. Taking the $x$ component of the Eq.(7):

$$E_x + (\mathbf{V} \times \mathbf{B})_x = \eta J_x. \qquad (21)$$

Using Ampere's law and momentum equation:

$$J_x = -\frac{1}{\mu_0} \frac{\partial B_y}{\partial z}. \qquad (22)$$

$$\frac{\partial V_y}{\partial t} = \frac{B_0}{\mu_0 \rho_0} \frac{\partial B_y}{\partial z} - \frac{B_0}{\mu_0 \rho_0} \frac{\partial B_z}{\partial y}. \qquad (23)$$

Combining Eqs.(21-23), we obtain:

$$\frac{\partial^2 E_x}{\partial z \partial t} = 2\Gamma_A \frac{\partial B_y}{\partial t} - V_A^2 \frac{\partial^2 B_y}{\partial z^2} + V_A^2 \frac{\partial^2 B_z}{\partial y \partial z}. \quad (24)$$

Combining Eqs.(10,16,20,24) we obtain:

$$\frac{\partial^2 B_y}{\partial t^2} + 2\Gamma_A \frac{\partial B_y}{\partial t} - V_A^2 \frac{\partial^2 B_y}{\partial z^2} = \lambda_e^2 \frac{\partial^4 B_y}{\partial z^2 \partial t^2} - V_A^2 \frac{\partial^2 B_z}{\partial y \partial z}. \quad (25)$$

We use envelope approximation [Rankin et al., 1995]:

$$A(x,y,z,t) = Re[A(x,t)exp(i\omega t - ik_y y)]sin(k_z z), \quad (26)$$

and neglect the higher derivatives of the slowly varying component $b(x,t)$. Using Eq.(26) and keeping terms to lowest order only, we obtain an equation describing nonlinear evolution of the perturbed magnetic field b.

$$\dot{b} + 2\Gamma_A b = -i\left(\Delta\omega - \frac{1}{4}\omega n\right)b - i\frac{\omega}{2}\lambda_e^2 \frac{\partial^2 b}{\partial x^2} + \frac{1}{2}R\omega. \quad (27)$$

The amplitude equation for the perturbed density n is given as [Rankin et al., 1995]:

$$\ddot{n} + \Omega^2 n = -\frac{1}{2}\omega^2 b^2. \quad (28)$$

Here, R is the strength of the driver:

$$R = \frac{k_y b_c}{k_z B_0}, \quad (29)$$

where $b_c$ is a constant [Rankin et al., 1995]. The other symbols are as follows:

$$b(t) = \frac{b_y}{B_0}; \frac{\delta\rho}{\rho_0} = n\cos(2k_z z), \text{ and } \Omega = 2k_z C_s. \quad (30)$$

$$\Gamma_A = -\frac{1}{2}\frac{\eta}{\mu_0}\nabla^2, \text{ and } \Delta\omega(x) = \omega(x_c - x)/2L. \quad (31)$$

The Eq.(27) assumes an Alfvén wave profile of the following form,

$$\frac{1}{V_A^2} \sim \frac{k_z^2}{\omega^2}\left(1 - \frac{x-x_c}{L} + \frac{\delta\rho}{\rho_0}\right). \quad (32)$$

Here, L is the length scale characterizing the radial variation of the Alfvén speed $(V_A)$ around $x_c$. The point $x_c$ is determined by the resonance condition $V_A^2(x_c) = \omega^2/k_z^2$. The quantity $\delta\rho$ (Eq. 32) denotes the density fluctuations about the density $\rho_0$.

It is clear from Eq.(27) that the ponderomotive forces and electron inertia result in the frequency shifts of the SAWs (which form FLRs) by $\frac{1}{4}\omega n$ and by $\frac{\omega}{2}\lambda_e^2\frac{\partial^2 b}{\partial x^2}$, respectively [Rankin et al., 1995]. The frequency shift (due to the ponderomotive forces) is proportional to the density perturbation (n) of the magnetosonic wave of frequency $\Omega$. The density perturbation (Eq.28) is driven by the ponderomotive forces resulting from the large amplitude of the FLRs.

In order to study the nonlinear evolution of FLRs in the presence of space-dependent Pedersen conductivity, the above equations should be coupled with the boundary conditions on the FAC associated with FLRs. The boundary conditions at the two ionospheric ends are obtained by equating the FAC of FLRs with the FAC which closes the Pedersen current. Using Eqs.(9,13) we obtain:

$$\frac{1}{\mu_0}\frac{\partial B_y}{\partial x} = -\Sigma_p V_A \frac{\partial B_y}{\partial x} - \frac{\partial \Sigma_p}{\partial x}V_A B_y. \quad (33)$$

The Eqs.(27,28) along with the boundary condition (Eq.33) can be studied numerically to examine the nonlinear evolution of FLRs in the presence of space-dependent ionospheric conductivity.

### 4.2. Nonlinear evolution of FLRs in the presence of time-dependent Pedersen conductivity

In this section, we study the case when the Pedersen conductivity is independent of "x", but is a function of the "t" co-ordinate only. We note that the time scale of variation of the ionospheric conductivity is much larger than Alfvén bounce period. Therefore, the steady state is not reached during the time scale of variation in the conductivity. Assuming that the time-dependent conductivity does not provide additional sources of charges, we use current continuity equation to obtain boundary conditions at the two ionospheric ends:

$$\frac{1}{\mu_0}\frac{\partial B_y}{\partial x} = -\Sigma_p V_A \frac{\partial B_y}{\partial x} \quad (34)$$

Here, $\Sigma_p$ is time-dependent.

The boundary condition (Eq.34) is coupled with Eqs.(27,28) to study the nonlinear evolution of FLRs in the presence of time-dependent ionospheric conductivity. Efforts are in progress to solve Eqs.(27-28) numerically using the boundary conditions given by Eqs.(33,34) with appropriate profiles for the ionospheric conductivity [Richmond, 1995]. The time-dependent conductivity can modify the dynamical evolution of the auroral arc observed during the substorms.

We list the plasma sheet parameters [Hughes, 1995] as follows: $T_e$ =1keV; $T_i$ =0.5keV; $\rho_i$ =100km; $B_0$ =20$\gamma$;

$n_0 = 1$cm$^{-3}$; $\beta = 0.2$. The wave and plasma parameters [*Wahlund et al.*, 1994] in the auroral zone at the Freja altitude are given as: $T_e = 5$eV; $T_i = 1$eV; $B_0 = 3 \times 10^4$ nT; $n_0 = 100$ cm$^{-3}$; $C_s = 16$ km/s; $V_A = 15 \times 10^3$ km/s. Here, $\gamma = 10^{-5}$ Gauss. In addition to the cold ion component, the ions with energies in the range 5-30 eV have also been reported at the Freja altitudes [*Wahlund et al.*, 1998].

In the cold plasma with low plasma $\beta$ value, the absence of thermal pressure fails to prevent the concentration of plasma into the nodes of the Alfvén waves. Therefore, large magnitudes of density perturbations are possible, which can lead to structuring of the auroral arc. It has been reported that the nonlinear effects resulting from the ponderomotive forces are strong in cold plasma, $\beta << b^2$ [*Rankin et al.*, 1994]. Therefore, with b of one per cent, we conclude (based on the above parameters) that the nonlinear effects can play a significant role in structuring the auroral arc in the cold component of the ions observed at the Freja altitudes. The effects due to the ponderomotive forces will be less prominent in the hot component of the auroral plasma observed at the Freja altitudes as well as near the plasma sheet.

## 5. SUMMARY AND CONCLUSIONS

We have derived amplitude equations and boundary conditions to study the nonlinear evolution of FLRs in the presence of electron inertia and the space-time dependent ionospheric conductivity. Efforts are in progress to solve these equations numerically to study the structuring of the FLRs and of the auroral arc. The outcome of these studies will be compared with data available from the CANOPUS ground based network. The theoretical and data analysis studies will shed some light on the mechanism of the small scale structures of the auroral arcs observed during substorms. We note that the space-time dependent conductivity can modify the phase mixing lengths of the FLRs. These effects and the feedback of the currents (associated with FLRs) on the ionospheric conductivity will be examined in future. The Eqs.(27-28;33,34) provide the starting point to carry out extensive studies relating to these effects.

Our studies ignore the role of space-time dependent conductivity on the magnetosonic wave describing the density perturbations. These effects are essentially collisional and will result in the linear damping of the wave. These effects will not significantly modify the nonlinear evolution of the FLRs. We have also ignored the effects arising from the non-uniform Alfvén speed and the temperature changes due to the inhomogeneous conductivity [*Lysak*, 1990]. The space-time dependent conductivity can modify the growth rate of the feedback instability as well as the reflection of Alfvén waves from the auroral acceleration region [*Lysak*, 1991; *Trakhengertz and Feldstein*, 1984]. The effects of the Hall current are significant in a dipole geometry and can lead to the coupling between the toroidal and poloidal modes [*Allan and Knox*, 1979]. These studies on the Hall current are also important in interpreting the ground signals of the FLRs. We will report the role of these effects in our future publications.

*Acknowledgments.* Manju Prakash would like to thank Prof. John Samson for his hospitality during her visit to the University of Alberta.

## REFERENCES

Allan, W., and F. B. Knox, A dipole field model for axis-symmetric Alfvén waves with finite ionosphere conductivities, *Planet. Space Sci., 27,* 79, 1979.

Bellan, P. M., Alfvén 'resonance' reconsidered: Exact equations for wave propagation across a cold inhomogeneous plasma, *Phys. Plasmas, 1,* 3523, 1994.

Bellan, P. M., Mode-conversion into non-MHD waves at the Alfvén layer: The case against the field line resonance concept, *J. Geophys. Res., 101,* 24,887, 1996.

Bellan, P. M., and K. Stasiewicz, Fine scale cavitation of ionospheric plasma caused by inertial Alfvén wave ponderomotive force, *Phys. Rev. Lett., 8,* 3523, 1998.

Buchert, S. C., and F. Budnik, Field-aligned current distributions generated by a divergent Hall current, *Geophys. Res. Lett., 24,* 297, 1997.

Ellis, P., and D. J. Southwood, Reflection of Alfvén waves by non-uniform ionosphere, *Planet. Space Sci., 31,* 107, 1983.

Frycz, P., R. Rankin, J. C. Samson, and V. T. Tikhonchuk, Nonlinear field line resonances: dispersive effects, *Phys. Plasmas, 5,* 3565, 1998.

Glassmeier, K. H., Reflection of MHD-waves in the PC4-5 period range at ionosphere with non-uniform conductivity distributions, *Geophys. Res. Lett., 10,* 678, 1983.

Greenwald R. A. and A.D.M. Walker, Energetics of long period resonant hydromagnetic waves, *Geophys. Res. Lett., 7,* 745, 1980.

Hasegawa, A., Particle acceleration my MHD surface wave and formation of aurora, *J. Geophys. Res., 81,* 5083, 1976.

Hughes, W. J., The magnetopause, magnetotail, and magnetic reconnection, in *Introduction to Space Physics*, edited by M. G. Kivelson and C. T. Russell, 227, Cambridge University, 1995.

Kan, J. R., Synthesizing a global model of substorms, *Magnetospheric substorms, AGU monograph, 64,* eds. J. R. Kan, T. A. Potemra, S. Kokubun, and T. Iijima, 74, 1991.

Kivelson, M. G., and D. J. Southwood, Coupling of global magnetospheric MHD eigen modes to field line resonances, *J. Geophys. Res., 91,* 4345, 1986.

Li., X., and M. Temerin, Ponderomotive effects on ion acceleration in the auroral zone, *Geophys. Res. Lett.*, *20*, 13, 1993.

Louarn, P., J. E. Wahlund, T. Chust, H. de Feraudy, A. Roux, B. Holback, P. O. Dovner, A. I. Eriksson, and G. Holmgren, Observations of kinetic Alfvén waves by FREJA spacecraft, *Geophys. Res. Lett.*, *21*, 1847, 1994.

Lysak, R. L., and C. T. Dum, Dynamics of magnetosphere-ionosphere coupling including turbulent transport, *J. Geophys. Res.*, *88*, 365, 1983.

Lysak, R. L., Electrodynamic coupling of the magnetosphere and the ionosphere, *Space Sci., Rev.*, *52*, 33, 1990.

Lysak, R. L., Feedback instability of the ionospheric resonant cavity, *J. Geophys. Res.*, *96*, 1553, 1991.

Prakash, M., and R. L. Lysak, Anomalous resistivity due to double layers, model for auroral arc thickness, *Geophys. Res. Lett.*, *19*, 2159, 1992.

Rankin, R., J. C. Samson, P. Frycz, Simulations of driven field line resonances in the Earth's magnetosphere, *J. Geophys. Res.*, *98*, 21,341, 1993.

Rankin, R., P. Frycz, V. T. Tikhonchuk, and J. C. Samson, Nonlinear standing shear Alfvén waves in the Earth's magnetosphere, *J. Geophys. Res.*, *99*, 21,291, 1994.

Rankin, R., P. Frycz, V. T. Tikhonchuk, J. C. Samson, Ponderomotive saturation of magnetospheric field line resonances, *J. Geophys. Res.*, *22*, 1741, 1995.

Rankin, R., P. Frycz, J. C. Samson, and V. T. Tikhonchuk, Shear flow vortices in magnetospheric plasmas, *Phys. Plasmas*, *4*, 829, 1997.

Rankin, J. C. Samson, V. T. Tikhonchuk, and I. Voronkov, Auroral density fluctuations on dispersive field line resonances, *J. Geophys. Res.*, *104*, 4399, 1999.

Richmond, A. D., Ionospheric Dynamics, *Handbook of atmospheric electrodynamics*, Hans Volland (Ed.) CRC press, 1995.

Samson J. C., R. Rankin, P. Frycz, V. T. Tikhonchuk, and L. L. Cogger, Observations of a detached, discrete arc in association with field line resonances, *J. Geophys. Res.*, *96*, 15,683, 1991.

Samson J. C., D. D. Wallis, T. J. Hughes, F. Creutzberg, J. M. Ruohoniemi, and R. A. Greenwald, Substorm intensifications and field line resonances in the nightside magnetosphere, *J. Geophys. Res.*, *97*, 8495, 1992a.

Samson J. C., L. R. Lyons, P. T. Newell, F. Creutzberg, and B. Xu, Proton aurora and substorm intensification, *Geophys. Res. Lett.*, *19*, 2167, 1992b.

Samson, J. C., L. L. Cogger, and Q. Pao, Observations of field line resonances, auroral arcs, and auroral vortex structures, *J. Geophys. Res.*, *101*, 17,373, 1996.

Stasiewicz, K., G. Gustafsson, G. Marklund, P.-A. Lindqvist, J. Clemmons, L. Zanetti, Cavity resonators and Alfvén resonance cones observed on Freja, *J. Geophys. Res.*, *102*, 2565, 1997.

Southwood, D. J., and W. J. Hughes, Theory of hydromagnetic waves in the magnetosphere, *Space Sci. Rev.*, *35*, 301, 1983.

Trondsen, T. S., L. L. Cogger, and J. C. Samson, Asymmetric auroral arcs and inertial Alfvén waves, *Geophys. Res. Lett.*, *24*, 2945, 1997.

Trakhtengertz, V. Y., and A. Y. Feldstein, Quiet auroral arcs: ionospheric effect of magnetospheric convection stratification, *Planet. Space Sci.*, *32*, 127, 1984.

Voronkov, I., R. Rankin, P. Frycz, V. T. Tikhonchuk, and J. C. Samson, Coupling of shear flow and pressure gradient instabilities, *J. Geophys. Res.*, *102*, 9639, 1997.

Wahlund, J. E., P. Louarn, T. Chust, H. de Feraudy, A. Roux, B. Holback, P. O. Dovner, and G. Holmgren, Ion-acoustic turbulence and the nonlinear evolution of kinetic Alfvén waves in aurora, *Geophys. Res. Lett.*, *21*, 1831, 1994.

Wahlund, J. E., A. I. Eriksson, B. Holback, M. H. Boehm, J. Bonnell, P. M. Kintner, C. E. Seyler, J. H. Clemmons, L. Eliasson, D. J. Knudsen, P. Norqvist, L. J. Zanetti, Broadband ELF plasma emission during auroral energization, 1. slow ion-acoustic waves, *J. Geophys. Res.*, *103*, 4343, 1998.

Walker, A. D. M., J. M. Ruohoniemi, K. B. Baker, and R. A. Greenwald, Spatial and temporal behaviour of ULF pulsations observed by the Goose Bay HF radar, *J. Geophys. Res.*, *97*, 12,187, 1992.

Wei, C. Q., J. C. Samson, R. Rankin, and P. Frycz, Electron inertial effects on geomagnetic field line resonances, *J. Geophys. Res.*, *99*, 11,265, 1994.

Yoshikawa, A., and M. Itonaga, Reflection of shear Alfvén waves at the ionosphere and the divergent Hall current, *Geophys. Res. Lett.*, *23*, 101, 1996.

---

M. Prakash, Department of Physics and Astronomy, SUNY at Stony Brook, Stony Brook, New York 11794-3800, USA.

R. Rankin, Canadian Network for Space Research, Department of Physics, University of Alberta, Edmonton, Canada T6G 2J1.

# FAST Observations of Upward Accelerated Electron Beams and the Downward Field-Aligned Current Region

R. C. Elphic[1], J. Bonnell[1,3], R. J. Strangeway[2], C. W. Carlson[3], M. Temerin[3],
J. P. McFadden[3], R. E. Ergun[3], and W. Peria[4]

Auroral processes result from the exchange of energy and momentum between the magnetosphere and ionosphere, and the current systems arising from this exchange. Field-aligned potential drops in the upward field-aligned current (FAC) region of the auroral zone maintain a required current density in the face of opposing mirror forces. But field-aligned potential drops occur in downward current regions as well, and result in upgoing accelerated electron beams. We investigate the occurrence of upward accelerated electrons with altitude, magnetic local time and season using FAST. By choosing orbits having perigee over the equator, northern-southern hemisphere differences in altitude coverage are largely removed. For an interval near solstice, no electron beams were observed over the summer auroral zone at FAST altitudes, while over the winter hemisphere, 71% of the crossings had upgoing beams, a clear seasonal effect. During the equinox interval the distribution between hemispheres was much more balanced, but the overall occurrence was less (~46%) than during winter solstice. There is a clear tendency for upgoing accelerated electron beam occurrence at FAST altitudes to maximize when dipole tilt places the field line footpoint deep in the nightside ionosphere, where background ionospheric density is lowest. In other words, as ionospheric density decreases, the potential structure tends to move to lower altitudes.

## INTRODUCTION

The auroral zone is the most dramatic and energetic manifestation of the inter-regional processes coupling the hot,

[1]Los Alamos National Laboratory, Los Alamos, New Mexico
[2]Institute of Geophysics and Planetary Physics, University of California, Los Angeles, California
[3]Space Sciences Laboratory, University of California, Berkeley, California
[4]University of Washington, Seattle, Washington

Magnetospheric Current Systems
Geophysical Monograph 118
Copyright 2000 by the American Geophysical Union

tenuous magnetospheric plasma and the cold, dense, collisional ionospheric plasma. Auroral processes are thus intimately linked to the driving mechanisms of magnetospheric current systems. To maintain current continuity, large potential drops develop along auroral zone field lines. Field-aligned potential drops develop in regions of upward FACs to accelerate the tenuous plasma sheet electrons into the ionosphere, in effect widening the loss cone to provide sufficient current flow in the circuit [*Knight*, 1973]. In general, no such potential was thought to be necessary for the downward FACs, because the density of ionospheric electrons would be sufficient to supply all the necessary current.

But sporadic evidence for field-aligned potential drops and upward accelerated electron beams in downward FAC regions has been observed over the years [*Shelley et al.*,

1976; *Sharp et al.,* 1980; *Klumpar and Heikkila,* 1982; *Johnstone and Winningham,* 1982; *Burch et al.,* 1983; *Gorney et al.,* 1985; *Hultqvist et al.,* 1988; *Marklund et al.,* 1994; *Boehm et al.,* 1995]. Recent observations by the Fast Auroral SnapshoT Explorer (FAST) mission have demonstrated that such potential drops are a very common feature in the auroral zone during winter [*Carlson et al.,* 1998; *McFadden et al.,* 1998; *Ergun et al.,* 1998]. Regions of downward current thus appear to sometimes require a parallel potential structure to maintain current flow, evidently when the ionospheric plasma distribution lacks sufficient charge carriers to do so otherwise.

Preliminary analysis suggests that the characteristic scale size of downward current regions, and their associated potential structures, is smaller than that of the typical inverted-V/upward current region. Current densities in the downward current regions are thus correspondingly larger. Early analysis of FAST upgoing electron beam observations indicated that they are found between 2000 and 4000 km altitude over the winter auroral zone on about 78% of the passes [*Carlson et al.,* 1998]. The occurrence probability drops quickly below 2500 km, and the overall beam occurrence varies with season. (See also *Peria et al.,* this volume). It has been demonstrated that the observed electron beams have number fluxes that agree closely with the current density inferred from magnetometer measurements [*Carlson et al.,* 1998; *Elphic et al.* 1998]. Thus these beams can carry most of the field-aligned current density in a downward current region: hence their importance.

To place this study in the broader context of inter-regional processes (the exchange of energy and momentum between the magnetosphere and ionosphere) and current-driving mechanisms, we consider the following simple picture. In the spirit of *Parker* [1996; and this volume] and *Iijima* [this volume], the observed magnetic deflections at low altitudes can be thought of as manifestations of Maxwell stresses applied to the ionosphere, $B_{dip} \cdot \Delta B/\mu_o$ (see also *Strangeway et al.,* this volume). These stresses are balanced in steady state in the ionosphere by friction due to ion-neutral collisions: $\nabla \cdot (B_{dip} \cdot \Delta B/\mu_o) = \rho(\mathbf{v}-\mathbf{U})\nu_{in}$, where $B_{dip}$ is the ionospheric magnetic field, $\Delta B$ is the perturbation field component, $\rho$ is the ionospheric plasma density, $\mathbf{v}$ is the ion drift velocity, $\mathbf{U}$ is the neutral drift velocity and $\nu_{in}$ is the ion-neutral collision frequency. The low-altitude Maxwell stresses should map to the motions of plasma in the magnetosphere, as illustrated in Figure 1 for the premidnight auroral zone. The top panel shows the stressed, tilted magnetic field lines and their connection to the ionosphere; here the high latitude stresses are applied in the anti-sunward/eastward direction, while at lower latitudes the stress is sunward/westward. The bottom panel shows the magnetospheric plasma motions needed to maintain the magnetic stress against friction in the ionosphere, namely antisunward convection in the outer magnetosphere and sunward convection in the inner magnetosphere. Field-aligned currents are present where there are shears in the magnetic field, in this case with downward FACs at the poleward and equatorward edges, and upward FACs in the

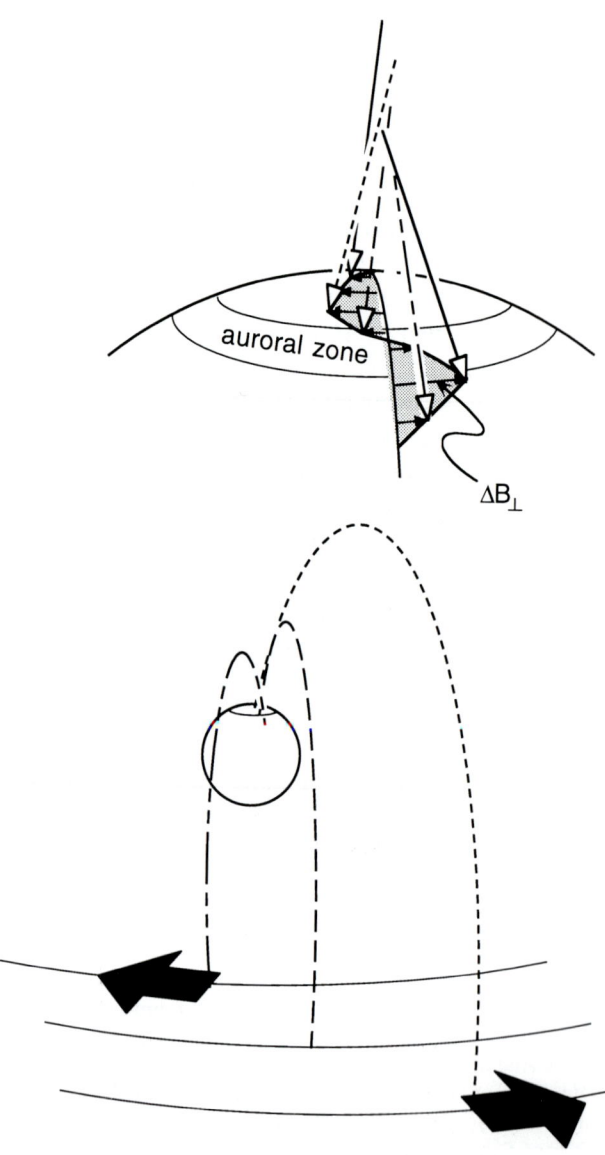

**Figure 1.** (top) Schematic diagram of sheared magnetic field lines applying Maxwell stress to the ionospheric plasma in the premidnight region. The horizontal perturbation field is shown by black arrows, which is opposite to the sense of the stress. (bottom) The magnetospheric motion of the flux tubes that is consistent with the stress applied to the ionosphere.

middle. Each sense of FAC is associated with electrostatic potential structure necessary to supply the current density required by magnetic field shears that are imposed by motions.

Using FAST observations, we have studied the altitude, invariant latitude (ILAT) distribution, seasonal dependence, and activity dependence of upward accelerated electron beams in the near-midnight magnetic local time (MLT) regions. We shall see that, for constant plasma sheet conditions and downward current density, a lower-density ionospheric profile will correspond to a lower-altitude accelerating potential, whereas for a higher-density ionospheric profile the potential structure will appear at higher altitudes.

## APPROACH

We make judicious use of the rapidly precessing FAST polar orbit to study beam dependence on altitude, MLT, and seasonal or dipole tilt effects. By picking out periods when perigee is at the equator, and apogee is in the midnight sector, we can confine our study to near midnight MLTs with similar altitude coverage in the north and south. Then by dividing these intervals into equinox and solstice subsets, we can study seasonal effects (which primarily affect the ionospheric density distributions, and hence possibly the altitudes at which the accelerating potentials form). For the equinox cases, we expect the ionospheric density distributions to be similar between the two hemispheres since the field line footpoints in the auroral zone have similar solar zenith angles. For the solstice cases one hemisphere will have auroral footpoints in the depleted, dark winter ionosphere, the other in the denser, sunlit summer ionosphere. We focus on two intervals in the FAST data meeting the foregoing conditions: the equinox interval 15 - 29 September, 1997, and the solstice interval 25 December, 1997 - 8 January, 1998. The FAST orbit provides roughly 11 northern and southern auroral passes per day, so each 15-day interval could potentially provide as many as 165 auroral passes through the near-midnight region for each hemisphere.

Plate 1 shows magnetic field and electron-related quantities that illustrate some of the features of the pre-midnight winter auroral zone. The top panel shows the east-west perturbation magnetic field, the second panel shows the field-aligned current density inferred from the slope of the field perturbation, the third panel shows the inferred height-integrated Pedersen conductivity $\Sigma_P$ based on precipitating electron energy fluxes [*Robinson et al.*, 1987], and the electron energy and pitch angle spectrograms are shown at the bottom. This pass illustrates that downward currents are often found at the edges of the inverted-V electron structures, and these downward current regions tend to be smaller and carry higher current densities than the inverted-V regions. Moreover, they are associated with minima in the inferred $\Sigma_P$, not at gradients in $\Sigma_P$ as would be expected if a constant electric field is being maintained in the ionosphere. The upward electron flux in these downward FAC regions is at times so high that the entire column of winter polar ionospheric plasma could be severely depleted within tens of seconds to minutes [*Elphic et al.*, 1998]. The downward FACs are found on the poleward side of the antisunward/eastward Maxwell stress region ($\Delta B_{EAST} < 0$), and on the equatorward side of the sunward/westward Maxwell stress region ($\Delta B_{EAST} > 0$). The downward current is carried here by upgoing electrons that have been accelerated, in some cases achieving energies above 1 keV. For this survey, we identify an electron beam event/downward FAC region using upgoing electrons.

The events were selected using FAST summary data, which have a 5-sec temporal resolution. We identified the accelerated upgoing electron beams on the basis of both pitch angle (confined to a few degrees within the loss cone) and energy (E > 20 eV), and energy flux greater than $10^7$ eV cm$^{-2}$ s$^{-1}$ sr$^{-1}$ eV$^{-1}$ (most beams have much higher energy fluxes, typically in the range of $10^9$ eV cm$^{-2}$ s$^{-1}$ sr$^{-1}$ eV$^{-1}$). The data were scanned and beam events were logged by orbit and time, and we also noted if the peak beam energy exceeded 1 keV. The corresponding altitude, magnetic local time (MLT), invariant latitude (ILAT), activity level (Kp) were also determined for each case.

## ALTITUDE AND ILAT DISTRIBUTIONS

The solstice cases in our study are confined to within a few hours local time of midnight, though orbital coverage extends outside this range. There is a slight asymmetry in MLT sampling between the north and south cases during the equinox interval, with northern observations biased to earlier local times and southern to later local times. In altitude, the solstice cases can be found as low as about 1600 km with many cases below 2000 km, whereas the equinox cases are mostly found above 2000 km. No beams were found below about 1500 km. These results are similar to those discussed by *Carlson et al.*, [1998]; there is an apparent cutoff of occurrence at altitudes above roughly 3500 km is due to orbital coverage for the intervals used here (the orbit passes through altitudes above 3500 km mainly at invariant latitudes equatorward of the auroral zone). The northern and southern altitude - ILAT distributions for the equinox cases are very similar. Most beams are found within the range of 60° - 80° ILAT, the normal range of auroral phenomena near midnight MLT.

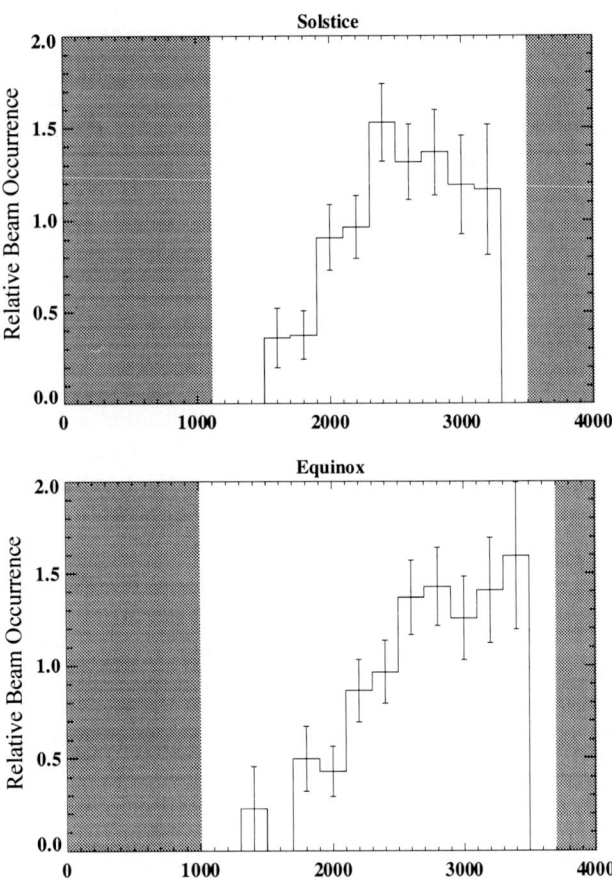

**Figure 2.** Histograms of number of upgoing beams divided by the number of orbits within each bin. Error bars denote the corresponding $N^{1/2}$ uncertainty, where N is the number of events in a bin. Upgoing electron beam occurrence drops below about 2000 km altitude, and reaches a fairly steady occurrence above 2500 km. Shaded regions denote no coverage for these intervals.

Figure 2 shows the fraction of electron beam cases divided by the fraction of orbital samples in that bin with altitude for the two seasons (the error bars denote the uncertainty based on $N^{1/2}$, where N is the number of cases in the bin). This ratio is proportional to the total number of beam events in a bin divided by the total number of orbital opportunities to observe a beam within that bin (in other words, fractional occurrence). For both the solstice and equinox cases, the beam occurrence drops to zero at low altitudes before the orbital coverage does, meaning that the incidence of upgoing electron beams really decreases below about 1600 km. Above about 3300-km altitude, the orbital coverage for the selected time intervals does not permit an estimate of beam occurrence. For both seasons, the beam fractional occurrence increases above 1600-km altitude and levels off at a roughly constant occurrence above about 2500-km altitude. This may imply that the entire field-aligned accelerating potential structure tends to lie within the altitudes between 1600 and 2500 km, with possibly a slightly higher altitude distribution for the equinox cases.

## SEASONAL AND ACTIVITY EFFECTS

Table 1 presents a list by day of the number of FAST near-midnight auroral passes with upgoing accelerated electron beams and the number of passes without beams. For solstice, the northern winter passes with beams outnumbered the passes without beams (71% had beams), whereas southern summer passes yielded no beams at all. Note that the overall coverage is similar, 142 passes in the north and 133 in the south. FAST telemetry coverage during the equinox period was somewhat lower than during solstice, with data from a total of 117 northern passes and 126 southern passes returned to the ground (out of 165 possible). Of the data returned, accelerated upgoing electron beams were observed on 59% of the northern and 33% of the southern passes (the bottom line of Table 1). This difference in occurrence between the northern and southern equinox passes is puzzling because we chose the orbital configuration to remove any biases in altitude, MLT, or ILAT between the two hemispheres.

Table 1. Upgoing Accelerated Electron Beam Occurrence

| Solstice (number of passes) | | | Equinox (number of passes) | | |
|---|---|---|---|---|---|
| | North | South | | North | South |
| Date | Yes/No | Yes/No | Date | Yes/No | Yes/No |
| 12/25/1997 | 7 / 3 | 0 / 10 | 9/15/1997 | 5 / 3 | 3 / 4 |
| 12/26/1997 | 5 / 4 | 0 / 10 | 9/16/1997 | 7 / 3 | 4 / 5 |
| 12/27/1997 | 3 / 7 | 0 / 10 | 9/17/1997 | 4 / 4 | 4 / 3 |
| 12/28/1997 | 6 / 4 | 0 / 10 | 9/18/1997 | 5 / 2 | 3 / 5 |
| 12/29/1997 | 5 / 5 | 0 / 9  | 9/19/1997 | 1 / 5 | 2 / 4 |
| 12/30/1997 | 5 / 2 | 0 / 8  | 9/20/1997 | 7 / 1 | 4 / 4 |
| 12/31/1997 | 6 / 4 | 0 / 9  | 9/21/1997 | 4 / 5 | 3 / 6 |
| 1/ 1/1998  | 4 / 2 | 0 / 6  | 9/22/1997 | 4 / 5 | 2 / 8 |
| 1/ 2/1998  | 10 / 1| 0 / 11 | 9/23/1997 | 3 / 3 | 4 / 5 |
| 1/ 3/1998  | 7 / 3 | 0 / 11 | 9/24/1997 | 4 / 4 | 2 / 6 |
| 1/ 4/1998  | 7 / 2 | 0 / 9  | 9/25/1997 | 5 / 2 | 1 / 3 |
| 1/ 5/1998  | 9 / 1 | 0 / 10 | 9/26/1997 | 5 / 1 | 3 / 5 |
| 1/ 6/1998  | 8 / 1 | 0 / 11 | 9/27/1997 | 5 / 3 | 2 / 6 |
| 1/ 7/1998  | 9 / 1 | 0 / 9  | 9/28/1997 | 6 / 4 | 4 / 6 |
| 1/ 8/1998  | 10 / 1| 0 / 11 | 9/29/1997 | 4 / 3 | 1 / 6 |
| Total      | 101 / 41 | 0 / 133 | Total | 69 / 48 | 42 / 84 |
|            | 71%   | 0%     |           | 59%    | 33%    |

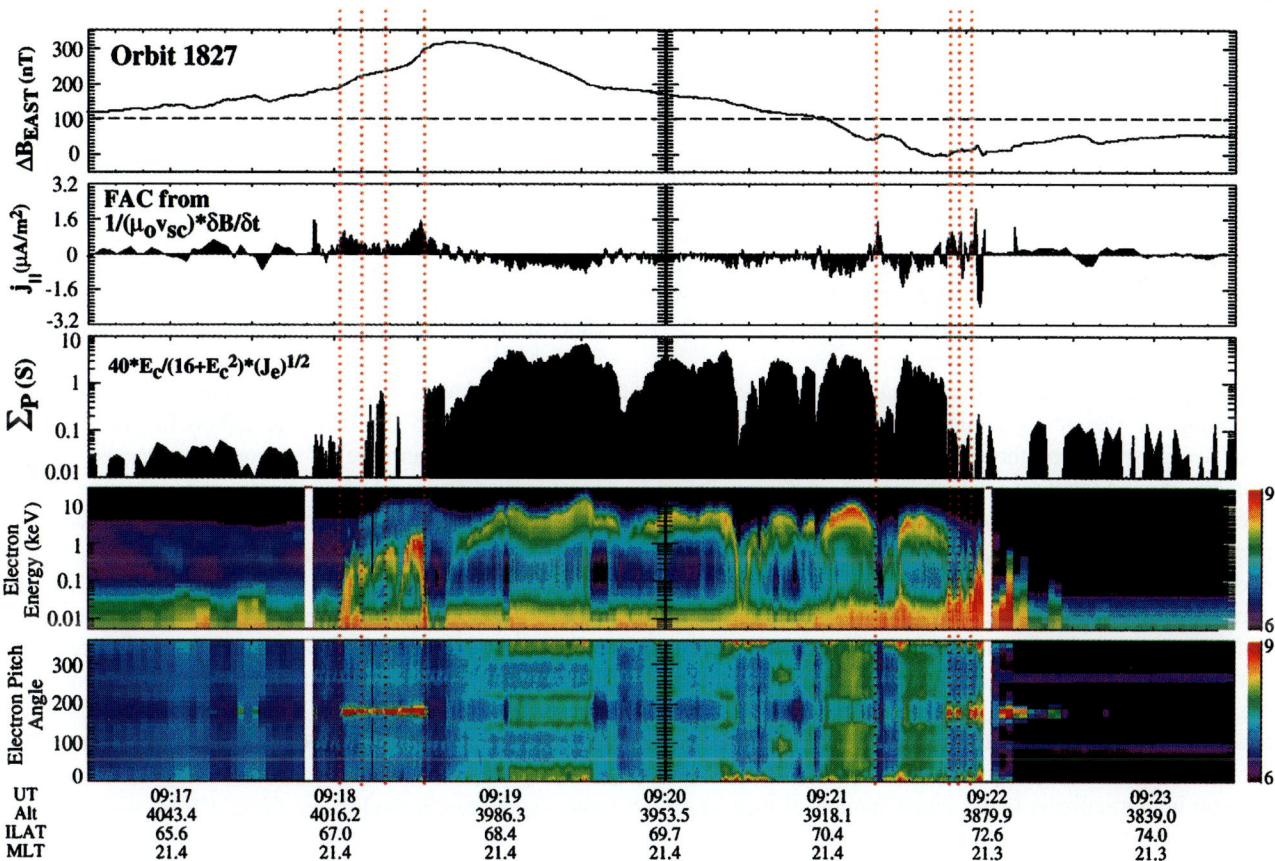

**Plate 1.** Data from a FAST auroral pass on February 6, 1997. The panels show the perturbation east-west magnetic field component, the inferred sheet current density (downward currents are positive), the inferred height-integrated Pedersen conductance based on electron precipitation, and electron energy and pitch angle spectrograms (color coded for log eV cm$^{-2}$ s$^{-1}$ sr$^{-1}$ eV$^{-1}$). The pass is at about 21.4 MLT, at an altitude of about 4000 km.

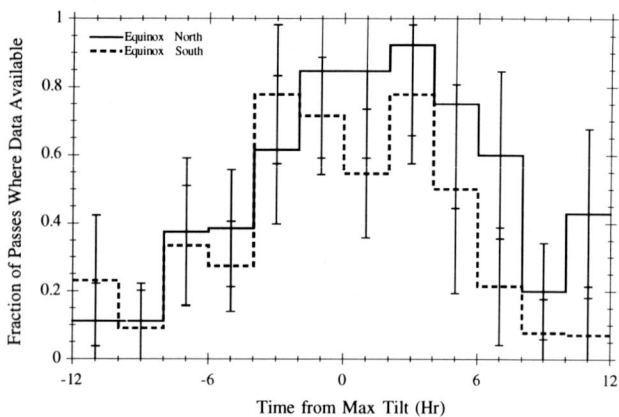

**Figure 3.** (top) Fractional occurrence of northern (southern) equinox passes with beams plotted as solid (dotted) lines versus the time of maximum away dipole tilt, 0440 UT for northern hemisphere cases, 1640 UT for southern hemisphere cases. The beam occurrence is peaked at the time of maximum away tilt for each corresponding hemisphere.

The apparent difference between equinox north and south occurrence is due to systematics in orbital coverage for FAST. Downlink of the data depends on availability of suitable ground stations, which occurs preferentially at about the same times every day. By the same token, outages due to lack of ground station availability also tend to occur systematically at other times every day. During the equinox season studied here, these outages just happen to occur at times far from the maximum dipole tilt for the north equinox cases, but are centered on the southern maximum tilt time. Thus the coverage bias tends to favor times when northern away tilt is maximized, and vice versa for the southern tilt conditions. We can take this bias into account by normalizing to the actual amount of time observations were possible at a given dipole tilt. After correcting for these biases, we see that the fractional occurrence shown in Figure 3 is essentially identical between the two hemispheres, and clearly demonstrates that dipole tilt (hence solar zenith angle) affects the observed beam occurrence for equinox conditions at FAST altitudes.

The upgoing electron beam energy is a rough indicator of the potential drop. We have tabulated the occurrence of electron beams with a peak energy below 1 keV, and those having peak energies above 1 keV. We have calculated the fraction of cases of each category as a function of Kp, together with the distribution of all Kp values found during the solstice and equinox intervals. For both solstice and equinox, there is a dearth of high-energy beams for the quietest conditions, Kp ~ 0. There is some suggestion that more energetic beams are found as Kp increases, but this tendency is far from clear. It is also of interest to determine if beam occurrence is higher during active times. We have compiled the fraction of beam cases in bins of Kp, together with the overall fraction of Kp values for the solstice and equinox periods. The histograms (not shown) are very similar (and the uncertainties generally overlap), indicating that beam occurrence does not appear to increase significantly for higher Kp values.

## SUMMARY AND DISCUSSION

We have studied the occurrence of upgoing accelerated electron beams by using precession of the FAST orbit at times when perigee (and apogee) is over the equator, providing comparable coverage in altitude, ILAT, MLT and dipole tilt over each hemisphere. We then selected intervals near northern winter solstice and equinox to clarify the role of ionospheric density in beam occurrence. We find that beams are found starting at altitudes near 1600 km, and rise to a constant occurrence frequency around 2500 km. This may imply that the accelerating potentials form in the neighborhood of 1600 - 2500 km, and not higher.

The overall occurrence of upgoing accelerated electron beams around midnight MLT is greatest near winter solstice (71% of the passes) and least near summer solstice (0%). The equinox occurrence was lower, with 59% and 33% of passes exhibiting beams in the north and south, respectively. The lower occurrence and the apparent asymmetry appears to be related to both dipole tilt and orbital coverage. The highest occurrence of beams is found when the magnetic pole is most tilted away from the sun, centered on 04:40 UT in the north and 16:40 UT in the south. This geometry means that the ionospheric footpoints have their largest solar zenith angles, are farthest from the terminator and hence have the lowest ionospheric densities. Moreover, these results imply that the altitude of the accelerating potential is lowest when the dipole tilt away from the sun is greatest. These accelerating potentials are thought to form closed banana-shaped surfaces, as shown in Figure 4. The figure shows that, for steady conditions, these surfaces would rise and fall in altitude as the dipole rocks during equinox.

Surprisingly, beam occurrence does not show a significant increase with geomagnetic activity. The occurrence of electron beams with peak energies greater than 1 keV does not clearly increase with Kp. These results imply that higher potentials are not found as activity (and the net field-aligned current) increases. This may indicate that during high Kp larger net field-aligned currents are carried by broader regions of modest current density, rather than in the nature of how magnetospheric stresses are accomodated by the ionosphere. In the winter nightside ionosphere the background density is low, and enhancements to this density occur only where significant electron precipitation occurs (see Plate 1). The frictional force $\rho(\mathbf{v}-\mathbf{U})\nu_{in}$ is thus higher in inverted-V regions, and lower outside these regions. An applied magnetospheric stress can cause the ionosphere to "slip" more easily in the low-density, low-friction regions away from the inverted-V precipitation. In other words, the low-density ionosphere can more easily reduce the applied stress, which results in a larger magnetic field shear between the high-density, high-friction inverted-V region and the low-density, low-friction region just outside of it. The enhanced magnetic shear is just another way of describing a high, localized downward field-aligned current density. In sunlit or higher-density ionospheric conditions, where inverted-V precipitation does not add much additional ionization to the background density, we would not necessarily expect to see such localized, high-current density downward FACs.

*Acknowledgments.* This work was supported by NASA order number S-57795-F. We are grateful to the FAST team for comments and suggestions. This paper is dedicated to the memory of Alan Johnstone.

## REFERENCES

Boehm, M. H., J. Clemmons, J.-E. Wahlund, A. Eriksson, L. Eliasson, L. Blomberg, P. Kintner, and H. Höfner, Observations of an upward-directed electron beam with the perpendicular temperature of the cold ionosphere, *Geophys. Res., Lett., 22*, 2103-2106, 1995.

Burch, J. L., P. H. Reiff, and M. Sugiura, Upward electron beams measured by DE-1: a primary source of dayside region-1 Birkeland currents, *Geophys. Res. Lett., 10*, 753, 1983.

Carlson, C. W., J. P. McFadden, R. E. Ergun, M. Temerin, W. Peria, F. S. Mozer, D. M. Klumpar, E. G. Shelley, W. K. Peterson, E. Möbius, R. Elphic, R. Strangeway, C. Cattell, and R. Pfaff, FAST observations in the downward auroral current region: Energetic upgoing electron beams, parallel potential drops, and ion heating, *Geophys. Res. Lett., 25*, 2017-2020, 1998.

Elphic, R. C., et al., The auroral current circuit and field-aligned currents observed by FAST, *Geophys. Res. Lett., 25*, 2033-2036, 1998.

Ergun, R. E., C. W. Carlson, J. P. McFadden, F. S. Mozer, G. T. Delory, W. Peria, C. C. Chaston, M. Temerin, R. Elphic, R. Strangeway, R. Pfaff, C. Cattell, D. M. Klumpar, E. G. Shelley, W. K. Peterson, E. Möbius, and L. Kistler, FAST satellite observations of electric field structures in the auroral zone, *Geophys. Res. Lett., 25*, 2025-2028, 1998.

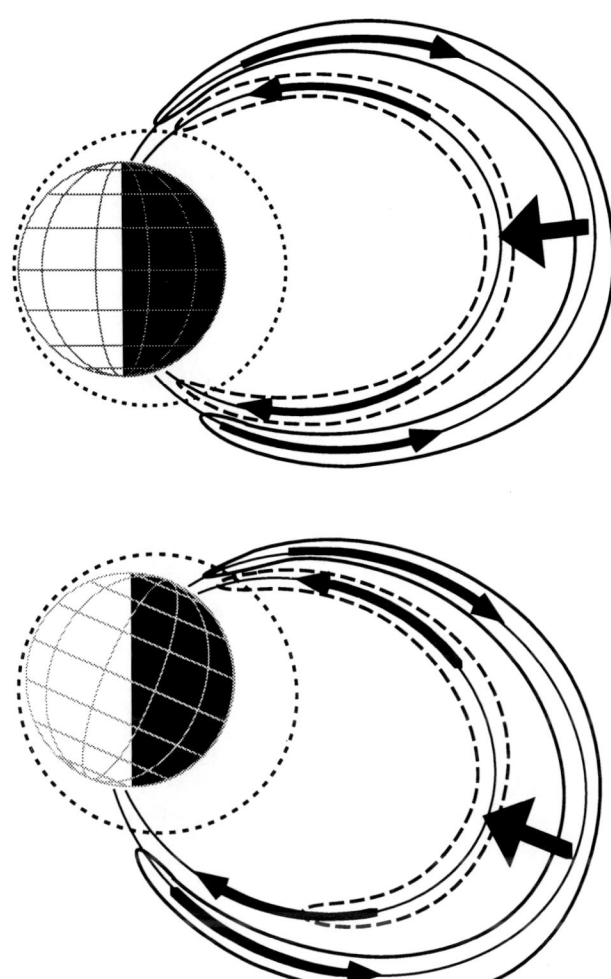

**Figure 4.** Schematic showing that the potential structure that accelerates the electrons sloshes up and down the field lines as the dipole rotates. (top) For the equinox case the potential structure associated with upgoing electron beams (dashed contour) drops down below the FAST orbit altitudes only in the hemisphere where the dipole rocks away from the sun. (bottom) For the solstice case, the potential structure only drops to FAST altitudes in the winter hemisphere, but stays there throughout the dipole rotation. The potential structures associated with inverted-V's may also drop to lower altitudes in the winter hemisphere. Current flow in both cases is shown by the solid arrows.

smaller regions of more intense current density with higher associated potential drops.

We are left with the question of why downward currents in the winter premidnight auroral zone form thin sheets of high current density instead of spreading out into broader layers of lower current density. The answer may lie

Gorney, D. J., Y. T. Chiu, and D. R. Croley, Jr., Trapping of ion conics by downward parallel electric fields, *J. Geophys. Res.*, *90*, 4205, 1985.

Hultqvist, B., R. Lundin, K. Stasiewicz, L. Block, P.-A. Lindqvist, G. Gustafsson, H. Koskinen, A. Bahnsen, T. A. Potemra, and L. J. Zanetti, Simultaneous observations of upward moving field-aligned energetic electrons and ions on auroral zone field lines, *J. Geophys. Res.*, *93*, 9765-9776, 1988.

Johnstone, A. D., and J. D. Winningham, Satellite observations of suprathermal electron bursts, *J. Geophys. Res.*, *87*, 2321-2329, 1982.

Klumpar, D. M., and W. J. Heikkila, Electrons in the ionospheric source cone: Evidence for runaway electrons as carriers of donward Birkeland currents, *Geophys. Res. Lett.*, *9*, 873-876, 1982.

Knight, L., Parallel electric fields, *Planet. Space Sci.*, *21*, 741, 1973.

Lin, C. S., J. L. Burch, J. D. Winningham, and J. D. Menietti, DE-1 observations of counterstreaming electrons at high altitudes, *Geophys. Res., Lett.*, *9*, 925-928, 1982.

Lundin, R., and L. Eliasson, Auroral energization processes, *Ann. Geophysicae*, *9*, 202-223, 1991.

Marklund, G., L. Blomberg, C.-G. Falthammar, and P.-A. Lindqvist, On intense diverging electric fields associated with black aurora, *Geophys. Res. Lett.*, *21*, 1859, 1994.

McFadden, J. P., C. W. Carlson, R. E. Ergun, C. C. Chaston, F. S. Mozer, M. Temerin, D. M. Klumpar, E. G. Shelley, W. K. Peterson, E. Möbius, L. Kistler, R. Elphic, R. Strangeway, C. Cattell, and R. Pfaff, Electron modulation and ion cyclotron waves observed by FAST, *Geophys. Res. Lett.*, *25*, 2045-2048, 1998.

Parker, E. N., The alternative paradigm for magnetospheric physics, *J. Geophys. Res.*, *101*, 10,587, 1996.

Robinson, R. M., R. R. Vondrak, K. Miller, T. Dabbs, and D. Hardy, On calculating ionospheric conductances from the flux and energy of precipitating electrons, *J. Geophys. Res.*, *92*, 2565, 1987.

Sharp, R. D., E. G. Shelley, R. G. Johnson, and A. G. Ghielmetti, Counterstreaming electron beams at altitudes of 1 $R_E$ over the auroral zone, *J. Geophys. Res.*, *85*, 92-100, 1980.

Shelley, E. G., R. D. Sharp, and R. G. Johnson, Satellite observations of an ionospheric acceleration mechanism, *Geophys. Res., Lett.*, *3*, 654, 1976.

Temerin, M., and C. W. Carlson, Current-Voltage relationship in the downward auroral current region, *Geophys. Res. Lett.*, *25*, 2365-2368, 1998.

---

J. Bonnell, C. W. Carlson, R. E. Ergun, J. P. McFadden, M. Temerin, Space Sciences Laboratory, University of California, Berkeley, CA 94720 USA

R. C. Elphic, MS D466, Los Alamos National Laboratory, Los Alamos, NM 87545 USA

R. J. Strangeway, Institute of Geophysics and Planetary Physics, University of California, 405 Hilgard, Los Angeles, CA 90024 USA

W. Peria, University of Washington, Geophysics Program, Box 351650, Seattle, WA 98195 USA

# Characteristics of Field-Aligned Currents near the Auroral Acceleration Region: FAST Observations

W. J. Peria, C. W. Carlson, R. E. Ergun, J. P. McFadden

*Space Sciences Laboratory, University of California, Berkeley*

J. Bonnell, R. C. Elphic

*Los Alamos National Laboratory, Los Alamos, NM*

R. J. Strangeway

*Institute of Geophysics and Planetary Physics, University of California, Los Angeles*

A pilot statistical study of field-aligned currents (FACs), using an automated FAC-finding technique, has been performed on the magnetometer data from the Fast Auroral Snapshot (FAST) spacecraft. The strategy of this survey is to eliminate all intervals during which the magnetometer data are not consistent with flight through a stationary sheet-like current flowing parallel to the geomagnetic field, and then examine the statistical properties of those data that remain. This survey is sensitive to currents with thicknesses from roughly 10- 1000 km, over an altitude range of 300- 4300 km, with densities of greater than 0.1 $\mu A/m^2$. We are able to reproduce the familiar statistical location and polarity pattern of large-scale currents, but we emphasize its high variability. The net current (that part which closes along the auroral zone or across the polar cap) comprises not only the large-scale currents, but the more numerous, finely-structured currents as well, implying that the fine structure is, in fact, an integral part of the global current system. FACs show a marked tendency to align themselves with the statistical auroral zone, and an even greater tendency, on each pass, to align themselves with each other, as expected for a population of sheet-like auroral currents. The scale sizes of currents are found to be symmetric with respect to current polarity, but downward currents contain larger spatial gradients of current density, particularly at low values of $K_p$. The likelihood of finding an intense upgoing electron beam within a downward FAC is found to be well correlated with the amount of time its magnetic footpoint has spent in darkness, suggesting suppression of the beams by ionospheric photoelectrons.

## 1. INTRODUCTION

The presence of field-aligned current (FAC) is the most fundamental signature of the auroral zone. The presence of FACs is inferred, in this study, from a satellite-borne 3-axis magnetometer, and FACs are seen on essentially every pass.

FACs couple electromagnetic energy and stresses from the outer magnetosphere to the ionosphere, and cause significant Ohmic dissipation in the ionosphere.

The existence of FACs, and their association with aurora, was first proposed by *Birkeland* [1908], but FACs were not detected by spacecraft magnetometers until much later [*Cummings and Dessler*, 1967]. Two landmark statistical studies were performed by *Zmuda and Armstrong*, [1970], and *Iijima and Potemra*, [1976], using Triad data at 800 km altitude, and subsequent similar studies include *Zanetti et al.*, [1983]. These studies focused on the location, polarity, intensity, and closure patterns of FACs, and the dependence of these on global geomagnetic conditions. *Yamauchi et al.*, [1998] performed a thorough multi-event study, examining the relationship between large-scale and meso-scale FACs, and how FACs are carried in various local time sectors, as well as reviewing recent FAC studies. For a thorough review of earlier space-based studies of FACs, see *Potemra*, [1985].

The present study reproduces some of the results of previous space-based studies of FACs, and also performs a high-resolution comparison of FAC magnetometer signatures with energetic electron data, using an automated database technique. This study examines all seasons and local times, and uniformly covers the altitude range within and below the auroral acceleration region, using data selected entirely at random from the first two years of the FAST mission.

## 2. METHODOLOGY

All spacecraft studies of FACs to date, including this one, suffer from the same rather serious shortcoming: they cannot directly measure currents! Either the current is inferred, using appropriately scaled time derivatives of the magnetometer data as a proxy for the curl of the local magnetic field, or some part of the current is measured directly (counting the current-carriers), under conditions where the unmeasured (typically low-energy) particles are assumed to be insignificant.

When using a single spacecraft magnetometer to measure currents, there is, of course, the usual space-time ambiguity. But even if it could be ascertained, somehow, that a set of magnetic perturbations was caused by the traversal of time-stationary currents, one still would not know if the current sheets were moving or at what rate; thus there would still be systematic uncertainty about how to scale the time derivative of the magnetometer signal into an estimate of the current density. At higher altitudes, *Chun and Russell*, [1991] have shown, using a two-point measurement (ISEE-1 and 2), that the velocity of current systems can be significant compared to that of orbiting spacecraft, making this systematic uncertainty a serious problem. At the relatively low altitudes visited by FAST, however, the spacecraft velocity (5-8 km/s) is typically considerably larger than that of the drift motions of auroral arcs (< 1 km/s), which is probably a reasonable upper limit for the motions of current systems. The only way to be safe from this uncertainty, of course, is with a multi-spacecraft measurement.

### 2.1. A multiscale automated FAC-finder

We use an automated FAC-finding routine to catalog FACs. The FAC-finder eliminates all intervals during which FAST is clearly not traversing a sheet of unipolar field-aligned current; what remains are the FACs. The sheer size of the FAST data set makes automated FAC-finding seem necessary. However, automation has the additional advantage that, once the FAC criteria have been established, investigator bias is excluded from the event selection process.

The FAC-finding algorithm is complex and difficult to describe. We will first describe the "multiscale" concept: the idea that since we do not know what size of FACs to search for, we will test for FACs of all sizes. Then, we will describe briefly the information that is recorded about each FAC. Finally, we will explain in detail the FAC criteria: the necessary conditions that an interval of magnetometer data must meet in order to be called a FAC.

For each data point, the FAC criteria are applied to several intervals which are centered on the point in question, and which have a range of different lengths. This procedure is repeated for all data points. The shortest interval which we can sensibly consider contains 3 data points. Since we sample the FAST magnetometer once per second, and since the minimum spacecraft speed is 5 km/s, there is a lower limit of 15 km for the thickness of FACs encountered at normal incidence. The longest interval we consider is set by the apparent thickness of the auroral zone on each pass: it is quite variable but typically substantially less than 1000 km. A particular point at a particular interval width will be consistent with the traversal by the spacecraft of either upward FAC, downward FAC, or no FAC at all. Intervals of different lengths, centered on the same point, will not necessarily agree with each other. For example, a long interval about a particular point might be consistent with upward FAC, a shorter interval with downward FAC. When this occurs, the shorter scale is given priority, and in this example the point in question would be said to belong to a downward FAC.

Each such suitable interval produces a single entry in our database. A database entry consists of the location of the FAC, its total size (km), characteristic scale size (km), intensity (mA/m), peak current density ($\mu A/m^2$), the roughness of its spatial profile ($\mu A/m^2/km$), its apparent orientation, and several "quality factors". We also compute, for each FAC, the current due to energetic electrons as seen by the FAST electrostatic analyzers (ESAs), and furthermore, an indication of whether or not the interval contains an upgoing electron beam. In this study, we have recorded over 15,000 FACs, representing roughly 10 % of the FAST data taken (poleward of ±50° invariant latitude) between 2 October, 1996, and 7 January, 1998.

The following criteria must be met by all the points in an interval of magnetometer data, in order for that interval to be recorded as a FAC.

- The vector difference $\delta\vec{B}$ between the measured magnetic field and the IGRF95 model $\vec{B}_0$ must be within 1° of the plane perpendicular to $\vec{B}_0$. (Thus, two components suffice to describe $\delta\vec{B}$; these are called "cross-track" and "along-track", with the obvious meaning.)

- The absolute value of the correlation coefficient between the two components of $\delta\vec{B}$ must exceed 0.8. (This is similar to, though not precisely the same as, demanding that the minimum-to-maximum variance ratio be smaller than a critical value. In practice, the correlation coefficient criterion is somewhat more selective.)

- The angle between the orbit track and the current sheet normal, in the plane perpendicular to $\vec{B}_0$, must not exceed 60°.

- The current estimated from the time derivatives of $\delta\vec{B}$, taking proper account of the apparent sheet orientation, must exceed 0.1 $\mu A/m^2$.

- The sign of the time derivative of the cross-track component of $\delta\vec{B}$, which indicates the polarity of the current, must not change.

The application of these criteria, to every data point, and to a variety of interval widths about every data point, yields a 2-dimensional array (the number of data points by the number of intervals sampled), and this must then be collapsed into the start and stop times of FACs. First, at each time point, we find whether current is indicated at any scale. Then, we find the shortest scale at which some amount of current is indicated, and record whether it is upward or downward at that scale. Finally, we find groups of time points where a consistent direction of current is indicated: these are the FACs.

Plate 1 shows an example of the operation of the FAC-finder. Plate 1A is the "cross-track" magnetometer, the derivative of which is the usual indicator of field-aligned currents. Plate 1B shows the "along-track" magnetometer. These two components of $\delta\vec{B}$, and the spacecraft ephemerides, are the only inputs to the FAC-finder.

Plate 1C shows the 2-dimensional output referred to above. The vertical axis shows the width of the windows, about each point, to which the FAC criteria are applied. If the data at a particular time, at a particular scale, are consistent with upward current, the corresponding pixel is colored red. Green indicates downward current, and black indicates that the data are inconsistent with the traversal of a sheet-like current. The polarity of each FAC in this example is indicated by the letters "U" (for upward) and "D" (for downward) in Plate 1A. The final output of the FAC-finder is indicated by the green and red vertical lines which span Plates 1A-E. These mark the start (green) and stop (red) times of FACs.

Plates 1D-E are the electron data. Unfortunately, the electron data are not suitable for FAC-finding, since the electron instrument is not always activated, and since the primary current carriers in some FACs are, in fact, unmeasurable thermal electrons. We can, however, compare the FAC-finder output to the electron data whenever it is available. Note how between 2043:35 and 2045:30 there is a net negative deflection of the magnetometer: a net upward field-aligned current. The two upward FACs which comprise it are, however, carried by electrons of rather different character. The downward FAC which interrupts the two upward FACs is apparently carried by the very low energy electrons seen at 0 and 360° pitch angle. (The green stripe at 90° is due to an instrumental effect.) A similar configuration (a net current of one polarity interrupted by a FAC of opposite polarity) is seen in the three FACs beginning just after 2041.

The arrow in the third FAC from the left in Plate 1A points at a small fluctuation which might be expected to result in another (very small) FAC. This does not occur however, because this fluctuation is not sufficiently highly correlated with the fluctuation in the along-track signal. Furthermore, the along-track fluctuation at this time is actually larger than the cross-track, indicating that if the fluctuations are due to a current sheet, it is being traversed at very oblique incidence, which makes estimates of current density unreliable. For these reasons, our algorithm rejects the fluctuation indicated by the arrow.

## 2.2. The sheet-current assumption

It is not possible in principle to determine, from a single spacecraft, whether or not a particular magnetic signature is produced by a sheet-like current. In this study, however, we will make an assumption about the geometry of the current systems, since we wish to evaluate the current density within each FAC. We have chosen, as have many other investigators before us, sheet geometry. This choice is not entirely arbitrary, and we attempt to justify it here.

First, a sheet-like geometry is difficult to miss. This is consistent with the ubiquitous and repeatable nature of FACs as viewed from FAST. A relatively small and regular number of currents is encountered on nearly every pass, suggesting that the currents are extended along the auroral zone. Furthermore, the visible aurora, corresponding to regions of upward current, has a generally sheet-like structure; the radius of curvature of all but the most active arcs is much greater than the thickness of the arcs themselves. This is precisely the sort of "sheet-likeness" required for a reasonably accurate estimate of current density to be made.

Finally, the distribution of apparent current orientations (see Figure 1) appears to contain a substantial sheet-like component. In the spacecraft-centered "cross-track, along-track" system, one finds that the most frequently occurring angle $\theta_c$ between the sheet direction and the cross-track direction is roughly 20° (see Figure 1A). The angle $\theta_c$ can be transformed to an angle $\theta_z$, between the apparent sheet direction and the tangent to the statistical auroral zone [Holzworth and Meng, 1975]. The most frequently occurring value of $\theta_z$ is zero (see Figure 1B), and the majority of currents are aligned within 30° of the statistical auroral zone. On individual passes, however, the sheet orientations are grouped even more tightly than this. Figure 1C shows the distribution of angles between each current sheet and the median orientation on the pass where each was encountered. The majority of current sheets lie within 15° of each other on any particular pass of the auroral zone. These tendencies toward alignment give important support to the current-sheet assumption.

The sort of time dependence which might produce the observed magnetic fluctuations (which we and many others interpret as spatial structure) is also strongly constrained by these tendencies toward alignment. Any sort of wave would have to be linearly polarized, and with a preferred orientation aligned along the auroral zone. This is perhaps not too hard to accept, but in the next section of the paper, we will show several results which also constrain the phase of this hypothetical wave relative to the passing of FAST, rendering a temporal interpretation less plausible still.

## 3. STATISTICAL PROPERTIES OF FACs

### 3.1. Regions 1 and 2

The "Region 1- Region 2" structure of auroral zone FACs [Iijima and Potemra, 1978] can be extracted from our database (see Plate 2), though doing so is not entirely straightforward. The standard picture leads one to expect that, depending on magnetic local time (MLT), any particular pass will yield at most 3 FACs. But fully 90 % of the ~ 1500 auroral zone passes (4 per orbit) contained in our database have more than three FACS, and the average number is ~ 10. Plate 2 is produced by selecting the most intense upward and

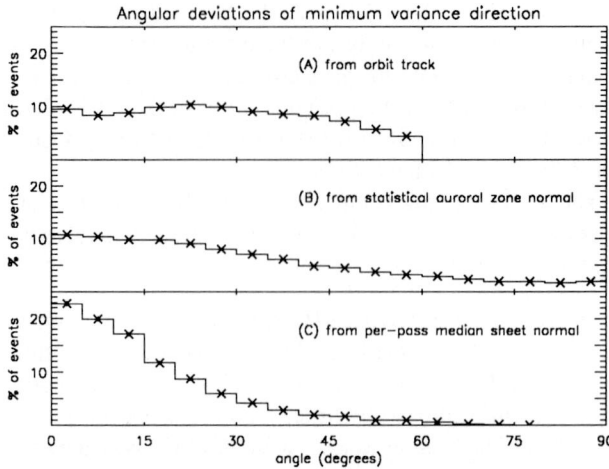

**Figure 1.** (A) Current sheet normals have a fairly broad distribution with respect to FAST's orbit. (B) The most frequently occurring current sheet normal is perpendicular to the statistical auroral zone. (C) On any particular pass of the auroral zone, the majority of current sheets are aligned to within 15° of the median orientation.

most intense downward current from each pass. These opposing pairs are then binned according to MLT. The poleward current of each pair is identified as Region 1, and the equatorward as Region 2. The dominant polarities for Regions 1 and 2 are then found, in each MLT bin. The median invariant latitude (ILAT) of each region is then plotted, colored according to polarity, with a latitudinal width equal to the median width of the corresponding FACs. Plate 2 shows the resulting pattern, for two different ranges of $K_p$.

The appearance of a "Region 1- Region 2" pattern under this method is not a foregone conclusion. We are showing that the polarity distribution for the most intense FACs agrees qualitatively with results obtained previously. Our method cannot, of course, reproduce the familiar three overlapping current sheets in the pre-midnight sector, nor at noon. Plate 2 is presented only as a point-of-contact with well-established results. When the differences between the two methods are taken into consideration, our result is consistent with that of [Iijima and Potemra, 1978]; this is a crucial internal consistency check of our methods.

We would, however, like to emphasize that the width of the colored rings in Plate 2 is not a measure of their statistical spread in latitude. Either ring has a typical thickness of only 1 degree, but, at any particular time, may be found as much as ±5° from its median location; i.e. the statistical spread in latitude of each ring is larger than the difference between the median latitudes of the inner and outer rings. It is sometimes suggested that the instantaneous global current pattern might be parameterized by a single number, such as $K_p$ or IMF $B_z$. Figure 2, a scatter plot of the invariant latitude of Region 1

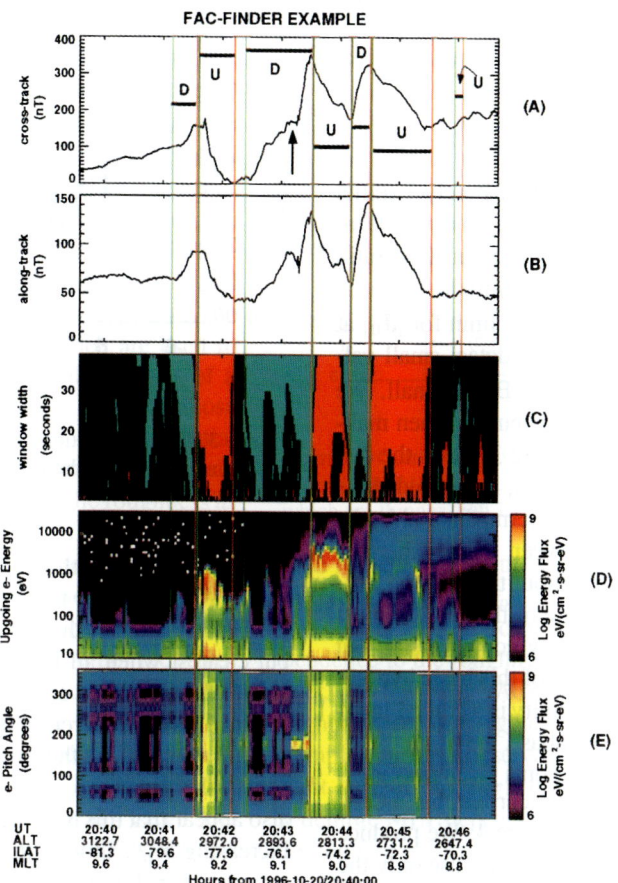

**Plate 1.** The output of the FAC-finding algorithm for an auroral zone pass. Also shown are energetic electron data, which provide further confidence in the FAC-finder output.

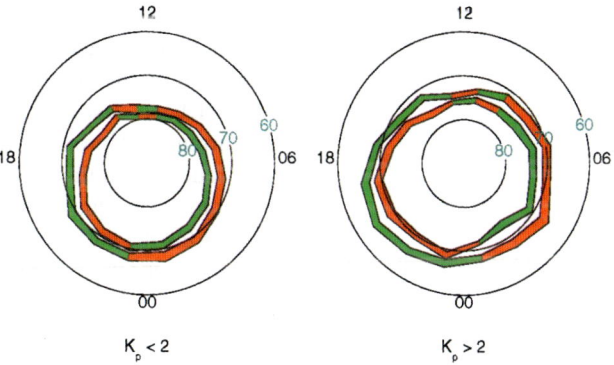

**Plate 2.** For the two most intense currents from each pass, the median ILAT for each MLT bin is shown. Green indicates predominantly downward currents, while red indicates predominantly upward. The latitudinal width of the bands indicates the median width of the currents in each bin, not their variance in ILAT.

the length of time a FAC's footpoint has been in darkness. The 5- 6 hour "rise-time" of the beam probability, following sunset, is qualitatively consistent with the characteristic dissociative recombination time for the height-integrated ionospheric electron density [*Chamberlin*, 1974], ignoring ionospheric convection entirely. This result complements the work of *Newell et al.*, [1996], which discusses the effects of a sunlit ionosphere on auroral electron precipitation.

## 5. DISCUSSION

Our automated study is capable of reproducing past results, as well as going beyond them. The emergence of the familiar Region 1- Region 2 pattern provides reassurance that our FAC-finding algorithm is operating in a sensible way. There is, however, much more going on than the familiar pattern would suggest. The currents used to produce the familiar polarity pattern in Plate 2 comprise about 20 % of the currents in our database, and represent roughly the same amount of total charge transport as the remaining currents combined. The net current pattern seen in Figure 4, on the other hand, is much less clear when the figure is made from only the intense currents of Plate 2. This indicates that the less intense currents are part of the global current system, and are not merely small-scale structures which close locally. Indeed, there is no evidence of substantial local current closure anywhere in our altitude range (see Figure 3).

Downward FACs tend to be "rougher" than upward FACs, and this tendency is more pronounced under quiet geomagnetic conditions ($K_p < 2$). The significance of this is not presently understood. *Hoffman et al.*,[1988] reported an increase in the structure of electron precipitation during quiescent auroral periods. The enhanced roughness of downward

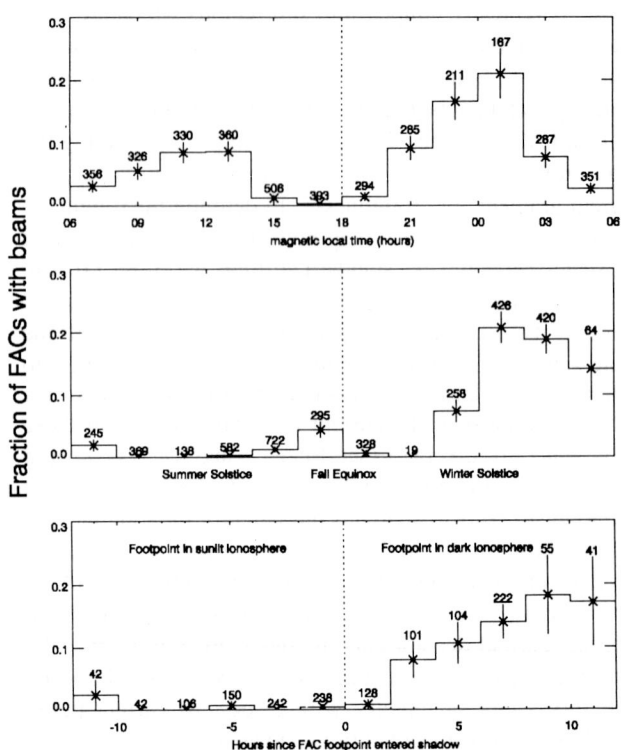

**Figure 7.** Upgoing electron beams are much more likely to occur on flux tubes which connect to a dark ionosphere. Flux tubes in daylight for 24 hours or more show a beam probability of less than 0.05 % . (The numbers indicate how many FACs are in each bin.)

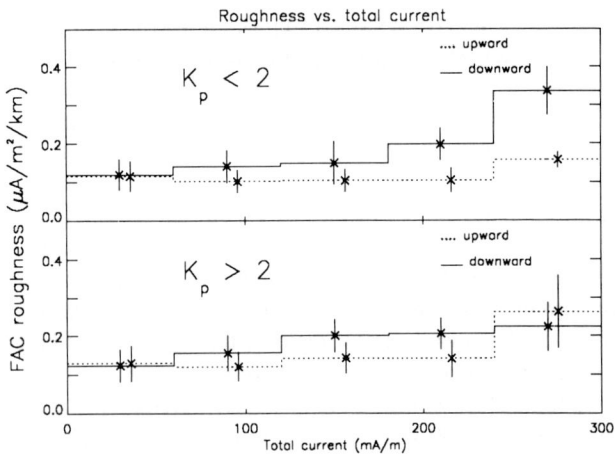

**Figure 6.** At low $K_p$, the roughness of downward currents depends more strongly on intensity than does the roughness of upward currents.

currents is not simply a manifestation of the structuring of FACs by upgoing electron beams, since it lacks the significant seasonal or solar zenith angle dependence displayed by the beams. There is however, an MLT dependence, with FACs near noon being some 40- 50 % rougher, suggesting that the inhomogeneity is determined somehow in the outer magnetosphere, and not in the ionosphere at all.

The ability, provided by FAST, to examine the microscopic particle distribution functions within individual FACs, and to compile statistics on these using an automated procedure, can provide evidence regarding the relative importance of the ionosphere and outer magnetosphere in a given physical process. In this study, the clear sorting of upgoing electron beams, according to the recent illumination history of their ionospheric footpoints, strongly suggests that it is the ionosphere that determines precisely how the return current is to be carried.

*Acknowledgments.* The authors gratefully acknowledge useful discussions with T. D. Phan and L. M. Peticolas. This research was conducted under NASA grant NAG5-3596.

# REFERENCES

Birkeland, Kr., *The Norwegian Aurora Polaris Expedition 1902-1903, Volume I*, H. Aschehoug and Co., Christiania, 1908.

Carlson, C. W., J. P. McFadden, R. E. Ergun, M. Temerin, W. Peria, F. S. Mozer, D. M. Klumpar, E. G. Shelley, W. K. Peterson, E. Moebius, R. Elphic, R. Strangeway, C. Cattell, and R. Pfaff, FAST observations in the downward auroral current region: Energetic upgoing electron beams, parallel potential drops, and ion heating, *Geophys. Res. Lett., 25,* p. 2017-2020, 1998.

Chamberlain, J. W., *Theory of Planetary Atmospheres: An Introduction to Their Physics and Chemistry,* p. 174, Academic, New York, 1978.

Chun, F. K. and C. T. Russell, The evolution of field-aligned currents as a function of substorm phase, *J. Geophys. Res., 96,* 15801-15810, 1991.

Cummings, W. D. and A. J. Dessler, Field-aligned currents in the magnetosphere, *J. Geophys. Res., 72,* 1007-1013, 1967

Elphic, R. C., J. W. Bonnell, R. J. Strangeway, L. Kepko, R. E. Ergun, J. P. McFadden, C. W. Carlson, W. Peria, C. A. Cattell, D. M. Klumpar, F. S. Mozer, M. Temerin, D. Klumpar, E. Shelley, W. Peterson, E. Moebius, L. Kistler, and R. Pfaff, The auroral current circuit and field-aligned currents observed by FAST, *Geophys. Res. Lett., 25,* p. 2033-2036, 1998.

Hoffman, R. A., M. Sugiura, N. C. Maynard, R. M. Candey, and J. D. Craven, Electrodynamic patterns in the polar region during periods of extreme magnetic quiescence, *J. Geophys. Res., 93,* 14515-14541, 1988.

Holzworth, R. H. and C. -I. Meng, Mathematical representation of the auroral oval, *Geophys. Res. Lett., 2,* 377-380, 1975.

Iijima, T. and T. A. Potemra, The amplitude distribution of field-aligned currents at northern high latitudes observed by TRIAD, *J. Geophys. Res., 81,* 2165-2174, 1976.

Iijima, T. and T. A. Potemra, Large-scale characteristics of field-aligned currents associated with substorms, *J. Geophys. Res., 83,* 599, 1978.

McFadden, J. P., C. W. Carlson, R. E. Ergun, F. S. Mozer, M. Temerin, W. Peria, D. M. Klumpar, E. G. Shelley, W. K. Peterson, E. Moebius, L. Kistler, R. Elphic, R. Strangeway, C. Cattell, and R. Pfaff, Spatial structure and gradients of ion beams observed by FAST, *Geophys. Res. Lett., 25,* p. 2021-2024, 1998.

Newell, P. T., C.-I. Meng, and K. M. Lyons, Suppression of discrete aurorae by sunlight, *Nature, 381,* 766, 1996.

Potemra, T. A., Field-aligned (Birkeland) currents, *Spa. Sci. Rev., 42,* 295-311, 1985.

Sugiura, M. and T. A. Potemra, Net field-aligned currents observed by Triad, *J. Geophys. Res., 81,* 2155-2164, 1976.

Taguchi, S., M. Sugiura, J. D. Winningham, and J. D. Slavin, Characterization of the IMF By-dependent field-aligned currents in the cleft region based on DE-2 observations, *J. Geophys. Res., , 98,* 1393-1407, 1993.

Yamauchi, M., R. Lundin, L. Eliasson, S. Ohtani, and J. H. Clemmons, Relationship between large-, meso-, and small-scale field-aligned currents and their current carriers, in *Polar Cap Boundary Phenomena,* edited by J. Moen, A. Egeland, and M. Lockwood, pp. 173-188, Kluwer, Dordrecht, The Netherlands, 1998.

Zanetti, L. J., T. A. Potemra, and W. Baumjohann, Ionospheric and Birkeland current distributions inferred from the MAGSAT magnetometer data, *J. Geophys. Res., 88,* 4875-4884, 1983.

Zmuda, A. J., and J. C. Armstrong, The diurnal flow pattern of field-aligned currents, *J. Geophys. Res., 79,* 4611-4619, 1974.

---

W.J. Peria, Space Sciences Laboratory, UC Berkeley, Berkeley, CA 94720

# Auroral Surge Currents and Electrodynamics with FAST and VIS

S. A. Cummer,[1] R. R. Vondrak,[1] R. F. Pfaff,[1] J. W. Gjerloev,[2] C. W. Carlson,[3] R. E. Ergun,[3] W. J. Peria,[3] R. C. Elphic,[4] R. J. Strangeway,[5] J. B. Sigwarth,[6] L. A. Frank[6]

We analyze data from two FAST satellite passes in the vicinity of auroral surges (as seen in the VIS auroral imager on the Polar satellite) to study the large scale ionospheric electrodynamics of the surge, focusing in particular on the configuration of field-aligned and ionospheric currents. While most previous ground and satellite observations have found the surge to be a region of predominantly upward field-aligned currents (FACs), some satellite observations have found evidence for equal downward and upward FACs near the surge, suggesting that the FACs close locally as opposed to globally through the substorm current wedge. Our observations show that while regions of both upward and downward FACs are present near the surge, there is a significant net upward FAC in the northern portion of meridians passing through the surge and just east of the surge. Current continuity requires that these upward currents be fed by downward currents traditionally thought to be in the eastern portion the auroral substorm bulge. Such a current configuration requires a westward ionospheric current to connect the two FAC regions, and this is supported by our observations of a significant southward electric field in the high Hall conductance surge region, which drives a strong westward Hall electrojet connecting the upward surge FACs and the presumed downward FACs east of the surge.

## INTRODUCTION

The auroral surge is arguably the most visually spectacular and electrodynamically complicated auroral form. However, due to its usually short-lived nature, ground-based and satellite electrodynamic observations of this phenomena are rather sparse. Surge electric fields, currents, and precipitation-produced ionization have been observed from the ground with all-sky cameras, magnetometers, and the STARE radar [*Inhester et al.*, 1981; *Opgenoorth et al.*, 1983] and with the Chatanika incoherent scatter radar [*Robinson and Vondrak*, 1990]. Satellite observations of the surge include measurements of the magnetic field deviations and energetic particles along a pass through an active auroral surge with DMSP satellites [*Bythrow and Potemra*, 1987], and detailed particle and field observations from the Dynamics Explorer (DE) [*Weimer et al.*, 1994] and Freja [*Marklund et al.*, 1998] satellites. Surge electrodynamics have also been studied statistically from a careful analysis of DE-1 images and DE-2 particle and field observations [*Fujii et al.*, 1994; *Gjerloev and Hoffman*, 1998].

---

[1] Laboratory for Extraterrestrial Physics, NASA/Goddard Space Flight Center, Greenbelt, Maryland
[2] Danish Space Research Institute, Copenhagen, Denmark
[3] Space Sciences Laboratory, University of California, Berkeley, California
[4] Los Alamos National Laboratory, Los Alamos, New Mexico
[5] Institute of Geophysics and Planetary Physics, University of California Los Angeles, California
[6] Department of Physics and Astronomy, University of Iowa, Iowa City, Iowa

Magnetospheric Current Systems
Geophysical Monograph 118
Copyright 2000 by the American Geophysical Union

The variability of individual auroral surges makes it difficult to compare directly the above-listed observations of different events. However, some features are common to most of these surge observations and we take these to be general characteristics of auroral surges. The surge head is a region of high ionospheric Hall conductivity typically produced by energetic inverted-V electron precipitation, and it contains broad and intense upward field-aligned currents (note that this does not preclude additional regions of downward field-aligned current). Radar and satellite observations also show that the electric field in the vicinity of the surge generally converges to the surge head, indicating negative space charge associated with this region which is consistent with models of surge electrodynamics [*Inhester et al.*, 1981; *Opgenoorth et al.*, 1983]. Lastly, the satellite electric field measurements consistently show a "stagnation region" of low electric fields (and therefore low ionospheric convection) south of the surge and extending somewhat to the east.

Despite these similarities, there remain some open questions regarding the structure of the field-aligned currents (FACs), ionospheric electric field, and ionospheric currents in the vicinity of the surge. Freja observations [*Marklund et al.*, 1998] showed an intense downward current northwest of a surge which balanced the upward current in the surge, from which it was concluded that the upward currents in the surge may be fed locally by equal downward currents surrounding the surge. Other observations [*Bythrow and Potemra*, 1987; *Weimer et al.*, 1994] and models [*Opgenoorth et al.*, 1983] do not produce equal downward currents nearby, suggesting that the upward currents in the surge are closed through downward currents elsewhere in the auroral bulge. Also, radar [*Inhester et al.*, 1981; *Opgenoorth et al.*, 1983] and satellite [*Fujii et al.*, 1994] observations have found significant southward electric fields near and to the east of the surge that drive strong westward Hall currents in the ionosphere, thereby feeding the net upward FACs associated with the surge head from downward currents distant from the surge. The same Freja observations [*Marklund et al.*, 1998], however, showed weak electric fields east of the surge, implying weak westward Hall currents and necessitating local closure of surge FACs through ionospheric currents in the immediate vicinity of the surge. The surge current configuration has important implications, as local field-aligned current closure precludes the presence of a substorm current wedge [*McPherron et al.*, 1973], which is generally thought to play an important role in substorm magnetosphere-ionosphere coupling.

To try to resolve these inconsistencies regarding surge electrodynamics, we have examined data from the FAST satellite [*Carlson et al.*, 1998] and the VIS instrument on the Polar satellite [*Frank et al.*, 1995] during the approximately three month period of late 1997 and early 1998 when FAST was in local time sectors (∼18–24 MLT) where surges are expected to occur and the VIS Low Res visible wavelength camera was in operation. We found two FAST passes in the vicinity of significant auroral surges, and we analyze the electrodynamic parameters observed on these two passes to compare them to models and previous observations, focusing in particular on the relationship of ionospheric and field-aligned currents.

## FAST AND VIS OBSERVATIONS

Possible substorm auroral surges were identified by bright x-ray auroral emissions (corresponding to intense, >3 keV precipitation) seen by the Polar Ionospheric X-ray Imaging Experiment (PIXIE) [*Imhof et al.*, 1995] in the ∼18–24 local time sector. The FAST passes through these possible surge regions were then examined for the precipitating electron signature of the surge, namely intense and energetic ($\gtrsim$5 keV) inverted-V precipitation at the northern edge of the auroral zone [*Meng et al.*, 1978]. Lastly, VIS images at a wavelength of 557.7 nm were examined to see whether the FAST footprint passed through or near a region with the optical signature of a surge, namely a bright emission at the northern edge of the auroral zone containing a noticeable curl [*Akasofu*, 1964]. We purposely avoid auroral morphologies containing multiple surges or spirals [e.g., *Johnson et al.*, 1998], as the electrodynamics of these cases may be quite different than for the single surges considered here. The two events that met these criteria are discussed below.

*Event 1: 17 Jan 98*

The top of Plate 1 shows an image of 557.7 nm auroral emissions taken by VIS on 17 January, 1998 at 17:04:46 UT with a 44 second integration time. The global aurora can be described as a typical well-developed substorm, with a bulge shape and a well-defined surge at the northwestern edge of the bulge. The image is overlayed with a grid of corrected geomagnetic (CGM) latitude and local time, and with the footprint of the southward-moving FAST satellite. FAST did not pass through the surge head but rather passed through the bright arc connecting to the surge about 1.5 hours of local time to the east.

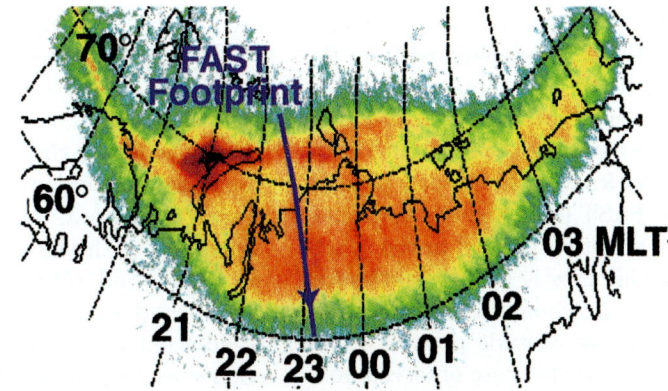
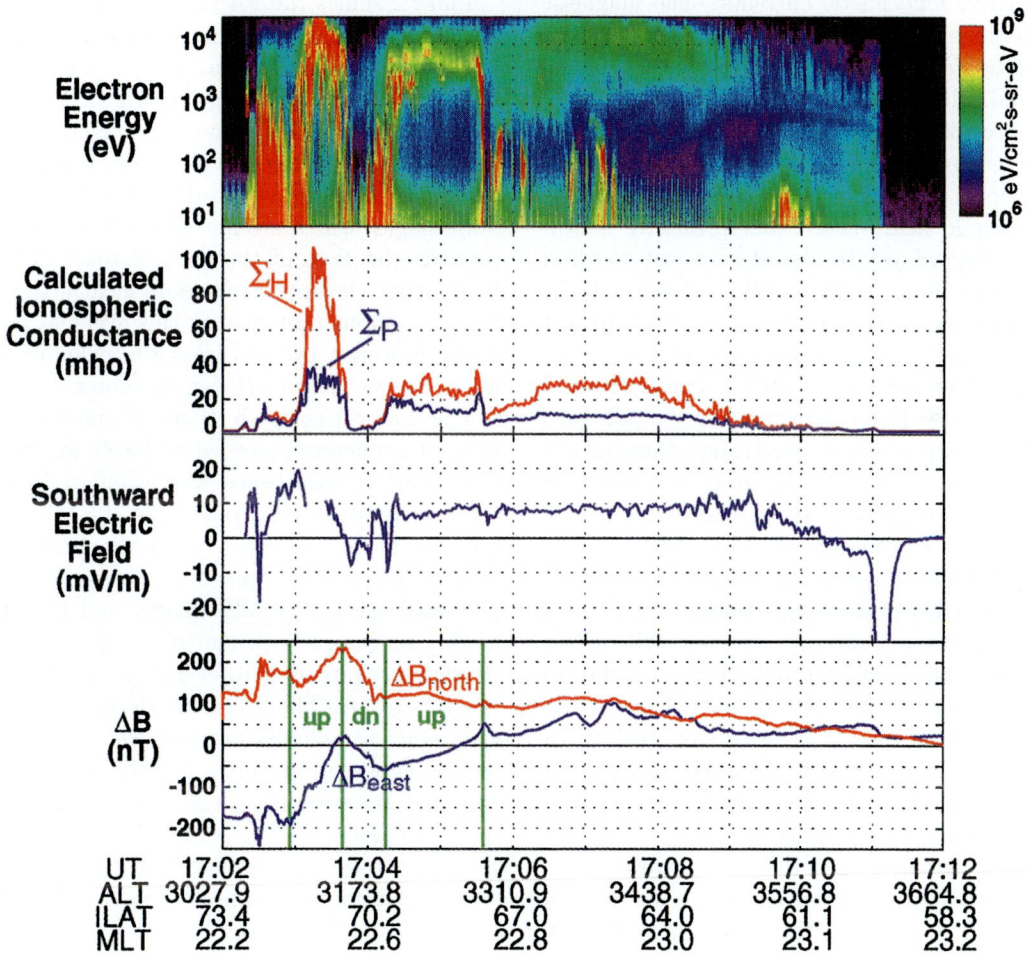

Plate 1. Observations from VIS and FAST on 17 Jan 98 at 17:05 UT. Top: 557.7 nm VIS image of the aurora. The grid coordinates are CGM local time and magnetic latitude. The magnetic footprint of the FAST satellite is overlaid in blue. Bottom: the observed and calculated electrodynamic parameters along the FAST footprint through the surge-connected arc. The panels show, from top to bottom, electron differential energy flux averaged over the loss cone, ionospheric conductances calculated from the precipitating electrons, observed southward electric field, and observed magnetic field deviations. Regions of upward and downward field-aligned currents are marked on the magnetic field deviations.

The lower panels in Plate 1 show the electrodynamic parameters observed by FAST and calculated from the FAST data along its path through the auroral bulge. The electrons in the loss cone, shown in the time-energy spectrogram, were used to calculate the Hall and Pedersen ionospheric conductivities with an ion-pair production model [*Rees*, 1963]. FAST measured the horizontal electric field only in the direction parallel to the spacecraft velocity, which in this case is almost southward. The electric field data gap corresponds to a period when FAST passed through the low altitude portion of the auroral acceleration region and therefore observed horizontal electric fields associated with parallel particle acceleration which are not present in the ionosphere and thus do not drive ionospheric currents. The magnetic field perturbations observed by FAST, which we assume are produced by FACs, are plotted in the bottom panel.

As expected, the intense and energetic (>10 keV) inverted-V precipitation corresponds to the bright surge-connected arc in the VIS image and produces high Hall and Pedersen conductances in the ionosphere. The meridional electric field varies from primarily southward in the poleward arc, to almost zero just south of the surge-connected arc, and to southward through the rest of the bulge. It turns very strongly northward just south of the auroral oval in association with a subauroral ion drift [*Anderson et al.*, 1993]. The southward electric field and high Hall conductance in the poleward arc produce a strong westward electrojet of magnitude $\sim 1.2 \times 10^5$ A that connects to the surge, in agreement with previous observations of strong westward currents east of the surge [e.g., *Opgenoorth et al.*, 1983].

The magnetic field observations show the expected strong upward current in the surge-connected arc and a region of net downward currents equatorward of this arc. The significant perturbations in the northward magnetic field indicate that the assumed FAC sheets are significantly tilted from their usual east-west alignment [*Fung and Hoffman*, 1992]. Comparing the integrated intensities of the upward and downward currents north of the Harang discontinuity that can connect to the westward electrojet, we find that the downward current is 56% of the upward current, leaving a net upward current of $\sim 0.25$ A/m along the northern portion of this meridian. Therefore, since 44% of the upward current does not close locally, current continuity requires that it must close through an ionospheric current that is connected to a net downward current that in turn is either localized or distributed elsewhere in the auroral oval. The strongest electrical connection from the upward current region to elsewhere in the oval is through the high Hall conductance channel at the northern edge of the bulge, and given the observation of southward electric fields in this channel which drive westward ionospheric currents, it is most likely that the net upward current is fed by these westward Hall currents that are connected to net downward currents in the eastern portion of the auroral bulge. This observation is in general agreement with the statistical observations of *Fujii et al.* [1994] and *Gjerloev and Hoffman* [1998] in which the upward surge FACs were closed by a combination of adjacent downward FACs and ionospheric currents connected to downward FACs in more distant regions.

*Event 2: 18 Jan 98*

Plate 2 shows the FAST and VIS observations from 18 January 98 in the same format as in the previous case. The lack of a latitudinally expanded auroral oval in the 557.7 nm VIS image shown at the top of Plate 2 taken at 13:04:51 UT shows that this is not a bulge-type substorm as was the previous case, but there is a clear and relatively static (as inferred from the sequence of VIS images, which are not shown here) surge structure towards the west of the auroral zone. The 13:08 UT image (not shown) shows more clearly the curling auroral emissions that qualify this event as a surge. The FAST footprint passes through a region of bright auroral emissions close to the surge center.

The lower panels in Plate 2 show the electrodynamic parameters observed by FAST for this pass. The ionospheric conductance was calculated directly from the precipitating electrons while the other parameters are directly measured. As in the previous case, intense inverted-V precipitation corresponds to the bright poleward arc in the VIS image, and the electric field in this surge arc is primarily southward, producing a strong westward electrojet in the surge of magnitude $\sim 2.0 \times 10^5$ A. The apparent electric field data gap is again a result of FAST entering the auroral acceleration region. However, in contrast to the other pass, the electric field is extremely weak and slightly northward in the region south of the surge. This difference between the two passes is consistent with the presence of a Harang discontinuity south of the surge that divides the regions of northward and southward electric fields.

The magnetic field perturbations are also slightly different than in the previous case. There is the expected strong upward current in the surge, but there is less downward current along the portion of this meridian north of the Harang discontinuity. Quantitatively, only 27% of the upward current is closed by adjacent downward currents, leaving a net upward current of 0.53

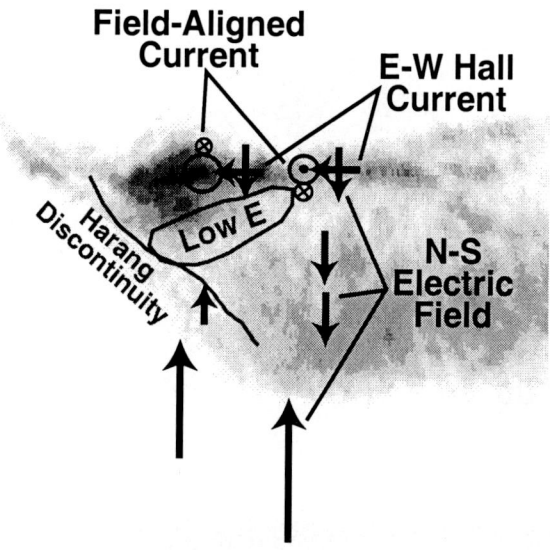

**Figure 1.** A schematic summary of the field-aligned currents, ionospheric Hall currents, and north-south electric fields observed in the two FAST passes presented herein. The surge and surge-connected arc contain strong upward FACs and adjacent, smaller downward FACs. The southward electric field in the high Hall conductance surge drives a strong westward electrojet which connects to the upward FACs.

A/m in the surge meridian. The non-local closure implies that these upward currents in the surge are closed primarily through downward currents elsewhere in the auroral oval.

## DISCUSSION

Figure 1 shows schematically the large-scale electrodynamics of the auroral surge observed in these two passes overlaid on a generic substorm auroral image. The intense upward current and high ionospheric conductivity in the surge and weak electric fields south of the surge, which have been consistently observed in previous surge studies, are found in these measurements as well, and we suggest that these are standard elements of the electrodynamics of the surge.

We find in both passes discussed above a significant southward electric field in the surge and surge-connected arc to the east, which when combined with the high Hall conductance in these regions produces a strong westward electrojet. The FAST FAC observations show that while both upward and downward FACs are present near the surge, the downward FAC is smaller by a factor of 2-4, leaving a significant net upward current in the surge and surge-connected arc. These two facts paint a picture that is generally consistent with the concept of the substorm current wedge [McPherron et al., 1973], in which a downward current in the eastern portion of the auroral bulge passes through the ionosphere as a westward current and leaves the ionosphere as an upward current through the auroral surge. However, our observations suggest that the upward wedge FACs are distributed in local time to a large degree and that significant downward currents are present in the surge and vicinity, as was seen in the statistical DE studies of Fujii et al. [1994] and Gjerloev and Hoffman [1998]. Our observations are in disagreement to the observations of Marklund et al. [1998] in which upward surge FACs were balanced by equal downward FACs in the immediate vicinity and in which weak westward ionospheric currents were observed east of the surge. It should be emphasized that since the surge is such an electrodynamically complicated and variable region, there is likely considerable variability from event to event. Nevertheless, the FAST observations here support the majority of the previous observations and show that the surge is a region of net upward FACs which are fed by a strong westward Hall current.

## CONCLUSIONS

We have analyzed the ionospheric electrodynamics observed by the FAST satellite on two passes near auroral surges seen by the VIS Low Res visible imager on the Polar satellite. The southward electric field and vector magnetic field perturbations were directly measured, and the ionospheric Hall and Pedersen conductances were calculated from the precipitating electron measurements. In agreement with most previous surge observations, we find that the surge is a region of strong upward field-aligned currents and high ionospheric conductance, while to the south of the surge is a region of weak electric fields. Our primary goal was to study the configuration of field-aligned and ionospheric currents, as previous surge observations have differed as to the details of this relationship. We find that the surge contains significant net upward currents and that a significant southward electric field in the high Hall conductance region drives a westward current which feeds this upward current. This configuration is consistent with the concept of the substorm current wedge [McPherron et al., 1973] in which a downward current in the eastern portion of the auroral bulge passes through the ionosphere as a westward current and leaves the ionosphere as an upward current through the auroral surge. These observations are consistent with many previous surge observations [Inhester et al., 1981; Opgenoorth et

196 AURORAL SURGE CURRENTS AND ELECTRODYNAMICS

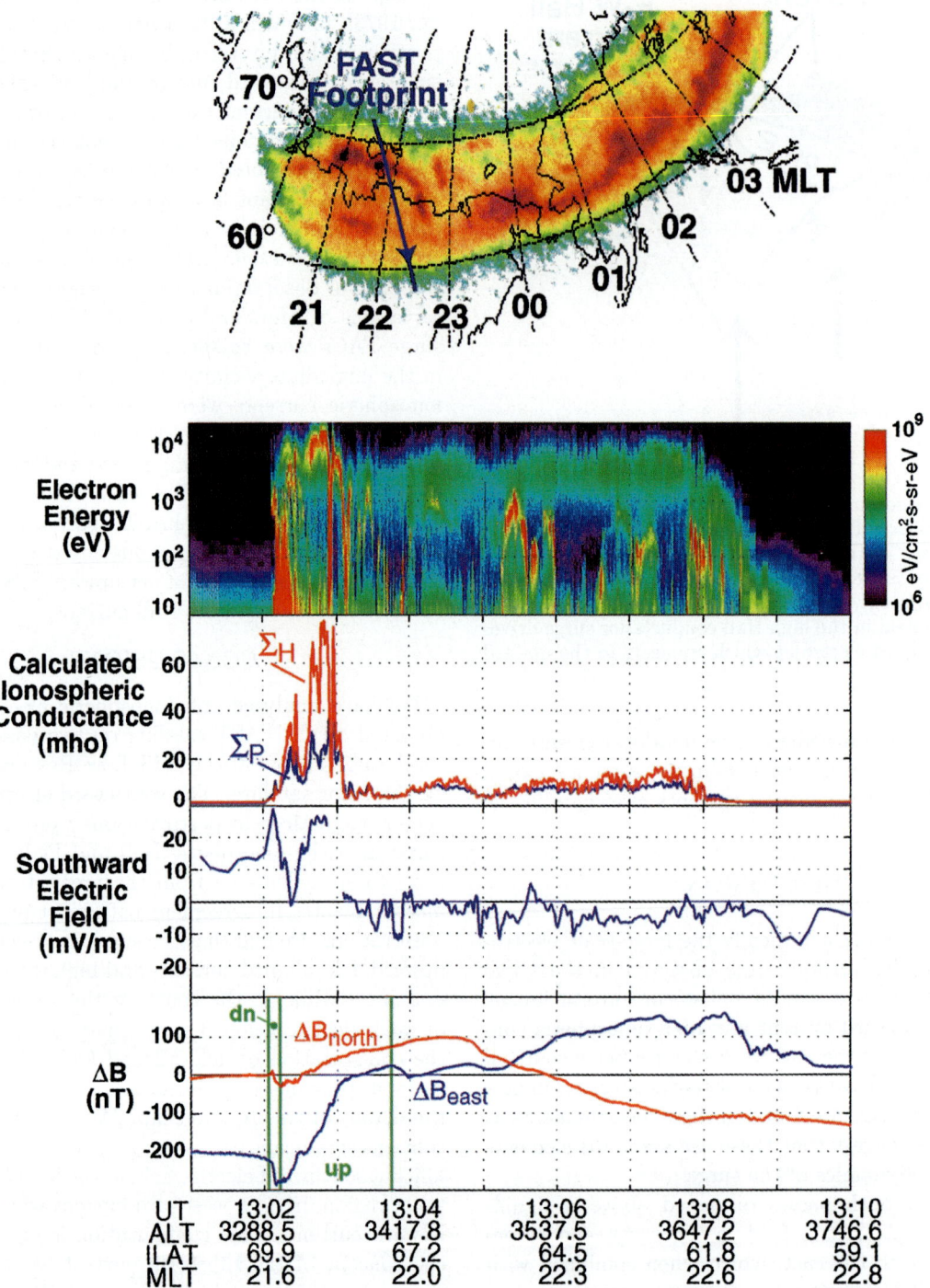

**Plate 2.** Observations from VIS and FAST on 18 Jan 98 at 13:05 UT in the same format as in Plate 1. Top: 557.7 nm VIS image of the aurora. Bottom: The observed and calculated electrodynamic parameters along the FAST footprint through the surge-connected arc.

al., 1983; *Weimer et al.*, 1994; *Fujii et al.*, 1994; *Gjerloev and Hoffman*, 1998] but are in disagreement with recent Freja surge observations [*Marklund et al.*, 1998] in which it was suggested that the upward surge currents close locally through equal downward currents in the immediate vicinity of the surge. We should emphasize that observations of the auroral surge are still rather sparse, and further simultaneous auroral imaging and electrodynamic observations of the surge will help fill in the gaps that remain in our understanding of the electrodynamics of this complicated and dynamic auroral form.

*Acknowledgments.* S. A. Cummer was supported by a National Research Council fellowship for this work. At the University of Iowa this research was supported in part by NASA under NAS5-30316. FAST efforts at LANL are supported by NASA order number S-57795-F.

## REFERENCES

Akasofu, S.-I., The development of the auroral substorm, *Planet. Space Sci., 12*, 273, 1964.

Anderson, P. C., W. B. Hanson, R. A. Heelis, J. D. Craven, D. N. Baker, and L. A. Frank, A proposed production model of rapid subauroral ion drifts and their relationship to substorm evolution, *J. Geophys. Res., 98*, 6069, 1993.

Bythrow, P. F., and T. A. Potemra, Birkeland currents and energetic particles associated with optical auroral signatures of a westward traveling surge, *J. Geophys. Res., 92*, 8691, 1987.

Carlson, C. W., R. F. Pfaff, and J. G. Watzin, The Fast Auroral SnapshoT (FAST) mission, *Geophys. Res. Lett., 25*, 2013, 1998.

Frank, L. A., J. B. Sigwarth, J. D. Craven, J. P. Cravens, J. S. Dolan, M. R. Dvorsky, P. K. Hardebeck, J. D. Harvey, and D. W. Muller, The Visible Imaging System (VIS) for the Polar spacecraft, *Space Sci. Rev., 71*, 297, 1995.

Fujii, R., R. A. Hoffman, P. C. Anderson, J. D. Craven, M. Sugiura, L. A. Frank, and N. C. Maynard, Electrodynamic parameters in the nighttime sector during auroral substorms, *J. Geophys. Res., 99*, 6093, 1994.

Fung, S. F., and R. A. Hoffman, Finite geometry effects of field-aligned currents, *J. Geophys. Res., 97*, 8569, 1992.

Gjerloev, J. W., and R. A. Hoffman, Electrodynamics in the high latitude nighttime sector during auroral substorms, in *Substorms-4*, p. 51, edited by S. Kokubun and Y. Kamide, Kluwer Academic Press, Boston, 1998.

Imhof, W. L., K. A. Spear, J. W. Hamilton, B. R. Higgins, M. J. Murphy, J. G. Pronko, R. R. Vondrak, D. L. McKenzie, C. J. Rice, D. J. Gorney, D. A. Roux, R. L. Williams, J. A. Stein, J. Bjordal, J. Stadsnes, K. Njoten, T. J. Rosenberg, L. Lutz, and D. Detrick, The Polar ionospheric x-ray imaging experiment (PIXIE), *Space Sci. Rev., 71*, 385, 1995.

Inhester, B., W. Baumjohann, R. A. Greenwald, and E. Nielsen, Joint two-dimensional observations of ground magnetic and ionospheric electric fields associated with auroral zone currents, *J. Geophys., 49*, 155, 1981.

Johnson, M. L., J. S. Murphree, G. T. Marklund, and T. Karlsson, Progress on relating optical auroral forms and electric field patterns, *J. Geophys. Res., 103*, 4271, 1998.

Marklund, G. T., T. Karlsson, L. G. Blomberg, P.-A. Lindqvist, C.-G. Fäthammar, M. L. Johnson, J. S. Murphree, L. Andersson, L. Eliasson, H. J. Opgenoorth, and L. J. Zanetti, Observations of the electric field fine structure associated with the westward traveling surge and large-scale auroral spirals, *J. Geophys. Res., 103*, 4125, 1998.

McPherron, R. L., C. T. Russell, and M. P. Aubry, Satellite studies of magnetospheric substorms on August 15, 1968, Phenomenological model for substorms, *J. Geophys. Res., 78*, 3131, 1973.

Meng, C.-I., A. L. Snyder, Jr., and H. W. Kroehl, Observations of auroral westward traveling surges and electron precipitations, *J. Geophys. Res., 83*, 575, 1978.

Opgenoorth, H. J., R. J. Pellinen, W. Baumjohann, E. Nielsen, G. Marklund, and L. Eliasson, Three-dimensional current flow and particle precipitation in a westward traveling surge (observed during the Barium-GEOS rocket experiment), *J. Geophys. Res., 88*, 3138, 1983.

Rees, M. H., Auroral ionization and excitation by incident energetic electrons, *Planet. Space Sci., 11*, 1209, 1963.

Robinson, R. M., and R. R. Vondrak, Electrodynamic properties of auroral surges, *J. Geophys. Res., 95*, 7819, 1990.

Weimer, D. R., J. D. Craven, L. A. Frank, W. B. Hanson, N. C. Maynard, R. A. Hoffman, and J. A. Slavin, Satellite measurements through the center of a substorm surge, *J. Geophys. Res., 99*, 23639, 1994.

---

C. W. Carlson, R. E. Ergun, W. J. Peria, Space Sciences Laboratory, University of California, Berkeley, CA 94720

S. A. Cummer, R. R. Vondrak, R. F. Pfaff, NASA/Goddard Space Flight Center, Code 690, Greenbelt, MD 20771. (email: steve.cummer@gsfc.nasa.gov)

R. C. Elphic, Los Alamos National Laboratory, D438, Los Alamos, NM 87545

J. W. Gjerloev, Danish Space Research Institute, Juliane Maries Vej 30, 2100 Copenhagen Oe, Denmark

J. B. Sigwarth, L. A. Frank, Department of Physics and Astronomy, University of Iowa, Iowa City, IA 52242

R. J. Strangeway, IGPP, University of California, Los Angeles, CA 90095

# A Synthetic View of the Magnetospheric-Ionospheric Current System Associated With Substorms

Gang Lu

*High Altitude Observatory, National Center for Atmospheric Research, Boulder, Colorado*

Electric currents play an important role in the energy coupling between the magnetosphere and ionosphere. This paper examines in detail the distribution and dynamics of ionospheric electric currents during an isolated substorm. In particular, the relative contributions of ionospheric Pedersen and Hall currents to field-aligned currents are evaluated in hoping to shed some new light on the ionospheric closure of the substorm current wedge. It is found that the substorm-related westward electrojet is dominated by Hall current, which is 2~3 times the Pedersen current in magnitude. Under winter-time substorm conditions, a significant portion (up to more than 1 MA) of field-aligned currents is closed in the ionosphere through Hall current. Although the contribution of Hall current to field-aligned currents is about 30~50% compared to the contribution from Pedersen current, it is Hall current, not Pedersen current, that appears to be the main contributor to the substorm current wedge.

## INTRODUCTION

Despite the controversy over its generation mechanisms, it is widely accepted that the substorm current system is composed of a pair of field-aligned currents (FACs) that flow into the ionosphere on the dawnside of the tail current sheet and flow out of the ionosphere on the duskside as a result of the disruption/reduction of the cross-tail current, forming the so-called substorm current wedge [*McPherron et al.*, 1973]. The FACs are closed in the ionosphere through a westward electrojet in the midnight sector. Understanding the development of the substorm current wedge is one of the key issues in substorm studies as it plays a critical role in the energy coupling between the magnetosphere and the ionosphere during substorms.

The 3-dimensional ionospheric current system associated with substorms has been studied extensively over the past 2 decades via a variety of techniques. Using ground-based magnetograms, *Rostoker and Hughes* [1979] and *Tighe and Rostoker* [1981] have proposed that the substorm related ionospheric current system is the superposition of an east-west aligned current wedge with relatively wider longitudinal extension and a north-south aligned current wedge that is confined to a longitudinally narrow region at the head of the westward traveling surge due to equatorward ionospheric currents. More sophisticated studies of the ionospheric current system have been presented by combining radar measurements of ion drifts or electric fields and ground magnetometer observations [e.g., *Inhester et al.*, 1981; *Baumjohann et al.*, 1981; *Opgenoorth et al.*, 1983], which show an east-west aligned substorm current wedge, with upward FACs at the surge head and downward FACs at the eastern trailing edge, and closed through enhanced westward electrojets attributed to the Cowling channel effect. Recently, *Fujii et al.* [1994]

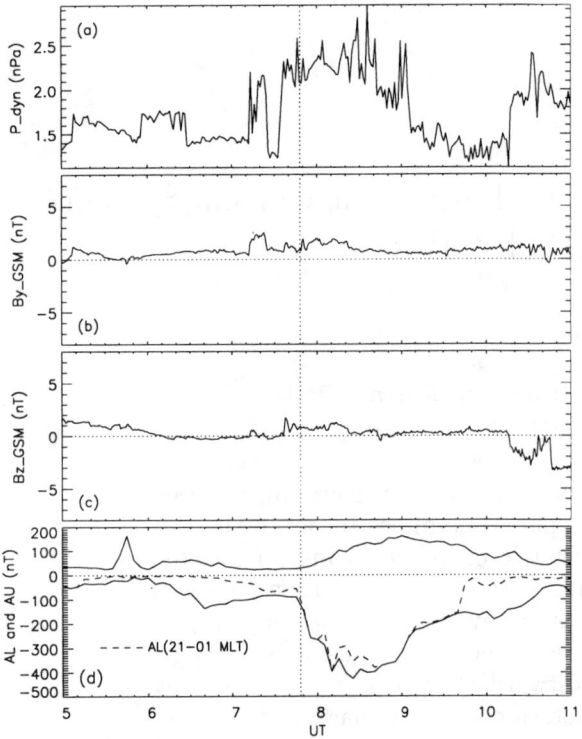

**Figure 1.** (top to bottom) Solar wind dynamic pressure, the IMF $B_y$ and $B_z$ components in GSM coordinates, and the $AU$ and $AL$ indices between 0500 and 1100 UT on January 9, 1997. A 21-min time delay has been applied to the Wind data. The vertical dotted line marks the substorm onset at 0748 UT.

and *Weimer et al.* [1994] have found a similar east-west aligned 3-D current wedge based on satellite observations.

The global magnetospheric currents associated with substorms are much less understood due to insufficient numbers of satellites that could physically cover the enormous spatial domain of the magnetosphere. Local observations at geosynchronous altitude have indicated that the substorm-associated magnetospheric current system consists of a pair of field-aligned currents with region-1 polarity, that is, they flow earthward on the post-midnight side and tailward on the pre-midnight side [e.g., *Nagai*, 1982, 1987; *Nagai et al.*, 1987; *Ohtani et al.*, 1988]. The spatial and temporal variation of the substorm current wedge in the magnetosphere, however, is yet to be established, and so is their corresponding ionospheric closure.

In this paper, we intend to provide a synthetic view of the magnetospheric-ionospheric current system through the study of an isolated substorm event on January 9, 1997. We first present the global distributions of ionospheric horizontal currents and FACs at the different phases of the substorm. We then show the evolution of the substorm current system in the equatorial magnetosphere through magnetic field line mapping. The main goal of this study is to evaluate the relative contributions of Pedersen and Hall currents to the substorm wedge currents both in the ionosphere and at the equatorial magnetosphere.

## RESULTS

Figure 1 shows, from top to bottom, the solar wind dynamic pressure, the $y$- and $z$-components of the interplanetary magnetic field (IMF), and the auroral electrojet indices of $AU$ and $AL$ for the period of 0500–1100 UT on January 9, 1997. The 1-min averaged solar wind and IMF data were obtained from the Wind satellite, which was located at $(78, -60, -3)R_E$ in GSE $(X, Y, Z)$ coordinates at 0700 UT, with the average bulk speed of about 400 km/s. A time delay of 21 min has been applied to the Wind data in Figure 1 to account for the solar wind propagation from the satellite location to the magnetopause. The solid curves of the $AU$ and $AL$ indices shown in the bottom panel were derived from the 5-min averaged north-south component of the magnetic perturbations measured by 68 ground magnetometer stations located in the auroral zone between $|55°|$ and $|76°|$ magnetic latitudes in all magnetic local times (MLTs). The dashed curve, on the other hand, is the $AL$ index but estimated from only those auroral stations located between 2100 and 0100 MLT. It therefore represents the westward auroral electrojet near the midnight sector. The good agreement between the solid and dashed $AL$ curves indicates that the enhanced westward electrojets during the substorm were mainly confined in the midnight region. The midnight $AL$ index (the dashed curve) began to decrease gradually at ∼0600 UT, and then dropped abruptly at ∼0745 UT. Sudden enhancement of Pi 2-type pulsations observed by the Canopus magnetometer chain (not shown) indicated that the onset of the substorm expansion phase was at ∼0748 UT, which was marked by the vertical dotted line. The Polar UVI camera, on the other hand, observed the initial auroral brightening at ∼0747 UT. This 1-min discrepancy in timing the substorm onset, however, should not affected the conclusions of this study since we focus mainly on the relative variations during the different phases of the substorm. After 0840 UT the substorm went into its recovery phase as indicated by the gradual increase in $AL$. During the entire substorm interval the

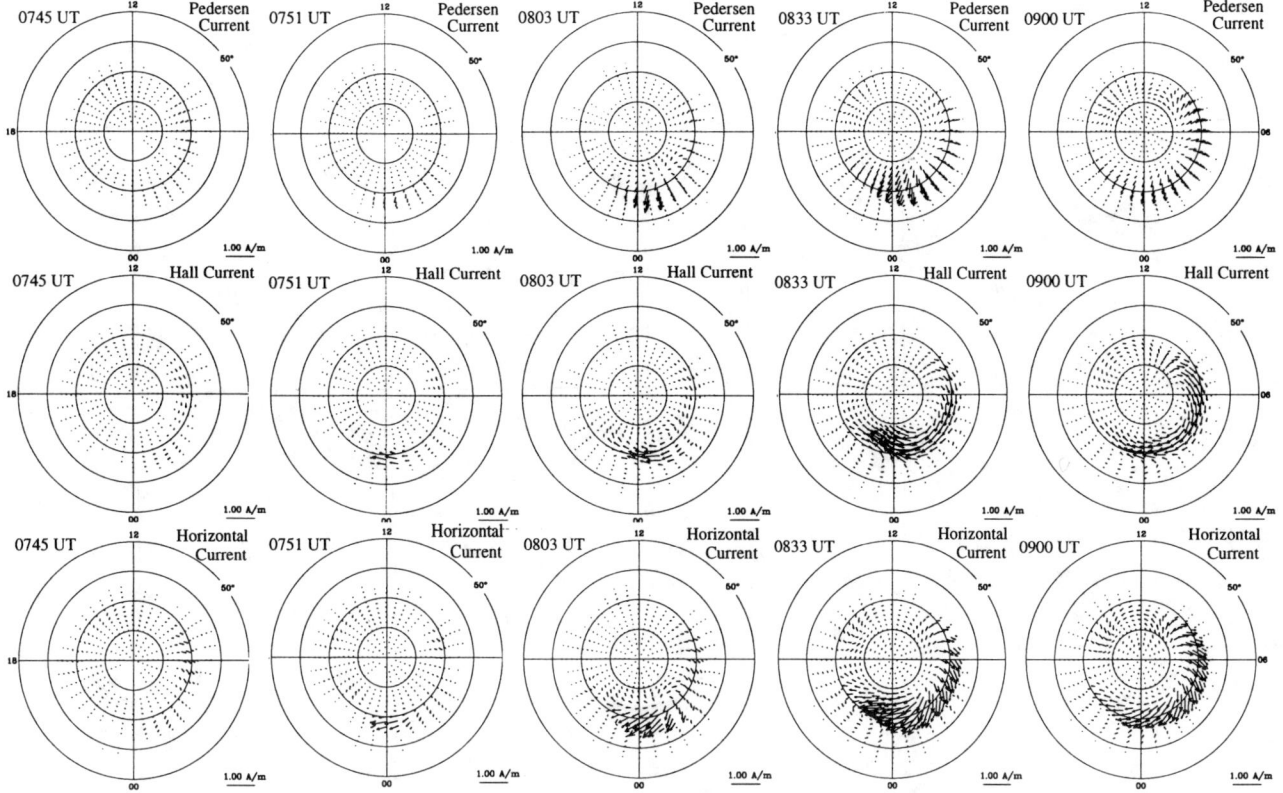

**Figure 2.** Distributions of (top row) Pedersen current, (middle row) Hall current, and (bottom row) total ionospheric horizontal currents at selected UT times during the substorm.

IMF remained nearly steady. But about 10 min prior to the substorm onset, there were a small northward turning (~2 nT) in the IMF $B_z$ and a 1-nPa increase in the solar wind dynamic pressure which may have induced the substorm. The stable and small magnitude of the $Dst$ index (not shown) with an average value of 2 nT indicated there were no magnetic storms during the period.

The assimilative mapping of ionospheric electrodynamics (AMIE) procedure [*Richmond and Kamide*, 1988] was used to derive the global distributions of ionospheric electric currents. Data inputs to AMIE for this study were the 1-min averaged ground magnetic perturbations recorded by a worldwide network of 123 magnetometer stations (among them, 88 were located in the northern hemisphere above 50° magnetic latitude), and the global auroral images from the Polar UVI instrument. Height-integrated Pedersen and Hall conductivities in the ionosphere were estimated from a pair of Polar UVI images [*Lummerzheim et al.*, 1997] at about every 3 min.

The representative patterns of ionospheric electric currents during the isolated substorm interval are illustrated in Figure 2. Prior to the substorm onset at 0745 UT both Pedersen (top row) and Hall (middle row) currents were rather weak over the entire polar region. After the onset Pedersen and Hall currents were intensified near the midnight sector, with Pedersen current pointing mainly toward south in the direction of electric field and Hall current pointing primarily westward. Since the ratio of Hall to Pedersen conductances was about 2~3 (as shown later in Figure 5) in the nightside auroral zone, the magnitude of Hall current was 2~3 times the magnitude of Pedersen current. As a result, the total ionospheric horizontal currents shown in the bottom row were dominated by Hall current during the substorm expansion phase. During the recovery phase at 0900 UT the enhanced Pedersen and Hall currents became concentrated towards the dawnside.

The divergence of ionospheric horizontal currents gives rise to field-aligned currents (FACs) $j_\parallel^{total}$. For simplicity, we refer hereinafter to the contribution of Pedersen

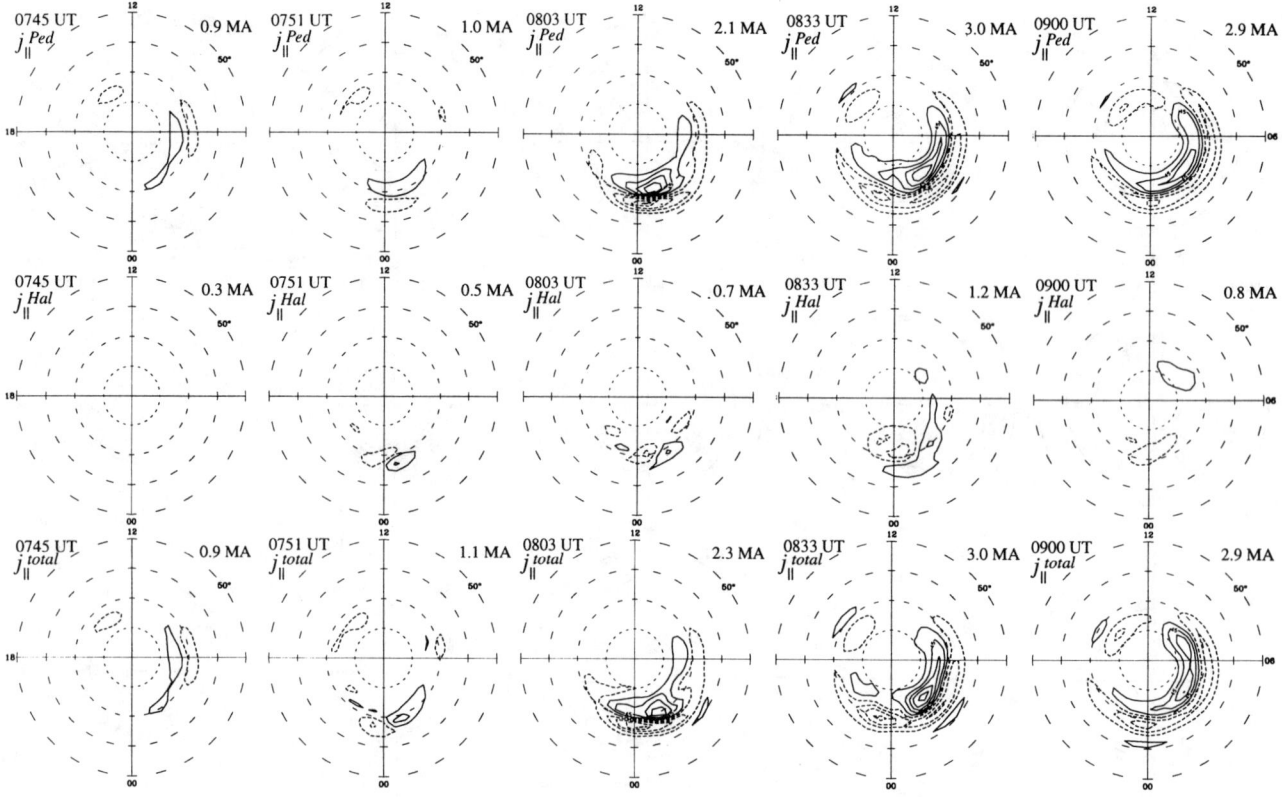

**Figure 3.** Distributions of (top row) the divergence of Pedersen current $j_\parallel^{Ped}$, (middle row) the divergence of Hall current $j_\parallel^{Hal}$, and (bottom row) the divergence of total horizontal currents $j_\parallel^{total}$ at the same selected UT times as in Figure 2. Solid contours represent downward currents and dashed contours for upward currents. The contour intervals are $0.3\mu A/m^2$, starting at $\pm 0.15\mu A/m^2$. The total hemisphere-integrated downward field-aligned current is given at the upper right of each pattern.

current to FACs as $j_\parallel^{Ped}$, and the contribution of Hall current to FACs as $j_\parallel^{Hal}$. Figure 3 presents the corresponding distributions of $j_\parallel^{Ped}$, $j_\parallel^{Hal}$ and $j_\parallel^{total}$ at the same selected UT times as in Figure 2. Before the onset $j_\parallel^{Ped}$ was comprised of a pair of current sheets with the polarity of the normal dawnside region 1 and region 2 currents. On the duskside there was only the upward region 1 current, and the downward region 2 current was too weak to show up in this $0.3\mu A/m^2$ contour interval. $j_\parallel^{Hal}$ was very small over the entire polar region. The distribution of $j_\parallel^{total}$ therefore was nearly identical to that of $j_\parallel^{Ped}$, indicating that FACs were closed predominantly through ionospheric Pedersen current prior to the substorm expansion phase. At 0751 UT (3 min after the onset) a pair of $j_\parallel^{Ped}$ as well as a pair of $j_\parallel^{Hal}$ evolved near midnight. The pair of $j_\parallel^{Ped}$ current sheets were oriented in the north-south direction. The pair of $j_\parallel^{Hal}$, on the other hand, were not exactly sheet-like but their peaks were oriented in the east-west direction. Consequently, the peaks of $j_\parallel^{total}$ in the midnight sector were oriented in the northeast-to-southwest direction. The total $j_\parallel^{Ped}$, defined as one-half the hemispheric integral of the absolute value of the current density over the area poleward of $50°$, was 1.0 MA, and the total $j_\parallel^{Hal}$ was 0.5 MA. Both $j_\parallel^{Ped}$ and $j_\parallel^{Hal}$ were intensified as the expansion phase progressed, with $j_\parallel^{Ped}$ aligned mainly in the north-south direction and the peaks of $j_\parallel^{Hal}$ aligned in the east-west direction. At 0833 UT, the peak of the substorm expansion phase, the total $j_\parallel^{Ped}$ had increased to 3.0 MA and the total $j_\parallel^{Hal}$ rose up to 1.2 MA. Although the magnitude of $j_\parallel^{Hal}$ was 2~3 times smaller than $j_\parallel^{Ped}$, the peaks of $j_\parallel^{total}$ near midnight oriented in the direction of northeast-to-southwest due to the contribution of $j_\parallel^{Hal}$. $j_\parallel^{Hal}$ in the midnight sector often had the opposite polarity as $j_\parallel^{Ped}$, and thus they tended to counteract with $j_\parallel^{Ped}$. During the recovery phase both $j_\parallel^{Ped}$ and $j_\parallel^{Hal}$ gradually decayed in the midnight sector. One interesting noticeable feature shown in Figure 3 is that both pairs of $j_\parallel^{Ped}$ and $j_\parallel^{Hal}$ evolved near midnight shortly after the substorm onset and then gradually expanded westward as well as eastward, with the eastward expansion being more pronounced; by 0833 UT $j_\parallel^{Ped}$ was centered at about 0200 MLT and $j_\parallel^{Hal}$ centered at about 0100 MLT. At the same time, the peaks of $j_\parallel^{Hal}$ also appeared to be moving poleward. For instance, the peak of the downward $j_\parallel^{Hal}$ had moved from about 67 MLAT to about 70 MLAT.

Figure 4 shows the mapped ionospheric FACs in the equatorial magnetosphere using a modified, time-evolving magnetic field model described by *Lu et al.* [1999]. Note that when mapping FACs from the ionosphere to the equatorial magnetosphere, we have assumed that there is no current leakage across the magnetic field lines so the projection of ionospheric FACs in the equatorial plane is scaled by a geometric factor of $B_e/B_i$, where $B_e$ and $B_i$ are the magnetic field strength at the equator and in the ionosphere, respectively. However, one should keep in mind that FACs may not remain exactly magnetic field-aligned when away from the low-altitude ionosphere. The solid (dashed) contours represent currents that flow out of (into) the equatorial plane. The contour interval is 0.1 nA/m² for the maps of $j_\parallel^{Ped}$ and $j_\parallel^{total}$, and 0.06 nA/m² for the maps of $j_\parallel^{Hal}$ (about 3 orders of magnitude smaller than in the corresponding ionospheric maps shown in Figure 3). After the substorm onset at 0751 UT the mapped $j_\parallel^{Ped}$ showed a pair of nightside region 2-sense currents located around 4 $R_E$, with currents flowing out of the equatorial plane on the west side of the midnight meridian and flowing into the equatorial plane on the east side of the midnight meridian. Tailward of 4 $R_E$ and near the midnight meridian $j_\parallel^{Ped}$ was predominantly flowing into the equatorial plane, and $j_\parallel^{Ped}$ flowing out of the equatorial plane was too weak to show up due to the fact that it was located further poleward in the ionosphere so that the mapping factor became relatively small. $j_\parallel^{Hal}$, on the other hand, showed a pair of currents with the nightside region 1-sense, consistent with the substorm current wedge model proposed by *McPherron et al.* [1973]. During the late expansion phase at 0803 and 0833 UT the mapped $j_\parallel^{Ped}$ also showed a pair of currents tailward of about 10 $R_E$. However, the wedge-shaped $j_\parallel^{Ped}$ had a larger extent in the Y-direction than what would be expected by the traditional substorm current wedge model; by 0833 UT they extended near the dawn and dusk flanks of the magnetopause. The pair of $j_\parallel^{Hal}$ were confined much closer to the X-axis, and their spatial extent was within 10 $R_E$ in the Y-direction and between -6 ~ -25 $R_E$ in the X-direction. Similar to its ionospheric counterpart, the mapped $j_\parallel^{Hal}$ also expanded longitudinally with the center rotating slowly towards the post-midnight sector. But unlike its ionospheric counterpart which showed a poleward motion, the mapped $j_\parallel^{Hal}$ did not propagate tailward during the expansion phase due to the dipolarization of the magnetic fields. During the recovery phase both $j_\parallel^{Ped}$ and $j_\parallel^{Hal}$ gradually diminished in the nightside central plasma sheet. The mapped $j_\parallel^{total}$ was somewhat similar to the mapped $j_\parallel^{Ped}$ during the substorm because the magnitude of the mapped $j_\parallel^{Ped}$ was larger than the mapped $j_\parallel^{Hal}$. The mapped $j_\parallel^{total}$ did not show the substorm current wedge structure at the early expansion phase, but showed an expanded wedge shape only during the late stage of the expansion phase.

## CONCLUSIONS AND DISCUSSION

The AMIE results presented above have reiterated that ionospheric westward electrojets during substorms are predominately Hall current, which is about 2~3 times the Pedersen current in terms of magnitude. Enhanced westward electrojets near midnight are attributed to the generation of a Cowling channel inside the highly conducting westward traveling surge. The westward currents or Cowling currents have been proposed as the combination of Pedersen current associated with the westward primary electric field and Hall current associated with the southward polarization southward electric field (see *Baumjohann* [1983] and references therein).

To further explore the electrodynamics concerning the substorm current system, Figure 5 shows the distributions of the various ionospheric parameters during the expansion phase of the substorm at 0809UT on January 9, 1997. Intense auroral precipitation was manifested by the enhancement of height-integrated ionospheric conductances over the nightside auroral zone. Both Pedersen and Hall conductances peaked near 70° magnetic latitude at about 2300 MLT, with a maximum value of about 13 S for Pedersen conductance and 26 S for Hall conductance, yielding a Hall-to-Pedersen conductance ratio of 2 at the surge head at this particular UT time.

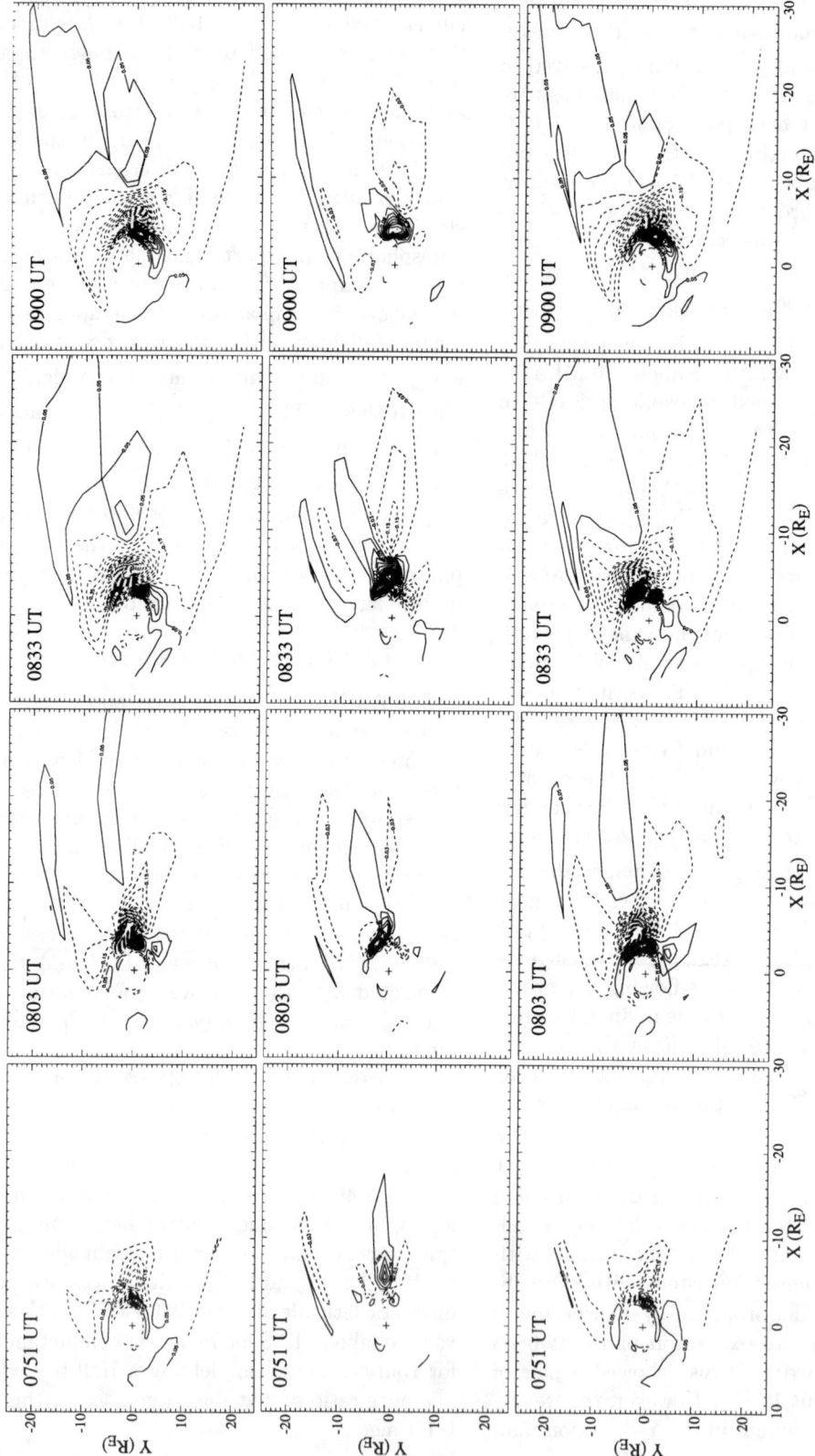

**Figure 4.** Distributions of field-aligned currents mapped to the magnetospheric equatorial plane: (top) mapped $j_\parallel^{Ped}$; (middle) mapped $j_\parallel^{Hal}$; and (bottom) mapped $j_\parallel^{total}$. Solid (dashed) contours indicate the currents that flow out of (into) the equatorial plane. The contour intervals are 0.1 nA/m² for the mapped $j_\parallel^{Ped}$ and $j_\parallel^{total}$, and 0.06 nA/m² for the mapped $j_\parallel^{Hal}$.

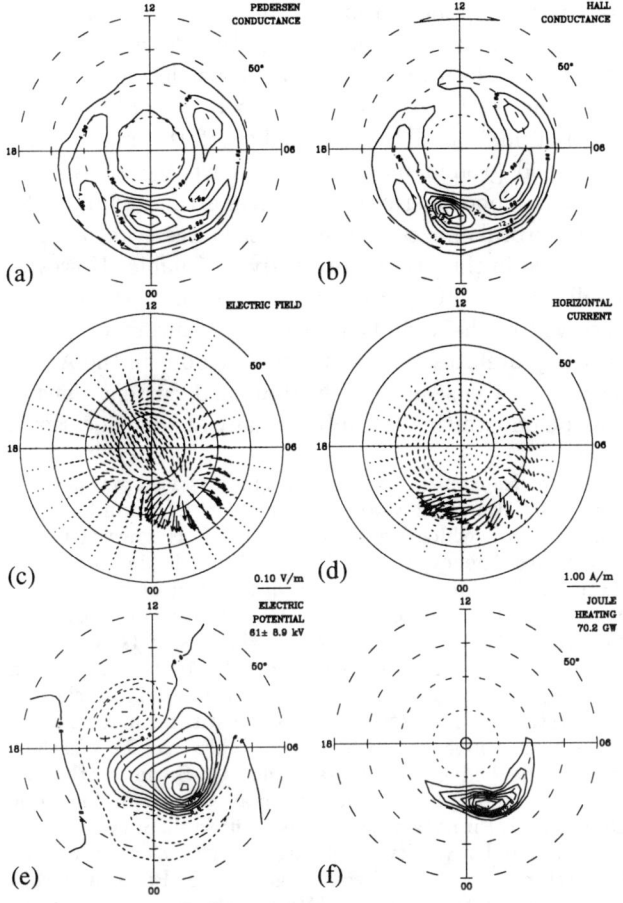

**Figure 5.** Distributions of various ionospheric parameters at 0809 UT on January 9, 1997: (a) Pedersen conductance; (b) Hall Conductance; (c) Electric fields; (d) total ionospheric horizontal currents; (e) electric potential; and (f) Joule heating. The contour intervals are 2 S for Pedersen conductance, 3 S for Hall conductance, 5 kV for electric potential, and 4 mW/m² for Joule heating rate.

The electric field in the pre-midnight sector was southward in the surge and northward to the south of the surge. Such a southward-to-northward change of the electric field is consistent with the previous observations [Inhester et al., 1981; Opgenoorth et al., 1983; Fujii et al., 1994]. East of the surge the electric field pointed southwestward in the post-midnight region, and the electric field strength was much larger than that inside the surge in the pre-midnight sector. Since the magnitude of Hall current was larger than the magnitude of Pedersen current by a factor of 2 (e.g., the Hall-to-Pedersen conductance ratio), the total ionospheric horizontal currents were dominated by Hall current and directed westward near the midnight sector. Compared with the distribution of electric field, it appears that the enhanced westward electrojets over the nightside auroral zone consisted of two components: one associated the enhancement of ionospheric conductance in the pre-midnight surge region and the other due to the enhancement of electric field in the post-midnight region to the east of the surge. This two-component concept has been postulated by Kamide et al. [1994] to interpret the ionospheric convection changes during substorms.

The southwest-directed electric field in the post-midnight sector shown in Figure 5c seems to be consistent with the Cowling channel suggested by Baumjohann [1983]. Unfortunately, using the AMIE procedure, we are unable to distinguish between the primary and polarization electric fields. However, the fact that $j_\parallel^{Hal}$ only partially cancels $j_\parallel^{Ped}$ indicates that the Cowling effect would be rather imperfect in this case. In addition, the Cowling channel appears to form at the trailing edge of the surge rather than within the surge head where the conductances are much higher, consistent with the modeling results of Inhester et al. [1981] and Baumjohann et al. [1981]. It should also be pointed out that, even though the substorm westward electrojets are dominated by the non-dissipative Hall current, $\mathbf{J} \cdot \mathbf{E}$ or Joule heating (when neglecting the neutral winds, as is the case in AMIE) is non-zero in the midnight sector. As a matter of fact, Joule heating tends to concentrate in that region (Figure 5f).

Figure 5e shows a 3-cell ionospheric convection configuration, with the regular 2 cells on the dawn and dusk sides and an isolated negative potential cell (dashed contours) near the midnight. In some previous substorm studies the deformation of convection patterns was found to be caused by the penetration of the dawnside cell across the midnight meridian into the duskside cell [Kamide et al., 1994; Weimer et al., 1994]. But in this case, an isolated nightside convection cell first evolved after the onset of substorm expansion phase; it then intensified and gradually merged with the pre-existing dusk cell as the substorm progressed [Lu et al., 1998]. It is interesting to note that the Harang discontinuity corresponding to the convection reversal of the nightside cell was further equatorward of the westward traveling surge.

In a relatively uniformly conductive ionosphere, such as that under sunlit condition, FACs are generally closed through ionospheric Pedersen current in the polar region. However, during substorms enhanced auroral precipitation can alter dramatically local ionospheric conductances so that their gradient is no longer negligible.

Consequently, part of FACs are fed through the divergence/convergence of Hall current. Up to now, the contribution of Hall current to FACs has only been modeled in localized regions [e.g., *Amm*, 1997, 1998; *Untiedt and Baumjohann*, 1993, and the references therein]. Using the AMIE procedure, we are able to evaluate Pedersen and Hall currents globally. We show quantitatively the relative contributions of Pedersen and Hall currents to FACs. We found that the magnitude of $j_\parallel^{Hal}$ amounted to more than 1 MA during this moderate substorm event, which was about 30–50% of the total $j_\parallel^{Ped}$. But more importantly, Hall current, rather than Pedersen current, played the dominant role in the formation of the substorm current wedge. During the substorm expansion phase the peaks of $j_\parallel^{Hal}$ aligned in the east-west direction, along with the enhanced substorm westward electrojet dominated by Hall current, they formed a 3-D current wedge aligned in the east-west direction as depicted by the traditional substorm current wedge model [*McPherron et al.*, 1973]. This study has demonstrated that the substorm wedge currents, although they have the same polarity as the nightside region 1 FACs, map to the region between -6 and -25 $R_E$ in the X-direction in the central equatorial plane, which obviously is not the same region where the normal region 1 FACs are considered to be generated. Our mapping results are in qualitative agreement with the MHD simulation of *Birn and Hesse* [1996] which shows that the substorm FACs are generated in the region from -5 to -13 $R_E$.

*Friedrich and Rostoker* [1998] have recently proposed an alternative configuration for the substorm current system. They suggest that the current system immediately after the expansion phase onset is the east-west distortion of the pre-existing north-south aligned region 1 and region 2 currents from the dawn side due to the westward shift of the upward region 2 current into the midnight sector, and the resultant northeast-to-southwest orientated FACs are closed in the ionosphere through Pedersen current. Although the distribution of $j_\parallel^{total}$ shown in Figure 3 bears a strong resemblance to the model of Friedrich and Rostoker, the AMIE results show clearly that the $j_\parallel^{total}$ in the midnight sector are the superposition of the north-south aligned $j_\parallel^{Ped}$ and the east-west aligned $j_\parallel^{Hal}$. This current system is not merely the distortion of the pre-existing region 1 and region 2 currents; rather, it evolves locally after the substorm onset.

It should be emphasized that the accuracy in our quantitative assessment of ionospheric Pedersen and Hall currents as well as their contributions to FACs is constrained by the spatial resolution of the AMIE algorithm which currently has an effective distance scale of about 1° in latitude and 5° in longitude (correspondingly, the AMIE grid size is ~1.7° in latitude and 10° in longitude), and by the quality of the input data. The good ground magnetometer coverage along with the global ionospheric conductances estimated from the Polar auroral images ensure that the large-scale features presented in this study are relatively reliable. However, small-scale (< 100 km) structures, such as those often observed in the vicinity of a discrete aurora by polar-orbiting satellites, are mostly smeared out in the AMIE patterns. Consequently, field-aligned current density may be underestimated substantially at some locations.

*Acknowledgments.* The auroral indices used in this study were calculated from a worldwide network of ground magnetometer data, which were provided by T. Hughes at Canadian Space Agency, L. Hakkinen at Finnish Meteorological Institute, T. Moretto at Danish Meteorological Institute, L. Morris at National Geophysical Data Center of NOAA, K. Yumoto and the STEL at Nagoya University, G. van Beek of the Geological Survey of Canada, J. Posch at Augsburg College, D. Milling at University of York, O. Troshichev at the Arctic and Antarctic Research Institute in Rassia, K. Hayashi at University of Tokyo in Japan, A. T. Weatherwax at the University of Maryland, G. Burns at the Atmospheric and Space Physics group at Australian Antarctic Division, M. Pinnock at the British Antarctic Survey, V. Papitashvili at University of Michigan, C. Maclennan at Bell Laboratories of Lucent Technologies, A. S. Potapov and S. I. Nechaev at the Institute of Solar-Terrestrial Physics at Irkutsk Observatory, and A. Zaitzev and V. Odintzov at IZMIRAN in Russia. The Polar UVI data were processed by D. Lummerzheim, and G. Parks is the PI of the Polar UVI instrument. The WIND magnetometer data were retrieved from the NASA CDAWeb database. This work was supported in part by the NSF Space Weather program, and by the NASA Space Plasma Theory and ISTP/SEC Guest Investigator programs. The author thanks A. D. Richmond for many helpful discussions.

## REFERENCES

Amm, O., Ionospheric elementary current systems in spherical coordinates and their application, *J. Geomag. Geoeletr.*, *49*, 947, 1997.

Amm, O., Method of characteristics in spherical geometry applied to a Harang-discontinuity situation, *Ann. Geophysicae*, *16*, 413, 1998.

Baumjohann, W., Ionospheric and field-aligned current systems in the auroral zone: A concise review, *Adv. Space Res.*, *2*, 55, 1983.

Baumjohann, W., R. J. Pellinen, H. J. Opgenoorth, and E. Nielsen, Joint two-dimensional observations of ground magnetic and ionospheric electric fields associated with auroral zone currents: Current systems associated with local auroral break-ups, *Planet. Space Sci.*, *29*, 431, 1981.

Birn, J., and M. Hesse, Details of current disruption and diversion in simulations of magnetotail dynamics, *J. Geophys. Res.*, *101*, 15,345, 1996.

Friedrich, E., and G. Rostoker, Reconfiguration of the directly driven currents during a substorm expansive phase: Implications for the substorm current, in *Substorms-4*, pp. 83-86, edited by S. Kokubun and Y. Kamide, Kluwer Academic Publishers, 1998.

Fujii, R., R. A. Hoffman, P. C. Anderson, J. D. Craven, M. Suguira, L. A. Frank, and N. C. Maynard, Electrodynamic parameters in the nighttime sector during auroral substorms, *J. Geophys. Res.*, *99*, 6093, 1994.

Inhester, B., W. Baumjohann, R. A. Greenwald, and E. Nielsen, Joint two-dimensional observations of ground magnetic and ionospheric electric fields associated with auroral zone currents, 3, Auroral zone currents during the passage of a westward traveling surge, *J. Geophys.*, *49*, 155, 1981.

Kamide, Y., et al., Ground-based studies of ionospheric convection associated with substorm expansion, *J. Geophys. Res.*, *99*, 19,451, 1994.

Lu, G., N. A. Tsyganenko, A. T. Y. Lui, H. J. Singer, T. Nagai, and S. Kokubun, Modeling of time-evolving magnetic fields during substorms, *J. Geophys. Res.*, *104*, 12,327, 1999.

Lu, G., A. D. Richmond, Y. Kamide, D. Lummerzheim, M. Brittnacher, and G. Parks, Global ionospheric convection during substorm expansion, in *Substorms-4*, edited by S. Kokubun and Y. Kamide, pp. 617-622, Kluwer Academic Publishers, 1998.

Lummerzheim, D., M. Brittnacher, D. Evans, G. A. Germany, G. K. Parks, M. H. Rees, and J. F. Spann, High time resolution study of the hemispheric power carried by energetic electrons into the ionosphere during the May 19/20, 1996 auroral activity, *Geophys. Res. Lett.*, *24*, 987, 1997.

McPherron, R. L., C. T. Russell, and M. P. Aubry, Satellite studies of magnetospheric substorms on August 5, 1968, 9, Phenomenological model for substorms, *J. Geophys. Res.*, *78*, 3131, 1973.

Nagai, T., Observed magnetic substorm signatures at synchronous altitudes, *J. Geophys. Res.*, *78*, 4405, 1982.

Nagai, T., Field-aligned currents associated with substorms in the vicinity of synchronous orbit, 2, GOES 2 and GOES 3 observations, *J. Geophys. Res.*, *92*, 2432, 1987.

Nagai, T., H. J. Singer, B. G. Ledley, and R. C. Olsen, Field-aligned currents associated with substorms, 1, The July 5, 1979, substorm observed by SCATHA, GOES 3 and GOES 2, *J. Geophys. Res.*, *92*, 2425, 1987.

Opgenoorth, H. J., R. J. Pellinen, W. Baumjohann, K. Nielsen, G. Marklund, and L. Eliasson, Three-dimensional current flow and particle precipitation in a westward traveling surge (observed during the Barium-GEOS rocket experiment), *J. Geophys. Sci.*, *88*, 3138, 1983.

Ohtani, S., S. Kokubun, R. C. Elphic, and C. T. Russell, Field-aligned current signatures in the near-tail region, 1, ISEE observations in the plasma sheet boundary layer, *J. Geophys. Res.*, *93*, 9709, 1988.

Richmond, A. D., and Y. Kamide, Mapping electrodynamic features of the high-latitude ionosphere from localized observations: Technique, *J. Geophys. Res.*, *93*, 5741, 1988.

Rostoker, G., and T. J. Hughes, A comprehensive model current system for high-latitude magnetic activity, II, The substorm component, *Geophys. J. R. Astr. Soc.*, *58*, 571, 1979.

Tighe, W. G., and G. Rostoker, Characteristics of westward traveling surges during magnetospheric substorms, *J. Geophys.*, *50*, 51, 1981.

Untiedt, J., and W. Baumjohann, Studies of polar current systems using the IMS Scandinavian magnetometer array, *Space Sci. Rev.*, *63*, 245, 1993.

Weimer, D. R., J. D. Craven, L. A. Frank, W. B. Hanson, N. C. Maynard, R. A. Hoffman, and J. A. Slavin, Satellite measurements through the center of a substorm surge, *J. Geophys. Res.*, *99*, 23,639, 1994.

---

G. Lu, High Altitude Observatory, National Center for Atmospheric Research, 3450 Mitchell Lane, Boulder, CO 80301. (e-mail: ganglu@ucar.edu)

# The Harang Discontinuity in Auroral Substorms

J.W.Gjerloev and E.Friis-Christensen

*Danish Space Research Institute, Juliane Maries Vej 30, DK-2100 Copenhagen Oe, Denmark.*

R.A.Hoffman and S.A.Cummer

*Laboratory for Extraterrestrial Physics, NASA/Goddard Space Flight Center, Greenbelt,*

We have investigated the electrodynamics of the pre-midnight convection reversal during substorm expansion phase through early recovery phase. This is done with measurements of the magnetic field, electric field, ion drift and electron precipitation made by the Dynamics Explorer 2 (DE-2). The measured electron precipitation has been used to calculate the height integrated Pedersen and Hall conductivity (conductance) which combined with the electric field measurements enables us to calculate the ionospheric current intensities. The field aligned current (FAC) density is then calculated from the divergence of the horizontal meridional current and from the measured magnetic field perturbations. The meridional electric field was used to identify the position of the Harang discontinuity or Harang region (HR) as we prefer to address it. We find that in general 3 different types of HRs exist: (1) A well defined point separating regions of poleward and equatorward meridional electric fields. This type is typically seen far west of the surge and is located near the high latitude boundary of the enhanced conductance region. Consequently, the HR is located at the poleward boundary of the eastward electrojet while the westward electrojet is very narrow or non-existent. In all cases, we find that the zonal component is vanishing at the HR. (2) A region of depleted meridional electric fields with a typical width of ~100 km, typically located just west of the surge. The electric field and the conductance exhibit a clearly inverse relationship. (3) A wide region (~500 km) located in the surge and bulge. The three dimensional current system is filamentary within the region separating the low latitude eastward electrojet and the intense high latitude westward electrojet.

## 1. INTRODUCTION

*Harang* [1946] used a Scandinavian magnetometer chain in a study of ground magnetic field perturbations. He found that in the pre-midnight sector a discontinuity exists, which separates regions of positive and negative ΔH (local north component). Later *Heppner* [1972] named it the Harang discontinuity and *Maynard* [1974a] used OGO-6 electric field data and found that it is located in higher latitudes at earlier local times thereby making an angle with constant invariant latitude. *Maynard* [1974b] reported that the Harang discontinuity is present even during extremely quiet times and that the latitudinal width of the discontinuity becomes smaller during disturbed conditions. It is obviously

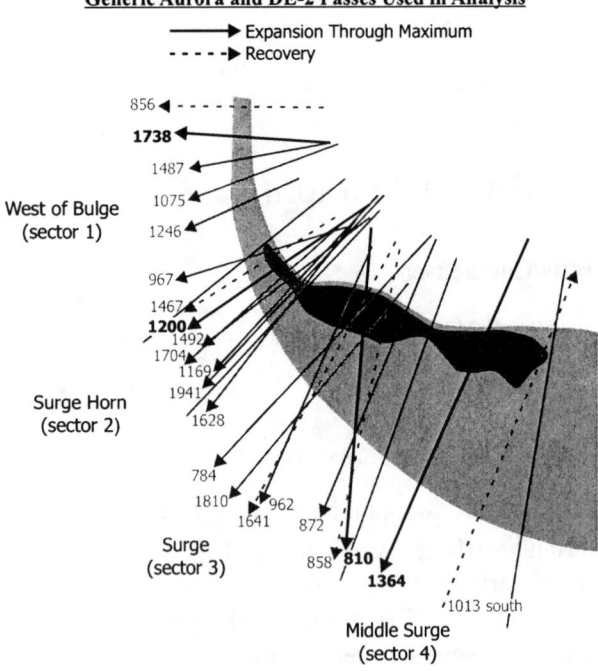

**Figure 1.** All passes used in the analysis superimposed on the generic bulge-type aurora with sectors indicated. Selected passes shown in Figure 2-5 are highlighted (after *Fujii et al., 1994*).

difficult to use ground based magnetometers to identify the discontinuity during quiet times since they require an ionospheric electrojet system of sufficient strength to produce perturbations that are observable on ground. Further, the eastward and westward electrojets are required to overlap in longitude and consequently enhanced conductance must be present poleward as well as equatorward of the electric field reversal. Hence, it is not obvious that the discontinuity determined by the use of ground magnetic field data is coincident with the discontinuity determined from electric field data. This was pointed out by *Kamide* [1978] who discussed various ways of identifying the Harang discontinuity (ground magnetic fields, ionospheric currents, auroral features and electric fields). He suggested that the discontinuity is a narrow region in which the electric field rotates counterclockwise from poleward over westward to equatorward. *Kamide and Vickrey* [1983] and *Kunkel et al.* [1986] found that the magnetic Harang discontinuity typically is located 1°-2° equatorward of the electric field Harang discontinuity.

In this study, we have investigated the electrodynamic properties of the convection reversal during substorms using data from DE-2. Emphasis has been put on the characteristics of the measured convection electric field; the calculated height integrated Pedersen and Hall conductivity; the calculated FACs and the ionospheric currents at the reversal and in its near vicinity. The paper consists of four sections: in section 2 we present four selected passes, section 3 is a discussion, and finally in section 4 we summarize and present our conclusions.

## 2. DATA AND OBSERVATIONS

Using the method by *Fujii et al.* [1994], global auroral images obtained by DE-1 were used to select DE-2 substorm crossings, which occurred during bulge-type auroral substorms. We use 22 DE-2 passes of which 17 occured during the expansion phase through maximum and the remaining 5 are early recovery phase. In Figure 1 is shown the selected DE-2 passes superimposed onto the generic aurora as defined by *Fujii et al.* [1994]. Notice that the passes are not equally distributed, e.g. only two passes are located in the Middle Surge sector and, unfortunately, none in the head of the Surge.

Figures 2-5 show four different substorm crossings from the West of Bulge sector through the Middle Surge sector. In all panels, we use the satellite coordinate system: x positive southward, y positive up and z positive westward. In panel 1 is shown the measured electric field component along the satellite trajectory (*Maynard et al.*, 1981) and the horizontal component perpendicular to the trajectory (if available). The electric field has been converted into a corotational frame by subtraction of the $\overline{V} \times \overline{B}$ electric field ($\overline{V}$ is the spacecraft velocity in the corotational frame and $\overline{B}$ is the magnetic field) and any offset has been removed assuming that the integrated potential difference across the high latitude region from 45° to 45° ILAT is zero. Unfortunately, DE-2 had a malfunction, which prevented measurements of the zonal electric field component. This problem can be partly overcome using data from the Retarding Potential Analyzer (RPA, see *Hanson et al.*, 1981) which measures the ion RAM drift. Assuming the ions $\overline{E} \times \overline{B}$ drift, the zonal electric field can be estimated by the use of the measured magnetic field. The DC electric field instrument provided data with a time resolution of 1/16 s while the RPA instrument had a time resolution of about 1 s. In panel 2 is shown the measured magnetic field perpendicular to the satellite trajectory after subtraction of the NASA 7/81 main field model with epoch 1980. Pedersen and Hall conductances are shown in panel 3. These were calculated from measurements of the electron precipitation provided by the Low Altitude Plasma Instrument (LAPI, *Winningham et al.*, 1981) and the monoenergetic conductance model by *Reiff* [1984] (see *Gjerloev and Hoffman*, 1998a and 1998b for a thorough study on the conductance in auroral substorms). LAPI provided a 32-step energy spectrum covering 5eV to 32 keV making a complete sweep every second, which enables us to calculate values of the Pedersen and Hall conductance with a 1 s time resolution (equivalent to a spatial resolution of ~7 km). Because the conductance has been deduced solely from electron precipitation data, we indicate the Boundary Plasma Sheet (BPS) and Central Plasma Sheet (CPS) type precipitation regions (*Winningham et al.*, 1975). However, to avoid a discussion of the morphology of the magnetosphere, it should be noted that we simply use the BPS-CPS terms as an indication of structured and diffuse precipitation respectively. In panel 4 is shown the height-integrated horizontal meridional current ($I_x$) calculated from the current equation: $I_x = (\Sigma_P E_x - \Sigma_H E_z)$, where $\Sigma_{P,H}$ is the Pedersen and Hall conductance, $E_{x,z}$ is the horizontal ionospheric convection electric field and spacecraft coordinates are used (x along trajectory or meridional, z perpendicular to trajectory or zonal and y is up). Finally, in panel 5 is shown the field aligned current density calculated by two independent methods. The FAC density (in $\mu A/m^2$) can be calculated directly from the measured magnetic field perturbations assuming infinite vertical FAC sheets oriented perpendicular to the satellite orbit: $j_\parallel \approx 0.8 \cdot \Delta b/D$, where $\Delta b$ is the perturbation across the infinite current sheet (in nT) of thickness D (in km). However, the FAC density can also be calculated from the divergence of the ionospheric currents, requiring the current to be divergence free: $j_\parallel = -\nabla \cdot \overline{I_\perp}$, where $\overline{I_\perp}$, is the horizontal ionospheric current intensity vector. This expression can be simplified if we assume the infinite FAC sheets close purely as ionospheric meridional currents: $j_\parallel = -d(\Sigma_P E_x - \Sigma_H E_z)/dx$. Consequently, we are able to calculate the FAC density using two different methods, which provide valuable information of the three-dimensional current system and serves as an independent verification of our results. The correlation coefficient between these two FAC results is also indicated in panel 5. Finally, it should be noted that in both methods a simple boxcar low pass filter was applied before calculating the FAC.

Since this paper primarily is concentrated on the electrodynamic parameters in or at the convection reversal we only display data from its near vicinity, i.e. ±90 s or a total of 3 minutes of data from all passes. As it will become apparent the pre-midnight convection reversal is generally a region of low fields rather than a discontinuity and consequently we will address it the Harang Region (HR).

### 2.1. West of Bulge Pass #1738

In Figure 2 is shown a DE-2 substorm crossing, which by the use of DE-1 images was placed in the West of Bulge sector. The HR can easily be identified as a point separating a broad region of poleward electric field and a small narrow region of equatorward electric field. Due to the classical in phase variations of the electric and magnetic field perturbations the HR is located in what can be classified as the large-scale region 1 upward FAC region. The HR is approximately coincident with the high latitude precipitation boundary, and consequently also with the poleward flowing $I_x$. It is important to notice that since the enhanced conductance region is embedded in the poleward E-field only an eastward electrojet is present. Consequently, the Harang discontinuity defined by the ground $\Delta H$ perturbations does not exist since this current system only produces a positive H-perturbation. Finally, despite the misfortune that no RPA data were available in this pass, it should be noticed that the FAC calculated from the two different methods is in good agreement with a correlation coefficient of 0.70. Consequently, our simplistic assumptions are actually valid: the FAC can be approximated as infinite current sheets, which are closed mainly by ionospheric Pedersen currents and there is no or little divergence of the eastward electrojet in this sector. As mentioned, the HR is nearly coincident with the high latitude BPS boundary and consequently the high latitude boundary of the enhanced conductance region and $I_x$. However, it is important to notice that other than this indirect indication, no changes in the other fundamental parameters would enable a determination of its location. No clear signature of the convection reversal is seen in any of the other parameters (e.g. an abrupt increase in electron precipitation as reported by *Wedde et al.*, 1977 or intense upward FACs as shown by *Kisabeth and Rostocker*, 1973). In three West of Bulge passes RPA data were available (not shown) but these do not provide a clear picture of the zonal electric field component at or near the HR. In one pass it is clearly eastward while it is vanishing in the other two passes.

### 2.2. Surge Horn Pass #1200

Pass #1200 located in the Surge Horn sector is shown in Figure 3. The HR is fundamentally different from the previous West of Bulge pass. In an approximately 85 km wide region, both electric field components vanish, separating regions of equatorward and poleward electric fields. In such a case, positioning the HR at a specific point becomes a matter of definition rather than physics. This is one example of why we prefer to address the convection reversal as the Harang region instead of the Harang discontinuity. Within the HR the zonal field component is vanishing although this is not typical for this sector since there is a clearly westward electric field in the other three passes where RPA data were available. It is important to notice that the depressed electric field is coincident with the enhanced conductance region. Near the low latitude boundary of the HR $\Sigma_P$ drops from ~30 mho at 16:54:31 UT to ~3

212 THE HARANG DISCONTINUITY IN AURORAL SUBSTORMS

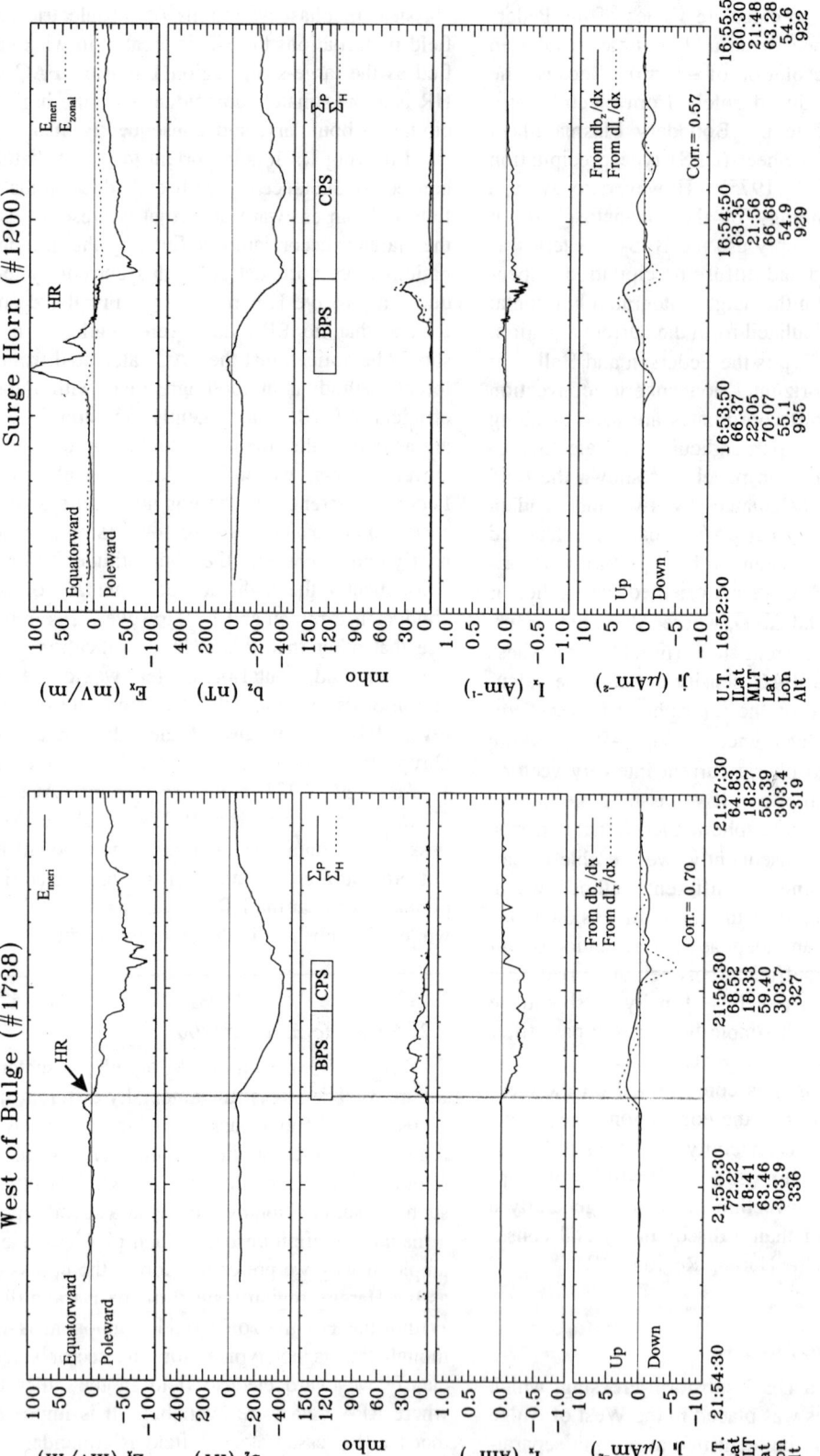

**Figure 2.** A typical West of Bulge pass #1738. The panels show: panel 1, the electric field; panel 2, the magnetic field perturbations; panel 3, the calculated Pedersen and Hall conductance with the BPS-CPS precipitation regions indicated; panel 4, the height integrated meridional current; panel 5, the FAC density calculated from magnetic field perturbations and from divergence of the meridional current. Further, the Harang region (HR) is indicated and the correlation coefficient between the two FAC results.

**Figure 3.** Surge Horn pass #1200, same format as in Figure 2.

mho only 3 s later (equivalent to ~21 km). However, simultaneously the electric field increases from ~10 mV/m to ~65 mV/m thereby keeping $I_x$ fairly constant across this region. These observations indicate that a polarization electric field is responsible for the decrease in the ambient electric field, which is imposed on the ionosphere by the magnetospheric convection pattern (*Marklund*, 1984). This polarization electric field provides the observed meridional current continuity by decreasing the ambient electric field inside the enhanced conductance region. Within the HR the weak poleward electric field drives a poleward ionospheric current. This current is steadily decreasing in magnitude toward higher latitudes and consequently an upward FAC is present. As in the West of Bulge pass we find a fairly good correlation between the two FAC methods again indicating that the FACs are closed by horizontal meridional currents with little or no divergence of the electrojets. Although, in general there is a good correlation within the HR, we find a very poor correlation poleward of the HR in the 4 Surge Horn passes where the enhanced conductance region is coincident with the equatorward electric field region. In these cases a narrow but well-defined westward electrojet is present. This was not seen in any of the West of Bulge passes and consequently a considerable divergence of the westward electrojet must be present in these passes. This violates our simple current assumptions hence a poor correlation between the FAC results is expected. However, several other possible explanations exist. For example: non-infinite FAC sheets which violates our assumptions (see *Hoffman et al.*, 1994 for an extensive discussion); if the observed variations are due to temporal variations rather than spatial gradients; additional ionospheric ionization not produced by precipitating electrons. Finally, as in the West of Bulge, we do not find any clear indication in the other parameters, which would enable a determination of the HR, although it should be mentioned that in all passes the HR is embedded in the upward region 1 FAC.

### 2.3. Surge Pass #810

In Figure 4 is shown a Surge sector pass, which exhibits an extremely broad HR, ~450-500 km. It is important to notice that this region is not coincident with the highest conductances, which are found in the BPS region. Instead it is located in a region with smoothly varying moderately high conductances, while on the other hand the highly enhanced conductances are coincident with a region of fluctuating but mostly equatorward electric fields. This does not support the previously mentioned inverse relationship between the conductance and the electric field and consequently $I_x$ is highly varying. Obviously, care must be taken in assuming that large gradients in $I_x$ necessarily result in intense FACs since horizontal diversion of the electrojet current can ensure the fulfillment of $\nabla \cdot \vec{j} = 0$ without any FAC at all. Since the electric field is vanishing over a broad region in which the conductance is moderate, the horizontal currents are too weak to feed any large scale intense FACs. Consequently, the FACs within the HR are filamentary and relatively weak. The zonal electric field is westward in the high latitude part of the HR and eastward in the low latitude part. RPA data were available in 4 other Surge passes and in 3 of these a westward electric field was found within the HR while in the remaining pass the zonal field was eastward. A westward electric field within the HR drives a poleward Hall current which can connect the eastward and westward electrojets (*Kamide*, 1978). However, as seen in this example the conductance is on the order of ~10-15 mho's, and the westward electric field component is on the order of a few mV/m, resulting in low Hall current intensities of ~0.05 A/m, which are not sufficient to drain the eastward electrojet. Hence, we find it intrinsically more likely that the eastward electrojet is closed through upward FACs located near midnight.

### 2.4. Middle Surge Pass #1364

Unfortunately, only two passes were found to pass through the Middle Surge sector of which pass #1364 is shown in Figure 5. Although two events do not provide a sound statistical basis, it should be mentioned that these two passes displayed a striking similarity in the electric field pattern. While the high latitude boundary of the HR is well defined, a broad region of weak fields is located equatorward of the HR making it difficult to identify a clear separation point. As in the Surge pass weak filamentary FACs are present in the wide HR and the depletion of the field is not coincident with the highest conductances which instead are found poleward of the HR. The two FAC results are remarkably poorly correlated in the BPS region. The structured electric field that result in a highly variable ionospheric current is one possible explanation, in addition to the previously mentioned. The FACs shown in panel 5 obviously does not resemble this structure due to the boxcar filter. However, it should be noted that the small-scale unfiltered data locally can be highly correlated. For example, the correlation coefficient is 0.85 between $b_z$ and $I_x$ in the 10 s interval from 19:48:57 to 19:49:07.

### 3. DISCUSSION

*Kamide* [1978] suggested that the Harang discontinuity is a region of weak meridional electric fields in which the

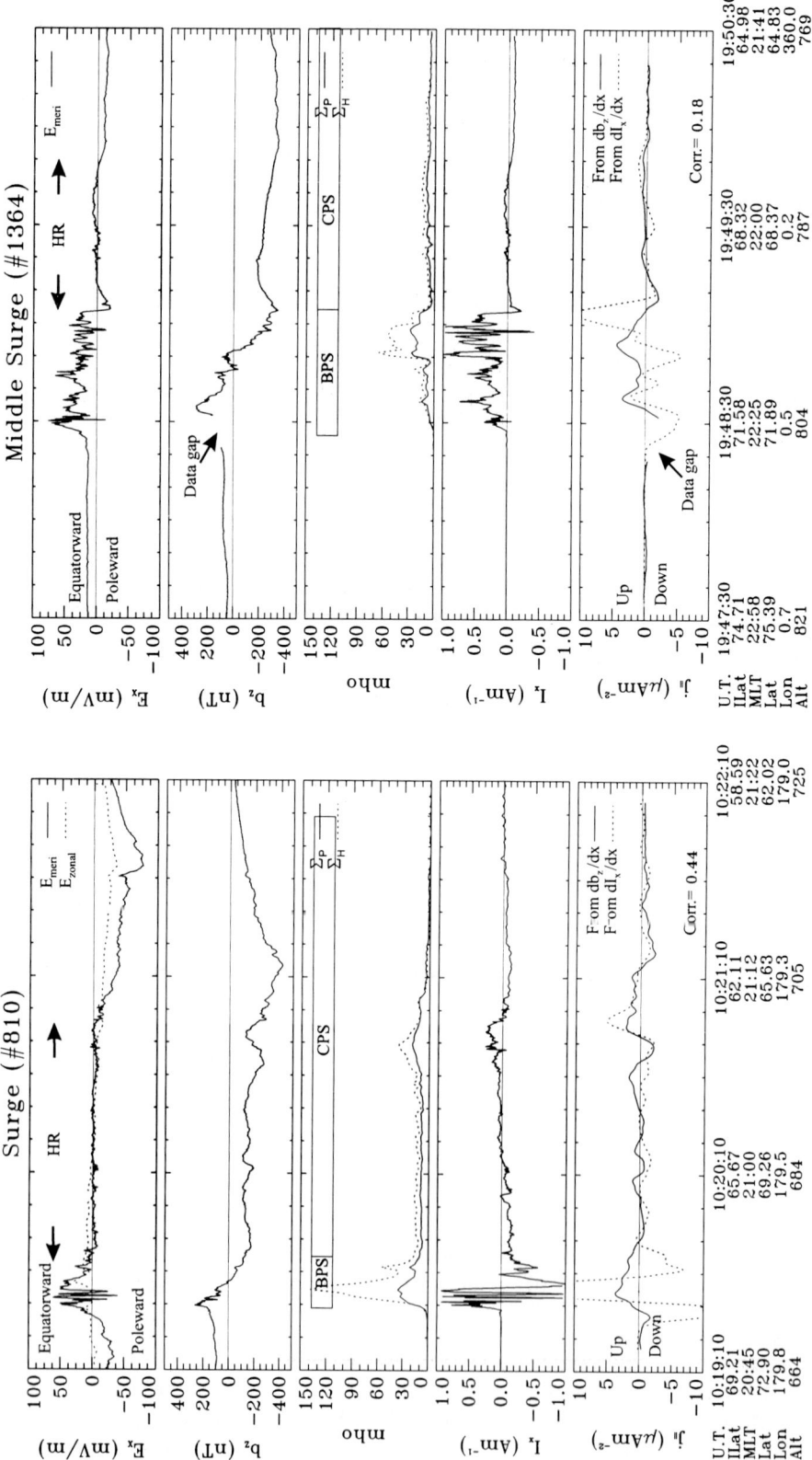

**Figure 5.** Middle Surge pass #1364, same format as in Figure 2.

**Figure 4.** Surge pass #810, same format as in Figure 2.

westward electric field dominates. *Marklund* [1984] explained the meridional electric field pattern for auroral arcs as consisting of a polarization electric field and an electric field, which is the low altitude signature of an inverted-V potential. This polarization electric field is then responsible for the depletion of the electric field in the HR where the conductance is highly enhanced and thereby providing current continuity across the arc (hereafter named Current Continuity Model, CCM). Although the Surge Horn pass shown in Figure 3 supports this simple CCM it is important to point out that we do not in general find this to be the case. Actually, we find that in only 3 of 22 passes the CCM seems to adequately describe the electrodynamics within the HR. On the other hand, relatively often we find regions, outside the HR, in which a decrease in the electric field strength is associated with increased conductance. Typically, the HR is embedded in diffuse CPS type precipitation (Surge and Middle Surge) where the conductance is moderately high and smoothly varying. Within the region of maximum conductances, located poleward of the HR, the electric field is erratic with large-scale fluctuations superimposed onto a mostly equatorward field. Further, it is important to note that in the 3 cases where the HR can be described by the CCM the width of the HR was 80 km, 110 km and 85 km respectfully. Although it is wider than the examples shown by *Marklund et al.* [1982] (20 km) it supports their suggestions. In contrast, the broad HR shown in Figure 4 is seen in 15 passes with a typical width of several hundreds of kilometers.

In apparent disagreement with our observations, *Maynard* [1974a & 1974b] concluded that the latitudinal width of the HR becomes smaller during disturbed conditions. *Maynard* [1974a] used AE(10) to bin his data (AE≤100, 100<AE<500 and AE≥500), but it is important to remember that a disturbed AE does not necessarily mean substorm conditions. Further, the magnitude of AE does not provide any substorm phase information, without including its temporal evolution. On the other hand, our database only consists of substorm expansion phase through early recovery phase passes, hence the databases are not obtained during similar conditions and therefore any direct comparison must be made with great caution. It should be noted, however, that of 21 HR crossings occurring during disturbed conditions, AE>100 nT, *Maynard* [1974a] found 8 HRs with a latitudinal width of 1°-2° and 6 with a width of 2°-7.5° (see his Figure 10). Hence, although our observations appear to be in contrast to the conclusions made by *Maynard* [1974a and 1974b] there are clearly similarities in the observations. Finally, it should be mentioned that *Marklund et al.* [1998] and *Cummer et al.* (this issue) used satellite passes through a substorm surge (Freja and FAST respectively) and found a wide region of weak electric fields in good agreement with our observations.

A classical point reversal of the meridional electric field is seen in three passes located in the West of Bulge and one in the Surge Horn. In each of our four events, the reversal is coincident with an upward FAC region (region 1). *Rostoker et al.*, [1975] suggested a schematic picture in which the HR separates poleward and equatorward electric fields and therefore converging ionospheric Pedersen currents (also see *Kamide*, 1978). However, we find that in all point reversal events, the high latitude boundary of the enhanced conductance region is approximately coincident with the HR and consequently no significant ionospheric currents are present poleward of the HR. This means that in all cases where the HR can be described as a point reversal no westward electrojet is present, hence, a determination of the HR location using the ΔH component from ground based magnetometers is not possible. The zonal component does not provide any insight since in the 3 events where RPA data were available since it was westward, vanishing and eastward respectively.

*Kamide* [1978] presented a schematic model of the current flow in the vicinity of the HR. He suggested that the HR was a belt and showed that no upward FAC was necessary within this belt. The westward electric field component within the belt resulted in a poleward Hall current connecting the low latitude eastward electrojet with the high latitude westward electrojet. This requires a westward component within the HR which we do not find in any of the 3 CCM passes since the zonal component is vanishing in all 3 passes. In the wide region type HR, which is primarily located in the Surge and Middle Surge we do, in general, find a weak westward component. However, the current pattern within this wide region is highly structured with filamentary FACs, which does not support the schematic model by *Kamide* [1978].

## 4. SUMMARY AND CONCLUSIONS

We have investigated the electrodynamics associated with the Harang discontinuity during substorm expansion phase through early recovery phase. Characteristics of the electric field, FACs, Pedersen and Hall conductances and ionospheric meridional currents have been discussed. We find that in general three types of HR exist. (1) A well-defined point separating regions of poleward and equatorward meridional electric fields typically located far west of the surge. The HR is nearly coincident with the high latitude boundary of the enhanced conductance region and consequently no westward electrojet is present. The HR is located in the high latitude part of an upward region 1 FAC, which is fed by the poleward ionospheric current component. The zonal electric field component within the HR is vanishing in all passes. (2) A relatively wide region (~100 km) typically located just west of the Surge. The electric field and the conductance exhibit a clearly inverse relationship. The zonal component does not exhibit any systematic trend. (3) A very wide region (~500km) located equatorward of the maximum conductance region which is typically seen in the Surge and east of the Surge. The zonal component is, in general, weak and westward and the three dimensional current system is filamentary.

*Acknowledgements.* The work was done at the Danish Space Research Institute and at Goddard Space Flight Center under USRA contract NAS5-32484.

## REFERENCES

Baumjohann, W., and G. Haerendel, Magnetic convection observed between 0600 and 21000 LT: Solar wind and IMF dependence, *J.Geophys.Res., 90,* 6370, 1985.

Cummer, S. A., R. R. Vondrak, R. F. Pfaff, J. W. Gjerloev, C. W. Carson, R. E. Ergun, W. J. Peria, R. C. Elphic, R. J. Strangeway, J. B. Sigwarth, and L. A. Frank, Auroral surge currents and electrodynamics with FAST and VIS, this issue.

Fujii, R., R. A. Hoffman, P. C. Anderson, J. D. Craven, M. Sugiura, L. A. Frank, and N. Maynard, Electrodynamic parameters in the nighttime sector during auroral substorms, *J.Geophys.Res., 99,* 6093,1994.

Gjerloev, J. W., and R. A. Hoffman, Height integrated conductivity in auroral substorms, Part I: Data, *J.Geophys.Res.,* 1999a, (accepted by JGR).

Gjerloev, J. W., and R. A. Hoffman, Height integrated conductivity in auroral substorms, Part II: Modeling, *J.Geophys.Res.,* 1999b, (accepted by JGR).

Hanson, W. B., R. A. Heelis, R. A. Power, C. R. Lippincott, D. R. Zuccaro, B. J. Holt, L. H. Harmon, and S. Sanatani. The retarding potential analyzer for Dynamics Explorer-B, *Space.Sci.Instrument., 5,* 503-510, 1981.

Harang, L. The mean field of disturbance of polar geomagnetic storms, *Terr.Mag., 51,* 353, 1946.

Heppner, J. P., The Harang discontinuity in auroral belt ionospheric currents, *Geofys.Publ., 29,* 105, 1972.

Hoffman, R. A., R. Fujii, and M. Sugiura, Characteristics of the field-aligned current system in the nighttime sector during auroral substorms, *J.Geophys.Res., 99,* 21,303, 1994.

Kamide, Y., On current continuity at the Harang discontinuity, *Planet.Space.Sci., 26,* 237, 1978.

Kamide, Y., and J. F. Vickrey, Variability of the Harang discontinuity as observed by the Chatanika radar and the IMS Alaska magnetic chain, *Geophys.Res.Lett., 10,* 159, 1983.

Kamide, Y., Electrodynamic processes in the Earth's ionosphere and magnetosphere, Kyoto Sangyo University Press, Kyoto, Japan, 1988.

Kisabeth, J. L., and G. Rostocker, Current flow in auroral loops and surges inferred from ground-based magnetic observations, *J.Geophys.Res., 78,* 5573, 1973.

Kunkel, T. W., W. Baumjohann, J. Untiedt, and R. A. Greenwald, Electric fields and currents in the Harang discontinuity: A case study, *J.Geophys., 59,* 73, 1986.

Marklund, G. I., I. Sandahl, and H. Opgenoorth, A study of the dynamics of a discrete auroral arc, *Planet.Space.Sci., 30,* 179, 1982.

Marklund, G. I., Auroral arc classification scheme based on the observed arc-associated electric field pattern, *Planet.Space.Sci., 32,* 193, 1984.

Marklund, G. T., T. Karlson, L. G. Blomberg, P.-A. Lindqvist, C.-G. Falthammer, M. L. Johnson, J. S. Murphree, L. Andersson, L. Eliasson, H. J. Opgenoorth, and L. J. Zanetti, Observations of the electric field fine structure associated with the westward travelling surge and large-scale auroral spirals, *J.Geophys.Res., 103,* 4125, 1998.

Maynard, N. C., Electric field measurements across the Harang discontinuity, *J.Geophys.Res., 79,* 4620, 1974a.

Maynard, N. C., The Harang discontinuity as defined by electric field (abstract), *EOS.Trans.,* Amer.Geophys.Union, 55, 1003, 1974b.

Maynard, N. C., E. A. Bielecki, and H. F. Burdick, Instrumentation for electric field measurements from DE-B, *Space Science Instrument, 5,* 523, 1981.

Reiff, P. H., Models of auroral-zone conductances, in Magnetospheric Currents, *Geophysical monograph, 28,* American Geophysical Union, Washington D.C., 1984.

Rostoker, G., J. C. Armstrong, and A. J. Zmuda, Field-aligned current flow associated with the intrusion of the substorm intensified westward electrojet into the evening sector, *J.Geophys.Res., 80,* 3571, 1975.

Wedde, T., J. R. Doupnik, and P. M. Banks, Chatanika observations of the latitudinal structure of electric fields and particle precipitation on November 21, 1975, *J.Geophys.Res., 82,* 2743, 1977.

Winningham, J. D., F. Yasuhara, S.-I. Akasofu, and W. J. Heikkila, The latitudinal morphology of 10 eV to 10 keV electron fluxes during magnetically quite and disturbed times in 2100-0300 MLT sector, *J.Geophys.Res., 80,* 3148, 1975.

Winningham J. D., J. L. Burch, N. Eaker, V. A. Blevins, and R. A. Hoffman, The low altitude plasma instrument (LAPI), *Space.Sci.Instrument., 5,* 465-475, 1981.

---

J. W. Gjerloev, and E. Friis-Christensen, Danish Space Research Institute, Juliane Maries Vej 30, DK-2100 Copenhagen Oe, Denmark.

R. A. Hoffman, and S. A. Cummer, Laboratory for Extraterrestrial Physics, NASA/Goddard Space Flight Center, Greenbelt, MD 20771, USA.

# The Effect of the January 10, 1997, Pressure Pulse on the Magnetosphere-Ionosphere Current System

E. Zesta[1], H. J. Singer[2], D. Lummerzheim[3], C. T. Russell[4], L. R. Lyons[1], M. J. Brittnacher[5]

On January 10, 1997, a strong pressure pulse, observed by the WIND spacecraft between 1030 and 1055 UT, hit the magnetosphere after about a one-half hour delay, causing the strengthening and widening of the auroral electrojet at all local times. The duration of the electrojet perturbation was the same as the duration of the solar wind pressure pulse. The pulse occurred during the well-studied January 10-11, 1997, magnetic storm and during strong geomagnetic activity. We study the effect of the pressure pulse on the ionospheric current using a global network of more than 100 ground magnetometers, images from the POLAR spacecraft, and solar wind measurements from the WIND and Geotail spacecraft. We find that the magnetospheric and ionospheric response is directly driven by the solar wind conditions and clearly related to the onset, duration and end of the pressure pulse. In addition, it appears that the enhancement of the Region 1 currents opposed the effect of the enhancement of the magnetopause current for locations near noon. These responses are not characteristics of a typical substorm.

## INTRODUCTION

The response of Earth's magnetosphere to solar wind pressure enhancements has been studied primarily for interplanetary shocks and sudden pressure increases.

---

[1] Department of Atmospheric Sciences, University of California Los Angeles, Los Angeles, CA 90095-1565
[2] Space Environment Center, National Oceanic and Atmospheric Administration, Boulder, CO, 80303
[3] Geophysical Institute, University of Alaska, Fairbanks, AK 99775
[4] Institute of Geophysics and Planetary Physics, UCLA, Los Angeles, CA 90095
[5] Geophysics Program, University of Washington, Seattle, WA, 98195

Magnetospheric Current Systems
Geophysical Monograph 118
Copyright 2000 by the American Geophysical Union

Sudden Impulses (SI) and Storm Sudden Commencements (SSC) are due to solar wind pressure enhancements usually associated with interplanetary shocks, the latter accompanied by a magnetic storm event. The effect of SI and SSC events has been studied by *Araki* [1977, 1994], and for both northward and southward IMF by *Russell et al.* [1994a, 1994b] and others. Sudden enhancements in the solar wind dynamic pressure are known to compress the magnetosphere and enhance the magnetopause current [*Kauffman and Konradi*, 1969], as well as the tail currents [*Russell et al.*, 1994a]. The response can be very different for northward and southward IMF. For northward IMF the magnetic field observed at low, mid- and sub-auroral latitudes, ≤55° geomagnetic, responds with a positive perturbation in the *H* component due primarily to the increased magnetopause current [*Araki*, 1994, and references therein; *Russell et al.*, 1994a; *Le et al.*, 1993; *Russell and Ginskey*, 1995]. At geomagnetic latitudes greater than 60° the response is due primarily to increased Region 1 currents that result from the magnetospheric compression. *Russell et al.* [1994b] showed that for SI

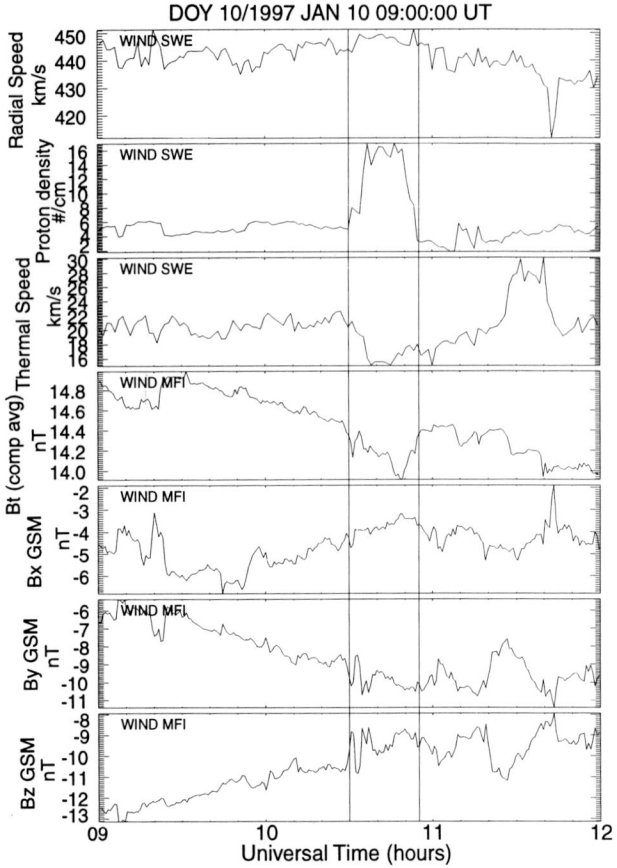

**Figure 1.** Solar wind plasma and interplanetary magnetic field data from the WIND spacecraft.

events during southward IMF the response of the dayside sub-auroral latitudes to the magnetopause currents is significantly reduced by the enhanced Region 1 currents that result from increased dayside reconnection.

The January 10-11, 1997, magnetic cloud event, and the magnetic storm resulting from it, is a well-documented event with extensive data coverage. There were two substorm events at ~0337 and 0645 UT [Li et al., 1998], before the pressure pulse hit the magnetosphere, followed by steady southward IMF and a continuous increase in geomagnetic activity between 0730 and 1030 UT. Shue and Kamide [1998] studied the correlation between the solar wind dynamic pressure and the strength of the westward electrojet, mostly in the vicinity of local midnight, for the 0 to 12 UT time period on January 10, 1997. They concluded that the solar wind density strongly affects the strength of the westward electrojet when the IMF has been southward for some time but the coupling is weak for northward IMF. Kamide et al. [1998] examined the same pressure pulse that we examine here and argued that there were two substorm onsets, the first at ~1035 UT, internally triggered, and a second around 1050 UT triggered by the solar wind dynamic pressure pulse. Both Shue and Kamide [1998] and Kamide et al. [1998] interpret the influence of the solar wind dynamic pressure on the magnetosphere in terms of substorm triggering. There are several studies that suggest that substorms can be triggered by SSC events when the IMF is southward [Burch, 1972; Iijima, 1973; Kokubun et al., 1977].

In the present paper we analyze the effect of the January 10, 1997, pressure pulse. We find an auroral enhancement at ~1035 UT, likely a substorm onset, preceding the pressure pulse; but we see no clear evidence for a substorm triggered by the pressure pulse. Rather the effects of the pulse are intense and global with a rapid response at all local times simultaneously. These include a dramatic increase of the morning-side westward auroral electrojet, during all local times from 0000 to 1200 MLT, and the poleward expansion and widening of the auroral oval. We should clarify here that in our use of the term "auroral westward electrojet" we refer to the global ionospheric electrojet associated with the morning convection cell (the global electrojet is westward in the morning sector and eastward in the afternoon) as opposed to the substorm westward electrojet, that is localized around 00 MLT and is associated with auroral surges and the substorm current wedge. The increase of the auroral electrojet during the pressure pulse event implies the global and dramatic increase of the Region 1 currents (see Siscoe et al. [1991] for a description of the circuit that connects the boundary layer, the R1 currents and the ionospheric currents). The poleward expansion of the auroral oval implies significant increase in closed flux in the magnetotail.

## THE PRESSURE PULSE IN THE SOLAR WIND

The pressure pulse examined herein was observed by the WIND spacecraft between 1030 and 1055 UT and is shown in Figure 1. At the time WIND was located at GSE (88.5,-58.6,-3.9) $R_E$. From top to bottom, Figure 1 shows the solar wind bulk velocity, proton density, thermal speed, total magnetic field strength, and the $B_x$, $B_y$, $B_z$ magnetic field components in the GSM coordinate system. The two vertical lines at 1030 and 1055 UT indicate the duration of the solar wind density pulse. The speed is steady at 445 km/s. The dynamic pressure pulse then is entirely due to the density pulse. The IMF remains strongly southward at about –10 nT with a strong negative $B_y$ component and it is slowly rotating northward. There is a small disturbance in $B_z$ and $B_y$, lasting only a couple of minutes and with a magnitude less than 2 nT, associated with the onset of the pressure pulse, but there are no major magnetic perturbations associated with the pulse. The sum of the perpendicular thermal and magnetic pressure is constant across the discontinuity and the velocity is constant. These are the properties of a tangential discontinuity [*Burgess*, 1995]. This non-propagating (in the solar wind frame) density structure causes a dynamic pressure pulse when it encounters the magnetosphere.

Figure 2 shows the plasma and magnetic field data from the Geotail spacecraft, which was at GSE (6.3,9.3,0.5) $R_E$ during this period and mostly within the afternoon magnetosheath. Geotail observes the same density pulse that WIND observes with a similar rise and decay and 25 min duration, shown between the two dashed lines in the Figure 2. The onset of the pulse at Geotail is at 1053 UT, 23 min later than WIND. From the magnetic field data we deduce that Geotail was continuously in the magnetosheath from 0950 until 1113 UT, when it crossed the magnetopause boundary. The dotted vertical lines indicate transitions from the magnetosphere to the magnetosheath and back. The two bars on the top of the figure indicate the time periods that Geotail was in the magnetosphere. The crossing of the boundary at 1113 UT coincides with the sharp drop of the pressure pulse, implying that the magnetopause expanded beyond the position of Geotail as the magnetosphere relaxed after being compressed by the pulse.

We determine the size and orientation of the solar wind structure associated with the pressure pulse in order to accurately determine its effect on the magnetosphere. By comparing the magnetic and thermal pressure in the structure we find that it is in thermal balance in the frame of reference of the moving plasma. Specifically, the total pressure was $8.58 \times 10^{-2}$ nPa before the pressure pulse, and $8.24 \times 10^{-2}$ nPa during the pressure pulse. This is only a 4% change. Both the total magnetic field, $B_t$, and the thermal speed (and therefore the plasma temperature), decrease during the pressure pulse to balance the strong increase in density (see Figure 1). Such a structure would more than likely be a tangential discontinuity aligned with the magnetic field and more than likely cylindrical with its axis along the IMF. If this were a rotational discontinuity, which could propagate with a normal at an oblique angle to the field, the density would be constant across the surface (e.g. *Burgess*, 1995), and this is not the case here. For a tangential discontinuity aligned with the IMF, the diameter of the cylinder should be at least 89 $R_E$, which is the distance determined by the duration of the pressure pulse projected in the direction perpendicular to the IMF. We also know that the length of the cylinder is at least 84 $R_E$, which is the Y separation of WIND and Geotail projected along the IMF. Therefore the solar wind structure would be at least 89 $R_E$ in the X direction and 84 $R_E$ in the Y direction

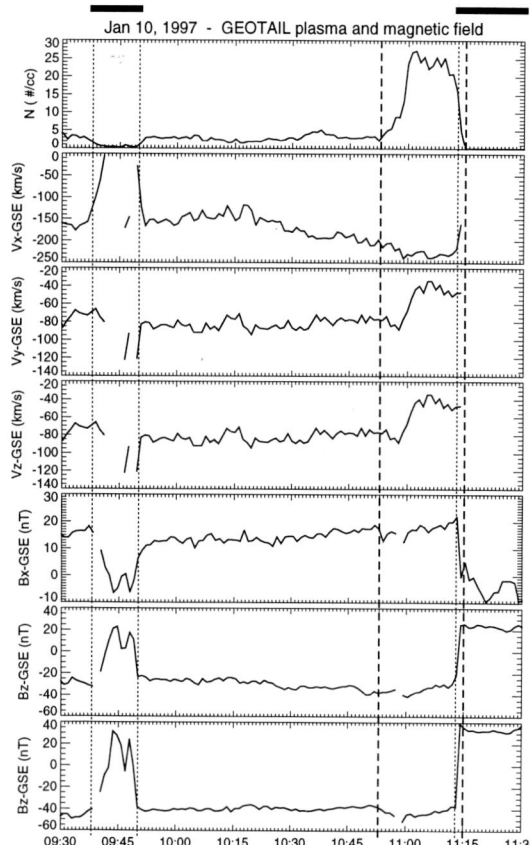

**Figure 2.** Plasma and magnetic field data from Geotail, located in the magnetosheath.

and therefore large enough that when it encounters the magnetosphere it engulfs it all from the dayside to the far nightside. The above analysis also implies that we can be certain that when the pressure pulse is observed by Geotail at 1053 UT it should also encounter the magnetopause at approximately the same time.

## THE MAGNETOSPHERIC AND IONOSPHERIC RESPONSE

We investigate the magnetospheric and ionospheric response by examining the aurora observed by the UVI instrument on Polar [*Torr et al.*, 1995], and by inferring the equivalent ionospheric current patterns from a global network of ground magnetometers.

Plate 1 shows a series of 25 auroral images, from 1020 to 1130 UT, observed with the POLAR UVI imager. Each image shows the two-dimensional distribution of the precipitating particle energy flux in a MLT–magnetic latitude coordinate system. The UVI instrument has several

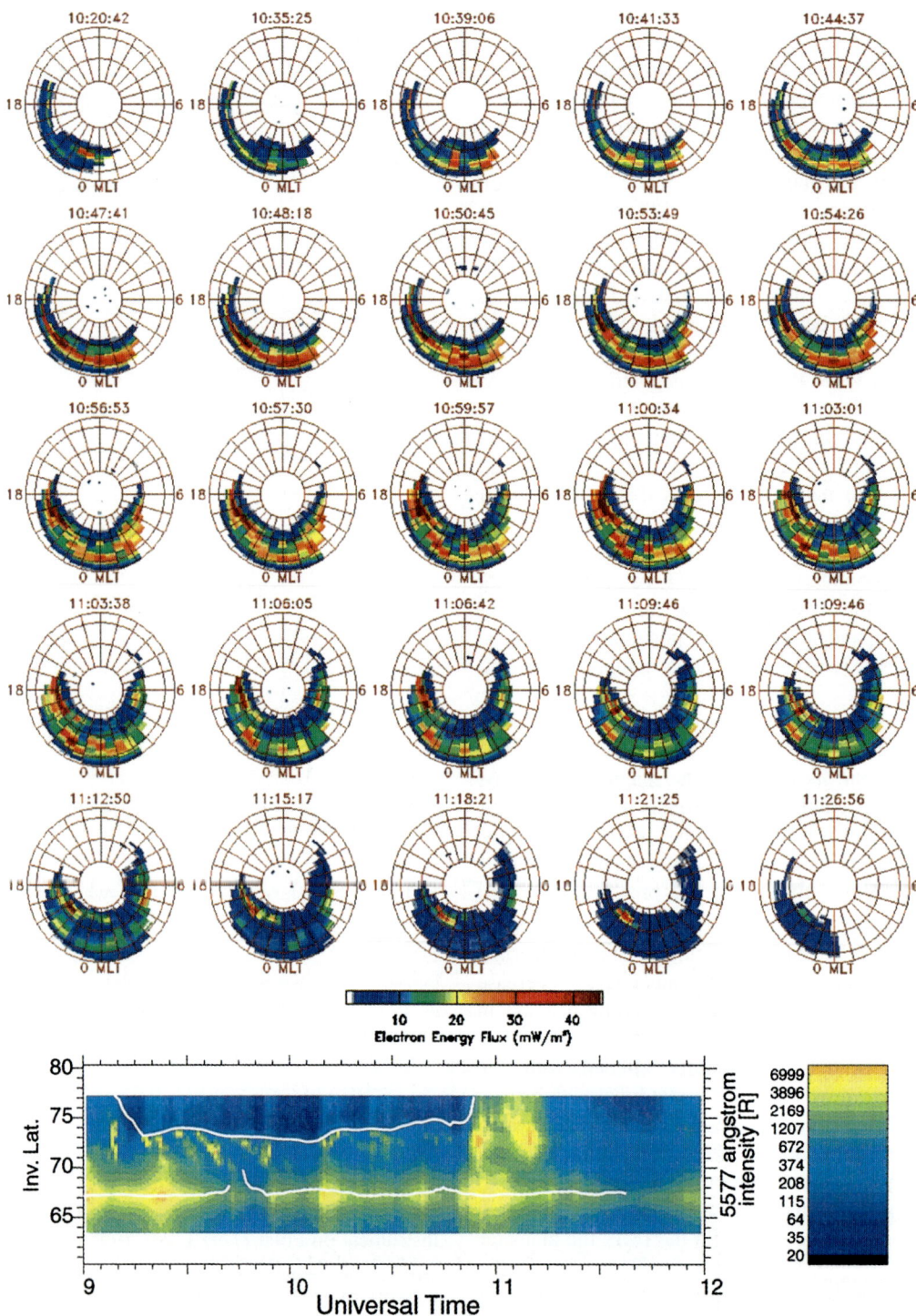

**Plate 1.** Top: A series of 25 images of the precipitating particle energy flux, from the POLAR UVI instrument. Bottom: 5577 Å emission line from the Rankin and Gillam meridional scanning photometers.

filters. We used images taken with the LBH-long filter. The brightness of the images in this passband is proportional to the incident energy. Each image is calibrated and pixels are binned into a regular grid (1° by 10°) in latitude-longitude geomagnetic coordinates. The energy flux is then obtained from a parameterization of the image brightness based on the *Lummerzheim and Lilensten* [1994] auroral model. Between 1000 and 1030 UT there is moderate auroral activity extending from the evening side to beyond midnight. Between 1035 and 1040 UT there is a strong auroral enhancement around 2400 MLT that by 1047 UT spreads towards both dusk and dawn and becomes a global enhancement, at 63° geomagnetic latitude. *Kamide et al.* [1998] identified this auroral enhancement at 1035 UT from ground magnetometers and auroral images as a substorm onset. *Li et al.* [1998] (their Figure 4) identified a particle injection at geosynchronous around 1035 UT. This intensification occurs more than 15 minutes before the pressure pulse hits the magnetosphere and is therefore unrelated to it. Kamide et al. also argue that there is a second substorm onset that is triggered by the pressure pulse. In our analysis the series of images in Plate 1 are consistent with a continuous strengthening of the aurora from 1050 UT onwards, rather than a clear second substorm onset.

The first signatures of the pressure pulse arriving at the magnetopause are seen shortly after 1050 UT, possibly 1-2 minutes earlier than at Geotail, due to the orthospiral orientation of the IMF and Geotail's location in the afternoon region. The image at 1050:45 UT shows a widening of the auroral oval as it continues to intensify. Until 1048 UT the oval extends from 55° to 71° geomagnetic latitude, while at 1050:45 UT it extends for the first time poleward of 71°. We believe this is the response of the magnetosphere to the initial compression as the pressure front hits the magnetopause boundary, resulting in an increase of the dayside magnetic field and the Chapman-Ferraro current. As a consequence, the Region 1 current circuit also strengthens. Around 1056:53 UT the poleward expansion of the oval accelerates. From 1047:41 UT to 1053:49 UT, a ~6-minute interval, the poleward boundary of the oval moved very little and remained at ~71°. However between 1054:26 and 1100:34 UT, another 6-minute interval, the poleward boundary moves from 71° to 78°. As can be seen from Figure 2, this corresponds to the time when the solar wind density at the dayside magnetopause began to increase at a faster rate. The poleward expansion of the auroral oval ended near 1103 UT, very soon after the time that the peak solar wind density hit the magnetopause. The 1106:05 UT image in Plate 1 shows the oval extending to almost 80° geomagnetic latitude, implying that a significant amount of nightside magnetic flux in the lobes has become closed. The continuous poleward expansion of the auroral oval from 1050 to 1103 UT suggests a continuous increase in nightside reconnection. Auroral intensities start to decrease at ~1103 UT and seem to follow the decrease in the solar wind density. By 1121:25 UT, seven minutes later, the intensity is dramatically decreased and the poleward boundary of the oval starts to move back to lower latitudes. By this time the end of the pressure enhancement region in the solar wind has reached far downtail. These observations show a very fast and direct response of the magnetosphere to the phases of the pressure pulse and therefore we interpret the intensification not as a new onset but as a directly driven response of the magnetosphere to the pressure pulse affecting ongoing activity. Similar conclusions have been reached using observations from the POLAR VIS camera (John Sigwarth, personal communication, 1999).

The bottom panel in Plate 1 shows the 5577Å-emission line from the meridional scanning photometers (MSP) from Rankin Inlet and Gillam. The two stations are located at the same MLT, separated only in latitude, so the data from the two stations are merged to increase the latitudinal coverage. The stations are located at 0430 MLT during the pressure pulse event. The data show very clearly the widening and strengthening of the auroral oval between 1050 and 1115 UT, exactly the duration of the pressure pulse that encounters the magnetosphere. The top white line in the keogram identifies the separatrix boundary to within 2° [*Blanchard et al.*, 1997]. This is yet another indication for the directly driven response of the magnetosphere-ionosphere to the pressure pulse.

Figure 3 complements the Plate 1 images and is a representative sample of the interpolated global equivalent ionospheric current patterns determined from the global network of magnetometers. The currents are plotted in an MLT–magnetic latitude coordinate system. The locations of the ground magnetometers are marked with solid triangles. Magnetic local noon is at the top of each dial and the UT time is indicated at the top right corner. The stations create an irregular grid of data points; therefore we applied a spherical interpolation technique to produce the two-dimensional equivalent ionospheric current patterns in a 10° by 1° longitude-latitude grid. To convert the magnetic perturbations in nT to height-integrated ionospheric current in A/m we used the conversion scale of 500 nT per A/m total ionospheric current, based on the measured geomagnetic effects of the Hall and Pedersen currents by *Araki et al.* [1989]. It should be noted that the *Araki et al.* [1989] conversion scale is accurate only under the well-developed auroral electrojets, in our case between latitudes 65° and 75°. Beyond those latitudes the approximation may

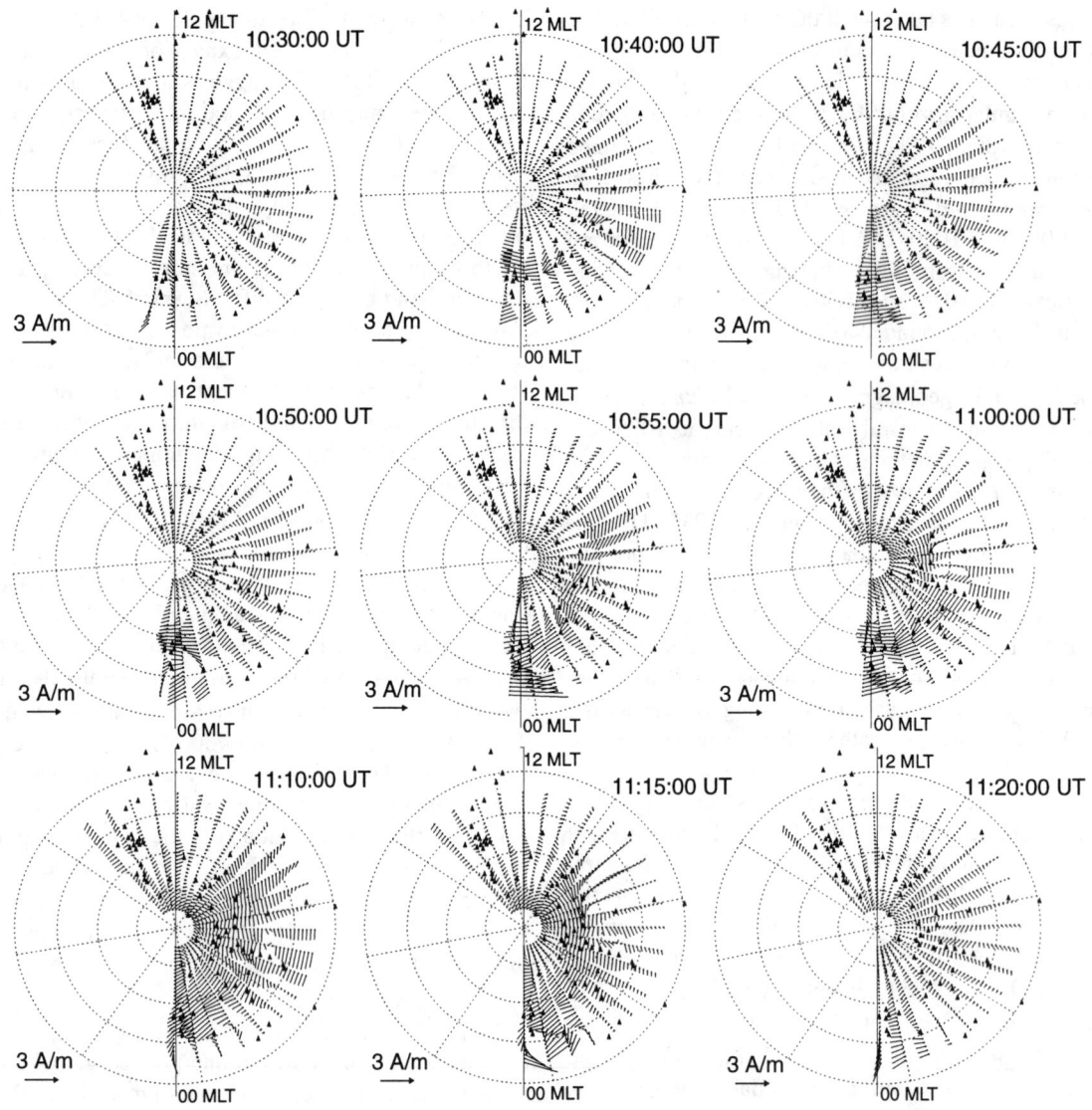

**Figure 3.** Interpolated equivalent current patterns from a global network of magnetometers.

not work as well. We corrected for the ground-induced currents by multiplying the equivalent current by a factor of 2/3. The current at the edges of our grid should be ignored, because of large edge effects, in particular the region around 00 MLT and at latitudes less than 60°.

At 1030 UT the morning convection cell is seen with the auroral westward electrojet located between 60° and 70° at local times from 0000 to 1300 MLT. At 1040 UT a small strengthening of the electrojet, localized in the midnight region, coincides with the strengthening in the auroral images in Plate 1. At 1050 UT, when the pressure pulse hits the magnetosphere, we see the first indication of a far more global strengthening of the ionospheric current system. By 1055 UT the region of significant current has widened, now located between 75° and 60° geomagnetic latitude, and reaching more than 3 A/m. During the next 15 minutes the auroral electrojet extends from 50° to 80° in excellent agreement with the very wide oval observed in the auroral images in Plate 1. The strengthening and widening of the ionospheric current system is indirect evidence for the significant strengthening of the Region 1 field-aligned currents (e.g. *Siscoe et al.*, 1991). Previously this has been indirectly inferred by *Russell et al.* [1994b], who found that the dayside response to the SI compression, namely the increase of the magnetopause current, is more than 25% smaller when the IMF is southward than when it is

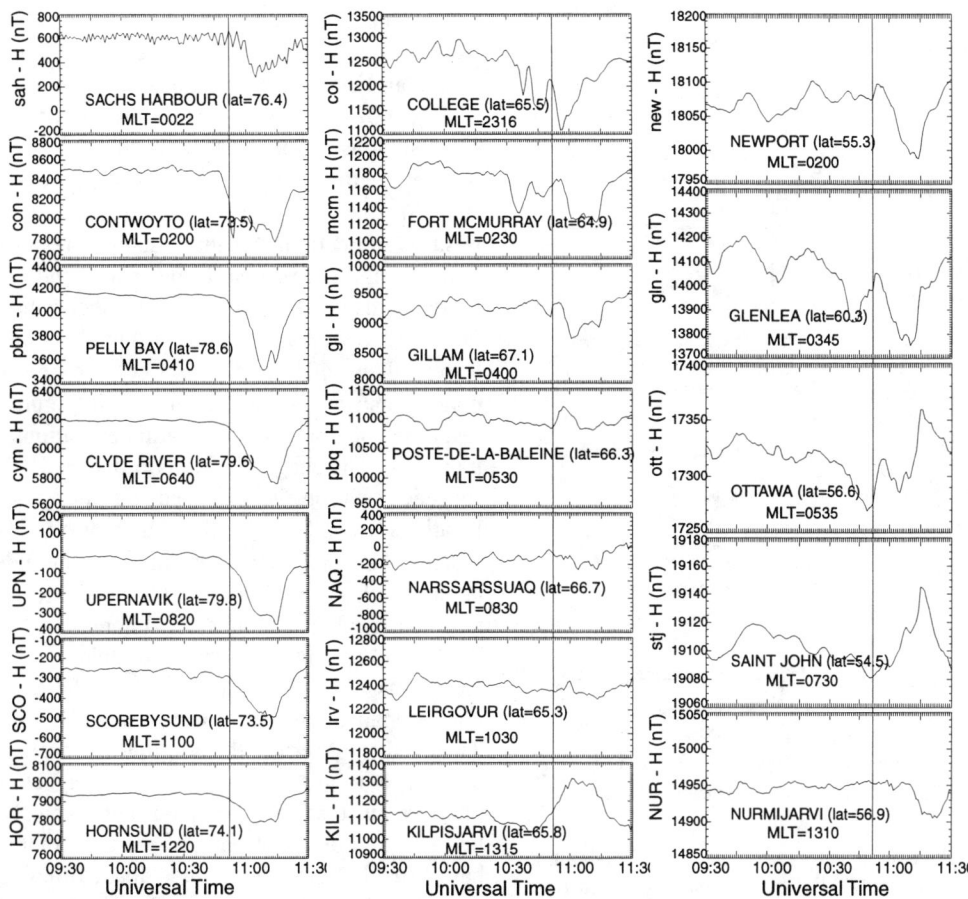

**Figure 4.** The H component of the magnetic field from stations at many local times and latitudes >70° on the left panel, between 60° and 70° in the middle panel, and between 50° and 60° in the right panel.

northward. They attribute this to the increase of the Region 1 currents associated with increased dayside reconnection.

Further evidence for the driven effect of the pressure pulse on the magnetosphere is shown in Figure 4. The $H$ component from ground magnetometers covering 0000 to 1300 MLT are plotted for stations at greater than 70° geomagnetic latitude in the left panel, stations with latitudes between 60° and 70° in the middle panel, and stations with latitudes between 50° and 60° in the right panel. The vertical line in all three panels indicates the time that Geotail observed the onset of the pressure pulse, 1053 UT. All the stations at higher than 70° latitude, except for Contwoyto, observe a large negative bay starting around 1050 UT (Contwoyto observes a strengthening of a pre-existing bay at that time). These stations are poleward of the auroral oval before the pressure pulse (see Plate 1) and observe the transient strengthening and widening of the westward auroral electrojet between 1050 and 1120 UT. All these stations (except Sachs Harbour) show the onset of a strong reduction in bay intensity at 1115 UT, the time of the end of the pressure pulse. The auroral stations in the middle panel also see the negative bay after 1050 UT, except for Kilpisjarvi that sees the strengthening of the afternoon eastward electrojet. The stations closer to 00 MLT however, College and Fort McMurray, observe the prior substorm intensification between 1035 and 1050 UT. The sub-auroral stations in panel 3 observe more complicated signatures. Some observe both the substorm and the pressure pulse effect (Glenlea, Ottawa) and others only the pressure pulse (Newport, Saint John, Nurmijarvi).

In summary, it is evident from Figures 3, and 4, and from Plate 1 that the response of the magnetosphere to the pressure pulse is almost instantaneous and global, and appears to be directly driven by the impact and propagation of the pressure pulse in the magnetosphere. We observe a possible substorm onset between 1035 and 1040 UT, also described by *Kamide et al.* [1998], that occurs before the pressure pulse arrives at the magnetosphere. The initial

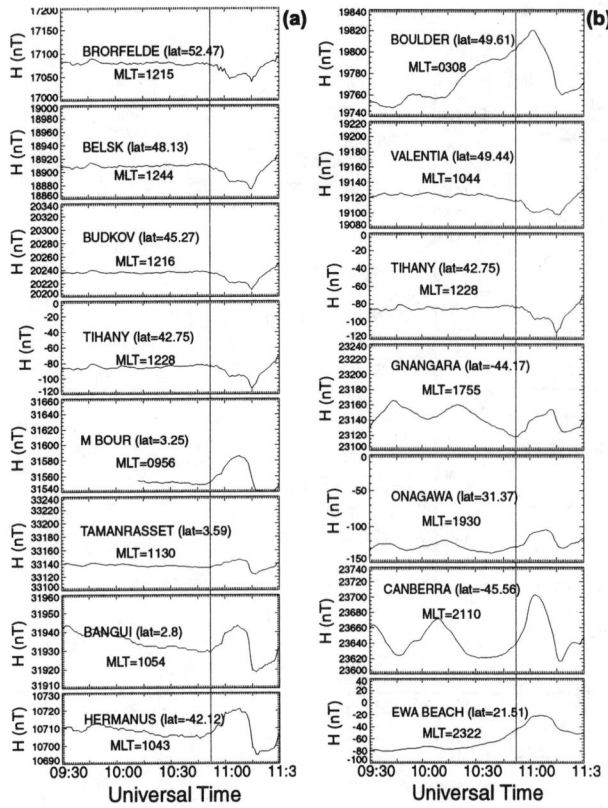

**Figure 5.** The H component of mid- and low-latitude magnetometers in the 11 to 13 MLT region, panel (a), and from all local times, panel (b).

response to the compression, occurring around 1050 UT, is the strengthening of the Region 1 currents evidenced by the global strengthening of the electrojet [*Siscoe et al.*, 1991]. About 7 minutes later, and for the next 15 minutes, we observe a significant increase in the closed flux in the tail indicated by the broadening of the electrojet and the auroral oval. The increase in closed flux implies that the pressure pulse caused an enhancement in nightside reconnection.

## COMPARISON WITH TYPICAL SUDDEN IMPULSE EVENTS

The response of the magnetosphere to pressure increases associated with shocks, SI and SSC events, has been investigated for all latitudes and both northward and southward IMF conditions [*Araki*, 1977, 1994; *Russell et al.*, 1994a, 1994b; *Le et al.*, 1993; *Russell and Ginskey*, 1995; and others]. The general consensus is that SI events have a sharp signature at all latitudes, but at the dayside sub-auroral mid- and low-latitudes the ground magnetometers all observe a positive main signature in the H component, as a direct response to the increased Chapman-Ferraro current at the magnetopause.

Figure 5, panel (a), is a stack plot of the H component of eight stations in the 1000 to 1300 MLT region ranging in magnetic latitude from 52° to -42°. Stations as low as 42° (L=1.5) show a negative bay instead of the customary positive deflection. All these stations respond to the very strong auroral westward electrojet and the Region 1 currents that cancel out and overshoot the effect of the increased Chapman-Ferraro current. It is only at latitudes lower than 40° that we see the response to the magnetopause current. Panel (b) of Figure 5 shows the response of mid- to low-latitude stations at all MLT locations. The magnetic latitude and exact MLT location for each station is indicated on the plot. Notice that locations away from local noon show the response to the increased magnetopause current at mid-latitudes around 40° geomagnetic, e.g. Boulder, Canberra. It is only at the region a few hours around local noon that the effect of the Region 1 currents penetrates to mid- and low-latitudes. Hermanus in the southern-hemisphere observes the magnetopause current, even though it is conjugate to Tihany that observes the Region 1 currents instead. The difference in the response of the magnetosphere to the pressure pulse studied here and more typical SI events can be explained by the strong pre-existing southward IMF that has already stretched and loaded energy into the magnetosphere. Also the fact that the perturbation studied here is a pulse with an extended distinct rise and decay time, as opposed to the SI and SSC events that are step like changes in the pressure, where the pressure rises fast and remains high for a longer period. As a result we do not observe the initial transient signatures customary in SI events [*Araki*, 1994].

## SUMMARY AND CONCLUSIONS

We have studied the effect of the January 10, 1997, dynamic pressure pulse on the magnetospheric currents. The pressure front hit the magnetosphere at 1050 UT and was large enough to engulf and constrict the entire magnetosphere within the 25 minutes of its duration. The response of the magnetosphere was practically instantaneous, global and directly driven by the propagation of the solar wind density enhancement region through the magnetosphere. It started with the onset of the pressure pulse and ended with the end of the pressure pulse. These are not the characteristics of a typical substorm. About 10 minutes before the pressure pulse encountered the magnetosphere an auroral enhancement did occur, likely a substorm onset [*Kamide et al.*, 1998]. The arrival of the pressure pulse had the profound effect of strengthening and widening the oval, the auroral electrojet and the Region 1 currents. We interpret this as the driven response of the magnetosphere to the pressure pulse affecting ongoing activity and not as a new substorm onset. The initial compression created a global rapid strengthening of the magnetopause, tail and Region 1 currents. A few minutes later and by the time the pressure front reached well into the magnetotail, the response was the closing of more flux in the nightside. We believe the effect of the pressure pulse is so profound and global because of the prior extended period of strongly southward IMF that stretched and loaded the magnetosphere with energy. Therefore when the pressure pulse hits, the magnetosphere has been "preconditioned" to respond strongly. This is in agreement with the conclusions of *Shue and Kamide* [1998]. *Li et al.* [1998] also report evidence for the "preconditioning" of the magnetosphere. Li et al. observed a rapid enhancement of 0.4 to 1.6 MeV electrons in the magnetosphere at L=4.2-6, immediately following the pressure pulse of January 10, 1997. They attributed that enhancement to the pressure pulse quickly energizing a source population of electrons that was present in the magnetosphere from the intense activity during the previous hours.

Finally, we found some basic differences between the magnetospheric response to the January 10, 1997, pressure pulse and typical SI events. There was no preliminary transient response to the pressure front, as has been reported at the beginning of other SI events [*Araki et al.*, 1994]. Most important the dayside magnetometers at mid and sub-auroral latitudes did not observe the positive bay characteristic of the increase in the magnetopause current, instead they observed the effect of the increased Region 1 currents. This was because, under conditions of southward IMF, the strengthening of the Region 1 currents from the compression was so strong that it completely canceled out the effect of the magnetopause current at dayside L-shells as low as 1.5.

*Acknowledgements*: This work was supported by the National Science Foundation grants OPP-9619733 and ATM98-03431 and NASA grants NAG5-6243 and NAG5-7962. D. Lummerzheim was supported by NASA grant NAG5-7683. We thank Drs K. Ogilvie and R. Lepping of Goddard Space Flight Center for the WIND plasma and magnetic field data, S. Kokubun for the Geotail magnetic field data, L. Frank and B. Patterson of U. of Iowa for the Geotail plasma data, and J. Samson of U. of Alberta for the MSP data. We also thank J. Hughes for the MACCS magnetometer data, the Danish Meteorological Institute for the Greenland data, the Finish Meteorological Institute for the IMAGE data, the Canadian Space agency for the CANOPUS data, the Geological Survey of Canada for their magnetometer data, and Les Morris of the National Geophysical Data Center for providing a useful source of ground magnetometer data from a global network of stations. The first author thanks the NOAA Space Environment Center and the National Research Council for the portion of this work completed at SEC while an NRC Associate.

## REFERENCES

Araki, T., Global structure of geomagnetic sudden commencement, *Plan. Space Sci.*, 25, 373, 1977.

Araki, T., K. Schlegel, and H. Lühr, Geomagnetic effects of the hall and pedersen current flowing in the auroral ionosphere, *J. Geophys. Res.*, 94, 17185, 1989.

Araki, T., A physical model of the geomagnetic sudden commencement, in *the Solar Wind Sources of Magnetospheric Ultra-Low-Frequency Waves* edited by M. J. Engebretson, K. Takahashi, and M. Scholer, 183, 1994.

Blanchard, G. T., L.R, Lyons, and J. C. Samson, Accuracy of 6300 Å auroral emission to identify the separatrix on the night side of the Earth, *J. Geophys. Res.*, 102, 9697, 1997.

Burch, J. L., Preconditions for the triggering of polar magnetic substorms by storm sudden commencements, *J. Geophys. Res.*, 77, 5629, 1972.

Burgess, D., Collisionless shocks, in *Introduction to Space Physics*, edited by M. G. Kivelson and C. T. Russell, 129-163, Cambridge U. Press, New York, 1995.

Iijima, T., Interplanetary and ground magnetic conditions preceding ssc-trigged substorms, *Rep. Ionos. Space Res. Japan*, 27, 205, 1973.

Kamide, Y., J.-H. Shue, X. Li, G. Lu, M. J. Brittnacher, G. K. Parks, and G. D. Reeves, Internally and externally triggered substorms: a case study of the January 10, 1997 events, in *SUBSTORMS-4* edited by S. Kokubun and Y. Kamide, 305, 1998.

Kaufmann, R. L., and A. Konradi, Explorer 12 magnetopause observations: Large-scale nonuniform motion, *J. Geophys. Res.*, 74, 3609, 1969.

Kokubun, S., R. L. McPherron, and C. T. Russell, Triggering of substorms by solar wind discontinuities, *J. Geophys. Res.*, 82, 74, 1977.

Le, G., C. T. Russell, S. M. Petrinec, and M. Ginskey, Effect of sudden solar wind dynamic pressure changes at subauroral latitudes: Change in magnetic field, *J. Geophys. Res.*, 98, 3982, 1993.

Li, X., D. N. Baker, M. Temerin, T. Cayton, G. D. Reeves, T. Araki, H. Singer, D. Larson, R. P. Lin, and S. G. Kanekal, Energetic electron injection into the inner magnetosphere during the Jan. 11-11, 1997 magnetic storm, *Geophys. Res. Let.*, 25, 2561, 1998.

Lummerzheim, D. and J. Lilensten, Electron Transport and Energy Degradation in the Ionosphere: Evaluation of the Numerical Solution, Comparison with Laboratory Experiments and Auroral observations, *Ann. Geophys.*, 12, 1039, 1994.

Russell, C. T., M. Ginskey, and S. M. Petrinec, Sudden impulses at low latitude stations: Steady state response for northward interplanetary magnetic field, *J. Geophys. Res., 99*, 253, 1994a.

Russell, C. T., M. Ginskey, and S. M. Petrinec, Sudden impulses at low latitude stations: Steady state response for southward interplanetary magnetic field, *J. Geophys. Res., 99*, 13403, 1994b.

Russell, C. T., and M. Ginskey, Sudden impulses at subauroral latitudes: Response for northward interplanetary magnetic field, J. Geophys. Res., 100, 23695, 1995.

Shue, J.-H., and Y. Kamide, Effects of solar wind density on the westward electrojet, in *SUBSTORMS-4* edited by S. Kokubun and Y. Kamide, 1998.

Siscoe, G. L., W. Lotko, and B. U. Ö. Sonnerup, A high-latitude boundary layer model of the convection current system, *J. Geophys. Res., 96*, 3487-3495, 1991.

Torr, M. R., D. G. Torr, M. Zukic, R. B. Johnson, J. Ajello, P. Banks, K. Clark, K. Cole, C. Keffer, G. Parks, B. Tsurutani, J. Spann, A far ultraviolet imager for the international solar-terrestrial physics mission, *Space Sci. Rev., 71*, 329, 1995.

---

M. J. Brittnacher, Geophysics Program, University of Washington, Seattle, WA 98195

D. Lummerzheim, Geophysical Institute, University of Alaska, Fairbanks, AK 99775

Larry R. Lyons, UCLA, Dept. of Atmospheric Sciences, 405 Hilgard Avenue, Los Angeles, CA 90095-1565, larry@atmos.ucla.edu

C. T. Russell, Institute of Geophysics and Planetary Physics, UCLA, Los Angeles, CA 90095

Eftyhia Zesta, UCLA, Atmospheric Sciences, 7127 Math Sciences, Box 951565, Los Angeles, CA 90095-1565, ezesta@atmos.ucla.edu

# Ionospheric Shear Flow Situations Observed by the MIRACLE Network, and the Concept of Harang Discontinuity

O. Amm, P. Janhunen, H. J. Opgenoorth[1], T. I. Pulkkinen, and A. Viljanen

*Finnish Meteorological Institute, Geophysical Research Division, Helsinki, Finland*

Ionospheric shear flow regions connected by magnetic field lines to corresponding magnetospheric shear zones contain a substantial portion of the large-scale field-aligned currents (FACs) in the ionosphere-magnetosphere system. Therefore, these regions play a key role in the interregional coupling mechanism. The most prominent shear zone, leading to the transition from positive to negative values of the northward (X) ground magnetic disturbance component in evening sector auroral latitudes to around magnetic midnight, is named as Harang discontinuity. Using the ground-based MIRACLE network in northern Fennoscandia, we focus on two aspects of shear flow regions: First, we show that two topologically different types of Harang discontinuities often confused in the literature exist, one associated with the Earth's rotation under a stable potential structure during quiet and moderately disturbed times, and another associated with the equatorward expansion of the evening cell potential minimum during intensifications and substorms. The latter typically appears in an earlier MLT sector than the former. Secondly, for five instantaneous shear flow situations observed in summer 1998, we infer the distributions of ionospheric conductances, true ionospheric currents, and field-aligned currents using the method of characteristics. The shear flow situations studied are associated with relatively small ionospheric conductances, which show large gradients somewhat equatorward of the shear regions. Hence, two latitudinally separated layers of FACs of the same sign are present at these zones: One directly at the shear zone associated with the divergence of the electric field, and another about 1-2 degrees of latitude southward, caused by the conductance gradient. These results are discussed in terms of the magnetospheric regions possibly connected with both areas.

## INTRODUCTION

The shear flow regions both in the ionosphere and in the magnetosphere are key areas for the interregional coupling as they contain much of the large-scale field-aligned currents that provide the electrodynamic connection between the two domains.

Most of the earlier studies on shear flow regions have concentrated on the Harang Discontinuity (HD) in the late afternoon to midnight sector. The term "Harang Discontinuity" was first introduced by *Heppner* [1972], with respect to a much earlier work of *Harang* [1946]. Heppner defined the HD as the transition of the ground magnetic disturbance from positive to negative X values around magnetic midnight measured by a magnetometer located in the auroral zone. We use the term HD for this magnetic field transition only, whereas the corresponding transition in the ionospheric convection from westward to

[1] also at: Swedish Institute of Space Physics, Uppsala Division, S-75591 Uppsala, Sweden (opg@irfu.se)

Magnetospheric Current Systems
Geophysical Monograph 118
Copyright 2000 by the American Geophysical Union

## 228 IONOSPHERIC SHEAR FLOW SITUATIONS

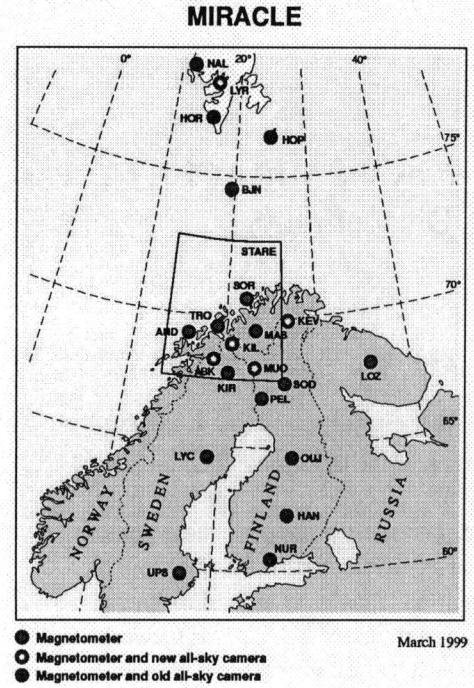

**Figure 1.** Overview of the MIRACLE network of ground-based instruments; the field of view of the STARE radar is marked by the box over northern Fennoscandia.

eastward flow is called "shear flow region", "shear zone", or "convection reversal boundary". These two transitions are typically not exactly colocated [e.g., *Untiedt and Baumjohann*, 1993; *Koskinen and Pulkkinen*, 1995; *Amm*, 1998]. It was found later that the HD region plays a crucial role during substorms when the initial breakup is often located at the HD [e.g. *Baumjohann et al.*, 1981; *Robinson and Vondrak*, 1990]. This observation has led to theories that the substorm expansion is initiated by some instability of the current system at the magnetospheric counterpart of the HD [e.g., *Kan*, 1993]. Several instantaneous event studies are available that infer spatial distributions of the ionospheric electrodynamic parameters at the HD and shear flow region [*Kunkel et al.*, 1986; *Inhester et al.*, 1992; *Amm*, 1998]. These studies agree in that the horizontal ionospheric current is not continuous over the HD, as was earlier suggested by *Kamide* [1978], but the currents of the electrojets are diverged by FACs at the HD. Moreover, they found a minimum of the ionospheric conductance associated with the shear flow region itself. Especially in the studies by *Inhester et al.* [1992] and *Amm* [1998], the parts of the electrojets were not equal: While most of the eastward electrojet current is diverged immediately south of the ionospheric shear flow because of the negative conductance gradient towards that region, the conductances immediately north

of the shear region were found to be low so that the westward electrojet was located somewhat north of it, outside of the area under study. *Amm* [1998] has shown that during the event chosen for his study, plasma flow over the shear zone adjacent to the HD was present. During events in earlier studies, like the six STARE radar observations of shear flows studied by *Koskinen and Pulkkinen* [1995], the question of the plasma flow direction in the center of the shear zone could not be answered because of lacking radar backscatter. *Koskinen and Pulkkinen* [1995] mapped the shear flow locations observed by the radar to the magnetospheric equatorial plane using the Tsyganenko89 model, and resulted in widely scattered locations between about 10 and 30 $R_E$.

To our knowledge, so far no detailed studies of instantaneous, spatial ionospheric electrodynamic situations, as mentioned above for the HD proper, have been carried out for the other ionospheric shear flow regions associated with the potential minima or maxima of the evening or morning convection cells.

In this paper, by using ground magnetometer data of two representative days selected from a large data set, we first point out that there are two types of ionospheric electrodynamic situations that can cause the HD signature according to its original definition, called "rotation-type" and "expansion-type" HD. These situations are topologically different with respect to their occurrence in the ionospheric potential pattern, and should therefore also map to different magnetospheric regions. Secondly, we study five shear flow situations from the morning side to infer spatial distributions of ionospheric conductances and currents with the method of characteristics inversion technique. We discuss the common features of all these events with a particular focus on the conductance and FAC distribution at the shear regions.

## INSTRUMENTATION

We study the shear flow regions by means of ground-based data of the MIRACLE network (Figure 1). This network consists of the IMAGE magnetometer network [*Lühr et al.*, 1998], the STARE coherent scatter radar [*Greenwald et al.*, 1979], and five digital all-sky cameras. The densest part of the network is located in northern Fennoscandia, where the fields of view of STARE and the all-sky cameras overlap with an array of magnetometers. In this part, it is therefore possible to infer spatial distributions of ionospheric electrodynamic parameters from the data by using an appropriate inversion technique.

## ROTATION-TYPE AND EXPANSION-TYPE HARANG DISCONTINUITY

In the first part of this paper, we use magnetometer data only to point out that there are two topologically different situations that may cause the ground magnetic signature of the Harang

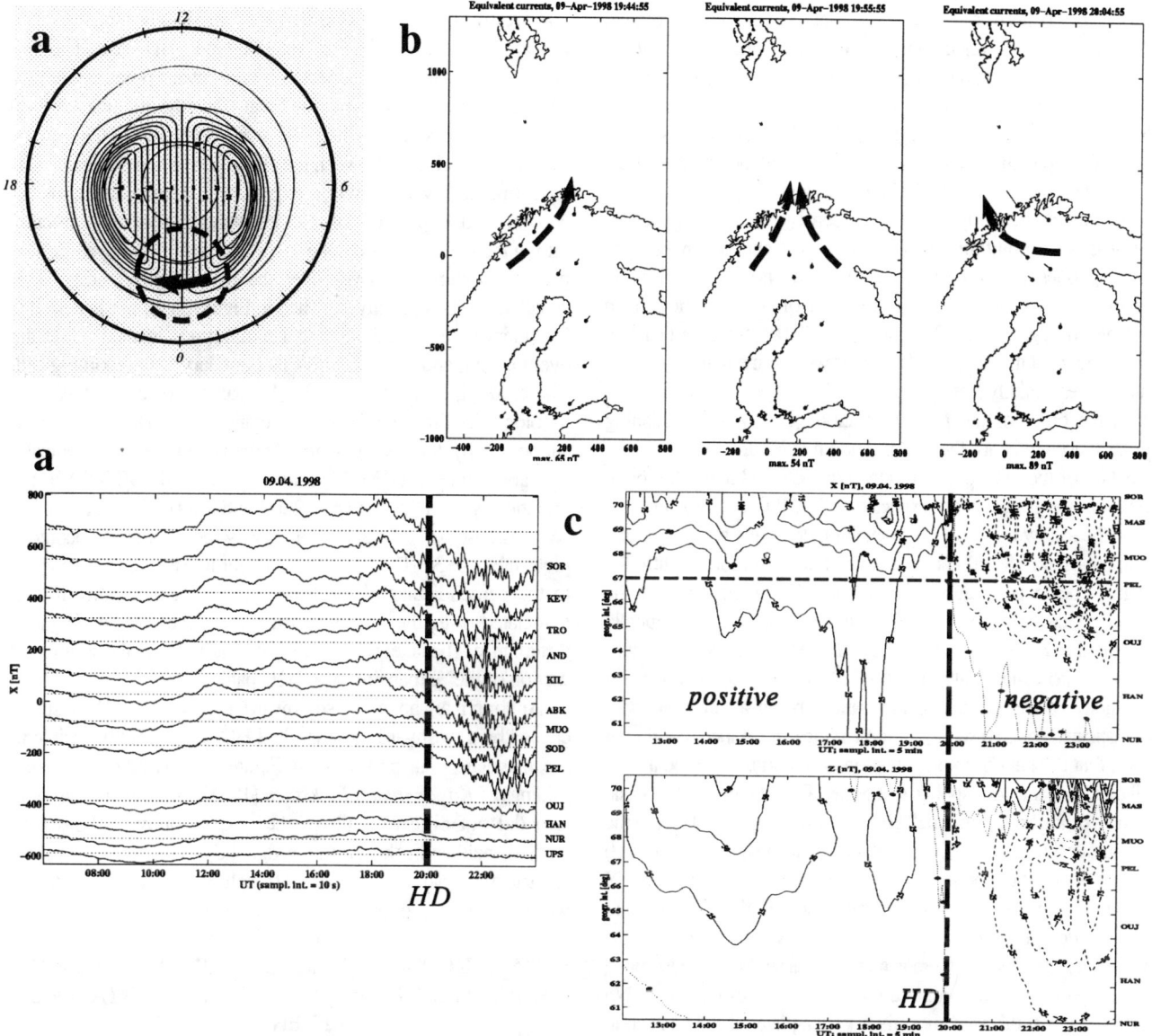

**Figure 2.** "Rotation-type" Harang Discontinuity (HD); (a) upper panel: Schematic representation of the potential structure and its movement causing the HD signature in the ground magnetic field disturbance; lower panel: IMAGE magnetograms for April 9, 1998, a representative day for an "Rotation-type" HD; (b) Evolution of by 90 degrees clockwise rotated ground magnetic disturbance vectors; (c) Isocontour lines of X and Z ground magnetic disturbance with time.

Discontinuity (HD) according to its original definition. The two different HD situations are termed "rotation-type" and "expansion-type" here. While the former is caused by the rotation of the magnetometer below the transition between the evening and morning side potential cells with the Earth's rotation velocity, the latter is caused by an equatorward expansion of the evening cell potential minimum. From many sets of magnetometer data, we have selected one representative day for both of these situations.

As an example for the "rotation-type HD", Figure 2a (lower panel) shows the X (geographic north) component of several IMAGE magnetometers located in Fennoscandia on April 9, 1998. Neglecting small differences due to the different longitudinal positions of the stations, all magnetometers show simultaneously the transition from positive to negative X values, i.e., from the eastward to the westward electrojet domain, at about 2000 UT. The quite stable eastward electrojet before the HD leads to ground magnetic disturbances of less than 100 $nT$, whereas the westward electrojet following the HD reaches about 150 $nT$. Such moderate values are typical for weakly disturbed periods without any substorm activity [e.g., *Kamide and Kokubun*, 1996]. Figure 2b shows the pattern of the by 90 degrees clockwise rotated horizontal magnetic disturbance vectors during the passage of the HD over the magnetometer array [note that these rotated vectors equal approximately, but not exactly the ground equivalent currents, cf. *Amm*, 1998]. At 19:44:55 UT, the vectors are pointing northeastward in the western part of the area and nearly northward in its eastern part. At 19:55:55 UT, the discontinuity is located over the center of the array, as can be seen by the convergence of northeastward pointing vectors in the west and northwestward pointing ones in the east, while the small northward pointing vectors in its center mark the location of the transition from positive to negative X . The pattern at 20:04:55 UT is nearly a mirror image of that of 19:44:55 UT. If we assume that the current pattern is essentially stationary, its westward speed is estimated to be in the order of the Earth's rotation velocity. For vanishing conductance gradients, the vectors shown in Figure 2b point in the direction of the ionospheric electric equipotential lines. Although conductance gradients are expected at the HD, as will be pointed out in the following section, the development of the rotated ground magnetic field disturbance vectors is in accordance with the magnetometers co-rotating with the Earth below the border of the evening side and morning side potential cells as schematically illustrated in Figure 2a (upper panel). Note that we took this "sun-aligned" potential pattern from *Heppner* [1977] only for simplicity of the schematic illustration, being aware that more sophisticated patterns were provided in the same paper. We do not claim that the pattern in our situation necessarily has to be a two-cell one, or that it corresponds to a certain IMF direction. The basic features of the "rotation-type" HD are most easily visible from the isocontour lines of the X and Z (vertical, positive downward) ground magnetic components with time (Figure 2c): The X contours show a "top on top" structure of the positive and negative domains, and no minimum of Z is associated with the HD. This indicates that there is no real shear in the currents, but a broad northward equivalent current region passes overhead. No northward or southward movement of the patterns can be seen in X or Z. We note that we do not exclude the possibility that this "rotation-type" HD signature shows a tilt further north in the auroral belt, as similarly seen in the "modified model A" convection pattern of *Heppner* [1977].

As an example for the "expansion-type" HD, we select August 26, 1998, as a representative day. From the IMAGE magnetograms (Figure 3a, lower panel), it can be seen that here the HD is associated with an intensification of the overall ground magnetic disturbance from values around 200 $nT$ before the HD to magnitudes up to 400 $nT$ after it. These still medium disturbed values increase even more later on that day during several intense substorms. The occurrence of the HD in the Fennoscandian magnetometers is seen first in the north, and then progresses subsequently to the south. The evolution of the clockwise rotated ground magnetic disturbance (Figure 3b) shows that the eastward and westward electrojet domains are latitudinally separated, and not longitudinally as in the "rotation-type" HD case. The two electrojet domains are connected by northward pointing vectors. This pattern appears to move southward, and the HD reaches each magnetometer station when the curl structure in the rotated magnetic disturbance between the electrojets, where X changes its sign, moves over it. We interpret this developement as an equatorward expansion of the ionospheric potential structure during the intensification. Due to this expansion, the evening side potential minimum moves over the magnetometer array. This situation is schematically sketched in Figure 3a (upper panel). It is important here that this potential minimum is located inside the evening cell, and does not mark its boundary to the morning cell. Therefore, the situation in the "expansion-type" HD case is topologically different from the "rotation-type" HD. This is also supported by the fact that the "expansion-type" HD typically appears in an earlier MLT sector (around 2100 MLT) than the "rotation-type" HD (between 2200 and 2400 MLT). The basic "expansion-type" HD features are easily seen from the isocontour plots in Figure 3c: The zero X isocontour line is tilted with respect to the perpendicular of the time axis, and a minimum in Z is associated with the curl-like structure in the rotated magnetic disturbance. The latter signature most clearly shows the southward motion of the "expansion-type" HD pattern.

Although an apparent southward movement could also be caused by a westward moving tilted HD, no such tilt is visible in the magnetic data presented in Figure 3. This agrees with the findings of *Koskinen and Pulkkinen* [1995] and *Amm* [1998] who also did not observe a tilt of the "expansion-type" HDs they studied, neither in the ground magnetic nor in the ionospheric electric field data. This does not exclude the possibility that a westward and a southward movement coexist.

## SPATIAL CONDUCTANCE DISTRIBUTIONS ALONG SHEAR FLOW REGIONS FOR FIVE INSTANTANOUS EVENTS

For the second part of this paper, five different non-HD shear flow events observed by MIRACLE have been selected for a detailed analysis with the method of characteristics [*Inhester et al.*, 1992; *Amm*, 1995, 1998]. With this method, spatial distributions of the ionospheric Hall ($\Sigma_H$) and Pedersen ($\Sigma_P$) conductances, currents $\vec{J}$, and field-aligned currents $j_l$ (FACs) can be inferred from spatial input data of the ground magnetic disturbance $\vec{B}_G$ and the ionospheric electric field $\vec{E}$. For the calculation, a rough estimate of the Hall to Pedersen conductance ratio $\alpha$ has to be given which can be assessed from the ground magnetic disturbance level [e.g., *Schlegel*, 1988; *Lester et al.*, 1996]. The influence of a possibly wrong assumption of $\alpha$ on the results of the method of characteristics has been studied in detail by *Amm* (1995). The basic result is that while the magnitude of the FACs is sensitive to the

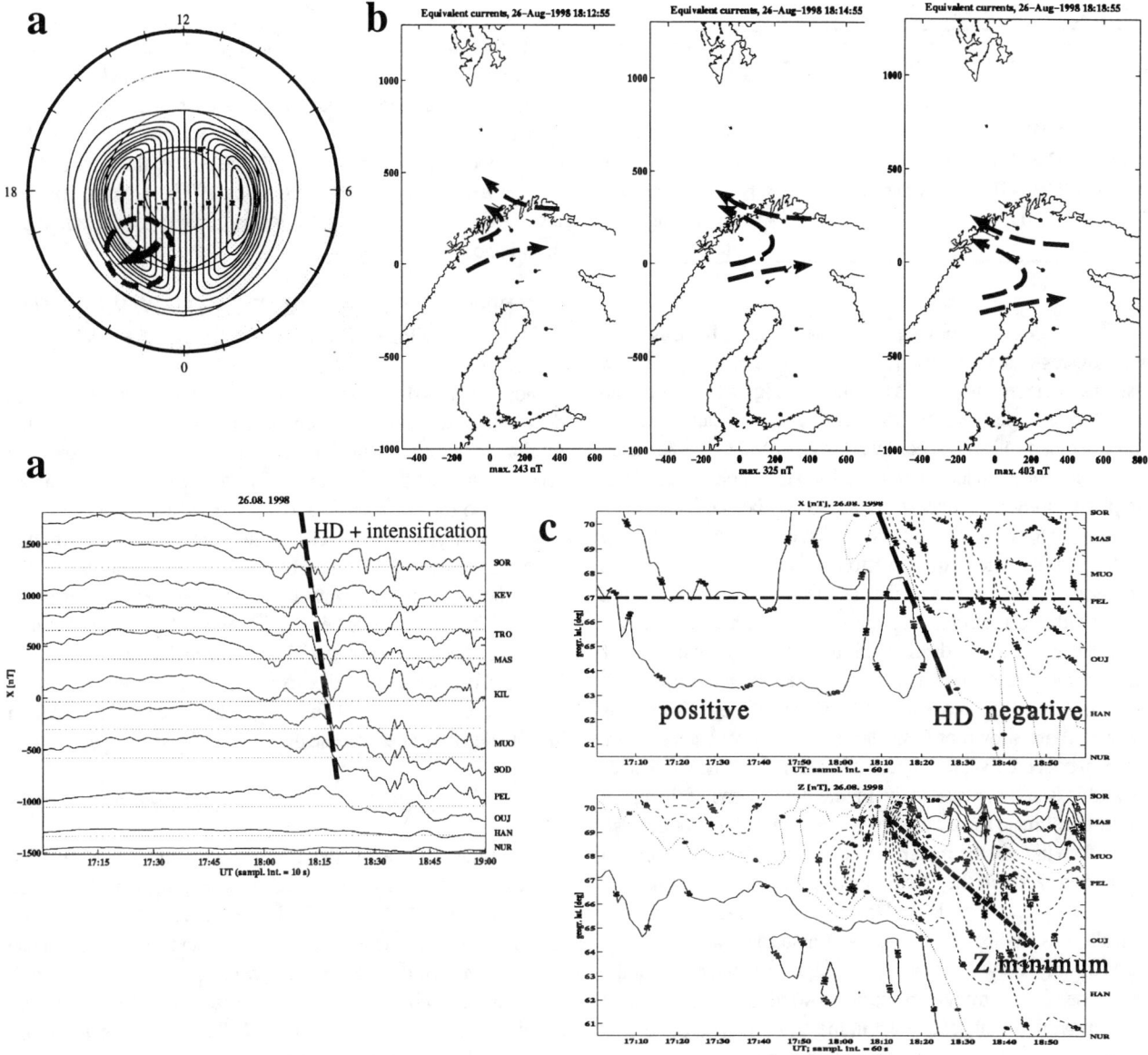

**Figure 3.** Same as Figure 2, for "expansion-type" Harang Discontinuity; representative day is August 26, 1998.

assumed α distribution, the location and general distribution of the FAC pattern is rather unsensitive to it.

As has been most clearly shown by *Lester et al.* [1996], the ratio α shows a much closer correlation to the ground magnetic disturbance value than the individual conductances that have to be assumed separately for other methods that infer ionospheric electrodynamic parameters from similar sets of ground-based data [cf. *Amm*, 1998]. Roughly following the statistics presented by *Schlegel* [1988], for the event studied in this paper we assume uniform distributions of α with a value of 1.2, except for the August 26, 1998, event for which we employ a value of 2. We emphasize that for the reasons stated above, none of the conclusions drawn below will be critically affected by the special value of α chosen.

While the magnetometers produce data nearly continuously, the strongest limiting factor regarding the input data for our inversion method is the presence of sufficient backscatter in the STARE radar to obtain a reasonable spatial distribution of ionospheric electric field measurements inside the radar's field of view (compare Figure 1). As the most important driver of E region instabilities which cause the radar backscatter is the ionospheric convection electric field [e.g., *Haldoupis*, 1989], strong backscatter

**Table 1.** Overview of shear zone events studied.

| Event | Location of shear zone | General level of activity |
|---|---|---|
| 26.6.1998, 05:36 UT | noonward edge of morning cell | medium, recovery after very intense storm |
| 11.7.1998, 22:50 UT |  | medium, steady |
| 22.8.1998, 23:45 UT | midnight edge of morning cell | medium, intensification |
| 26.8.1998, 23:36 UT |  | very high, recovery |
| 27.8.1998, 23:30 UT |  | medium to high, recovery |

is usually not detected from regions of a low electric field, as in the shear flow regions. One reason for the selection of five shear flow situations in the morning cell is that they belong to the events with the strongest backscatter observed during such situations so far. More afternoon cell and especially HD events will be studied later. An overview of the selected events is found in Table 1. All events take place during moderately to strongly disturbed periods.

Figure 4 shows the five input data distributions on the left side of the panels, where thick vectors relate to by 90 degrees clockwise rotated ground magnetic disturbances, and STARE electric field measurements are marked by thin vectors. In the right panels, the resulting $\Sigma_H$ distributions as inferred by the method of characteristics are shown. The calculation is carried out approximately on the area of the STARE field of view. The results in Figure 4 are shown on a smaller region to avoid areas where unknown boundary values affect the solution. In some small areas where this is still the case in the Figure 4, the mean of the upper and lower estimate for $\Sigma_H$ is shown [see *Amm*, 1998, for details]. The broken lines in the right panels of Figure 4 mark the location of the shear zone in the electric field data, i.e., the approximate latitude of $E_{north} = 0$, as derived from STARE data.

In all cases studied, the shear zone is associated with small $\Sigma_H$ values of only a few $S$. Typically 1-2 degrees equatorward of the shear zone, $\Sigma_H$ is strongly increasing towards south. In all cases the main ground magnetic effect in the STARE field of view is caused by this area equatorward of the shear zone which is governed by the westward electrojet. Note that also the largest ground magnetic disturbances are detected at or south of the southern boundary of the STARE area. Directly from the measurements, it is most visible for the most disturbed event on August 26, 1998 (Figure 4d) that the conductances at and immediately north of the shear zone must be low: While STARE shows a large north to northwestward directed electric field north of the shear zone, the rotated ground magnetic disturbance vectors in the same area still point westwards with some northward deflection. In case of uniform conductances, they would have to point nearly exactly in the opposite direction. The same effect, although less pronounced, is seen in the July 7 (Figure 4b) and August 22 (Figure 4c) cases. Only in the June 26 (Figure 4a) and August 27 (Figure 4e) events, the by 90 degrees clockwise rotated $\vec{B}_G$

vectors point eastward immediately north of the shear zone, but with small magnitudes as compared to the westward electrojet domain in the south. However, it is clear that the conductance has to increase again north of our analysis area, because the relatively large magnetic disturbances over Svalbard cannot be explained as equivalent return currents caused by the ends of the electrojet [cf. *Untiedt and Baumjohann*, 1993; *Amm*, 1997]. Also the effect of the FACs at the shear zone is insufficient to explain these disturbances.

## FAC DISTRIBUTION AT THE SHEAR FLOW REGION ON THE EXAMPLE OF THE AUGUST 27, 1998, EVENT

For a detailed analysis of the field-aligned current (FAC) distribution which we can derive from the divergence of the resulting real ionospheric current flow (data not shown here), we select the August 27, 1998, event. The total FACs (Figure 5a) are pointing mostly downwards, as to be expected at a shear flow zone where the divergence of the electric field is positive. However, the largest downward FACs are not located at the shear zone itself where the FAC density is between 1 and 1.5 $\mu A/m^2$, but somewhat south of it where $j_l$ reaches values around 2 $\mu A/m^2$. Also it is noticable that the downward FAC region has a much larger latitudinal extent than would be expected from the divergence of $\vec{E}$ alone (compare Figure 4e). In order to study the FAC distribution in more detail, we decompose the total FACs with respect to

$$j_l = \Sigma_P \ div\vec{E} + \nabla\Sigma_P \cdot \vec{E} + \nabla\Sigma_H \cdot (\hat{z} \times \vec{E}) \quad (1)$$

which results from taking the divergence of Ohm's law. The first term essentially describes the FAC associated with the divergence of the electric field ("term 1"), the second one ("term 2") those associated with gradients of the Pedersen, and the third one those with gradients of the Hall conductance. In earlier studies, term 1 has occasionally been named as "magnetospheric FACs" and the remaining terms as "ionospheric FACs" [e.g., *Sofko et al.*, 1995; *Sato et al.*, 1995]. We do not use this terminology here since we do not believe such a separation is meaningful in the coupled magnetosphere-ionosphere system. The FACs associated with term 1 are shown in Figure 5b. In this distribution, the downward FACs are restricted to the shear zone, whereas in the southern part of the analysis area term 1 leads to upward FACs which is a specific feature of this particular event. However, the largest downward FACs are caused by the conductance gradient towards the south and represented by term 2 (Figure 5c). These FACs form a second latitudinally aligned sheet of FACs, the maximum of which is located about 2 degrees equatorward of the shear zone. The third term of equation (1) is sensitive to the conductance gradient perpendicular to the electric field, i.e., here in east-west direction, which is more event-dependent and less important for our study.

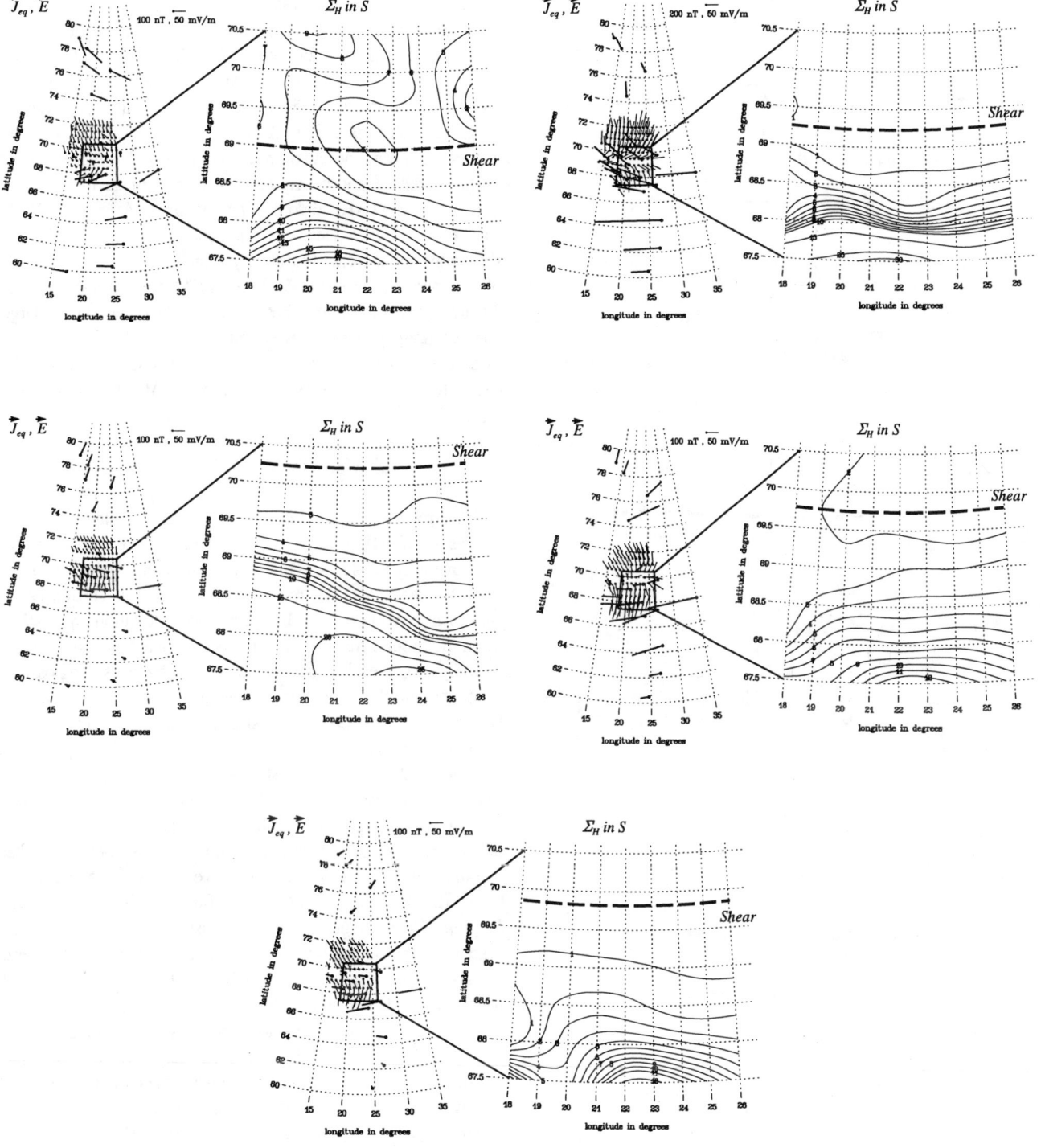

**Figure 4.** Analysis of the Hall conductance distribution at shear zones for five different events; left panels: Input data of by 90 degrees clockwise rotated ground magnetic disturbance vectors (thick) and STARE ionospheric electric field vectors (thin); right panels: calculated Hall conductance distribution; the broken line indicates the shear zone as observed in the electric field data; (a) June 26, 1998, 0536 UT; (b) July 11, 1998, 2250 UT; (c) August 22, 1998, 2345 UT; (d) August 26, 1998, 2336 UT; (e) August 27, 1998, 2330 UT.

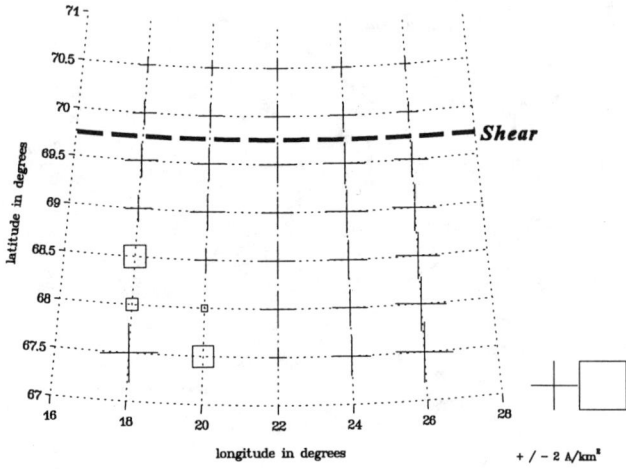
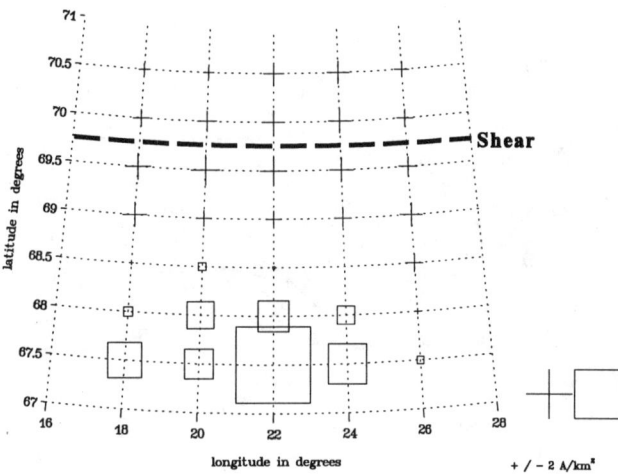
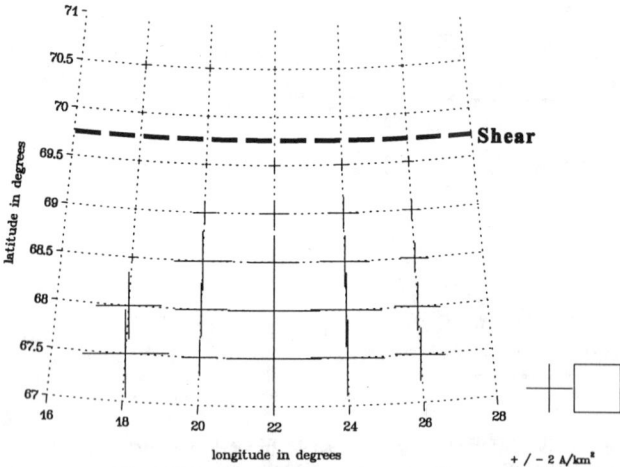

**Figure 5.** Field-aligned currents (FACs) analysis for the August 27, 1998, 2330 UT event; (a) total FACs; (b) FACs due to $\Sigma_H \, div \vec{E}$; (c) FACs due to $\nabla \Sigma_H \cdot \vec{E}$.

Since a substantial increase of the conductances equatorward of the shear zone is a common feature of all events studied, this is consequently true also for the double sheet structure of FACs as shown in Figure 5b and c. A schematical synthesis of the common features for all situations with respect to conductances, divergence of $\vec{E}$, and FACs is shown in Figure 6: The divergence of $\vec{E}$ is large at the shear zone and decreases equatorward of it. Depending on the special event, it may become negative near the southern border of the STARE field of view like in the August 27, 1998, case, or take values near zero. The conductances are small at the shear zone and increase first slowly, then drastically equatorwards of that zone. This leads to two sheets of FACs, one associated with the divergence of $\vec{E}$ at the shear zone, and the other with the large conductance gradient 1-2 degrees equatorwards of it. The FACs in these two sheets always have the same direction which, for the cases studied, corresponds to the region 1 FAC orientation in the morning sector.

## DISCUSSION AND CONCLUSIONS

In terms of our initial classification, all instantaneous event studies on the HD listed in the introduction were carried out on "expansion-type" HDs. This is not surprising, because those studies focused on the role of the HD during substorms or substorm growth phases. Moreover, during "rotation-type" HDs, often the ground magnetic disturbance is quite weak, and radars typically do not receive any backscatter which makes it impossible to carry out the inversion techniques used in these works. However, it is important to notice that statistical studies of the ionospheric potential structure [e.g., *Heppner*, 1977; *Heppner and Maynard*, 1987] or of the distribution of FACs [e.g., *Iijima and Potemra*, 1978] will result in the average location of the "rotation-type" HD because of its much more frequent occurrence. This difference is likely to be the explanation for the inconsistence that *Koskinen and Pulkkinen* [1995] observed between the location of their HD events and the HD as defined by the *Heppner and Maynard* [1987] patterns. This inconsistency became even more dramatic after the mapping of the patterns to the magnetospheric equatorial plane. This shows that, while the Heppner and Maynard patterns are certainly valuable for many other applications, care has to be taken if they are applied to single substorm period events.

Although a corrobation by a statistical study is not available yet, we suggest the following relation between the two HD types and the two-component electrojet classification of *Kamide and Kokubun* [1996]: While the "rotation-type" HD is observed in cases when solely the "convection electrojet" is present, i.e., in the geomagnetic disturbance range from quiet times to moderately disturbed times with enhanced convection but without substorm activity [e.g., *Sergeev et al.*, 1996], the "expansion-type" HD is observed during the presence of the "substorm electrojet", usually during geomagnetically disturbed periods. Note that also the

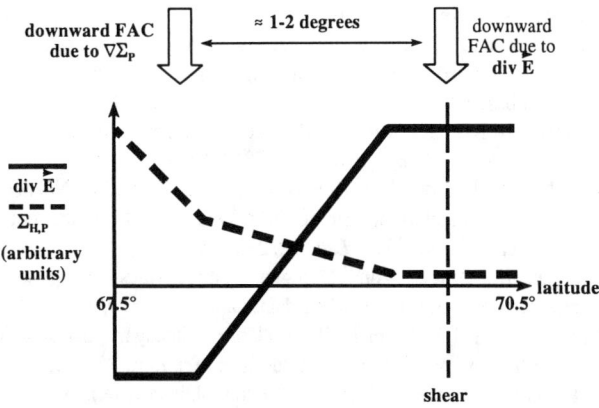

**Figure 6.** Schematic representation of common features for all events studied, regarding conductance, divergence of $\vec{E}$, and FACs at the shear zones.

potential structure proposed by *Kamide and Kokubun* [1996] as being associated with the "substorm electrojet" (their Figure 12) agrees very well with the observed latitudinal transition between the eastward and westward electrojets (Figure 3).

The five shear flow events in the morning potential cell that we studied in detail are not a large enough number to derive very general results. More situations, also from the evening cell have to be added later. Moreover, additional support by all sky camera observations from the MIRACLE network (not available because of the polar summer for our events), and possibly by satellites should be used to support the analysis and interpretation. In particular, this could provide a more detailed estimation of the conductance ratio $\alpha$ which may show substantial gradients especially in the morning sector due to energetic particle precipitation. Still, it is remarkable that all five events tally in that there are small conductance values at the shear zone as observed by the radar, and a strong conductance increase about 1-2 degrees equatorward of that zone, leading to two separate latitudinally aligned sheets of FACs of the same sign, one at the shear zone due to the divergence of $\vec{E}$, and the other due to the conductance gradient equatorward of it. In none of the events, a remarkable electrojet is observed north of the shear region inside our analysis area, although the Svalbard magnetometers indicate that there must be a substantial current further in the north. While such small conductances in downward FAC regions like the morning cell shear zones are not surprising, it is interesting that a somewhat similar conductance behaviour as we derived for our events was also observed in the *Kunkel el al.* [1986], *Inhester et al.* [1992] and *Amm* [1998] HD cases in the evening sector, where the shear regions are governed by upward FACs. Further work has to be carried out to study if this behaviour is typical for shear zones in general.

Finally, it would be most interesting to obtain satellite particle spectra of the two FAC sheets. Since one is associated with a low conductance and the other with a medium or high conductance ionospheric region, it might be suspected that they are connected to magnetospheric source regions of different plasma temperature. An identification of these source regions would lead to a considerable improvement in the understanding of the ionosphere-magnetosphere coupling mechanism. Although no such satellite data is available for our events, the DE2 measurements presented by *Coley et al.* [1987] closely agree with what would be expected to be observed by a satellite for our events: For the events shown in their Plates 1 and 2, it is clearly seen that the trapped particles precipitation boundary is located 1-3 degrees equatorward of the convection reversal boundary. *Coley et al.* [1987] conclude that the part of the sunward flow (corresponding to the southward electric field region in Figure 4e, left panel) equatorward of that reversal is located on open field lines in a lobe cell. The weak precipitation in this region and poleward of it would lead to small ionospheric conductances. In contrast to that, the region equatorward of the precipitation boundary is located on closed field lines and connected to the plasma sheet. The more energetic electron precipitation would lead to enhanced ionospheric conductances, in correspondence with our results (Figure 4e, right panel).

*Acknowledgments.* The work of O.A. was supported by a DAAD fellowship HSP III, financed by the German Federal Ministry for Education and Science. We would like to thank K. Kauristie (Helsinki) for comments on the manuscript.

## REFERENCES

Amm, O., Direct determination of the local ionospheric Hall conductance distribution from two-dimensional electric and magnetic field data: Application of the method using models of typical ionospheric electrodynamic situations, *J. Geophys. Res., 100*, 24173, 1995.

Amm, O., Ionospheric elementary current systems in spherical coordinates and their application, *J. Geomag. Geoelectr., 49*, 947, 1997.

Amm, O., Method of characteristics in spherical geometry applied to a Harang discontinuity situation, *Ann. Geophys., 16*, 413, 1998.

Baumjohann, W., R.J. Pellinen, H.J. Opgenoorth, and E. Nielsen, Joint two-dimensional observations of ground magnetic and ionospheric electric fields associated with auroral zone currents: Current system associated with local auroral break-ups, *Planet. Space Sci., 29*, 431, 1981.

Coley, W.R., R.A. Heelis, W.B. Hanson, P.H. Reiff, J.R. Sharber, and J.D. Winningham, Ionospheric convection signatures and magnetic field topology, *J. Geophys. Res., 92*, 12352, 1987.

Greenwald, R.A., W. Weiss, E. Nielsen, and N.R. Thomson, STARE: A new radar backscatter experiment in northern Scandinavia, *Radio Sci., 13*, 1021, 1978.

Haldoupis, C., A review on radio studies of auroral E-region ionospheric irregularities, *Ann. Geophysicae, 7*, 239, 1989.

Harang, L., The mean field of disturbance of polar geomagnetic storms, *Terrest. Magn. Atmospheric Electr., 51*, 353, 1946.

Heppner, J.P., The Harang discontinuity in auroral belt ionospheric currents, *Geofys. Publ., 29,* 105, 1972.

Heppner, J.P., Empirical models of high-latitude electric fields, *J. Geophys. Res., 82,* 1115, 1977.

Heppner, J.P., and N.C. Maynard, Empirical high-latitude electric field models, *J. Geophys. Res., 92,* 4467, 1987.

Iijima, T., and T.A. Potemra, Large-scale characteristics of field-aligned currents associated with substorms, *J. Geophys. Res., 83,* 599, 1978.

Inhester, B., J. Untiedt, M. Segatz, and M. Kürschner, Direct determination of the local ionospheric Hall conductance distribution from two-dimensional electric and magnetic field data, *J. Geophys. Res., 97,* 4073, 1992.

Kamide, Y.: On current continuity at the Harang discontinuity, *Planetary Space Sci., 26,* 237, 1978.

Kamide, Y., and S. Kokubun, Two-component auroral electrojet: Importance for substorm studies, *J. Geophys. Res., 101,* 13027, 1996.

Kan, J.R., A global magnetosphere-ionosphere coupling model of substorms, *J. Geophys. Res., 98,* 17263, 1993.

Koskinen, H.E.J., and T.I. Pulkkinen, Midnight velocity shear zone and the concept of Harang discontinuity, *J. Geophys. Res., 100,* 9539, 1995.

Kunkel, T., W. Baumjohann, J. Untiedt, and R.A. Greenwald, Electric fields and currents at the Harang discontinuity: a case study, *J. Geophys., 59,* 73, 1986.

Lester, M., J.A. Davies, and T.S. Virdi, High-latitude Hall and Pedersen conductances during substorm activity in the SUNDIAL-ATLAS campaign, *J. Geophys. Res., 101,* 26719, 1996.

Lühr, H., A. Aylward, S.C. Buchert, A. Pajunpää, K. Pajunpää, T. Holmboe, and S.M. Zalewski, Westward moving dynamic substorm features observed with the IMAGE magnetometer network and other ground-based instruments, *Ann. Geophys., 16,* 425, 1998.

Robinson, R.M., and R.R. Vondrak, Electrodynamic properties of auroral surges, *J. Geophys. Res., 95,* 7819, 1990.

Sato, M., Y. Kamide, A.D. Richmond, A. Brekke, and S. Nozawa, Regional estimation of electric fields and currents in the polar ionosphere, *Geophys. Res. Lett., 22,* 283, 1995.

Schlegel, K., Auroral zone E region conductivities during solar minimum derived from EISCAT data, *Ann Geophys., 6,* 129, 1988.

Sergeev, V.A., R.J. Pellinen, and T.I. Pulkkinen, Steady magnetospheric convection: A review of recent results, *Space Sci. Rev., 75,* 551, 1996.

Sofko, G.J., R. Greenwald, and W. Bristow, Direct determination of large-scale magnetospheric field-aligned currents with SuperDARN, *Geophys. Res. Lett., 22,* 2041, 1995.

Untiedt, J. und W. Baumjohann, Studies of polar current systems using the IMS Scandinavian magnetometer array, *Space Sci. Rev., 63,* 245, 1993.

---

O. Amm, P. Janhunen, H. J. Opgenoorth[1], T. I. Pulkkinen, and A. Viljanen, Finnish Meteorological Institute, Geophysical Research Division, P.O. Box 503, FIN-00101 Helsinki, Finland. (Olaf.Amm@fmi.fi, Pekka.Janhunen@fmi.fi, Tuija.Pulkkinen@fmi.fi, Ari.Viljanen@fmi.fi)

# Statistical Characteristics of Field-Aligned Currents in the Earth's Inner Magnetosphere

Francis K. Chun

*Department of Physics, U.S. Air Force Academy CO*

Christopher T. Russell

*Institute of Geophysics and Planetary Physics, Los Angeles, CA*

Field-aligned currents (FACs) play a major role in the solar wind-magnetospheric interaction by electrodynamically coupling magnetospheric and ionospheric plasmas, so that stresses applied to the outer magnetospheric plasma are transmitted to the ionosphere and ultimately to the upper atmosphere. FACs have been studied extensively since 1967 and most of our statistical knowledge and understanding of this coupling mechanism is based on data from low-altitude polar orbiting satellites. In this paper, we present results from a statistical study of the inner magnetosphere FACs. Specifically, we examine their current intensity, current density, current sheet thickness, and current sheet velocity. Our data extends over the first 360 orbits of ISEE-1 and 2, and ranges over all local times and from altitudes between 2-9 $R_E$. We extrapolate our data down to ionospheric altitudes for comparison with low-altitude FACs. We also determine the location of the FACs as well as how closely they resemble sheets of current. We find that quasi-stationary FACs (FACs which move very slowly compared to transient events) are magnetospheric extensions of the low-altitude region 1 and 2 currents. Inner magnetosphere FAC characteristics are consistent with the region 1 and 2 currents, with densities on the order of a few $\mu A/m^2$. Velocities and thickness extrapolated to the ionosphere are also consistent with auroral structures. Current sense and location also resemble region 1 and 2 currents, with inner magnetosphere FACs flowing into the ionosphere on the dawn side poleward of the outward flowing currents, and opposite for the dusk side. Finally, we find that the currents are most sheet-like in the dawn and dusk sectors and least two dimensional at noon and midnight.

## 1. INTRODUCTION

Early statistical studies of spacecraft magnetometer data at low altitudes revealed a pattern of field-aligned currents (FACs) coincident with the auroral oval termed region 1 and 2 [*Iijima and Potemra*, 1976]. Region 1 currents consist of current flow into the ionosphere on the dawnside and outward flow on the duskside. Region 2 lies equatorward of region 1 and has the opposite sense current flow. Region 1 and 2 have been studied extensively with respect to geomagnetic activity and solar wind/interplanetary magnetic field (IMF) conditions [Iijima and Potemra, 1976; 1978; 1982]. One reason why many studies have centered on low-altitude FACs is because at low altitudes, one can assume that current structures are moving quite slowly compared to the speed of the observing spacecraft. Thus the magnetic perturbation measured by a single spacecraft can be confidently attributed to the motion of the spacecraft through a "stationary" current sheet. The assumption of stationary current sheets unfortunately does not apply to field-aligned currents at higher altitudes, and in fact current structures at high altitudes can sometimes be moving as fast or faster than the observing spacecraft.

Any large-scale study of field-aligned currents in the earth's inner magnetosphere must overcome the problem of determining the relative motion of the FAC and the satellite in order to accurately determine current density and thickness. At middle and high altitudes, the only feasible means of doing so is to use two spacecraft and compare their signatures. The International Sun-Earth Explorer (ISEE) 1 and 2 satellites are ideal for this task. Chun and Russell [1997] provided a detailed description of the ISEE orbits and the advantages of using them for such a statistical study. Our previous studies presented data on how inner magnetospheric FACs respond to substorm phase [Chun and Russell, 1991] and geomagnetic activity [Chun and Russell, 1997]. In this paper, we complete our earlier studies by presenting overall statistical characteristics of field-aligned currents in the earth's inner magnetosphere, to include current intensity, current density, current sheet velocity, and current sheet thickness. In addition, we provide data on the location of these currents and how sheet-like they appear.

## 2. DATA

Our data extends over the first 360 orbits of ISEE-1 and 2, and ranges over all local times and from altitudes between 2-9 $R_E$ (Figure 1). Examples of ISEE orbital coverage in magnetic latitude, radial distance, local time, and $L$ value can be found as Figures 1 and 2 in Chun and Russell [1997]. Figure 2 in this paper (from Chun and Russell [1997]), is an example of field-aligned currents taken from orbit 43. This plot

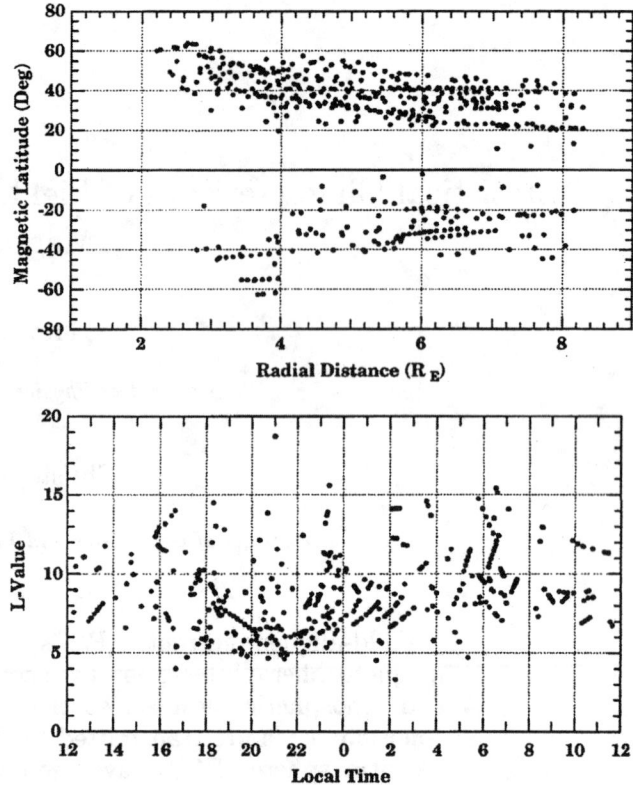

**Figure 1.** Location (radial distance vs. magnetic latitude and local time vs. L-value) of the field-aligned currents used in this study.

shows magnetic signatures from ISEE-1 and 2 which have been detrended by the Tsyganenko [1987] magnetic field model and rotated into a field-aligned coordinate system, where BZ DIP is field-aligned, BX DIP is the outward normal, and BY DIP is the azimuthal normal. The FAC signatures in Figure 2 have been previously discussed [Kelly et al., 1986; Chun and Russell, 1991; 1997], and one can observe that some signatures are easily separated by ISEE-1 and 2 (e.g., interval 1), while others appear almost simultaneously at both spacecraft (e.g., interval 4).

The near-simultaneous signatures are attributed to fast-moving, impulsive waves encountering both spacecraft at the same time [Chun and Russell, 1997], or a consequence of temporal changes in a current structure seen simultaneously at both spacecraft [Strangeway et al., 1997]. In either case, since it is extremely difficult to unambiguously determine the properties of the current sheet (such as current density, thickness and velocity), events such as interval 4 have been discarded

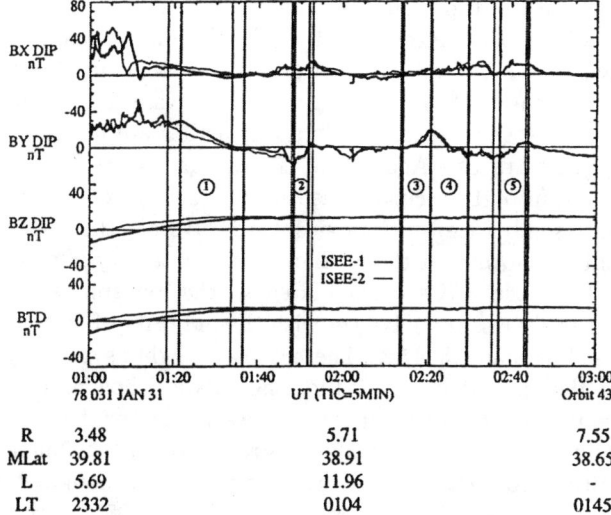

**Figure 2.** Time series of ISEE 1 and 2 detrended and field-aligned magnetic field data for orbit 43 (from *Chun and Russell* [1997]), showing outward normal (BX DIP), azimuthal normal (BY DIP), and field-aligned (BZ DIP) detrended components, and difference between the total measured field strength and total model field strength (BTD). ISEE-1 data are shown by the thick traces, while ISEE-2 are the thin traces. Field-aligned currents are indicated by the numbered labels (1-5), and the start and stop times of each current are shown for both ISEE 1 (thick line) and ISEE 2 (thin line).

from this study. Instead, we focus our analysis on quasi-stationary field-aligned currents. Additionally, in order to have more confidence in our determination of current sheet thickness and hence density, we limited our events to only those in which ISEE-1 and 2 intersect the current sheet at a greater than 10° grazing angle.

The determination of current intensity (current per length), current density (current per area), velocity, and thickness has been discussed in depth before [*Kelly et al.*, 1986; *Chun and Russell*, 1991]. Briefly though, we first calculate the current intensity of each FAC assuming that the FAC is a current sheet aligned along $L$ shells. Thus magnetic signatures of the current sheet will be seen primarily in the BY DIP component. Ampere's law allows us then to calculate the intensity. Current sheet velocities are determined from the difference in the time signatures of the current structures (e.g., interval 1 from Figure 2). A negative velocity indicates motion radially inward or to lower invariant latitudes. Once we know the current sheet velocity, we can estimate the thickness of the current sheet by the product of velocity and time, with the appropriate modification to the velocity due to the spacecraft motion. Finally, the current sheet density can be calculated from the current intensity and thickness. All FAC properties have been extrapolated to an altitude of 110 $km$ in order to mitigate any differences due to where in the magnetosphere the FAC was observed. This also allows comparison to low-altitude studies. The extrapolation technique is a simple dipole extrapolation and was previously discussed in detail [*Chun and Russell*, 1991].

## 3. RESULTS AND DISCUSSION

Figures 3 through 9 are local time distributions of various current properties. Figure 3 (top panel) is the median location (in $L$ value determined from the Olson-Pfitzer model) of the quasi-stationary FAC events divided into their current sense. FACs flowing into the

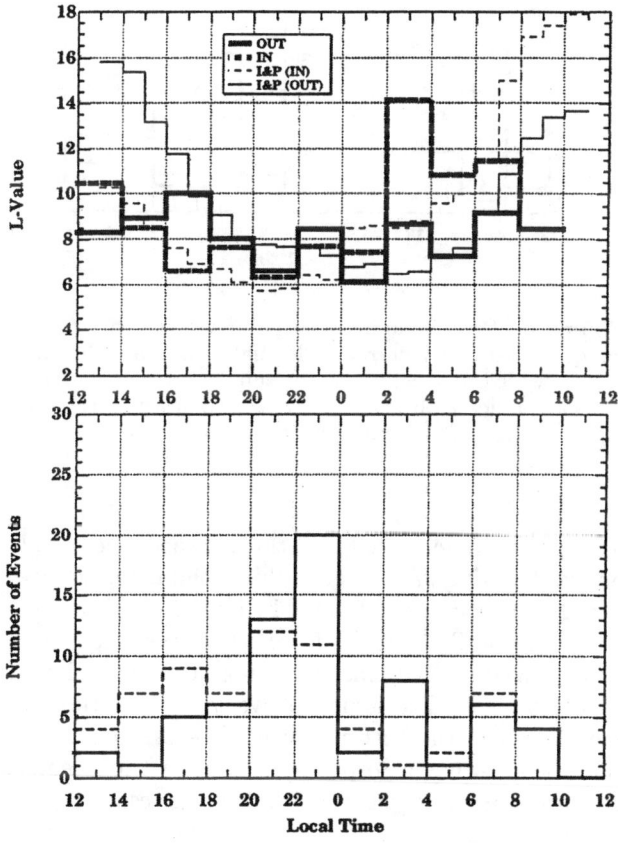

**Figure 3.** Local time distribution of the quasi-stationary FAC median $L$ values (thick traces). The location of region 1 and 2 currents (thin traces) are also plotted for comparison (from *Chun and Russell* [1997]). Outward flowing currents are indicated by solid traces, while inward flowing currents are dashed traces. The number of events per local time bin per current sense is plotted below.

240 INNER MAGNETOSPHERIC FACS

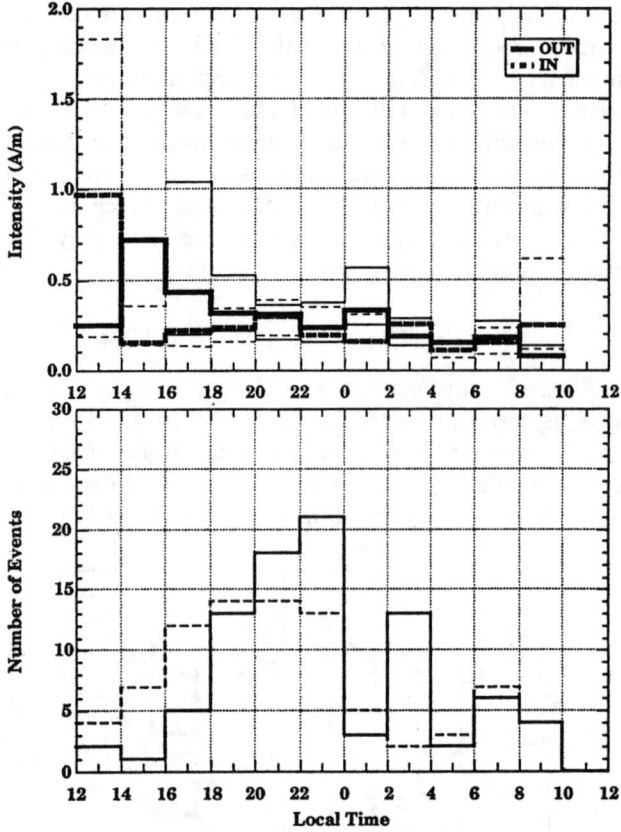

**Figure 4.** Local time distribution of the intensity ($A/m$) for quasi-stationary currents. Median values are the thick traces, while the quartiles are the thin traces. Outward currents are indicated by solid traces, while inward currents are dashed traces. The number of events per local time bin are plotted below.

ionosphere are indicated by a thick dashed trace, while outward flowing FACs are the thick solid trace. The region 1 and 2 locations, denoted by I&P(IN) (thin dashed trace) and I&P(OUT) (thin solid trace), where I&P is an abbreviation for Iijima and Potemra, are also plotted for comparison. The number of events per local time bin are plotted in the bottom panel. One can readily see that the inner magnetospheric FACs appear to resemble the low altitude region 1 and 2 currents, especially on the night side. The difference in $L$ value location near noon and the lack of ISEE cases are due to ISEE not routinely reaching high latitudes to observe those currents. Most of the noon ISEE events are probably due to current systems migrating to lower latitudes, which appears to be reflected in the negative velocity medians (or motion toward lower invariant latitudes) shown in Figure 7.

Figure 4 (top panel) is the local time distribution of median current intensity ($A/m$) for the quasi-stationary events, with the number of events per local time plotted in the bottom panel. The median values are indicated by thick traces while the quartiles are indicated by thin traces. Again, the current intensities are divided into their current sense, in (dashed trace) and out (solid trace) of the ionosphere. Overall, the distribution shows that the median current intensity of magnetospheric FACs extrapolated to the ionosphere are generally less than 1 $A/m$ with the majority of intensities less than 0.5 $A/m$. There is also a slight asymmetry in the current intensities for the outward flowing currents, with the dusk currents appearing larger than the dawn currents. This could be a reflection of the contribution from the partial ring current that is thought to form on the dusk side [*Crooker and Siscoe*, 1981; *Clauer et al.*, 1983; *McPherron*, 1991].

Figure 5 is the local time distribution of median current density ($\mu A/m^2$) for the quasi-stationary currents. Since the number of events per local time bin for Figures 5-8 are the same as for Figure 4 (current intensity), those panels have been eliminated from the figures. The region 1 and 2 current densities (connected black and white circles) are plotted for a comparison [*Iijima and Potemra*, 1976]. Statistically, our estimate of the median current densities based on magnetospheric FACs appear two to three times that of the low-altitude FACs. However, global features of the region 1 and 2 current densities are reflected in the magnetospheric FACs in that the current densities are normally lowest throughout the night side and strongest around the 1400 and

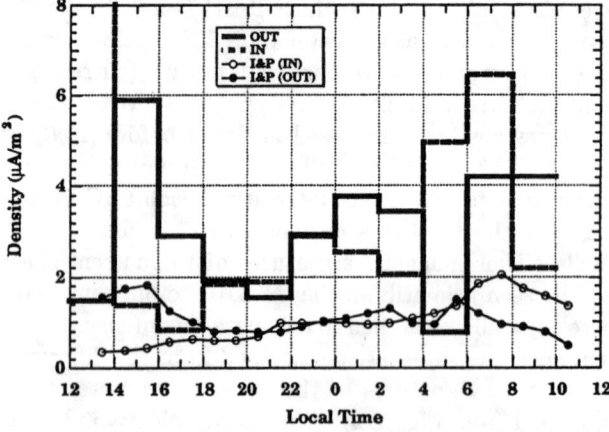

**Figure 5.** Local time distribution of the median density ($\mu A/m^2$) for quasi-stationary currents (outward is solid, inward is dashed). Region 1 and 2 currents (I&P (IN) and I&P (OUT)) are also plotted for comparison.

0800 local time sectors. The density imbalance between the outward and inward flowing currents are also reflected in both the ionospheric (region 1 and 2 currents) and magnetospheric FACs in those two local time sectors, while balancing one another between dusk and midnight. The density imbalance between outward and inward flowing currents could be in part due to poor statistics. However, although ISEE density estimates are larger than Iijima and Potemra's values [*Iijima and Potemra*, 1976], the imbalance is consistent with their findings, and is perhaps due to some closure of the field-aligned currents through traverse currents in either the magnetosphere or ionosphere, consistent with a three-dimensional current system.

The difference in the quasi-stationary magnetospheric current density values and the ionospheric region 1 and 2 current density values could be due to several factors. First, and perhaps most significant, ISEE current density estimates are highly dependent on an accurate determination of current sheet thickness. The ideal situation would be for ISEE-1 and 2 to intersect the current sheet at a normal angle. This scenario would provide the best estimate of current sheet thickness, and anything less would lead to a lower estimate. Since current sheet thickness varies as the sine of the grazing angle, current density would be inversely proportional to the sine of that angle. Thus since most of our events have grazing angles between 40° and 10°, magnetospheric current density estimates could be 1.6 to 6 times greater, respectively.

Secondary factors which could affect current density estimates include uncertainties in estimating current sheet velocities, use of a simple dipole model to extrapolate magnetospheric values to ionospheric altitudes, and a bias to measuring stronger currents in the magnetosphere. Current sheet thickness is also dependent on an *in situ* current sheet velocity estimate, which in turn is based on the average of the velocities determined at the beginning, middle, and end of the current sheet. Temporal evolution of a current sheet would thus lead to different velocity estimates depending on what part of the current structure one references. Although the extrapolation technique uses a simple dipole model, at the middle altitudes where we have made our observations, the magnetosphere is to first order a dipole configuration. Therefore, any errors associated with the extrapolation should be minimal. Finally, our events could be biased toward stronger currents measured in the magnetosphere since those currents are inclined to be more noticeable and observable. This would then lead to our statistical medians being biased toward stronger currents.

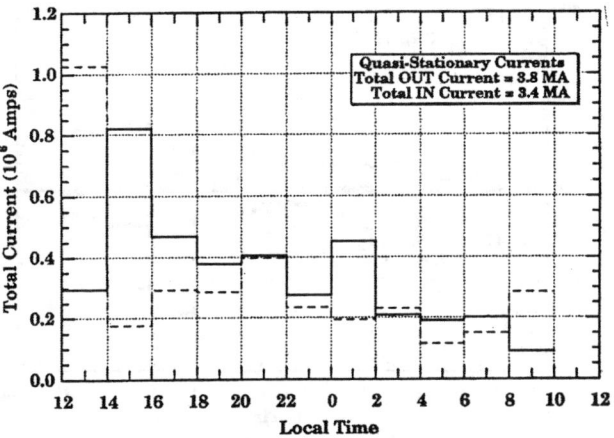

**Figure 6.** Local time distribution of the total current ($MA$) for quasi-stationary currents based on median values of current intensity (outward is solid, inward is dashed).

Total current can be estimated from the intensities assuming that the FACs are current sheets which extend across each two-hour local time sector. We account for the variation in the median $L$ value location of the currents when estimating the total current. Figure 6 is a local time distribution of the total current flowing in and out of the ionosphere, and as expected, resembles the intensity distribution (Figure 4). Except for a few local time sectors, the total current flowing out of the ionosphere is greater than that flowing in. This imbalance between the inward and outward flowing currents could be due to current flow across the polar cap or current flow around the auroral oval in the electrojets. Globally however, the currents appear to balance, with 3.8 million amperes of current flowing out of the ionosphere compared to 3.4 million amperes flowing inward, consistent with low-altitude FACs (3.5 and 3.4 million amperes, respectively) [*Iijima and Potemra*, 1976].

One of the primary reasons for conducting this study with ISEE-1 and 2 was the unique opportunity to determine for the first time the statistical velocity characteristics of field-aligned currents. The dual spacecraft configuration is ideal to determine the motion of the FAC structures in their outward normal direction. Figure 7 is the local time distribution of velocities ($m/s$) for the quasi-stationary current sheets. The dusk currents appear to always move equatorward regardless of their current flow sense, however there are certainly occasions when currents move poleward. Because we are examining quasi-stationary currents, the velocities in which we observed are generally less than 200 $m/s$, very much consistent with auroral structures [*Akasofu*, 1968; *Nakai et al.*, 1986].

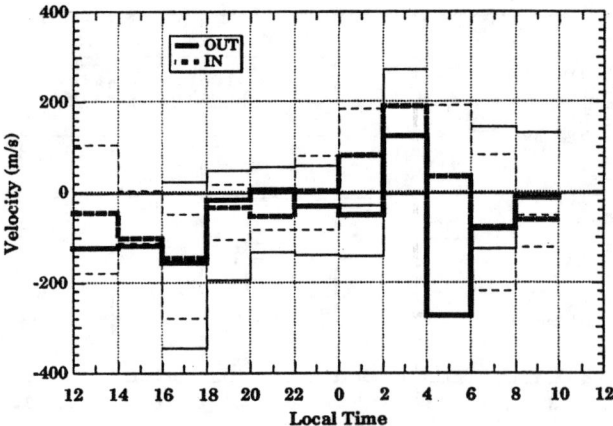

**Figure 7.** Local time distribution of the velocity $(m/s)$ for quasi-stationary currents (outward is solid, inward is dashed). Median values are the thick traces, while the quartiles are the thin traces.

Figure 8 is the local time distribution of current sheet thickness $(km)$ in the north-south direction for the quasi-stationary currents. Most of the median values of current sheet thickness are less than 200 $km$, approximately 2 to 3 times smaller than what one can infer from the region 1 and 2 currents [*Iijima and Potemra*, 1976], and consistent with our current densities being larger than region 1 and 2 by the same factors. There are some instances (as seen in the quartile values) of current sheet thicknesses upward to 600 $km$ or so. These appear reasonable since some auroral structures can range from a few kilometers to over 1000 $km$ in size (such as spirals) [*Davis*, 1979].

Finally, one last piece of information that can be determined from this study is an indication of how sheet-like magnetospheric field-aligned currents appear based on a maximum variance analysis. The eigenvalues contain useful information as to how sheet-like the magnetospheric currents appear. The ratio of the maximum eigenvalue to the intermediate eigenvalue describes whether the current structure is sheet-like or filamentary in nature. The eigenvalue ratio is an indication of how successful the maximum variance analysis was in rotating all of the magnetic field change into one principal axis. A large ratio (roughly on the order of 100) means that the rotation was successful, and that the resulting magnetic field profile resembles that of a current sheet. A small ratio (an order of magnitude less) is just the opposite, meaning that comparable magnetic field changes are present in both transverse directions (maximum and intermediate principle axes) resulting in a profile similar to a filamentary current structure or current sheets with embedded small-scale structures.

Figure 9 (top panel) is a local time distribution of the median (thick trace) and quartile (thin traces) eigenvalue ratios for all quasi-stationary currents, with the number of events per local time bin plotted in the bottom panel. The quasi-stationary currents near the dawn and dusk meridian appear more sheet-like than at other local times. The ratios near the dawn-dusk meridian are generally greater than 70, while the ratio at other local times are less than 40. There is also an asymmetry in the dawn-dusk ratios with the median dawn ratio considerably higher than the dusk ratio (approximately 150 to 100). This observation of sheet-like FACs near the dawn-dusk meridian is consistent with the findings of *Kelly et al.* [1986] who showed more examples of large-scale, sheet-like currents away from midnight towards both dawn and dusk. Additionally, the finding of filamentary-like currents elsewhere is consistent with *Robert et al.* [1984] who observed many of their fast-moving FAC tubes around midnight, especially during substorm onset.

## 4. CONCLUSION

In this paper, we presented the statistical characteristics of large-scale, quasi-stationary field-aligned currents in the earth's inner magnetosphere. We find that these FACs, which are between altitudes of 2-9 $R_E$, appear to be extensions of the low-altitude region 1 and 2 currents. The current properties (intensity, density,

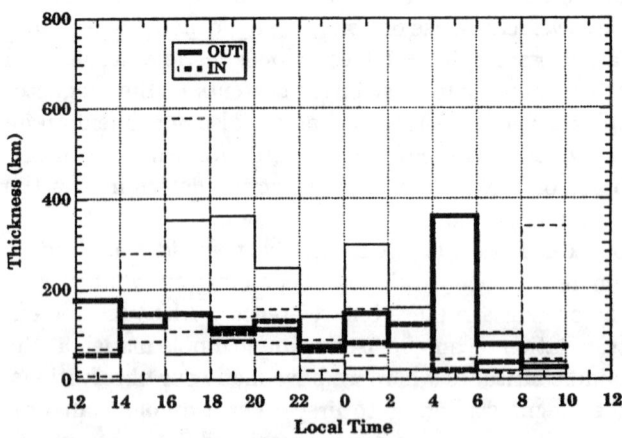

**Figure 8.** Local time distribution of the thickness $(km)$ in the north-south direction for quasi-stationary currents (outward is solid, inward is dashed). Median values are the thick traces, while the quartiles are the thin traces.

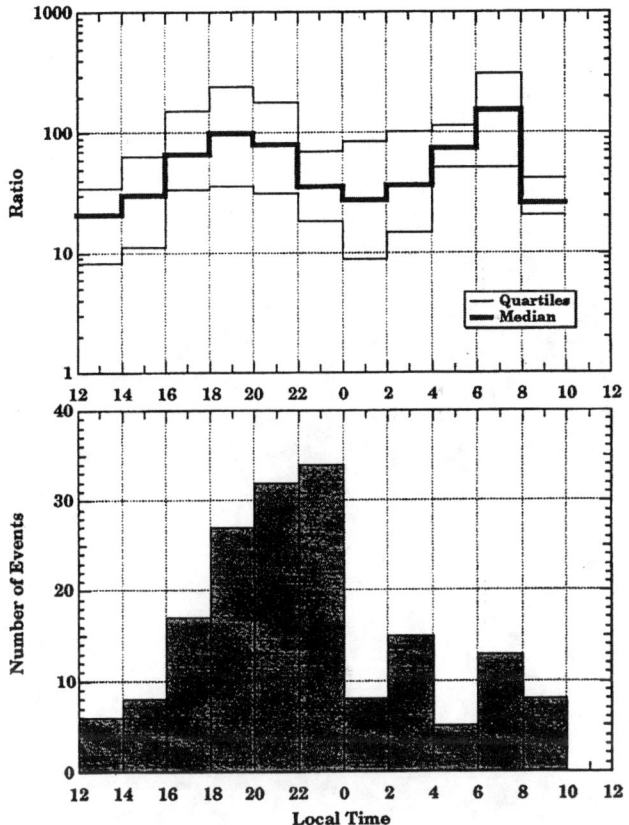

**Figure 9.** Local time distribution of eigenvalue ratios for all quasi-stationary currents. Median values are the thick traces, while the quartiles are the thin traces. The number of events are plotted below.

velocity, and thickness), when extrapolated to the ionosphere are consistent with the strength, dimension, and dynamics of region 1 and 2 currents and auroral structures. Additionally, inner magnetospheric FACs appear to be more sheet-like near the dawn-dusk meridian region compared to the noon-midnight region.

*Acknowledgments.* F. K. Chun would like to thank the Department of Physics, USAF Academy for their sponsorship.

The Editor would like to thank the reviewers of this manuscript.

## REFERENCES

Akasofu, S.-I., *Polar and Magnetospheric Substorms*, p. 44, D. Reidel, Norwood, Mass., 1968.

Chun, F.K., and C.T. Russell, The evolution of field-aligned currents as a function of substorm phase, *J. Geophys. Res.*, *96*, 15,801-15,810, 1991.

Chun, F.K., and C.T. Russell, Field-aligned currents in the inner magnetosphere: Control by geomagnetic activity, *J. Geophys. Res.*, *102*, 2261-2270, 1997.

Clauer, C.R., R.L. McPherron, C. Searls, Solar wind control of the low-latitude asymmetric magnetic disturbance field, *J. Geophys. Res.*, *88*, 2123-2130, 1983.

Crooker, N.U., and G.L. Siscoe, Birkeland currents as the cause of the low-latitude asymmetric disturbance field, *J. Geophys. Res.*, *86*, 11,201-11,210, 1981.

Davis, T.N., Observed microstructure of auroral forms, in *Auroral Processes*, edited by C.T. Russell, pp. 171-180, Center for Academic Publications Japan, Tokyo, Japan, 1979.

Iijima, T., and T.A. Potemra, The amplitude distribution of field-aligned currents at northern high latitudes observed by TRIAD, *J. Geophys. Res.*, *81*, 2165-2174, 1976.

Iijima, T., and T.A. Potemra, Large-scale characteristics of field-aligned currents associated with substorms, *J. Geophys. Res.*, *83*, 599-615, 1978.

Iijima, T., and T.A. Potemra, The relationship between interplanetary quantities and Birkeland currents, *Geophys. Res. Lett.*, *9*, 442-445, 1982.

Kelly, T.J., C.T. Russell, R.J. Walker, G.K. Parks, and J.T. Gosling, ISEE 1 and 2 observations of Birkeland currents in the earth's inner magnetosphere, *J. Geophys. Res.*, *91*, 6945-6958, 1986.

McPherron, R.L., Physical processes producing magnetospheric substorms and magnetic storms, *Geomagnetism*, *4*, 593, 1991.

Nakai, H., Y. Kamide, D.A. Hardy, and M.S. Gussenhoven, The dynamics of the equatorward boundary of the auroral oval, in *Solar Wind-Magnetosphere Coupling*, edited by Y. Kamide and J.A. Slavin, pp. 633-641, Terra Scientific, Tokyo, 1986.

Robert, P., R. Gendrin, S. Perraut, A. Roux, and A. Pedersen, GEOS 2 identification of rapidly moving current structures in the equatorial outer magnetosphere during substorms, *J. Geophys. Res.*, *89*, 819-840, 1984.

Strangeway, R.J., T.C. Meseroll, and C.T. Russell, The variability of magnetic field perturbations and IMF control of field-aligned currents in the inner magnetosphere, *Adv. Space Res.*, *20* (3), 469-472, 1997.

Tsyganenko, N.A., Global quantitative models of the geomagnetic field in the cislunar magnetosphere for different disturbance levels, *Planet. Space Sci.*, *35*, 1347-1358, 1987.

---

F. K. Chun, Department of Physics, 2354 Fairchild Drive, Suite 2A31, U.S. Air Force Academy, CO 80840. (e-mail: Francis.Chun@usafa.af.mil)

C. T. Russell, Institute of Geophysics and Planetary Physics, 3845 Slichter Hall, Box 951567, University of California, Los Angeles, CA 90095-1567. (e-mail: ctrussell@igpp.ucla.edu)

# Independency of the Dayside Field-Aligned Current System: A Restriction to Cusp Models

M. Yamauchi[1], R. Lundin[1], L. Eliasson[1], S. Ohtani[2], P.-A. Lindqvist[3], and R. P. Lepping[4]

Data from the Freja satellite, with its unique longitudinal cusp traversals during southward IMF due to 63° inclination, reinforced past evidences that the field-aligned currents (FACs) in the cusp region are separated from and independent of the dayside region 1 FAC outside the cusp. The results are consistent with the traditional concept of the cusp (open due to non-global mechanism such as the weak magnetic field), and indicate that the FAC generation is directly related to local plasma injection in the cusp and independent of the global magnetospheric convection during southward IMF. Existing global 3-D MHD models do not explain the observed independency without essential modifications.

## 1. INTRODUCTION

To avoid confusion in terminology (see *Yamauchi et al.* [1993] for detail), we call the dayside region 1 field-aligned currents (FAC) in the cusp region "cusp region 1" (or "C-1") FAC, the dayside region 1 FAC outside the cusp "auroral oval region 1" (or "R-1") FAC, and the region 0 FAC in the cusp region (i.e., the traditional cusp FAC or the mantle FAC) "cusp region 0" (or "C-0") FAC. We here consider only southward interplanetary magnetic field (IMF) conditions.

There is a fundamental discrepancy between the FAC observations and the magnetohydrodynamic (MHD) models of the convection (and FACs) in the cusp region. Although all satellite observations during the past decade have repeatedly emphasized that the cusp-part (region 1 and region 0) FACs are isolated from the auroral oval region 1 FAC with completely different location and plasma populations [e.g., *Potemra*, 1994; see next section for other references], many models and theories on the cusp still employ a so-called "unified" concept of MHD which predicts the cusp region 0 FAC as an extension of the auroral oval region 1 FAC [e.g., *Cowley et al.*, 1991; *Onsager and Elphic*, 1996; *Lockwood*, 1998]. Because of its one-fluid treatment, any global MHD model predicts a unified source no matter how we change the IMF condition [*Crooker*, 1988].

This misunderstanding stems partly from the observed global ionospheric convection (or potential) pattern, a quantity which is equivalent to an integration of FAC distribution but can also be directly measured by ground observation. Integration of real data means a smoothing of patterns. Although the FAC pattern in 50 km resolution contradicts the global MHD prediction, the convection pattern of 50 km resolution agrees with the MHD prediction. Therefore, we are not yet able to distinguish the cusp region 1 FAC and auroral-oval region 1 FAC using ground-based instruments such as the SuperDarn radar system. Furthermore, the convection pattern and the FAC pattern have a one-to-one relation only in MHD theory, while in reality many different FAC patterns produce the same convection pattern within observational errors or variations.

There have, however, been some attempts to model two independent sources for the cusp-part FACs and the dayside region 1 FAC outside the cusp [*Burch et al.*, 1985; *Taguchi et al.*, 1993; *Ohtani et al.*, 1995; *Yamauchi and Lundin*, 1997]. Most of these non-global attempts assume a direct dynamo mechanism for the cusp-part FAC

[1]Swedish Institute of Space Physics, Kiruna, Sweden
[2]JHU/APL, Johns Hopkins Road, Laurel, MD, USA
[3]Royal Institute of Technology, Alfvén Laboratory, Stockholm, Sweden.
[4]NASA/Goddard Space Flight Center, Greenbelt, MD, USA

generation. A careful examination of these models will reveal that two types of openness are assumed in these models, namely, a global openness, which allows solar wind access through the dayside magnetospheric boundary (for R-1) and a local openness which allows the access only near the cusp (for C-1/C-0). We here call the former a "semi" openness and the latter an "extra" openness because the global openness does not necessarily mean completely open compared to the local openness in the cusp region. The semi-open mechanism may include the viscous-like interaction as well as the merging.

The "double openness" concept is not new; i.e., the cusp singularity (weak magnetic field and intense turbulence) has long been considered independent of the global solar wind-magnetosphere interaction [e.g., *Spreiter and Summers*, 1967; *Heikkila and Winningham*, 1971; *Haerendel et al.*, 1978]. This concept (the extra-open cusp in the semi-open magnetosphere) is also theoretically feasible within a large-scale fluid scheme by using a multi-component plasma [*Wallis*, 1973; *Yamauchi and Lundin*, 1997]: adding a new plasma component (e.g., mass-loading oxygen or a local obstacle) means an additional freedom in the equation system and hence a new wave mode and a boundary such as the mass-loading boundary. Global simulations already exist for the mass loading near Venus and Mars [*Kallio*, 1996, and references therein], although the real cause of the double openness could of course be different.

Double openness implies independent drift motions for two major plasma populations: a poleward background convection driven by the global semi-open mechanism and an east west deflected injection driven by the cusp "extra-open" mechanism. In fact such multiple flow velocities for different plasma populations have been observed in the Terrestrial and Jovian magnetospheres [*Lundin et al.*, 1987; *Krupp et al.*, 1993]. However, the double openness concept has been somewhat forgotten in recent 3-D global simulations of the Terrestrial magnetosphere.

Let us mention the flux transfer event (FTE) models [e.g., *Smith and Lockwood*, 1990]. The existing FTE cusp models are the same as the global MHD model in the sense that a unified source of FAC is simply divided into a meso-scale newly open flux to the already-open polar cap [*Cowley et al.*, 1991], so its large-scale physics is the same as that of traditional global MHD. Thus, introducing the FTE concept does not change the essence of the present discussion.

With this background (cf. second paragraph), it is important to re-emphasise the independence between the cusp (C-1/C-0) FACs and the auroral oval (R-1) FAC. Therefore, it is worth adding even minor contributions to support the past observations of this independency. We show Freja observations in this paper.

## 2. BRIEF REVIEW OF PAST EVIDENCES

The history of the FAC studies is briefly summarized by *Potemra* [1994] (see also *Yamauchi et al.* [1998]). The large-scale current system at high-latitude was established in mid 1970s from both ground [*Friis-Christensen and Wilhjelm*, 1975] and in-situ observations [*Iijima and Potemra*, 1976]. Soon after this finding many in-situ observations showed evidences that cusp (C-1/C-0) FACs and auroral oval (R-1) FAC are independent.

*1.* Seasonal and IMF control: The intensity of C-1/C-0 FACs depends linearly on IMF $B_Y$ [*Iijima et al.*, 1978; *Wilhjelm et al.*, 1978; *Yamauchi and Araki*, 1989] while this is not the case for R-1 FAC [*Woch et al.*, 1993]. Furthermore, the intensity of C-1/C0 FACs depends strongly on season (order of a magnitude difference between the winter and summer) whereas R-1 FAC changes only by a factor of two between summer and winter [*Fujii et al.*, 1981; *Yamauchi and Araki*, 1989].

*2.* Co-existing plasma population: The plasma domain is different between C-1/C-0 FACs and R-1 FAC [*Klumpar*, 1979; *Bythrow et al.*, 1982; *Potemra et al.*, 1987; *Erlandson et al.*, 1988; *Woch et al.*, 1993; *Yamauchi et al.*, 1993, 1998]. This is why and how we distinguish the "cusp" part and "auroral oval" part of the dayside large-scale FACs. The auroral oval region 1 and 2 FACs flow in the so-called BPS and CPS regions [*Winningham et al.*, 1975], respectively, whereas the cusp region 1 and 0 FACs flow in the cusp/cleft/mantle region.

*3.* Four sheet FACs: DE-2 particle and field data showed the simultaneous existence of two separate convection cells [*Burch et al.*, 1985] or equivalently four FAC sheets [*Taguchi et al.*, 1993] at the same meridian at different latitudes: the region 2 in CPS (upward in prenoon), R-1 in BPS (downward in prenoon), and C-1 (upward for negative IMF $B_Y$) and C-0 (downward for negative IMF $B_Y$) in the cusp region. The four-FAC system was also observed in the afternoon sector. *Ohtani et al.* [1995] showed an example of the four-FAC system simultaneously observed by low-altitude DMSP and mid-altitude Viking satellites, proving that it is a spatial structure.

*4* Meso-scale characteristics: Although the cusp (C-1/C-0) FACs contain many pairs of minor FACs which seem to be related to cusp meso-scale multiple injections [e.g., *Lundin et al.*, 1991], these meso-scale FACs simply add complexity to the background large-scale FACs [*Yamauchi et al.*, 1998]. Thus the cusp (C-0 or C-1) FAC is not formed as a simple summation of the monopolar signatures. Multiple injections, especially overlapping ones, are more often observed during high solar wind dynamic pressure and northward IMF than during other conditions [*Yamauchi and Lundin*, 1994; *Yamauchi et al.*, 1996; see also *Newell and Meng*, 1994]. These facts contradict the existing FTE cusp models.

*5.* Cusp-mantle interface: Excluding coincidental cases, the cusp proper is clearly distinguished from the mantle cusp by the particle, wave, and field data [*Kremser and Lundin*, 1990; *Yamauchi and Blomberg*, 1997]. The mantle cusp is not a simple downstream of the cusp proper, but they are rather two semi-independent regions as is expected by the extra-open scenario (cf. Introduction).

Table 1. IMF $B_Y$ control of the cusp

|  | IMF $B_Y > 0$ | IMF $B_Y < 0$ |
|---|---|---|
| $\Delta Be > 0$ | 4 | 1 |
| $\Delta Be < 0$ | 2 | 11 |

*6. Asymmetry*: During southward IMF the cusp region 1 FAC is generally stronger than the cusp region 0 FAC and vice versa for northward IMF [*Potemra, 1994*; and references therein]. Since the strongest FAC (of the adjacent FACs) is most likely connected to a dynamo, the dynamo is probably located equatorward of the cusp during southward IMF contrary to prediction of the existing MHD models [*Yamauchi et al., 1998*].

## 3. FREJA DATA

The 63° inclination of Freja [*Lundin et al., 1994*] enables a longitudinal examination of the cusp and the dayside FACs during southward IMF. The advantages of Freja are, in addition to providing further statistics to the above evidences, (1) it traverses both the cusp and the dayside auroral oval simultaneously at different magnetic local times; (2) it can observe the cusp regardless of its longitudinal shift (due to IMF $B_Y$); and (3) it can distinguish transient (= FTE type) and semi-stationary (= large-scale) FAC sheets if the sheet lies longitudinally.

Since the cusp has a quite variable morphology [e.g., *Yamauchi and Lundin, 1994*] and is the site of very active waves and turbulence [*Lundin et al., 1991*, and references therein], we must examine all the Freja cusp traversals to extract common features of the cusp FACs. Unfortunately, Freja had some attitude determination problems (no baseline for the magnetic field data) for nearly half the traversals. Hence, we have only 20 traversals with useful magnetic field data and IMP-8 IMF data out of more than 100 clear cusp traversals.

To increase the database, we estimate the solar wind condition from the Freja data. One can safely assume southward IMF for the Freja encounters of the cusp because of the 63° inclination, as is demonstrated by *Yamauchi et al.* [1996] (Figure 8). Figure 8 also confirmed that the cusp multiple injections observed by Freja (i.e. during southward IMF) indicate a high solar wind dynamic pressure condition.

The polarity of the IMF $B_Y$ is the same as that of the east-west magnetic deviations ($\Delta Be$) in the cusp region [*Iijima et al., 1978*], as is also demonstrated in Table 1 which is obtained from the 20 Freja cusp traversals mentioned above. A $\Delta Be < 0$ observation most likely indicates dawnward IMF ($B_Y < 0$), and vice versa. Statistics is not enough for $\Delta Be > 0$ or IMF $B_Y > 0$. The relation can also be applied to meso-scale FACs because the perpendicular magnetic deviation ($\Delta Be$) penetrates along any open geomagnetic field lines as an Alfvén wave.

## 4. OBSERVATIONS AND ANALYSES

### 4.1. Four-sheet FACs and plasma domain

Figure 1 shows a Freja cusp traversal during southward and dawnward IMF ($B_X = +3$ nT, $B_Y = -3$ nT, $B_Z = -4$ nT). As marked in the figure, one can identify BPS (1753:15-1753:25 UT and 1759:20-1759:30 UT; almost no proton injection with intense and structured electrons), the boundary cusp or "cleft" (1753:25-1754:15 UT and 1755:35-1759:20 UT; moderate or weak proton injection with either low intensity or highly structured electrons), and the cusp proper (1754:15-1755:35 UT; intense and continuous proton and electron injection). The noon part of the boundary cusp (1755:40-1758:15 UT) is somewhat mantle-like with faint proton injection and laminar poleward convection (eastward electric field) but we yet classify it as the (duskward) boundary cusp because of the existence of abrupt electron bursts. This ambiguity is not essential for our FAC analyses.

We first derive east-west aligned FAC sheets from $\Delta Be$. There are three intense FACs in this category (indicated in the bottom of the Figure): a downward FAC at 1753:15-1753:25 UT (the alignment is in the northwest-southeast direction because $\Delta Bn$ also jumped) in the morning BPS, an upward FAC at 1753:55-1754:10 UT inside the boundary cusp, and an upward FAC at 1759:20-1759:30 UT in

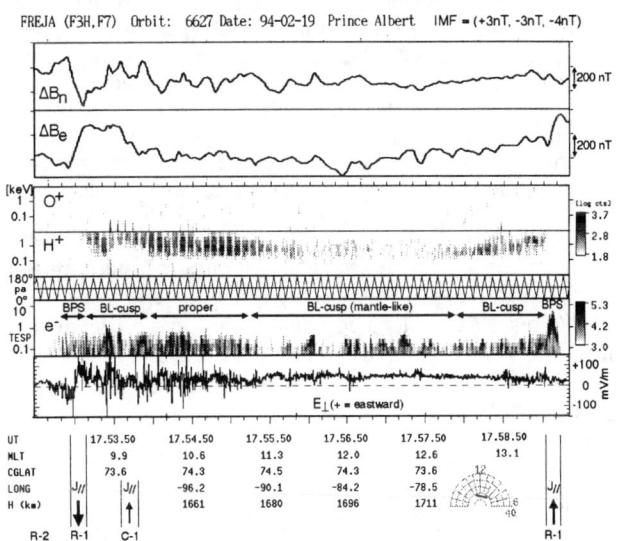

**Figure 1.** Freja particle and field data for orbit 6627 (1994-2-19). From the top to bottom: northward ($\Delta Bn$) and eastward ($\Delta Be$) magnetic deviation (subtracted by the model field), energy-time spectrograms of heavy ions (mainly oxygen), light ions (mainly proton), and electrons, and eastward electric field. Large-scale FACs indicated in the bottom (R-1 and C-1) are derived from the large-scale slope of $\Delta Be$ assuming east west aligned FAC sheets.

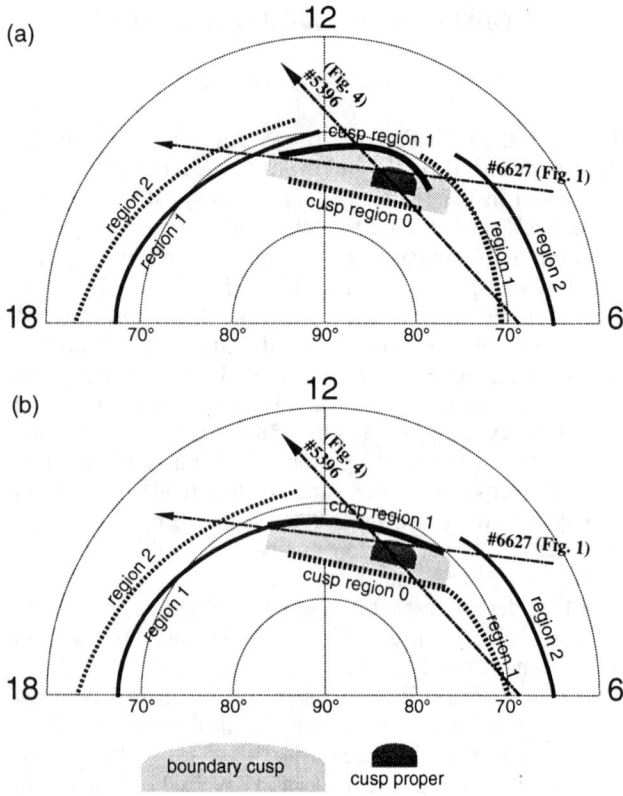

**Figure 2.** Two major models of the FAC distributions are compared with the present data. FACs poleward of these regions (e.g., polar arc FACs [*Yamauchi et al.*, 1993]) are not included in the Figure. (a) DE2 model [*Taguchi et al.*, 1993] agrees with data from both orbit 6627 and orbit 5396, (b) the unified MHD model [*Cowley et al.*, 1991] is not consistent with the data from orbit 6627 because the auroral oval region 1 FAC is connected to the cusp region 0 FAC in the model.

the afternoon BPS. The first and last ones are morning and afternoon R-1 FACs whereas the second one is C-1 or C-0 FAC. This FAC is located equatorward and/or dawnward of the cusp where a direct penetration of $\Delta Be<0$ is seen. We also see a wide-spread positive trend of $\Delta Be$ (gradual exit from the $\Delta Be<0$ penetration region) during 1556:45-1559:20 UT in the afternoon boundary cusp, suggesting a sheet FAC nearly aligned to the traversal.

All the three FAC sheets and the gradual increase of $\Delta Be$ are in good agreement with the model by *Taguchi et al.* [1993] (see also *Burch et al.* [1985] and *Ohtani et al.* [1995]) as is illustrated in Figure 2a. Here we interpret the second FAC sheet inside the boundary cusp (1753:55-1754:10 UT) as the cusp region 1 (C-1) FAC, and this FAC is clearly isolated from the auroral oval region 1 (R-1) FAC at its northward-noonward side. On the other hand, it is impossible to fit this observation to the "unified" FAC model by the existing global MHD (Figure 2b) because the model inevitably connects C-0 and the morning R-1 FACs during IMF $B_Y<0$ and does not allow a downward FAC (such as R-1) equatorward-dawnward of C-1 FAC. Note that *Erlandson et al.* [1988] showed another model which is also consistent with this observation, but we do not go in detail because this model is in principle the same as Figure 2a (their model adds an extra pair of FACs to Figure 2a).

These topological comparisons are not affected by the semi-steady assumption (it is inherent to the FAC analyses) because the essential information for this topology comes from the large plateau of positive $\Delta Be$ (1753:25-1753:55 UT) between R-1 and C-1 FACs. The only concern is if the entire positive $\Delta Be$ plateau were the result of a pulsed injection. However, the pulsed injection is simply a high-resolution limit of the continuous injection (cf. Introduction), and hence the above conclusion is still valid if we always have such pulsed "large-scale" injections equatorward of the cusp proper. This question ("if always") is already answered positively by past statistics.

The other transient scenario is that the strongest downward FAC at 1753:15-1753:25 UT were not R-1 FAC but merely the result of a temporal change. But this is most unlikely. If so, the entire plasma population should suddenly be accelerated toward the northwest direction (opposite direction to the magnetic deviation) only in this region, but no such signature is seen in the particle data. We instead observe an electron burst without protons, indicating a spatial structure inherent to BPS [*Potemra et al.*, 1987].

Finally, let us examine the north-south aligned meso-scale FACs using the $\Delta Bn$ data. In addition to the morning region 1 FAC, we see two large gradients of $\Delta Bn$: an upward FAC at 1753:25-1753:55 UT in the boundary cusp (this is continuous to C-1 FAC mentioned above) and a downward FAC at 1754:15-1754:25 UT in the cusp proper. These FACs are also consistent with the convection shear seen in the electric field data. The rather low $\Delta B/E$ ratio (about 500 km/s, i.e., 0.6 mho) indicates that the observed structure is a spatial one [e.g., *Sugiura*, 1984]. The first FAC (upward) fits Figure 2a but not Figure 2b, whereas the second FAC (downward) does not fit either model. Therefore, we believe that these FACs (especially the second one) are the meso-scale FACs which adds fine structures to the large-scale FACs, and these additional FACs do not alter the above conclusion on the large-scale FACs

### 4.2. Statistics

According to *Taguchi et al.* [1993], nearly half the traversals (they examined 27 cases for IMF $B_Y<0$) have such an extra downward (i.e., morning R-1) FAC equatorward of the upward cusp region 1 FAC. Table 2 shows the Freja statistics (with useful magnetic field data) of the morning R-1 FAC equatorward of C-1 FAC during IMF $B_Y<0$ (i.e., $\Delta Be<0$ inside the cusp). Statistics for $\Delta Be>0$ is too little

Table 2. Non-cusp region 1 FAC when $\Delta Be$ (cusp) <0

| between BPS and cusp | < 11 MLT | 12 MLT | > 13 MLT |
|---|---|---|---|
| with $\Delta Be>0$ plateau | 9 | 5 | 0 |
| without $\Delta Be>0$ plateau | 4 (+1)* | 1 (+3)* | 1 (+3)* |

* Inside () is the cases when $\Delta Be>0$ is not related to region 1 FAC.

(cf. Table 1). We eliminated traversals with extremely dispersed multiple injections during high solar wind dynamic pressure [*Yamauchi et al.*, 1996] because the IMF is then no longer controlling the cusp (we cannot guess the IMF polarity).

In Table 2 we used $\Delta Be$ instead of FAC (= spatial derivative of $\Delta Be$) because the Freja's longitudinal traversal is no longer perpendicular to the (assumed) current sheet, and because the total $\Delta Be$ is a good estimate of the latitude-integrated intensity of FAC in [A/m]. The four-sheet FAC system by *Taguchi et al.* [1993] means the existence of a narrow $\Delta Be>0$ region in the morning sector equatorward of a $\Delta Be<0$ region in the cusp when IMF $B_Y<0$. This is what we examined in Table 2. Another necessary caution in Table 2 is to make sure that the satellite really traverses the cusp (not the boundary cusp) where the IMF is eventually penetrating. This extra caution inherently requires identification of regions by the particle data, and in this sense the present study is the first statistics of extra FAC sheet with clear separation between BPS (R-1 FAC) and cusp/cleft (C-1 FAC).

Table 2 demonstrates that a majority of the traversals belongs to the upper row ($\Delta Be>0$ observations), which means an additional (downward) region 1 FAC in BPS equatorward of the cusp region 1 (upward) FAC. Inside the parenthesis in the lower row is the case when we do not see any BPS signatures and we regard them as the region 2 FAC or a signature of simple return flow. The other cases in the lower row (absence of the region 1 FAC) simply mean that the morning R-1 FAC does not extend there. However, this does not necessarily mean the absence of the afternoon R-1 FAC which flows in the same direction as C-1 FAC. The 27 examples of Table 2 add significant statistics to the Taguchi's result (Figure 2a) on the separation of R-1 FAC and C-1/C-0 FACs (cf. §2.2 and §2.3).

### 4.3. FAC distribution

Since Freja may simultaneously traverse both the cusp and the dayside auroral oval, we can correlate the FAC intensities (total change of $\Delta Be$) in the cusp region (C-1/C-0) and the auroral oval region (R-1). Figure 3a shows the result. Most of these traversals are over the cusp in the morning sector and over the BPS in the afternoon sector because we have more data with IMF $B_Y<0$ than IMF $B_Y>0$ (cf. Table 1).

The figure demonstrates that (1) the FAC intensity in the cusp region varies more than that in the auroral oval region,

(2) the average intensity of both FACs are about the same, and that (3) there is no correlation between these FACs' intensities. Thus the cusp (C-1/C-0) FACs and the auroral oval (R-1) FAC behave independently (cf. §2.1) although their average intensities are similar to each other. The first two points (the intensity deviation and average intensity) are also demonstrated in Figure 3b, a statistics from the Viking meridional traversals. The same results repeat here. An immediate consequence of these results is that for half of the cases the auroral oval FAC is more intense than the cusp FACs during southward IMF.

### 4.4. Meso-scale FACs

We now move to the forth point of section 2: the roles of the meso-scale fluctuations in the large-scale FACs. A Freja example is shown in Figure 4 (orbit #5396). The particle data shows extremely structured BPS-like electrons up to

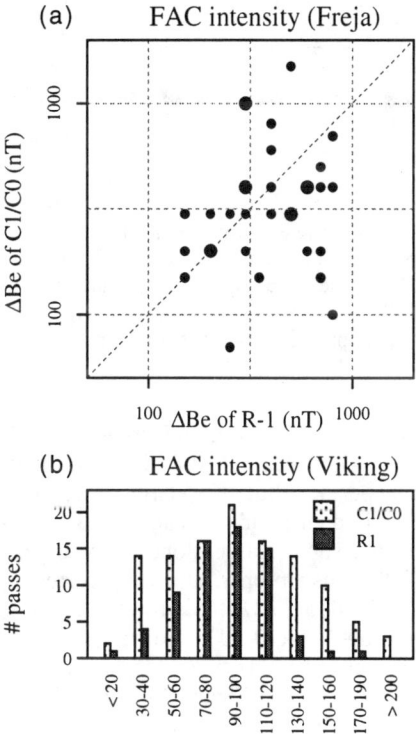

**Figure 3.** (a) Scatter plot of the latitude-integrated FAC intensities (in total change of $\Delta Be$) of simultaneously observed auroral oval region 1 FAC and cusp region 1 FAC. Low inclination of Freja made it possible to observe both FAC in the same traversal. A large dot means two traversals whereas a small dot means one traversal. (b) Distribution of the FAC intensities observed by Viking meridional traversals. The data for the morning LLBL is taken from *Woch et al.*, [1993].

# 250 DOUBLE OPENNESS OF THE DAYSIDE MAGNETOSPHERE

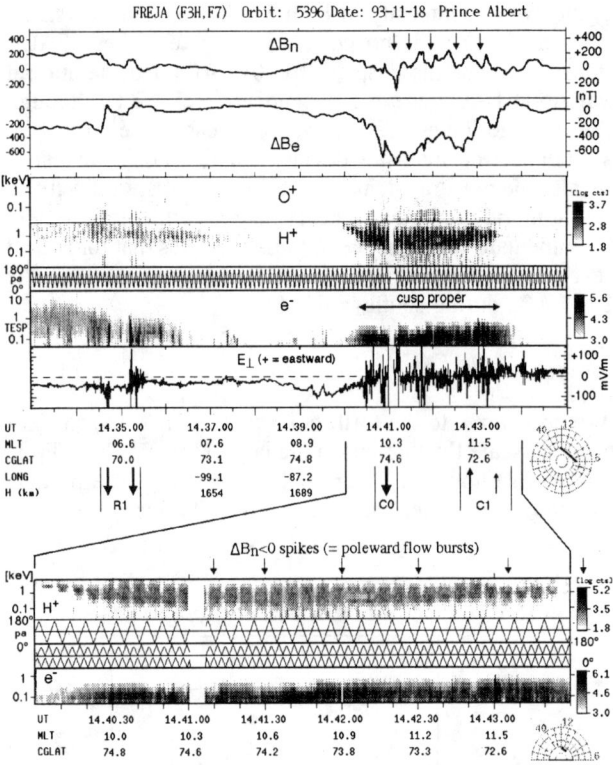

**Figure 4.** Freja particle and magnetic field data for orbit 5396 (1993-11-18) with the same format as Figure 1. The large negative $\Delta Be$ in the cusp (800 nT) indicates extremely strong IMF $B_Y < 0$. The electric field data has non-zero offset due to the satellite attitude problem, but yet we can use its slope and spikes.

1435:20 UT and dense plasma injection of the cusp proper at 1439:50-1443:40 UT. IMP-8 does not measure IMF during this period, but the large $\Delta Be<0$ inside the dawnward-shifted cusp indicates an IMF $B_Y<0$ condition.

The $\Delta Be$ data shows large slopes only in the BPS-like region and the cusp proper, and they correspond to R-1 FAC (1434:40-1435:25 UT; downward) and C-0 and C-1 FACs (1440:00-1441:10 UT; downward, and 1442:40-1443:45 UT; upward). They are topologically consistent with both Figure 2a and Figure 2b, and hence this case is not included in the statistics of Table 2. The $\Delta Bn$ data does not alter this conclusion because the $\Delta Bn$ variation is smaller than the $\Delta Be$ variation except for some meso-scale bipolar signatures inside the cusp. Therefore we must use another clue to answer whether C-0 FAC of Figure 4 is the simple extension of R-1 FAC. We here point out the difference in the meso-scale characteristics.

The auroral oval FAC in the BPS-like region (R-1) is composed of steps of sharp downward FACs associated with the electron bursts at 1434:45 UT (+400 nT jump), at 1435:10 (+50 nT jump) and at 1435:25 UT (+50 nT jump). Past observations also show the same: the majority of the morning R-1 FAC is carried by spiky electron bursts [*Potemra et al.,* 1987; *Yamauchi et al,* 1998]. On the contrary, C-0 FAC at 1440:00-1441:10 UT is scattered and not in the form of the summation of the jumps. From this difference and the current carrier difference, we conclude that C-0 FAC and R-1 FAC are two different FACs as is illustrated in Figure 2a

Let us further examine the meso-scale FACs. Many meso-scale monopolar signatures of $\Delta Bn$ (localized pairs of up and down FACs, e.g., at 1441:10 UT) are found inside the cusp in Figure 4. However, they do not constitute the large-scale FACs contrary to predictions by the existing FTE cusp models (which is essentially the same as Figure 2b). The same result is obtained with Viking data (cf. §2.4). The other obvious feature in Figure 4 is the discontinuity between the cusp proper and the polar cap (cf. §2.5), but we do not discuss further in this paper.

## 4.5. The other important features

We have shown many observations that support the past evidence of independency summarized in section 2. In this subsection we note some overlooked features. One is the meso-scale monopolar $\Delta Bn$ signatures in Figure 4. They can be interpreted as either poleward flow channels (i.e., paired FACs) or large-amplitude standing waves between the ionosphere and the exterior cusp (simple oscillation [e.g., *Maynard et al.,* 1991]). Both causalities predict the same relation of $\mathbf{u} \parallel -\Delta\mathbf{B}$ (where $\mathbf{u}$ is the localized horizontal flow) in the northern hemisphere because a standing wave of $\Delta Bn$ causes a localized convection or clustering of injecting particles, and vice versa. To distinguish this chicken-and-egg problem, we must compare the $\Delta Bn$ dips (poleward flow bursts) and the meso-scale proton injections.

The bottom panels of Figure 4 show an expanded view for the cusp data. No good correlation is found between the $\Delta Bn$ dips (arrows in the figure) and the meso-scale proton injections, suggesting that the monopolar $\Delta Bn$ dips in Figure 4 are probably caused by a standing wave. In fact, the $\Delta Bn$ dip repeats every 30 seconds, supporting the wave scenario. Thus we have at least one example when the $\Delta Bn$ dips inside the cusp are not "caused" by the plasma injection. In general we observe both cases, i.e., well-correlated case and independent case for $\Delta Bn$ dips and meso-scale injections in the cusp, and the former is often seen when the cusp is composed of dispersed multiple injections (not shown here). This phenomenon needs future investigations.

The other feature is the strong east-west deflection of the convection inside the cusp, seen as the large $\Delta Be$ deviation over the $\Delta Bn$ dips (we again use the $\mathbf{u} \parallel -\Delta\mathbf{B}$ relation). The eastward convection exceeds the poleward convection even inside the flow channels ($\Delta Bn<0$ dips) in Figure 4. The strongest poleward convection is instead found at the boundaries of the cusp. This feature is also recognized in

Figure 1. In other words, the most intense proton injection (seen in particle data) seems to block and deflect the background poleward convection. Such stagnancy of the flow at the strongest cusp injection during southward IMF was already found by Viking [*Woch and Lundin, 1992*] and Søndre Strømfjord radar [e.g., *Nilsson et al., 1996*].

## 5. OBSERVATION SUMMARY AND DISCUSSION

Using longitudinal cusp traversals of Freja, we showed supporting evidences for the independency between the cusp (region 1 + region 0) FACs and the auroral oval region 1 FAC outside the cusp: (1) The intensities of these FACs are not correlated at all (Figure 3). (2) We often observe an extra (downward-flowing) region 1 FAC in the BPS region equatorward/dawnward of the (upward-flowing) cusp region 1 FAC in the cusp/cleft region when IMF $B_Y < 0$ (Figure 1 and Table 2). (3) The morning auroral oval region 1 FAC is mostly divided into a few narrow and intense FACs associated with electron bursts, whereas the meso-scale FACs in the cusp region does not form the large-scale cusp FACs (Figure 4).

In addition, we noted the following features in Figure 4: (4) The meso-scale $\Delta Bn$ dips (paired FACs) are probably not caused by the injections but by standing Alfvén waves inside the cusp. (5) The cusp proton injection seems to block and deflect the poleward background convection instead of driving the poleward convection.

One might consider the possibility that the global semi-openness is caused by a viscous-like interaction and the cusp extra-openness is caused by the merging. However, this is not likely. More than half of the cases of Figure 3a show that the FAC due to the global openness is more intense than the FAC due to the cusp local openness during southward IMF. The blocking effect in the cusp (5th point in the summary) also contradicts the merging scenario for the local cusp.

It is not wise to seek the cusp formation mechanism (FAC and particles) within the existing global MHD scheme. One should instead consider a local source for the cusp region 1 and region 0 FACs (and cusp particles). One way for such modelling is to forget MHD and start from the above observational facts. The other way is to add non-MHD effects to the global MHD to see if such a modification allows a local and independent open region in the cusp, which blocks the poleward convection. As mentioned in the introduction, adding one more plasma population (e.g., ionospheric escaping oxygen) to MHD allows us to have a new boundary (e.g., mass loading boundary), and hence the open/closed regions are no longer divided by a single boundary (magnetopause).

## 6. CONCLUSION

The cusp (region 1 and region 0) FACs are disconnected from and independent of the auroral oval region 1 FAC.

The cusp is most likely a local "extra" open region in a semi-open magnetosphere during southward IMF, and the cusp FACs must have its own energy source (dynamo) within the cusp. This makes it hard to describe the cusp by existing global MHD models or drift models starting from global MHD.

*Acknowledgments.* The IMP 8 plasma data are provided by the space research group at MIT, Cambridge. The Freja project is supported by Swedish National Space Board and German Space Agency. Authors thank J. H. Clemmons for helpful suggestions.

## REFERENCES

Burch, J.L., P.H. Reiff, J.D. Menietti, R.A. Heelis, W.B. Hanson, S.D. Shawhan, E.G. Shelley, M. Sugiura, D.R. Weimer, and J.D. Winningham, IMF BY dependent plasma flow and Birkeland currents in the dayside magnetosphere 1. Dynamics Explorer observations, *J. Geophys. Res. 90*, 1577-1593, 1985.

Bythrow, P.F., T.A. Potemra, and R.A. Hoffman, Observations of field-aligned currents, particles, and plasma drift in the polar cusps near solstice, *J. Geophys. Res. 87*, 5131-5139, 1982.

Cowley, S. W. H., J.P. Morelli, and M. Lockwood, Dependence of convective flows and particle precipitation in the high-latitude dayside ionosphere on the X and Y components of the interplanetary magnetic field, *J. Geophys. Res., 96*, 5557-5564, 1991.

Crooker, N.U., Mapping the merging potential from the magnetopause to the ionosphere through the dayside cusp, *J. Geophys. Res., 93*, 7338-7344, 1988.

Erlandson, R.E., L.J. Zanetti, T.A. Potemra, P.F. Bythrow, and R. Lundin, IMF BY dependence of region 1 Birkeland currents near noon, *J. Geophys. Res., 93*, 9804-9814, 1988.

Friis-Christensen, E., and J. Wilhjelm, Polar cap currents for different directions of the interplanetary magnetic field in the Y-Z plane, *J. Geophys. Res., 80*, 1248-1260, 1975.

Fujii, R., T. Iijima, T.A. Potemra, and M. Sugiura, Seasonal dependence of large-scale Birkeland currents, *Geophys. Res. Lett., 8*, 1103-1106, 1981.

Haerendel, G., G. Paschmann, N. Sckopke, H. Rosenbauer, and P.C. Hedgecock, The frontside boundary layer of the magnetosphere and the problem of reconnection, *J. Geophys. Res., 83*, 3195-3216, 1978.

Heikkila, W.J., and J.D. Winningham, Penetration of magnetosheath plasma to low altitudes through the dayside magnetospheric cusps, *J. Geophys. Res. 76*, 883, 1971.

Iijima, T., and T.A. Potemra, The amplitude distribution of field-aligned currents at northern high latitudes observed by Triad, *J. Geophys. Res., 81*, 2165-2174, 1976.

Iijima, T., R. Fujii, T.A. Potemra, and N.A. Saflekos, Field-aligned currents in the south polar cusp and their relationship to the Interplanetary magnetic field, *J. Geophys. Res., 83*, 5595-5603, 1978.

Kallio, E., Mars-solar wind interaction: ion observations and their interpretation, *PhD Thesis, 57pp*, Finnish Meteorological Institute, Helsinki, 1996.

Klumpar, D.M., Relationships between auroral particle distributions and magnetic field perturbations associated with field-aligned currents, *J. Geophys. Res.* 84, 6524-6532, 1979.

Kremser, G., and R. Lundin, Average spatial distributions of energetic particles in the mid-altitude cusp/cleft region observed by Viking, *J. Geophys. Res.*, 95, 5753-5766, 1990.

Krupp, N., E. Keppler, A. Korth, M. Fränz, A. Balogh, and M.K. Dougherty, Three dimensional particle anisotropies in and near the plasma sheet of Jupiter observed by the EPAC-experiment on board the Ulysses spacecraft, *Planet. Space Sci.*, 41, 953-966, 1993.

Lockwood, M., Identifying the open-closed field line boundary, in *Polar Cap Boundary Phenomena*, edited by J. Moens, et al., Kluwer Acad. Pub., 73-90, 1998.

Lundin, R., K. Stasiewicz, and B. Hultqvist, On the interpretation of different flow vectors of different ion species in the magnetospheric boundary layer, *J. Geophys. Res.*, 92, 3214-3222, 1987.

Lundin, R., J. Woch, and M. Yamauchi, The present understanding of the cusp, *ESA SP 330*, 83-95, 1991.

Lundin, R., G. Haerendel, and S. Grahn (eds), The Freja Special Issue, *Space Sci. Rev.*, 70, 505-602, 1994.

Maynard, N.C., T.L. Aggson, E.M. Basinska, W.J. Burke, P. Craven, W.K. Peterson, M. Sugiura, and D.R. Weimerr Magnetospheric boundary dynamics: DE-1 and DE-2 observations near the magnetopause and cusp, *J. Geophys. Res.*, 96, 3505-3522, 1991.

Newell, P.T., and C.-I. Meng, Ionospheric projections of magnetospheric regions under low and high solar wind pressure conditions, *J. Geophys. Res.*, 99, 273-286, 1994.

Nilsson, H., M. Yamauchi, L. Eliasson, O. Norberg, and J. Clemmons, The ionospheric signature of the cusp as seen by incoherent scatter radar, *J. Geophys. Res.*, 101, 10947-10963, 1996.

Ohtani, S., T.A. Potemra, P.T. Newell, L.J. Zanetti, T. Iijima, M. Watanabe, L.G. Blomberg, R.D. Elphinstone, J.S. Murphree, M. Yamauchi, and J. Woch, Four large-scale field-aligned current systems in the dayside high-latitude region, *J. Geophys. Res.*, 100, 137-153, 1995.

Onsager, T.G., and R.C. Elphic, Is magnetic reconnection intrinsically transient or steady state? The Earth's magnetopause as a laboratory, *EOS*, 77, 241, 1996.

Potemra, T.A., L.J. Zanetti, R.E. Erlandson, P.F., Bythrow, G. Gustafsson, M.H. Acuna, and R. Lundin, Observations of large-scale Birkeland currents with Viking, *Geophys. Res. Lett.*, 14, 419-422, 1987.

Potemra, T.A., Sources of large-scale Birkeland currents, in J.A. Holtet and A. Egeland (eds.), *Physical signatures of magnetospheric boundary layer process*, Kluwer Academic Publishers, Dordrecht, pp. 3-27, 1994.

Smith, M.F., and M. Lockwood, The pulsating cusp, *Geophys. Res. Lett.*, 17, 1069-1072, 1990.

Spreiter, J.R., and A.L. Summers, On conditions near the neutral points on the magnetosphere boundary, *Planet. Space Sci.*, 15, 787-798, 1967.

Sugiura, M., A fundamental magnetosphere-ionosphere coupling mode involving field-aligned currents as deduced from DE-2 observations, *Geophys. Res. Lett.* 11, 877-880, 1984.

Taguchi, S., M. Sugiura, J.D. Winningham, and J.A. Slavin, Characterization of the IMF $B_Y$-dependent field-aligned currents in the cleft region based on DE 2 observations, *J. Geophys. Res.*, 98, 1393, 1993.

Wallis, M.K., Weakly shocked flows of the solar wind plasma through atmospheres of comets and planets, *Planet. Space Sci.*, 21, 1647-1660, 1973.

Wilhjelm, J., E. Friis-Christensen, and T.A. Potemra, The relationship between ionospheric and field-aligned currents in the dayside cusp, *J. Geophys. Res.*, 83, 5586-5594, 1978.

Winningham, D.J., F. Yasuhara, S.-I. Akasofu, and W.J. Heikkila, The latitudinal morphology of 10 eV to 10 keV electron fluxes during quiet and disturbed times in the 2100-0300 MLT sector, *J. Geophys. Res.*, 80, 3148, 1975.

Woch, J., and R. Lundin, Magnetosheath plasma precipitation in the polar cusp and its control by the interplanetary magnetic field, *J. Geophys. Res.*, 97, 1421-1430, 1992.

Woch, J. M. Yamauchi, R. Lundin, T.A. Potemra, and L.J. Zanetti, The low-latitude boundary layer at mid-altitudes: Relation to large-scale Birkeland currents, *Geophys. Res. Lett.*, 20, 2251-2254, 1993.

Yamauchi, M., and T. Araki, The interplanetary magnetic field $B_Y$-dependent field-aligned current in the dayside polar cap under quiet conditions, *J. Geophys. Res.*, 94, 2684-2690, 1989.

Yamauchi, M., and R. Lundin, Classification of large-scale and meso-scale ion dispersion patterns observed by Viking over the cusp-mantle region, in *Physical Signatures of Magnetospheric Boundary Layer Processes*, edited by J. A. Holtet, and A. Egeland, pp. 99-109, Kluwer academic press, Dordrecht, 1994.

Yamauchi, M., and L. Blomberg, Problems on mappings of the convection and on the fluid concept, *Phys. Chem. Earth*, 22, 709-714, 1997.

Yamauchi, M., and R. Lundin, The wave-assisted cusp model: Comparison to low-altitude observations, *Phys. Chem. Earth*, 22, 729-734, 1997.

Yamauchi, M., R. Lundin, and J. Woch, The interplanetary magnetic field $B_Y$ effects on large-scale field-aligned currents near local noon: Contributions from cusp part and noncusp part, *J. Geophys. Res.*, 98, 5761-5767, 1993.

Yamauchi, M., H. Nilsson, L. Eliasson, O. Norberg, M. Boehm, J.H. Clemmons, R.P. Lepping, L. Blomberg, S. Ohtani, T. Yamamoto, T. Mukai, T. Terasawa, and S. Kokubun, Dynamic response of the cusp morphology to the solar wind: A case study during passage of the solar wind plasma cloud on February 21, 1994, *J. Geophys. Res.*, 101, 24675-24687, 1996.

Yamauchi, M., R. Lundin, L. Eliasson, S. Ohtani, and J.H. Clemmons, Relationship between large-, meso, and small-scale field-aligned currents and their current carriers, in *Polar Cap Boundary Phenomena*, edited by J. Moens, et al., Kluwer Acad. Pub., 173-188, 1998.

---

M. Yamauchi, R. Lundin, and L. Eliasson, Swedish Institute of Space Physics, Box 812, SE-981 28 Kiruna, Sweden.

S. Ohtani, The Johns Hopkins University, Applied Physics Laboratory, Johns Hopkins Road, Laurel, MD 20723-6099, USA.

P.-A. Lindqvist, Royal Institute of Technology, Alfvén Laboratory, SE-100 44 Stockholm, Sweden.

R.P. Lepping, NASA/Goddard Space Flight Center, Greenbelt, MD 20771, USA.

# Disappearance of Large-scale Field-aligned Current Systems: Implications for the Solar Wind–Magnetosphere Coupling

S. Ohtani

*The Johns Hopkins University Applied Physics Laboratory, Laurel, Maryland, U.S.A.*

T. Higuchi

*The Institute of Statistical Mathematics, Tokyo, Japan*

T. Sotirelis and P. T. Newell

*The Johns Hopkins University Applied Physics Laboratory, Laurel, Maryland, U.S.A.*

Large-scale field-aligned current (FAC) systems occasionally disappear in the midday sector (9 < MLT < 13), even though this is the local time sector where FACs tend to be most intense. The present study investigates 26 such events observed by the DMSP-F7 satellite. It is found that the events tend to occur in the winter hemisphere in a certain UT range when the dipole axis was inclined most antisunward and therefore the contribution of solar illumination to the ionospheric conductivity is minimum. Before most events, IMF $B_z$ was positive, and the solar wind bulk flow momentum was below the average. Although one might expect to see such events when the sign of $B_x$ was unfavorable for merging at the high-latitude tail magnetopause in the winter hemisphere, no clear preference for IMF $B_x$ was found. It was also found that the asymmetry of the ionospheric conductivity between the summer and winter hemispheres suppresses the IMF-dependent characteristics of particle precipitation in the polar cap. These results suggest that ionospheric conductivity not only determines the intensity of FACs for the convection electric field imposed by the solar wind–magnetosphere interaction, but also affects how the magnetosphere interacts with the solar wind.

## 1. INTRODUCTION

Dayside large-scale field-aligned currents (FACs) are classified into three systems, that is, region 2 (R2), region 1 (R1), and region 0 (R0) systems from equatorward to poleward [*Iijima and Potemra*, 1976]. Good correlation between the intensity of R1 and R0 currents and solar wind parameters [*Iijima and Potemra*, 1982] strongly suggests that these two current systems are generated by interaction between the solar wind and the magnetosphere. The ionospheric distribution of R1 and R0 FACs depends on the IMF orientation, and these FACs are collocated with precipitation of magnetosheath particles (see review article by *Potemra* [1994]). There is little doubt that features related to these current systems can be used for a diagnosis of the solar wind–magnetosphere (SW–MS) interaction.

The ionospheric condition is another important factor for the formation of FAC systems, which is demonstrated by strong dependence of the intensity of R1 FACs on the solar zenith angle [*Fujii and Iijima*, 1987]. This dependence suggests that the generation mechanism of FACs is a voltage source, and therefore the current intensity is proportional to ionospheric conductivity. Although this idea is perhaps correct in the first approximation, it is based on the (implicit) assumption that the generation mechanism itself is not affected by the ionospheric condition. The question of whether or not this is truly the case has rarely been addressed observationally.

The issue can be addressed using the following simple considerations. The electric field **E** imposed to the polar cap by the SW–MS interaction should depend on $-\mathbf{v} \times \mathbf{B}$ where **v** and **B** are the magnetosheath flow velocity and the magnetic field in the interaction region. Energy transfer from the solar wind to the magnetosphere is accomplished through the deceleration of the magnetosheath flow, which is attributed to the Lorentz force, $\mathbf{J} \times \mathbf{B}$; **J** is the current density in the interaction region. Thus, we infer that **E** depends on **J**, which is then controlled by ionospheric conductivity $\sigma$, provided that the relevant FAC is closed in the interaction region. If this is the case, the role of ionospheric conductivity is more than a parameter that determines the intensity of FACs for given **E**. Is there any observation that can test this idea?

In a separate study [*Higuchi and Ohtani*, 1999] we developed an automatic procedure to identify spatial structures of FACs from satellite magnetometer data and applied it to the entire set of the DMSP-F7 magnetometer data. An unexpected result of that study was that the satellite occasionally observed no large-scale FAC signature in the midday sector. As is shown later, the occurrence of such extraordinary events strongly depends on ionospheric conductivity. By comparing particle and magnetic field signatures between the winter and summer hemispheres and examining solar wind conditions, the present study addresses the role of the ionosphere in the SW–MS interaction.

## 2. DATA ANALYSIS

We used 1-s magnetometer and particle precipitation data from the DMSP-F7 satellite. DMSP-F7 is a Sun-synchronous satellite with a nearly circular polar orbit at about 835 km altitude, with its ascending and descending nodes at 1030 and 2230 local time (LT), respectively. The orbital period is about 101 min.

We applied the procedure to identify FAC structures [*Higuchi and Ohtani*, 1999] to the entire set of DMSP-F7 magnetometer data acquired from December 1983 to January 1988 (1339 days). The procedure fits line segments to magnetic variations and recognizes a line segment as a FAC sheet if its amplitude is more than 50 nT (47 mA/m if projected to 110 km in altitude). No FAC was detected for 7,956 dayside orbits. It is possible, however, that for skimming orbits, the satellite did not reach a latitude high enough to detect large-scale FACs. Based on orbits for which large-scale FAC systems were detected, we examined the locations of the equatorward-most points of large-scale FACs in the range 09 < MLT < 13 and determined the latitude of the 95 percentile from equatorward as a function of MLT. For most of the remaining 5 percent of events, we found that the boundary was misidentified due to apparent artificial signals. As a result, the magnetic latitude of the 95-percentile point is higher than 75° at MLT = 9 and is as high as 79° at MLT = 12. Orbits that did not reach that latitude were removed from the present analysis.

Focusing on 185 events for which the IMP-8 satellite measured the solar wind at $x > +10\ R_E$ and $\sqrt{(y^2 + z^2)} < 30\ R_E$, we visually examined a plot of the east–west magnetic component. If there was a signature that could be associated with large-scale FACs (e.g., a triangle-shaped variation), however small the amplitude, we excluded such an orbit. The final event list contains 26 events.

One may suspect that the satellite did not observe any large-scale FAC signature in these events because the satellite trajectory happened to thread through the demarcation between prenoon and postnoon FAC systems. However, we found consistent dependence of the event occurrence on the dipole tilt angle, which excludes such possibility. Four out of these 26 events occurred in the northern hemisphere between November and January. The remaining 22 events were observed in the southern hemisphere and were also concentrated in winter months: 20 events during May to August and one event in each of March and April. Furthermore, 16 (2) out of 22 (4) events in the southern (northern) hemisphere took place at 12–15 (07–08) UT, when the dipole axis was further inclined antisunward in the corresponding hemisphere. The geographic latitude and longitude of the southern and northern magnetic poles are (–74.2°, 126.1°) and (81.0°, –81.7°), respectively, in PACE coordinates based on the IGRF 85 model (Wing, private communication, 1999). Considering that the southern magnetic pole is more offset from the geographic pole than the northern magnetic pole, the preference of the event occurrence for the southern hemisphere can also be explained in terms of the dipole tilt. We conclude that ionospheric conductivity was extremely asymmetric between the northern and southern hemispheres when the events took place.

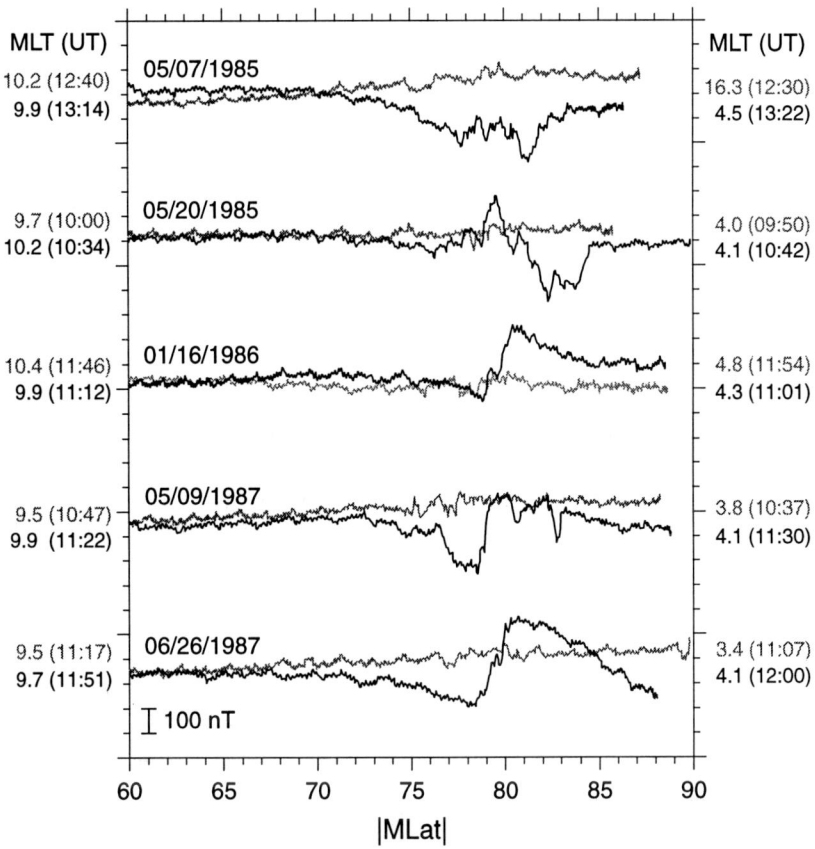

**Figure 1.** The azimuthal magnetic field component plotted against magnetic latitude for the winter (gray lines) and summer (black lines) hemispheres for five representative events.

Figure 1 plots the azimuthal magnetic field (cross-track) component, $B_z$, (positive eastward) against |MLat|, the absolute value of magnetic latitude, for five examples. The gray lines represent data from the winter hemisphere, whereas the black lines represent data from the closest pass of the summer hemisphere. Superposed high-frequency variations and quasi-periodic undulations are common with the DMSP-F7 magnetometer data and are presumably related to the satellite operation; the instrument is mounted on the satellite body. However, the amplitude of these artificial signals is a few tens of nanotesla at most, smaller than the typical amplitude of large-scale FAC signatures by an order of magnitude.

No large-scale magnetic variation can be seen for the winter passes. In contrast, for the summer hemisphere, a three-sheet structure is clear for the events of May 20, 1985; January 16, 1986; and June 26, 1987; and at least two large-scale FAC sheets can be identified for the events of May 7, 1985, and May 9, 1987. The present result interestingly suggests that the intensity of region 2 currents is also strongly controlled by the ionospheric condition like other FAC systems, although it is not a common idea that the region 2 system is driven by a voltage source.

Because the occurrence of events is rather rare, it is expected that there are some external conditions for the disappearance of large-scale FACs. Figure 2 examines the IMF orientation and the solar wind bulk flow momentum $nV$ measured by IMP 8 before the events. Both solar wind plasma and IMF data were averaged over 30 min before the events, and the travel time from the satellite position to the subsolar point was taken into account. Solar wind plasma data were not available for four events, for which the values of $B_z$ are plotted on the vertical axis. All IMF components are given in GSM coordinates, and in Figure 2b both vertical and horizontal scales are reversed for the events that were observed in the northern hemisphere (NH), which are represented by the open circles.

Three points are clear. First, the distribution of $B_z$ is biased positively; there are no events with $B_z < -1$ nT. Second, the values of $nV$ are distributed mostly below its median (the vertical line in Figure 2a). Finally, there is no clear preference for the IMF sector structure, that is, the

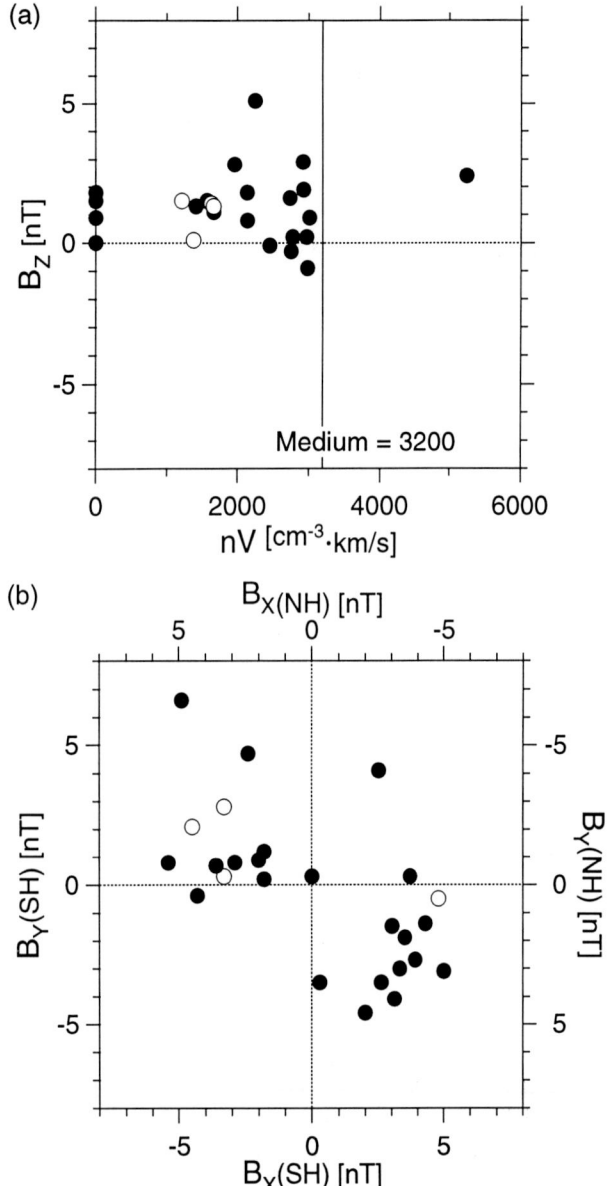

**Figure 2.** (a) IMF $B_z$ vs. $nV$ and (b) IMF $B_y$ vs. $B_x$ before events. For events observed in the northern hemisphere (NH, open circles) both vertical and horizontal scales are inverted in Figure 2b.

and magnetospheric field lines takes place at the tail magnetopause antisunward of the dayside cusp [*Maezawa*, 1976; *Crooker*, 1979], where magnetospheric field lines are open to the solar wind even before merging with the IMF. In fact, the ionospheric distribution of FACs is well explained in terms of this hypothesis [*Potemra et al.*, 1984; *Burch et al.*, 1985; *Cowley et al.*, 1991]. Therefore, the associated FAC is inferred to have separate sources in the winter and summer hemispheres. In contrast, for southward IMF $B_z$, the current circuit driven by the SW–MS coupling may be envisioned as a single source at the dayside magnetopause wired to two resistivities in the winter and summer ionospheres. For such a circuit, it would be more difficult to examine the ionospheric feedback to the source mechanism.

When IMF $B_z$ is positive, positive IMF $B_x$ is more favorable for merging in the southern hemisphere than for merging in the northern hemisphere, and vice versa for negative IMF $B_x$. Thus it is expected that the events tend to take place when IMF $B_x$ is negative and positive for the southern and northern hemispheres, respectively. The fact that there is no clear preference for the IMF sector structure suggests that the ionospheric condition is the dominant factor for the occurrence of events and the IMF orientation plays, if at all, a secondary role.

Let us examine the winter–summer asymmetry in terms of particle precipitation. An event selected for this purpose took place on January 16, 1986, one of the five events shown in Figure 1. The northern hemisphere was in winter for this event. Figure 3 plots the IMF data acquired by IMP-8 at (25–27, 16–17, 15) $R_E$ for 0900–1200 UT. Although fluctuations are noticeable for each component, the IMF continued to point approximately in the negative $X$ direction throughout the interval. For this orientation, the tail magnetopause in the northern, therefore winter, hemisphere is the preferable site for the merging.

Plate 1a shows two DMSP-F7 trajectories, one in the southern summer (S) hemisphere and the other in the northern winter (N) hemisphere, along with a DMSP-F6 trajectory in the northern winter hemisphere for this event. Plate 1c–1d show the DMSP-F7 measurements of the east–west magnetic component (top panel) and particle precipitation (color-coded panels) in the southern and northern dayside high-latitude regions, respectively. Note that the satellite orbited from poleward to equatorward in the southern hemisphere and vice versa in the northern hemisphere. No noticeable large-scale magnetic variation was observed in the winter hemisphere, whereas three large-scale FAC systems were observed in the summer hemisphere, which are presumably R2, R1, and R0 currents.

sign of IMF $B_x$ or $B_y$. The first and second points are important not only because these tendencies are favorable for weak FACs [*Iijima and Potemra*, 1982] but also because under such conditions, the dayside cusp tends to be at higher magnetic latitudes, which is favorable for lower ionospheric conductivity.

The hypothesis of the antiparallel merging predicts that when the IMF is northward, the merging between the IMF

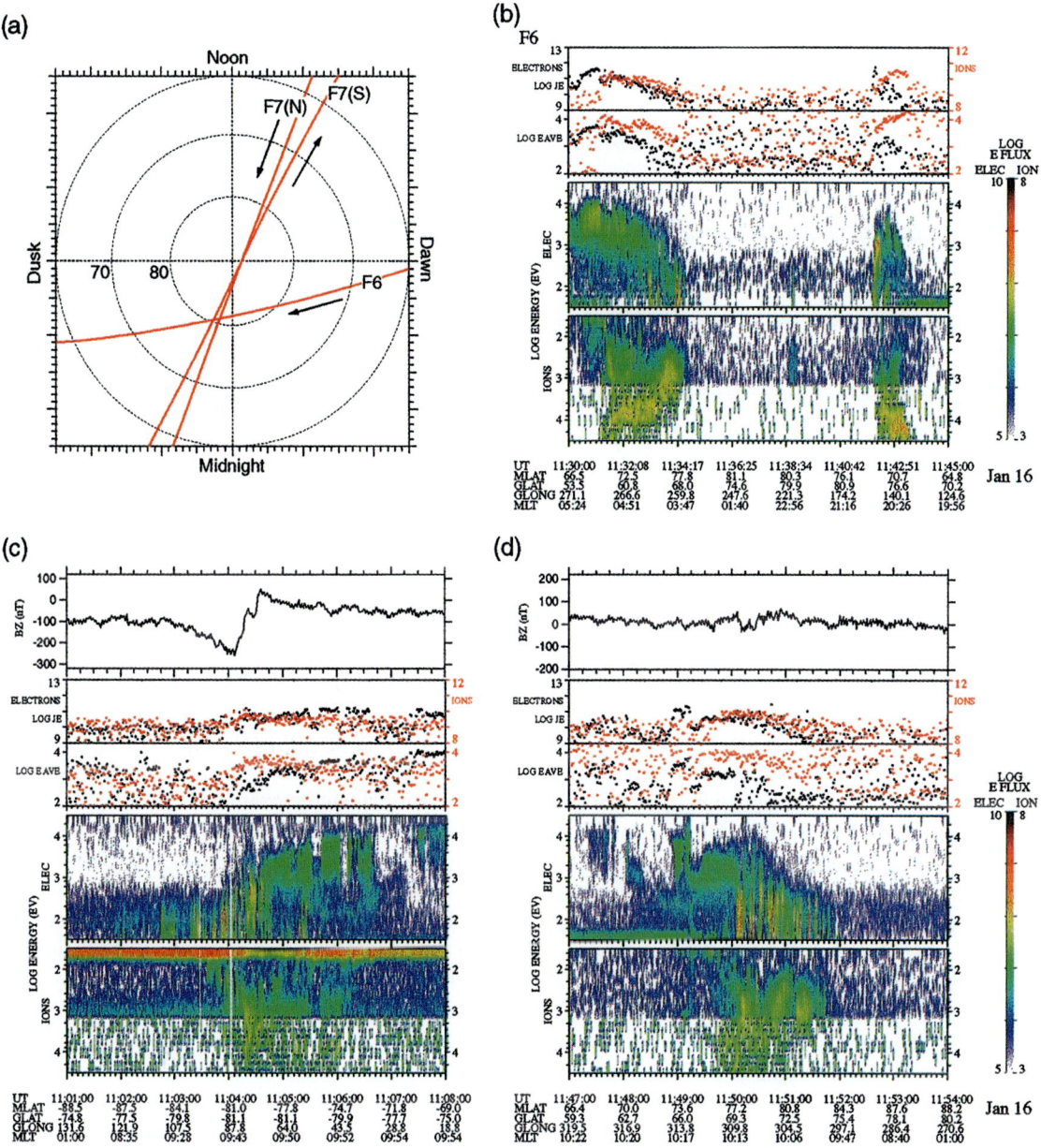

**Plate 1.** (a) DMSP-F6 and –F7 trajectories in the polar diagram, (b) DMSP-F6 particle precipitation data in the northern (winter) hemisphere, DMSPS-F7 azimuthal magnetic component ($B_z$) and particle precipitation data in the (c) southern (summer) and (d) northern (winter) hemisphere. Shown for DMSP particle measurements are energy fluxes, $J_E$ (eV/cm² s sr) and the average energies $E$ AVE (eV) for ions (red) and electron (black), and electron and ion differential energy fluxes (eV/cm² s sr eV). The energy scale is inverted for the ion $E$–$t$ diagram.

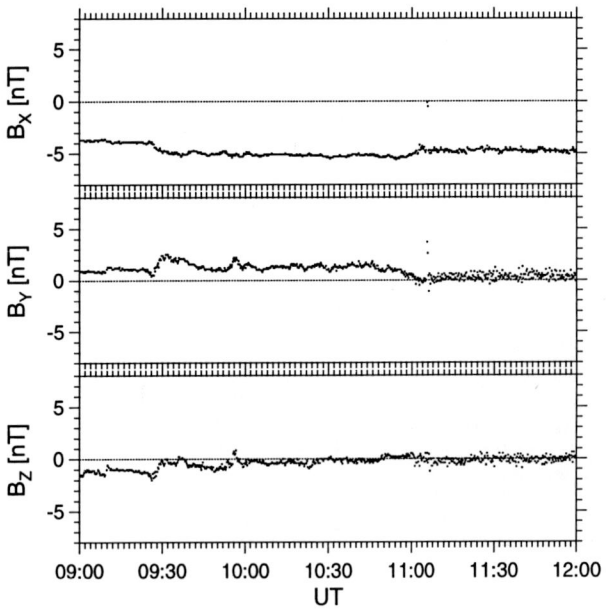

**Figure 3.** Three IMF components in GSM meassured by IMP 8 at (25–27, 16–17, 15) $R_E$ for the January 16, 1986, event.

At first glance, precipitation signatures look similar between the summer and winter hemispheres. However, there is a noticeable difference in features at the poleward boundary. In the winter (Plate 1d) hemisphere, the termination of low-energy (<1 keV) ion and electron precipitation was coincident. In contrast, in the summer (Plate 1c) hemisphere, bursty electron precipitation extended farther poleward of the termination of ion precipitation. The previous study [*Veselovsky et al.*, 1995] shows that such bursty electron precipitation is better described by the total flux enhancement than by field-aligned acceleration that can be attributed to the magnetosphere–ionosphere coupling; though, this may depend on flux intensity [Figure 7 of *Shinohara and Kokubun*, 1996]. It was also reported that the IMF dependence and other statistical characteristics of such bursty electron precipitation not accompanied by ion precipitation are well explained in terms of the merging process at the high-latitude tail magnetopause [*Shinohara and Kokubun*, 1996]. However, the observed asymmetry is just opposite to what is expected from negative IMF $B_X$.

Plate 1b shows particle precipitation data from DMSP-F6, which crossed the northern (winter) polar cap during an interval between the summer and winter DMSP-F7 passes; unfortunately, the satellite only skimmed the dayside oval in the opposite hemisphere. No magnetic field data are available from this satellite. The polar cap is void of precipitation except for the isolated low-energy ion and electron precipitation observed around 1139 UT, although this is the hemisphere favorable for observing polar rain for negative IMF $B_X$ [*Fairfield and Scudder*, 1985].

DMSP-F7 particle data were available for 20 out of the 26 events, and we visually examined the E–t diagram for both winter and summer orbits for each event. It is sometimes difficult to uniquely determine the poleward boundary of precipitation. Nevertheless the coincident termination of ion and electron precipitation similar to or even sharper than the January 16, 1986, event, was observed for six other events. All of them took place in the southern winter hemisphere. For three out of the six events, IMF $B_X$ was positive, which is favorable for the merging in the southern winter hemisphere. For those three events, the satellite observed in the northern summer hemisphere, not in the southern winter hemisphere, bursty electron precipitation to extend farther poleward of the termination of ion precipitation. That is, the winter–summer asymmetry was just opposite to what is expected from the sign of IMF $B_X$, as was the case for the January 16, 1986, event. IMF $B_X$ was negative for another event and was not clear for the other two events. The result suggests that the ionospheric conductivity controls particle precipitation in the open field line region, possibly overturning the preference expected from the IMF orientation. We also found that the energy flux associated with bursty electron precipitation tends to be higher in the summer hemisphere than in the winter hemisphere (not shown).

## 3. DISCUSSION AND SUMMARY

The present study examined 26 events in which no large-scale FAC signature was observed by DMSP-F7 in the dayside high-latitude region. The events took place in the winter hemisphere in the UT range when the dipole axis was inclined most antisunward and therefore the ionospheric conductivity due to solar illumination was extremely low. In contrast, large-scale FAC systems were clearly observed in the summer hemisphere. The events provide the most ideal opportunity to address the role of the ionosphere in the SW–MS interaction from a winter–summer comparison.

Most of events took place when IMF $B_Z$ was positive and the solar wind bulk momentum was below the average, suggesting that the energy and momentum transfer from the solar wind to the magnetosphere was small. For northward IMF $B_Z$, it is inferred that field line merging takes place at the tail magnetopause tailward of the dayside cusp, which allows us to assume separate sources of FACs for different hemispheres. Although it is expected that large-scale FACs disappear when the IMF orientation is not favorable for the merging in that hemisphere, no clear preference for the sign of IMF $B_X$ was found for the event occurrence, indicating that the low ionospheric conductivity is the dominant control factor.

No clear preference for the IMF orientation was found for particle precipitation, either. Instead, the most clear tendency for the examined events is that the bursty electron precipitation in the polar cap is observed farther poleward and with higher energy fluxes in the summer hemisphere than in the winter hemisphere. The result suggests that the ionospheric conductivity is the dominant factor and the preference for the IMF orientation, which is generally the most important parameter for the merging, appears to be suppressed by the extraordinary interhemispheric asymmetry of the ionospheric conductivity.

The role that the ionosphere plays in the magnetospheric electrodynamics may have been underestimated. The results of the present study suggest that the ionospheric conductivity even affects the way the magnetosphere interacts with the solar wind. It is possible, however, that such an active role of the ionosphere is noticeable only when the interhemispheric asymmetry of conductivity is extraordinary and there are separate current sources for different hemispheres, as inferred to be the case for the examined events. Further investigation is required for addressing the mechanism of the ionospheric feedback to the SW–MS interaction.

*Acknowledgments.* The author (S.O.) thank J. M. Ruohoniemi for fruitful discussions. The DMSP-F7 particle data were provided by C.-I. Meng and D. A. Hardy. The DMSP-F7 magnetometer data were provided by F. J. Rich. The IMP 8 magnetometer data were provided by R. P. Lepping and the National Space Science Data Center through the World Data Center-A for Rockets and Satellites. The IMP 8 solar wind data were provided by the MIT Space Plasma Physics Group. Work at APL was supported by NASA, NSF, and the Office of Naval Research.

## REFERENCES

Burch, J. L., P. H. Reiff, J. D. Menietti, R. A. Heelis, W. B. Hanson, S. D. Shawhan, E. G. Shelley, M. Sugiura, D. R. Weimer, and J. D. Winningham, IMF By-dependent plasma flow and Birkeland currents in the dayside magnetosphere, 1, Dynamics Explorer observations, *J. Geophys. Res.*, 90, 1577, 1985.

Cowley, S. W. H., J. P. Morelli, and M. Lockwood, Dependence of convective flows and particle precipitation in the high-latitude dayside ionosphere on the X and Y components of the interplanetary magnetic field, *J. Geophys. Res.*, 96, 5557, 1991.

Crooker, N. U., Dayside merging and cusp geometry, *J. Geophys. Res.*, 84, 951, 1979.

Fairfield, D. H., and J. D. Scudder, Polar rain: Solar coronal electrons in the Earth's magnetosphere, *J. Geophys. Res.*, 90, 4055, 1985.

Fujii, R., and T. Iijima, Control of the ionospheric conductivities on large-scale Birkeland current intensities under geomagnetic quiet conditions, *J. Geophys. Res.*, 92, 4505, 1987.

Higuchi, T., and S. Ohtani, Automatic identification of large-scale field-aligned current structures and its application to night-side current systems, this volume, 1999.

Iijima, T., and T. A. Potemra, Field-aligned currents in the dayside cusp observed by Triad, *J. Geophys. Res.*, 81, 5971, 1976.

Iijima, T., and T. A. Potemra, The relationship between interplanetary quantities and Birkeland current densities, *Geophys. Res. Lett.*, 9, 442, 1982.

Maezawa, K., Magnetospheric convection induced by the positive and negative z components of the interplanetary magnetic field: Quantitative analysis using polar cap magnetic records, *J. Geophys. Res.*, 81, 2289, 1976.

Potemra, T. A., Sources of large-scale Birkeland currents, in *Physical Signatures of Magnetospheric Boundary Layer Processes*, ed. by J. A. Holtet and A. Egeland, pp. 3–27, Kluwer Academic Publishers, Netherlands, 1994.

Potemra, T. A., L. J. Zanetti, P. F. Bythrow, A. T. Y. Lui, and T. Iijima, $B_Y$-dependent convection patterns during northward interplanetary magnetic field, *J. Geophys. Res.*, 89, 9753, 1984.

Shinohara, I., and S. Kokubun, Statistical properties of particle precipitation in the polar cap during intervals of northward interplanetary magnetic field, *J. Geophys. Res.*, 101, 69, 1996.

Veselovsky, I. S., P. T. Newell, and A. T. Y. Lui, Pervasive small-scale enhancements in mantle and polar rain precipitation, *Geophys. Res. Lett.*, 22, 3263, 1995.

---

T. Higuchi, The Institute of Statistical Mathematics, Tokyo 106, Japan.

P. T. Newell, S. Ohtani, and T. Sotirelis, The Johns Hopkins University Applied Physics Laboratory, 11100 Johns Hopkins Rd., Laurel, MD 20723-6099.

# Storm-Time Energetic Particle Penetration Into the Inner Magnetosphere as the Electromotive Force in the Subauroral Ion Drift Current Circuit

J. De Keyser

*Belgian Institute for Space Aeronomy, Brussels, Belgium*

Subauroral ion drift layers (SAID) with strong westward ion flows are frequently observed in the ionosphere during substorms. They are interpreted as the ionospheric footprint of a current sheet that separates the cold magnetospheric plasma from the energetic particles penetrating into the inner magnetosphere during the substorm. We present numerical simulations of this current sheet and the corresponding SAID based on observed injected particle energy spectra. We find an SAID signature that is not very sensitive to the precise form of the spectra.

## INTRODUCTION

Sub-auroral ion drifts (SAID) are observed mainly in the F region at latitudes ranging from 50° to 70° [e.g., *Galperin et al.*, 1973; *Smiddy et al.*, 1977; *Spiro et al.*, 1979]. They are narrow (~1° latitude) and show westward drifts typically in excess of 1 km/s. SAID are observed from about half an hour after substorm onset and last up to a few hours. SAID appear in the evening to midnight local time sector, at a latitude that is close to the plasmapause projection and that varies with it as $K_p$ changes [*Karlsson et al.*, 1998]. The ionospheric processes in SAID are well understood [*Anderson et al.*, 1991, 1993] but the magnetospheric driver remains unclear. One explanation is based on the energy-dependent separation of the guiding-center trajectories of ions and electrons as they drift inward from the tail, which produces a space charge layer and the associated shielding currents [e.g., *Southwood and Wolf*, 1978; *Harel et al.*, 1981a, b; *Spiro et al.*, 1981].

*De Keyser et al.* [1998] and *De Keyser* [1999] recently proposed another driving mechanism. They regard SAID as the ionospheric footprint of an inward moving magnetospheric current sheet that interfaces hot injected particles and the cold magnetospheric plasma. They use a finite gyroradius model (henceforth called FGM) to compute the electric field across this interface. Figure 1 sketches the current system. The electromotive force is due to the differential motion of the injected particles and the (partially corotating) cold plasma in the terrestrial magnetic field, leading to a potential difference $\Delta\Phi$ across the interface of a few to tens of kiloVolts; strong radial electric fields are produced in the thin interface. The conducting magnetic field lines project this potential difference onto the ionosphere. Downward field-aligned currents flow equatorward of the SAID and upward currents flow on the poleward side. The SAID acts like a resistive load as heat is dissipated by ion-neutral collisions. The ionospheric magnetic field $B_{iono}$ and the poleward electric field $E_{iono}$ result in a westward drift $V_{iono} = E_{iono} \times B_{iono}/B_{iono}^2$ in both hemispheres. The power deposited in the ionosphere is typically 1 kiloWatt per meter along the westward drift band [*Smiddy et al.*, 1977].

The purpose of the present paper is to examine the relationship between the properties of the injected particles and the observed characteristics of SAID. In particular, we present a detailed SAID simulation using observed injected particle energy spectra, thereby confirming the conclusions of the FGM model.

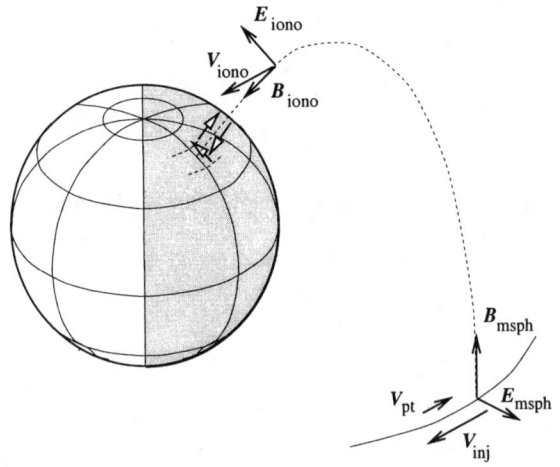

**Figure 1.** Geometry of the ionospheric magnetic and electric fields $B_{iono}$ and $E_{iono}$, and the drift $V_{iono}$. Westward drift corresponds to poleward electric field. The arrows indicate the currents closing through the SAID. The SAID is connected by field lines to a current sheet in the equatorial magnetosphere. The electromotive force is generated by the difference between the plasmatrough and injected plasma velocities $V_{pt}$ and $V_{inj}$.

## INJECTED PARTICLE PROPERTIES

Substorms are associated with the penetration or injection of energetic particles into the inner magnetosphere [e.g., *Ejiri et al.*, 1980]. The "substorm injection boundary" concept introduced by *McIlwain* [1974] describes the time-evolving injected plasma front. This front can be identified with the current sheet considered in the FGM.

*Ejiri et al.* [1980, Figure 11] present energy spectra observed during a substorm with particle penetration up to the plasmapause ($L = 3.5 - 4$). This inspired us to repeat the FGM calculations with these observed spectra rather than Maxwellian populations. The ion and electron spectra were each approximated by a sum of two Maxwellians. We added a third component to mimic the high energy tails seen in plasmasheet spectra. Bearing in mind plasma neutrality, we adopt electron populations with densities of 3.8, 0.1995, and 0.0005 cm$^{-3}$, and energies of 0.2, 3, and 15 keV, respectively, and proton distributions with densities of 2, 1.95, and 0.05 cm$^{-3}$ and energies of 0.3, 5, and 15 keV. Figure 2 shows the corresponding fluxes. These spectra are consistent with a ring current or plasmasheet origin of the particles.

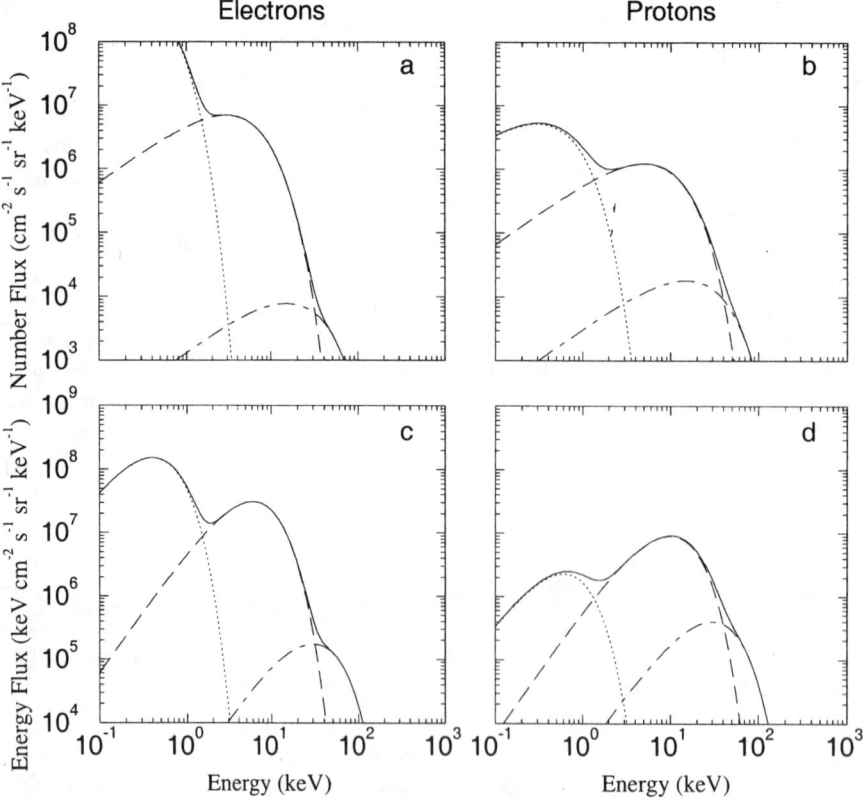

**Figure 2.** Injected proton and electron fluxes. (a) Electron number flux. (b) Proton number flux. (c) Electron energy flux. (d) Proton energy flux.

The energy fluxes of Figures 2c and 2d quantitatively resemble the precipitation spectra observed poleward of SAID [*Shiokawa et al.*, 1997, Plate 1] with a significant ion energy flux above about 1 keV and an important electron flux in the broad 0.1 – 10 keV range. The equatorward electron precipitation boundary is never located equatorward of the SAID, which the FGM explains in terms of the different traveling times of Alfvén waves and energetic particles along field lines [*De Keyser*, 1999].

For SAID on the $L = 4$ shell a width $\Delta\lambda = 1°$ corresponds to a thickness of 1600 km in the equatorial magnetosphere. Since the FGM current sheet thickness is a few to a few hundred injected ion gyroradii $\rho^+_{\rm inj}$, the particles responsible for the formation of the current sheet must have energies of 1 keV to 1 MeV ($\rho^+_{\rm inj} = 10 - 300$ km) which agrees with the observed spectra.

## SIMULATION

Finite gyroradius effects are responsible for charge separation in the inward moving interface layer. Field-aligned currents flow as a consequence of the thermo-electric and convection electric fields, and try to neutralize the charge separation on a time scale of $10^2 - 10^3$ s [*De Keyser*, 1999]. The charges, however, are continuously replenished as plasma flows toward the current sheet from both sides (as seen in a frame co-moving with the sheet). The current sheet therefore is in a state of dynamic equilibrium. Ignoring the effects of discharge and replenishment (which cancel each other), the current sheet can be regarded as a tangential discontinuity (TD). We assume that this TD is locally planar, with its normal in the radial direction ($x$ direction). A particle (mass $m$, charge $Ze$, velocity $v$) is fully characterized in such a configuration by its invariants: energy $H = mv^2/2 + Ze\phi$ and canonical momenta $p_{y,z} = mv_{y,z} + Zea_{y,z}$ ($\phi$ and $a_{y,z}$ are the electric and magnetic potentials). If the magnetic field is unidirectional ($\boldsymbol{B} \parallel \mathbf{1}_z$) and perpendicular to the mean velocities ($\boldsymbol{V} \parallel \mathbf{1}_y$) $p_z$ plays no role. We choose velocity distribution functions (VDFs)

$$f(H, p_y) = N\left(\frac{m}{2\pi k_B T}\right)^{3/2} e^{-(H + mV^2/2 - p_y V_y)/k_B T}$$
$$\times \frac{1}{2} \operatorname{erfc} \frac{\pm p_y}{ZeB_0\sqrt{\mathcal{L}^2 - \rho^2}},$$

that is, a shifted Maxwellian times a smooth transition with length scale $\mathcal{L}$, where the sign selects the halfspace to the left or right of the interface centered at $x = 0$; $B_0$ is the mean ambient field and $\rho$ is the gyroradius of a thermal particle in this mean field. VDFs in terms of invariants automatically satisfy the Vlasov equations. Densities and currents are

$$n = \frac{N}{2} e^{-Ze(\phi - a_y V_y)/k_B T} \operatorname{erfc} \frac{\pm a_y}{B_0 \mathcal{L}},$$
$$j_y = ZenV_y \mp \frac{|Z|eN}{\mathcal{L}/\rho}\sqrt{\frac{k_B T}{2\pi m}} e^{-Ze\phi/k_B T - a_y^2/B_0^2 \mathcal{L}^2}.$$

The magnetic field $B_z = da_y/dx$ follows from Ampère's Law $d^2 a_y/dx^2 + \mu_0 j_y = 0$. The electric field is obtained from the quasi-neutrality condition $d(\sum_i Z_i n_i)/dx = \sum_i Z_i(B_z dn_i/da_y - E_x dn_i/d\phi) = 0$, where the sum is over all populations [*Roth et al.*, 1996].

We model a plasma front that has reached the plasmapause at $L = 4$, where the density is 10 cm$^{-3}$ (1.5 eV protons, 1 eV electrons [*Comfort*, 1996]). We adopt a velocity shear across the interface of $\Delta V_{\rm msph} = 10$ km/s in the pre-midnight sense. We use the injected particle distributions of Figure 2. We choose ion transition lengths $\mathcal{L}^+ = 50\rho^+$ in order to match the observed SAID thickness, and electron transition lengths $\mathcal{L}^- = \mathcal{L}^+/5$ (large enough to avoid unstable TD configurations [*Roth et al.*, 1996]).

Figure 3 shows the computed SAID structure. The abscissa gives the radial distance; cold plasma is to the left, injected plasma to the right of the interface. The kinetic pressure of the injected plasma is small, leaving the magnetic field essentially unaffected (Figure 3a). Density, temperature, and azimuthal bulk velocity vary considerably across the interface (Figures 3b–3d). While the model assumes constant plasma states on either side of the current sheet, the magnetic field and the density profiles in reality rise earthward of the interface. Figure 3b shows that a sharp (cold plasma) density peak can exist inside the layer; such density fluctuations at the plasmapause are known to exist. A similarly strong peak does not appear in the simulation in [*De Keyser et al.*, 1998]; its presence depends, for instance, on the density of the injected plasma. As the reference frame is chosen to co-move with the plasmatrough, the electric field is zero there; the convection electric field is positive in the injected plasma region. A radially outward electric field peak of 10 mV/m is formed, with a width of 700 km (Figure 3e). When mapped down into the ionosphere, the poleward electric field is found to peak at about 150 mV/m (Figure 3f); Figure 3g shows that the corresponding ion drift exceeds 2 km/s and is directed westward. The peak is 0.5° wide. This profile is similar to the one presented by *De Keyser et al.* [1998, Figure 3]. Energetic particle precipitation, slightly lagging the inward moving SAID, strongly enhances the ionospheric conductivity immediately poleward of the SAID and reduces the electric field there;

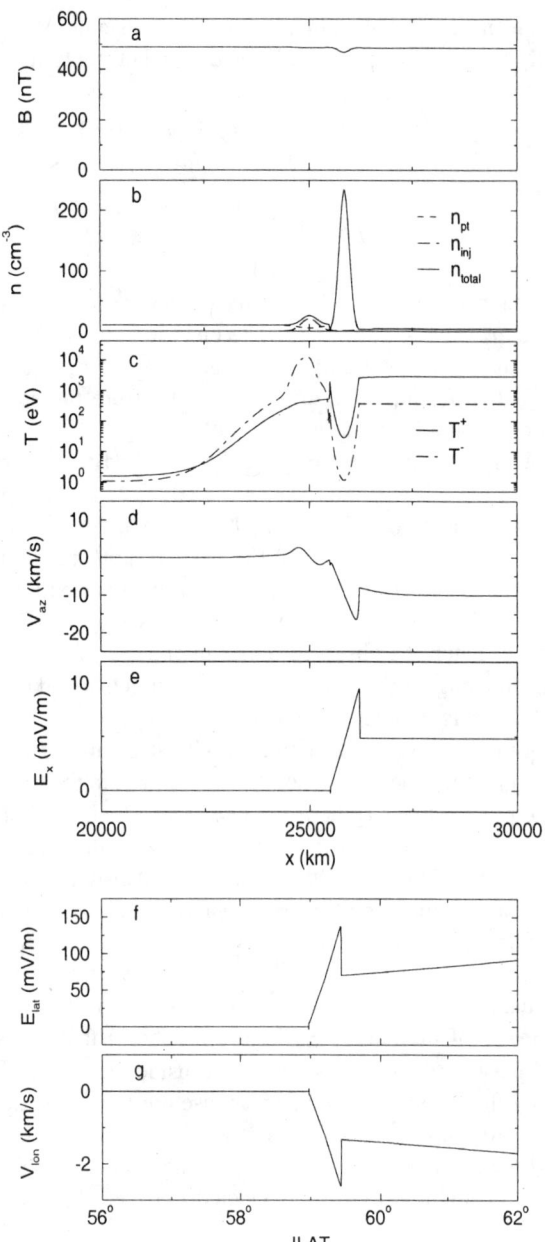

**Figure 3.** Computed structure of a magnetospheric TD at $L=4$ in the pre-midnight sector, in a frame corotating with the plasmatrough/plasmasphere: (a) magnetic field strength; (b) plasmatrough density $n_{pt}$, injected particle density $n_{inj}$, and total density $n_{total}$; (c) mean electron and proton temperatures $T^-$ and $T^+$; (d) azimuthal velocity; (e) magnetospheric electric field (outward); (f) ionospheric electric field (poleward); (g) ionospheric drift (eastward).

the magnetospheric flow is altered there such that the assumption of constant azimuthal velocity tailward of the interface is no longer justified. The electric field poleward of the SAID, about 75 mV/m in this simulation, is therefore overestimated.

The ionospheric drift can be estimated by $V_{iono,peak} \approx 2\Delta\Phi/\Delta\lambda R_E B_{iono} \approx 1-4$ km/s when $\Delta\Phi = 5-20$ kV. If there are no field aligned potential drops, the same potential difference exists across the magnetospheric current sheet and the magnetospheric shear flow is $\Delta V_{msph} = 8-30$ km/s (for $L=4$). If there are field aligned potential drops, the shear flow must be even stronger. Such shear flows correspond to the thermal velocity of the cold ions (1-2 eV), which is the threshold for the onset of velocity induced current sheet instabilities. Note that the cold plasma velocity is a few km/s at most and has almost no influence on $\Delta V_{msph}$.

## DISCUSSION AND CONCLUSIONS

We have considered the role of the injected plasma that penetrates the inner magnetosphere in the course of a substorm. We have approximated observed injected ion and electron spectra by a sum of Maxwellians. Associating the "substorm injection boundary" with the magnetospheric interface in the FGM model of SAID [*De Keyser et al.*, 1998; *De Keyser*, 1999] we have demonstrated that such energetic particles can indeed produce SAID.

The model used here remains crude: the geometry is simplified and the field-aligned currents have not been modeled. We have also ignored the shielding electric field in the inner magnetosphere; this is justified in a first approximation since the shielding Alfvén layer has a characteristic thickness in the order of 1 $R_E$ [*Karlson*, 1971], while the interface discussed here is only 0.1 $R_E$ wide. In spite of its shortcomings, the simulation corroborates the scenario that SAID are formed as the ionospheric footprint of the inward moving injected plasma front.

*Acknowledgments.* The author acknowledges many valuable discussions with M. Roth, J. F. Lemaire, Y. I. Galperin, S. Sazykin, and B. Fejer. Part of the TD model development was supported by a PRODEX contract with ESA. The support of the Belgian Federal Services for Scientific, Technological and Cultural Affairs is acknowledged.

## REFERENCES

Anderson, P. C., W. B. Hanson, and R. A. Heelis, The ionospheric signatures of rapid subauroral ion drifts, *J. Geophys. Res.*, 96, 5785, 1991.

Anderson, P. C., W. B. Hanson, R. A. Heelis, J. D. Craven, D. N. Baker and L. A. Frank, A proposed production model of rapid subauroral ion drift and their relationship to substorm evolution, *J. Geophys. Res.*, 98, 6069-6078, 1993.

Comfort, R. H., Thermal structure of the plasmasphere, *Adv. Space Res.*, 17, 175-184, 1996.

De Keyser, J., M. Roth, and J. Lemaire, The magnetospheric driver of subauroral ion drifts, *Geophys. Res. Lett.*, 25, 1625-1628, 1998.

De Keyser, J., Formation and evolution of subauroral ion drifts in the course of a substorm, *J. Geophys. Res.*, 104, 12,339-12,349, 1999.

Ejiri, M., R. A. Hoffman, and P. H. Smith, Energetic particle penetrations into the inner magnetosphere, *J. Geophys. Res., 85*, 653–663, 1980.

Galperin, Y. I., V. N. Ponomarev, and A. G. Zosimova, Direct measurements of ion drift velocity in the upper atmosphere during a magnetic storm, *Kosm. Issled., 11*, 273, 1973.

Harel, M., R. A. Wolf, P. H. Reiff, R. W. Spiro, W. J. Burke, F. J. Rich, and M. Smiddy, Quantitative simulation of a magnetospheric substorm, 1, Model logic and overview, *J. Geophys. Res., 86*, 2217–2241, 1981a.

Harel, M., R. A. Wolf, R. W. Spiro, P. H. Reiff, C.-K. Chen, W. J. Burke, F. J. Rich, and M. Smiddy, Quantitative simulation of a magnetospheric substorm, 2, Comparison with observations, *J. Geophys. Res., 86*, 2242–2260, 1981b.

Karlson, E. T., Plasma flow in the magnetosphere, *Cosmic Electrodynamics, 1*, 474–495, 1971.

Karlsson, T., G. T. Marklund, L. G. Blomberg, and A. Mälkki, Subauroral electric fields observed by the Freja satellite: A statistical study, *J. Geophys. Res., 103*, 4327–4341, 1998.

McIlwain, C. E., Substorm injection boundaries, in *Magnetospheric Physics*, edited by B. M. McCormack, pp. 143–154, D. Reidel Publishing Company, 1974.

Roth, M., J. De Keyser and M. M. Kuznetsova, Vlasov theory of the equilibrium structure of tangential discontinuities in space plasmas, *Space Sci. Rev., 76*, 251–317, 1996.

Shiokawa, K., C.-I. Meng, G. D. Reeves, F. J. Rich, and K. Yumoto, A multievent study of broadband electrons observed by the DMSP satellites and their relation to red aurora observed at midlatitude stations, *J. Geophys. Res., 102*, 14,237–14,253, 1997.

Smiddy, M., M. C. Kelley, W. Burke, F. Rich, R. Sagalyn, B. Shuman, R. Hays, and S. Lai, Intense poleward directed electric fields near the ionospheric projection of the plasmapause, *Geophys. Res. Lett., 4*, 543, 1977.

Southwood, D. J., and R. A. Wolf, An assessment of the role of precipitation in magnetospheric convection, *J. Geophys. Res., 83*, 5227–5232, 1978.

Spiro, R. W., R. H. Heelis, and W. B. Hanson, Rapid subauroral ion drifts observed by Atmospheric Explorer C, *Geophys. Res. Lett., 6*, 657–660, 1979.

Spiro, R. W., M. Harel, R. A. Wolf, and P. H. Reiff, Quantitative simulation of a magnetospheric substorm, 3, Plasmaspheric electric fields and evolution of the plasmapause, *J. Geophys. Res., 86*, 2261–2272, 1981.

---

J. De Keyser, Belgian Institute for Space Aeronomy, Ringlaan 3, B-1180 Brussels, Belgium. (e-mail: Johan.DeKeyser@oma.be)

# Structured Currents Associated with Tail Bursty Flows During Turbulent Plasma Sheet Conditions

L. R. Lyons[1], T. Nagai[2], J. C. Samson[3], E. Zesta[1], T. Yamamoto[4], T, Mukai[4], A. Nishida[4], S. Kokubun[5]

Flow in the tail often consists of highly structured bursts. Here we use Geotail spacecraft data to show that bursty flows in the tail are associated with significant (~5 nT) magnetic structure indicating structured currents with estimated densities of $\geq 1 \times 10^{-10}$ A/m$^2$. When mapped along field lines to the auroral ionosphere, these currents appear to be sufficiently intense ($\geq 1 \times 10^{-6}$ A/m$^2$) to account for auroral poleward boundary intensifications, which are nightside geomagnetic disturbances having an auroral signature that moves equatorward from the poleward boundary of the auroral oval. Our analysis suggests that there is a dramatic difference between two states of the tail plasma sheet: a stable state with a minimum of bursty flow activity and associated structured currents and a turbulent state with considerable bursty flow activity and structured currents. The transition between these two states can be quite abrupt; however intermediate states also exist. Structured currents appear to be just as important a part of the turbulent plasma sheet as are the bursty flows, so that the concept of a stable tail current distribution may not be appropriate when the plasma sheet is in its turbulent state. We furthermore find that significant flows and structured currents, when present, appear to exist throughout the entire height of the tail plasma sheet during the turbulent periods, suggesting that it may generally be inappropriate to separate plasma sheet flows within the central plasma sheet from those within

---

[1]Department of Atmospheric Sciences, University of California, Los Angeles, Los Angeles, California
[2]Department of Earth and Planetary Sciences, Tokyo Institute of Technology, Meguro, Tokyo, Japan
[3]Department of Physics, University of Alberta, Edmonton, Alberta, Canada
[4]Institute of Space and Astronautical Science, 3-1-1 Yoshinodai, Sagamihara, Kanagawa 229-8510, Japan
[5]Solar-Terrestrial Environment Laboratory, Nagoya University, Toyokawa, Aichi, Japan

plasma sheet boundary layer. We also find that dipolarizations of the magnetic field do not accompany the majority of flow burst events that we have examined. However, dipolarizations are occasionally seen, primarily during substorm associated bursty flow activity.

## INTRODUCTION

Plasma flow in the geomagnetic tail often exhibits large temporal variations. Individual flow bursts often have peak speeds of ~250-1000 km/s [*Baumjohann et al.*, 1990; *Angelopoulos et al.*, 1992]. Peak velocities are in the ±x direction, whereas plasma sheet turbulence at lower ($\lesssim$ 250 km/s) velocities appear to be isotropic[*Borovsky et al.*, 1997]. Bursty plasma sheet flows are believed to have significant spatial structure, so that there should be significant shears and currents associated with the flows.

Periods of bursty flows in the tail have been found to correspond to periods when disturbances (here referred to as "poleward boundary intensifications" or "PBIs" ) having an auroral signature that moves equatorward from the magnetic separatrix are frequently observed in the ionosphere [*Kauristie et al.*, 1996; *Yeoman and Lühr*, 1997; *Lyons et al.*, 1999]. Periods without PBIs are found to correspond to periods when the plasma sheet is far more stable. Discrete auroral arcs are generally associated with upward currents that connect along field lines to the magnetosphere, and PBIs are presumable associated with such currents. Thus the association of PBIs with bursty flows in the tail further suggests that significant currents are associated with the bursty flows within the plasma sheet.

Here we examine Geotail measurements which indicate that structured currents in the tail are indeed associated with bursty flows in the tail and that such currents are generally absent when bursty flows are absent. Studies of bursty flows [e.g., *Baumjohann et al.*, 1990; *Angelopoulos et al.*, 1992, 1994] have been careful to distinguish central plasma sheet flows that are perpendicular to the ambient magnetic field from the field-aligned flows observed within the plasma sheet boundary layer (PSBL) [*Eastman et al.*, 1984, 1985]. We find evidence that the flows within the central plasma sheet and those within the PSBL may not be distinct phenomena. We also find a significant distinction between periods when the plasma sheet is highly turbulent with considerable bursty flow activity and associated structured currents and periods when the plasma sheet is stable with a minimum of such flows and currents.

## GEOTAIL OBSERVATIONS

We first examine Geotail magnetic field and plasma moment observations (Figure 1) from a 3 hr conjunction on November 13, 1996 between Geotail in the tail plasma sheet and the Canadian Auroral Network for the OPEN Program Unified Study (CANOPUS) ground array in central Canada. Figure 1 and subsequent figures show the x, y, and z component of magnetic field $B_x$, $B_y$, and $B_z$, the total magnetic field $B_t$, the x, y, and z components of ion velocity $V_x$, $V_y$, and $V_z$, density, ion temperature, and plasma $P_{plasma}$ and total $P_{tot}$ (magnetic plus plasma, heavier line in pressure panel) pressures. The heavier lines in the velocity panels give $V_{perp,x}$, $V_{perp,y}$, and $V_{perp,z}$, which are the x, y, and z components, respectively, of the velocity component perpendicular to the measured magnetic field. Spacecraft locations in units of $R_E$ are given in the bottom of the figure. GSM coordinates are used.

The time interval shown in Figure 1 included a substorm with an onset time indicated by the arrow at the bottom of the figure. Except for a few minutes near the time of the substorm onset, $V_{perp,x} \approx V_x$ and $B_x$ was small, implying that the spacecraft was near the center of the tail current sheet. The substorm was identified as a global substorm using data from the CANOPUS ground magnetometers and

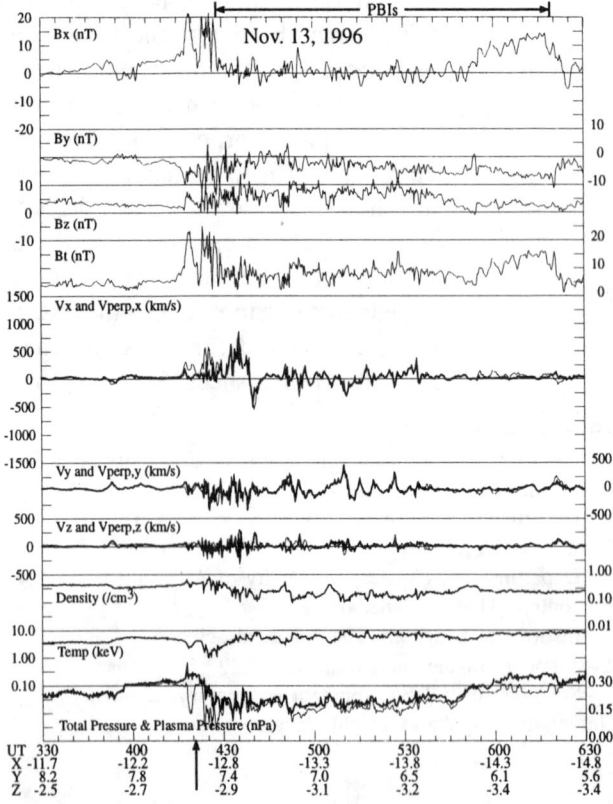

**Figure 1.** Geotail magnetic and plasma data for 0330 to 0630 UT on November 13, 1996. Vertical arrow indicates time of global substorm onset.

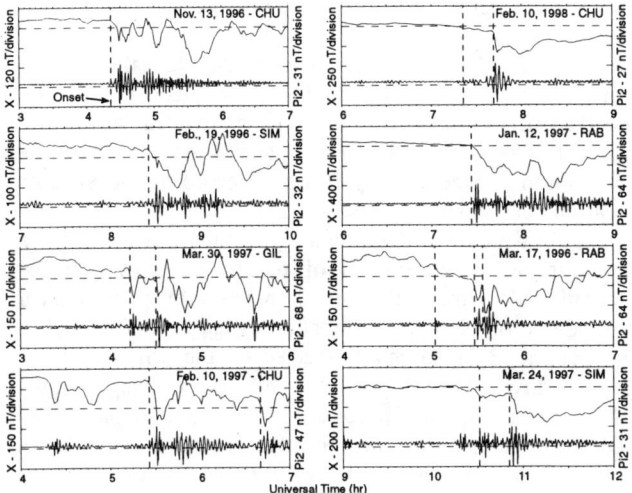

**Figure 2.** Magnetic X-component and Pi 2 data from a representative CANOPUS station for each example used in the present paper. Vertical dashed lines indicate the times of global substorm onsets identified using all CANOPUS magnetic, Pi 2, and MSP data. Stations in this figure are Fort Churchill (CHU), Fort Simpson (SIM), Rabbit Lake (RAB), and Gillam (GIL).

meridian scanning photometers (MSPs). The magnetic X component and Pi 2 pulsation data from a representative CANOPUS station (Fort Churchill) are shown in the upper left panel of Figure 2 (see *Rostoker et al.*, [1995] for CANOPUS instrument descriptions and station locations). As indicated at the top of Figure 1, PBIs were observed in the MSP data for an approximately two hour period following the substorm onset (see Figures 5 and 6 of *Lyons et al.*, 1999). It has not yet been determined whether the above relation of PBIs to substorms is common. Such a relation is at least occasionally seen; however, PBIs are also often seen at other times, including the growth phase of substorms, quiet times, and convection bays [*Lyons et al.*, 1998]. Figure 1 clearly shows that bursty flows were more prevalent during the period when PBIs where observed than during the ~ 50 min period before the onset when PBIs were not observed. Notice that the magnetic field data show significantly less structure during the period without PBIs and structured flows than during the ensuing period. This implies that there were structured currents throughout the period of the bursty flows, and that such currents were generally not present during the period without the flows.

Many of the rapid magnetic changes seen in $B_y$ and $B_z$ were of magnitude ~5 nT. Bursty flows in the tail are believed to be longitudinally localized structures, and *Angelopolous et al.* [1994, 1996] have used ~1-3 $R_E$ as representative scale sizes for the structures. Since the magnetic structures are associated with gradients in the flows, their spatial scale size should be similar in magnitude to, or perhaps smaller than, the scale size of the bursty flows. Taking the maximum spatial scale size for a "localized structure" to be 4 $R_E$ (1/10 of the tail diameter), we assume that the scale sizes for the magnetic changes are $\lesssim 4\ R_E$. With this assumption, we find that ~5 nT magnetic changes correspond to a current density $J \gtrsim 1.6 \times 10^{-10}$ A/m$^2$. Mapping this along field lines to the ionosphere from a tail region of ~10 nT total field would give $J \gtrsim 1 \times 10^{-6}$ A/m$^2$. This is sufficiently intense to require a field-aligned potential drop, and to thus account for the auroral enhancements associated with PBIs. (For a magnetospheric electron population of density 1 cm$^{-3}$ and thermal energy a few tenths of keV, a significant field-aligned potential drop will form when the upward field-aligned current density from the ionosphere exceeds ~0.5 $\times 10^{-6}$ A/m$^2$ [*Lyons*, 1981]).

To determine whether the association between bursty flows and structured currents may be a general feature of the tail, we have examined Geotail data from a data set that we are currently using for a study of plasma sheet dynamics in association with substorms. Each example is for a time interval when observations from CANOPUS allowed the identification of global substorm onsets. Magnetic X-component and Pi 2 data from a representative CANOPUS station are shown in Figure 2 for each example used in the present paper. Optically, substorm onsets show most clearly in the CANOPUS MSPs (at Gillam and Fort Smith) that are located at ~67° magnetic latitude, whereas PBI identification requires MSP observations from the more poleward CANOPUS station (Rankin Inlet at 73° magnetic latitude). Unfortunately, except for the example in Figure 1, PBI identifications were not possible for events selected for the present study due to either poor longitudinal conjunctions between Geotail and Rankin Inlet or poor ground viewing conditions at Rankin Inlet. Since studies have associated some bursty flows with substorms [e.g., *Nagai et al.*, 1998], our use of this data allows possible associations with substorms to be considered. However, this is not a primary goal of the present study. Here, we have selected clear examples of periods with and without bursty flow activity and examples showing a well-defined series of flows bursts. The selected examples are representative of our entire data set, other than we have not emphasized examples with weak flows for which the transitions between intervals with and without flows are less definitive than in the examples presented.

Figure 3 shows Geotail data from a 1 hr interval on February 10, 1998. Throughout this interval, the magnetic and plasma pressures were about equal, implying that Geotail was well within the plasma sheet but away from the current sheet. During some time intervals, we find that the plasma sheet is extremely stable and has almost no detectable bursty flows, and the data in Figure 3 is from such a period. It can be seen that the magnetic field

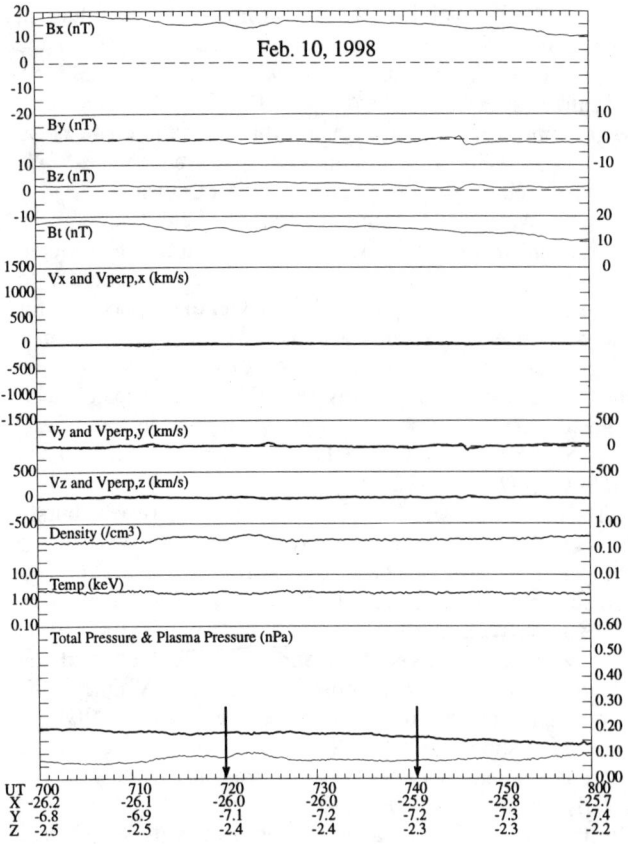

**Figure 3.** Geotail magnetic and plasma data for 0700 to 0800 UT on February 10, 1998. Vertical arrows indicate times of global substorm onsets.

over the spacecraft. However much of the variations in $B_y$ and $B_z$ are not associated with specific variations in $B_x$, and must be related to structured currents. Much of the changes in $B_y$ and $B_z$ are in the range of ~2-10 nT, as is the case for the data in Figure 1 and the data shown in subsequent figures. The data in Figures 3 and 4 do not show an association between bursty flows observed at the satellite location and substorm activity.

The flow bursts in Figure 4 are generally largest in the x direction, which is typical of the flow bursts seen by Geotail. At times, the bursts have little perpendicular component of **V** in the x-direction ($V_{perp,x} \ll V_x$). When this occurs, $|B_x|$ is large as expected for the primarily field-aligned flows of the PSBL. At other times, $V_{perp,x} \approx V_x$. When this occur, the spacecraft can be seen to have been near the center sheet of the current as indicated by $|B_x| \approx 0$. Note that the flows appear to be nearly continuous at times when the spacecraft moved from/to the PSBL to/from the current sheet and that there is a continuous increase/decrease in $V_{perp,x}$ relative to $V_x$ at these times. (For example, this

measured by Geotail was quite smooth, consistent with an absence of large field-aligned current densities in the tail in the vicinity of the spacecraft.

During other time intervals, Geotail detected a nearly continuous sequence of bursty flow activity, such as during the one hour interval on February 19, 1996 shown in Figure 4. Geotail was at approximately the same y (~-7 $R_E$) during this time interval as during the time interval shown in Figure 3, but was ~10 $R_E$ closer to the Earth. Figure 4 includes periods when Geotail was quite near the current sheet, as indicted by $P_{plasma} \approx P_{tot}$, and some periods when Geotail moved away form the current sheet but was still within the plasma sheet. A highly structured magnetic field, which is very different from the smooth magnetic field seen in Figure 3, can be seen during the time interval of bursty flows shown in Figure 4. Rapid fluctuations, some on time scales < 1 min, can be seen, though it is not possible to determine the extent to which any particular fluctuation is spatial or temporal. $B_x$ variations can be related to motions of the tail current sheet

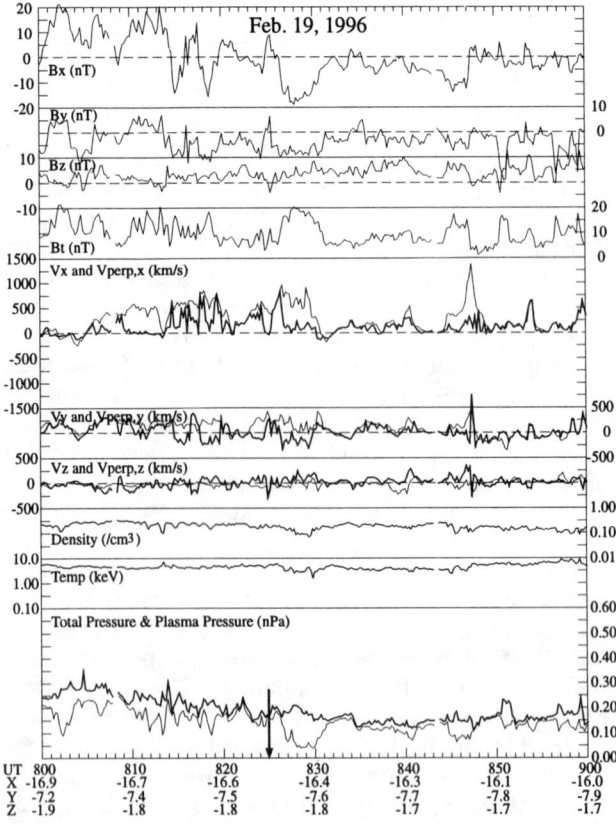

**Figure 4.** Geotail magnetic and plasma data for 0800 to 0900 UT on February 19, 1996. Vertical arrow indicates time of global substorm onset.

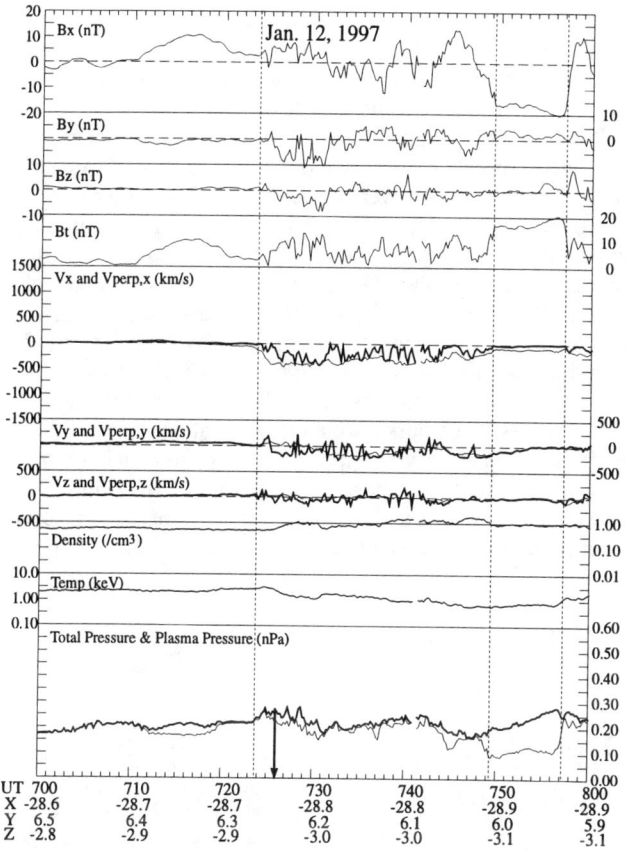

**Figure 5.** Geotail magnetic and plasma data for 0700 to 0800 UT on January 12, 1997. Vertical arrow indicates time of global substorm onset.

occurred a number of times between 0805 and 0830 UT.) It thus appears that the bursty flows and associated structured currents engulf the entire plasma sheet, and are not confined to either the PSBL or the central plasma sheet.

The data in Figures 3 and 4 show that the plasma sheet can be in a stable state without significant bursty flows and magnetic perturbations due to structured currents or can be in a very turbulent state with considerable bursty flow activity and associated structured currents. Figures 5 and 6 show data that illustrate the transition from one state to the other, transitions being identified in each figure by vertical short-dashed lines. In both examples, the magnetic field is relatively stable during the time periods without bursty flows but shows considerable structure of the type expected from structured currents during the intervals with significant bursty flow activity. It can be seen that the transitions between stable and turbulent plasma sheet are quite abrupt and that they occur at nearly the same time in both the flow and magnetic field data.

Figure 5 shows a transition from a stable to a turbulent plasma sheet that occurred at ~0724 UT on January 12, 1997 during a prolonged interval when Geotail was very near the center of the cross-tail current sheet ($P_{plasma} \approx P_{tot}$ and $|B|_x$ is small), and Figure 6 shows a such transition at ~0420 UT on March 30, 1997 when Geotail was near the outer edge of the plasma sheet ($P_{plasma}$ considerable less than $P_{tot}$ and $|B|_x$ is large). This supports the inference above that when the bursty flows exist, they engulf the entire plasma sheet and are not confined to either the PSBL or the central plasma sheet. The data in Figure 6 before 0420 UT also suggest that significant flows along the outer portion of the plasma sheet may not exist when the plasma sheet is in a stable mode.

The above two transitions to a turbulent plasma sheet were observed at positive values of y where *Nagai et al.* [1994, 1998] have reported plasma sheet flows associated with substorms, and both of these transitions may be associated with substorms. The transition near the center of the current sheet in Figure 5 occurred ~2 min before a substorm onset was observed using the CANOPUS ground observations (See Figure 2). Note that none of the flow

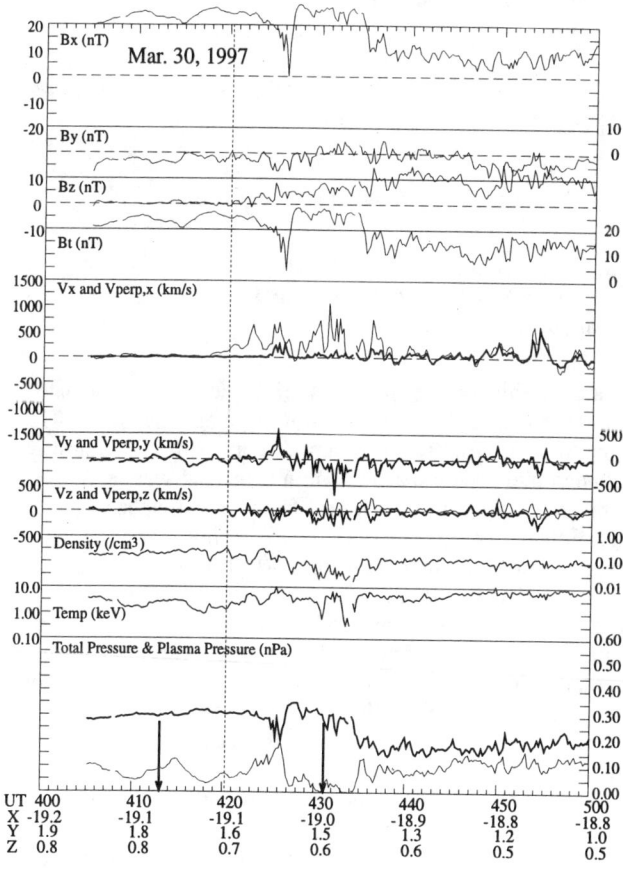

**Figure 6.** Geotail magnetic and plasma data for 0400 to 0500 UT on March 30, 1997. Vertical arrows indicate times of global substorm onset.

between bursty flows, substorms, and dipolarizations is needed. Bursty flows associated with dipolarizations and those not associated with dipolarizations are both associated with a similar magnetic field structure.

The data in Figures 1 and 2-6 show an absence of structured currents in the absence of the bursty flows and the presence of such currents when bursty flows are present. However, it is difficult to identify individual flow bursts and the direct association between the shears associated with the bursts and currents. Occasionally, however, a series of distinct flow bursts can be identified. Several such series are identified in the Geotail examples shown in Figures 7 (1 hr interval on March, 17, 1996) and 8 (0.5 hr intervals on Feb. 10, 1997 and March 24, 1997). Figures 7 and 8 include cases when bursts were observed at positions within the plasma sheet varying from very near the center of the current sheet (e.g., the bursts between 0525 and 0533 UT on March 17, 1996) and to the outer edge of the plasma sheet (e.g., the bursts seen in the March 24, 1997 example,

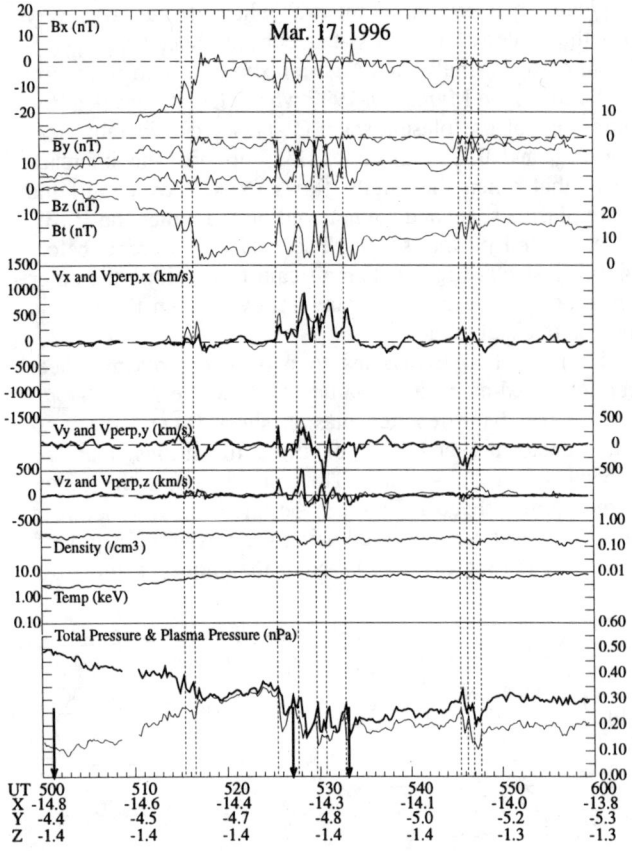

**Figure 7.** Geotail magnetic and plasma data for 0500 to 0600 UT on March 17, 1996. Vertical arrows indicate times of global substorm onset.

bursts in this or the previous examples were associated with identifiable dipolarizations of the magnetic field. The transition in Figure 6 occurred approximately midway between two onsets spaced 18 min apart and was associated with a temporal transition from a stretched to a significantly more dipolar configuration of the tail magnetic field ($B_z$ increased from ~0 nT at 0421 UT to ~10 nT at 0435 UT, and $B_x$ was ~25 nT at the beginning and near the end of this interval). This dipolarization and associated period of bursty flows looks very much like dipolarizations and bursty flows that *Nagai et al.* [1999] found closer to the Earth within the longitude range of substorm onset. In addition to the case in Figure 6, a dipolarization can be seen following a short, substorm-associated, period of flow bursts that occurred from 0525 to 0533 UT on March 17, 1996 (See Figure 7). While such dipolarizations and associated flows are likely an important aspect of substorms, the majority of bursty flows examined here are not clearly associated with substorm onsets and are not associated with significant dipolarizations of the tail magnetic field. Clearly a more detailed study of the relation

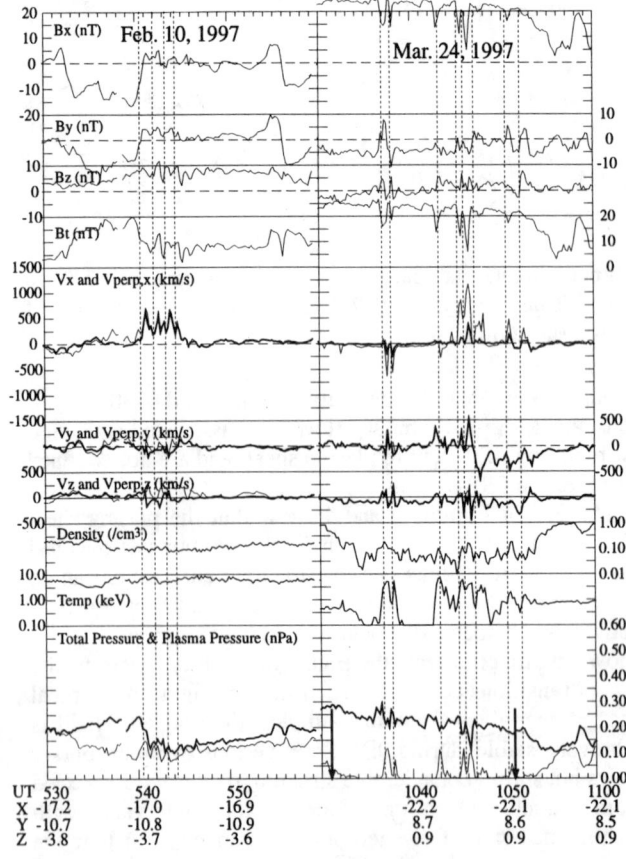

**Figure 8.** Geotail magnetic and plasma data for 0530 to 0600 UT on February 10, 1997 and for 1030 to 1100 UT on March 24, 1997. Vertical arrows indicate times of global substorm onset.

some of which occurred as the plasma sheet traversed the boundary between the lobes and the plasma sheet). The vertical small-dashed lines in Figures 7 and 8 identify the beginning of 23 distinct bursts of flow in the x-direction. It can be seen that the rapid increase in $|V_x|$ immediately following each vertical line is associated with a corresponding rapid ~2-10 nT change in $B_y$ or $B_z$, and generally with a rapid change in both $B_y$ or $B_z$. This correspondence is equally good for the rapid decreases in $|V_x|$ the follows the peak $|V_x|$ of each burst. These correspondences show quite clearly that individual bursts are accompanied by a pair of, presumably oppositely directed, currents. This is just what is expected if the shears associated with individual flow bursts are directly associated with significant current structures.

## DISCUSSION AND CONCLUSIONS

Bursty flows in the tail are an important dynamical feature of the magnetosphere-ionosphere system. Here we have presented evidence that bursty flows are associated with equivalently important structured currents in tail. Current densities are estimated to often be $\geq 1 \times 10^{-10}$ A/m². This would map along field lines to current densities of $\geq 1 \times 10^{-6}$ A/m² in the auroral ionosphere, which is sufficient to account for PBIs.

The data considered here indicates that there is a dramatic difference between two states of the tail: a stable state with a minimum of bursty $\geq 250$ km/s flow activity, and structured currents and a turbulent state with considerable $\geq 250$ km/s bursty flow activity and associated structured currents. The transition between these two states can be quite abrupt. These two distinct states are often clearly identifiable in the data we have examined; however intermediate states also exist (e.g., 0533 to 0600 UT on March 17, 1996; see Figure 7). We have not determined the frequency of occurrence of different states. The data suggests that, when present, significant flows and structured currents exist throughout the height of the tail plasma sheet during the turbulent periods. On the other hand, such flows and currents appear to not be present anywhere from the center of the tail current sheet to the edge of the plasma sheet during the stable periods. While a difference clearly exists between the current sheet region where flows are primarily perpendicular to **B** and the PSBL region where flows are primarily parallel to **B**, flows appear to be continuous from one region to the other. Such a continuity of flows can be explained as being the result of Speiser motion [e.g., *Speiser*, 1965] of particles in a local current sheet region with enhanced electric fields, and the subsequent ejection of particles along field lines from the current region [*Lyons et al., 1999*].

We have also found that dipolarizations of the magnetic field did not occur in association with the majority of flow burst events we have examined. We do see some dipolarizations, and these occur primarily in association with substorm-associated flow bursts. A superposed epoch analysis by *Angelopolous et al.* [1992] shows some dipolarization of the field in association with flow bursts, and this is often interpreted as indicating that flow bursts are generally associated with dipolarizations. We do not see such a general association. But, since *Angelopolous et al.* [1992] included both substorm-onset related and non-onset related flow bursts in their analysis, their superposed epoch analysis would show a dipolarization even if dipolarizations do not occur for a majority of events. Thus their results are consistent with our results.

Finally, since magnetic perturbations associated with current structures within the turbulent plasma sheet are often of the same magnitude as the ambient field and are up to ~1/2 of the lobe field strength, the concept of a stable tail current distribution may not be applicable when the plasma sheet is in its turbulent state.

*Acknowledgments* This research was supported at UCLA by NSF Grant OPP-9619733 and NASA grants NAG5-7962 and NAG5-6243. CANOPUS data have been obtained with support of the Canadian Space Agency. Research by J. C. Samson was supported in part by the Natural Sciences and Engineering Research Council of Canada.

## REFERENCES

Angelopoulos, V., W. Baumjohann, C. F. Kennel, F. V. Coroniti, M. G. Kivelson, R. Pellat, R. J. Walker, H. Lühr, and G. Paschmann, Bursty bulk flows in the central plasma sheet, *J. Geophys. Res.*, 97, 4027, 1992.

Angelopoulos, V., C. F. Kennel, F. V. Coroniti, M. G. Kivelson, R. Pellat, M. G. Kivelson, R. J. Walker, C. T. Russell, W. Baumjohann, W. C. Feldman, and J. T. Gosling, Statistical characteristics of bursty bulk flow events, *J. Geophys. Res.*, 99, 21,257, 1994.

Angelopoulos, V., et al., Multipoint analysis of a bursty bulk flow event, *J. Geophys. Res.*, 101, 4967, 1996.

Baumjohann, W. J., G. Paschmann, and H. Lühr, Characteristics of high-speed flows in the plasma sheet, *J. Geophys. Res.*, 95, 3801, 1990.

Borovsky, J. E., R. C. Elphic, H. O. Funsten, and M. F. Thomsen, The Earth as a laboratory for flow turbulence in high-β MHD, *J. Plasma Physics*, 57, 1, 1997.

Eastman, T. E., L. A. Frank, W. K. Peterson, and W. Lennartsson, The plasma sheet boundary layer, *J. Geophys. Res.*, 89, 1553, 1984.

Eastman, T. E., L. A. Frank, and C. Y. Huang, "The boundary layers as the primary transport regions of the Earth's magnetotail, *J. Geophys. Res.*, 90, 9541, 1985.

Kauristie, K., V. A. Sergeev, T. I. Pulkkinen, R . J. Pellinen, V. Angelopoulos, and W. Baumjohann, Study of the ionospheric signatures of the plasma sheet bubbles, in *Substorms 3*, p. 93, Eur. Space Agency Publ. Div., Noordwijk, 1996.

Lyons, L. R., The field-aligned current versus electric potential relation and auroral electrodynamics, in *Physics of Auroral Arc Formation*, Geophys. Monogr. Ser., vol. 25, edited by J. R. Kan and S.-I. Akasofu, p. 252, AGU, Washington, D. C., 1981.

Lyons, L. R., G. T. Blanchard, J. C. Samson, J. M. Ruohoniemi, R. A. Greenwald, G. D. Reeves, and J. D. Scudder, Near Earth plasma sheet penetration and geomagnetic disturbances in *New Perspectives on the Earth's Magnetotail*, edited by A. Nishida, S. W. H. Cowley, and D. N. Baker, p. 241, AGU, Washington, D. C., 1998.

Lyons, L. R., T. Nagai, G. T. Blanchard, J. C. Samson, T. Yamamoto, T. Mukai, A. Nishida, and S. Kokubun, Association Between GEOTAIL Plasma Flows and Auroral Poleward Boundary Intensifications Observed by CANOPUS photometers, *J. Geophys. Res.*, *104*, 4485, 1999.

Nagai, T., M. Fujimoto, Y. Saito, S. Machida, T. Terasawa, R. Nakamura, T. Yamamoto, T. Mukai, A. Nishida, and S. Kokubun, Structure and dynamics of magnetic reconnection for substorm onsets with GEOTAIL observations, *J. Geophys. Res.*, *103*, 4441, 1998.

Nagai, T., H. Singer, T. Mukai, T. Yamamoto, and S. Kokubun, Development of substorms in the near-Earth tail, *Adv. Space Res.*, 1999 (in press).

Rostoker, G., et al., CANOPUS-A ground-based instrument array for remote sensing the high latitude ionosphere during the ISTP/GGS program, *Space Sci. Rev.*, *71*, 743, 1995.

Speiser, T. W., Particle trajectories in model current sheets, 1, Analytical solutions, *J. Geophys. Res.*, *70*, 4219, 1965.

Yeoman, T. K., and H. Lühr, CUTLASS/IMAGE observations of high-latitude convection features during substorms, *Ann. Geophys.*, *15*, 692, 1997.

----------

S. Kokubun, Solar-Terrestrial Environment Laboratory, Nagoya University, Honohara 3-13, Toyokawa, Aichi 442-8507, Japan

L. R. Lyons and E. Zesta, Department of Atmospheric Sciences, University of California, Los Angeles Los Angeles, CA 90095-1565

T, Mukai, A. Nishida, and T. Yamamoto, Institute of Space and Astronautical Science, 3-1-1 Yoshinodai, Sagamihara, Kanagawa 229-8510, Japan

T. Nagai, Department of Earth and Planetary Sciences, Tokyo Institute of Technology, Ookayama 2-12-1, Meguro, Tokyo 152-8551, Japan

J. C. Samson, Department of Physics, University of Alberta, Edmonton, Alberta, Canada T6G 2E9

# Substorm Associated Tail Current Changes Inferred From Lobe Magnetic Field Observations

Christian Jacquey

*Centre d'Etude Spatiale des Rayonnements, Toulouse, France*

In this paper, we present two case studies of the magnetic field changes observed during substorms in the tail lobe at geocentric distances ranging from 20 to 36 Re. For the first one, we study the 3 components of the lobe magnetic field recorded by ISEE-1 during a double onset event which occurred on March 30, 1978. An earlier study has shown that each of these successive substorm expansion phases was associated with a disruption of the cross-tail current which first occurred close to the Earth, and then propagated tailward. The present study suggests that the disrupted cross-tail current was diverted through the ionosphere via field aligned currents, consistent with the current wedge model. For the second case, we study an isolated substorm observed simultaneously on November 24, 1996 by the INTERBALL and IMP-8 spacecraft located respectively at radial distances of 26 and 36 Re. Data analysis and simple 2D current sheet model simulations suggest that the observed signatures are consistent with the superimposition of a tailward propagating cross-tail current disruption and of a Traveling Compression Region (TCR). Both phenomena appear to be initiated on the earthward side of the satellites (i.e., at R < 26 Re). Timing analysis and model simulation suggest that the TCR speed is larger than the disruption propagation one.

## 1. INTRODUCTION

The current wedge model (Akasofu, 1972, McPherron et al., 1973) is an attempt to organize substorm signatures detected both on the ground and in the near-Earth and middle tail. In this model, the substorm expansion is associated with the disruption of the cross-tail current and its diversion through the ionosphere via field aligned currents. The cross-tail current disruption and the associated field aligned currents are expected to be initiated in the near-Earth tail and then to expand both tailward and longitudinally. The analysis of the field and particle data recorded at geosynchronous (e.g. Ohtani, 1998) and AMPTE (e.g. Lui et al., 1992) orbits supports the current wedge model. Moreover, statistical studies of the field aligned currents detected inside the plasma sheet boundary layers by the ISEE-1/2 satellites showed that their observed longitudinal distribution was as predicted by the current wedge model (Ohtani et al., 1988). Furthermore, the cross-tail current redistribution associated with substorms inferred from the magnetic field changes obtained in the lobe at radial distances ranging from 10 to 36 Re is also consistent with a tailward expanding cross-tail current disruption (Jacquey et al., 1991, 1993, Ohtani et al., 1992, Jacquey and Sauvaud, 1994, Sauvaud et al., 1996a,b).

# 276 SUBSTORM ASSOCIATED TAIL CURRENT CHANGES

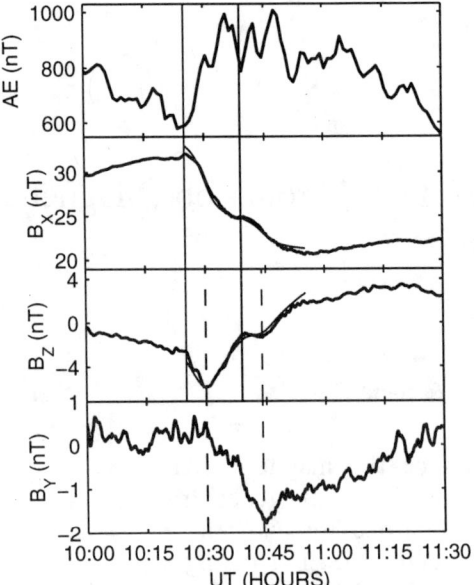

**Figure 1**. Changes of the GSM BX, BZ and BY components of the magnetic field measured by the ISEE-1 satellite in the North lobe on March 30, 1978. The thin lines superimposed on the BX and BZ profiles represent the predictions of the disruption model obtained with the parameter values used in this study.

The plasmoid model (Hones, 1976) provides an interpretation of the observations performed in the deep-tail by the ISEE-3 and GEOTAIL missions. In this model, a neutral line forms at radial distances of about 15 Re and ejects a plasmoid in the tailward direction. The signature of plasmoids in the lobe consists of a peak of BX, main component of the magnetic field and a bipolar change of BZ which have been interpreted as Traveling Compression Regions (e.g. Slavin et al., 1993).

In the lobes, the plasma is tenuous and cold (e.g., Lui, 1987). Thus in such regions, there is no significant local electric current and the plasma diamagnetic effect is negligible. The magnetic field variations in the lobes are the signatures of changes of the large scale tail current system. Both statistical analysis (e.g., Nakai et al., 1991) and case studies (e.g. Fairfield et al., 1981, Jacquey and Sauvaud, 1994) have identified three principal causes of the lobe magnetic field changes: (i) disturbances of the solar wind pressure, (ii) energy input supplied by the solar wind, and (iii) internal dissipation processes.

In this paper, we report two case studies based on lobe magnetic field data obtained in the near-Earth and middle tail. In section 2, we take advantage of a double substorm onset event which occurred on March 30, 1978. The analysis of the 3 components of the magnetic field measured onboard ISEE-1 and detailed ground data shows that the observed redistribution's of both tail currents and low altitude currents are consistent with the predictions of the current wedge model. In section 3, we present simultaneous observations performed by the INTERBALL and IMP-8 satellites located in the lobe at radial distances of 26 and 36 Re respectively. The event consists of an isolated substorm which occurred during a long period of quiet solar wind pressure conditions. It is shown that the observed magnetic field changes can be interpreted as the superposition of a tailward propagating cross-tail current disruption and of a traveling compression region.

## 2. DISRUPTION AND DIVERSION OF THE CROSS-TAIL CURRENT: MARCH 30, 1978 EVENT

### 2.1. Lobe magnetic field observations

Figure 1 exhibits lobe magnetic field observations obtained by ISEE-1 during a substorm on March 30, 1978. The top panel represents the AE index and the others display the three GSM magnetic field components measured onboard ISEE-1 (Russell, 1978). The satellite was located at geocentric distance of about 20 Re (Xgsm = -20.7, Ygsm = -0.2 and Zgsm = 8.3 Re), in the North lobe and close to the noon-midnight meridian plane. The distance to the neutral sheet was close to 8 Re. The two solid vertical lines indicate the onset times (10:25 and 10:39 UT) of two successive substorm expansions, as inferred from detailed ground data analysis.

In the course of the two successive substorm expansions, the BX and BZ components of the lobe magnetic field observed by ISEE-1 exhibit similar signatures. These consist of a decrease of BX and a V-shaped profile of BZ. The BX decrease suggests that the cross-tail current is continuously reduced during the event. The initial BZ decrease indicates that the cross-tail current reduction first occurs on the earthward side of the spacecraft. Conversely, the following BZ increase shows that the cross-tail current reduction has been displaced on the tailward side. Such changes have been interpreted as the signature of partial cross-tail current reduction starting close to the Earth and then propagating tailward (Jacquey et al., 1991, Ohtani et al., 1992). For the March 30, 1978 case, the BX and BZ profiles have been regarded as due to two successive tailward propagating cross-tail current reductions. The observed BX and BZ changes were successfully reproduced with the help of a simple 2D uniform current sheet model (see Jacquey et al., 1991). The comparison of the data with this model indicated that the two cross-tail current disruptions started at geocentric distance of about 7 Re and propagated tailward with a velocity of the order of 300 km/s. It was also found that the cross-tail current was only partially reduced by $J_1 \sim 18$ mA/m (30 % of the initial current) and $J_2 \sim 6$ mA/m (10%) respectively for the two successive disruptions. In Figure 1, the thin lines superimposed on the BX and BZ profiles represent the simulation corresponding to these parameter values.

In this model, the minimum of BZ is reached when the current disruption front passed under the satellite; the corresponding times $T_{Z,1}$ (10:30) and $T_{Z,2}$ (10:44) are marked by the vertical dashed lines in Figure 1. During the present period, the BY profile exhibits three clearly distinct trends. From the beginning of the period up to the time $T_{Z,1}$ of the passage of the first disruption front at the ISEE location, BY was constant on average. Then, at the time $T_{Z,1}$, the BY slope became suddenly negative and BY decreased quasi-linearly until the time of the passage of the second disruption front, $T_{Z,2}$. At that time, again, the BY slope suddenly changed and BY then increased in a roughly linear way. There are three features to be interpreted: (i) the sign of the BY changes, (ii) the coincidence of the BY slope changes and the passage of the disruption fronts under the satellite and (iii) the quasi-linearity of the BY variations.

The BY variations are generally interpreted as due to field aligned currents, i.e., flowing approximately along the X-direction. Although they are weak, the BY changes exhibit a clearly structured profile and the timing of their slope changes appears to be closely related to the one of the two successive disruption expansions. It is thus assumed that the observed BY variations are the signature of the field aligned current system associated with the disruptions. The first BY change (10:30 - 10:44) was a decrease indicating the dominating effect of an earthward directed field aligned current. According to the current wedge model, this suggests that the central axis of the cross-tail current disruption was located westward of the satellite. The second BY variation (10:44 - 11:30) consisted of an increase and is interpreted as due to the signature of the dominating effect of a tailward field aligned current. This indicates that the central axis of the second disruption was located in the eastward side of ISEE-1. As the satellite was located close to the noon-midnight meridian plane, the BY changes suggest that the first cross-tail current disruption developed mainly in the pre-midnight sector and the second one in the post-midnight sector.

Both timing and quasi-linearity of the BY changes can be interpreted in terms of the tailward propagating diversion of cross-tail current. Let us consider a cross-tail current disruption which starts at the location $X_o$ and which expands tailward thereafter in the equatorial plane as illustrated in Figure 2a. Let us also assume that the disruption propagates at a constant velocity V and that the current density reduction $\Delta J$ is uniform. If we assume that the disrupted current completely closes through the field aligned currents, their total intensity $I_o$ remains constant:

$$I_o = |I_{FAC,T}| = |I_{FAC,E}| = |I_{DIS}| = \Delta J . V . (t - T_o) \quad (1)$$

where $T_o$ represents the onset time of the disruption and the subscripts T and E designate the tailward and earthward polarities of the field aligned currents. Thus, the total field aligned current intensity $I_o$ is a linearly increasing function of the time. Note that this intensity is carried by pairs of

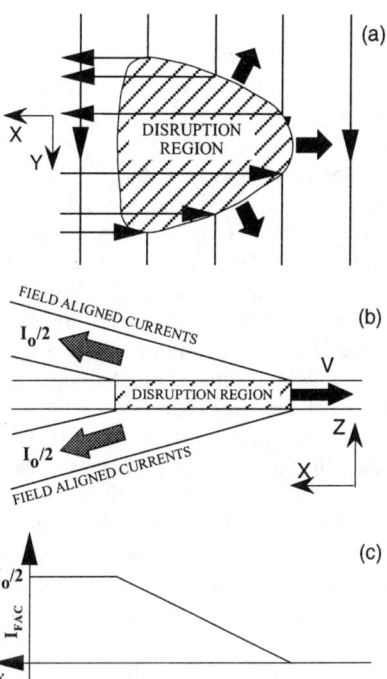

**Figure 2.** Schematical illustration of the model of the diversion of the disrupted current.

field aligned currents flowing in both North and South plasma sheet boundary layers, as displayed in Figure 2b. In this model, the field aligned current system expands tailward simultaneously with the disruption. At a given time, the local field aligned current intensity I(X) flowing at a particular location X corresponds to the amount of the cross-tail current intensity disrupted in its tailward side:

$$|I(X)| = \Delta J . (X - X_F) = \frac{X - X_F}{X_o - X_F} I_o \quad (2)$$

where $X_F$ indicates the instantaneous location of the disruption front. Thus, there is a field aligned current intensity gradient along the X direction as schematically represented in Figure 2c. The BY component of the magnetic field resulting from a tailward expanding field aligned current at the satellite location $(X_S, Z_S)$ can be expressed by:

$$B_Y = \frac{\mu_o}{4\pi} \frac{I_o}{2} G \quad (3)$$

In this formula, we only consider the field aligned current flowing in the North plasma sheet boundary layer carrying the half of the disrupted current $I_o/2$. G represents a geometric factor depending on the instantaneous spatial distribution of the field aligned currents relative to the spacecraft location. During the period preceding the passage of the disruption front under the satellite, G is weak, close to zero. This is due to the fact that the field aligned currents are still located far away from the satellite. When the disruption front arrives in the vicinity of the spacecraft location, the field aligned currents also become closer and we expect a sharp increase of the G factor. After the disruption front has passed the satellite location, BY remains dominated by the field aligned currents flowing close to the spacecraft. Their expansion continues but now occurs far away. Thus, the G-factor should not change very much after the passage of the disruption front. By multiplying the G-factor, interpreted as a step function of the time and the field aligned current intensity which is a linearly function of the time, the BY changes are expected to be weak until the time of the passage of the disruption front under the satellite and then to increase in a quasi-linear way, consistently with the observations. In order to test this interpretation, the results of a simple model are given in Figure 3. In this model, we compute the BY changes due to a field aligned current which is created at $X_o$ and then expands tailward with a constant velocity. The field aligned current flows along the X-direction and is considered to be filamentary. Its total intensity is the half of the disrupted cross-tail current intensity given by expression (1) and its gradient by the formula (2). The BY changes are obtained by integrating the Biot and Savart law:

$$B_Y = \frac{\mu_o}{4\pi} \frac{I_o}{2} \frac{1}{L.\Delta Z} \left\{ \sqrt{\Delta X_F^2 + \Delta Z^2} - \frac{\Delta X_F . \Delta X_o + \Delta Z^2}{\sqrt{\Delta X_o^2 + \Delta Z^2}} \right\} \quad (4)$$

where L represents the length of the current filament and $\Delta X_o$, $\Delta X_F$ and $\Delta Z$ designate its coordinates relatively to the satellite. The separation along the Z-direction between the satellite and the current filament and its earthward edge location $\Delta X_o$ are arbitrarily chosen. The location of its tailward edge $\Delta X_F$ and its length are computed by time-integration of the velocity. Figure 3b shows the BY values computed with the help of this model. Here, we considered a field aligned current of earthward polarity in order to compare the computed BY changes to the ones observed during the first disruption. The parameter values used are the ones derived from the disruption analysis. As ISEE-1 was located at ~8 Re above the neutral sheet and as the current filament is supposed to flow inside the North plasma sheet boundary layer, the values of the $\Delta Z$ parameter have been taken between 4 and 8 Re. The computed values exhibit a profile similar to the observed one. BY is weak before the time of the passage of the disruption under the

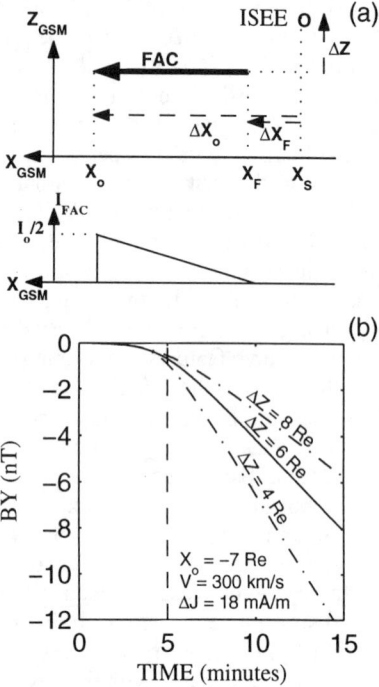

**Figure 3.** (a) Schematic illustration of the current filament model. (b) BY changes computed at the ISEE-1 location. The parameter values are the ones inferred from the disruption analysis. The vertical dashed line indicates the passage of the disruption front under the satellite.

satellite and then decreases linearly. However, the values of the computed BY are larger than the observed ones by a factor of 2~6, depending on the separation along the Z-direction between the field aligned current and the satellite. This discrepancy may be attributed to the following reasons: (i) in this simple model, the whole disrupted current is diverted in filamentary field currents. We may expect slighter effect on BY by considering that field aligned currents are distributed in wide longitudinal range; (ii) according to the current wedge model, a field aligned current of opposite polarity is expected to flow in the westward side of the current disruption region. This field aligned current should reduce the effect of the first one; (iii) The formula (4) is valid only if both the field aligned current and the observation point have the same location along the Y-axis. If it is not the case, the distance separating the field aligned current and the satellite becomes larger. This will result in a decrease of the absolute value of the BY changes.

This analysis shows that a model consisting of a field aligned current system expanding in association with a tailward propagating uniform cross-tail current disruption is qualitatively consistent with the observed changes of the three components of the lobe magnetic field.

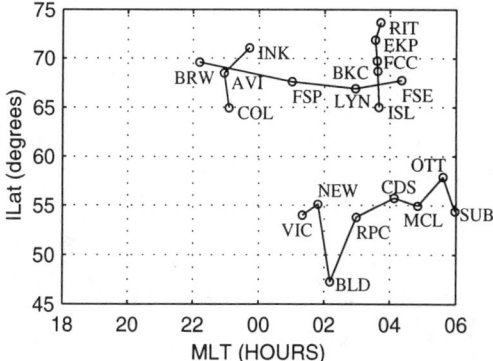

**Figure 4.** Distribution of the ground stations in the MLT,Ilat frame on March 30, 1978, 10:30 UT.

## 2.2. Ground observations

The nightside ionospheric electrojet and the associated field aligned current have been analyzed with the help of ground magnetograms coming from both auroral and mid-latitude stations. Their locations in the MLT,Invariant Latitude frame are represented in Figure 4. At auroral latitudes, we used one longitudinal chain from MLT ~ 22 hours to MLT ~ 05 hours and two meridian chains: the Alaska one at MLT ~ 23-00 hours and Fort Churchill one at MLT ~ 03.5 hours. We also used a longitudinal chain at mid-latitude located in North America for MLT from 01 to 06 hours.

In Figure 5, the circles represent the negative bays observed by each auroral station. Their diameter is proportional to the maximum negative H-component enhancement recorded during the two successive disruption periods. The changes of both the D and H components of the horizontal magnetic field observed at the mid-latitude stations are represented by normalized vectors.

During the first event (10:25 - 10:39 UT), auroral latitude magnetograms show that negative bays were observed in the whole longitudinal sector covered by the available stations (22 - 04 MLT). They were maximum in the pre-midnight sector (MLT = 22~00) reaching 500 nT and weaker (~200 nT) in the morning sector. This suggests that a first electrojet took place in a wide longitudinal sector centered in the pre-midnight sector, as illustrated in the middle panel of Figure 5a. At mid-latitude, all the stations recorded a negative D component variation (D is eastward directed). This is the signature of a large scale downward field aligned current. The H-component increased at the mid-latitude stations located at MLT ~ 1-2 hours. Conversely, H decreased in the far morning sector, for local times later than 4 hours. Interpreting the H-changes as due to the edge effect of the field aligned currents, the observed D and H variations suggest that a downward field current developed in the morning sector within a large local time sector of more than 3 hours, as displayed in the top panel of Figure 5a.

During the second disruption event (10:39 - 11:00), the negative bays were less important in the pre-midnight sector, reaching a maximal value of about 300 nT. In the morning sector, the Fort Churchill chain detected strong negative bays reaching 450 nT. This suggests that a second westward electrojet developed mainly in the morning sector. At mid-latitude, the stations located at local times later than 4 hours detected a negative D-variation, indicating a downward field aligned current. Conversely, the stations located at MLT < 4 hours recorded a positive D-variation, interpreted as due to a field aligned current of opposite polarity, i.e. upward. The H variations were negative for MLT < 02 hours and for MLT > 04 hours and positive for 02 < MLT < 04 hours. These changes are consistent with a pair of field aligned currents of opposite polarities. One, downward directed is centered at MLT ~ 04 hours. The other one is upward directed and centered at MLT ~ 02 hours.

## 2.3. Comparison of lobe and ground observations

The data obtained both on the ground and onboard ISEE-1 during this double onset event can be consistently

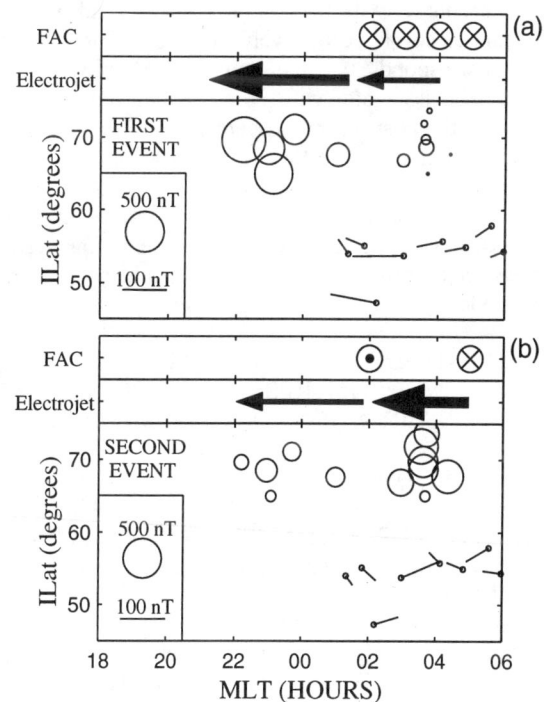

**Figure 5.** Magnetic field changes observed on the ground during the two successive disruptions. The two top pannels of each figure illustrate the inferred low altitude current distribution.

interpreted as due to the contribution of the three main components of the current wedge model (Akasofu, 1972, McPherron et al., 1973): tailward propagating disruption of the cross-tail current, its diversion via field aligned currents and the westward electrojet. The magnetic field measurements obtained onboard ISEE-1 in the tail lobe are consistent with a model which includes (i) two successive cross-tail current disruptions initiated close to the Earth and then propagating tailward and (ii) two successive field aligned current systems expanding tailward in association with the disruptions. According to the current wedge model, the analysis of the polarity of the field aligned current inferred from measurements in the lobe suggests that the first disruption developed mainly in the pre-midnight sector and the second one in the post midnight sector. For the first event, the auroral ground data can be interpreted as the signature of a westward electrojet stretching out a large longitudinal sector centered in the pre-midnight sector. A wide downward field aligned current was observed in the post-midnight sector at the mid-latitude stations. According to the current wedge model, this low altitude current system is the one expected from a cross-tail current disruption developing in the pre-midnight sector, consistent with the ISEE data analysis. For the second event, ground auroral data show that a new westward electrojet developed in the post-midnight sector. Downward and upward field aligned currents were detected in the eastward and westward sides of the electrojet respectively. This set of low altitude currents corresponds to the current diversion system expected to be associated with a cross-tail current disruption occurring in the post-midnight sector. Thus, for each event, the longitudinal redistribution of the low altitude current system observed on the ground corresponds to the one of high altitude tail currents inferred from the lobe observations. Both ground and lobe observations can be interpreted in a consistent way with the help of the current wedge model.

## 3. CROSS-TAIL CURRENT DISRUPTION AND TRAVELING COMPRESSION REGIONS (TCR): DUAL OBSERVATIONS ON NOVEMBER 24, 1996.

Figure 6 displays the GSM X and Z components of the magnetic field measured by the INTERBALL/ASPI-MIFM (Klimov et al., 1997) and IMP-8/MAG (Mish and Lepping, 1976) on November 24, 1996. The vertical line indicate the time of the main onset of an isolated substorm, as inferred from ground, geosynchronous and AKR data (not shown, see Petrukovich et al., 1998, Jacquey et al., 1998). Both INTERBALL and IMP-8 satellites were located close to the noon-midnight meridian in the middle magnetotail at Xgsm = -26.4 Re, Ygsm = 1.3 Re, Zgsm = 9.3 Re and Xgsm = -36.2 Re, Ygsm = -3 Re, Zgsm = 10.4 Re respectively. Thus, they were nearly aligned along the X-axis and separated by about 10 Re. Electron spectrograms

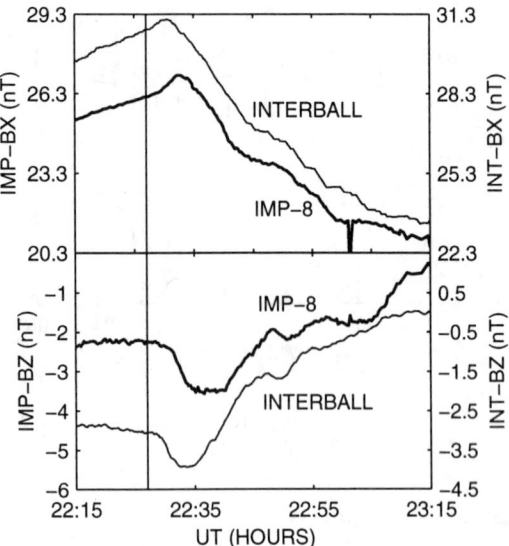

**Figure 6**. Changes of the BX and BZ GSM components of the magnetic field measured by the INTERBALL and IMP-8 satellites in the north lobe on November 24, 1996. The vertical dashed line indicates the substorm onset time.

(not shown) show that the INTERBALL satellite was inside the North tail-lobe during the whole 1 hour period. The IMP-8 satellite was also located inside the North tail-lobe between 21:40 UT and 24:00 UT as suggested by the weak variance of the magnetic field (not shown). During this period, the WIND measurements (not shown, see Jacquey et al., 1998) indicate that the solar wind pressure was roughly constant. Thus, this event provides an opportunity to analyze the large scale tail current redistribution associated with substorm and undisturbed by solar wind pressure variations.

The average trends observed by both satellites are similar to those observed at the ISEE orbit and described in section 2. Both spacecraft recorded a decrease of the BX component and a V-shaped profile of the BZ one. These changes suggest that the large scale redistribution of the tail currents globally consisted of a cross-tail current reduction having started initially on the earthward side of both satellite (i.e., R < 26.4 Re) and then having propagated beyond their locations (i.e., to R > 36.2 Re). Consistently, the BZ signature recorded by IMP-8 is delayed relative to the INTERBALL one.

According to the model of a simple tailward propagating current disruption, BX should start to decrease at the onset time and BZ should reach its minimum by exhibiting a negative peak. The observed signatures show some second order deviations from this prediction: (i) a positive peak of the BX component was measured by both satellites a few minutes after the substorm onset and (ii) BZ did not exhibited a clear minimum, but a plateau: at the IMP-8

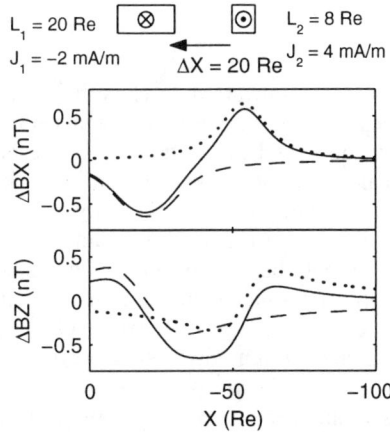

**Figure 7.** Example of the distribution along the X-direction of the ΔBX and ΔBZ components of the magnetic field resulting from a couple of opposite polarity current slabs for an observation point located 9 Re above their symetry plane. The dashed and dotted lines displays the contribution of the duskward current slab (representing the disruption) and the dawnward one (representing the excess current) respectively. The solid line shows the effects of both superimposed currents.

location, it had a duration of about ~330 seconds. The plateau was less pronounced at INTERBALL and lasted only ~110 seconds. Although small, these discrepancies relative to the disruption model are significant and need to be interpreted. Similar signatures, BX peaks associated with bipolar BZ changes observed by IMP-8 or ISEE-3 have been previously interpreted in terms of Traveling Compression Regions. The present BX peak was first observed by INTERBALL and later by IMP-8. The time delay of 94~168 seconds can be interpreted if the source of the BX peaks is an excess current slab traveling tailward at speed of 370~660 km/s (Jacquey et al., 1998). This value is consistent with the ones found for the TCR and plasmoids (Slavin et al., 1993). However, a travelling compression region can not account for the BX decrease following the peak and the plateau of BZ observed at IMP-8.

To attempt to account for both the general trends and second order changes depicted above we combined the current wedge and the TCR models. To illustrate the prediction of this model, Figure 7 displays an example of the distribution along X of the magnetic field components ΔBX and ΔBZ due to two current slabs of opposite polarity (solid lines). The TCR and the current wedge are represented by an excess (J>0) and reduction (J<0) current slabs respectively. Here, we consider a static situation which includes an excess current slab (dotted lines) centered at X = -54 Re and a reduction one (dashed lines) centered at X = -20 Re. The values of ΔBX and ΔBZ have been computed for an observation point located in the North lobe at Z = 9 Re above the neutral sheet. In the region between the current slabs, ΔBZ is constant, due to the superposition of the edge effects of both current slabs. Note that the constant ΔBZ region can appear only if the separation ΔX between the two current slabs is sufficiently large compared to the distance to the neutral sheet of the observation point. ΔBX is positive above the excess current and negative above the reduction. Superposing this contribution on the main tail magnetic field results in a compression region on the tailward side and a depression region on the earthward one. If such a current structure moves tailward under a satellite, the magnetic field signature should be similar to the one observed during the November 24, 1996 event. Figure 8 displays the temporal changes of the magnetic field computed with the help of a simple 2-D uniform current sheet model. The thin and thick lines represent the changes expected at the INTERBALL and IMP-8 locations respectively. The shape and the timing of the BX and BZ changes observed at both IMP-8 and INTERBALL are successfully reproduced. This model includes two uniform current sheets. One (J>0), moving tailward with a constant speed and having a constant width corresponds to the excess current slab. The other (J<0) represents to the cross-tail current disruption expanding tailward, i.e. its width increases linearly with time. Note that the model depends on a large number of free parameters and is only qualitative. By varying the parameters, it appears that the shape and the timing of the observed changes can be reproduced under two conditions: (i) both the initial disruption and the centre of

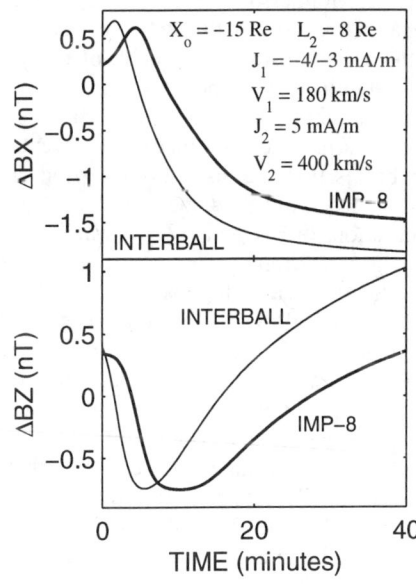

**Figure 8.** Example of the BX and BZ profiles computed at the INTERBALL (thin line) and IMP-8 (thick line) locations from a model including a tailward expanding uniform cross-tail current disruption and an excess uniform current slab ejected tailward.

the excess current slab have to be initially located in the earthward side of the satellite. Otherwise, the BX peak is not be observed; (ii) the speed of the excess current slab has to be significantly greater than that of disruption expansion. To fit the observed timing, their values were 400 and 180 km/s respectively. If this condition is fulfilled, the BZ plateau is predicted to be longer at IMP-8 than at INTERBALL, consistently with the observations.

## 4. DISCUSSION AND CONCLUSION

The analysis of the BX and BZ components of the lobe magnetic field data obtained by ISEE-1 at radial distance of about 20 Re on March 30, 1978 by ISEE-1 can be interpreted as due to two successive cross-tail current disruptions starting in the near-Earth tail and propagating tailward at velocity of the order of 150~350 km/s (Jacquey et al., 1991). This interpretation was supported by other case studies using both differential timing analysis of dual satellite observations and 2D current sheet model simulation which provided consistent results (Ohtani et al., 1992, Jacquey et al., 1993). In this paper we show that the combined analysis of the three components of the lobe magnetic field and detailed ground data obtained on the March 30, 1978 event provides an evidence of the diversion of the disrupted current through the ionosphere via field aligned currents observed at both low and high altitude. We finally conclude that the lobe magnetic field observations obtained at the ISEE orbit support the current wedge model (Akasofu, 1972, McPherron et al., 1973). Furthermore, they provide quantitative estimates of the spatio-temporal features of the current disruptions capable of constraining on the current destabilization model.

Assuming that they propagate at a constant velocity of a few hundreds of km/s, the duration of these events (20 ~ 30 minutes) indicates that the disruptions expand far into the tail, beyond 40 Re. This prediction has been confirmed by IMP-8 observations of BX and BZ (Sauvaud et al., 1996a) which show general trends similar to those detected by ISEE 1/2. The dual satellite observations obtained from INTERBALL and IMP-8 at radial distances of 26 and 36 Re on November 24, 1996 enable us to study the redistribution of the tail current during roughly constant solar wind pressure conditions. The data revealed small deviations from the model which can be interpreted as second order features superimposed on the tailward propagating cross-tail current disruption signatures. Their analysis suggests that the BX and BZ changes observed at both locations can be interpreted by a model which includes the simultaneous tailward ejection of a TCR/plasmoid and tailward propagating disruption of the cross-tail current. This scenario has been already suggested by a very large scale multi-satellite study (Sauvaud et al., 1996b) who used simultaneous data obtained in the mid-tail lobe by IMP-8 and in the deep tail by ISEE-3. Data and timing analysis of the November 24, 1996 event suggest (i) that the TCR/plasmoid speed is significantly larger than the disruption expansion speed and (ii) that both phenomena are initiated on the earthward side of both satellite, i.e., at radial distances less than 26 Re. A key question concerns the initial conditions and the initial process responsible for the large scale current redistribution associated with substorm expansion. One possibility is that the TCR/plasmoid forms initially at the same location where the cross-tail current disruption starts. This suggests that the TCR/plasmoid formation and ejection result from the disruption development. Alternatively, if the TCR/plasmoid and current disruption are not initiated at the same location, both phenomena would be the consequences of another process developing prior the main substorm onset, as suggested by GEOTAIL observations (Nagai et al., 1998, Ohtani et al., 1999).

*Acknowledgments.* The author thanks S.I. Klimov and S.A. Romanov for providing INTERBALL ASPI/MIFM data, R.P. Lepping for producing the IMP-8 data and C.T. Russell for supplying the ISEE data. The AL data were provided by the World Data Center A for Solar Terrestrial Physics. The author benefited from helpful discussion with D.G. Sibeck and R. Anderson in Bern, Switzerland, as member of visiting scientist team of the International Space Science Institute (ISSI) during 1997 and 1998.

## REFERENCES

Akasofu, S.I., Magnetosphere substorm: a model, in *Solar Terrestrial Physics/1970: Part III*, edited by D. Dyer, p. 131, D. Reidel, Norwell, Mass., 1972.

Fairfield, D.H., R.P. Lepping, E.W. Hones Jr., S.J. Bame and R.J. Asbridge, Simultaneous measurements of magnetotail dynamics by IMP spacecraft, *J. of Geophys. Res.*, 86, 1396, 1981.

Hones, E.W. Jr., The magnetotail: Its generation and dissipation, Physics of Solar Planetary Environnements, edited by D.J Williams, 559, AGU, Washington, D.C., 1976.

Jacquey, C., J.A. Sauvaud, and J. Dandouras, Location and propagation of the magnetotail current disruption during substorm expansion: Analysis and simulation of an ISEE multi-onset event, *Geophys. Res. Lett.*, 18, 389-392, 1991.

Jacquey, C., J.A. Sauvaud, J. Dandouras, and A. Korth, Tailward propagating cross-tail current disruption and dynamics of the near-Earth tail: a multi-point measurement, *Geophys. Res. Lett.*, 20, 983-986, 1993.

Jacquey, C., and J.A. Sauvaud, Magnetosphere-ionosphere response to enhanced energetic coupling between the solar wind and the magnetosphere, *J. Geophys. Res.*, 99, 11331, 1994.

Jacquey, C., J.A. Sauvaud, D. Popescu, H. Rème, D.G. Sibeck, S.I. Klimov, S.A. Romanov, R.P. Lepping, G.D. Reeves, Large scale response of the magnetotail to a substorm expansion: INTERBALL and IMP-8 observations on November 24, 1996, International Conference on Substorms ICS-4, Lake Hamano, Japan, ASSL, 155, 1998.

Klimov, S.I., et al., ASPI experiment: Measurements of fields and waves onboard the Interball-1 spacecraft, *Ann. Geophys.*, 15, 514, 1997.

Lui, A.T.Y., Road map to magnetotail domains, in *Magnetotail Physics, the Johns Hopkins University Press*, p. 3, 1987.

Lui, A.T.Y, R.E. Lopez, B.J. Anderson, K. Takahashi, L.J. Zanetti, R.W. McEntire, T.A. Potemra, D.M. Klumpar, E.M. Greene and R. Strangeway, Current disruptions in the near-Earth neutral sheet region, *J. Geophys. Res.*, 97, 1461, 1992.

McPherron, R.L., C.T. Russel and M.P. Aubry, Satellite studies of magnetospheric substorms on August 15, 1968. 9: Phenomenological model for substorm., *J. Geophys. Res.*, 78, 3131, 1973.

Mish, W.H. and R.P. Lepping, *Rep. X-694-76-158, NASA GSFl, Greenbelt, Md., Aug.* 1976.

Nagai, T., M. Fujimoto, Y. Saito, S. Machida, T. Teresawa, R. Nakamura, T. Yamamoto, T. Mukai, A. Nishida and S. Kokubun, Structure and dynamics of magnetic reconnection for substorm onsets with Geotail observations, *J. Geophys. Res.*, 103, 4419, 1998.

Nakai, H., Y. Kamide and C.T. Russell, Influences of solar wind parameters and geomagnetic activity on the tail lobe magnetic field: a statistical study, *J. of Geophys. Res.*, 96, 5511, 1991.

Ohtani, S., Earthward expansion of tail current disruption: dual-satellite study, *J. of Geophys. Res.*, 103, 6815, 1998.

Ohtani, S., S. Kokubun, R.C Elphic and C.T. Russell, Field-aligned current signatures in the near-Earth region 1. ISEE observations in the plasma sheet boundary layer, *J. Geophys. Res.*, 93, 9709, 1988.

Ohtani, S., S. Kokubun and C.T. Russell, Radial expansion of the tail current disruption during substorm: a new approach to the substorm onset region, *J. Geophys. Res.*, 97, 3129, 1992.

Ohtani, S., T. Mukai, H. Singer, A.T.Y. Lui, M. Nakamura, P. Prikryl, K. Yumoto and G. Rostoker, Substorm onset timing: The December 31, 1995 event, *J. Geophys. Res.*, in press, 1999.

Petrukovich, A.A., S.A. Romanov, L.M. Zelenyi, T. Mukai, Y. Siato, T. Yamamoto, S. Kokubun and O.A. Troshichev, Substorm-associated pressure variations in the magnetotail, International Conference on Substorms ICS-4, Lake Hamano, Japan, ASSL, 199, 1998.

Russell, C.T., The ISEE 1 and 2 fluxgate magnetometers, *IEEE Transactions on Geoscience Electronics*, GE16, 239, 1978.

Sauvaud, J.A., C. Jacquey, T.Beutier, R.P. Lepping, C.T. Russell, R.J. Belian, A.T.Y. Lui, Dynamics of the Magnetospheric Mid-Tail Induced by Substorms: a Multisatellite Study, *Adv. Space Res.*, 18, 35, 1996a.

Sauvaud, J.A., C. Jacquey, T.Beutier, C. Owen, R.P. Lepping, C.T. Russell, R.J. Belian, Large Scale Dynamics of the Magnetospheric Tail Induced by Substorms: a Multisatellite Study, *J. Of Geomag. Geoelectr.*, 48, 675, 1996b.

Slavin, J.A., E.J. Smith, E.L. Mazur, D.N. Baker, E.W. Hone, T. Iyemori and E.W. Greenstadt, ISEE-3 observations of traveling compression regions in the Earth magnetotail, *J. Geophys. Res.*, 98, 15425, 1993.

---

Christian Jacquey, Centre d'Etude Spatiale des Rayonnements, 9, avenue du Colonel Roche, 31400 Toulouse, France

# The Current Disruption Myth

### Joachim Birn
*Los Alamos National Laboratory, Los Alamos, New Mexico, U.S.A.*

### Michael Hesse
*Electrodynamics Branch, NASA Goddard Space Flight Center, Greenbelt, Maryland, U.S A.*

We investigate signs and symptoms of current disruption and diversion on the basis of resistive MHD simulations of magnetotail dynamics. The simulations generally support the view that the local occurrence of anomalous dissipation (here modeled by a resistive term) leads to the reduction (disruption) and diversion of cross-tail current. However, the dynamic relation between the occurrence of resistivity and its effects differs from views originally derived from circuit analogies. The primary myth addressed here is the expectation that current disruption is the direct consequence of anomalously enhanced dissipation, so that the current disruption region and the dissipation region should be collocated. We find that the initial consequence of enhanced resistance indeed is a reduction of the local cross-tail current threading the region of enhanced resistance. However, this local reduction is accompanied by a diversion of current around that region, so that there are no global consequences. The later, more global, current changes result from the plasma dynamics (flows) rather than from the local dissipation. Hence the region of reduced cross-tail current is distinct (located earthward) from the region of enhanced resistivity. The region of dipolarization, the most commonly considered indicator of current disruption, again is distinct from the region of current reduction. This region expands in all three space directions, which might affect the timing between satellites. The energy changes are consistent with the expected conversion of magnetic into kinetic and thermal energy. However, these processes are not highly localized and the release or conversion regions also are not collocated with the region of enhanced resistivity.

## 1. INTRODUCTION

As pointed out by *Alfvén and Carlquist* [1967], many electric current circuits (including laboratory plasma devices) have a tendency to "explode." That means that they undergo an instability that leads to the dissipation of the magnetic energy associated with the circuit when the current is forced to exceed a certain threshold. Using this circuit analogy, Alfvén and Carlquist proposed a solar flare model based on the interruption or disruption of the electric current in a magnetic flux loop. A similar concept was also applied to magnetospheric substorms [see, e.g., *Akasofu*, 1977]. Although it is rather difficult to measure electric currents in space directly, indirect measurements provide strong support for the crucial role that changes in the magnetospheric current systems play in magnetospheric activity. The magnetic field strength in the lobes of the magnetotail provides a good measure of the total current flowing across the tail at that particular distance along the tail. Therefore, the increase of the lobe field prior to substorm onset and its subsequent decrease is good evidence of current changes. Similarly, the stretching of the magnetic field in the near tail during the substorm growth phase and the subsequent dipolarization during the expansion phase are plausible consequences of a build-up and subsequent reduction of the cross-tail current [e.g., *Thomas and Hedgecock*, 1975; *McPherron*, 1979; *Lopez et al.*, 1988; *Baker et al.*, 1981, 1993; *Pulkkinen et al.*, 1992; *Sergeev et al.*, 1993].

Magnetospheric Current Systems
Geophysical Monograph 118
Copyright 2000 by the American Geophysical Union

Observations of the formation of a thin current sheet in the near-tail region in the late substorm growth phase [e.g., *Mitchell et al.*, 1990; *Sergeev et al.*, 1993; *Pulkkinen et al.*, 1994; *Sanny et al.*, 1994] indicate a possibly crucial role of current intensification in the initiation of substorms [e.g., *McPherron et al.*, 1987; *Lui et al.*, 1990; *Schindler and Birn*, 1993]. Furthermore, combining the idea of a disruption of cross-tail current with the diversion of currents to the Earth and a closure through the ionosphere ("substorm current wedge" [*McPherron et al.*, 1973]) provides a plausible and generally accepted model of magnetosphere-ionosphere coupling, consistent with ground magnetic signatures of substorms.

Current disruption or interruption and magnetic tearing or reconnection are frequently seen as alternative models of magnetospheric [see, e.g., *Akasofu*, 1977] as well as solar activity [e.g., *Kaburaki*, 1975; see, also, *Kahler et al.*, 1980]. However, simulations of magnetotail dynamics [e.g., *Birn and Hesse*, 1991, 1996] have demonstrated both features as aspects of the same dynamic evolution, described by the same large-scale model.

Although there is little doubt that current disruption and diversion are an intrinsic part of substorm activity in the magnetotail, there are several beliefs surrounding this topic, including its expected association with anomalous dissipation (leading to an effective resistance), that are not verified in self-consistent models. This view, termed here "the current disruption myth," makes a spatial association between the signs and symptoms of current reduction or disruption and the underlying dissipation mechanism, as in the original *Alfvén and Carlquist* [1967] model. Specifically, in the myth it is expected that the region of anomalous dissipation is identical to the region of current reduction (disruption) and thus to the region from which the disrupted current gets diverted to field-aligned currents connected with the ionosphere. Consequently, a tracing of the field-aligned currents from the ionosphere along magnetic field lines back to the equatorial plane is expected to indicate that source region. It is the purpose of this paper to address these topics on the basis of self-consistent, nonlinear, large-scale simulations, providing an overview of the changes of the current systems and their consequences. We address, particularly, the connection between a region of anomalous resistance and the reduction and diversion of currents, the three-dimensional view of the current changes and their consequences, such as the dipolarization of the magnetic field, and the global changes in energy contributions.

## 2. APPROACH AND BASIC ASSUMPTIONS

The simulations that form the basis of our investigations have been described earlier in more detail [e.g., *Birn and Hesse*, 1996]. Here we summarize some basic properties and assumptions. The initial state for the dynamic tail evolution is a section of the tail extending from $x = -5R_E$ to $-65R_E$ downtail, and to $\pm 10 R_E$ in the east-west ($y$) and south-north ($z$) directions, with symmetry around $y = 0$ and $z = 0$ imposed. The configuration thus includes the transition region from a more dipolar field to a stretched tail field. It also includes a thin current sheet in the near tail, formed dynamically by imposing a driving electric field at the boundary.

The dynamic energy release phase is initiated by imposing finite (anomalous) resistivity $\eta$ (while the external driving is discontinued). *Birn and Hesse* [1996] used several models for $\eta$, uniform resistivity as well as models based on the expected thresholds for current driven microinstability or a collisionless tearing mode. The numerical values for the magnitude of the resistivity were chosen arbitrarily, but turn out to be consistent with estimates of expected effects of cross-field current instabilities [e.g., *Lui*, 1996]. Since the large-scale dynamic evolution was found to be quite similar for the different resistivity models, Birn and Hesse primarily presented results from the run with uniform resistivity. Localized resistivity, however, more closely reflects the original idea of *Alfvén and Carlquist* [1967], and we will here focus on that model. It is based on the expected threshold for a current-driven instability, given by

$$j > \alpha n e v_{ti} \qquad (1)$$

where $v_{ti}$ is the ion thermal speed and $\alpha$ some numerical factor of the order of $(m_e/m_i)^{1/2}$ [*Mikhailovskii and Timofeev*, 1963; *Gary and Sanderson*, 1978]. Fast growth may be expected when the current density becomes comparable to $n e v_{ti}$ [e.g., *Lui et al.*, 1990]. Our adopted resistivity model is given by

$$\eta = \eta_o \max[0, \tanh(j - \bar{\alpha} n \sqrt{T})] \qquad (2)$$

Dimensionless quantities are used in (2), based on a length scale of, say, $1 R_E$, a characteristic magnetic field strength of 80 nT (lobe field strength at the inner boundary), and a characteristic velocity of 1000 km/s, corresponding to the Alfvén speed calculated with the lobe field strength and the plasma density at the inner boundary. In view of the fact that (2) is rather ad hoc and the parameters have not been confirmed theoretically for realistic tail conditions yet, a value of $\alpha = 3.4(m_e/m_i)^{1/2}$ was chosen (corresponding to $\bar{\alpha} = 3.0$) to be consistent with an onset at an observed thickness of $\sim 0.2 R_E$ of a thin current sheet at the end of the growth phase. This corresponds to a current density threshold of approximately 20-30 nA/m$^2$, at the lower end of the range estimated by *Lui* [1996]. The maximum resistivity $\eta_o$ chosen here corresponds to a Lundquist number (magnetic Reynoldsnumber) of 500. This corresponds to a maximum anomalous resistivity of $1.5 \times 10^4 \, \Omega$ m, consistent with the lower range of estimates cited by *Lui* [1996].

The boundary conditions at the near-Earth boundary and at the maximum $y$ and $z$ boundaries correspond to highly-conducting solid walls, while the distant boundary in $x$ is open to outflow. The initial plasma sheet profile, and even more the profile after the formation of the thin current sheet, exhibit a reduction of the plasma/current sheet thickness near midnight, and correspondingly an increase of $B_z$ from midnight toward the flanks of the tail. As a consequence, the reconnection effects start in, and remains confined to, a region of finite width around $y = 0$. Therefore the boundary conditions for $y = y_{max}$ have little influence on the evolution. Similarly, the boundary conditions at $z = z_{max}$ have little effect (when the driving is discontinued). In contrast, the near-Earth boundary may have a stronger impact, because closure of field-aligned currents in the ionosphere

[1999], the diversion to field-aligned currents is the consequence of flow braking and shear (rather than associated with the locally increased resistance), which occur much closer to the Earth than the enhancement of resistivity. In addition to the reduction of cross-tail currents associated with the region 1 type field aligned currents fo the substorm current wedge, we find enhanced westward current father earthward. This current may be interpreted as a contribution to the partial ring current, connected with region 2 type field aligned currents [*Cowley*, 1999].

We expanded an earlier discussion of $B_z$ signatures, showing global views in the equatorial plane as well as in the midnight meridian plane. These views again demonstrated at first a weak earthward propagating positive $\Delta B_z$ signal, then a more significant increase of $B_z$, starting out in the near tail, expanding both tailward and earthward, longitudinally, and (as a new result) to higher latitude. For a timing of dipolarization signals between different satellites, therefore, the separation in all three space directions may play a role.

As demonstrated earlier [*Birn and Hesse*, 1996], the peaks of the field-aligned current density do not follow field lines and cease at finite distance from the equatorial plane (being converted from perpendicular currents) when traced form closer to the Earth. The simulation results agree qualitatively with conclusions from the mapping of ground signatures, namely, that a brightening auroral arc, which one might associate with the peak field-aligned current density, indicates current diversion in the near tail inside the closed field line region. However, they also show that a simple mapping of near-Earth or ground features of substorms via magnetic field lines into the equatorial plane does not necessarily reveal the source region or the onset location.

An investigation of the changes in the various contributions to the energy density showed that overall the simulations are consistent with the conversion of magnetic into kinetic and thermal energy, as expected from a reconnection model. However, the release and conversion process occurs predominantly not coincident with the region of enhanced resistivity, which is contrary to expectations by *Alfvén and Carlquist* [1967]. Although the evolution is very dynamic and non-steady, it more closely resembles the reconnection model of *Petschek* [1964]. Again the major conversion involves the plasma dynamics and takes place earthward of the region of enhanced resistivity.

In summary, the simulation results are consistent with the logical sequence that a microscopic process, triggered by a local increase in current density, leads to an effective increase in anomalous resistance, which subsequently causes current disruption and diversion (substorm current wedge). The simulations, however, also suggest that the logical connection differs from what one might conclude by just considering local Ohmic dissipation. In the simulations, the local breakdown of ideal MHD leads to a rise of a large-scale tearing type instability, which also causes neutral line formation and plasmoid ejection. The flow pattern associated with this instability causes magnetic field changes (earthward of the region of enhanced resistivity) that are consistent with those interpreted as signs of current disruption and diversion.

The simulations hence support the logical sequence of the "neutral line model" of substorms [e.g., *Baker et al.*, 1996], in contrast to the "current disruption model" [*Lui et al.*, 1990], in which reconnection is a possible consequence of current disruption rather than vice versa. The myth addressed here, however, is not unique to the current disruption model. In earlier versions of the neutral line model as well, current disruption was attributed to the anomalous dissipation region, necessary for reconnection, and hence the field-aligned currents of the current wedge were believed to be magnetically connected to the westward and eastward edges of the reconnection site. The simulations discussed in this paper, consistent with results by *Scholer and Otto* [1991], demonstrate that this expectation is not verified and that current reduction and diversion, involving the plasma dynamics with induced electric fields and plasma motions, is quite distinct (and spatially separated) from the dissipation process.

The simulations discussed here use the one-fluid MHD approximation. Observations of the late growth phase and the early expansion phase indicate that the current sheet thickness in the near tail may become quite small during these phases so that Hall effects and finite Larmor radius corrections are no longer negligible. In this situation, the plasma flows responsible for the magnetic field changes are replaced by the electron flows. While the dynamic time and length scales might well be affected by this change, the qualitative conclusions about the role of the induced electric field and the plasma dynamics should remain valid. Indeed, fully three-dimensional particle simulations of dynamic changes in the magnetosphere show quite similar associations between plasma flows and current changes as the MHD simulations [*Nishikawa*, private communication].

The major difference between the simulations discussed here and the model suggested by *Lui et al.* [1990] lies in the fact that the effects of anomalous resistance occur farther tailward than postulated by Lui et al. To investigate the possible influence of the location of anomalous resistance, we first tried to generate a thin current sheet closer to Earth. While an inner edge of enhanced current density could be produced dynamically somewhat closer than in the present simulations, the main effects on resistance were still found tailward of the region considered by Lui et al. We then arbitrarily placed localized resistance very close to the earthward boundary. Depending on the resistivity model and on the properties of the preexisting configuration, we found two possible dynamical responses. Either the large scale configuration remains stable. In this case, the locally enhanced current gets dissipated without global consequences. Or the large scale configuration becomes unstable. In that case the dynamic evolution is qualitatively similar to the one described in this paper, involving reconnection (with a neutral line in the region of enhanced resistivity) and fast flows. Although the region of earthward flow is more limited in $x$, it nevertheless is responsible for the magnetic field and current changes in the inner region in the same way as described here. We therefore conclude that the location of the occurrence of nonideal MHD effects, causing anomalous dissipation, is irrelevant for the mechanism of current disruption and diversion in the magnetotail.

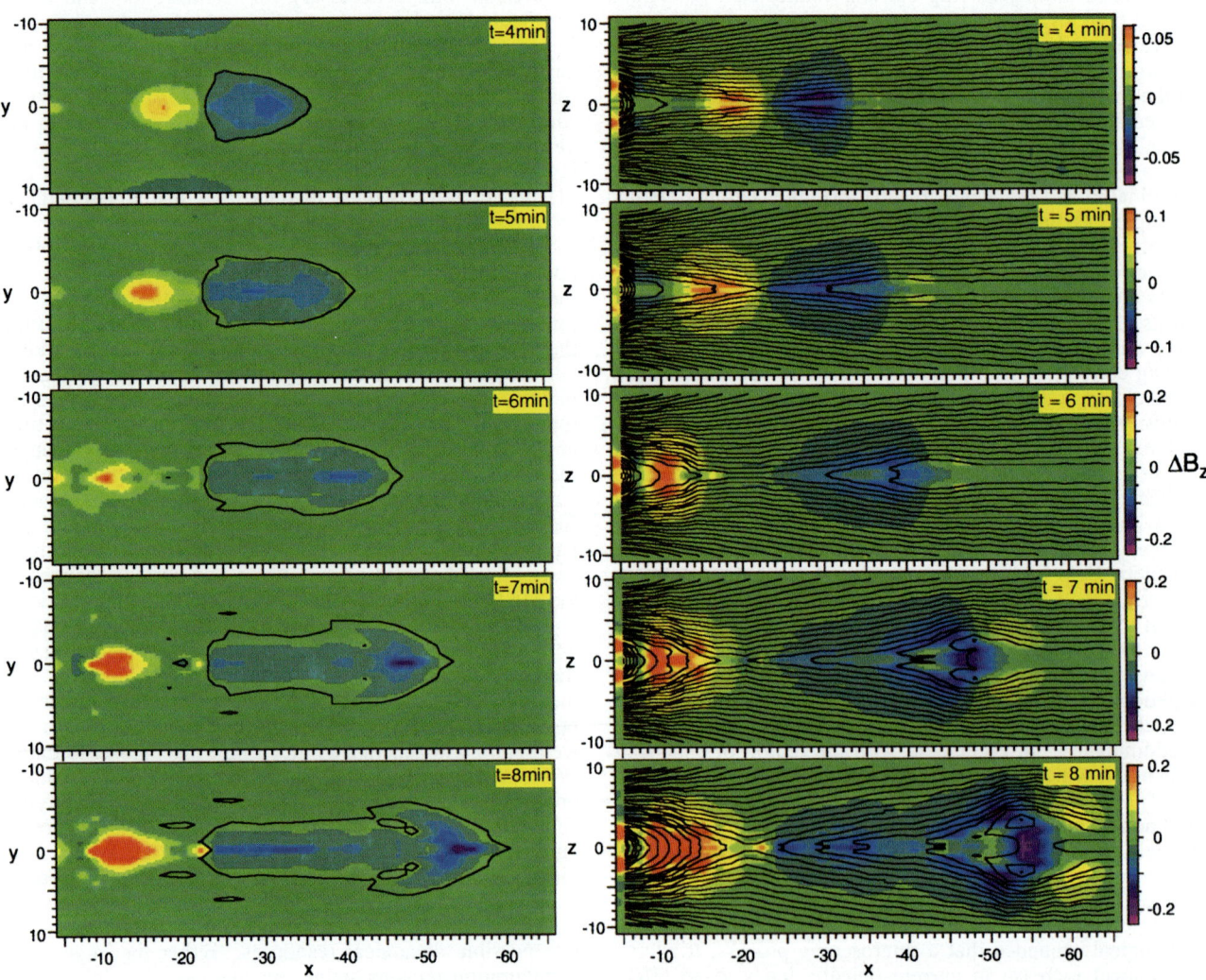

**Plate 3.** Color-coded changes of the magnetic field component $\Delta B_z = B_z(t) - B_z(0)$, (left) in the equatorial plane with the solid line representing the magnetic neutral line, $\mathbf{B} = 0$, and (right) in the midnight meridian plane with solid lines representing magnetic field lines.

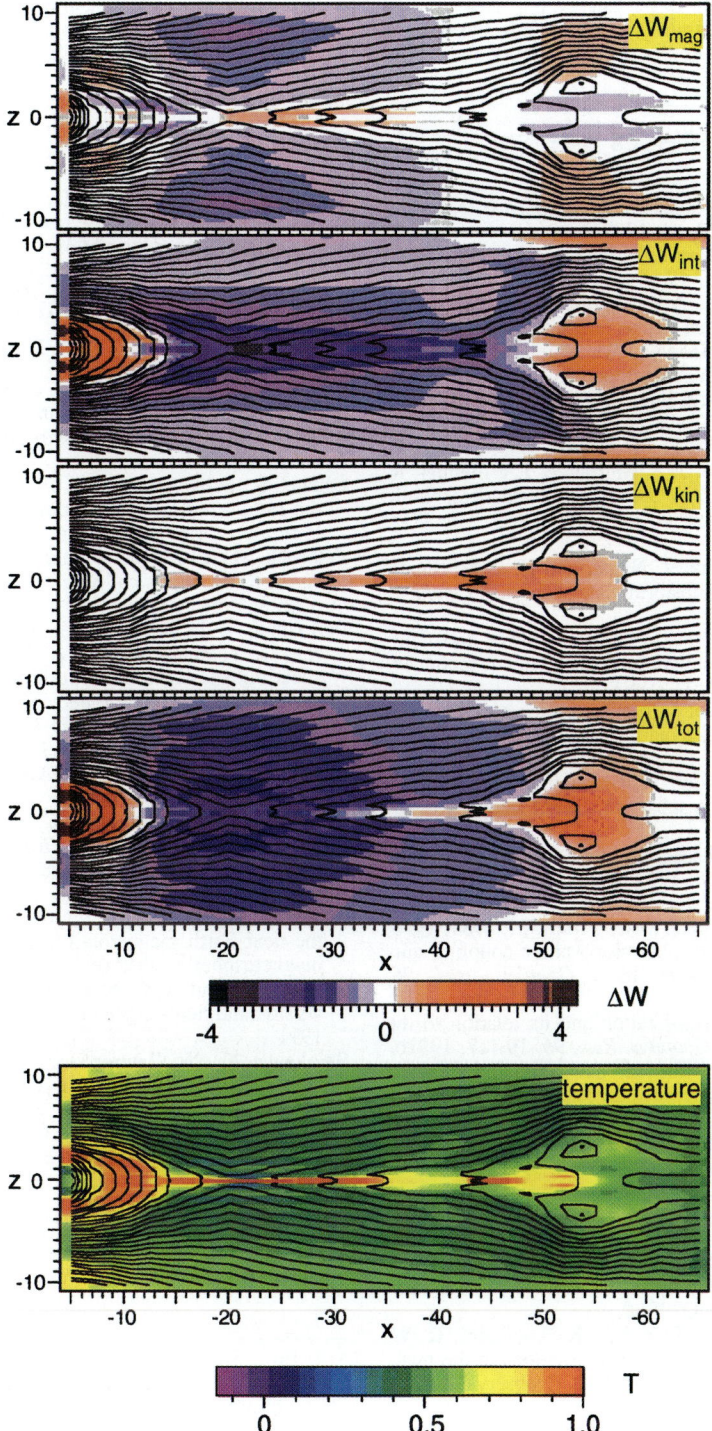

**Plate 4.** Energy losses and gains as functions of $x$ and $z$, integrated from $y = -10$ to $y = 10$, and (bottom) temperature at $y = 0$, for $t = 8$ min. $\Delta W = W(t) - W(0)$ represents the changes of the energy density contributions, (from top) magnetic energy, thermal (internal) energy, kinetic energy, and total energy. Solid lines represent magnetic field lines at $y = 0$.

*Acknowledgments.* This work was supported by the U.S. Department of Energy's Office of Basic Energy Sciences through its Geosciences Research Program and by NASA through its Sun Earth Connection Theory Program. The authors thank Karl Schindler for valuable discussions.

## REFERENCES

Akasofu, S.-I., *Physics of Magnetospheric Substorms*, D. Reidel Publ. Comp., Dordrecht-Holland, 1977

Alfvén, H., and P. Carlquist, Currents in the solar atmosphere and a theory of solar flares, *Solar Phys.*, *1*, 220, 1967.

Baker, D. N., E. W. Hones Jr., P. R. Higbie, R. D. Belian, and P. Stauning, Global properties of the magnetosphere during a substorm growth phase, *J. Geophys. Res.*, *86*, 8941, 1981.

Baker, D. N., T. I. Pulkkinen, V. Angelopoulos, W. Baumjohann, and R. L. McPherron, Neutral line model of substorms: Past results and present view, *J. Geophys. Res.*, *101*, 12,975, 1996.

Baker, D. N., T. J. Pulkkinen, R. L. McPherron, J. D. Craven, L. A. Frank, R. D. Elphinstone, J. S. Murphree, J. F. Fennel, R. E. Lopez, and T. Nagai, CDAW 9 analysis of magnetospheric events on May 3, 1986: Event C, *J. Geophys. Res.*, *98*, 3815, 1993.

Baumjohann, W., G. Paschmann, T. Nagai, and H. Lühr, Superposed epoch analysis of the substorm plasma sheet, *J. Geophys. Res.*, *96*, 11,605, 1991.

Birn, J., and M. Hesse, The substorm current wedge and field-aligned currents in MHD simulations of magnetotail reconnection, *J. Geophys. Res.*, *96*, 1611, 1991.

Birn, J., and M. Hesse, Details of current disruption and diversion in simulations of magnetotail dynamics, *J. Geophys. Res.*, *101*, 15,345, 1996.

Birn, J., M. Hesse, G. Haerendel, W. Baumjohann, and K. Shiokawa, Flow braking and the substorm current wedge, *J. Geophys. Res.*, in press, 1999.

Cowley, S. W. H., Magnetosphere-ionosphere interaction - A tutorial review, *this monograph*, 1999.

Gary, S. P., and J. J. Sanderson, Density gradient drift instabilities: Oblique propagation at zero beta, *Phys. Fluids*, *21*, 1181, 1978.

Hesse, M., and J. Birn, Magnetosphere-ionosphere coupling during plasmoid evolution: First results, *J. Geophys. Res.*, *96*, 11,513, 1991a.

Hesse, M., and J. Birn, On dipolarization and its relation to the substorm current wedge, *J. Geophys. Res.*, *96*, 19,417, 1991b.

Hesse, M., and J. Birn, Three-dimensional MHD modeling of magnetotail dynamics for different polytropic indices, *J. Geophys. Res.*, *97*, 3965, 1992.

Hesse, M., and J. Birn, MHD simulations of magnetotail instability for localized resistivity, *J. Geophys. Res.*, *99*, 8565, 1994.

Kaburaki, O., Thermodynamical approach to current interruption model of solar flares, *Publ. Astron. Soc. Japan*, it 27, 45, 1975.

Kahler, S., D. Spicer, Y. Uchida, and H. Zirin, 3. Primary energy release, in *Solar Flares*, edited by P. Sturrock, p. 83, Colorado Ass. Univ. Press, Boulder, CO, 1980.

Lopez, R. E., D. G. Sibeck, A. T. Y. Lui, K. Takahashi, R. W. McEntire, and T. A. Potemra, Substorm variations in the magnitude of the magnetic field: AMPTE/CCE observations, *J. Geophys. Res.*, *93*, 14,444, 1988.

Lui, A. T. Y., Current disruption in the Earth's magnetosphere: Observations and models, *J. Geophys. Res.*, *101*, 13,067, 1996.

Lui, A. T. Y., A. Mankofsky, C.-L. Chang, K. Papadopoulos, and C. S. Wu, A current disruption mechnism in the neutral sheet: A possible trigger for substorm expansions, *Geophys. Res. Lett.*, *17*, 745, 1990.

McPherron, R. L., Magnetospheric substorms, *Rev. Geophys.*, *17*, 657, 1979.

McPherron, R. L., C. T. Russell, and M. A. Aubry, Satellite studies of magnetospheric substorms on August 15, 1968, 9, Phenomenological model for substorms, *J. Geophys. Res.*, *78*, 3131, 1973.

McPherron, R. L., A. Nishida, and C. T. Russell, Is near-Earth current sheet thinning the cause of auroral substorm onset?, in *Quantitative Modeling of Magnetosphere-Ionosphere Coupling Processes*, edited by Y. Kamide and R. A. Wolf, p. 252, Kyoto Sangyo Univ., Kyoto, Japan, 1987.

Mikhailovskii, A. B., and A. V. Timofeev, Theory of cyclotron instability in a non-uniform plasma, *Sov. Phys. JETP*, Engl. Transl., *17*, 626, 1963.

Mitchell, D. G., D. J. Williams, C. Y. Huang, L. A. Frank, and C. T. Russell, Current carriers in the near-Earth cross-tail current sheet during substorm growth phase, *Geophys. Res. Lett.*, *17*, 583, 1990.

Ohtani, S., S. Kokobun, and C. T. Russell, Radial expansion of the tail current disruption during substorms: A new approach to the substorm onset region, *J. Geophys. Res.*, *97*, 3129, 1992.

Petschek, H. E., Magnetic field annihilation, *AAS-NASA Symposium on the Physics of Solar Flares*, NASA Spec. Publ. SP-50, p. 425, 1964.

Pulkkinen, T. I., D. N. Baker, D. G. Mitchell, R. L. McPherron, C. Y. Huang, and L. A. Frank, Global and local current sheet thickness estimates during the late growth phase, in *Substorms 1*, Eur. Space Agency Spec. Publ., SP-335, 131, 1992.

Pulkkinen, T. I., D. N. Baker, D. G. Mitchell, R. L. McPherron, C. Y. Huang, and L. A. Frank, Thin current sheets in the magnetotail during substorms: CDAW 6 revisited, *J. Geophys. Res.*, *99*, 5793, 1994.

Sanny, J., R. L. McPherron, C. T. Russell, D. N. Baker, T. I. Pulkkinen, and A. Nishida, Growth-phase thinning of the near-Earth current sheet during the CDAW 6 substorm, *J. Geophys. Res.*, *99*, 5805, 1994.

Schindler, K., and J. Birn, On the cause of thin current sheets in the near-Earth magnetotail and their possible significance for magnetospheric substorms, *J. Geophys. Res.*, *98*, 15,477, 1993.

Scholer, M., and A. Otto, Magnetotail reconnection: Current diversion and field-aligned currents, *Geophys. Res. Lett.*, *18*, 733, 1991.

Shiokawa, K., W. Baumjohann, and G. Haerendel, Braking of high-speed flows in the near-Earth tail, *Geophys. Res. Lett.*, *24*, 1179, 1997.

Sergeev, V. A., D. G. Mitchell, C. T. Russell, and D. J. Williams, Structure of the tail plasma/current sheet at $\sim 11\,R_E$ and its changes in the course of a substorm, *J. Geophys. Res.*, *98*, 17,345, 1993.

Thomas, B. T., and P. Hedgecock, Substorm effects in the neutral sheet inside 10 $R_e$, in *Magnetospheres of the Earth and Jupiter*, edited by P. Formisano, p. 55, D. Reidel, Norwell, Mass., 1975.

---

J. Birn, Space and Atmospheric Sciences Group, M.S. D466, Los Alamos National Laboratory, Los Alamos, NM 87545. (e-mail: jbirn@lanl.gov)

M. Hesse, Electrodynamics Branch, Code 696, NASA Goddard Space Flight Center, Greenbelt, MD 20771. (e-mail: hesse@gsfc.nasa.gov)

# Near- and Mid-Tail Current Flow During Substorms: Small- and Large-Scale Aspects of Current Disruption

Michael Hesse

*Electrodynamics Branch, NASA Goddard Space Flight Center, Greenbelt, Maryland*

Joachim Birn

*Los Alamos National Laboratory, Los Alamos, New Mexico*

We investigate kinetic processes of relevance to the re-arrangement of magnetotail current systems during the substorm growth and expansion phases. The studies address microphysical processes occurring in the magnetotail current layer with clear macroscopic consequences. They are based on fully self-consistent, electromagnetic, particle-in-cell simulations. The first focus of the analyses is on the pre-onset formation of a thin current sheet. The additional current, brought about by the lobe magnetic field increase associated with solar wind-like driving electric fields, appears to be carried by the electrons, in a thin current sheet of substantially enhanced current density. Thin current sheet formation leads to a reduction in the normal magnetic field, which is shown to be sufficient for magnetic reconnection to initiate. This onset is analyzed in detail. Last, an investigation of the kinetic kink instability, which can lead to strong current sheet warping, and potentially to turbulence, is presented. The results here support earlier analytical results indicating that kinetic kinking growth rates are strongly reduced if realistic ion-electron mass ratios are considered.

## 1. INTRODUCTION

Current sheets are a ubiquitous feature of plasmas, occurring anywhere from laboratory plasmas [e.g., *Yamada et al.*, 1997], solar coronal plasmas [e.g. *Priest*, 1984], to astrophysical systems [*Kahn and Brett*, 1993]. An important example in the neighborhood of the Earth is found in the Earth's magnetotail, where a plasma and current sheet separate regions of oppositely directed magnetic flux in the lobe regions [e.g., *Baker et al.*, 1996]. This current sheet often undergoes rapid reconfiguration during the dynamics of the nightside magnetosphere.

Satellite observations have fostered the creation of a phenomenological picture of the most frequent mode of magnetotail dynamics, the magnetospheric substorm. During the growth phase of a substorm, solar wind electric fields lead to a thinning of the plasma sheet, or the formation of a thinner embedded current sheet [*Sanny et al.*, 1994]. In the process, the overall magnetic configuration of the nightside magnetosphere becomes more stretched, with reduced magnetic flux threading the equatorial current sheet [e.g., *Tsyganenko*, 1989].

This growth phase evolution terminates at the onset of fast dynamics, commonly attributed to a yet undetermined instability. It is generally agreed, however, that the magnetic field configuration on the nightside of the Earth reverts to a more dipolar configuration, with reduced current flow across the magnetotail [*McPherron et al.*, 1973]. It is also agreed that magnetic reconnection plays an important role in this dynamical phase [e.g., *Fairfield*, 1992].

The immediate cause of substorm onset, in tail observations identified by the dipolarization of the magnetic field, remains a contentious issue. One school alleges that the dipolarization be a consequence of a process operating locally in the inner magnetospheric region, probably associated with a kinetic, current driven instability [*Lui et al.*, 1990]. This contention is supported by observations of fast magnetic field fluctuations during dipolarization [e.g., *Ohtani et al.*, 1992]. Magnetic reconnection is seen as a consequence, whose action is triggered by the tailward propagation of a rarefaction wave [*Lui et al.*, 1990]. The argument brought forth against the action of magnetic reconnection at, or prior to, substorm onset is based on the apparent stability of collisionless tearing in typical, magnetotail-like, configurations [e.g., *Pellat et al.*, 1991].

The other school, backed by energy, and flux balance arguments [*Hesse*, 1995; *Hesse et al.*, 1996], magnetohydrodynamic simulations [e.g., *Birn et al.*, 1996], and recent observations of reconnection signatures prior to substorm onset [*Nagai et al.*, 1998], favors a scenario in which magnetic reconnection, as well as features such as fast flows, and dipolarization, are part of the same, large-scale instability [e.g., *Hesse and Birn*, 1994]. Here MHD simulations have been successful in explaining the dipolarization of the magnetic field by magnetic flux pile-up [*Hesse and Birn*, 1991], the formation of the substorm current wedge [*Birn and Hesse*, 1991; *Birn et al.*, 1996; 1999], the acceleration of energetic particles [*Birn et al.*, 1998], and the fast earthward and tailward flows seen in actual observations [e.g., *Raeder*, 1994].

Despite the successes of MHD modeling in this area, a still outstanding issue revolves around the actual onset of magnetic reconnection itself. Observations as well as recent modeling results have shown that the old paradigm of fast processes initiating in a thick current sheet needs to be replaced by models including thin current sheets, generated by the growth phase evolution [*Sanny et al.*, 1994]. The changed initial configuration, with the potentially much thinner current sheet and reduced normal magnetic field, merits further stability investigations with respect to collisionless tearing.

Furthermore, kinetic models of thin current sheet formation [e.g., *Pritchett and Coroniti*, 1995] as well as recent Geotail observations [T. Mukai, this issue] have shown that the current densities in these thin current sheets are supported by electrons, rather than ions, thereby putting into question the applicability of ion current based instabilities as causes of substorm onset. Accordingly, one should consider kinetic instabilities which operate in systems where the electrons provide current carriers. Of these, the most commonly discussed is the kinetic kink instability [e.g., *Zhu and Winglee*, 1996].

Many studies of this instability have been performed and shown it to operate in thin current sheets [e.g., *Lapenta and Brackbill*, 1998]. The small ion-to-electron mass ratio mandated by most particle-in-cell simulations, however, has been some cause for concern. Furthermore, while some numerical experiments have indicated stronger growth for more realistic mass ratios [*Lapenta and Brackbill*, 1998], a semi-analytic investigation comes to the opposite conclusion [*Daughton*, 1999]. Thus further numerical work is required to shed light onto the viability of the kinetic kink instability as a mediator of substorm onset.

In this paper, we present results of fully electrodynamic particle-in-cell simulations addressing the above issues. In particular, we will analyze in detail the structure of a pre-onset current sheet model, with focus on the current changes brought about by the increase of the lobe magnetic field during the growth phase. Furthermore, we will use this model to study the stability of this configuration with respect to collisionless tearing. After onset of the fast dynamics, we will again revisit the structure of the reconnecting current sheet with the goal of analyzing current carriers for the purpose of comparison with recent Geotail observations.

We will also address the issue of current driven instabilities by studying the growth rate dependence of the kinetic kink instability in an electron current sheet. Through this study, we hope to shed light on the role of this mode in magnetospheric dynamics, as well as on the importance of three-dimensional effects in the reconnection process itself.

## 2. THE MODEL

### 2a. Simulation Code

All investigations which are part of the present study are based on our fully electromagnetic, particle-in-cell, simulation code [*Hesse and Winske*, 1998; *Hesse et al.*, 1999]. Only the 2.5 dimensional version of the code is utilized in the present studies, neglecting variations either in the cross-tail ($y$) direction (reconnection study), or in the $x$ direction (kink mode study).

Ions are assumed to be protons in the following investigations. Further, we normalize lengths to the ion inertial length $c/\omega_i = c(e^2 n_0/\epsilon_0 m_i)^{-1/2}$ using the maximum value of the current sheet density $n_0$, and times are normalized to the inverse of the ion cyclotron frequency $\Omega_i = eB_0/m_i$ in the asymptotic magnetic field $B_0$ (maximum lobe field) unless noted otherwise. The system dimensions and initial conditions are detailed below. A time step of an inverse electron plasma frequency $\omega_e \Delta t = 1$ is used.

Reflecting wave boundary conditions for the electromagnetic fields are used at nonperiodic and nondriven boundaries. This choice of boundary conditions is possible because of the wave damping in the implicit scheme. At driven boundaries, the $y$ component of the electric field is prescribed. Particles are specularly reflected at all nonperiodic boundaries.

### 2b. Initial Conditions, Tail Model

The magnetotail study initiates in a tail-like equilibrium of the class developed by [*Birn and Schinder*, 1983]. Here the magnetic field is derived from a flux function

$$A = -\ln\cosh(\nu(x)z) + \ln(\nu) \quad (1)$$

The function $\nu(x)$ regulates the current sheet width and its dependence on the $x$ coordinate. For the purpose if this study, we adopt the following form

$$\nu = \left(1 + \frac{b_n x}{\gamma x_{max}}\right)^{-\gamma} \quad (2)$$

which resembles the average properties of the magnetotail tailward of some 10-15 Earth radii [*Birn and Schindler*, 1983]. Here the simulation region size is $x_{max} = 60$, the exponent $\gamma = 0.6$, and the normal magnetic field component $B_z(x = 0, z = 0) = b_n = 0.03$. The plasma sheet half-width at the earthward edge of the simulation is one ion inertial length, corresponding to some $720 km$ for a plasma density of $0.1 cm^{-3}$.

This configuration in integrated in a box of size $L_x = 60$, and $L_z = 20$. Clearly, this simulation box size is considerably smaller than the actual magnetotail. This smaller size, is at present, unavoidable due to the numerical effort involved in these simulations. Nevertheless, results from simulations such as presented here can help to shed light on stability properties of current sheets, such as found in the real magnetosphere.

The number of grid points is 600 in the $x$, and 300 in the $z$ direction. $12.5 \times 10^6$ ions and electrons macro particles each are used in the simulation. We utilize an ion-electron mass ratio of $m_i/m_e = 100$, and will comment on the effect of a realistic mass ratio. In order to avoid the finite-grid instability. discussed, e.g., by *Brackbill and Lapenta* [1994], the Debye length has to be sufficiently large compared to the grid spacing. In order to satisfy this requirement, we set the ion-electron temperature ratio to unity.

Runs without further modifications show this configuration to be tearing stable, at least for times less than about 400 ion cyclotron times, where no indication of instability could be seen. In order to study the impact of thin current sheet formation, however, we apply driving electric fields at the boundaries in the manner of *Hesse et al.* [1997], for times less than 20 ion cyclotron times. These driving electric fields increase the lobe magnetic field strength, and the flaring angle of the lobe field direction similar to what is found in actual observations. As a result, previous simulations have shown the formation of thin current sheets [e.g., *Hesse et al.*, 1997], but none so far has been undertaken in a fully kinetic model with such large a mass ratio.

### 2c. Initial Conditions, Kink Mode Study

The initial configuration was chosen as a Harris-sheet equilibrium in the $y$-$z$ plane,

$$B_x = \tanh(z) \qquad (3)$$

Four particle species, two of ions and electrons each, were integrated in each run, in a sqare simulation domain of size $L_y = L_z = 16$. The first set of ion and electron species establishes the pressure and currents demanded by (2). The second set of species constitute a constant density level background $n_b = 0.2$. Here the system is set up such that foreground electrons only carry the current, by the inclusion of an electric field in the inital condition. Background temperatures are identical to the temperatures of the current carrying species $T_i + T_e = 0.5$, with $T_e/T_i = 0.2$. The mass ratio is varied in four steps, between $m_i/m_e = 9$ to $m_i/m_e = 100$. Particle numbers range from a total of $2.1 \times 10^6$ for mass ratio 9, to $2.1 \times 10^7$ for mass ratio 100. The number of grid cells varies between $n_y = n_z = 120$, and $n_y = n_z = 800$, dependent on the mass ratio. The current sheet helf-thickness of one ion inertial length is chosen for simplicity and to obtain a rapid growth of the kink instability.

## 3. ONSET OF MAGNETIC RECONNECTION

### 3a. Formation and Structure of a Thin Current Sheet

The pressure in the center of the plasma sheet, and hence its gradient along $x$, are determined largely by the balance

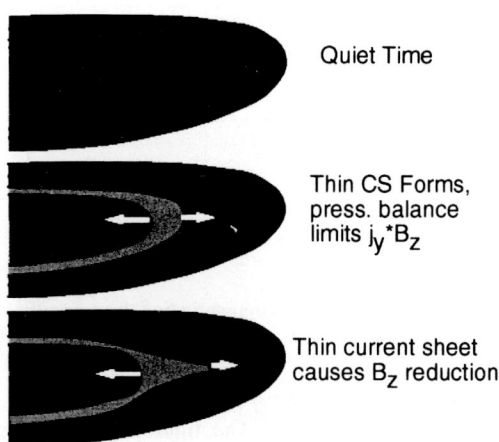

Figure 1. Sketch of the formation of a thin current sheet (light) within the thicker plasma sheet (dark). The enhanced current density requires a reduction of the north-south magnetic field, with a weak divergent flow.

with the lobe magnetic pressure, which increases moderately during the growth phase, driven by the external electric field. For a slowly driven evolution, force balance inside the plasma sheet should be maintained. A significant increase in the current density inside the plasma sheet hence must be accompanied by a reduction of $B_z$ and therefore to a change of the radial distribution of $B_z$. This scenario is illustrated in Figure 1, which schematically depicts the formation of a thin current sheet (light) within a thicker plasma sheet (dark). External driving leads to a transition from an unperturbed state (top panel) to a state with an enhanced and localized current density (center panel). Now force balance requires a reduction of the magnetic field $B_z$ brought about by an expansion of the field lines in the regions of enhanced current density in the equatorial plane. Both of these effects are expected to destabilize the tail current sheet.

Plate 1 shows the evolution of the magnetic field and the current density in the $x, z$ plane, demonstrating the formation of a thin current sheet and its subsequent breakup. Along with an increase in lobe magnetic field strength, the current density in the plasma sheet region increases strongly and disproportionally, in a localized region almost aligned with a magnetic flux tube. After $t = 30$, the current sheet becomes unstable and reconnection begins to operate, by X line formation, and a significant rearrangement of magnetic flux and current density is evident. We will study the onset of the reconnection process below. At later times, the ion flow velocity in the central current sheet approached the Alfvén speed.

The increase of magnetic flux through the $x = 0$ plane is evident in the graph of Figure 2. The graph shows the effects of the driving boundary conditions, which generate a flux increase by about 65%. After termination of the driven phase ($t > 20$), the total magnetic flux remains constant, bearing evidence to the flux conservation quality of the simulation code.

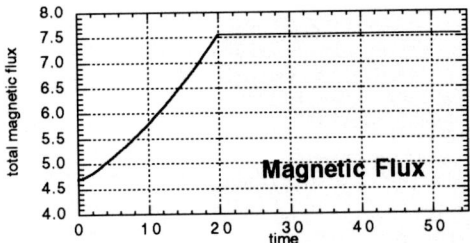

Figure 2. Evolution of the total magnetic flux threading the $x=0$, $z>0$ half-axis.

The evolution of the total $y$ momentum of ions and electrons is shown in Figure 3. The figure demonstrates that the total plasma momentum in the $y$ direction is very well conserved in the numerical simulation. The total ion and electron currents, obtained by integration over the simulation plane, show a different behavior. The additional current produced by the electric field driving is virtually exclusively carried by the electrons. This fact is seen in Figure 4, and can be demonstrated by analytical theory [Hesse et al., 1997].

Thus the increase of the total current during the driven phase can be attributed to the electrons. This leaves open the question, though, of whether the current density in the thin current sheet proper is also supplied by the electrons. This question is answered by Figure 5, which displays the profiles along the $x$ axis of ion and electron current densities, for three different times of the simulation. While the initial current densities, shown in the top panel, are equal by virtue of the equal initial temperature, the evolution has produced, by $t=30$, a large enhancement of the electron current density, locally exceeding a factor of three of its original value. Similar conclusions can be drawn from an analysis of the entire two-dimensional distribution of the $y$-directed current densities. Interestingly, the dynamic evolution after the onset of magnetic reconnection still leads to significant electron current density enhancements, as evidenced by the bottom panel of fig. 5. Electron current enhancements even after onset of substorms have recently been observed by GEOTAIL [T. Mukai, private communications].

The reduced electron distribution function, integrated over the $z$ coordinate in velocity space, is shown in Plate 2, for $t=30$. A total of 20500 particles in the area $8<x<13$, and $-1<z<1$ were used in its construction. The distribution features a noticeable drift in the $y$ direction of an essentially isotropic structure. Similar conclusions can be drawn for the other two reduced distribution functions (not shown). Thus we conclude that electron phase space densities in the thin current sheet region are likely described by drifting isotropic distributions.

### 3b. Onset of Magnetic Reconnection

The onset of magnetic reconnection in the present geometry requires that $B_z$ vanishes at the prospective reconnection site. Electron stabilization prevents this from occurring, unless a suitable diffusion mechanism becomes active. Thus the onset of magnetic reconnection requires a diffusive electric field $E_D$ in the electron Ohm's law

$$\mathbf{E} + \mathbf{v}_e \times \mathbf{B} = \mathbf{E}_D \qquad (4)$$

which we shall investigate in the following, for the time $\Omega_i t = 30$, just prior to $B_z < 0$ transition. This diffusive electric field has to be of sufficient magnitude to overcome the stabilization by a finite $B_z$ magnetic field. A reduction in $B_z$, as evidenced in Figure 6, is expected to lead to a destabilization.

Previous investigations [e.g., Hesse and Winske, 1998] have shown that the pressure-based contribution

$$E_{Dy} = -\frac{1}{ne}\left(\frac{\partial P_{xye}}{\partial x} + \frac{\partial P_{yze}}{\partial z}\right) \qquad (5)$$

provides the dominant dissipation in slowly evolving systems. In the present situation, inspection of the simulation results indicates that the most dominant contribution of (5) stems from the $x$ derivative of $P_{xye}$. In a situation with nonvanishing magnetic field, this term can be evaluated to an excellent approximation [e.g., Kuznetsova et al., 1998]

$$P_{xye} \approx -\frac{p_e}{2\Omega_{ze}}\frac{\partial v_{xe}}{\partial x} \qquad (6)$$

Here $p_e$ denotes the isotropic part of the electron pressure tensor, $\Omega_{ze}$ the electron cyclotron frequency in the normal magnetic field, and $v_{xe}$ the $x$ component of the electron flow velocity.

Insertion of expression (6) into (5) requires taking derivatives in with respect to $x$. Inspection of fig. 6 suggests that derivatives of the electron cyclotron frequency are dominant, a result that is supported by a detailed analysis of the other quantities in (6). Thus we find

$$E_{Dy} \approx \frac{p_e}{2ne\Omega_{ze}^2}\frac{\partial \Omega_{ze}}{\partial x}\frac{\partial v_{xe}}{\partial x} \qquad (7)$$

This expression is readily evaluated from the actual simulation results. To this extent, Figure 7 displays the behavior of the electron flow velocity in the vicinity of the $B_z$ minimum, in units of the Alfvén velocity. While featuring a high fluctuation level, which locally leads to very large gradients, a conservative estimate of the trend of the velocity

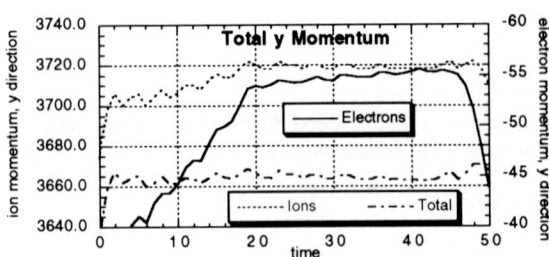

Figure 3. Time evolution of the total ion and electron momenta in the $y$ direction, and of their sum.

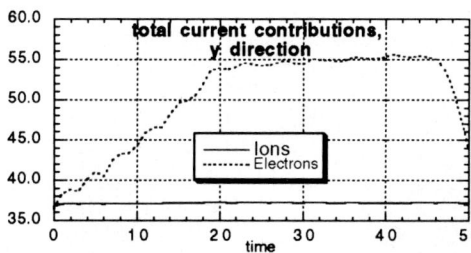

Figure 4. Time evolution of the total ion and electron currents in the $y$ direction.

distribution leads to a slope of $\frac{\partial v_{xe}}{\partial x} \approx 8 \times 10^{-3}$. If no significant temperature change is assumed, this number can be inserted into (7) to provide an estimate for the $y$ component of the diffusive electric field $E_{Dy} \approx 2 \times 10^{-3}$ in units of the product of Alvén velocity and lobe magnetic field magnitude, which matches closely the magnitude of $E_y$ at the site of X point formation. Inspecting fig. 6, we estimate a scale length of $\Delta x \approx 1$, such that the temporal evolution of $B_z$ in the region of enhanced current density can be estimated as $\delta B_z \approx -2 \times 10^{-3} \delta t$. This time evolution leads to a reduction of $B_z$ to zero in about five ion cyclotron times, from $\Omega_i t = 30$. We like to point out that this estimate is conservative, since the diffusive electric field will increase nonlinearly with decreasing magnetic field strength by virtue of (6). We point out that this latter, "diffusive" time scale is much faster than the formation time scale of the thin current sheet, which leads to the inital reduction of $B_z$. This time scale is set by typical wave travel times across the system, or prescribed by the etxernal driving, which are, in reality, more comparable to time scales of the growth phase. Once the reduction of $B_z$ through this process has occurred, a rapid diffusive mechanism is responsible for reversing the sign of $B_z$, a process which is impossible in the absence of dissipation.

We also note that, by virtue of (7), the efficiency of this process is a function of the electron-ion mass ratio. While this implies that a realistic mass ratio requires a further reduction of $B_z$ by thin current sheet formation, we are confident that such reductions will actually occur. In fact, MHD modeling of thin current sheet formation [*Birn et al.*, 1998b] show that current density enhancements by more than a factor of ten are easily achieved by moderate driving. Such current density enhancements are impossible to observe in the present model, because dissipation starts earlier due to the higher electron mass. Thus our results strongly suggest that thin current sheet formation can destabilize collisionless tearing in an otherwise stable magnetotail configuration.

## 4. THE KINETIC KINK MODE

The small scale nature of these current sheets, down to or even below the ion Larmor radius, lends itself to large current densities and thus becomes the likely site of dissipation processes. While thin current sheets may support the onset of magnetic reconnection, another important agent, however, may be provided by instabilities with wave vectors in the current direction. While instabilities of this nature have been invoked to support the current disruption phenomenology (mostly assuming ion-based current densities), the kinetic kink instability is unstable also in electron current sheets. The kink mode has also been shown in recent three-dimensional particle simulations to destabilize the collisionless tearing instability in a situation where the latter would otherwise be stable [*Lapenta and Brackbill*, 1998].

The kink mode appears to grow rapidly, with growth times of a few ion cyclotron times, in current sheets with thicknesses of the ion inertial length [e.g., *Zhu and Winglee*, 1996]. While this mode in itself is unlikely to provide dissipation for a magnetic reconnection process, it might destroy the thin current sheet equilibrium and generate substructure thin enough to enable collisionless reconnection.

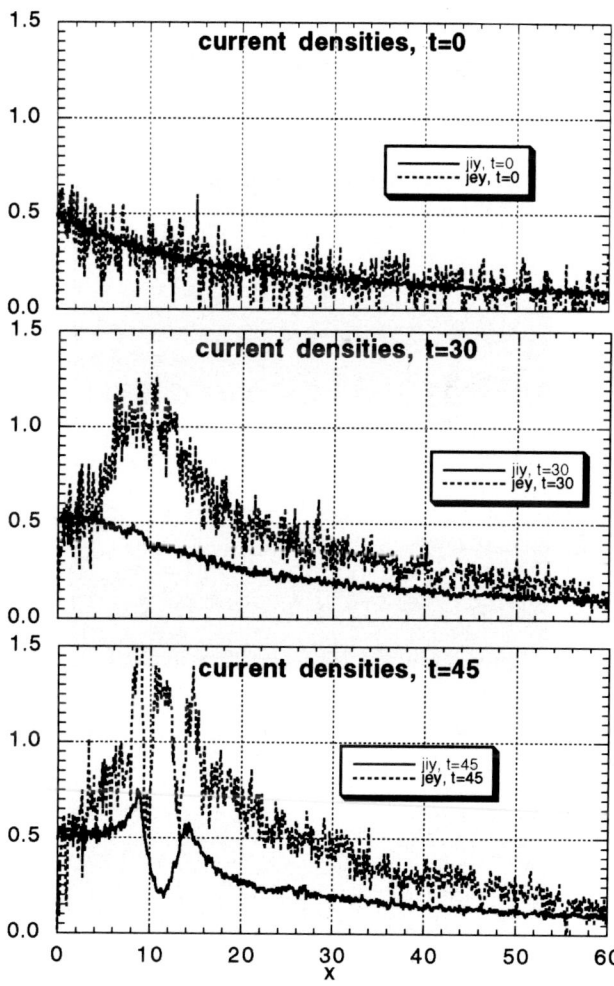

Figure 5. Ion (solid) and electron (dotted) current densities along the $x$ axis of the simulation for three different times.

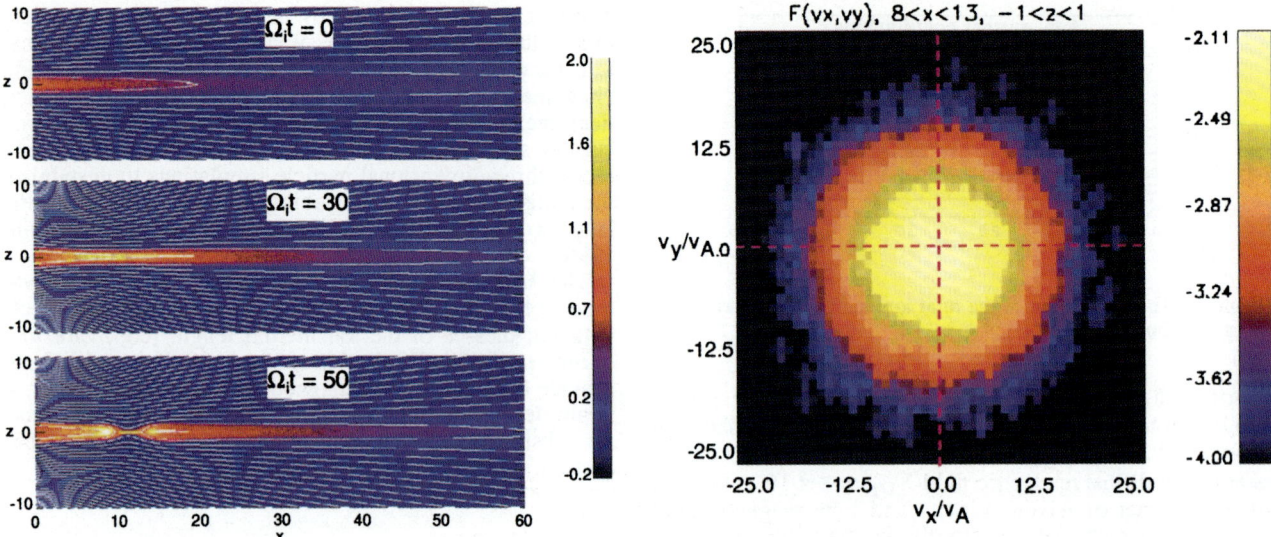

Plate 1. Magnetic field evolution and current density (color coded) evolution. The figure shows the effects of driving, as well as strong changes brought about by magnetic reconnection.

Plate 2. Reduced electron distribution function in the range $8 < x < 13$, and $-1 < z < 1$.

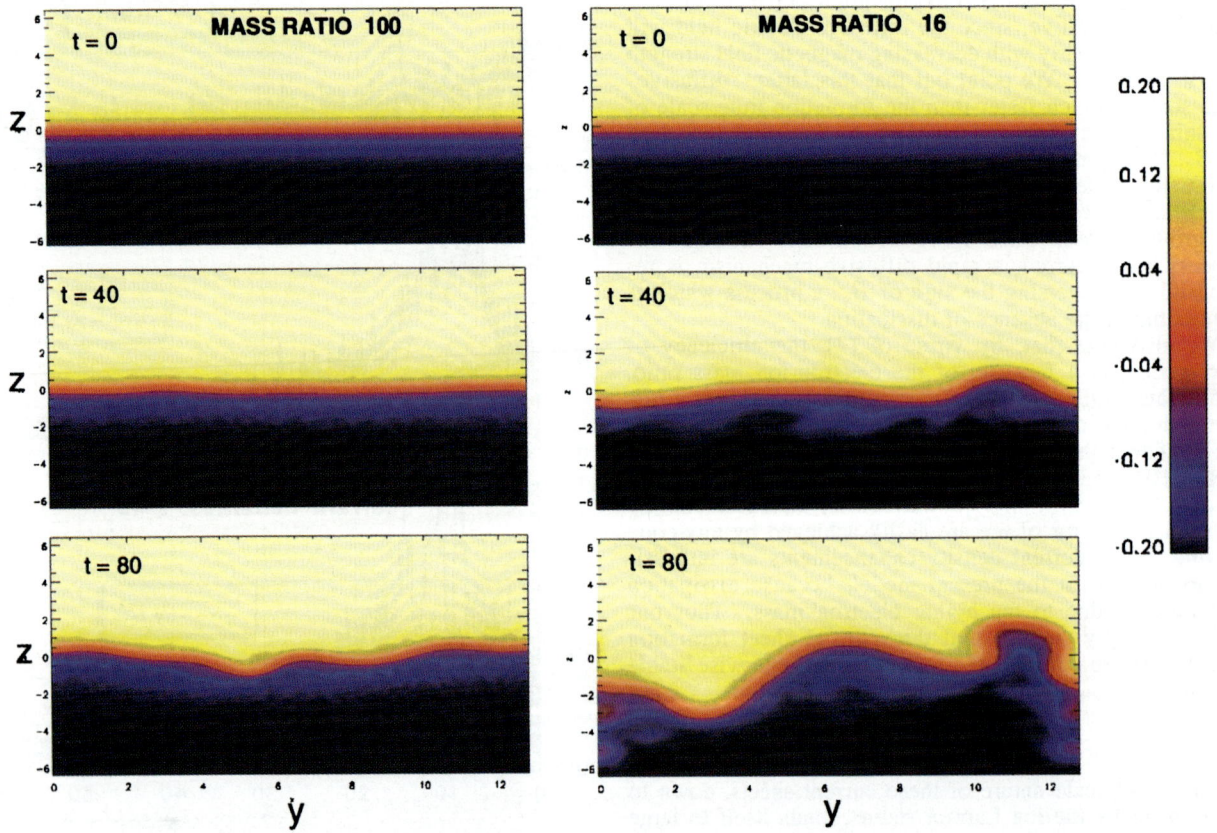

Plate 3. Time evolution of the $B_x$ component of the magnetic field for ion-electron mass ratios of 100 (left panels), and 16, right panels. The panels are stretched in the $y$ direction.

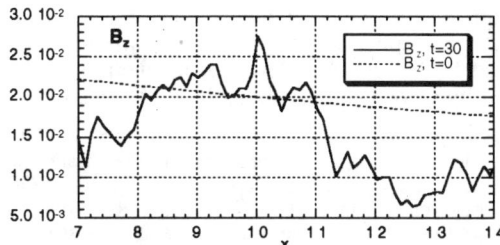

Figure 6. North-south magnetic field component $B_z$ in the vicinity of the forming X point, at $\Omega_i t = 30$ (solid line). The initial profile (dotted line) is shown for comparison.

A recent controversy focusses on the dependence of the growth rate of the kink instability on the ion-electron mass ratios. Based on analytical theory and numerical modeling, *Lapenta and Brackbill* [1998] conclude that kink mode growth increases with ion/electron mass ratio. *Daughton* [1999], however, uses a detailed analysis of the linear stability of the kink mode to arrive at the opposite conclusion. We will here use out particle-in-cell algorithm to clarify the issue.

Plate 3 shows the evolution of the $B_x$ magnetic field component in the $y - z$ plane of two simulations, selected for $m_i/m_e = 16$, and $m_i/m_e = 100$. Comparison of the two columns shows a clearly faster growth for lower mass ratios. This effect can be quantified further by Fourier-analyzing in $y$ the displacement in $z$ of the neutral sheet, i.e., the location where $B_x = 0$. This is performed for all simulations, and the growth rates of the two most important long wavelength modes are shown in Figure 8. Here the $m = 1$ mode fits one full wave length into the simulation box, and the $m = 2$ mode exhibits two full wavelengths.

The figure shows a clear decrease of growth rates for increasing ion-electron mass ratio, consistent with the conclusions of *Daughton* [1999]. While our simulations at present cannot be extended beyond mass ratio 100, it appears therefore likely that *Daughton's* [private communication] calculation of a realistic mass ratio growth rate of $\gamma \approx 10^{-3} \Omega_i$ is correct.

Accordingly, we are forced to conclude that the kinetic kink instability may be too slow to be involved in substorm onset. It might still, however, contribute to the onset of magnetic reconnection itself, if it had sufficient time to grow, and the stability threshold for magnetic reconnection and tearing has not been reached already. A corollary of the above result is related to the validity of three-dimensional kinetic simulations, which are often and out of need performed with small ion-electron mass ratios. It is to be expected that kink modes will grow in such models with unrealistic growth rates, limiting the validity of the simulation results. Unfortunately, there is no obvious way to avoid this problem, except perhaps for numerical filtering, such as performed by *Ozaki et al.* [1996].

## 5. SUMMARY AND CONCLUSIONS

In this paper, we have investigated kinetic processes in the magnetotail plasma sheet, which give rise to the reorganization of the cross-tail current density. All studies presented here are based on the application of our self-consistent fully electromagnetic particle-in-cell simulation code.

Like the natural procession of events in a magnetospheric substorm, our first focus was on the formation of thin current sheets during the substorm growth phase. Here we found, consistent with earlier analyses, that thin current sheets form if a driving electric field is applied to a magnetotail model, similar to what would be expected for the effects of the solar wind.

In accordance with momentum conservation, the ion current increased only by a small fraction, so that essentially the entire additional cross-tail current required by the lobe magnetic field enhancement is carried by the electrons, a result consistent with earlier predictions based on analytic theory.

Furthermore, a stronger, more localized current density enhancement was also carried by the electrons. This current density enhancement was accompanied by a reduction in the north-south magnetic field component $B_z$ and of the current sheet thickness. Both effects are expected to destabilize the current sheet, and eventually initiate reconnection.

The simulation indeed led to the onset of reconnection and the typical features commonly associated with substorm expansion, within the limits of the 2.5 dimensional model. We then conducted a study of the diffusive electric field in the spatial region which becomes the locus of the X point in the subsequent evolution. We found the existence of a weak, divergent electron flow in the $x$ direction, as required for the reduction of $B_z$. Such flows, together with gradients in the electron cyclotron frequency, can give rise to electron pressure anisotropies, and thereby generate diffusive electric fields.

Further, taking actual simulation output into analytical theory arguments led us to conclude that a negative $B_z$ should be reached in a very brief time period of a few ion cyclotron times. The fact that the actual onset of reconnection occurred even earlier results from the nonlinearity of the problem, which increases the diffusive electric field as $B_z$ is decreased further. Thus we concluded that electron pressure based dissipation is likely responsible for the onset of magnetic reconnection, as well as for its operation thereafter.

There still are a number of open questions associated with this result. First, one might wonder whether current sheets can become thin enough if a realistic ion-electron mass ratio, rather than $m_i/m_e = 100$, were employed. While

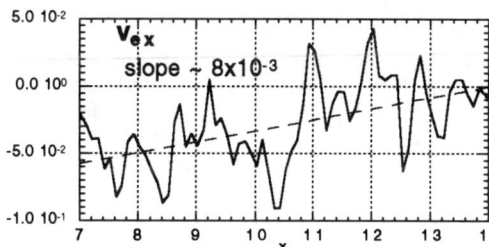

Figure 7. $x$ component of the electron flow velocity in the vicinity of the forming X point, at $\Omega_i t = 30$.

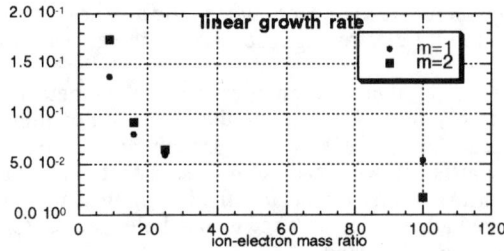

Figure 8. Mass dependence of the linear growth rate for the two most unstable normal modes of the kinetic kink instability.

computational capabilities presently prohibit answering this question, we refer to ideal MHD simulations, which showed that current density enhancements of some factor of ten can be expected for reasonable electric field driving, and in the absence of dissipation [*Birn et al.*, 1998b]. Therefore, we are led to conclude that, in reality, the current sheet would just continue to thin until a dissipation mechanims would be excited, initiating magnetic reconnection. Based on the present results, we would require the current sheet to be strong enough to reduce $B_z$ to the point where the local electron Larmor radius exceeds the gradient scale length of the reduced $B_z$.

A second question related to the above results addresses the role of instabilties with wave vector components in the $y$ direction. While we can safely discard such processes if they rely on ion currents, other current-driven instabilities may still play a role. The most prominent of these candidates is the collisionless kink instability.

Because of the obvious importance this instability might have on substorm evolution, or even the onset of magnetic reconnection, we performed a set of particle simulations to study the mass-ratio dependence of kink mode growth. Commensurate with our and other results on thin current sheet structure, we set up the simulated system such that electron constitute the major current carrier.

A Fourier analysis of current sheet displacement in the $z$ direction showed that the two longest wavelength modes grow the fastest. The linear growth rates strongly decrease for increasing ion-electron mass ratio, i.e., lighter electrons. This result is consistent with conclusions of *Daughton* [1999], based on linear stability analysis. Therefore, we concluded that the kinetic kink instability is an unlikely contributor to the fast magnetic field fluctuations often observed at the onset of fast dynamics in the magnetotail [e.g., *Ohtani et al.*, 1992]. It remains conceivable, however, that slow kinking of the current sheet might locally enhance electron currents and thereby speed up the onset of magnetic reconnection. A study of this interaction, however, will have to be relegated to future simulation efforts on more powerful computational platforms.

Our models have shown that current driven instabilities are not required to destabilize the magnetotail current sheet. Instead, a reduction of $B_z$ associated with thin current sheet formation plays a critical role in destabilizing collisionless tearing, and in enabling magnetic reconnection, even in the absence of other kinetic processes. Collisionless tearing and magnetic reconnection thus remain a crucial ingredient in magnetospheric dynamics, a fact which is strongly supported by the clear evidence of magnetic reconnection provided by Geotail observations [e.g., *Nagai*, 1998].

*Acknowledgments.* This work was supported by NASA's Sun-Earth-Connection Theory and Supporting Research and Technology Programs, and by the U.S. Department of Energy's Office of Basic Energy Sciences through its Geosciences Research Program. The authors thank Karl Schindler for valuable discussions.

## REFERENCES

Baker, D.N., T.I. Pulkkinen, V. Angelopoulos, W. Baumjohann, and R.L. McPherron, Neutral line model of substorms: Past results and present view, *J. Geopys. Res.*, *101*, 12,975, 1996.

Birn, J., and K. Schindler, Self-consistent theory of three-dimensional convection in the geomagnetic tail, *J. Geophys. Res.*, *88*, 6969, 1983.

Birn, J., and M. Hesse, The substorm current wedge and field-aligned currents in MHD simulations of magnetotail reconnection, *J. Geophys. Res.*, *96*, 1611, 1991.

Birn, J., M. Hesse, and K. Schindler, MHD simulations of plasmoid formation and magnetotail current disruption, *J. Geophys. Res.*, *101*, 12939, 1996.

Birn, J, M. F. Thomsen, J. E. Borovsky, G. D. Reeves, D. J. McComas, R. D. Belian, and M. Hesse, Substorm electron injections: Geosynchronous observations and test particle simulations, *J. Geophys. Res.*, *103*, 9235, 1998.

Birn, J., M. Hesse, and K. Schindler, Formation of thin current sheets in space plasmas, *J. Geophys. Res.*, *103*, 6843, 1998b.

Birn, J., M. Hesse, G. Haerendel, W. Baumjohann, and K. Shiokawa, Flow braking and the substorm current wedge, *J. Geophys. Res.*, in press, 1999.

Brackbill, J. U., and G. Lapenta, A method to suppress the finite grid instability in plasma simulations, *J. Comput. Phys.*, *114*, 77, 1994.

Daughton, W., Linear stability of a current sheet, *J. Geophys. Res.*, *104,*, 4657, 1999.

Fairfield, D. H., Advances in magnetospheric storm and substorm research, *J. Geophys. Res.*, *97*, 10,865, 1992.

Hesse, M., The magnetotail's role in magnetospheric dynamics: Engine or exhaust pipe, US National Report to International Union of Geodesy and Geophysics 1991-1994, *Rev. Geophys. Suppl.*, 675, 1995.

Hesse, M., and J. Birn, On dipolarization and its relation to the substorm current wedge, *J. Geophys. Res.*, *96*, 19,417, 1991.

Hesse, M., and J. Birn, MHD simulations of magnetotail instability for localized resistivity, *J. Geophys. Res.*, *99*, 8565, 1994.

Hesse, M., J. Birn, and T. Pulkkinen, Estimates of magnetic flux, and energy balance in the plasma sheet during substorm expansion, Substorms 3, p. 549, *Eur. Space Agency Spec. Publ.*, ESA SP, 1996.

Hesse, M., D. Winske, and J. Birn, On the ion scale structure of thin current sheets in the magnetotail, *Phys. Scr.*, *T74*, 63, 1997.

Hesse, M. and D. Winske, Electron dissipation in collisionless magnetic reconnection, *J. Geophys. Res.*, *103*, 26479, 1998.

Hesse, M., K. Schindler, J. Birn, and M. Kuznetsova, The diffusion region in collisionless magnetic reconnection, *Phys. Plasmas*, *6*, 1781, 1999.

Kahn, F. D., and L. Brett, Magnetic reconnection in the disc and halo, *Mon. Not. R. Astron. Soc.*, *263*, 37, 1993.

Kuznetsova, M., M. Hesse, and D. Winske, Kinetic quasi-viscous and bulk flow inertia effects in collisionless mag-netotail reconnection, J. Geophys. Res., 103, 199, 1998.

Lapenta, G., and J. U. Brackbill, 3D study of the interaction of the tearing and kink instability for a Harris equilibrium with realistic mass ratios, J. Geophys. Res., submitted, 1998.

Lui, A. T. Y., A. Mankofsky, C.-L. Chang, K. Papadopoulos, and C. S. Wu, A current disruption mechnism in the neutral sheet: A possible trigger for substorm expansions, *Geophys. Res. Lett.*, *17*, 745, 1990.

McPherron, R. L., C. T. Russell, and M. A. Aubry, Satellite studies of magnetospheric substorms on August 15, 1968, 9, Phenomenological model for substorms, *J. Geophys. Res.*, *78*, 3131, 1973.

Nagai, T., et al., Structure and dynamics of magnetic reconnection for substorm onsets with Geotail observations, J. Geophys. Res., 103, 4419, 1998.

Ohtani, S., S. Kokobun, and C. T. Russell, Radial expansion of the tail current disruption during substorms: a new approach to the substorm onset region, *J. Geophys. Res.*, *97*, 3129, 1992.

Ozaki, M., T. Sato, R. Horiuchi, and C. S. Group, Electromagnetic instability and anomalous resistivity in a magnetic neutral sheet, *Phys. Plasmas*, *3*, 2265, 1996.

Pellat, R., F. V. Coroniti, and P. L. Pritchett, Does ion tearing exist?, *Geophys. Res. Lett.*, *18*, 143, 1991.

Priest, E. R., Magnetic reconnection at the sun, in *Magnetic Reconnection in Space and Laboratory Plasmas, Geophys. Monogr. Ser.*, vol. 30, edited by E. W. Hones, p. 63, AGU, Washington, D. C., 1984.

Pritchett, P. L., and F. V. Coroniti, Formation of thin current sheets during plasma sheet convection, *J. Geophys. Res.*, *100*, 23,551, 1995.

Raeder, J., Global MHD simulations of the dynamics of the magnetosphere: Weak and strong solar wind forcing, in *Substorms 2*, edited by J. R. Kan, J. D. Craven, and S.-I. Akasofu, p. 561, Geophys. Inst., Univ. of Alaska Fairbanks, 1994.

Sanny, J., R. L. McPherron, C. T. Russell, D. N. Baker, T. I. Pulkkinen, and A. Nishida, Growth-phase thinning of the near-Earth current sheet during the CDAW 6 substorm, *J. Geophys. Res.*, *99*, 5805, 1994.

Tsyganenko, N. A., A magnetospheric field model with a warped tail current sheet, *Planet. Space Sci.*, *37*, 5, 1989.

Yamada, M., et al., Identification of Y-shaped and O-shaped diffusion regions during magnetic reconnection in a laboratory plasma, Preprint, Plasma Physics Laboratory, *Phys. Rev. Letts.*, *78*, 3117, 1997.

Zhu, Z., and R. M. Winglee, Tearing instability, flux ropes, and the kinetic current sheet kink instability in the Earth's magnetotail: A three-dimensional perspective from particle simulations, *J. Geophys. Res.*, *101*, 4885, 1996.

---

Michael Hesse, NASA GSFC, Code 696, Greenbelt, MD 20771; phone: 301-286-8224; fax: 301-286-1648; e-mail: hesse@gsfc.nasa.gov

Joachim Birn, Los Alamos National Lab, MS D466, Los Alamos, NM 87545; phone: 505-667-9232; fax: 505-665-3332; e-mail: jbirn@lanl.gov

# Intrinsic Variability in the Quiet-Time Magnetotail

Vahé Peroomian, Maha Ashour-Abdalla

*Institute of Geophysics and Planetary Physics, University of California, Los Angeles*

Lev M. Zelenyi

*Space Research Institute, Russian Academy of Sciences, Moscow, Russia*

This study investigates the evolution of the magnetotail's magnetic field with the aid of a self-consistent two-dimensional model in which the ion current periodically updates the magnetic field. The plasma mantle supplies particles continuously to the magnetotail, and the perturbation magnetic field is calculated from the ion current using the Biot-Savart law. The simulated magnetotail evolves into a quasi-steady state, characterized by the periodic motion of the near-Earth X-line in the model. This variability is caused by the nonadiabatic acceleration of ions in the current sheet and their rapid loss from the tail. Particularly noteworthy is the value found for the characteristic time scale of variability in the magnetotail, on the order of 4 – 5 minutes.

## 1. INTRODUCTION

Mobilis in mobili: Jules Verne, 19th Century French novelist, used these words to evoke the essence of one of his fantastic inventions, the submarine Nautilus. The words describe equally well the magnetotail, for surely no other phenomenon better epitomizes movement within a moving body than this extremely variable and dynamic windsock-like region in the highly variable supersonic solar wind plasma stream. Given that changeability is the most notable feature of the magnetotail, it is surprising that the obvious question of how the variability of the two regions affect each other is rarely asked. Instead, it is commonly assumed that variations in the magnetotail are caused by changes in the solar wind, in spite of the fact that the magnetotail's response to solar wind dynamics is complicated and nonlinear, and the tail retains a "memory" of solar wind history.

In this paper, we challenge the assumption that magnetotail dynamics are entirely determined by solar wind events. We show that, even for steady external solar wind conditions, the tail never attains a self-consistent equilibrium, but achieves only a quasi-steady state (QSS). In a QSS, internal magnetospheric conditions are steady on the average, but show pronounced intrinsic variability at sufficiently large amplitudes. This internal variability is caused by the properties of nonadiabatic particle dynamics and therefore cannot be reproduced in fluid models.

Numerous attempts have been made during the last decade to construct a self-consistent global MHD model of the magnetotail [*Birn and Hesse*, 1991; *Ogino et al.*, 1994; *Raeder et al.*, 1995; *Fedder and Lyon*, 1995]. Unfortunately, fluid descriptions are incapable of describing the most important region of the magnetotail, as the region around the current sheet proper and thin nonadiabatic current sheets are smeared when such an approach is employed. Attempts to overcome MHD limitations are documented in a series of interesting papers by *Cheng* [1992a, b], who addresses magnetotail equilibria in nonisotropic plasmas. To our knowledge, no attempt that includes full pressure tensors has been made thus far to develop a hydrodynamic model of the magnetotail.

Several interesting magnetotail models based on a Vlasov or hybrid description of plasma have been developed recently [*Hesse et al.*, 1996; *Pritchett and Coroniti*, 1996; *Büchner*, 1996]. These models realistically describe the system's particle dynamics and the plasma instability; their internal evolution is almost perfect for studies of local processes like reconnection, but is significantly less valuable when applied to a global magnetotail configuration. One difficulty is the choice of appropriate boundary conditions for the borders of the local simulation region within the global system. The external region provides both particles and energy to the simulated domain, but it also depends heavily on the processes that take place within it. Simple ballistic coupling takes place throughout the magnetotail, and thus, it is almost impossible to choose a region within the magnetotail to designate as the simulation box. When a simulation box is chosen, the region surrounding the box is treated as static plasma with fixed parameters and receives no feedback from or connection to the simulation box portion of the tail. The only logical solution to this problem is to choose the entire magnetotail as a simulation box. Recently, *Nishikawa* [1997, 1998a, b] and *Nishikawa and Ohtani* [1998] modeled the magnetosphere by using fully 3D kinetic simulations. These simulations have resulted in a better understanding of the interaction of the solar wind with the magnetosphere and yielded a self-consistent picture of the nightside magnetic field.

Our goal is to study the effect of particles on the magnetotail magnetic field in the absence of external stimuli. Our study employs a large-scale kinetic self-consistent (LSK-SC) model [*Peroomian et al.*, 1998]. The difference between this and our previous large-scale kinetic (LSK) models [*Ashour-Abdalla et al.*, 1993, 1994, 1995] is that the LSK-SC model, instead of using a prescribed magnetic field, uses a magnetic field that is continuously renewed according to the existing distribution of particle currents. The model reproduces many characteristic kinetic effects, including a thin current sheet, nongyrotropic pressure, and intermediate scale structures, or beamlets, that were produced by our previous models using calculations that are not self-consistent. In addition, though, the system exhibits several new global features of magnetotail equilibrium.

In section 2 we briefly describe our self-consistent kinetic model of the magnetotail. Results from the simulations are discussed in section 3. In section 4 we calculate auto-correlation functions for the basic magnetotail parameters to quantify the spatial and temporal aspects of its variability. In the last section we address the shortcomings of the model and outline our plans for resolving them. We also briefly consider implications of our results for in situ spacecraft observations.

## 2. THE MODEL

In this study, we use a modified version of the 2-D Birn-Zwingmann equilibrium [*Birn et al.*, 1975; *Zwingmann*, 1983] analytical magnetic field model as a zeroth order background magnetic field for our LSK calculations. We modified the original field by stretching the magnetotail field and including an X-line at a distance of 100 $R_E$ downtail. $B_y = 0$ in this model. We use a uniform dawn to dusk electric field of 0.1 mV/m in this simulation, which corresponds to a cross-tail potential drop of 17 kV. The inductive electric field is also calculated and added to the cross-tail electric field. Although the magnetic field is 2-D, ions are allowed to traverse the finite width of the magnetotail in the $y$-direction (25 $R_E$ in this study). Ion bulk parameters and the magnetic field are calculated in a box extending from $x = 0$ (at the Earth) to $x = 120$ $R_E$ (downtail) and $z = \pm 4$ $R_E$. We average over the $y$-direction when calculating bulk plasma parameters.

The first step in carrying out the simulation is to populate the magnetosphere. This is achieved by launching a continuous stream of ions from the plasma mantle in both hemispheres (defined in this model by $x = 15$ $R_E$, $3.0$ $R_E \leq |z| \leq 3.5$ $R_E$). We emulate steady solar wind conditions (constant n, V, southward IMF) and launch ~ 20,000 ions per minute, randomly distributed in phase space to make up a drifting Maxwellian distribution with 300 eV temperature and 200 km/s streaming velocity downtail (comparable to values found in shocked solar wind plasma). Because of the 2-D geometry of the magnetic field, we cannot launch ions from the low latitude boundary layer (LLBL), since these ions would remain trapped in the current sheet and unnaturally increase the ion density cross-tail current in the region. However, *Richard et al.* [1994] and *Peroomian and Ashour-Abdalla* [1995] have shown that the LLBL is a significant source of ions only during northward IMF. We plan to incorporate this source in future three-dimensional versions of the model. We launch ions throughout the two hours of the simulation presented in this paper. After one hour, we calculate the particle current. We then assume that the current in each 1 $R_E \times 0.1$ $R_E$ cell in the $x - z$ plane is carried by an infinite (in the $y$-direction) wire of 0.5 $R_E$ thickness and a circular cross-section. The current density within the wire is taken to be uniform, so that the current at the center of the wire is equal to zero. Wires are placed everywhere in the simulation box, not just in high-current regions. This allows us to use the Biot-Savart law to calculate the perturbation magnetic field resulting from the particle current everywhere in the magnetosphere. The calculated perturbation field is then added to the background magnetic field.

Since the perturbation magnetic field causes a significant change in the magnetic field's topology, we allow the initial transition from the background field to the new field to occur over 10 minutes, and every 30 seconds thereafter the particle current is calculated and the magnetic field is updated. In order to smooth the iteration process the magnetic field at any given time is taken to be a linear combination of the fields calculated from the particle current in the previous two time intervals.

To take into account the physical "inertia," or the resistance of the electrodynamic system to changes occurring within it, we calculate the inductive electric field ($E_{IND}$) caused by changes in the magnetic field topology. To do this, we solve Poisson's equation with Dirichlet boundary conditions over the northern half of the magnetosphere, and because of the symmetry of our system we apply the results to the entire magnetosphere. Because of the 2-D geometry of the magnetic field, the inductive electric field only has a non-vanishing $y$-component. The largest values of $E_{yIND}$ occur in the $z = 0$ plane in the vicinity of the X-line. The inclusion of $E_{IND}$ is of primary importance in our calculations, as without it we would lose the electromagnetic character of the system. Including $E_{IND}$ automatically accounts for the propagation of low frequency electromagnetic perturbations. In the strongly curved magnetic geometry under consideration, these perturbations may be characterized as coupled Alfvén–magnetosonic modes. Therefore, despite the use of the Biot-Savart law for calculating the perturbation magnetic field, a measurable time interval exists between the occurrence of the change in the local current in one place of the tail and the instant its magnetic effect is felt elsewhere.

## 3. RESULTS

### 3.1. Properties of the Quasi-Steady State: Merging Current Sheets

Equilibrium in the tail exists as a delicate balance between the source of the plasma populating it (i.e., the flux of solar wind particles) and the characteristics of plasma convection (i.e., the dawn-dusk electric field which is determined by the IMF direction). It is not evident a priori that tail equilibrium ever actually occurs for an arbitrary combination of these parameters. The original Birn-Zwingmann equilibrium is not compatible with the values of ion influx and the dawn-dusk electric field imposed on the system. The magnetic field begins to conform to the particle population approximately 20 minutes after we start updating the **B** field. Figure 1 shows 8 snapshots of the magnetic field taken approximately 40 minutes after we begin updating **B**. This figure shows many large- and small-scale changes to the magnetic field topology, among them the formation and disappearance of magnetic islands, thinning and thickening of the plasma sheet, and the motion of the X-line in the simulation.

One of the unexpected features of the LSK-SC simulation is the oscillation of the position of the X-line in the model. The dashed curve in Figure 2a indicates the position of the X-line closest to Earth as a function of time in the simulation, and the solid curve shows the equatorial crossing of the last closed field line in the model. At $t \leq 3600$ s, the X-line is held at $x = 100$ $R_E$. The next 10 minutes (until $t = 4200$ s) make up the initial transition period during which the field gradually changes to include the ion current contribution (gray shaded region in Figure 2a), and after which the **B** field is updated every 30 s. Both curves in Figure 2a show that the magnetotail evolves into a quasi-steady, periodic state in which the position of the near-Earth X-line oscillates between $x \sim 40$ $R_E$ and $x \sim 60$ $R_E$ downtail, and the location of the last closed field line oscillates between $x \sim 40$ $R_E$ and $x \sim 80$ $R_E$ downtail. There is no sign of saturation of these oscillations, and their characteristic period is on the order of 4 – 5 minutes. This appears to be an inherent feature of the newly achieved quasi-steady state of the magnetotail. The contour plot in Figure 2b shows the number of particles lost to the flanks of the simulation (at $y = -25$ $R_E$) as a function of $x$ (vertical axis) and of time (horizontal axis). We discuss particle loss from the model below.

To illuminate the physics underlying the periodic motion of the X-line, we analyze a single segment of this motion between $t = 5910$ s and $t = 6180$ s. To do this, we plot in Figure 3 the ion current profile in the $x - z$ plane for the 8 time intervals shown in Figure 1. We note that in both figures, the first 3 panels show **B** (Figure 1) and ion current (Figure 3) at 1-min intervals, after which the time elapsed between panels is 30 sec. The ion current sheet formed by the interaction of mantle plasma with the magnetic field reversal is relatively long at the beginning of the cycle (Figure 3a, $t = 5910$). Mantle ions encountering the thin magnetic reversal region of the neutral sheet experience strong nonadiabatic Speiser-type acceleration in the dawn-dusk electric field. Gaining energy, they move in the dusk direction and convect earthward. Because of this, the current-carrying ions are quickly lost to the dusk flank of our model (at $y = -25$ $R_E$), resulting in the collapse or disappearance of the current sheet (Figures 3b and 3c). The characteristic time scale for this collapse is on the order of

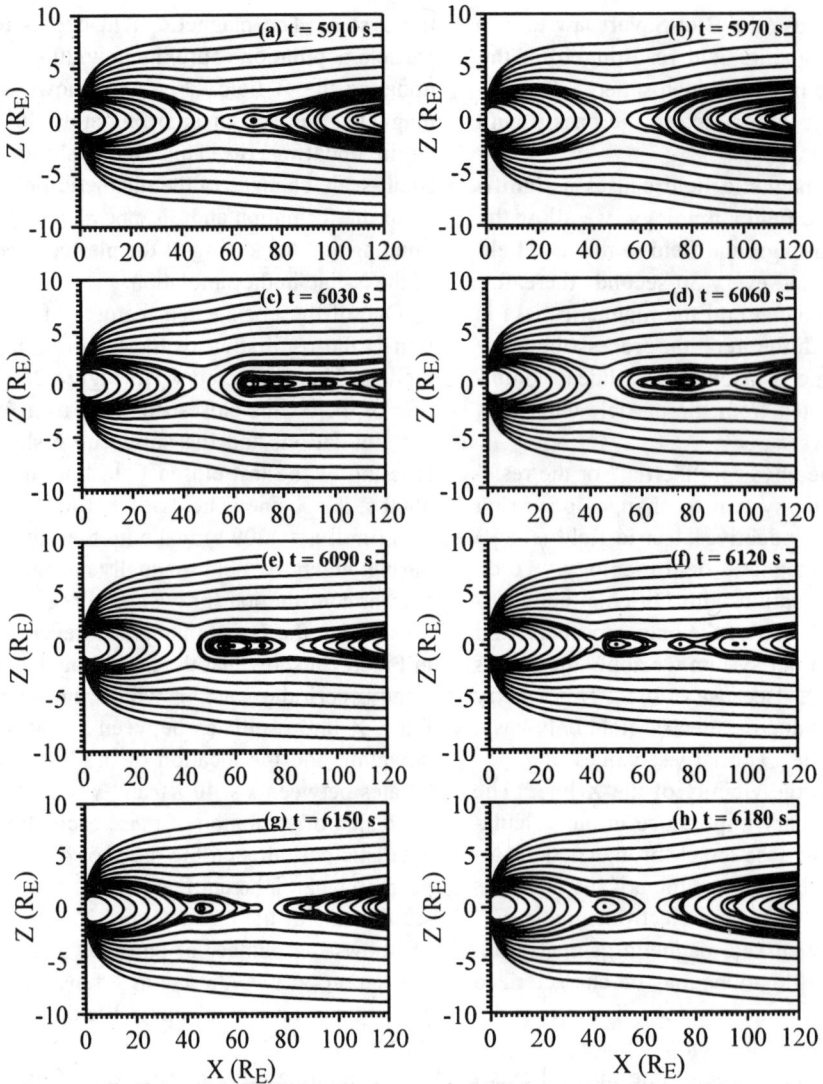

**Figure 1.** Snapshots of the magnetic field in the $x-z$ plane for $t = 5910$ s to $t = 6180$ s.

2 – 3 minutes in our model. However, 2 minutes later, at $t = 6030$ s, there is already evidence of new mantle ions entering the neutral sheet region and forming a new current sheet in the distant tail ($x > 80 \, R_E$). This process continues over the next two minutes, until $t = 6150$ (Figure 3g), when the newly formed current sheet reaches the near-Earth region. The rapid jump of the X-line downtail seen at the end of this cycle (Figures 1g and 1h) occurs when the new current sheet merges with the remnants of the old current sheet, causing the near-Earth X-line to coalesce and to be replaced by an X-line further downtail.

This sequence repeats itself, and snapshots of subsequent cycles are similar. The average cycle time is of the order of 5 minutes, and this peridiocity is manifested in the acceleration–escape–replenishment cycle of the current-carrying ions in the self-consistent thin current sheet. Evidence of this can be seen in the contour plot shown in Figure 2b, which shows periodic increases in the number of ions lost to the flanks corresponding to the motion of the X-line downtail. In fact, the majority of the ion loss occurs between the position of the near-Earth X-line and the last closed field line in the model (i.e. between $x = 40 \, R_E$ and $x = 80 \, R_E$).

It may be argued that these results are a direct consequence of using a two-dimensional magnetic field model. However, as we argue in section 5, these results can be applied on a local scale to the three-dimensional magnetotail.

### 3.2. Characteristic Scales of the Plasma Sheet Variability

It is currently customary to characterize the tail as a highly dynamic body. In one of the first papers specifically

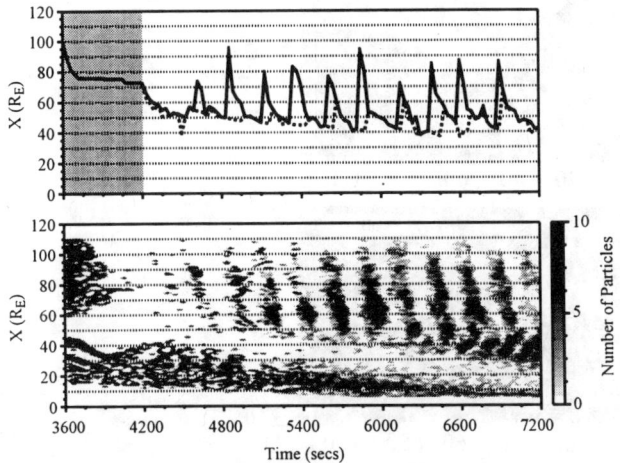

**Figure 2.** (a) (Dashed curve) Position of the near-Earth X-line as a function of time in the simulation. (Solid curve) Position of the equatorial crossing of the last closed field line as a function of time. (b) Contour plot of ion losses to the dusk flanks as a function of $x$ and time (scale on right of figure).

devoted to this subject, *Coroniti et al.* [1980] argued that the quiet-time plasma sheet could contain turbulent plasma flows and magnetic field structures which would not necessarily disrupt the global structure of the plasma sheet. Recent studies have confirmed that the flows in the plasma sheet are turbulent and transient rather than laminar [*Sergeev and Lennartsson*, 1988; *Baumjohann et al.*, 1990; *Baumjohann*, 1991; *Angelopoulos et al.*, 1994; *Borovsky et al.*, 1997]. Actual spacecraft observations routinely produce the impression of variability over a wide range of scales, with the cut-off on the lowest scale existing only because of limitations in the temporal resolution of measurement instruments.

To quantify the temporal variability we calculated the temporal autocorrelation functions for all locations within our magnetotail for the most characteristic parameter – the density of ions in the simulation

$$C(x_o, z_o, \tau) = \frac{\int n(x_o, z_o, t) n(x_o, z_o, t-\tau) dt}{\int n^2(x_o, z_o, t) dt}. \quad (1)$$

The integrals in (1) above are calculated from $\tau = 30$ s to $\tau = 900$ s. Figure 4 shows three profiles of $C(\tau)$ calculated at $x = 60$ $R_E$ and three $z$ values, $z = 0.0$ (dotted curve), $0.6$ $R_E$ (dashed curve) and $1.3$ $R_E$ (solid curve). To estimate the characteristic autocorrelation time, we take $C(\tau) = 0.5$ and compute the corresponding correlation time $\tau_o$. Figure 4 illustrates how the autocorrelation time increases from $\tau_o \sim 60$ s in the PSBL to $\tau_o \gg 15$ min in the CPS.

By calculating this profile for our entire $x - z$ spatial grid, we obtain the distribution of $\tau_o$ everywhere in our two-dimensional magnetotail. To automate this process, we have used the following criteria. If $C(\tau)$ becomes less than 0.5 at any time, then the time at which this occurs is the correlation time (as is the case for the $x = 60$ $R_E$, $z = 0.6$ $R_E$ and $z = 1.3$ $R_E$ curves in Fig. 4). If $C(\tau)$ remains above 0.5 (as is the case for $x = 60$ $R_E$, $z = 0$ curve in Fig. 4), then the correlation time is calculated by comparing the slope of the $C(\tau)$ curve at successive time intervals. When the ratio of slopes from successive 30-second intervals is less than 0.5, the correlation time is set equal to the median of these points. Thus, the curve for $x = 60$ $R_E$, $z = 0$ yields a correlation time of 459 seconds. The result from this process is shown in Figure 5. The smallest values of $\tau_o$ ($< 50$ s) occur at the outer edges of the PSBL. As expected, the CPS earthward of the nominal location of the X-line shows the least variability. This figure shows autocorrelation times of $200 - 400$ s in the neutral sheet at $x > 60$ $R_E$, in the region where the collapse and replenishment of the current sheet occurs. These time scales are consistent with the time scale of periodicity found in Figure 2. Clearly, the characteristic time scales of variability differ from one region to another in the magnetotail. This variability is intrinsic and is not caused by external large-scale motions of the tail.

## 5. CONCLUSIONS AND DISCUSSION

In this paper we presented the results of our large-scale kinetic self-consistent (LSK-SC) modeling of a two-dimensional magnetotail. We showed that the tail achieves a new self-consistent state that is nearly at equilibrium for the given external parameters (flux of solar wind plasma and dawn-dusk electric field). Though it hovers around equilibrium, the self-consistent state appears to be very dynamical (we call it a quasi-steady state). Periodic disruptions of the thin current sheet due to nonadiabatic acceleration of current-carrying particles occur in the system and are followed by the replenishment of the current sheet by new mantle plasma and its recovery (by the coalescence of its two disjoint parts) followed again by its disruption in the next cycle. The characteristic periodicity of this process is about 4-5 minutes.

Perhaps the chief limitation of the present study is its two-dimensional magnetic field geometry. Although ions are allowed to move in the $y$-direction, the magnetic field everywhere in $y$ is the same as that in the $y = 0$ plane. We expect that the dynamics of ions near an X-line occupying a local region in a 3D model will be similar to those in a 2D model. Ions leave this region in $1 - 3$ minutes, and the local current sheet in this region collapses at this location. However, just as *Bhattachargee et al.* [1998] found that a 2D thin current sheet is unstable to a 3D ballooning instability, the motion of the X-line in a 3D model may occur in both the $x$ and $y$ directions, rather than just in the $x$ direction as in the 2D model. We note here that previous studies ex-

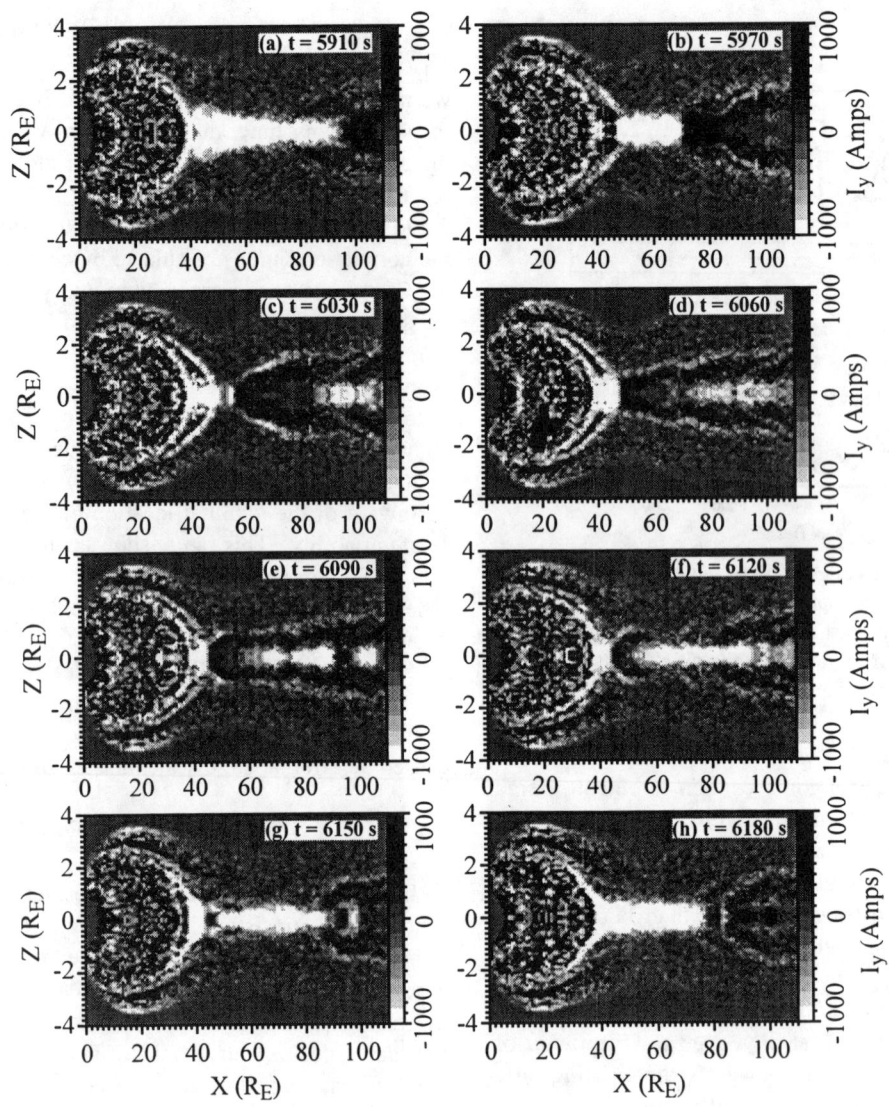

**Figure 3.** Snapshots of the transverse ion current in the $x-z$ plane for the 8 intervals shown in Fig. 1.

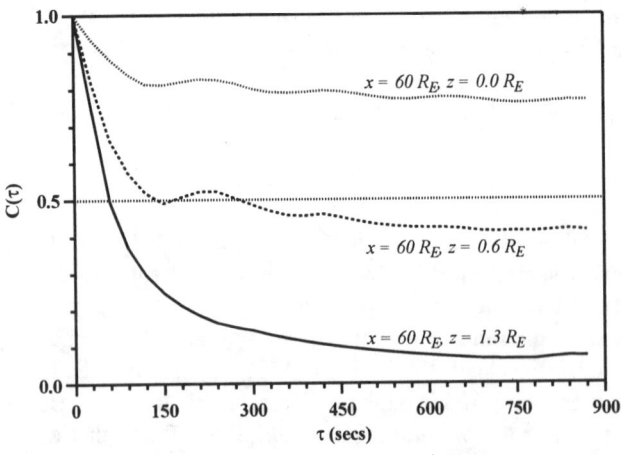

**Figure 4.** Profiles of $C(\tau)$, calculated at $x = 60\ R_E$ and $z = 1.3\ R_E$ (solid curve), $z = 0.6\ R_E$ (dashed curve), and $z = 0.0$ (dotted curve).

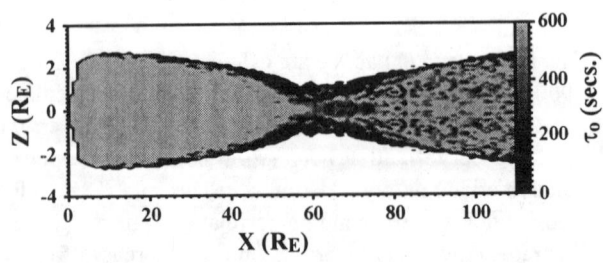

**Figure 5.** Profile of the autocorrelation time $\tau_o$ calculated for the magnetotail (color scale on right).

tending the results of 2D LSK calculations [e.g. *Ashour-Abdalla et al.*, 1993, 1994, 1995; *Bosqued et al.*, 1993] to three dimensions [*Peroomian*, 1994; *Peroomian and Ashour-Abdalla*, 1995, 1996] indicate that results obtained using 2D models can be surprisingly realistic when applied to local regions in the magnetotail. Specifically, the investigation of velocity dispersed ion structures (VDIS) [*Zelenyi et al.*, 1990] in 2D [*Bosqued et al.*, 1993] and in 3D [*Peroomian*, 1994; *Bosqued et al.*, 1994] showed that in 3D, the structures occupy narrow regions in local time and are similar in characteristics to the VDIS reproduced in the 2D model. We therefore expect the mechanism causing the X-line oscillations in this model to operate in the Earth's magnetotail as well, albeit in a more complicated manner. We will investigate this conjecture in a future study using three-dimensional magnetic and electric fields.

Choosing the entire magnetotail as the simulation box necessitated the neglect of electron dynamics in our model. The consistent orbit tracing (COT) analysis of particle trajectories carried out by *Larson and Kaufmann* [1996] and the subsequent studies by *Kaufmann et al.* [1997a, 1997b] show that the electron contribution to the cross-tail current is small. However, electrons have been shown to carry the bulk of the cross-tail current in some parts of the tail, especially near X-lines and in thin current sheets [*Hesse et al.*, 1997, 1998]. Preliminary calculations that include the contribution of electrons via the assumption of a Boltzmann distribution and charge neutrality indicate that the inclusion of electron dynamics does not destroy the variability of the system. Based on this result, we do not expect electrons to influence the variability drastically, since the effects of variability (and its scales) are related to the ion inertia, and it is difficult to imagine that electrons could fully compensate for the dropout of the ion current in the vicinity of the X-line after the ions' escape. If this region with a weak field (in the vicinity of the X-line) is local in $y$, the accelerated electrons will also quickly leave it, and the effect of the variability will remain.

*Acknowledgements.* The authors thank J. M. Bosqued, M. El-Alaoui V. Sergeev, and R. J. Walker for valuable comments and suggestions. This work was supported by the NASA ISTP grant NAG5-6689 and NASA grant NAGW-4683. Lev M. Zelenyi acknowledges contracts RFBR 96-15-96723, RFBR 97-02-16489, and INTAS 96-2346. Computing support was provided by the Office of Academic Computing at UCLA and by NPACI at the San Diego Supercomputer Center. UCLA-IGPP publication number 5342.

## REFERENCES

Angelopoulos, V., C. F. Kennel, F. V. Coroniti, R. Pellat, M. G. Kivelson, R. J. Walker, C. T. Russel, W. Baumjohann, W. C. Feldman, and J. T. Gosling, Statistical characteristics of bursty bulk flows events, *J. Geophys. Res.*, 99, 21257, 1994.

Ashour-Abdalla, M., J. Berchem, J. Büchner, and L. M. Zelenyi, Shaping of the magnetotail from the mantle: Global and local structuring, *J. Geophys. Res.*, 98, 5651, 1993.

Ashour-Abdalla, M., L. M. Zelenyi, V. Peroomian, and R. L. Richard, Consequences of magnetotail ion dynamics, *J. Geophys. Res.*, 99, 14,891, 1994.

Ashour-Abdalla, M., L. M. Zelenyi, V. Peroomian, R. L. Richard, and J. M. Bosqued., The mosaic structure of plasma bulk flows in the Earth's magnetotail, *J. Geophys. Res.*, 100, 19,191, 1995.

Baumjohann, W., Heating and fast flows in the near-Earth tail, in *Magnetospheric Substorms*, edited by J. R. Kan, T. A. Potemra, and T. Iijima, *Geophys. Monogr. Ser.*, vol. 64, 141, AGU, Washington, D. C., 1991.

Baumjohann, W., G. G. Paschmann, and H. Luhr, Characteristics of high-speed ion flows in the plasma sheet, *J. Geophys. Res.*, 95, 3801, 1990.

Bhattachargee, A., Z. W. Ma, and Xiaogang Wang, Dynamics of thin current sheets and their disruption by ballooning instabilities: A mechanism for substorm onset, in *Substorms-4*, ed. S. Kokubun and Y. Kamide, Kluwer Academic Pub., Dordrecht, p. 367, 1998.

Birn, J., and M. Hesse, The substorm current wedge and field-aligned currents in MHD simulations of magnetotail reconnection, *J. Geophys. Res.*, 96, 1611, 1991.

Birn, J., R. Sommer, and K. Schindler, Open and closed magnetospheric tail configurations and their stability, *Astrophys. Space Sci.*, 35, 389, 1975.

Borovsky, J. E., R. C. Elphic, H. O. Funsten, and M. F. Thomsen, The Earth's plasma sheet as a laboratory for flow turbulence in high-B MHD, *J. Plasma Phys.*, 57, 1, 1997.

Bosqued, J. M., M. Ashour-Abdalla, M. El-Alaoui, V. Peroomian, L. M. Zelenyi, and C. P. Escoubet, Dispersed ion structures at the poleward edge of the auroral oval: Low-altitude observations and numerical modeling, *J. Geophys. Res.*, 98, 19,181, 1993.

Bosqued, J. M., M. Ashour-Abdalla, V. Peroomian, R. L. Richard, and L. M. Zelenyi, Auroral precipitation patterns: Observations and 3D simulations, *Substorms 2*, edited by J. R. Kan, J. D. Craven, and S.-I. Akasofu, University of Alaska Fairbanks, p. 365, 1994.

Büchner, J., Three-dimensional current sheet tearing in the earth's magnetotail, *Adv. Space Res.*, 18, 267, 1996.

Cheng, C. Z., Kinetic extensions of magnetohydrodynamics for axisymmetric toroidal plasmas, *Phys. Rep.*, 211, 1, 1992a.

Cheng, C. Z., Magnetospheric equilibrium with anisotropic pressure, *J. Geophys. Res.*, 97, 1497, 1992b.

Fedder, J. A., and J. G. Lyon, The Earth's magnetosphere is 165 $R_E$ long: Self consistent currents, convection, magnetospheric structure and processes for northward interplanetary magnetic field, *J. Geophys. Res.*, 100, 3623, 1995.

Hesse, M., D. Winske, M. Kuznetsova, and J. Birn, Hybrid simulations of the formation of thin current sheets in magnetotail configurations, *J. Geomagn. Geoelectr.*, 1996.

Hesse, M., J. Birn, and D. Winske, On the ion scale structure of thin current sheets in the magnetotail, *Phys. Scr.*, T74, 63, 1997.

Hesse, M., J. Birn, and D. Winske, Formation and structure of thin current sheets: Dipolarization, *Substorms-4*, edited by S. Kokubun and Y. Kamide, Kluwer Acad. Pub., 727, 1998.

Kaufmann, R. L., D. J. Larson, I. D. Kontodinas, and B. M. Ball, Force balance and substorm effects in the magnetotail, *J. Geophys. Res.*, *102*, 22,141, 1997a.

Kaufmann, R. L., I. D. Kontodinas, B. M. Ball, and D. J. Larson, Nonguiding center motion and substorm effects in the magnetotail, *J. Geophys. Res.*, *102*, 22,155, 1997b.

Larson, D. J., and R. L. Kaufmann, Structure of the magnetotail current sheet, *J. Geophys. Res.*, *101*, 21,447, 1996.

Nishikawa, K.-I., Particle entry into the magnetosphere with a southward IMF as simulated by a 3-D EM particle code, *J. Geophys. Res.*, *102*, 17,631, 1997.

Nishikawa, K.-I., Reconnections at near-Earth magnetotail and substorms studied by a 3-D EM particle code, *Geospace Mass and Energy Flow: Results from the International Solar-Terrestrial Physics Program*, ed. J. L. Horwitz, W. K. Peterson, and D. L. Gallagher, AGU Geophys. Monogr., *104*, p. 175, 1998a.

Nishikawa, K.-I., Particle entry through reconnection grooves in the magnetopause with a dawnward IMF as simulated by a 3-D EM particle code, *Geophys. Res. Lett.*, *25*, 1609, 1998b.

Nishikawa, K.-I., and S. Ohtani, Particle entry through reconnection by a time-varying IMF as simulated by a 3-D EM particle code, in *Substorms-4*, ed. S. Kokubun and Y. Kamide, Kluwer Academic Pub., Dordrecht, p. 535, 1998.

Ogino, T., R. J. Walker, and M. Ashour-Abdalla, A global magnetohydrodynamic simulation of the response of the magnetosphere to a northward turning of the interplanetary magnetic field, *J. Geophys. Res.*, *99*, 11,027, 1994.

Peroomian, V., Large Scale Kinetic Modeling of Magnetospheric Plasma, PhD thesis, University of California at Los Angeles, March 1994.

Peroomian, V., and M. Ashour-Abdalla, Relative contribution of the solar wind and the auroral zone to near-Earth plasmas, in *Cross-Scale Coupling in Space Plasmas, Geophys. Monogr. Ser.*, vol. *93*, edited by J. Horwitz et al., pp. 213-217, AGU, Washington, D. C., 1995.

Peroomian, V., and M. Ashour-Abdalla, Population of the near-Earth magnetotail from the auroral zone, *J. Geophys. Res.*, *101*, 15,387, 1996.

Peroomian, V., M. Ashour-Abdalla, and L. M. Zelenyi, Self-consistent simulation of the magnetotail, *Substorms-4*, edited by S. Kokubun and Y. Kamide, Kluwer Acad. Pub., p. 165-168, 1998.

Pritchett, P. L., and F. V. Coroniti. The role of the drift kink mode in destabilizing thin current sheets, *J. Geomagn. Geoelectr.*, *48*, 833, 1996.

Raeder, J., R. J. Walker, and M. Ashour-Abdalla, The structure of the distant geomagnetic tail during long periods of northward IMF, *Geophys. Res. Lett.*, *22*, 349, 1995.

Richard, R. L., R. J. Walker, and M. Ashour-Abdalla, The population of the magnetosphere by solar wind ions when the interplanetary magnetic field is northward, *Geophys. Res. Lett.*, *21*, 2455, 1994.

Sergeev, V. A., and W. Lennartsson, Plasma sheet at $X \sim 20\, R_E$ during steady magnetospheric convection, *Planet. Space Sci.*, *36*, 353, 1988.

Zelenyi, L.M., R. A. Kovrazkhin and J.M. Bosqued, Velocity-dispersed ion beams in the nightside auroral zone: AUREOL 3 observations, *J. Geophys. Res.*, 95, 12,119, 1990.

Zwingmann, W., Self-consistent magnetotail theory: Equilibrium structures including arbitrary variation along the tail axis, *J. Geophys. Res.*, *88*, 9101, 1983.

---

Vahé Peroomian and Maha Ashour-Abdalla, UCLA-IGPP, Box 951567, Los Angeles, CA 90095-1567.

Lev M. Zelenyi, Space Research Institute, Russian Academy of Sciences, Moscow 117810, Russia.

# Self-Consistent Model of 1D Current Sheet: The Role of Drift, Magnetization and Diamagnetic Currents

Helmi V. Malova and Mikhail I. Sitnov

*Nuclear Physics Institute, Moscow State University, Moscow, Russia*

Lev M. Zelenyi

*Space Research Institute, Moscow, Russia*

Surja Sharma

*Department of Astronomy, University of Maryland, USA*

Thin anisotropic current sheets are important elements of the magnetospheric dynamics and structure. We develop an analytical theory of the self-consistent current sheet (CS) created by the ion streams impinging from the plasma mantle. The nonlocal analogue of the Grad-Shafranov equation is obtained and numerically solved in the quasi-adiabatic approximation, i.e. neglecting jumps of the sheet adiabatic invariant $I_z$ which is possible when the minimum curvature radius of the magnetic field is less than the thermal ion gyroradius. General solutions allow to consider both the cases of strong and weak anisotropy. Resulting self-consistent current is a sum of the drift cyclotron and magnetization currents flowing in opposite directions. In the immediate vicinity of the neutral plane the paramagnetic current arising from the meandering motion of ions on Speiser orbits becomes dominating. The maximum CS thickness is achieved in the case of weak plasma anisotropy and is of the order of the thermal ion gyroradius outside the sheet. Separately the CSs with nonadiabatic ions are considered, where the jumps of $I_z$ become essential. For that case of the so called "superstrong" anisotropy, when the ratio of the drift velocity outside CS to the thermal ion velocity exceeds the ratio of the magnetic field outside CS to its value inside CS ($v_D/v_T > B_0/B_n$), the assessment of the CS thickness is also performed. It is found in particular that the CS thickness can be reduced down only up to some finite value, which depends upon the parameter $B_n/B_0$. The model predicts essential reduction of the CS thickness under the influence of the convection electric field, which might have important implications for the substorm dynamics.

## 1. INTRODUCTION

Thin current sheets (TCSs) are found now to be a phenomenon of a common occurrence in the Earth magnetosphere. Contemporary space investigations demonstrate the existence of CS of the order of ion gyroradius (or less) at different locations in the magnetotail. Such thin sheets have been detected in the near-Earth and the midtail regions during the substorm growth phase [e.g., *Fairfield*, 1984; *Kaufmann*, 1987; *McPherron et al.*, 1987; *Mitchell et al.*, 1990; *Lui et al.*, 1992; *Sergeev et al.*, 1993; *Sanny et al.*, 1994; *Pulkkinen et al.*, 1994, 1998] and in the distant region of magnetotail [*Pulkkinen et al.*, 1993]. The most of studies revealed the thinning of CS during the growth phase of substorm before the dipolization [*Mitchell et al.*, 1990]. The important characteristic feature of the structure of TCS is that their cross-tail current density profile doesn't coincide and might be significantly more narrow than the profile of the plasma density [*Sergeev et al.*, 1993; *Sanny et al.*, 1994].

These experimental investigations stimulated the development of the mathematical models of TCSs in the magnetotail [*Pritchett and Coroniti*, 1994, 1995] The isotropic models where magnetic tension of curved field lines is balanced by the plasma pressure gradient [*Schindler*, 1974; *Kan*, 1973] were extensively exploited in magnetospheric research for more than two decades [e.g., *Birn*, 1987]. There is another principal opportunity to balance the magnetic field line tension by centrifugal force exerted on finite mass particles moving along curved field lines [*Eastwood*, 1972]. Such kind of balance implies the existence of the finite plasma anisotropy outside the sheet. The class of anisotropic CS models was based on the existence of so called Speiser orbits [*Speiser*, 1965] which were essentially nonadiabatic in the CS. Recent studies of anisotropic ion CSs made by [*Alexeev and Malova*, 1990; *Burkhart et al.*, 1992; *Pritchett and Coroniti*, 1992, *Holland and Chen*, 1993] demonstrated some of the characteristic features of the ion dominated anisotropic sheets, i.e. small ($\sim c/\omega_{pi}$) scales and embedded structure. [*Alexeev and Malova*, 1990, 1995] found that the CS supported by two counterstreaming almost field-aligned plasma flows has a resulting structure with the "diamagnetic wings" near the boundaries due to the properties of particle motion far from the region of magnetic field reversal and within it, which conforms with the results of numerical kinetic models by *Pritchett and Coroniti* [1995]. At last the investigations of *Holland and Chen* [1993], *Harold and Chen* [1996] clearly demonstrated the important role of magnetization currents in equilibrium CSs particularly in the case of weak anisotropy. *Kaufmann* [1997] tried to take into account the influence of non-guiding center particles on the structure of CS found that although the net currents carried by trapped particles is equal to zero, their existence could appreciably change the resulting CS structure.

First numerical estimates of ion CS thickness were made by *Eastwood* [1972]. Later *Francfort and Pellat* [1976] obtained analytical non-selfconsistent estimate of the thickness of the thin CS: $L \approx \rho_0 (v_T/v_D)^{4/3}$ where $v_T$ and $v_D$ are respectively the thermal ion velocity and the bulk speed of the ion flow outside the sheet. Later the same estimate have been done by *Burkhart et al.* [1992]. The particular case of super-strong anisotropy (ion motion almost along the magnetic field lines) was considered by *Pritchett and Coroniti* [1992] and *Burkhart et al.* [1992]. They found scaling of the CS in a form $L \approx \rho_0 (B_n/B_0)^{4/3}$, where $B_n$ and $B_0$ are, respectively, the magnetic field near the neutral plane (normal to the sheet plane) and far outside it.

The first entirely self-consistent thin anisotropic sheet model has been proposed by *Kropotkin et al.* [1997] for the region of the moderately strong plasma anisotropy outside the TCS $b_n \ll v_T/v_D \ll 1$, where $b_n = B_n/B_0$. They obtained the analytical solution for magnetic field and current density profiles similar to the well-known *Harris* [1962] solution with $B_n = 0$. In this paper we propose a more general approach for the cases of both weak and strong anisotropies, which enables to consider all the above mentioned regimes as a different limiting cases of the general equilibrium solution.

Our paper consists of two separate parts. In the first one in the frame of quasiadiabatic approximation we assume the conservation of adiabatic invariant $I_z$. This assumption allows us to obtain the equations of Grad-Shafranov type describing self-consistent ion CSs. In the second, more qualitative part, we explicitly take into account the nonconservation of adiabatic invariant $I_z$ and estimate the minimum thickness which could have self-consistent CS when the parameter $v_D/v_T$, characterizing the anisotropy is increasing unlimitedly. In the last section we tried to merge these two approaches and present the combined picture of CS scalings. This picture clearly demonstrates the essential tendency to CS thinning for large values of convection electric field.

## 2. BASIC THEORY

Let us first briefly describe the basics of the new approach to the self-consistent theory of anisotropic equilibria where the magnetic field line tension is balanced by the finite inertia of ions moving along the curved magnetic field lines. The details may be found elsewhere [*Sitnov et al.*, 1998, 1999a, 1999b]. The CS under study is homogeneous along the Earth-Sun ($X$) and dawn-dusk ($Y$) directions. The magnetic field is assumed to have two components: $\mathbf{B} = (B(z), 0, B_n)$ with $B_0 = |B(z \to \pm\infty)|$. The Maxwell equation $rot\mathbf{B} = (4\pi/c)\mathbf{J}$ may be reduced to the following scalar equation

$$dB/dz = (4\pi/c) \sum_{\alpha=i,e} q_\alpha \int v_y f_\alpha(z, \mathbf{v}) d\mathbf{v} \quad (1)$$

We assume that: (a) the current is supported mainly by the ion population, and (b) the dynamics of this population is quasi-adiabatic [*Büchner and Zelenyi*, 1989]. The latter assumption is valid under the condition

$$\kappa = \sqrt{R_c/\rho_z} \ll 1 \quad (2)$$

where $\rho_z$ is the maximum gyroradius of the ion in the magnetic field with the smallest curvature radius $R_c$.

We suppose also that the electron component may play the role of quasi-neutralizing background. Some investigators of TCSs [*Shay et al.*, 1998; *Hesse et al.*, 1996; *Nishikawa*, 1998] underline the essential role of electron currents in the vicinity of reconnection region near $X$-line. Contrary, our 1D model assumes the existence of the finite value of $B_n$ all over the sheet and we assume that electrons are magnetized. Therefore the condition of magnetic field being frozen into electron component still holds. And we do not expect strong electron currents in this model, similar to electron currents found in the dissipation region. The arguments about smalness of electron current have been given also by *Larsson and Kaufmann* [1996] in their consistent orbit tracing model.

Our model could be applied for the sheets satisfying the condition: $\kappa_i \ll 1 \ll \kappa_e$, so that the value of $B_n$ component of magnetic field is small enough to demagnetize the ions but large enough to keep electrons magnetized. For very small values of $B_n$ when electrons could also become nonadiabatic the whole sheet could be considered as representing the "dissipation region" and contribution of electron currents might be significant.

The assumption (b) allows us to use a new set of integrals of motion to avoid the explicit solving of Vlasov equation. We use in particular the first approximate integral of the "fast" $z-$ motion [*Speiser*, 1970; *Sonnerup*, 1971; *Büchner and Zelenyi*, 1989]

$$I_z = \frac{1}{2\pi} \oint mv_z dz \quad (3)$$

and the total particle energy $W = mv_0^2/2$. Conservation of quasi-adiabatic invariant (3) have been studied in a number of papers particularly by *Büchner and Zelenyi* [1989], *Zelenyi and Savenkov* [1993] who have shown that on average for $\varepsilon > B_n/B_0$ the jumps $|\Delta I_z| \ll I_z$ and could be neglected.

Based on the equations of motion and on the conservation of generalized moment of impulse $P_y$ the invariant (3) may be written explicitly in the form

$$I_z = \frac{2m}{\pi} \int_{z_0}^{z_1} \sqrt{v_y^2 + v_z^2 - \left[v_y + \frac{e}{mc}\int_z^{z'} B(z'') dz''\right]^2} dz' \quad (4)$$

where the limits of the integration are given by the equation

$$\frac{e}{mc}\int_z^{z_{0,1}} B(z'') dz'' = \mp\sqrt{v_y^2 + v_z^2} - v_y \quad (5)$$

with the additional restriction $z_0 = 0$ if the formal solution $z_0$ of (5) turns to be negative. The general structure of the distribution function could therefore be presented in a form

$$f(z, \mathbf{v}) = f(W(v_x, v_y, v_z), I_z(v_x, v_y, z, z_{1,2}(v_x, v_y, z))) \quad (6)$$

To obtain the self-consistent profile of the magnetic field from the system (1) - (5) it is sufficient to specify the form of the distribution function. We assume in particular that far outside the current sheet ($z \to \pm\infty$) it takes the form of the drifting Maxwellian [*Francfort and Pellat*, 1976; *Burkhart et al.*, 1992]

$$f \sim \exp\left\{-\left[(v_\parallel - v_D)^2 + v_\perp^2\right]/v_T^2\right\} \quad (7)$$

Here $v_\parallel$, $v_\perp$ are respectively the projections of the ion velocity along and across the magnetic field.

To connect the distribution (6) with its asymptotic form outside the sheet we use the relation $I = (2mc/e)\mu$ between the sheet invariant $I$ and the magnetic moment $\mu = mv_\perp^2/(2B)$ at the boundary between crossing and non-crossing parts of the transient orbit. This allows us to represent the distribution in the form

$$f = \frac{n_0 \exp\left\{-\varepsilon^{-2/3}\left[\left(\sqrt{w_0^2 - I} - \varepsilon^{-2/3}\right)^2 + I\right]\right\}}{\pi^{3/2} v_T^3 [1 + erf(\varepsilon^{-1})]} \quad (8)$$

where $\varepsilon = v_T/v_D$, $\mathbf{w} \equiv \varepsilon^{-2/3}\mathbf{v}/v_D$, $w_0 = |\mathbf{w}|$, $I = \varepsilon^{2/3}I_z\omega_0/(mv_T)$, $n_0$ is the plasma density outside the sheet which is created by two counterstreaming beams, $\omega_0 = eB_0/(mc)$. Here we also suggest the validity of Liuville theorem.

Using (8) and introducing the dimensionless variables $\zeta = z\varepsilon^{-4/3}\omega_0/v_D$, $\zeta_{0,1} = z_{0,1}\varepsilon^{-4/3}\omega_0/v_D$, and $b = B/B_0$, as well as their transforms $b(\zeta) \to b(a) \to \gamma(\eta)$ with

$$a = \int_0^\zeta b(\zeta')d\zeta', \quad \eta = \varepsilon^{2/3}a, \quad \gamma(\eta) = b(a),$$

$$\frac{db}{d\zeta} = \frac{db}{da}\frac{da}{d\zeta} = \frac{1}{2}\frac{db^2}{da} \quad (9)$$

we arrive at the following dimensionless equation for the sheet magnetic field $\gamma$ as a function of the nontrivial part $\eta \sim \int B(z')\,dz$ of the vector-potential in the region $\eta \geq 0$

$$\gamma^2(\eta) = \frac{8\varepsilon^{1/3}}{\pi^{3/2}}\left(\frac{v_D}{v_A}\right)^2 \frac{F_{(+)}(\eta) + F_{(-)}(\eta)}{1 + erf(\varepsilon^{-1})} \quad (10)$$

where

$$F_{(\pm)}(\eta) = \pm \int_0^\eta d\eta' \int_0^\infty dw_x \int_0^\infty w_y dw_y \int_0^\infty dw_\zeta$$
$$\exp\left\{-\varepsilon^{-2/3}\left[\left(\sqrt{w_x^2 + w_y^2 + w_\zeta^2 - I_\pm} - \varepsilon^{-2/3}\right)^2 + I_\pm\right]\right\}, \quad (11)$$

$$I_\pm = \frac{2}{\pi}\int_{\eta_{0\pm}}^{\eta_\pm}\sqrt{w_y^2 + w_\zeta^2 - (\pm w_y + \eta''' - \eta')^2}\,\frac{d\eta'''}{\gamma(\eta''')}, \quad (12)$$

$$\eta_{0\pm} = \max\left\{0, \eta' - \sqrt{w_y^2 + w_\zeta^2} \mp w_y\right\}, \quad (13)$$

$$\eta_\pm = \sqrt{w_y^2 + w_\zeta^2} + \eta' \mp w_y, \quad w_\zeta = d\zeta/dt. \quad (14)$$

The parameter $v_A/v_D$ must obey the boundary relation $\gamma(\eta \to \infty) = 1$, which corresponds to the pressure balance condition in a sheet and should be satisfied for self-consistent solution [Burkhart et al., 1992]:

$$\frac{v_A}{v_D} = \frac{2\sqrt{2}}{\pi^{3/4}}\frac{\varepsilon^{1/6}}{\sqrt{(1 + erf(\varepsilon^{-1}))}}$$
$$\sqrt{\int_0^\infty d\eta'(F_{(+)}(\eta', w_x) + F_{(-)}(\eta', w_x))} \quad (15)$$

## 3. LIMITING CASES OF THE BASIC THEORY

In the limiting case $\varepsilon \to 0$ due to the estimates $w_0 \sim \varepsilon^{-2/3} \to \infty$, $I_+ \sim \varepsilon^{2/3} \to 0$, $I_- \sim w_y^2 \to \infty$ the basic Eq. (10) may be reduced to the following parameterless form

$$b^2(a) = \frac{2\sqrt{2}}{\pi}\int_0^a da' \int_0^{\pi/2}(\cos\phi)^{3/2}\,d\phi \int_{a'}^\infty \frac{da''}{\sqrt{a'' - a'}}$$
$$\exp\left(-I_+^{(s)}\right)(\phi, a'') \quad (16)$$

with

$$I_+^{(s)} = \frac{2^{3/2}(\cos\phi)^{1/2}}{\pi}\int_0^{a''}\sqrt{a'' - a'''}\,\frac{da'''}{b(a''')} \quad (17)$$

which in fact coincides with the expression found by *Kropotkin et al.* [1997]. According to Eq. (10) and Eq. (11) the corresponding current and magnetic field profiles have the characteristic scaling $L \sim \varepsilon^{4/3}v_D/\omega_0$, which was predicted on the basis of non self-consistent approach by *Francfort and Pellat* [1976].

In the other limit $\varepsilon \to \infty$ the parameters $\eta$, $w_x$, $w_y$, $w_\zeta$, and $I_\pm^{1/2}$ scale as $\varepsilon^{1/3}$. As a result expanding exponent in the R.H.S. of (11) and using the new set of variables $\gamma(\eta) = \beta(\alpha)$, $\alpha = \varepsilon^{-1/3}\eta = \varepsilon^{1/3}a$, one can obtain the following parameterless equation like (16)-(17)

$$\beta^2(\alpha) = (16/\pi)\int_0^\infty r^2 \exp(-r^2)\,dr \int_0^{\pi/2}\sin\theta d\theta \int_0^{\pi/2} d\varphi$$
$$\left(\int_0^{2y} - \int_\alpha^{\alpha+2y}\right)da'\sqrt{r^2 - I_+^{(w)}(\alpha', y, z)} \quad (18)$$

where

$$I_+^{(w)}(\alpha', y, z) = \frac{2}{\pi}\int_{\max\{0,\alpha'-\sqrt{y^2+z^2}-y\}}^{\sqrt{y^2+z^2}+\alpha'-y} \frac{da''}{\sqrt{y^2+z^2-(y-\alpha'+\alpha'')^2}}\,\frac{da''}{\beta(\alpha'')} \quad (19)$$

and $x = r\sin\theta\cos\varphi$, $y = r\sin\theta\sin\varphi$, $z = r\cos\theta$. Contrary to the case of strong anisotropy the corresponding universal profiles of current density and magnetic field scale as $L \sim \varepsilon v_D/\omega_0 = \rho_0$, so that the width of the weakly anisotropic sheet is always of order of the thermal ion gyroradius. This result is consistent with the another earlier estimate of the CS thickness [Ashour-Abdalla et al., 1994]. Similar scaling $L \sim \rho_0$ have been mentioned earlier by *Chen et al.* [1990].

## 4. THE ROLE OF DIAMAGNETIC DRIFT CURRENTS IN CURRENT SHEET EMBEDDING STRUCTURE

The functions $F_{(+)}(\eta)$ and $F_{(-)}(\eta)$ with the opposite signs which determine the structure of (10) reflect the contributions of partial currents along $Y$-axis. These partial currents might have different physical origin in the vicinity of the CS. Far from the sheet particle motion might be considered roughly as magnetized and in terms of guiding center theory resulting currents are combination of gradient, drift, centrifugal and magnetization currents. The importance of the role of magnetization currents is underlined by investigations of *Holland and Chen* [1993], *Harold and Chen* [1996] who were found with the help of numerical models that the diamagnetism of plasma can play the significant role for the case $v_D \ll v_T$ but is quite small in the opposite case.

Three general kinds of drift currents can exist far from sheet, i.e. magnetization, curvature and gradient one. Their sum manifests the natural plasma diamagnetism and their total current is directed in negative $Y$-direction. Near the plane of the field reversal these currents exist together with the positive current, which is similar to the boundary current flowing along the walls of the box with plasma as a result of non-compensated Larmor circles of particles, reflected by walls. The plane of magnetic field reversal ($z = 0$) where particles accomplish the meandering motions plays the role of such "wall".

Fig. 1 demonstrate the self-consistent magnetic field at different $\varepsilon = v_T/v_D$ values. The results of calculations for arbitrary $\varepsilon$ demonstrate that in case of moderate and weak anisotropy the self-consistent current sheet equilibrium exists (contrary to paper by *Holland and Chen* [1993]). The sheet thickness is changing with parameter $\varepsilon$ and the profile of magnetic field converges to the limiting profiles for strong and weak anisotropies, assuming the corresponding scaling of *Francfort and Pellat* [1976] and *Ashour-Abdalla et al.* [1994].

The comparison of Fig. 2 and Fig. 3 demonstrates that the profiles of current and plasma densities are similar and have unique scaling in case of strong anisotropy. In the case of weak anisotropy completely different picture emerges. At large values of $\varepsilon$ profile of density tends to be homogeneous along $Z$-axis. At the same time the corresponding current is concentrated inside this plasma sheet (PS) with the characteristic thickness $L_{PS}$ about gyroradius. We show in Fig. 4 that the current at $\varepsilon \gg 1$ although having essential paramagnetic kernel ($z \approx 0$) has also quite long diamagnetic "wings".

The existence of these wings results in the difference of plasma and current structures of the sheet.

Fig. 4 shows that the structure of CS is determined by process of compensation of positive and negative components of the diamagnetic current, which can change the scaling of the CS at weak anisotropy and create therefore embedded structure. This result is in good agreement with papers by *Holland and Chen* [1993], *Harold and Chen* [1996] where the essential role of diamagnetic currents in the case of weak anisotropy is demonstrated. In the opposite case of strong anisotropy the influence of diamagnetic currents decreases and diamagnetic effects (proportional to $v_\perp^2$) disappear.

## 5. CURRENT SHEET STRUCTURE FOR SUPERSTRONG ANISOTROPY

We present in Fig. 5 the combined picture of thin ion CSs scaling $L/\rho^*$, where $\rho^* = v_D/\omega_0 \simeq v_0/\omega_0 \simeq \rho_0$, at different $\varepsilon$ values. We distinguish two kinds of thin ion sheets: quasi-adiabatic CSs ($\varepsilon > B_n/B_o$) and non-adiabatic CS ($\varepsilon < B_n/B_o$; which we refer as the superstrong anisotropy case). Why are these two regions appearing?

In general, magnetized ion, which crosses the reversal field plane, always experiences the jumps of quasi-adiabatic invariant $I_z$. Magnetic field in the field reversal configuration has very small radius of curvature near the midplane and dynamics of particles is controlled by centrifugal force. It's influence on conservation of CS invariant $I_z$ was first estimated by *Zelenyi and Savenkov* [1993]. Then in series of papers by *Delcourt et al.*, [1994, 1999] the theory of the jumps of magnetic invariant under the influence of sharply peaked centifugal impulses have been developed for the "chaotic" case at $\kappa = 1-3$. One could roughly estimate the characteristic value of the jump accordingly *Büchner and Zelenyi* [1989], *Zelenyi and Savenkov* [1993], *Vainshtein et al.* [1999] as

$$|\Delta I_z| \sim \kappa \tag{20}$$

where $\kappa = B_n/B_o\sqrt{L/\rho_0}$ is a parameter of adiabaticity. If the jump of $|\Delta I_z|$ is smaller than the value of $I_z$ itself then our analytical model (10) - (11) could more or less adequately describe thin ion CS neglecting the non-conservation of $I_z$. One could easily estimate the condition when this quasi-adiabatic approximation is violated. One could find the relation between parameter $\kappa$ and normalized $B_n$ component of the magnetic field for the self-consistent solution in the form $b_n = \kappa^{2/3}$ which gives the value of the self-consistent CS thickness $L$

$$L = b_n \rho^* \tag{21}$$

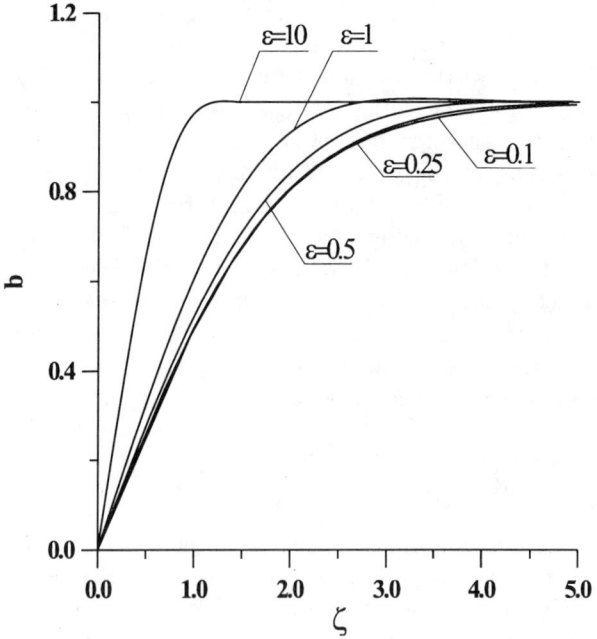

**Figure 1.** The self-consistent profile of the magnetic field $b$ as a function of the dimensionless Z-coordinate $\zeta = z\epsilon^{-4/3}\omega_0/v_D$.

So the marginal value of $v_\perp^0$ at the boundary $|z| \sim L$ below which one can not neglect the non-conservation of $I_z$ could be estimated as:

$$v_\perp^0 \leq v_0 b_n \quad (22)$$

Expression (22) is equivalent to $\varepsilon \leq b_n$ (as shown in Fig. 5).

As one can see from Fig. 5, the thickness of ion CS is not going to zero in the limit of super strong anisotropy but has finite value due to unavoidable jumps of the adiabatic invariant. Actually, in quasi-adiabatic case the thickness is determined by the initial cross-field velocity $v_\perp^0$ (or by parameter $\varepsilon$, as in estimation of *Francfort and Pellat* [1976]). In the non-adiabatic approximation where the initial velocity $v_\perp^0 \approx 0$ the dynamics of particle meander motion in the sheet is depending from the above mentioned finite gain of cross-field velocity $v_\perp^*$, which determines the effective value of adiabatic invariant $I_z$ at the moment of first encounter with the field reversal plane. It means that (21) gives the absolute minimal thickness of the CS. This estimate coincides with the estimation of the thickness of meandering in a step-like non self-consistent magnetic field [*Alexeev and Malova*, 1990] $\mathbf{B} = \{B_x sign(z), 0, B_z\}$.

One can also demonstrate that the estimation of CS thickness (21) does not contradict to the scaling $L^* = \rho^* b_n^{4/3}$. Moreover there is a one-to-one correspondence between both estimates. The thickness of super-strong anisotropic CS obtained by *Pritchett and Coroniti* [1992] and by *Burkhart et al.* [1992] refers to a different moment of meandering motion than (21). It is a height of ion trajectory element at the center of meander semi-circle. While we estimated the maximal thickness at the begining of meandering regime when particle just crosses the separatrix. The absolute minimal thickness of the current sheet is proportional to $b_n^{4/3}$, and the relation of scales is as follows: $L^*/L = b_n^{1/3}$.

## 6. THE COMBINED PICTURE OF THIN ADIABATIC AND NON-ADIABATIC CURRENT SHEETS.

The combined scheme of the self-consistent CS thickness $L/\rho^*$ as function of parameters $B_n/B_0$ and $\varepsilon$ is represented in Fig. 6. The region 1 corresponds to weak anisotropy approximation with scaling $L/\rho^* \approx 1$. Region 2 corresponds to the scaling obtained by *Francfort and Pellat* [1976] with $L/\rho^* \approx (v_T/v_D)^{4/3}$. Region 3 reflects the scaling of super-strong anisotropic CS ($L^* \approx \rho^* b_n^{4/3}$) and as we argued above at the same time is equivalent to (21). Its thickness doesn't depend already from parameter $\varepsilon$, but is entirely determined by the value of normal component of magnetic

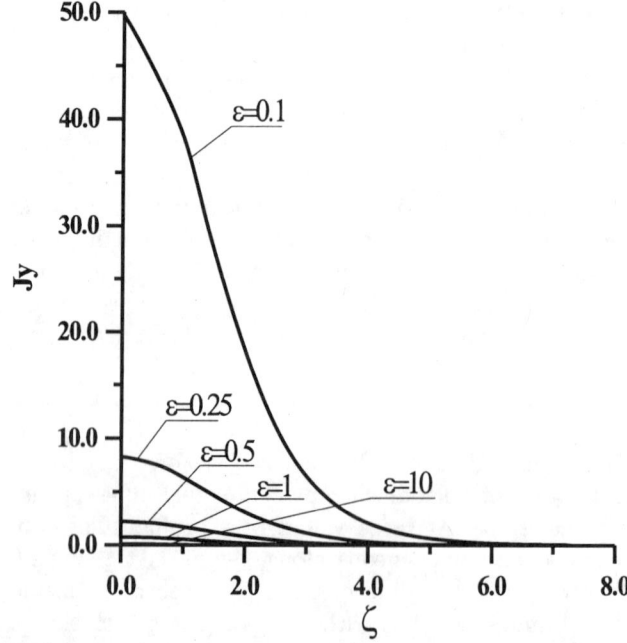

**Figure 2.** The current density $j_y$ as a function of the dimensionless Z-coordinate $\zeta$.

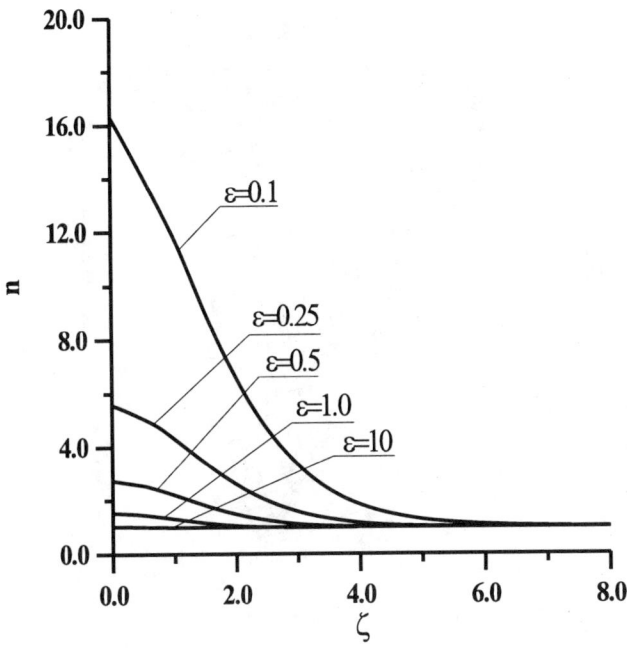

**Figure 3.** The plasma density $n$ as a function of the dimensionless $Z$-coordinate $\zeta$.

field $B_n$. The general thinning of CS due to increase of plasma flows anisotropy outside the sheet is in good agreement with the detailed consideration of *Harold and Chen*, [1996].

Although our model is applied to deHoffman-Teller system of coordinates where the electric field is transformed to zero, we also can represent our results in the magnetospheric coordinate system where exists a finite dawn-dusk $E_y$. Transferring from deHoffmann-Teller system having velocity $V_d = cE_y/B_n$ we represent the drift flow velocity as a function of the external electric field: $\tilde{V}_d = v_D + cE_y/B_n$. Here $v_D$ is a velocity of source ion beam in the magnetospheric frame, $E_y$ is a dawn-dusk (external) electric field. The dependence of CS thickness $L/\rho^*$ from $b_n$ and $\varepsilon^{-1} = v_D/v_T$ parameters at at two values of dimensionless electric field $E_y^* = cE_y/(v_DB_0)$ is shown in Fig. 7a,b. One can see that the thicknesses characteristic to the regimes of strong and super-strong anisotropy are prevailing for nonzeroth values of normalized electric field for arbitrary values of parameters $b_n$ and $\varepsilon = v_T/v_D$. This last result is consistent with the generally recognized concept of CS thinning during the beginning of substorm. The Fig. 7a,b support the general view of CS thinning presenting several "momentary" pictures of the TCS scaling for absent and present normalized "dawn-dusk" electric field $E_y^*$.

## 7. SUMMARY

Our model presented here is an analytical development of kinetic investigations of TCSs by *Speiser* [1965], *Eastwood* [1972], *Francfort and Pellat* [1976], *Burkhart et al.* [1992], *Pritchett and Coroniti* [1992], *Holland and Chen* [1993], *Kropotkin et al.* [1997] and other scientists. The evidence of such sheets have been provided recently by the in situ measurements in the tail of the Earth magnetosphere.

Two essential types of kinetic models of TCSs exist now. First, this is a full particle-in cell and hybrid simultaions by *Pritchett and Coroniti* [1994, 1995], *Hesse and Winske* [1996]; *Hesse et al.* [1996] and *Kuznetsova et al.* [1998] to describe very thin CSs in the magnetotail. Electrons constitute significant or even dominant part of the cross-tail current in such sheets. Analytical model presented in this paper is a part of another class of kinetic models, where the tension of magnetic field lines is balanced by the finite inertia of ions. We have shown that the self-consistent solution of Vlasov-Maxwell equations could be obtained for the anisotropic

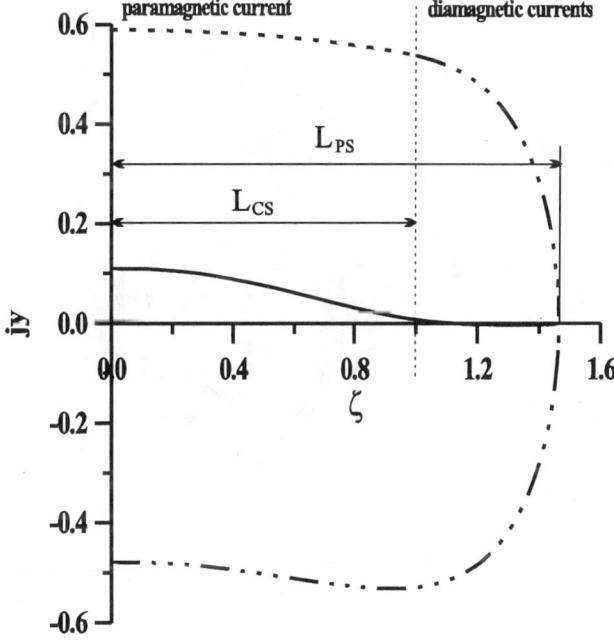

**Figure 4.** The total current density (solid line) at $\varepsilon = v_T/v_D = 10$ and corresponding partial positive and negative currents (dashed and dotted-dashed lines); the vertical dashed line conditionally divides two regions where only diamagnetic currents exist and both dia- and paramagnetic ones.

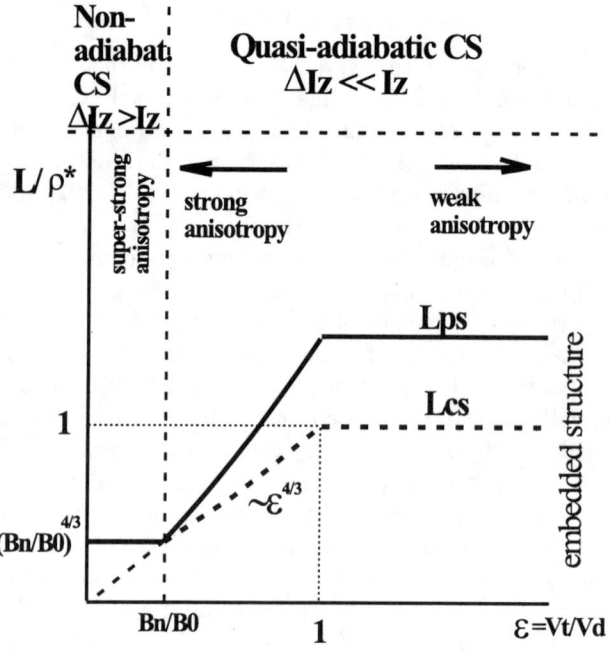

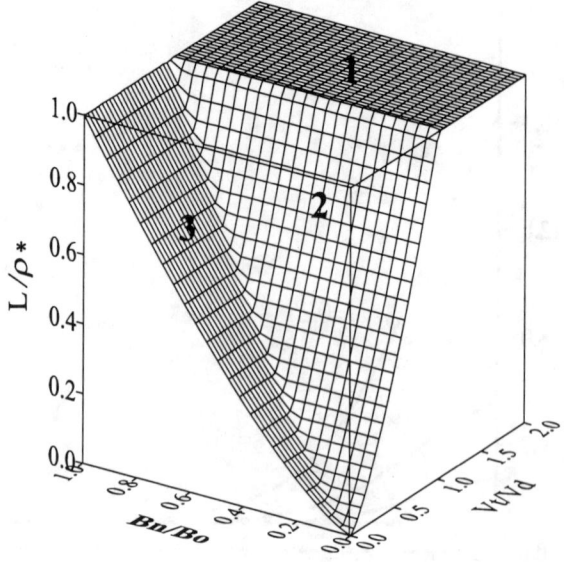

**Figure 5.** The combined picture of thin ion CSs scaling $L/\rho^*$ at different $\varepsilon$ values.

**Figure 6.** The combined scheme of the self-consistent CS thickness $L/\rho^*$ as function of parameters $B_n/B_0$ and $\varepsilon$. The region 1 corresponds to weak anisotropy scaling $L/\rho^* \approx 1$ (the explanations are in the text), the region 2 corresponds to the scaling with $L/\rho^* \approx (v_T/v_D)^{4/3}$, the region 3 reflects the scaling of super-strong anisotropic CS.

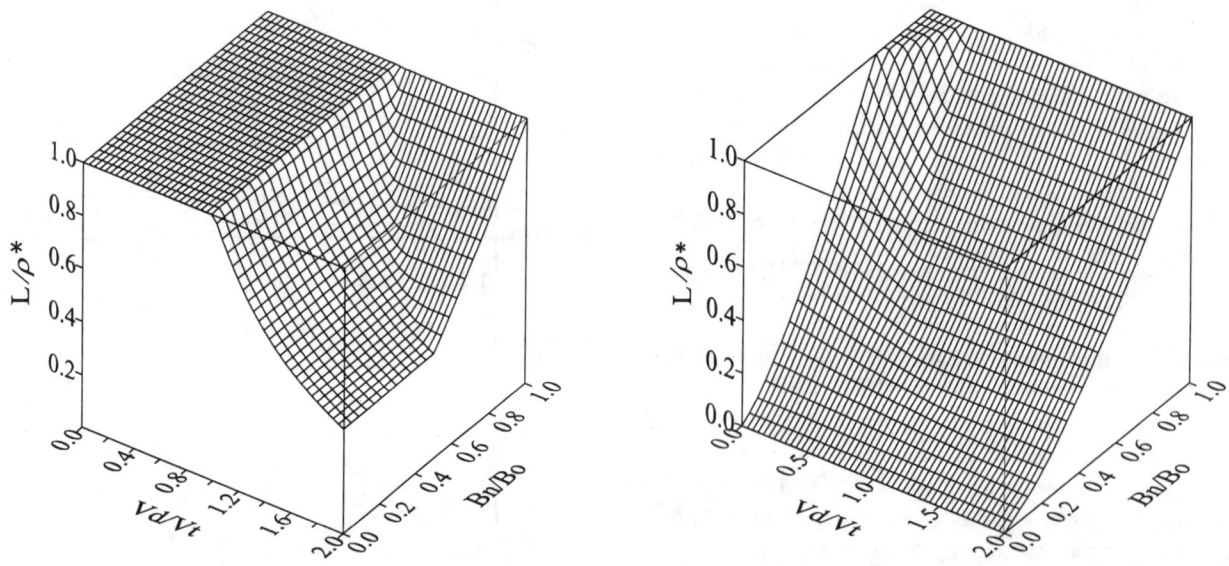

**Figure 7.** The dependence of CS thickness $L/\rho^*$ from $B_n/B_0$ and $\varepsilon^{-1} = v_D/v_T$ parameters at two values of dimensionless electric field $E_y^* = cE_y/(v_T B_0) = 0.0$ (a), 0.7 (b).

sheets on the basis of the new set of integrals of ion motion in the analytical form like the well known isotropic solutions before [Harris,1962] or [Schindler, 1972]. The characteristic property of self-consistent CS structure predicted by this model is the embedding of CS inside the PS due to compensation of dia- and paramagnetic currents at the sheet outer parts. It is demonstrated that the maximum thickness of the CS is achieved in the case of weak anisotropy and equals thermal gyroradius of ions outside the sheet. We have estimated also that the absolute minimal thickness of the sheet is achieved for super-strong anisotropy $v_T/v_D < b_n$ and lies within the limits $\rho^* b_n^{4/3} < L < \rho^* b_n$. In this regime the CS structure is determined by nonadiabatic effects ($I_z$ is essentially non conserved) and consists generally of paramagnetic currents. CS in our model thins with the growth of the dawn-dusk electric field and this effect might have important implications for the dynamics of substorms.

*Acknowledgments.* The authors are grateful to Drs. D. Delcourt and V. Sergeev for helpful discussions.

This work was supported by the Russian Basic Research Foundation grants 97-02-16489, 99-05-65567, INTAS grant 96-2346, 97-1612 and grant of Basic Research Competitive Center of Ministry of Education of Russian Federation.

## REFERENCES

Alexeev, I.I., and H.V. Malova, The plasma sheet structure in the magnetotail (in Russian), *Geomagn. Aeron.*, *30*, 407, 1990.

Alexeev, I.I., and H.V. Malova, On the model of current sheet in the magnetosphere tail, taking into account the interaction of transit and traped particles, *Advances in Space Research*, *16*, 205, 1995.

Ashour-Abdalla, M., L.M. Zelenyi, V. Peroomian, and R.L. Richard, Consequences of magnetotail ion dynamics, *J. Geophys. Res.*, *99*, 14,891, 1994.

Ashour-Abdalla, M., L.A. Frank, W.R. Paterson, V. Peroomian, and L. M. Zelenyi, Proton velocity distributions in the magnetotail: theory and observations, *J. Geophys. Res.*, *101*, 2587, 1996.

Birn, J., Magnetotail equilibrium theory: The general three-dimensional solution, *J. Geophys. Res.*, *92*, 11,101, 1987.

Büchner, J., and L.M.Zelenyi, Regular and chaotic charged particle motion in magnetotaillike field reversals 1. Basic theory of trapped motion, *J. Geophys. Res.*, *94*, 11821, 1989.

Burkhart, G.R., J.F. Drake, P.B. Dusenbery, and T.W. Speiser, A particle model for magnetotail neutral sheet equilibria, *J. Geophys. Res.*, *97*, 13,799, 1992.

Burkhart, G.R., and J. Chen, Differential memory in the Earth's magnetotail, *J. Geophys. Res.*, *97*, 6479, 1992.

Chen, J., G.R. Burkhart, and C.Y. Huang, Observations signatures of nonlinear particle dynamics, *Geophys. Res. Lett.*, *17*, 2237, 1990.

Delcourt, D.C., and R.F. Martin Jr., Application of the centrifugal impulse model to particle motion in the near-Earth plasma sheet, *J. Geophys. Res.*, *99*, 23,583, 1994.

Delcourt, D.C., and R.F. Martin Jr., Pitch angle scattering near energy resonances in the geomagnetic tail, *J. Geophys. Res.*, *104*, 383, 1999.

Eastwood, J.W., Consistency of fields and particle motion in the "Speiser" model of the current sheet, *Planet Space Sci.*, *20*, 1555, 1972.

Fairfield, D.H., Magnetotail energy storage and the variability of the magnetotail current sheet, in *Magnetic Reconnection in Space and Laboratory Plasmas, Geophys. Monogr. Ser.*, vol. 30, edited by E. W. Hones, p. 168, AGU, Washington, D. C., 1984.

Francfort, P., and R. Pellat, Magnetic merging in collisionless plasmas, *Geophys. Res. Lett.*, *3*, 433, 1976.

Galeev, A.A., and L.M. Zelenyi, Tearing instability in plasma configurations (in Russian), *Zh. Eksp. Teor. Fiz.*, *70*, 2133, 1976.

Harold, J.B., and J. Chen, Kinetic thinning in one- dimensional self-consistent current sheets, *J. Geophys. Res.*, *101*, 24,899, 1996.

Harris, E.G., On a plasma sheath separating regions of oppositely directed magnetic fields, *Nuovo Chimento*, *23*, 115, 1962.

Hesse, M., and D. Winske, Hybrid modeling of the formation and structure of thin current sheets in the magnetotail, in *Proceedings of the Third International Conference on Substorms (ICS-3), Spec. Publ. ESA SP-389* p. 231, Eur. Space Agency, Paris, 1996.

Hesse, M., D. Winske, M.M. Kuznetsova, J. Birn, and K. Schindler, Hybrid modeling of the formation of thin current sheets in magnetotail configurations, *J. Geomagn. Geoelectr.*, *48*, 749, 1996.

Holland, D.L., and J. Chen, Self-consistent current sheet structures in the quiet-time magnetotail, *Geophys. Res. Lett.*, *20*, 1775, 1993.

Kan, J.R., On the structure of the magnetotail current sheet, *J. Geophys. Res.*, *78*, 3773, 1973.

Kaufmann, R. L., Substorm currents: Growth phase and onset, *J. Geophys. Res.*, *92*, 7471, 1987.

Kaufmann, R.L., I.D. Kontodinas, B.M. Ball, and D. J. Larson, Nonguiding center motion and substorm effects in the magnetotail, *J. Geophys. Res.*, *102*, 22,155, 1997.

Kropotkin, A.P., H.V. Malova, and M.I. Sitnov, Self-consistent structure of a thin anisotropic current sheet, *J. Geophys. Res.*, *102*, 22,099, 1997.

Kuznetsova, M.M., M. Hesse, and D. Winske, Kinetic quasi-viscous and bulk flow inertia effects in collisionless magnetotail reconnection, *J. Geophys. Res.*, *103*, 199, 1998.

Larsson, D.J., R.L. Kaufmann, Structure of magnetotail current sheet, *J. Geophys. Res.*, *101*, 21, 447, 1996.

Lembege, B. and R. Pellat, Stability of a thick two-dimensional quasineutral sheet, *Phys. Fluids*, *25*, 1995, 1982.

Lui, A.T.Y., R.E. Lopez, B.J. Anderson, K. Takahashi, L.Z. Zanetti, R.W. McEntire, T.A. Potemra, D.M. Klumpar,

E.M. Greene, and R. Strangeway, Current disruptions in the near-Erath neutral sheet region, *J. Geophys. Res.*, *97*, 1461, 1992.

McPherron, R.L., A. Nishida, and C.T. Russell, Is near-Earth current sheet thinning the cause of auroral substorm onset? in *Quantitative Modeling of Magnetosphere-Ionosphere Coupling Processes*, edited by Y. Kamide and R. A. Wolf, pp.252-265, Kyoto Sangyo University, Kyoto, Japan, 1987.

Mitchell, D.G., G.J. Williams, C.Y. Huang, L.A. Frank, and C. T. Russell, Current carriers in the near-Earth crosstail current sheet during substorm growth phase, *Geophys. Res. Lett.*, *17*, 583, 1990.

Nishikawa, K.I, Reconnection at near Earth magnetotail and substorm studied by 3-D EM particle code. In: "Geospace mass and energy flow results from ISTP Program", ed. J. L. Horwitz, W. K. Peterson, and D. J. Gallagher, AGU Geophys. Monograph., 104, p.175, 1998.

Pritchett, P.L., and F.V. Coroniti, Formation and stability of the self-consistent one-dimensional tail current sheet, *J. Geophys. Res.*, *97*, 16,773, 1992.

Pritchett, P.L., and F.V. Coroniti, Convection and the formation of thin current sheets in the near-Earth plasma sheet, *Geophys. Res. Lett.*, *21*, 1587, 1994.

Pritchett, P.L., and F.V. Coroniti, Formation of thin current sheets during plasma sheet convection, *J. Geophys. Res.*, *100*, 23,551, 1995.

Pulkkinen, T.I., D.N. Baker, C.J. Owen, J.T. Gosling, and N. Murthy, Thin current sheets in the deep geomagnetotail, *Geophys. Res. Lett.*, *20*, 2427, 1993.

Pulkkinen, T.I., D.N. Baker, D.G. Mitchell, R.L. McPherron, C. Y. Huang, and L. A. Frank, Thin current sheets in the magnetotail during substorms: CDAW 6 revisited, *J. Geophys. Res.*, *99*, 5793, 1994.

Pulkkinen, T.I., D.N. Baker, L.L. Cogger, T. Mukai, and H.J. Singer, Coupling of inner and midtail processes, in: *SUBSTORMS-4*, Ed. by S. Kokubun and Y. Kamide, Terra Scientific Publishing Company/Kluwer Academic Publishers, p.749, 1998.

Sanny, J., R.L. McPherron, C.T. Russell, D.N. Baker, T.I. Pulkkinen, and A. Nishida, Growth phase thinning of the near-Earth current sheet during the CDAW-6 substorm, *J. Geophys. Res.*, *99*, 5805, 1994.

Savenkov, B.V., L.M. Zelenyi, M. Ashour-Abdalla, and J. Büchner, Regular and chaotic aspects of charged particle motion in a magnetotail-like field with a neutral line, *Geophys. Res. Lett.*, *18*, 1587, 1991.

Schindler, K., A self-consistent theory of the tail of the magnetosphere, in *Earth's Magnetospheric Processes*, edited by B. M. McCormac, p.200, D. Reidel, Norwell, Mass., 1972.

Schindler, K., A theory of the substorm mechanism, *J. Geophys. Res.*, *79*, 2803, 1974.

Sergeev, V.A., D.G. Mitchell, C.T. Russell, and D.J. Williams, Structure of the tail plasma/current sheet at 11 Re and its changes in the course of a substorm, *J. Geophys. Res.*, *98*, 17345, 1993.

Sergeev, V.A., T.I. Pulkkinen, and R.J. Pellinen, Coupled mode scenario for the magnetospheric dynamics, *J. Geophys. Res.*, *101*, 13,047, 1996.

Sergeev, V.A., V. Angelopoulos, C. Carlson, and P. Sutcliffe, Current sheet measurements within a flapping plasma sheet, *J. Geophys. Res.*, *103*, 9177, 1998.

Shay, M.A., J.F. Drake, R.E. Denton., D. Biscamp, Structure of dissipation region during collisionless magnetic reconnection *J. Geophys. Res.*, *103*, 9165, 1998.

Sitnov, M.I., H.V. Malova, and L.M. Zelenyi, Self-consistent structure of anisotropic current sheet with quasi-adiabatic ion dynamics, *Proc. of Int. Conf. "Problems of Geospace"*, St.Peterburg, 1998 (in press).

Sitnov, M.I., H.V. Malova, and A.S. Sharma, Linear stability of a tearing mode in a quasi-neutral current sheet, *Plasma. Phys. Rep. (transl. from Fizika Plazmy)*, *25*, 227, 1999a.

Sitnov, M.I., L.M. Zelenyi, H.V. Malova, and A.S. Sharma, Thin current sheet embedded within a thicker plasma sheet: Self-consistent kinetic theory, submitted to *J. Geophys. Res.*, 1999b.

Smets, R., and D. Delcourt, Ion and electron distribution functions in the distant magnetotail: modeling of Geotail observations, *J. Geophys. Res.*, *103*, 20,407, 1998.

Sonnerup, B.U.O., Adiabatic particle orbits in a magnetic null sheet, *J. Geophys. Res.*, *76*, 8211, 1971.

Speiser, T.W., Particle trajectories in model current sheets; 1. Analytical solutions, *J. Geophys. Res.*, *70*, 4219, 1965.

Speiser, T.W., Conductivity without collisions or noise, *Planet. Space Sci.*, *18*, 613, 1970.

Zelenyi, L.M., and B.V. Savenkov, Violating of quasiadiabaticity during the particle motion in magnetic field configurations with strongly curved field lines, *Plasma. Phys. Rep. (Transl.from Russian)*, *19*, 712, 1993.

Vainshtein, D.L., L.M. Zelenyi, A.I. Neishtadt, and B.V. Savenkov, The jumps of adiabatic invariant at its small initial values, *Plasma. Phys. Rep. (in press)*, 1999.

---

H. V. Malova and M. I. Sitnov, Nuclear Physics Institute, Moscow State University, 119899 Moscow, Russia. (e-mail: mlv@dec1.npi.msu.su; stn@dec1.npi.msu.su)

L. M. Zelenyi, Space Research Institute, 117810, Moscow, Russia. (e-mail: lzelenyi@iki.rssi.ru)

A. S. Sharma, Department of Astronomy, University of Maryland, USA (e-mail: ssh@astro.umd.edu)

# Pressure Anisotropy and $B_y$ in the Magnetotail Current Sheet

Richard L. Kaufmann, Bryan M. Ball

*Department of Physics, University of New Hampshire, Durham, New Hampshire*

W. R. Paterson and L. A. Frank

*Department of Physics and Astronomy, The University of Iowa, Iowa City, Iowa*

Ions were almost isotropic at the neutral sheet throughout the $(-31 < x < -7\ R_E)$, $(0 < |y| < 15\ R_E)$ region studied using Geotail data. Pressure anisotropies developed in the middle and outer current sheet, indicating the influence of nonguiding center motion or parallel electric fields. The pressure anisotropies were only large enough to provide from 10% to 30% of the force needed to balance the $\mathbf{j} \times \mathbf{B}$ force. The sign of the measured $B_y$ was used to find where magnetic field fluctuations and the influence of the IMF $B_y$ dominate over the long term averaged $B_y$. The fluctuation and IMF effects dominated throughout the current sheet near midnight and near the neutral sheet away from midnight. The average fluctuation plus IMF influence produced a 2 to 3 nT contribution to $B_y$ in the low and mid β regions near midnight and a 1 to 2 nT contribution to $B_y$ in the highest β region. Since shielding of these contributions to $B_y$ was found to be weak, this suggests that any IMF effect most likely distorts the current sheet rather than generating a steady field in the plane of the neutral sheet. The lack of shielding also suggests that fluctuations have scale lengths on the order of an ion gyroradius or less.

## 1. INTRODUCTION

This paper is part of a project to study the structure of the current sheet in the magnetotail. Calculations using a consistent orbit tracing (COT) method [*Larson and Kaufmann*, 1996] are being compared with distribution functions and plasma moments measured by the Geotail comprehensive plasma instrumentation [*Frank et al.*, 1994] and with data from the Geotail magnetic field experiment [*Kokubun et al.*, 1994]. An evaluation of each term in the momentum and energy equations will be presented elsewhere. The present report describes alternate ways to study several features of the plasma sheet. The goal is to get a better understanding of magnetotail structure by looking at observations from different viewpoints. The specific topics discussed here involve ion pressure anisotropy and the combined influence of fluctuations plus the IMF on measurements of $B_y$ in the plasma sheet.

One-minute averages of Geotail data taken between Jan. 31, 1995 and Mar. 1, 1996 were used. *Paterson et al.* [1998] averaged several fluid parameters for a period that included the interval used here. Only the 104,000 data points taken at $|y| < 15\ R_E$ in aberrated GSE coordinates were considered to assure that the measurements were inside the magnetosphere. A 4.5° aberration correction was made to account for the motion of the Earth around the Sun [*Troshichev et al.*, 1999]. Data points then were removed if the count rate was very low. This criterion eliminated many of the lobe obser-

Magnetospheric Current Systems
Geophysical Monograph 118
Copyright 2000 by the American Geophysical Union

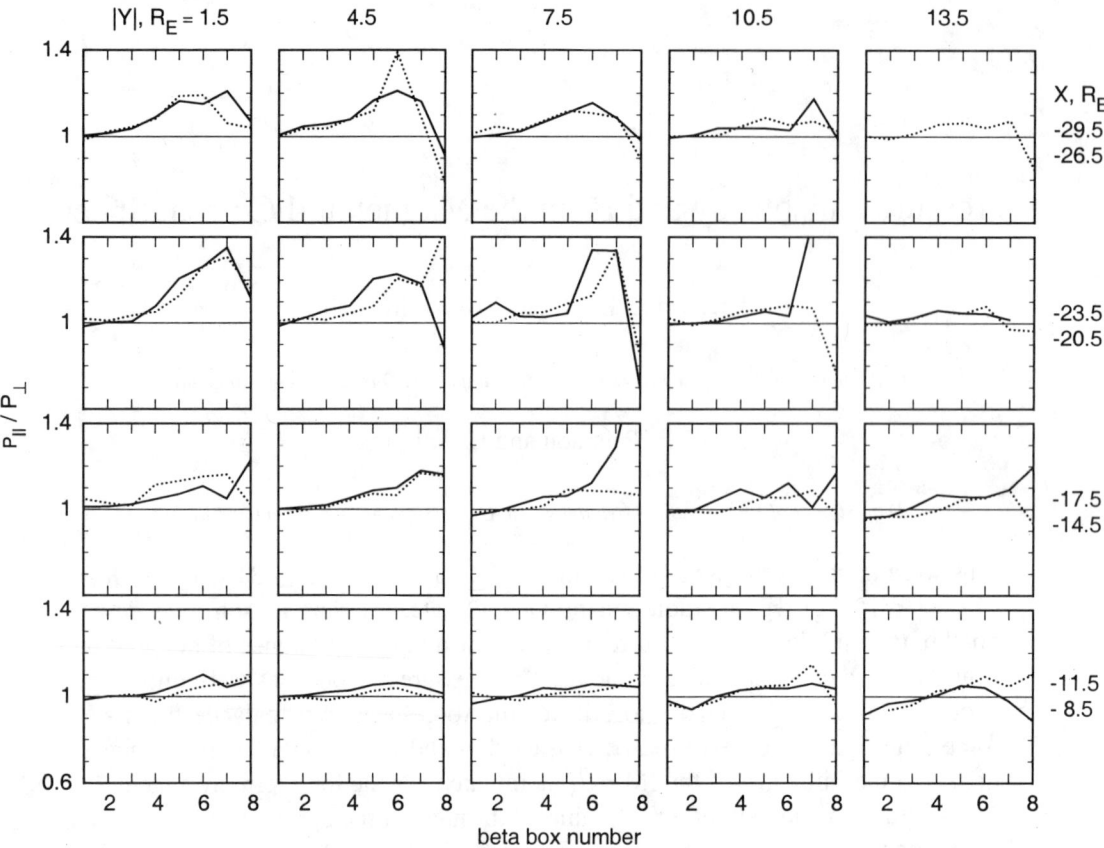

**Figure 1.** The ion pressure anisotropy ratio $P_\parallel/P_\perp$ is shown for the $(-31 < x < -7\ R_E)$, $(0 < |y| < 15\ R_E)$ region studied. Data were sorted into $3\ R_E \times 3\ R_E$ boxes in the $x$-$y$ plane. The center of each $y$-box is listed along the top of the figure. Data from two adjacent $x$-boxes are included in each panel. The $x$-box locations are listed along the right side of the figure. The solid curve shows data from the more distant of the two $x$-boxes. Data in each $x$-$y$ box was further sorted according to plasma beta. The horizontal axis in each panel is the $\beta$-box number. The ranges associated with each $\beta$-box are: 1: $\beta > 30$, 2: $10 < \beta < 30$, 3: $3 < \beta < 10$, 4: $1 < \beta < 3$, 5: $0.3 < \beta < 1$, 6: $0.1 < \beta < 0.3$, 7: $0.03 < \beta < 0.1$, 8: $\beta < 0.03$.

vations. Data points also were removed if the measured ion and electron densities differed by more than a factor of two, indicating a problem with determinations of the satellite potential or photoelectron corrections. Several other checks were made, but they removed only a small number of suspicious data points. The remaining 90,000 1-minute averages were used for the analysis.

## 2. ION PRESSURE ANISOTROPY

Two fluid parameters are examined in this section. The first shows that plasma within the current sheet does not behave as would be expected using the MHD or guiding center approximations. The second parameter provides some information about force balance.

Figure 1 shows how the ion pressure anisotropy varies in the middle magnetotail. The ion $P_\parallel/P_\perp$ measurements were sorted into $3\ R_E \times 3\ R_E$ $x$-$y$ boxes. Similar patterns were seen on the dawn and dusk sides so the results at $(x, y)$ and $(x, -y)$ were combined to reduce the number of plots and to increase the amount of data in each plot. Centers of the five different $y$-boxes are listed at the top of the Figure. There are eight $x$-boxes in the $-31 < x < -7\ R_E$ region studied. Two curves, showing data at the two adjacent $x$-locations listed along the right edge of Figure 1, are plotted in each of the four rows of panels. The solid line is for the $x$-box that is farther from Earth. For example, the solid curve in the top row represents $-31 < x < -28\ R_E$ and the dotted curve in the top row is for the $-28 < x < -25\ R_E$ box.

Data in each $x$-$y$ box also has been sorted according to plasma beta or the ratio of the thermal particle energy density to the magnetic field energy density. The horizontal axis of each panel in Figure 1 is the $\beta$-box number, with the beta ranges listed in the Figure caption. Sorting by $\beta$ is used to remove effects of tail distortion and motion such as flapping, warping, and twisting. The largest beta, which is expected to correspond to the smallest $|z|$ or distance from the neutral sheet, is at the left side of each panel. The smallest beta, corresponding to the largest $|z|$, is at the right side of each panel. The β-box number therefore is used as a nonlinear and presently uncalibrated measure of the instantaneous distance from the satellite to the neutral sheet.

There usually were fewer than 50 data points in β-boxes 1, 7, and 8. Beta-boxes 7 and 8 had few data points because the minimum count rate criterion eliminated 30% and 80% of the measurements in these boxes, respectively. The removal of so much data can produce systematic errors and the erratic behavior that is apparent in these boxes. Fewer than 2% of the measurements were removed from β-boxes 1 to 5 and 8% were removed from box 6 as a result of the low count rate cutoff. The small number of points in β-box 1 also can produce large uncertainties, but simply represents the fact that β rarely exceeds 30. Beta boxes 3 to 5 usually contained about 200 points on each the dawn and dusk sides.

Almost every curve throughout the region studied in Figure 1 shows $P_{\parallel}/P_{\perp}$ near 1.0 in β boxes 1 and 2, indicating that ions are almost isotropic near the neutral sheet. Figure 1 also shows that $P_{\parallel}/P_{\perp}$ averages about 1.1 to 1.2 in the outer current sheet beyond $x = -13\ R_E$, where the magnetic field structure is tail-like. The average $B_z$ is less than 5 nT beyond $x = -13\ R_E$ and $B_z$ increases rapidly to between 10 and 15 nT in the most earthward two $x$-boxes, indicating the transition to a dipolar geometry. The plots in this paper were prepared using the uncorrected Geotail magnetic field data. A revised data set with $B_z$ reduced by 0.5 ± 0.3 nT is expected to be available in the near future (S. Kokubun, personal communication, 1999). Sets of plots similar to those shown here have been run with approximate magnetic field corrections. The expected revision produced only minor changes in Figure 1.

Errors are difficult to estimate because the assumption that the 1-minute averages can be treated as a random sampling drawn from a normal distribution is invalid for many fluid parameters. The satellite entered each $3\ R_E \times 3\ R_E$ $x$-$y$ box only about 10 times per year. Individual β-boxes within each $x$-$y$ box therefore were entered only a few times. Some fluid parameters remain relatively steady while the satellite passes through a box. The standard deviation of individual data points was calculated for each $x$-$y$-β box in Figure 1. Most of these standard deviations of the $P_{\parallel}/P_{\perp}$ ratio were between 0.1 and 0.2 when β > 0.3, and they increased to an average of nearly 0.5 for β-box 8. An estimate of the standard error of the curves in Figure 1, derived by dividing the standard deviation by $(N-1)^{1/2}$ and using the number of 1-minute averages for $N$, gives an estimated error of between 0.01 and 0.02 for β > 0.3. This error estimate increases to an average of nearly 0.1 for β-box 8. An examination of the box-to-box fluctuations in Figure 1 shows that these estimates are much too small, and that something between the number of data points and the number of times a box was entered should be used for $N$ to estimate errors.

The best way to treat uncertainties is probably to compare separate plots made using different data sets. Lacking this, the box-to-box fluctuations seen in plots such as Figure 1 provide a reasonable estimate of the statistical errors.

One significant conclusion that can be drawn from Figure 1 is that nonguiding center effects or parallel electric fields are important in the inner current sheet. The fact that $P_{\parallel}/P_{\perp}$ increases from 1.0 near the neutral sheet to 1.1 or 1.2 in the outer current sheet throughout the large region studied shows that the pressure anisotropy increases as one moves along field lines away from the neutral sheet. Similar features have been noted in orbit tracing studies using nonguiding center particles. In contrast, the pressure would remain isotropic and constant along a field line in the guiding center approximation if pressure was isotropic at the equator and if $E_{\parallel} = 0$.

Figure 2 shows the closely related fire hose instability parameter $[P_{\parallel} - P_{\perp}]/[B^2/\mu_o]$. The fire hose parameter would be 1.0 at the outer edge of the current sheet if the tailward centrifugal force associated with the change in $v_x$ from tailward to earthward during each current sheet interaction was equal to the earthward $\mathbf{j} \times \mathbf{B}$ force [*Rich et al.*, 1972; *Cowley*, 1978; *Schindler*, 1979]. The parameter plotted in Figure 2 was obtained using β-box averages of $P_{\parallel}$, $P_{\perp}$ and $B^2$ in the above expression. This was done because a single 1-minute point can give an extremely large value of the fire hose parameter if $B$ is nearly zero. One or a few points therefore tended to dominate the averages in the first few β-boxes, where $B$ is smallest, when the parameter was calculated separately for each 1-minute interval and then averaged. The fire hose parameter also depends very sensitively upon small errors in the magnetic field data in weak field regions. Since both the numerator and denominator of the fire hose parameter approach zero in β-box 1, results for this box are especially unreliable. The standard deviation of the fire hose parameter decreased from 1 in β-box 2 to 0.1 in box 5 and 0.01 in box 8. The estimated errors of the curves in Figure 2 derived using the number of 1-minute averages for $N$ are 0.1 for β-box 2, 0.01 for box 4, and 0.003 for boxes 6 through 8. These estimates again are unrealistically small, and a better estimate is obtained by simply observing the box-to-box variations.

The fire hose parameter in Figure 2 peaks near 0.1 somewhere in the inner current sheet throughout the tail-like region. Excluding β-boxes 1 and 2, a plot similar to Figure 2 but using the average of the fire hose parameter calculated separately for each 1-minute data point peaks near 0.2. The anticipated magnetic field correction and the low ion count rates both suggest that the fire hose parameter may be underestimated at the outer edge of the principal current sheet. These results indicate that the earthward acceleration of ions as they cross the equator requires only 20% ± 10% of the $\mathbf{j} \times \mathbf{B}$ force. Pressure variations in the $x$ direction balance a

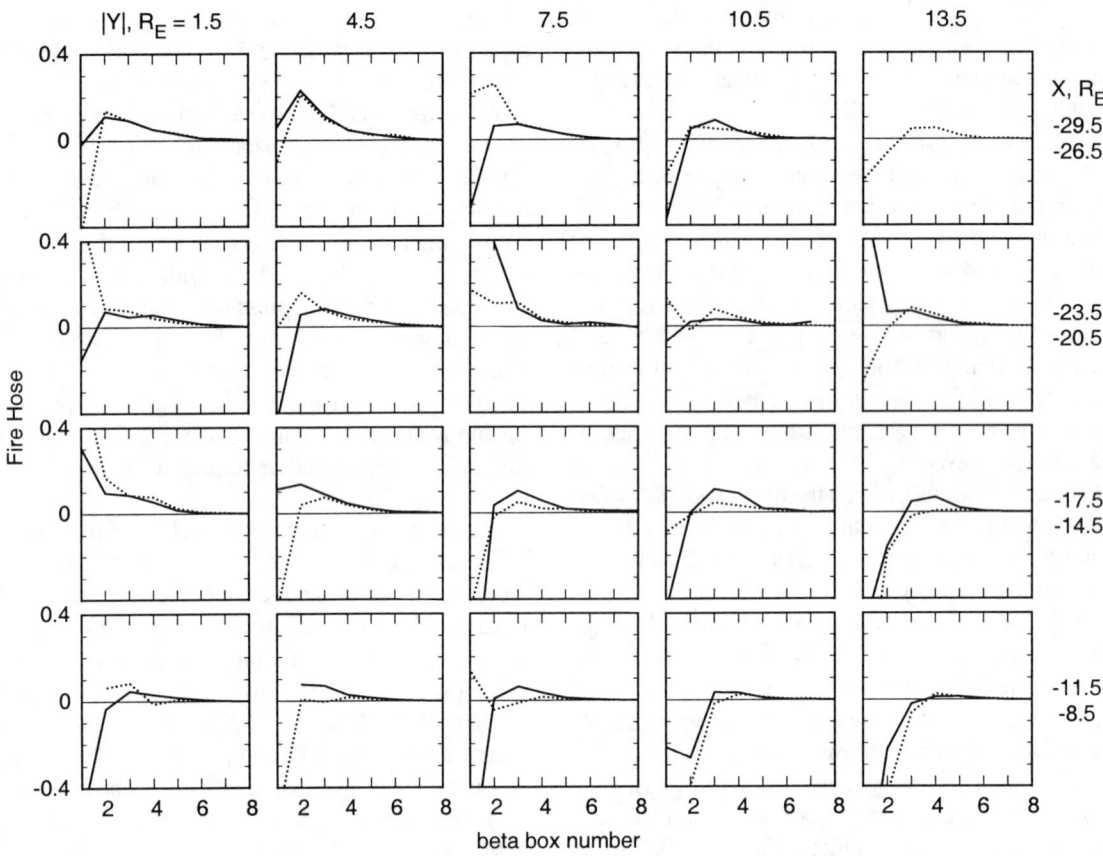

**Figure 2.** The fire hose instability parameter $[P_\parallel - P_\perp]/[B^2/\mu_o]$ is plotted in a format similar to that used in Figure 1.

larger portion of the earthward force and a small residual force imbalance is needed to explain the variations in average bulk velocity as plasma drifts earthward. A more complete study of force balance including effects of each element of the pressure tensor will be presented elsewhere.

## 3. INFLUENCE OF FLUCTUATIONS AND OF THE IMF ON THE MAGNETOTAIL $B_y$

Most studies of the influence of the interplanetary magnetic field (IMF) on the tail have involved direct correlations of measurements made in the solar wind or magnetosheath with measurements made in the magnetotail. *Kaymaz et al.* [1994] presented IMP 8 observations and summarized results from several earlier analyses. These studies revealed significant correlations between the IMF $B_y$ and the magnetotail field. The correlation is sometimes discussed in terms of the penetration of a fraction of the IMF $B_y$ into the plasma sheet. Large fluctuations also are commonly seen in the tail magnetic field [*Borovsky et al.*, 1997]. In some regions it is possible for fluctuations or effects of the IMF to dominate over the long term averaged field in determining the sign of the 1-minute averaged $B_y$. The long term averaged field includes contributions from the steady ring, tail, Birkeland, and magnetopause current systems as well as the Earth's dipole field [*Tsyganenko*, 1989; *Tsyganenko and Stern*, 1996]. This section presents a different way to study $B_y$ effects and is based only on Geotail data.

Figure 3 shows the 1-year averages of $B_x$ and $B_y$ measured when Geotail was at the outer edge of the current sheet. Magnetic field data were rotated by 4.5° so that $B_y$ approximately equals the field in the aberrated GSE y direction. A total of 12,000 data points fell into the $0.1 < \beta < 0.3$ boxes. The signs of both $B_x$ and $B_y$ were reversed whenever $B_x < 0$ in order to combine data from north and south of the neutral sheet. Since the observed $B_y$ is antisymmetric across the $y = 0$ plane, data taken at equal $|y|$ on the dawn and dusk sides also have been folded together for the following analysis by reversing the sign of $B_y$ for all data taken at $y < 0$. The $B_y$ effects therefore will be described as if all measurements were taken in the northern dusk quadrant. The average folded $B_y$ beyond $x = -13$ $R_E$ in Figure 3 is $-0.4$ nT in the $0 < |y| < 3$ $R_E$ boxes and $-5$ nT in the $12 < |y| < 15$ $R_E$ boxes. Averaging over a full year removes nearly all effects of fluctuations and the IMF $B_y$ because the sign of $B_y$ is reversed whenever the satellite is in two quadrants and because fluctuations and the IMF are roughly equally likely

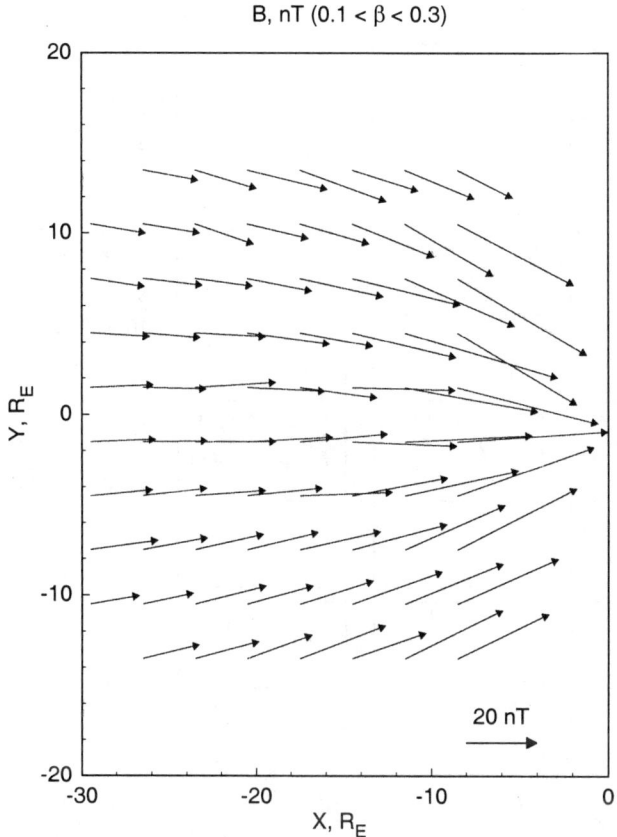

**Figure 3.** The vectors show the measured average $B_x$ and $B_y$ in each $x$-$y$ box in the outer current sheet. Data taken in the range $0.1 < \beta < 0.3$ were used to generate the averages.

to produce positive and negative $B_y$ effects. Figure 3 therefore gives an approximation of the steady portion of the tail field, and is similar to the empirical models [*Tsyganenko*, 1989; *Tsyganenko and Stern*, 1996].

Magnetization currents shield steady $x$ and $y$ components of the magnetic field from the high $\beta$ inner current sheet if the current sheet lies in the $x$-$y$ plane. In the MHD approximation magnetization currents are produced by the circular motion of particles about stationary guiding centers [*Parker*, 1957]. Such currents are associated with pressure gradients and magnetic field line curvature. Magnetization currents carried by nonguiding center particles are easiest to study in one-dimensional magnetic field models because trapped particles have no net drift in such a field. *Kaufmann and Lu* [1993] described the importance of magnetization currents in the construction of a model current sheet using nonguiding center particles. Magnetization current flows from dusk to dawn at the neutral sheet and from dawn to dusk in the outer current sheet in a magnetotail with $B_y = 0$. This current system reduces $B_x$ in the current sheet from its maximum value which is found in the lobes. Figure 4 shows how complete this shielding is for the $\beta > 10$ or neutral sheet region.

A total of 12,000 data points fell into the $\beta > 10$ boxes. The average folded $B_y$ beyond $x = -13\ R_E$ in Figure 4 is between $-0.5$ and $0.2$ nT in the various $|y|$ boxes. Note that the magnetic field scale changes by a factor of four between Figures 3 and 4. Since magnetization currents are generated by orbital motion in whatever field is present, any relatively steady contribution of the IMF to the magnetic field in the plane of the current sheet should also be shielded near the neutral sheet.

Figure 5 shows one way to examine the influence of fluctuations and of the IMF $B_y$. The ratio $-B_y/|B_y|$ for each 1-minute data point equals $+1$ whenever the measured $B_y$ is negative and $-1$ whenever $B_y$ is positive. Since all data has been folded into the northern dusk quadrant, each 1-minute point with $B_y$ pointing in the direction shown in Figure 3 produces $-B_y/|B_y| = +1$. The 1-year averages of this ratio would equal $+1$ if the long term averaged field always determined the sign of $B_y$. At the other extreme, if fluctuations and the influence of the IMF $B_y$ were always so strong that they dominated the measured $B_y$ then the above ratio would be $+1$ half the time and $-1$ the other half. The 1-year average of the $-B_y/|B_y|$ ratio therefore should be near zero in

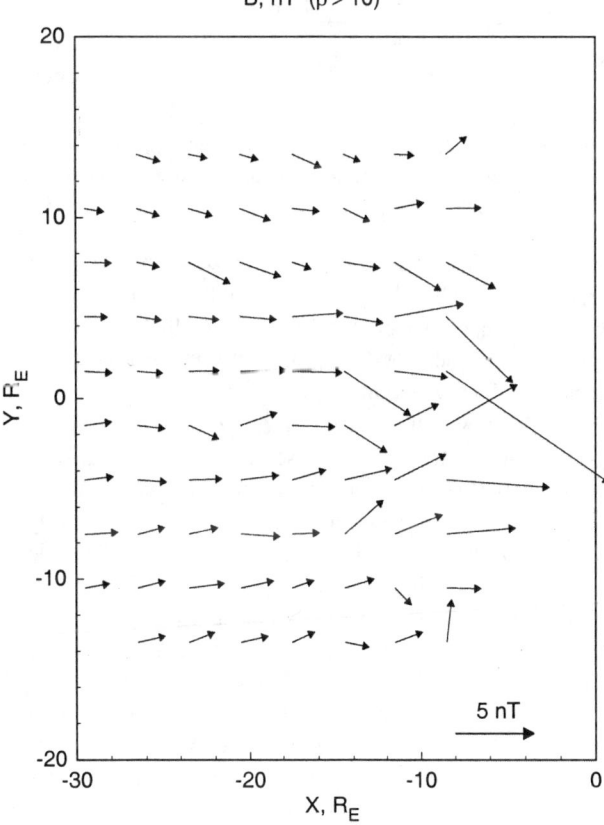

**Figure 4.** Similar to Figure 3 except for the neutral sheet. Data with $\beta > 10$ were used. Note that the scale is different from the scale in Figure 3.

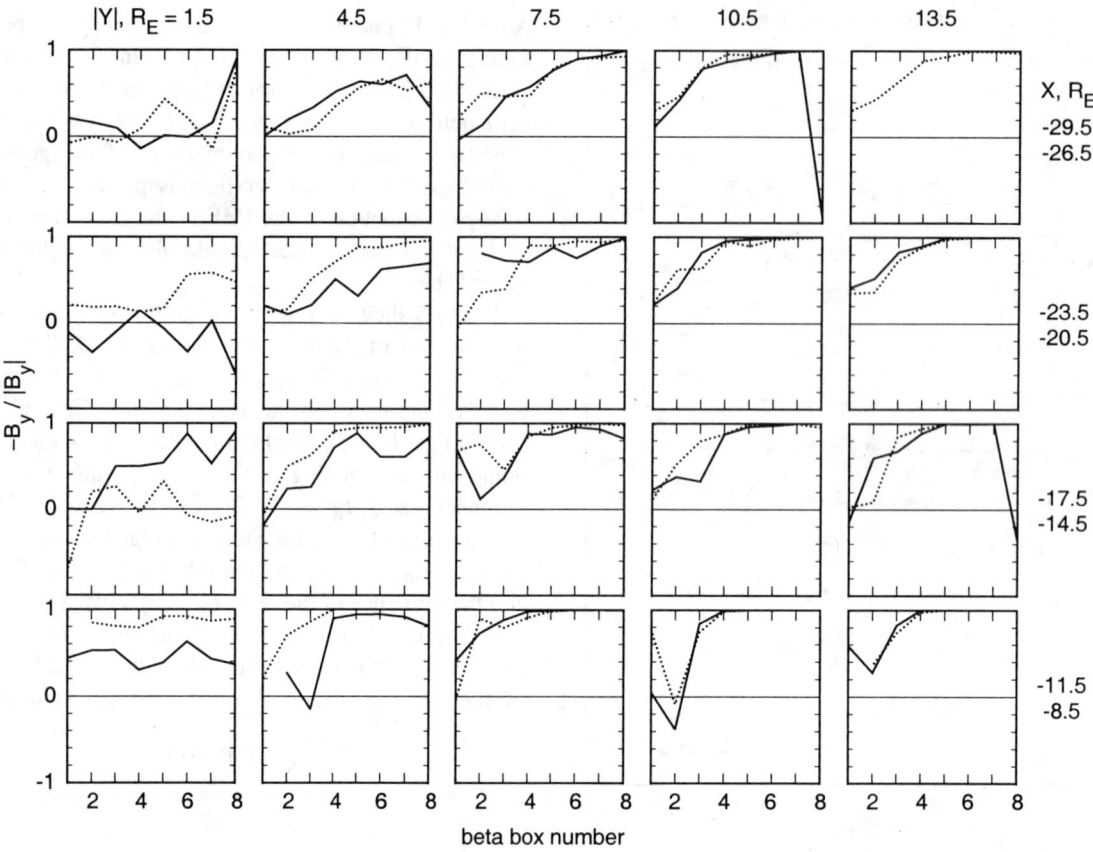

**Figure 5.** Average of the $-B_y/|B_y|$ ratio in the format used for Figure 1. This parameter should be near +1 when the long term averaged field dominates and near zero when either fluctuations or the IMF dominates in determining the sign of the measured $B_y$.

regions where fluctuations or the IMF $B_y$ dominates. The incorporation of other data would permit a separation of the effects of fluctuations from the effects of the IMF $B_y$, but the present analysis does not distinguish between these two possible causes of a reversal in the sign of $B_y$.

Figure 5 shows the 1-year average of the 1-minute $-B_y/|B_y|$ ratio. Most of the observations in the tail-like region show that the long term averaged field nearly always dominates in the outer current sheet and lobe well away from midnight ($|y| > 6 R_E$). Figure 5 also shows that the ratio fluctuates about zero in the tail-like region at $|y| < 3 R_E$, indicating that fluctuations or the IMF almost always dominates in determining the sign of $B_y$ near midnight. A comparison with Figure 3 shows that these qualitative results are not surprising. The long term averaged $B_y$ is nearly zero at midnight so it is easiest for fluctuations or the IMF to dominate.

Since the $-B_y/|B_y|$ ratio always equals ±1 its standard deviation has little significance. The standard deviation of the folded $B_y$ averages 1 nT in β-box 1, 2 nT in β-box 2, and 3 nT in β-boxes 3 through 7. The standard errors obtained if the number of 1-minute averages is used for $N$ is between 0.1 and 0.2 nT for β-boxes 1 through 6. One useful conclusion from Figure 5 is that the transition to dominance by fluctuations or the IMF when β < 1 (β-boxes 5 to 8) takes place in the $3 R_E < |y| < 6 R_E$ region. Another interesting feature of Figure 5 is that the $-B_y/|B_y|$ ratio is almost always small at the largest β (β-boxes 1 and 2) indicating that fluctuations or the IMF $B_y$ dominates in the neutral sheet throughout the region studied. Therefore, even though a comparison of Figures 3 and 4 showed that the long term averaged $B_x$ and $B_y$ are shielded from the high β region, effects of fluctuations or the IMF $B_y$ are not as well shielded.

While Figure 5 used only the sign of each 1-minute average of $B_y$ to determine which source dominates, Figure 6 shows a way to study the influence of fluctuations and of the IMF $B_y$ more quantitatively near midnight. The sum $B_y + |B_y|$ was calculated for each 1-minute period and this sum was then averaged over the 1-year study interval to create Figure 6. This parameter also has been folded to represent the northern dusk quadrant. For a single 1-minute point $B_y + |B_y|$ will be zero if $B_y$ is negative and will equal $2B_y$ if $B_y$ is positive. The average of this sum therefore

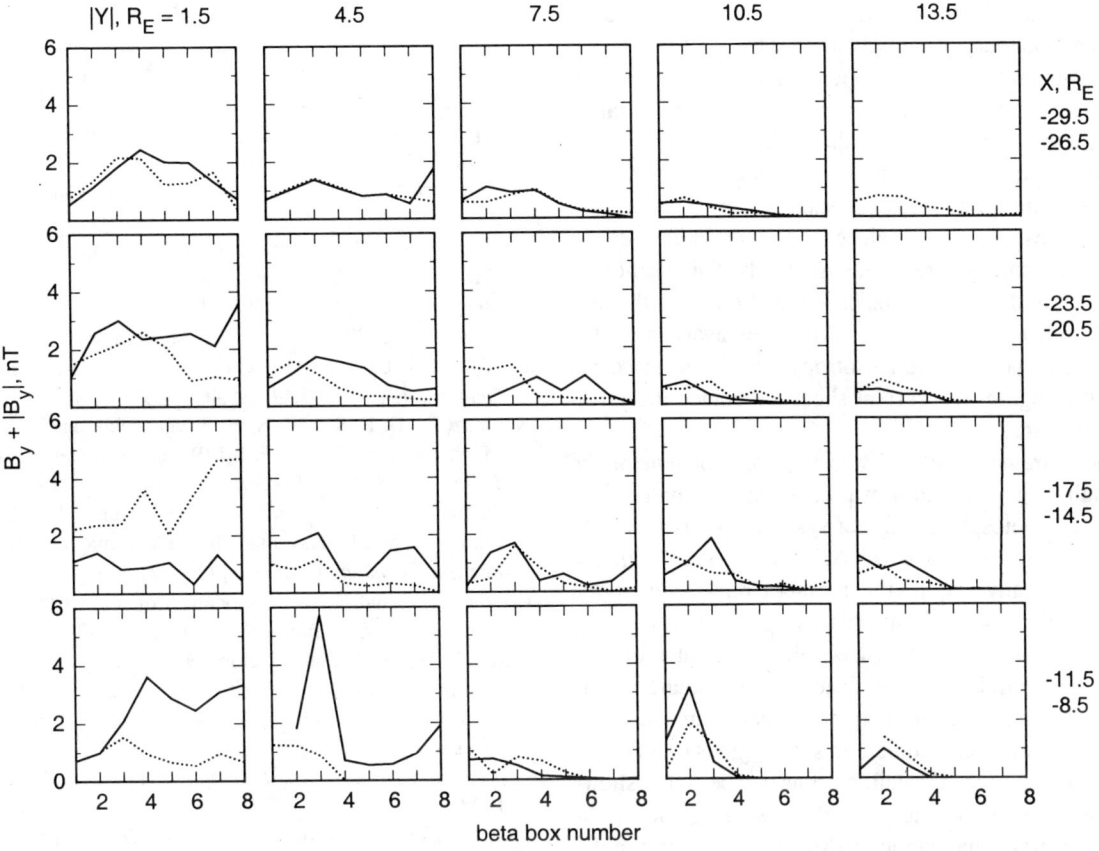

**Figure 6.** Average of $B_y + |B_y|$. This parameter should be near zero when the long term averaged field dominates and should equal the $|B_y|$ generated by fluctuations or the IMF when these influences determine the sign of $B_y$.

should equal zero in regions where fluctuations and the IMF never dominate because $B_y$ will always be negative. Near midnight, where fluctuations and the IMF effects always dominate, it will be assumed that the effects of fluctuations or the IMF are positive half the time and negative half the time. The average $B_y + |B_y|$ then should equal twice the IMF plus fluctuation field half the time and zero the other half. As a result, curves in the $0 < |y| < 3\ R_E$ boxes in Figure 6 provide an estimate of the fluctuation plus IMF associated contributions to $B_y$. The decrease away from midnight in the curves shown in Figure 6 reflect the fact that fluctuations are dominant less often in these $x$-$y$ boxes.

Figure 6 shows that this average fluctuation plus IMF effect is between 2 and 3 nT in most of the plasma sheet in the tail-like region near midnight. The average IMF or fluctuating $B_y$ appears to drop to between 1 and 2 nT in the highest $\beta$ region ($\beta$-boxes 1 and 2) indicating possible weak shielding. Fluctuations would not be shielded by organized magnetization currents if their scale size was comparable to or smaller than the radius of curvature of a typical ion. Such fluctuations are likely to be frozen into and carried with the plasma in the high $\beta$ region. If a relatively steady uniform IMF $B_y$ simply penetrated into the plasma sheet and altered the magnetic field in the plane of the neutral sheet, then this field would be expected to be shielded from the high $\beta$ region along with the Earth's field. To remain unshielded, any IMF contribution to $B_y$ therefore must primarily alter the shape of the plasma sheet through twisting or other distortions. The field normal to the neutral sheet is not well shielded by magnetization currents, which flow primarily in the plane of the current sheet. Any change in orientation will cause the normal field to have a $y$ component in the satellite or GSE coordinate system.

## 4. SUMMARY

This paper introduced a format for presenting some three-dimensional information about various parameters in a large section of the magnetotail. Separate panels were used to show measurements in each observation box in the $x$-$y$ plane. The measurements in each $x$-$y$ box were further sorted by $\beta$ which was used as a nonlinear indicator of the instantaneous $z$ location of the satellite relative to the neutral sheet. The satellite is assumed to be very close to the neutral sheet when $\beta > 10$ and to be in the outer current sheet or plasma sheet boundary layer when $0.1 < \beta < 1$.

The first measurements studied involved the ion pressure anisotropy or the closely related fire hose instability parameter. It was seen that pressure is almost fully isotropic at the neutral sheet and that anisotropies develop as one moves along a field line away from the equator. The development of anisotropies can be produced by nonguiding center motion of ions and by parallel electric fields. This result may be important to theoretical studies, simulations, and the projection of other measurements made well away from the neutral sheet. The pressure anisotropy was found to be too small at the edge of the current sheet to balance the earthward $\mathbf{j} \times \mathbf{B}$ force.

Another parameter studied was the $y$ component of the magnetotail magnetic field. It was seen that the contribution by steady magnetospheric current systems to this field was shielded by magnetization currents. This long term averaged field has a negative $B_y$ in the northern dusk side of the plasma sheet. It was found that positive and negative observed $B_y$ fields were almost equally likely at midnight and near the neutral sheet for all local times studied. This could either be an effect of fluctuations that are much larger than the long term averaged field in such regions or an effect of the IMF $B_y$. Individual satellite orbits can be found showing the temporary dominance of each $B_y$ source. The amplitude of the average fluctuation or IMF $B_y$ effect was about 2 to 3 nT near midnight throughout most of the plasma sheet and 1 to 2 nT very near the neutral sheet.

*Acknowledgments.* This material is based upon work supported by the National Science Foundation under grants ATM-9422056 and ATM-9730845 and by the National Aeronautics and Space Administration under grant NAG5-4453 at the University of New Hampshire. Research at the University of Iowa was supported by the National Aeronautics and Space Administration under grants NAG5-2371 and NAG5-7684. The authors would like to thank S. Kokubun who supplied full Geotail magnetic field measurements throughout the period studied. The reviewers also made a number of suggestions that are included in this paper.

## REFERENCES

Borovsky, J. E., R. C. Elphic, H. O. Funsten, and M. F. Thomsen, The Earth's plasma sheet as a laboratory for flow turbulence in high-$\beta$ MHD, *J. Plasma Phys.*, *57*, 1-34, 1997.

Cowley, S. W. H., The effect of pressure anisotropy on the equilibrium structure of magnetic current sheets, *Planet. Space Sci.*, *26*, 1037-1061, 1978.

Frank, L. A., K. L. Ackerson, W. R. Paterson, J. A. Lee, M. R. English, and G. L. Pickett, The comprehensive plasma instrumentation (CPI) for the GEOTAIL spacecraft, *J. Geomag.Geoelectr.*, *46*, 23-37, 1994.

Kaufmann, R. L., and C. Lu, Cross-tail current: Resonant orbits, *J. Geophys. Res.*, *98*, 15,447-15,465, 1993.

Kaymaz, Z., G. L. Siscoe, N. A. Tsyganenko, and R. P. Lepping, Magnetotail views at 33 $R_E$: IMP 8 magnetometer observations, *J. Geophys. Res.*, *99*, 8705-8730, 1994.

Kokubun, S., T. Yamamoto, M. H. Acuna, K. Hayashi, K. Shiokawa, and H. Kawano, The Geotail magnetic field experiment, *J. Geomagn. Geoelectr.*, *46*, 7-21, 1994.

Larson, D. J., and R. L. Kaufmann, Structure of the magnetotail current sheet, *J. Geophys. Res.*, *101*, 21,447-21,461, 1996.

Parker, E. N., Newtonian development of the dynamical properties of ionized gases of low density, *Phys. Rev.*, *107*, 924-933, 1957.

Paterson, W. R., L. A. Frank, S. Kokubun, and T. Yamamoto, Geotail survey of ion flow in the plasma sheet: Observations between 10 and 50 $R_E$, *J. Geophys. Res.*, *103*, 11,811-11,825, 1998.

Rich, F. J., V. M. Vasyliunas, and R. A. Wolf, On the balance of stresses in the plasma sheet, *J. Geophys. Res.*, *77*, 4670-4676, 1972.

Schindler, K., Theories of tail structures, *Space Sci. Rev.*, *23*, 365-374, 1979.

Troshichev, O., S. Kokubun, Y. Kamide, A. Nishida, T. Mukai, and T. Yamamoto, Convection in the distant magnetotail under extremely quiet and weakly disturbed conditions, *J. Geophys. Res.*, *101*, 10,249-10,263, 1999.

Tsyganenko, N. A., A magnetospheric magnetic field model with a warped tail current sheet, *Planet. Space Sci.*, *37*, 5-20, 1989.

Tsyganenko, N. A. and D. P. Stern, Modeling the global magnetic field of the large-scale Birkeland current systems, *J. Geophys. Res.*, *101*, 27,187-27,198, 1996.

---

B. M. Ball and R. L. Kaufmann, Department of Physics, University of New Hampshire, Durham, NH 03824-3568.

L. A. Frank and W. R. Paterson, Department of Physics and Astronomy, The University of Iowa, Iowa City, IA 52242-1479.

# Formation of the Storm-Time Ring Current and the Dst Field: Some Recent Topics

T. Iyemori

*Graduate School of Science, Kyoto University, Japan.*

Some recent topics related to the formation of storm-time ring current are reviewed and their mutual relationships are discussed along a scenario of the current generation associated with an enhanced magnetospheric convection.

In particular, the asymmetric distribution of disturbance magnetic fields on the ground and in the magnetosphere, their relation to the ring current structure and growth/decay mechanisms, the physical meaning of the Dst index, and the solar wind density effects on the current formation are discussed. It is stressed that (1) the asymmetry of disturbance magnetic field in the magnetosphere is essentially in noon-midnight direction, although it is in dawn-dusk on the ground or at low altitudes just over the ionosphere because of the field-aligned current effect, (2) the initial rapid decay of the ring current as observed in the Dst field is mainly caused by the flow-out of ring current particles from dusk or afternoon magnetopause, (3) the tail current also contributes to the Dst field, and (4) the generalized relationship between the magnetospheric energy (i.e., both kinetic and magnetic field energies) and the Dst field should be used instead of the original D-P-S relation when we discuss the ring current quantitatively with the Dst index.

## INTRODUCTION

Since extensive research on the ring current has been made for more than 35 years, we tend to assume that the fundamental structure and generation mechanism of it are well known (e.g., see Gonzalez et al., 1994 and the references therein). However, the consensus of our understanding on these problems is still not enough. For example, Campbell (1996) discussed that the ring current structure is far from a symmetric ring and questioned our traditional understanding on the meaning of the Dst index which has been widely used as a measure of the ring current intensity. In a traditional view, the Dst index represents the ring current intensity which is proportional to the kinetic energy of the particles that form the symmetric ring current. A theoretical base of this view is the Dessler-Parker-Sckopke (D-P-S) relationship (e.g., Sckopke, 1966).

The discussion on 'storm-substorm relation' (e.g., Kamide, 1992; Iyemori and Rao, 1996; Siscoe and Petschek, 1997; McPherron, 1997) concerns the generation mechanism of the ring current and the role of substorms in it. It is often assumed that the substorms are the source of particle injection to the inner magnetosphere and the injected particles form a storm-time ring current.

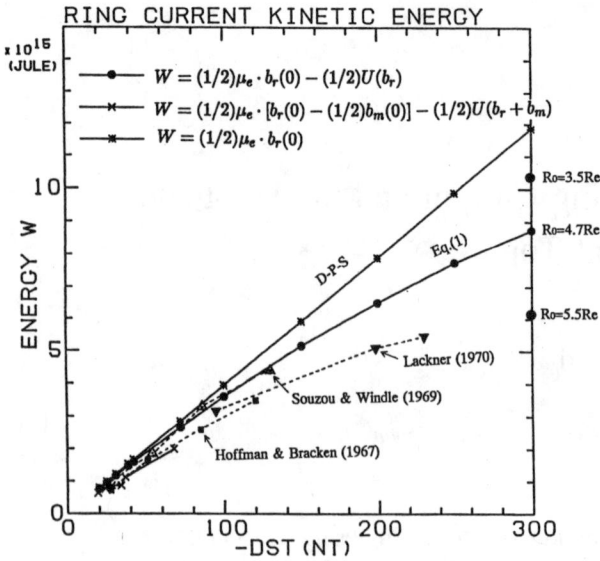

**Figure 1.** Relation between the Dst index and the total kinetic energy of plasma particles in the magnetosphere estimated from the Dessler-Parker-Sckopke relation and those estimated from the generalized theorem discussed by, for example, Siscoe (1970). The relation estimated from other methods is also plotted with broken likes. The estimation from eq. (1) in the text is shown for three cases, R0=3.5, 4.7 and 5.5 Re, where R0 indicate the location of the ring current with doughnut shape from the center of the Earth. The radius of the current is set to be 1 Re. For R0=3.5 and 5.5 Re cases, only the values at Dst=-300nT are shown. The symbol br and bm in the panel respectively means the ring current field and the magnetic field generated by the magnetopause and tail currents. The Tyganenko 1987 model fields in Xgsm>-15Re was used for the case where bm appears in the equation (i.e., x-symbols connected by a solid line). br(0) and bm(0) indicate the value at the center of the Earth.

However, this scenario was questioned based on the observational results obtained by Iyemori and Rao (1996), and a theoretical discussion which interprets their results was given by Siscoe and Petschek (1997) using a generalized energy theorem and a current circuit analogy.

A clear dependence of the plasma sheet plasma density on the solar wind density has been found by Borovsky et al. (1997) and Terasawa et al., (1997), and the dependence is discussed in conjunction with the ring current formation (e.g., Kozyra et al., 1998; Ebihara and Ejiri, 1998).

In this paper we review these recent topics to show their mutual relationship in understanding the formation and the decay of storm-time ring current. The term 'ring current' is used in this paper for an azimuthal current near magnetospheric equator in the inner magnetosphere typically inside 6 - 8 Re (Earth radii). That is, the ring current does not necessarily encircles the Earth.

## ASYMMETRIC RING CURRENT

From the OGO3 and 5 magnetic field observations in the inner magnetosphere, Sugiura (1972) obtained a statistical distribution of the delta-B which is the difference in magnitude between an observed field and the corresponding reference magnetic field of Earth's internal origin. The region with negative delta-B near the equatorial plane has been interpreted as an effect of the southward field generated by the ring current. The Figures 1 and 2 in his paper clearly indicate that the magnetic field distribution is strongly asymmetric even in the quiet period with Kp=0-1 both in noon-midnight and dawn-dusk meridional planes.

From ground based geomagnetic observation, an equivalent current pattern under disturbed condition has been obtained by Fukushima and Oguti (1953) by subtracting the axially symmetric disturbance field. The two vortices at the dawn and the dusk mid-latitudes in their Figure 4 have been attributed to a partial ring current which flows on the dusk side (e.g., Cummings, 1966; Fukushima and Kamide, 1973) and is caused by a substorm injection. However, such pattern of equivalent current was explained by Crooker and Siscoe (1981) as a net field-aligned current effect which arises from the ionospheric Hall conductance under the existence of a dawn-to-dusk polar cap potential drop. That is, the dawn-dusk asymmetry observed on the ground was explained by a pair of the net field-aligned currents which results from unbalanced Region-1 and Region-2 currents and flows into the ionosphere on the day side and out from the ionosphere on the night side.

In addition to the net field-aligned currents generated by the polar cap potential drop, an existence of the downward field-aligned current in the afternoon sector and the upward current in the post-midnight region in a storm main phase has been suggested from mid-latitude geomagnetic data (Iyemori, 1990). This current system is similar to the traditional partial ring current system centered around 18 MLT (Magnetic Local Time), although the proposed system in storm main phase is twisted in anti-sunward direction (see Figure 3).

From a computer simulation with the Rice convection model, a consistent result with the idea of the net

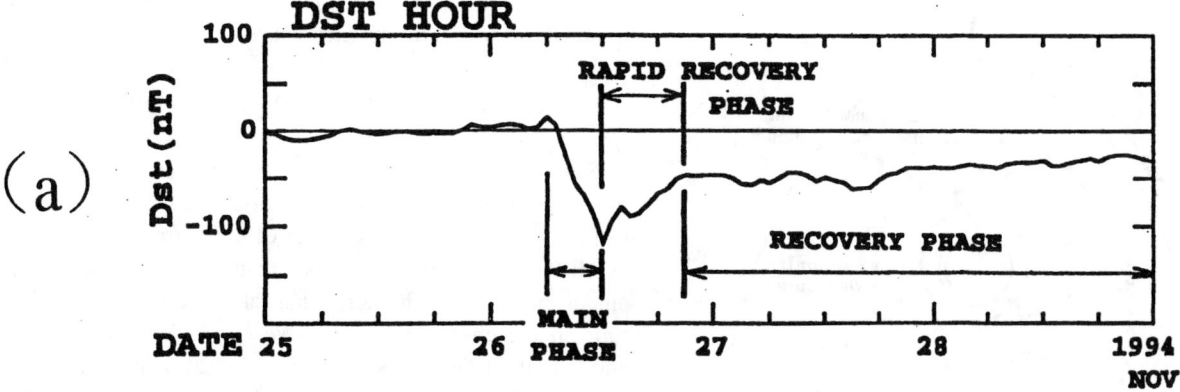

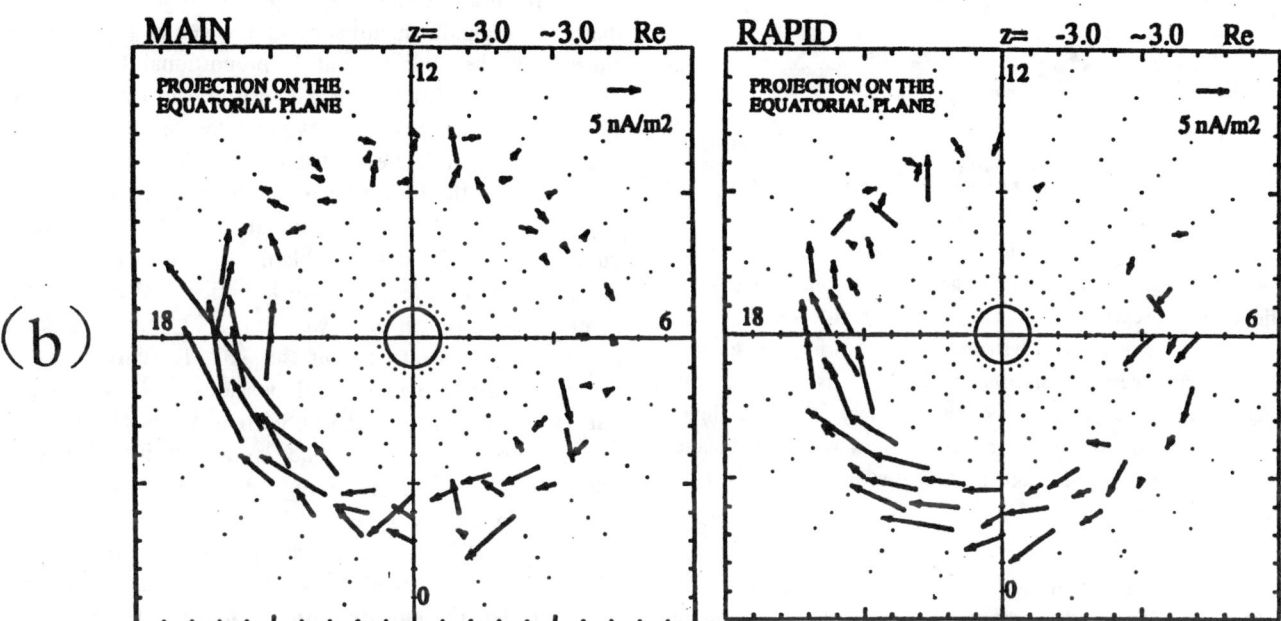

**Figure 2.** (a) A typical example of the Dst variation during a geomagnetic storm. The main phase, rapid recovery phase and recovery phase are defined as indicated. (b) A statistically obtained current vector by taking a curl of the averaged disturbance vector for each phase of geomagnetic storm. These panels are composed from Figure 1 and 7 in Terada et al. (1998).

field-aligned currents introduced in the previous section has been obtained (Chen et al., 1982). A Magsat data analysis (Suzuki and Fukushima, 1984) also shows the existence of the net current flowing from day side to night side below the Magsat orbit which was fixed in the dawn-dusk meridian.

On the other hand, the results obtained by Iijima et al. (1990), Nakabe et al. (1997) and Terada et al. (1998) from high-altitude satellite observations such as by AMPTE, DE-1 or ETS-VI indicate that the (partial) ring current flows mainly on the night side rather than on the dusk side. Such noon-midnight asymmetry of the ring current has been suggested also by numerical simulations in realistic magnetic field models of the magnetosphere (e.g., Takahashi and Iyemori, 1990). This is in contrast with the observation just over the ionosphere, as has been obtained by the OGO-2 (Langel and Sweeney, 1971) or by other low-altitude satellites such as the Magsat or the

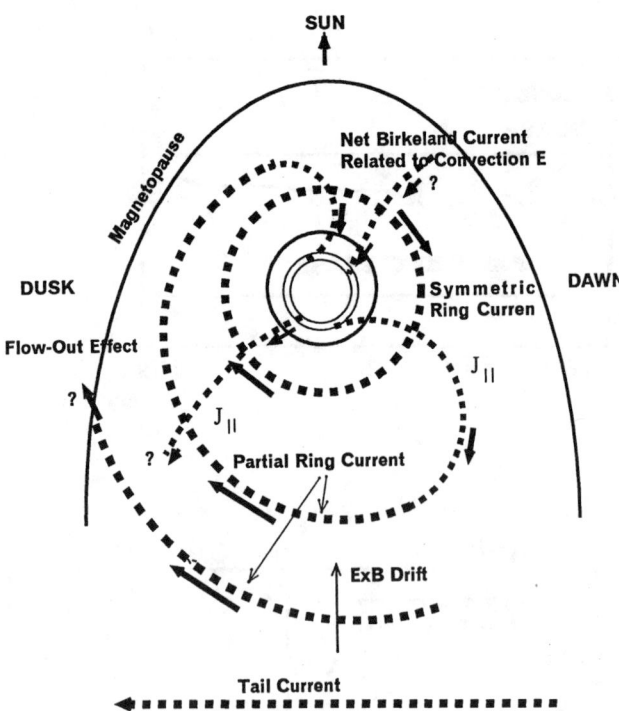

**Figure 3.** A schematic drawing of the storm time current systems which appear in the text. The Region-1/-2 field-aligned currents and the cusp current as well as the substorm wedge current system are not shown. Two types of "partial ring current" are shown, i.e., one is similar to the traditional partial ring current which close via field-aligned currents and ionospheric current and another is the continuation of the tail current, although it flows the inner magnetosphere near geosynchronous orbit.

DE-2 which shows dawn-dusk asymmetry similar to the characteristics on the ground. At low-altitudes or on the ground, the effect of the net field-aligned currents discussed above often dominates that of the distant currents in the magnetosphere such as the ring current.

Therefore we can summarize that the ground magnetic signature at middle and low latitudes in a storm main phase, that is, the dawn-dusk asymmetry on the ground is generated mainly by the net field-aligned currents which flow into the ionosphere on the day-side and out on the night-side including the field-aligned currents associated with the partial ring current in a storm main phase. On the other hand, the magnetospheric part of the partial ring current system mostly flows on the night-side equatorial plane. The currents mentioned above are schematically depicted in Figure 3. Here, we include in the term "partial ring current" the current which flow in the night-side inner magnetosphere near geosynchronous orbit and reach the dusk or afternoon magnetopause.

## RING CURRENT AND THE DST INDEX

Some essential features of the particles in the inner magnetosphere, where the ring current has been assumed to flow, were obtained by Frank (1967) from the OGO-3 satellite observations. His results show a clear enhancement of high-energy particle flux during a storm time in the energy range of 3 keV - 50 keV (proton) and radial distance inside 7Re. (See also Williams and Sugiura, 1985 for recent satellite observations).

It is no doubt that the energetic particles enhance in the inner magnetosphere during a storm, and if we adopt the D-P-S relation as the first approximation, which states that the magnetic disturbance at the center of the Earth caused by the ring current is proportional to the total kinetic energy of the particles drifting in a given dipole field, those enhanced particles must develop the Dst field.

However, the relative contribution of the particles to the Dst field, i.e., the quantitative significance of the energetic "ring current particles" in comparison with other source currents is a separate problem. As for the current systems which affect to the Dst field other than the symmetric ring current, we have (1) magnetopause current, (2) partial ring current, (3) tail current, (4) substorm current wedge, and possibly (5) ionospheric current. As for the partial ring current, we have two types of the current closure. One is the traditional system which closes via the field-aligned currents connected to the ionosphere and another is connected to some magnetopause boundary currents with particle flow out effect which we shall discuss later, although we have not yet identified the closure current for the latter type (see Figure 3).

The question on relative contribution of the ring current to the Dst index was also pointed out by Campbell (1996). In fact, Arykov and Maltsev (1996) discussed that the magnetic flux from the tail current can be a dominant source for the Dst field. An estimation of the contribution from different parts of the magnetospheric current system with the model by Alexeev et al. (1996) indicates that the effect from the tail current in a storm time is often 50 ~ 100 nT and sometimes dominates the ring current effect. On the other hand, the Tsyganenko 1987 model gives the effect to be about 30 nT for Kp=4 (see Figure 14 in Iyemori and Rao, 1996). The cause of the difference in quantitative estimation of the tail current effect should be clarified.

There exists a series of papers appeared in 1960s on the magnetospheric energy and its relation to the Dst field, and the discussion was summarized by Siscoe (1970) and also by Carovillano and Siscoe (1973). After some approximation to the generalized magnetospheric energy theorem, a relationship, which is an extension of the D-P-S relation obtained under the assumption of fixed magnetic field, has been deduced (e.g., Siscoe, 1970) as

$$\delta B \propto B0[2W+U]/3Em \qquad (1)$$

Here W is the total kinetic energy of plasma particles in the magnetosphere, U is the magnetic field energy generated by the magnetospheric currents (i.e., the self-energy) and $\delta B$ is the magnetic field generated at the center of the Earth by various magnetospheric currents, which is approximated with the Dst field. B0 is the magnetic field intensity at the Earth's equator and Em denotes the magnetic field energy of the Earth's main field integrated in the region outside of the Earth. If we neglect U, this equation is reduced to the D-P-S relation.

The term of magnetic energy U in eq.(1) has been neglected in most cases, and the D-P-S relation has been used to estimate the total kinetic energy of the ring current particles with the Dst index. However, if we calculate the magnetic field energy generated by a simple ring current (i.e., axially symmetric and doughnut shape with radius of the current 1 Re) and apply eq.(1), we realize that the effect (U) should not be neglected. In fact, if we compare the kinetic energy estimated from the D-P-S relation, which does not take into account the magnetic field distortion caused by the ring current itself, and that estimated from the generalized magnetospheric energy theorem or from some other methods, we see that the difference is considerably large under some conditions. Figure 1 shows the relation between the kinetic energy and the magnetic disturbance at the Earth (i.e., the Dst field) estimated with the theorem as well as those estimated by other methods (Hoffman and Bracken, 1967; Souzou and Windle, 1969; Lackner, 1970) in which the magnetic field distortion by the ring current is taken into account.

From Figure 1, we see that the difference between the estimation from the D-P-S relation and that from the generalized energy theorem becomes large and can not be neglected when the location of the ring current is more distant than 3-4 Re from the center of the Earth (see the dots on the right side edge of the figure). The development of the outer ring current near geosynchronous orbit in storm main phase has been suggested from observations or numerical simulations (e.g., Terada et al., 1998: Ebihara and Ejiri, 1998). The difference in the estimation of the kinetic energy should become more large if we take into account the magnetic effect from the tail current.

Iyemori and Rao (1996) found that the Dst field decays after substorm onset, and this finding seems to contradict with the traditional view that a storm is caused by successive occurrence of substorms. Siscoe and Petcheck (1997) discussed a theoretical possibility which explain the finding. That is, eq.(1) should be applied also for the discussion on storm-substorm relation, because we know that a large part of the magnetic energy deposited from the solar wind is released and dissipated during a substorm expansion phase. More than half of the energy deposited during a substorm growth phase is possibly lost as a plasmoid ejection from the magnetosphere (e.g., Ieda et al., 1998) and as a Joule heat in the polar ionosphere (e.g., Siscoe and Petschek, 1997). If this is the case, the storm field measured with the Dst index should weaken according to eq.(1), and the weakening after a substorm onset (Iyemori and Rao, 1996) is a natural consequence of the energy theorem.

That is, even if the total kinetic energy of the ring current particles is increased with a substorm injection (e.g., Kamide and McIlwein, 1974; Reeves, 1998), the Dst field can decay, because the magnetosphere loses much more magnetic energy than the particle energy which is converted from the magnetic energy and injected to the inner magnetosphere.

Therefore we have to distinguish the two concepts, i.e., a development of the storm field measured with the Dst index and that of the ring current field generated by the particles in the inner magnetosphere. The contribution from the tail current to the Dst field can be significant under some conditions as has been pointed out by Arykov and Maltsev (1996), Campbell (1996) or Alexseev et al. (1996). The development and decay of the Dst field during a substorm growth phase and an expansion phase are probably caused by an enhancement and a weakening of the tail current, respectively. The physical meaning of the Dst index should be given with eq.(1) rather than the traditional D-P-S relation.

## PARTICLE "FLOW-OUT" EFFECT ON THE RING CURRENT DECAY

Another concept which is important in understanding the ring current formation is the flow-out effect of particles from the magnetosphere. Because of the distortion of magnetospheric magnetic field from a dipole field, the drift velocity is higher on the night-side than that on the day-side (e.g., Takahashi and Iyemori, 1990) in realistic magnetospheric magnetic field models such as the Mead and Fairfield (1975) or the Tsyganenko (1987) model. This tendency is more clear for higher energy particles, and the ions with energy more than about 50keV tend to flow out from the dusk magnetopause. Because of the faster drift speed on the night side than that on the day side, the drift paths of particles are not symmetric to 6-18 MLT axis even in a steady state, and the energetic ions tend to reach dusk-side magnetopause. This effect causes a rapid decay of the ring current in simulation studies of the storm-time ring current with particle tracing method such as Takahashi et al. (1990, 1991), Ebihara and Ejiri (1998) or Kozyra et al. (1998). That is, the initial rapid decay of the Dst field, which is often observed just after the minimum Dst epoch in a storm main phase, is mainly caused by the particle flow out, and the slow decay in late recovery phase is due to the charge exchange with neutral atoms or some wave-particle interactions. In a storm main phase, the particles which drift inward by ExB drift and trapped in the magnetosphere after the reduction of the dawn-to-dusk electric field are energized through conservation of the first invariant, and they form more symmetric ring current.

It should be pointed out that these numerical simulations predict that the average energy of the particles carrying most of the current in a storm main phase is not so high as has been observed in the inner magnetosphere (cf., Williams and Sugiura, 1985). It is suggested that the location and energy of the current carrier are quite different in a storm main phase from those in a late recovery phase. The location is more inside in a late recovery phase as a result of flow-out of outer ring current particles in the main and early recovery phases, and the shape of the ring current becomes more symmetric, and the average particle energy is higher than that of the current carriers in the main phase.

Using the ETS-6 satellite observation of the magnetospheric magnetic fields on the transfer orbits, Terada et al. (1998) examined the current distribution in each phase of geomagnetic storms by taking the curl of averaged magnetic disturbance fields. From Figure 2, which is composed from their Figures 1 and 7, we see that the current on the equatorial plane near geosynchronous orbit flows mainly in the pre-midnight and evening sectors in storm main and rapid-recovery phases and that the current in the pre-noon sector is very weak. This observational result suggests that the outer part of the westward current (i.e., the outer ring current) does not encircle the Earth except for the late recovery phase but continues to the magnetospheric boundary region as depicted in Figure 3, and it is consistent with the concept of particle flow-out although the current continuation at the boundary region has not been examined yet. A part of the ring current seems to divert to the ionosphere through the field-aligned currents to form so-called 'partial ring current system' when the dawn-to-dusk electric field varies in time as has been shown by Takahashi et al. (1991).

The flow-out of energetic ring current particles from the magnetosphere or the Joule dissipation in the ionosphere associated with the partial ring current system should also contribute to the decay of the Dst field according to eq. (1), because it consumes the magnetospheric energy.

## RING CURRENT DEVELOPMENT AND PLASMA SHEET DENSITY

If the ring current is essentially formed by the drift of particles transported to the inner magnetosphere from the plasma sheet by the dawn-to-dusk electric field associated with a strong southward IMF, the strength of the current should directly depend on the density and temperature of the plasma sheet particles near the inner boundary. A clear correlation of the plasma sheet density with the solar wind density has been found by Borovsky et al. (1997) from the data obtained with the ISEE-2 and MPA on the LANL geosynchronous spacecrafts or by Terasawa et al., (1997) from those obtained with the Geotail satellite.

Taking the plasma sheet density dependence on the solar wind density into the boundary condition, Ebihara and Ejiri (1998) and Kozyra et al. (1998) made a computer simulation of the ring current with a charged particle tracing technique including the flow-out effect mentioned above. They calculated the magnetic effect at the Earth and obtained a good agreement with the observed Dst index supporting the above scenario of the ring current formation.

## SUMMARY

The ring current formation and its relation to the Dst field reviewed and discussed in this paper are summarized as follows:
(1) The main part of the ring current is formed by the earthward drift of the plasma sheet particles driven by an enhanced dawn-to-dusk electric field associated with a

strong southward IMF, and the structure has a clear noon-midnight asymmetry. The solar wind plasma density contributes to the efficiency of the ring current formation through the plasma sheet plasma density.

(2) The initial rapid decay of the ring current usually observed in the Dst index is mainly caused by the flow-out of ring current particles from the dusk or afternoon magnetopause.

(3) The tail current also contributes to the Dst field, although quantitative assessment has to be made.

(4) The generalized relationship between the magnetospheric energy and the Dst field, i.e., eq. (1), should be used instead of the original D-P-S relation when we discuss the ring current quantitatively with the Dst index.

(5) It is necessary to distinguish the concepts, 'ring current', 'Dst field' and 'high-energy particles' in understanding the physics of ring current formation.

In this paper, we did not discuss the effects of induced electric fields associated with a dipolarization (e.g., Fok et al., 1996) nor the effect of an injection (e.g., Kamide and McIlwain, 1974; Reeves, 1998) at substorm expansion phase on the ring current development. The O+ ion effect and the wave-particle interaction effect on the ring current decay (e.g., Li et al., 1993; Chen and Schulz, 1995) were also not discussed.

The relative significance of these effects compared with the process discussed in this paper should be evaluated quantitatively both from observations and theoretical calculations.

*Acknowledgments.* This study has been supported in part by "Ground Research for Space Utilization" promoted by NASDA and Japan Space Forum and also by grant 10640430 under the Ministry of Education, Japan.

## REFERENCES

Alexeev, I.I., E.S. Belenkaya, V.V. Kalegaev, Y.I. Feldstein and A. Grafe, Magnetic storms and magnetotail currents, *J. Geophys. Res.*, 101, 7737-7747, 1996.

Arykov, A.A. and Yu. P. Maltsev, Direct-driven mechanism for geomagnetic storms, *Geophys. Res. Lett.*, 23, 1689-1692, 1996.

Borovsky, J.E., M.F. Thomsen and D.J. McComas, The superdense plasma sheet: Plasmaspheric origin, solar wind origin, or ionospheric origin?, *J. Geophys. Res.*, 102, 22089-22097, 1997.

Campbell, W.H., Geomagnetic storms, the Dst ring-current myth and lognormal distributions, *J. atmos. terr. Phys.*, 58, 1171-1187, 1996.

Carovillano, R.L. and G.L. Siscoe, Energy and momentum theorems in magnetospheric processes, *Rev. Geophys. Space Phys.*, 11, 289-353, 1973.

Chen, C.-K., R.A. Wolf, M. Harel, and J.L. Karty, Theoretical magnetograms based on quantitative simulation of a magnetospheric substorm, *J. Geophys. Res.*, 87, 6137-6152, 1982.

Chen, M.W. and M Schulz, Ring current formation and decay: A review of modeling work, *Advances in Space Res.*, 17, 7-16, 1995.

Crooker, N.U. and G.L. Siscoe, Birkeland currents as the cause of the low-latitude asymmetric disturbance field, *J Geophys. Res.*, 86, 11201-11210, 1981.

Cummings, W.D., Asymmetric ring currents and the low-latitude disturbance daily variation, *J. Geophys. Res.*, 71, 4495-4503, 1966.

Ebihara, Y. and M. Ejiri, Modeling of solar wind control of the ring urrent buildup: A case study of the magnetic storms in April 1997, *Geophys. Res. Lett.*, 25, 3751-3754, 1998.

Fok, M.-C., T.E. Moore and M.E. Greenspan, Ring current development during storm main phase, *J. Geophys. Res.*, 101, 15311-15322, 1996.

Frank, L.A., On the extraterrestrial ring current during geomagnetic storms, *J. Geophys. Res.*, 72, 3753-3767, 1967.

Fukushima, N. and Y. Kamide, Partial ring current models for worldwide geomagnetic disturbances, *Rev. Geophys. Space Phys.*, 11, 795-853, 1973.

Fukushima, N. and T. Oguti, II, Polar magnetic storms and geomagnetic bays, Appendix I, A theory of Sd-field, *Rep. Ionos. Res.*, 7, 137, 1953.

Gonzalez, W.D., J.A. Joselyn, Y. Kamide, H.W. Kroehl, G. Rostoker, B.T. Tsurutani, and V.M. Vasyliunas, What is a geomagnetic storm?, *J. Geophys. Res.*, 99, 5771-5792, 1994.

Hoffman, R.A. and P.A. Bracken, Higher order ring currents and particle energy storage in the magnetosphere, *J. Geophys. Res.*, 72, 6039, 1967.

Ieda, A., S. Machida, T. Mukai, Y. Saito, T. Yamamoto, A. Nishida, T. Terasawa and S. Kokubun, Statistical analysis of plasmoid evolution with GEOTAIL observations, *J. Geophys. Res.*, 103, 4453-4465, 1998.

Iijima, T., T.A. Potemra and L.J. Zanetti, Large-scale characteristics of magnetospheric equatorial currents, *J. Geophys. Res.*, 95, 991-999, 1990.

Iyemori, T., Storm-time magnetospheric curents inferred from mid-latitude geomagnetic field variations, *J. Geomag. Geoelectr.*, 42, 1249-1265, 1990.

Iyemori, T., and D.R.K.Rao., Decay of the Dst component of geomannetic disturbance after substorm onset and its implication to storm substorm relation, *Ann. Geophys.*, 14, 608-618, 1996.

Kamide, Y. and C.E. McIlwain, Onset time of magnetospheric

substorms determined from ground and synchronous satellite records, *J. Geophys. Res.*, 79, 4787-4790, 1974.

Kamide, Y., Is substorm occurrence a necessary condition for a magnetic storm?, *J. Geomag. Geoelectr.*, 44, 109-117, 1992.

Kozyra, J.U., V.K. Jordanova, J.E. Borovsky, M.F. Thomsen, D.J. Knipp, D.S. Evans, D.J. McComas and T.E. Cayton, Effects of a high-density plasma sheet on ring current development during the November 2-6, 1993, magnetic storm, *J. Geophys. Res.*, 103, 26285-26305, 1998.

Lackner, K., Deformation of a magnetic dipole field by trapped particles, *J. Geophys. Res.*, 75, 3180, 1970.

Langel, R.A. and R.E. Sweeney, Asymmetric ring current at twilight local time, *J. Geophys. Res.*, 76, 4420-4427, 1971.

Li, X., M. Hudson, A. Chen and I. Roth, Loss of ring current O+ ions due to interaction with Pc5 waves, *J. Geophys. Res.*, 98, 215-231, 1993.

McPherron, E.L., The role of substorms in the generation of magnetic storms, in Magnetic Storms, 131-147, *Geophys. Monograph 98*, 1997.

Mead, G.D. and D.H. Fairfield, A quantitative magnetospheric model derived from spacecraft magnetometer data, *J. Geophys. Res.*, 80, 523-534, 1975.

Nakabe, S., T.Iyemori, M.Sugiura, and J.A.Slavin, A statistical study of the Magnetic field structure in the inner magnetosphere, J. Geophys. Res., 102, 17571-17582, 1997.

Reeves, G.D., New perspectives on substorm injections, in *SUBSTORM-4*, ed. S. Kokubun and Y. Kamide, Terra Sci. Pub. Co./Kluwer Academic Pub., p.785, 1998.

Sckopke, N., A general relation between the energy of trapped particles and the disturbance field near the Earth, *J. Geophys. Res.*, 71, 3125-3130, 1966.

Siscoe, G.L., The virial theorem applied to magnetospheric dynamics, *J. Geophys. Res.*, 75, 5340-5350, 1970.

Siscoe, G.L. and H.E. Petschek, On storm weakening during substorm expansion phase, *Ann. Geophys.*, 15, 211-216, 1997.

Sozou, C. and D.W. Windle, A self consistent ring current in the Earth's dipole field, *Planet. Space Sci.*, 17, 375-387, 1969.

Sugiura, M., The ring current, in Critical Ploblems of Magnetospheric Physics, p.195-210, *Proc. Joint COSPAR/IAGA/URSI Sympo.*, ed. by E.R. Dyer, Nat. Acad. Sci., Washington, D.C., 1972.

Suzuki, A. and N. Fukushima, Anti-sunward space current below the Magsat level during magnetic storms, *J. Geomag. Geoelectr.*, 36, 493-506, 1984.

Takahashi, S. and T. Iyemori, Simulation of charged particle motions in realistic model magnetosphere and the effect of corotating electric field, *Ann. Geophys.*, 8, 503-510, 1990.

Takahashi, S., T. Iyemori, and M. Takeda, A simulation of the storm-time ring current, *Planet. Space Sci.*, 38, 1133-1141, 1990.

Takahashi, T., M. Takeda, and Y. Yamada, Simulation of storm-time partial ring current system and the dawn-dusk asymmetry of geomagnetic variation, *Planet. Space Sci.*, 39, 821-832, 1991.

Terada, N., T.Iyemori, M.Nose, T.Nagai, H.Matsumoto, T.Goka, Storm-time magnetic field variations observed by the ETS-VI satellite, *Earth Planets Space*, 50, 853-864, 1998.

Terasawa, T., M. Fujimoto, T. Mukai, I. Shinohara, Y. Saito, T. Yamamoto, S. Machida, S. Kokubun, A.J. Lazarus, J.T. Steinberg and R.P. Lepping, Solar wind control of density and temperature in the near-Earth plasma sheet: WIND/GEOTAIL collaboration, *Geophys. Res. Lett.*, 24, 935-938, 1997.

Tsyganenko, N.A., Global quantitative models of the geomagnetic field in the cislunar magnetosphere for different disturbance levels, *Planet. Space Sci.*, 35, 1347, 1987.

Williams, D.J. and M. Sugiura, The AMPTE Charge Composition Explorer and the 4-7 September 1984 geomagnetic storm, *Geophys. Res. Lett.*, 12, 305-308, 1985.

---

T. Iyemori, Data Analysis Center for Geomagnetism and Space Magnetism, Graduate School of Science, Kyoto University, Kyoto 606-8502, Japan.

# Currents and Flows in Distant Magnetospheres

Margaret Galland Kivelson

*Institute of Geophysics and Planetary Physics and Department of Earth and Space Sciences, University of California, Los Angeles, CA 90095-1567*

Space scientists have explored, described, and explained the terrestrial magnetosphere for four decades. Rarely do they point out that the planetary and solar wind parameters controlling the size, shape, and activity of Earth's magnetosphere map out only a small portion of the space of dimensionless parameters that govern magnetospheric properties. With the discovery of Ganymede's magnetosphere, the range of parameters relevant to magnetospheric studies has grown by orders of magnitude. Consider the extremes of Ganymede's and Jupiter's magnetospheres. Jupiter's magnetosphere forms within a plasma flowing at super-Alfvénic speed, whereas Ganymede's forms in a sub-Alfvénic flow. The scale sizes of these magnetospheres, characterized by distances to the magnetopause of order $7\times10^6$ km and $5\times10^3$ km, respectively, differ by three orders of magnitude, ranging from 100 to 0.1 times the scale of Earth's magnetosphere. The current systems that control the structure and dynamics of a magnetosphere depend on specific plasma and field properties. Magnetopause currents at Ganymede differ greatly from the forms familiar for Earth and Jupiter, principally because the Mach number of the ambient plasma flow greatly influences the shape of the magnetosphere. A magnetodisk current, present at Jupiter because of its rapid rotation, is absent at Earth and Ganymede. The ring current, extensively investigated at Earth, is probably unimportant at Ganymede because the dynamical variations of the external flow are slow. The ring current is subsumed within the magnetodisk current at Jupiter. This paper describes and contrasts aspects of these and other current systems for the three bodies.

## 1. INTRODUCTION

Magnetospheres form when a magnetized planet or moon is embedded in a flowing plasma. If the planetary magnetic moment is sufficiently large and the temporal and spatial scales of the system are larger than the natural temporal and spatial scales of the incident plasma (so that the magnetohydrodynamic or MHD description is valid), the external plasma flows around the magnetized obstacle. A thin magnetopause boundary separates internal and external plasmas. The large-scale structure of the system can be described using MHD equations. The forces that act on the system are the Lorentz force, produced by electrical currents, and pressure gradients

$$\frac{d(\rho\mathbf{u})}{dt} = -\nabla p + \mathbf{j}\times\mathbf{B} \qquad (1)$$

Here $\rho$ is the mass density, $\mathbf{u}$ is the flow velocity, $d/dt = \partial/\partial t + \mathbf{u}\cdot\nabla$ is the convective time derivative, $p$ is thermal pressure, assumed isotropic, $\mathbf{j}$ is current density, and **B** is magnetic induction. It is critical to note that equation (1) does not single out the current for special emphasis. If currents (perpendicular to the field) are present, a $\mathbf{j} \times \mathbf{B}$ force acts on the plasma. But equally fundamental are flows or pressure gradients, whether steady or changing in time. In some parts of the system the flows and pressure gradients react to the $\mathbf{j} \times \mathbf{B}$ force; a current generated non-locally may flow into remote regions and impose motions on a plasma or modify its pressure. Elsewhere imbalances between pressure forces and flows generate currents in the plasma. The critical conclusion is that where forces are required to change the motion of a plasma or to change its energy density, currents are likely to be present.

Field aligned (or parallel) currents $j_\parallel = \mathbf{j} \cdot \mathbf{B}/B$ do not exert force. They flow unimpeded in an ideal plasma. In order to identify the source of field aligned currents, one must supplement equation (1) by the MHD requirement that current is divergenceless, $\nabla \cdot \mathbf{j} = 0$. Then one finds that

$$j_\parallel = -\int \frac{ds}{B} \nabla \cdot \mathbf{j}_\perp \qquad (2)$$

which implies that where flows and pressure gradients generate currents perpendicular to the background field ($\mathbf{j}_\perp = \mathbf{j} - j_\parallel \hat{\mathbf{B}}$ where $\hat{\mathbf{B}}$ is a unit vector along the background magnetic field), they may also drive parallel currents, although the integral in the above equation adds subtlety to the relationship.

In the material that follows, I assume that the reader is already familiar with the large-scale current systems, described by articles in *Potemra* [1984], that account for the structure and dynamics of the terrestrial magnetosphere. Therefore, following a brief enumeration of those current systems, I concentrate on the special cases of the magnetospheres of Jupiter and Ganymede. I introduce the dimensionless parameters that govern these two objects, noting how they differ from the parameters typical at Earth. I emphasize the ways in which these differences affect the large-scale currents in these very different magnetospheres.

## 2. FAMILIAR QUASI-STATIC AND TIME-VARYING CURRENT SYSTEMS IN THE TERRESTRIAL MAGNETOSPHERE

Magnetopause currents flow on the interface between a flowing external plasma and the magnetosphere proper. The currents flow across the field ($\mathbf{j}_\perp$). In a cross section of the magnetotail of Earth (or Jupiter), the principal currents form a theta pattern. Looking towards the Earth up the tail, the magnetopause currents flow clockwise from dusk to dawn in the northern hemisphere and counterclockwise in the southern hemisphere (as illustrated in Figure 9.4b in [*Hughes*, 1995]). They close from dawn to dusk through the magnetotail current sheet, which is thus not really a separate current.

Another major current system is the ring current, which flows around the Earth across the magnetic field, peaking inside of geostationary orbit (magnetic shell crossing the equator at 6.6 $R_E$). The ring current is most intense on the night side of the Earth. The region of enhanced current is referred to as the partial ring current [*Crooker and McPherron*, 1972]. The partial ring current links to Region 2 field aligned currents [*Crooker and McPherron*, 1972; *Spiro and Wolf*, 1984] which flow in sheets into the equatorial regions pre midnight and back out post midnight [*Iijima and Potemra*, 1976].

In addition to these quasi-static or slowly varying currents, important transient currents flow when the magnetosphere is disturbed. Familiar dynamic magnetospheric currents include the substorm current wedge in the magnetotail [*McPherron*, 1995], currents produced by reconnection on the dayside magnetopause such as FTE-associated currents/cusp currents, flux rope or plasmoid currents [*Hughes*, 1995]. Most of these current systems include both $\mathbf{j}_\perp$ and $j_\parallel$, and the latter couples the magnetosphere to the ionosphere.

In the most general case, changes of plasma flow in the magnetosphere compress and bend the magnetic field. The perturbations launch MHD waves and conversely, MHD waves produce flow perturbations. The allowed wave perturbations for a propagating plane wave (wave vector **k** at an oblique angle relative to **B**, the background magnetic field, assumed uniform) differ for the two possible wave polarization directions (see Figures 11.2 a and b in [*Kivelson*, 1995]). In compressional waves, the perturbation field **b** is oblique to **B** while in shear Alfvén waves **b** is perpendicular to **B**. The compressional wave produces perturbation current (**j**) perpendicular to the background field. Only the shear Alfvén wave carries parallel current between the magnetosphere and the ionosphere.

The magnetosphere is set into corotation by its interaction with the corotating ionosphere. Departures from corotation drive parallel currents that couple the magnetosphere to the ionosphere and act to restore corotation [*Hill*, 1983]. Typically these currents are overlooked in Earth's magnetosphere as they are small, but when we turn our attention to Jupiter, the currents that impose corotation will be significant.

**Table 1.** Some Planetary Magnetospheres (external and internal plasma properties) [1]

|  | Mercury | Earth | Jupiter | Ganymede[2] |
|---|---|---|---|---|
| Distance (from center) | ~0.4 AU[3] | 1 AU | 5.2 AU | 15 $R_J$ |
| $\rho_{wind}$ (amu cm$^{-3}$) [4] | 35-80 | 8 | 0.3 | 27 - 107 |
| $u_{wind}$ (km/s) | 400 | 400 | 400 | 140 |
| $R_{planet}$ (km) | 2,439 | 6,373 | 71,398 | 2634 |
| Rotation Period[5] | 59 days | 1 day | ~9.8 hours | 7.15 days |
| $B_{eq}$ (nT) | 330 | 31,000 | 428,000 | 750 |
| $\theta_M$ | ~14°(S) | 10.8°(S) | 9.6°(N) | ~10°(S)[6] |
| Magnetosphere Size | 1.5 $R_M$ | 10 $R_E$ | ~100 $R_J$ | 2 $R_G$[6] |

[1] From [Kivelson and Bagenal, 1999] except where indicated.
[2] From [Bagenal, 1998] except where indicated
[3] 1 A.U. = $1.5 \times 10^8$ km., $R_J$ = Jupiter's radius = 71,000 km.
[4] The density of the solar wind fluctuates by a factor of ~5 about typical values of $\rho_{SW}$ and varies with distance from the Sun as ~[(8 amu cm$^{-3}$) / (distance)$^2$]
[5] [Beatty and Chaikin, 1990]
[6] [Kivelson et al., 1997]

## 3. OTHER MAGNETOSPHERES

More than twenty years ago, [Siscoe, 1979] emphasized that properties of magnetospheres may differ dramatically depending on the properties of the flowing plasmas in which they are embedded and on pertinent features of the planet around which they form. Siscoe suggested that we think of planetary magnetospheres as a family of related objects. The emphasis in this discussion is principally on Ganymede and Jupiter, the two family members (among magnetospheres that have been explored in situ) that illustrate the extremes of pertinent properties. Some of the parameters that establish the scales of these two magnetospheres as well as those of Earth and Mercury are given in Table 1. Listed are the distances of the planetary orbits from the gravitational center (Sun or Jupiter), the mass density $\rho_{wind}$ and flow speed $u_{wind}$ of the plasma in which they are embedded, the planetary radius $R_{planet}$, rotation period, surface equatorial magnetic field $B_o$, the tilt of the dipole moment, and the scale size of the magnetosphere (characterized by the distance to the nose of the magnetosphere). Important properties of the plasma flowing onto the magnetospheres are provided in Table 2. Listed are the Alfvénic ($M_A$) and magnetosonic ($M_{ms}$) Mach numbers, the ratio of the dynamic pressure ($\rho u^2$) to the magnetic pressure ($p_B = B^2/2\mu_o$) of the incident plasma and the plasma beta $\beta = p_{thermal}/p_B$. The temporal variability of the flow is also noted. Finally, internal features including significant sources of plasma internal to the magnetosphere, the location of the plasmapause (if any), and a characteristic height-integrated conductivity of the planetary ionosphere are given.

The tabulated material makes it clear why Jupiter and Ganymede attract special interest. Compared with Earth, Jupiter is large: its radius ($R_J$) is ~11 $R_E$. It is highly magnetized with a surface field more than an order of magnitude larger than Earth's. Despite its size, Jupiter rotates in 10 hours, even faster than Earth; rotational forces dominate gravitational forces beyond ~2 $R_J$. At Jupiter, Io provides a plasma source, that injects fresh plasma into the magnetosphere at the rate of a ton per second (~3 × 10$^{28}$ ions/s). Ganymede and Mercury are small and rotate more slowly than Earth. They can trap very little plasma in their magnetospheres. Ganymede is of special interest because it is embedded in a sub-Alfvénic flow.

In other magnetospheres as at Earth, pressure gradients and flow disturbances relate to electrical currents, both transverse and field aligned. Differences among the current contributions in different magnetospheres relate to the relative importance of the principal current systems and to the scales over which they act.

## 4. CURRENTS IN JUPITER'S MAGNETOSPHERE

The quasi-static currents at Jupiter include familiar ones like magnetopause currents, which have been encountered

**Table 2.** Properties of the Ambient Plasma and Internal Features of Some Planetary Magnetospheres [1]

|  | Mercury | Earth | Jupiter | Ganymede[2] |
|---|---|---|---|---|
| $M_{ms,ambient}$ | 3-4 | 6-10 | 8-12 | 0.18-0.48 |
| $\beta_{ambient}$ | 0.5 | ~2 | 1-3 | 0.6 – 2.5 |
| $(\rho u^2/p_B)_{wind}$ | >>1 | >>1 | >>1 | 0.1 – 1.6 |
| Variable wind? | Yes | Yes | Yes | No |
| Internal plasma sources[1] | none | ionosphere | Io and other moons | minor |
| Plasmapause | none | ~4-5 $R_E$ | at $R_{MP}$ | none[3] |
| $\Sigma_{ion}(S)$ | 0.1[4] | 1-10[4] | 0.3-10[5] | ? |

[1] Inferred from [*Kivelson and Bagenal*, 1999] unless otherwise noted.
[2] Inferred from [*Bagenal*, 1998] unless otherwise noted.
[3] [*Kivelson et al.*, 1998].
[4] [*Glassmeier et al.*, 1999]
[5] [*Strobel and Atreya*, 1983]

a few times as spacecraft entered or left the Jovian system [*Smith et al.*, 1978; *Lepping et al.*, 1981; *Balogh et al.*, 1992; *Huddleston et al.*, 1998a]. At Earth, Region 1 field-aligned currents communicate to the ionosphere the effects of viscous and reconnection processes near the equatorial flanks of the magnetopause. One anticipates that similar processes occur at the magnetopause of Jupiter's magnetosphere. Thus Region 1 currents should exist at Jupiter, but these currents have yet not been directly investigated. The ring current is not readily definable and, in any case, unimportant relative to other azimuthal currents arising from rotational effects (see below). The existence of Region 2 type currents has been inferred from magnetometer data by Khurana (personal communication, 1999) who attributes them to day-night asymmetries in Jupiter's magnetosphere. Substorm currents and other dynamic current systems have not yet been identified.

Considerable attention has been paid to other important current systems of the Jovian magnetosphere. In particular, there is a strong near-equatorial current sheet flowing around the planet on both the day and night sides. This magnetodisk current flows azimuthally. It is confined near the equator and stretches magnetic field lines radially. The effect of stretching is to increase the curvature of the field near the equator. If the electromagnetic force is written in the form

$$\mathbf{j} \times \mathbf{B} = -\nabla B^2 / 2\mu_o + \mathbf{B} \cdot \nabla \mathbf{B} / \mu_o \qquad (3)$$

the second term on the right is the curvature force that acts inward and becomes increasingly important near the equator where the radius of curvature attains its minimum value. Acting outward in addition to the magnetic pressure gradient is the force of an inward thermal pressure gradient present over much of the equatorial plane. Inertial effects arising from the rotation of the relatively dense plasma sheet containing heavy ions of Iogenic origin are also important. The challenge is to understand how the plasma maintains the stretched configuration of the field. The dominant effect has not been unambiguously identified. Rotational effects and pressure gradients did not balance the curvature force of the magnetic field during the Voyager epoch [*Caudal*, 1986; *McNutt*, 1982; *Mauk and Krimigis*, 1987]. *Paranicas et al.* [1991] noted that the plasma pressure is anisotropic. For anisotropic pressure, equation (1) must be replaced by

$$\frac{d(\rho \mathbf{u})}{dt} = -\nabla^2(p_\perp + B^2/2\mu_o) + \mathbf{B} \cdot \nabla (p_\parallel \mathbf{B}/B^2 + \mathbf{B}/\mu_o) \qquad (4)$$

where the electromagnetic term has been expressed as in equation (3), $p_\parallel$ is the component of pressure along the field and $p_\perp$ is the component of pressure transverse to the field. Using Voyager particle data, Paranicas demonstrated that the field configuration near the equator is maintained principally by the pressure anisotropy. Between the Voyager and Galileo epochs, the field configuration did not change significantly in the regions where overlapping measurements are available (K. K. Khurana, personal communication, 1999). On the other hand, the plasma pressure inside of L = 10 decreased markedly [*Mauk et al.*, 1998], and lower pressures have been reported elsewhere as well [*Williams*, 1999]. Thus, at the present time, there is

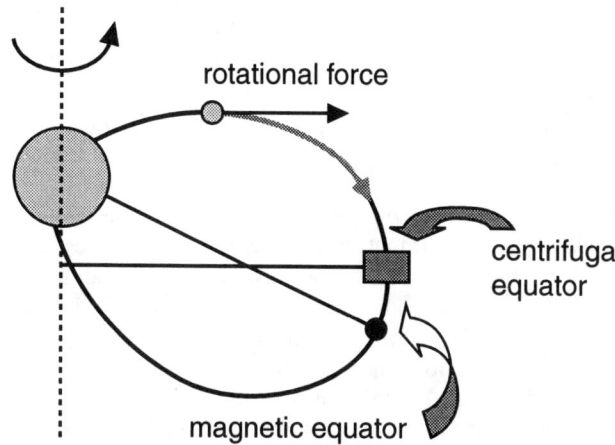

**Figure 1.** Schematic illustration of the locations of the magnetic equator (minimum B along a field line) and the centrifugal equator (location along the field line most distant from the planetary spin axis).

some question whether the pressure anisotropy can account for the field configuration during the Galileo epoch.

In either case, there is a near-equatorial current sheet that stretches the field into a disc-like configuration with plasma concentrated near the equator. Rotational forces accelerate plasma away from the spin axis as illustrated in Figure 1. A component of this acceleration along the field lines of the tilted dipole field is everywhere directed to the location on the field line that is farthest from the spin axis. This position is called the centrifugal equator. Because of the dipole tilt, the centrifugal equator coincides with the magnetic equator, the location of minimum field strength, only at two longitudes. Elsewhere, the two equators occur at different locations on a field line. Field aligned forces include the magnetic mirror force, directed towards the magnetic equator, pressure gradient forces, and the ambipolar electric field, directed away from the highest density region along a flux tube [Bagenal, 1994]. Close in, the current sheet lies between the centrifugal equator and the magnetic equator [Hill et al., 1974]. In the middle magnetosphere, the plasma is concentrated in a warped sheet referred to as the plasma sheet. The warp of the plasma sheet arises from two effects. A finite signal propagation time that communicates the changing orientation of Jupiter's field to the plasma at large distance from the planet is partially responsible [Kivelson et al., 1978]. Further warping results from distortion of the field out of meridian planes required by the currents that flow to impose approximate corotation on outward-moving plasma [Khurana and Kivelson, 1993], the process that we next consider.

Consider the fate of plasma introduced near the orbit of Io. It is transported outward by diffusion, interchange, and convective flows. In the absence of currents, plasma conserves angular momentum and lags corotation as it moves outward. The constancy of angular momentum, $\rho \omega r^2 =$ constant, implies that the angular velocity $\omega$ must decrease as the distance from the spin axis $r$ increases. Because the plasma is frozen to the field, a corotation lag twists (or curls) the near-equatorial portions of the field lines out of meridian planes, which means that the azimuthal component of the magnetic field ($B_\varphi$) varies with distance $s$ along the field line. Recalling that Ampére's law links the curl of the field and the radial current $j_r$ $\left(\partial B_\varphi / \partial s = \mu_o j_r\right)$, one recognizes that radial currents must arise wherever the field is bent out of meridian planes.

The twist of the field in the plasma sheet (identified in Voyager data [Khurana and Kivelson, 1993]) is illustrated in Figure 2 from a field model of Khurana [1997]. Selected field lines are viewed from above (north). The field-aligned currents associated with the bendback of the field lines flow towards the equator in the inner part of the plasma sheet and back into the ionosphere at large radial distances. Closure currents flow radially outward in the equatorial regions (and the associated $\mathbf{j} \times \mathbf{B}$ force accelerates the equatorial plasma, working to restore corotation) and from high to low latitude in Jupiter's ionosphere (decelerating ionospheric rotation). Neutrals freshly ionized in the torus link to an analogous current system that acts to accelerate them to corotation, a point that will be further elaborated below.

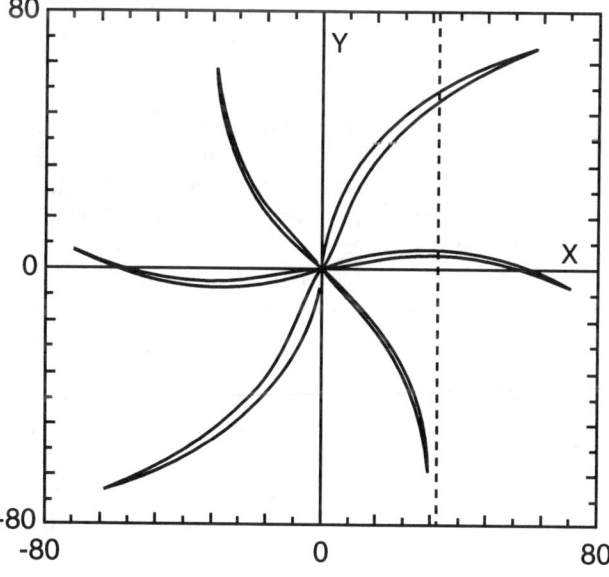

**Figure 2.** Selected field lines are viewed from above (north) in the field model of Khurana [1997].

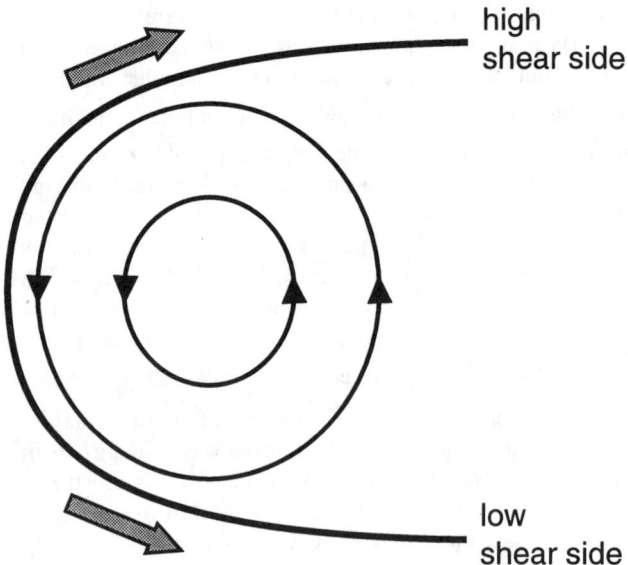

**Figure 3.** Schematic in the equatorial plane of the Jovian magnetosphere showing flows both inside and outside the magnetopause. Essential asymmetries arise from the different shears of flow on the dawn and dusk sides.

Indeed, the heavy ion plasma approximately (within a factor of 2) corotates with the ionosphere out to many tens of $R_J$. *Kane et al.* [1999], using Galileo measurements, found that although the azimuthal flow speed is variable, its average value decreases from near corotation at ~20 $R_J$ to roughly 0.5 times corotation in the outer magnetosphere (L > 50). The results are in agreement with earlier analysis based on Voyager measurements [*Kane*, 1995].

An important effect of Jupiter's corotation is that it breaks the symmetry of the Jovian magnetosphere as illustrated in Figure 3. Current patterns and the associated flows and fields on the dawn and dusk hemispheres differ greatly even if the solar wind has no dawn-dusk asymmetry. The broken symmetry is illustrated in Figure 4 (courtesy of R. J. Walker) showing characteristic flows and pressure in the Jovian magnetosphere [*Ogino et al.*, 1998]. Flow shears appear in the simulation, and there are strong dawn-dusk asymmetries both in the flow (note vectors) and the pressure distribution (gray scale). We cannot avoid recognizing this fact.

In studying the Earth, on the other hand, we are often tempted to impose unwarranted symmetry in theoretical formulations. Processes that are highly improbable or even totally impossible in less idealized cases may then seem reasonable. As an example of a result that may apply only in a limiting case and not more generally, consider the treatment of reconnection at Earth in the presence of a northward interplanetary magnetic field (IMF). Many simulations and even cartoon treatments of the problem adopt a configuration with a strictly northward IMF and an untilted dipole. Reconnection occurs simultaneously in the northern and southern lobes, thereby adding closed flux tubes to the day side (as illustrated on the left side of Figure 5). Yet a strictly northward IMF, in principle possible, is highly improbable. And this does matter because it is virtually an axiom that broken symmetry changes the physics. For example, for a less symmetric and more probable configuration of the interplanetary field (for example, an IMF with a significant component along the Earth-Sun direction as on the right of Figure 5) reconnection occurs only in one hemisphere. Newly formed flux tubes are not closed at both ends, no flux is added to the day side, and the net effect of reconnection is greatly modified.

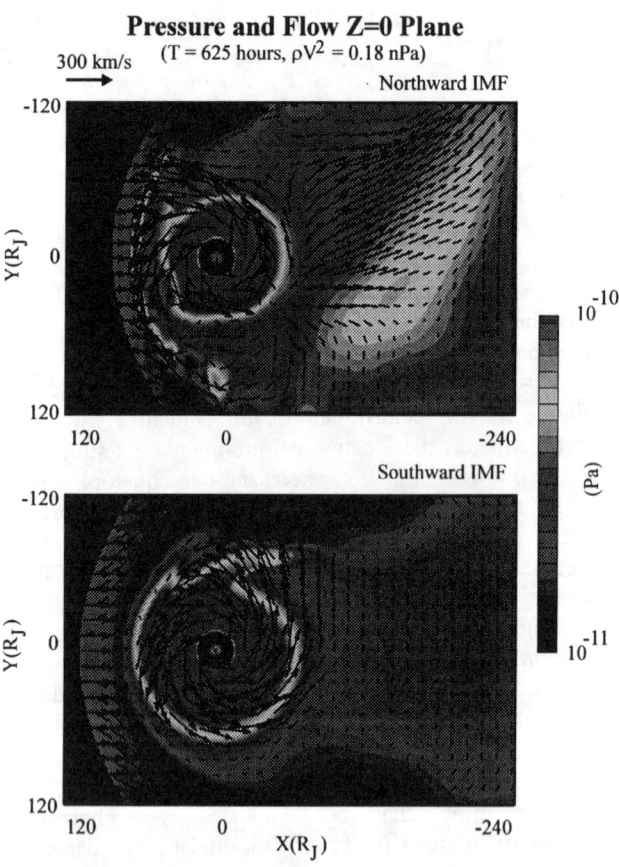

**Figure 4.** *Ogino et al.* [1998] simulation (see also *Walker et al.*, 1998). Arrows represent plasma flow velocity. Gray scale indicates plasma pressure (see color bar). The upper (lower) panel is run for a northward (southward) orientation of the interplanetary magnetic field. In both cases, the solar wind flow velocity is 300 km s$^{-1}$ and the solar wind dynamic pressure is 0.18 nPa.

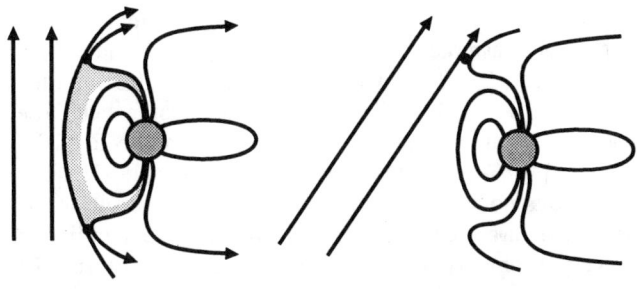

**Figure 5.** Reconnection for northward interplanetary magnetic field (IMF). Left: Reconnection with a strictly northward IMF occurs simultaneously in both hemispheres, creating a closed flux tube. Right: Reconnection for oblique IMF with a finite component along the solar wind flow direction occurs only in the northern lobe and produces no closed flux tubes.

Jupiter forces us to recognize asymmetries as fundamental. In studies of Earth as well as Jupiter, we would be wise to direct more attention to the different and challenging physics of systems with broken symmetry.

Let us return to further discussion of the way in which rotation is maintained in the plasma sheet at Jupiter. We have noted that magnetospheric drag slows the rotation of its ionosphere yet corotation of the ionosphere is approximately maintained. The reason is that the ionosphere gains momentum through collisions with the atmosphere whose large mass makes it an effective reservoir of momentum. Eddy viscosity is critical. *Kennel & Coroniti* [1975] described how this process acts at Jupiter. "The . . . torque is communicated to the ionosphere and atmosphere by a system of currents in at the dipole and spin axis, across the ionosphere, and out at the boundary of the polar cap. The $\mathbf{j} \times \mathbf{B}$ torque is exerted first on the atmosphere at ionospheric levels. This should be balanced by the diffusion of angular momentum upward from the ionospheric layers below the ionosphere." Their schematic of the coupling is shown in Figure 6.

*Kennel and Coroniti* [1975] suggested that "where the current flows out, an aurora borealis may be found, if Earth-like physics prevails." The aurora that they predicted has been identified. It is narrow and bright as illustrated in Figure 7 from [*Clarke et al.*, 1996]. It lies on field lines that link to the disk-like current sheet region. Using Ulysses magnetometer measurements, *Dougherty et al.* [1998] reported dusk sector field-aligned currents that they believed link to the aurora. However, the currents that they reported close through the equatorial plane near 12 $R_J$ whereas the field models require the currents that impose corotation, and which are likely to couple to the aurora, close at much greater distances.

The extent of the current system that imposes corotation can be inferred either from field models or from direct measurements of the azimuthal flow speed. As noted above, although full corotation is not maintained beyond about 20 $R_J$, the plasma continues to rotate at about half the corotation speed to large distances, requiring coupling to the ionosphere out to distances greater than ~50 $R_J$ [*Kane*, 1995]. However, at some distance, the required link between the ionosphere and the plasma sheet must break. Signal propagation between the distant magnetosphere and the ionosphere requires hours, a point that has been emphasized by *Vasyliunas* [1994]. While the signal propagates, plasma continues to move both azimuthally and radially. Field-aligned currents (carried by Alfvén waves as previously noted) may not be able to keep the ionosphere and the outer layers of the magnetosphere moving together because of the lengthy time required for signals to propagate from the ionosphere to the equatorial portion of a flux tube. This means that the return currents that bound the aurora at high latitude probably flow well within the magnetopause boundary.

## 5. AURORAS DIFFER AT EARTH & JUPITER

The aurora at Earth and Jupiter differ both in appearance and in generation mechanism. At Earth, the primary auroral zone is approximately annular in form with a center shifted antisunward of the magnetic pole by 5° and, when active, the aurora is brightest, broadest (in latitude), and most active near midnight. Earth auroras are dynamically related to substorms. At Jupiter, the latitude of the auroral oval is roughly fixed relative to Jovigraphic coordinates. The auroral oval width is narrow over part of its local time extent, being particularly narrow and quiescent near dawn, but extending far into the polar cap near dusk. In the dusk hemisphere, the Jovian aurora is

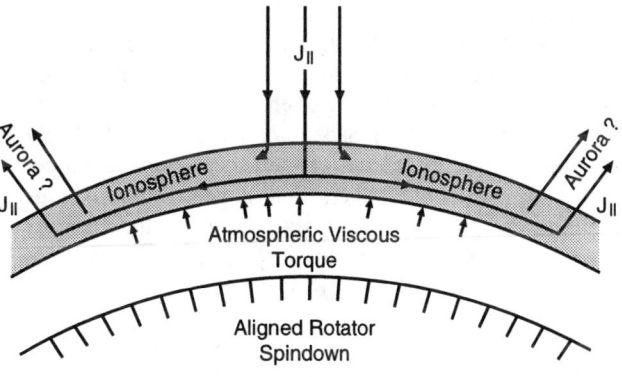

**Figure 6.** Schematic of Jupiter's auroral ionosphere [*Kennel and Coroniti*, 1975] showing that the drag exerted by the current system that imposes corotation on the magnetosphere is balanced by momentum carried upward from the atmosphere.

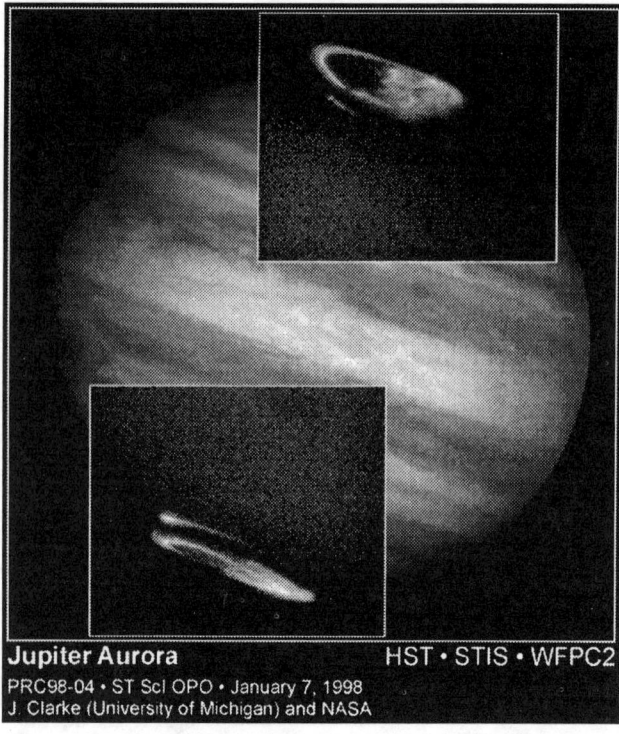

**Figure 7.** Jovian northern and southern auroras as captured by NASA's Hubble Space Telescope [*Clarke*, 1999].

dynamic with a strong dependence on Jovian longitude [*Ballester et al.*, 1996]. The brightest portion of the aurora, on the other hand, is observed near dawn [*Ballester et al.*, 1996] and the morning side oval lies well poleward of the flux tubes that link to Ganymede at ~15 $R_J$ [*Clarke et al.*, 1998]. According to reports based on Galileo imaging [*Ingersoll et al.*, 1998], the aurora is present at also on the night side. The nightside aurora is narrow and relatively quiescent.

It seems likely that different processes control the Jovian and the terrestrial auroras. For Earth, the aurora occurs at latitudes where field-aligned currents are associated with field-aligned potential drops. A similar situation may well account for the observations at Jupiter. On the day side of Jupiter, field-aligned currents must couple to currents that impose corotation on the outward-moving plasma. On the night side, outflow of heavy ion plasma down the magnetotail must relate to the observed luminosity, but it is not yet certain if the transport on the night side is relatively steady or extremely dynamic. Indeed, the relation between aurora and "activity" has not yet been established. However, the auroral emissions are most variable and intense in the dusk sector and link magnetically to regions where corotating flux tubes break free from the confinement of the dayside magnetopause.

In the Jovian magnetotail, plasma "bubbles" like terrestrial plasmoids or flux ropes have been identified [*Russell et al.*, 1998], with bursty outflow also reported [*Krupp et al.*, 1998]. The analysis is ongoing but these phenomena may relate to tail reconnection predicted by *Nishida* [1983.] and illustrated by *Vasyliunas* [1983]. Thus far there is no accepted evidence of global scale substorms or of the large-scale substorm current wedge familiar to terrestrial observers but the existence of substorm-like activity has been described [*Woch et al.*, 1999; *Louarn et al.*, 1999]. Direct links between the plasma bubbles in the magnetotail and the aurora have not been established.

## 6. GANYMEDE: THE OTHER EXTREME

Compared with Jupiter, or even with Earth, Ganymede is small with a radius $R_G \sim 0.4$ $R_E \approx 2,600$ km. It is weakly magnetized, yet its "surface" magnetic field magnitude (750 nT), ~50 times smaller than Earth's surface field, is large compared with the ~100 nT field of the Jovian magnetosphere at Ganymede's orbit. Because the Keplerian orbital speed of Ganymede is small compared with the roughly corotational flow speed of the ambient Jovian plasma, the corotating plasma of Jupiter's magnetosphere impinges on Ganymede from its trailing hemisphere (relative to orbital motion), hereinafter referred to as the upstream side. Relative to Ganymede, the flow speed is ~140 km/s. Table 2 shows that the magnetic pressure of the ambient plasma dominates both the dynamic pressure ($M_A < 1$) and the thermal pressure ($\beta < 1$). Thus the magnetic structure of Ganymede's magnetosphere is quite well approximated by superposition of a uniform external magnetic field and the field of a moon-centered dipole moment oriented southward. Figure 8 shows a cut through such a vacuum superposition model of Ganymede's magnetosphere, modified to include schematically some of the plasma interactions. The magnetosphere, defined as the volume within which at least one end of each field line enters Ganymede's surface, is about 5 $R_G$ across.

Ganymede is a slow rotator. Its rotation period of ~one week is the same as its orbital period. Because Ganymede's rotation is phase locked, it always presents the same face to the corotating plasma of Jupiter's plasma disk. As Ganymede remains fixed relative to its own magnetospheric reference frame, rotational forces do not act on the plasma within this mini-magnetosphere.

The orientation of Ganymede's internal field facilitates reconnection near the equatorial magnetopause where the internal and external fields are on average antiparallel as shown in Figure 8. With the reconnection centered on the upstream magnetosphere, conservation of magnetic flux requires secondary reconnection on the downstream side of the moon.

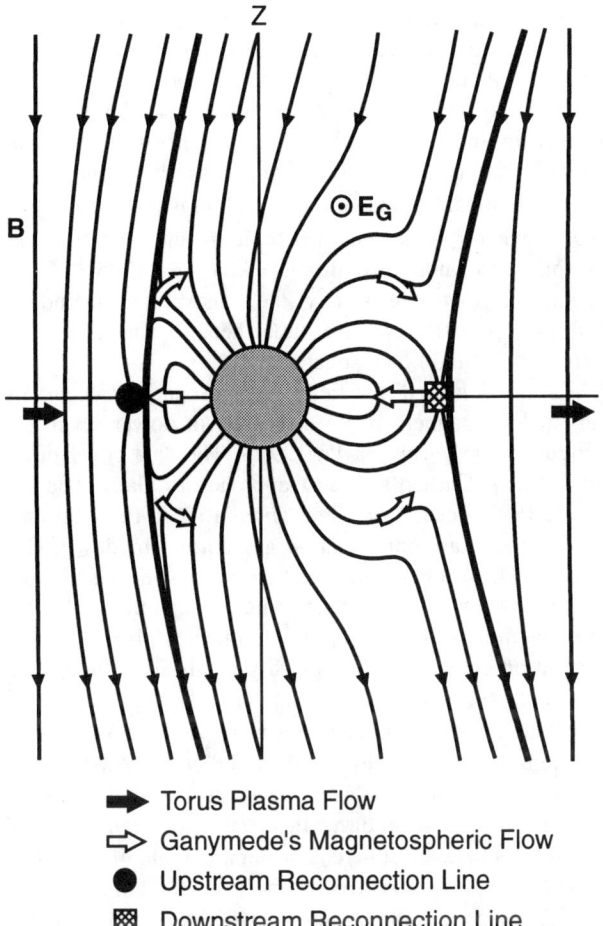

**Figure 8.** Schematic cross section of Ganymede's magnetosphere. Modified from *Volwerk et al.* [1999].

The magnetopause, the boundary between the flowing plasma of Jupiter's plasma sheet and the plasma on field lines linked to Ganymede is represented by heavy curves in Figure 8. Notice that there are lobe-like structures extending north and south. A cross section of the magnetosphere at a few $R_G$ above the equatorial plane would show a roughly circular magnetopause with the magnetopause current flowing around this boundary as it does around the northern or southern lobe cross section of the terrestrial magnetosphere. Now imagine folding the two lobes of Ganymede's magnetosphere down and to the right in the image. Following this modification, a cut across the plane of the image would reveal a structure similar both in terms of magnetic field orientation and current paths to that in a cut of the terrestrial magnetotail. This exercise for a magnetosphere in which the topology is familiar but the geometry is unfamiliar, helps us recognize that the bundles of field lines emerging from Ganymede's polar caps are the tail lobes! It also makes sense of the idea that at Earth the magnetotail sheet current is integral with the magnetopause current.

A subtle effect of the lack of a corotation electric field within Ganymede's magnetosphere is that the plasma density on closed field lines is low because of the flow patterns imposed by reconnection [*Volwerk et al.*, 1999]. Flux tubes reconnected upstream and near the equator flow over the polar cap and return to low latitudes downstream (see Figure 8). Flux conservation imposes a requirement for return flow on closed field lines, but little of the plasma present on flux tubes that reconnect downstream near the equator returns upstream. The reason is that, in the absence of a corotation electric field within Ganymede's magnetosphere, the cold plasma flows approximately along $-\hat{x}$ (Figure 8) and is lost on Ganymede's surface which blocks most of it from returning to the upstream side [*Volwerk et al.*, 1999]. Particles with energies greater than a few keV can avoid Ganymede because their drift is modified by magnetic gradient drift. Thus, they can drift on closed field lines from the downstream reconnection location into the upstream region. Indeed, energetic electrons distributions of the form expected for particles bouncing on closed field lines have been reported [*Williams et al.*, 1997]. It is not clear whether the electrons observed by Williams et al. were on drift orbits that encircle Ganymede. They may have been following trajectories draped around Ganymede that link a source region near the downstream reconnection line to a sink at the upstream magnetopause (see Figure 8). Because magnetospheric reconnection is likely to be rather steady in the slowly changing external flow, temporal fluctuations may be too weak to scatter energetic ions from such open trajectories to orbits closed to drift around Ganymede. Such orbits must be populated in a magnetosphere with a ring current.

## 7. WHAT ABOUT CURRENTS AT GANYMEDE?

Figure 8 shows no upstream shock because the flow of the surrounding plasma is sub-magnetosonic (see Table 2). Small upstream flow perturbations are deliberately illustrated because fast mode waves, moving faster than the flow, begin to divert the plasma upstream of the magnetospheric obstacle.

At the magnetopause, localized currents develop. They were observed by Galileo as rotations of the field [*Kivelson et al.*, 1998]. The magnetopause currents have not yet been fully described but some features are evident. For example, because the closed field lines in Figure 8 are roughly antiparallel to the external field, it is apparent that the low latitude magnetopause currents must close in a ring around the near-equatorial magnetopause, thus differing from the

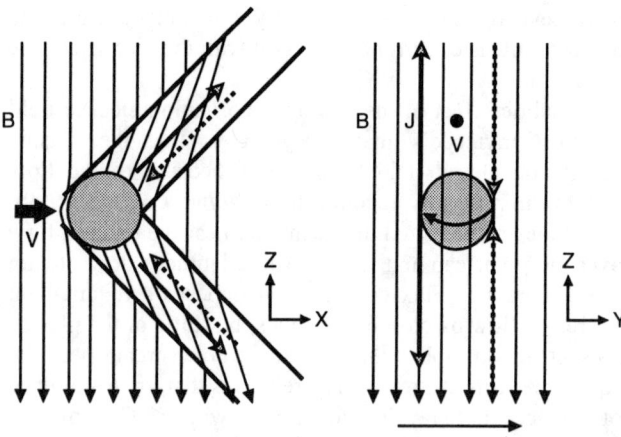

**Figure 9.** Schematic of the field (B), the currents (J), the flows (v) in an Alfvén wing interaction of a flowing magnetized plasma with a conducting obstacle. Plasma flows in from the left. The background field of Jupiter is oriented southward. The left panel shows the plane of the flow (v) and the field (B) through the moon's center. The right panel is the cross section orthogonal to the flow with Jupiter to the right. Alfvén wing currents (J) are shown flowing inward on the side near Jupiter and outward on the anti-Jupiter side. They cause the field to bend back in the Alfvén wing as illustrated on the left.

Chapman-Ferraro current system at Earth. (The latter currents close at high latitudes above the polar cusp.)

In addition, Alfvén wing currents, partially field-aligned, also flow on the magnetopause. Alfvén wing currents develop when the flow slows because of the presence of a conducting or magnetized moon as illustrated in Figure 9 [*Neubauer*, 1980]. The Alfvén wing currents cause the open "lobe" field lines to bend back in the flow direction. Because the perturbations propagate along **B** at the Alfvén speed while convecting downstream at the flow speed, **u**, the tilts occur behind a front illustrated in Figure 9. The front is tilted at an angle $\theta_A = \tan^{-1}(M_A)$.

The field rotations with $\Delta B_x = O(100\,\text{nT})$ observed by Galileo in crossings at high latitudes [*Kivelson et al.*, 1998] have been found to be consistent with predicted perturbations from Alfvén wing currents. The net current in each Alfvén wing can be estimated from Ampére's law by integrating the average change of the magnetic field, $0.5\Delta B_x = 50\,\text{nT}$, along the arc of the Alfvén wing (length $\sim 4\pi R_G$) giving $I = \Delta B_x \pi (2R_G)/\mu_o \approx 1\,\text{MA}$. This current flows into the Jovian ionosphere where a signature has been observed in auroral images [*Clarke et al.*, 1998].

## 8. STANDING WAVES AT GANYMEDE

The region of closed fields at Ganymede has been found to support "field line resonances". At Earth, field line resonances, familiar as oscillations of entire closed flux tubes, link to Earth at both ends through field-aligned currents. A poloidal field line resonance is illustrated schematically in Figure 10 in which the flux tube is shown to move in and out radially. The ionospheric end of the flux tube imposes boundary conditions, here shown as "fixed". Resonant frequencies depend on mass density: $f \propto B/\ell\rho^{1/2}$ where $\ell$ is length of field line and on the ionospheric boundary conditions [*Kivelson*, 1995]. Only eigenmodes that can satisfy the ionospheric boundary conditions persist on a flux tube, hence such waves are referred to as field line resonances.

Waves identified as field line resonances occur in Ganymede's magnetosphere as shown in power spectrum of Figure 11 [*Volwerk et al.*, 1999]. The data cover three minutes near Galileo's closest approach to Ganymede on May 7, 1997 (near $L = 1.6$ at about 1556 UT). Multiple peaks with a harmonic relation are normal modes of the plasma. They convey information about the plasma and about Ganymede's ionosphere which damps them. As the field magnitude is known, the frequency gives indirect information about the mass density on the flux tube. The mass density inferred on the closed flux tubes of Ganymede's upstream magnetosphere is $\sim 2\,\text{AMU cm}^{-3}$. This density is much lower than the $>100\,\text{AMU cm}^{-3}$ density of the ambient plasma torus, as anticipated in our discussion of the return flow at low latitudes in Ganymede's magnetosphere. The damping rate of field line resonances is quite sensitive to the Pedersen conductance, $\Sigma_P$, of Ganymede's ionosphere. Damping is smallest for extremes of ionospheric boundary conditions [*Newton et al.*, 1978] and maximum for intermediate values of $\Sigma_P$. The waves observed at Ganymede persisted for several cycles implying either than the ionospheric conductivity was relatively high or relatively low.

## 9. NEAR MOONS, PICKUP CURRENTS ARISE FROM NEWLY IONIZED MATTER

Energetic particles present in Jupiter's magnetosphere bombard at least portions of the atmospheres and surfaces of the moons and release neutrals into the surroundings. As the neutrals, on average at rest with respect to the moon, are ionized by photoionization or by impact ionization, they find themselves in the electric field of the flowing plasma.

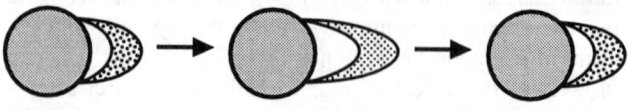

**Figure 10.** Schematic of a poloidal field line resonance for which the position of a flux tube oscillates radially.

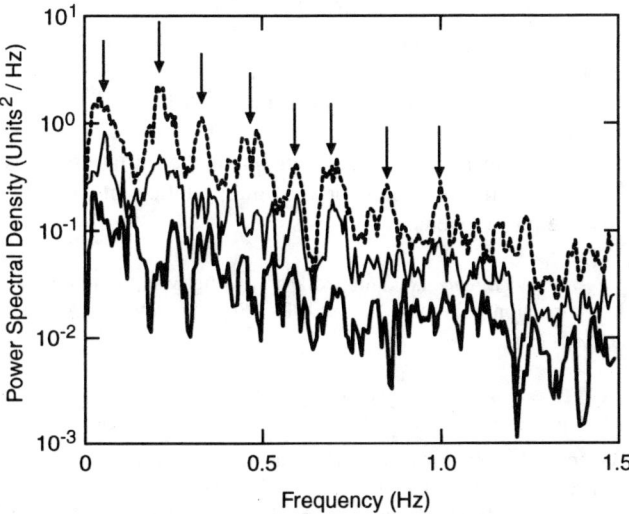

**Figure 11.** Power spectral density of waves observed on Ganymede's closed field lines. From *Volwerk et al.* [1999].

Their presence is most obvious near Io, a prodigious source of heavy ions.

The link between plasma pickup and currents is understood by examining the motion of a newly ionized neutral embedded in a flowing plasma as illustrated in Figure 12. Because the flowing plasma imposes an electric field in the frame of the pickup ion ($E \neq 0$), the ion is immediately accelerated. In the diagram, the initial motion is "up" for ions and "down" for electrons. The gyrocenters, initially superimposed, separate by roughly one ion gyroradius, $\sim \rho_{L,i}$. If the rate of ionization is $\dot{n}_{pu}$, the pickup current is $j_{pu} = q\dot{n}_{pu}\rho_{L,i}$ in the direction of the flow electric field. Following initial acceleration, both ions and electrons gyrate and drift to left at speed $u$, with a perpendicular thermal speed equal to the flow speed. Because the flow speed is typically large compared with the ambient thermal speed, pickup distribution are usually highly anisotropic, with a ring distribution in velocity space that is unstable to growth of ion cyclotron waves. Such waves have been identified in the surroundings of Io [*Warnecke et al.*, 1997; *Huddleston et al.*, 1998b]. *Huddleston et al.* were able to infer the rate of pickup of molecular $SO_2^+$ ions from the changing amplitude of the wave spectrum.

In a pickup plasma, current continuity requires that the cross field pickup current illustrated in the figure must close in field-aligned currents. Effectively, when a cloud of pickup ions exists around a moon, an Alfvén wing type current system extends over a spatial volume whose cross section is larger than the moon's. It is likely that in this case the field aligned currents are not confined to sheet-like structures of the sort illustrated in Figure 9.

Although of considerable theoretical interest, pickup currents are rarely important in Earth's magnetosphere. However, they contribute in an important way when artificial plasma seeding of the magnetosphere is undertaken [*Haerendel*, 1983].

## 10. INDUCED MAGNETIC FIELDS AND CURRENTS WITHIN THE MOONS

As noted above, Jupiter's moons are embedded in the magnetic field of Jupiter's magnetosphere. Because of the tilt of Jupiter's dipole moment relative to its spin axis, the orientation of the field imposed on a moon changes as Jupiter rotates. The dominant effect is a near ten-hour variation of the radial component of the ambient field. Faraday's law specifies the electric field induced by a time-varying magnetic field. If appropriate conducting paths exist within the moons, the induced electric field drives electrical currents that in turn perturb the local magnetic field. In Galileo passes by Europa and Callisto, perturbation fields consistent with expectations from an inductive response have been identified [*Khurana et al.*, 1998; *Kivelson et al.*, 1999]. The direction and magnitude of the observed perturbations are well approximated as the inductive response of a perfectly conducting spherical body, i.e., the field of a centered dipole moment whose surface field at the pole exactly cancels the externally imposed time-varying field, $B_{ext}(t)$. The observed signature implies that currents within the moons flow in shells near or not far below the surface.

These observations are evidently significant for understanding the internal structure of the moons. Near the surfaces, there must be layers of thickness $h$ that satisfy $h \lesssim 0.1 \, \delta$, where $\delta \equiv$ skin depth $= (2/\mu_0\omega\sigma)^{1/2}$, expressed in terms of $\omega$, the angular frequency of the temporal variation, and $\sigma$, the electrical conductivity. For $\omega \sim 2\pi/10$(hrs) and $\sigma = 2.5$ S/m, the conductivity of terrestrial seawater, $\delta$ is $\sim 60$ km and $h$ is about 6 km.

Although the Earth is not exposed to a periodically varying external magnetic field (as is Europa), the external

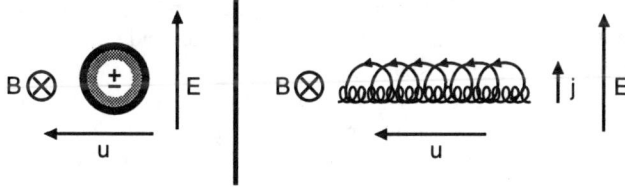

**Figure 12.** Illustration of ion pickup in a flowing plasma. On the left, the neutral particle is embedded in a flowing magnetized plasma. On the right, the ion and electron are gyrating in independent orbits, their gyrocenters having separated by one ion gyroradius in the direction of **E**, thus contributing to the current density, **j**.

field can change by of order 100 nT over intervals of several hours to days, for example during geomagnetic storms [*McPherron*, 1995]. These changes are large enough and Earth's conductivity is sufficiently great to produce an inductive response to the changing field [*Chapman and Bartels*, 1962]. The induced field opposes the imposed field perturbation at the pole where it may have a surface magnitude that equals the imposed perturbation field. At the equator, the perturbation at the surface is aligned with the perturbation field, with a maximum value of ~1/2 its amplitude. The inductive response is measured as part of the storm index Dst [*McPherron*, 1995] which is then 3/2 as large as it would be with no induced contribution.

The Dst index has long been used as a proxy for the intensity of the ring current at Earth. The link between the total energy of ring current particles, $W_{particles}$, and Dst is referred to as the *Dessler-Parker-Sckopke* relation [*Wolf*, 1995]. The relation tells us that a ring current depresses the field at Earth, giving $Dst = \Delta B_{particles} = -\mu_o W_{particles}/2\pi B_o R_E^3$. *Feldstein* [1992] pointed out that if inductive effects are present, the measured Dst exceeds the perturbation imposed by the ring current alone. Thus to infer the energy in the ring current from the Dessler-Parker Sckopke relation, one should reduce the measured Dst by a factor ~2/3, assuming that the maximum inductive field is present. Recent studies of the induced fields at the Galilean moons of Jupiter suggest that Feldstein's interesting idea deserves more thorough investigation.

## 11. CURRENTS AT JUPITER AND ITS MOONS

We have discussed some current systems in Jupiter's magnetosphere that differ considerably from the analogous current systems at Earth (such as the very different auroral current system). Other currents observed at Jupiter also exist at Earth but are not dominant there. For example, the effects of corotation-related currents on the terrestrial aurora are never discussed because they are probably too minor to be interesting. Substorm currents, central to our studies at Earth, have not yet been unambiguously identified at Jupiter and may not be present because rotational effects dominate plasma transport and convective processes differ in character.

In Ganymede's magnetosphere as at Earth and Jupiter, currents shape the flow and the flow drives currents. The difference between magnetospheres in super-Alfvénic and sub-Alfvénic flows is dramatically evident when one compares the global structure of planetary magnetospheres with the structure at Ganymede. The absence of a corotation electric field in Ganymede's magnetosphere leads to strong upstream-downstream asymmetry of the plasma distribution on closed flux tubes, as contrasted with the relatively symmetric distribution of plasma within the terrestrial magnetosphere where the corotation electric field is important.

The induced fields observed near Europa and Callisto suggest that inductive responses at Earth can be significant, depending on the frequency content of the variation. This feature has implications for the interpretation of Dst that deserve more extensive analysis.

*Acknowledgements.* This work was partially supported by the Jet Propulsion Laboratory under JPL 958694 and by NASA under NAG 5-7959.

## REFERENCES

Bagenal, F., Empirical model of the Io plasma torus: Voyager measurements, *J. Geophys. Res.*, *99*, 11043-62, 1994.

Bagenal, F., Plasma parameters near Galilean satellites and during satellite encounters, web page - http://dosxx.colorado.edu/Galileo/encounter.html, consulted August 14, 1998.

Ballester, G.E., J.T. Clarke, J. Trauger, W.M. Harris, K. Stapelfeldt, D. Crisp, R. Evans, E.B. Burgh, C.J. Burrows, S. Casertano, I. J. S. Gallagher, R.E. Griffiths, J.J. Hester, J.G. Hoessel, J.A. Holtzman, J.E. Krist, V. Meadows, J.R. Mould, R. Sahai, P.A. Scowen, A.M. Watson, and J.A. Westphal, Time-resolved observations of Jupiter's far-ultraviolet aurora, *Science*, *274*, 409-13, 1996.

Balogh, A., M.K. Dougherty, R.J. Forsyth, D.J. Southwood, E.J. Smith, B.T. Tsurutani, N. Murphy, and M.E. Burton, Magnetic field observations during the Ulysses flyby of Jupiter, *Science*, *257*, 1515-18, 1992.

Beatty, J.K., and A. Chaikin, *The New Solar System*, pp. 326, Sky Pub. Co., Cambridge, MA, 1990.

Caudal, J., A self-consistent model of Jupiter's magnetodisc including the effects of centrifugal force and pressure, *J. Geophys. Res.*, *91*, 4201, 1986.

Chapman, S., and J. Bartels, *Geomagnetism*, 1049 pp., Oxford at the Clarendon Press, New York, 1962.

Clarke, J.T., G.E. Ballester, J. Trauger, R. Evans, J.E.P. Connerney, K. Stapelfeldt, D. Crisp, P.D. Feldman, C.J. Burrows, S. Casertano, I. J. S. Gallagher, R.E. Griffiths, J.J. Hester, J.G. Hoessel, J.A. Holtzman, J.E. Krist, V. Meadows, J.R. Mould, P.A. Scowen, A.M. Watson, and J.A. Westphal, Far-ultraviolet imaging of Jupiter's aurora and the Io "footprint", *Science*, *274*, 404-9, 1996.

Clarke, J.T., G.E. Ballester, J. Trauger, J. Ajello, W. Pryor, K. Tobiska, J.E.P. Connerney, G.R. Gladstone, J. J. H. Waite, L.B. Jaffel, and J.-C. Gerard, Hubble Space Telescope imaging of Jupiter's UV aurora during the Galileo orbiter mission, *J. Geophys. Res.*, *103*, 20,217-36, 1998.

Clarke, J.T., Image STScI-PRC98-04, January 7, 1998, John Clarke (University of Michigan), from website: http://oposite.stsci.edu/pubinfo/pr/1998/04/ consulted July 2, 1999.

Crooker, N., and R.L. McPherron, On the distinction between the auroral electrojet and partial ring current systems, *J. Geophys. Res.*, *77*, 6886, 1972.

Dougherty, M.K., M.W. Dunlop, R. Prange, and D. Rego,

Correspondence between field aligned currents observed by Ulysses and HST auroral emission, *Planet. Space Sci.*, 46, 531-40, 1998.

Feldstein, Y. I., Modeling of the magnetic field of magnetospheric ring current as a function of interplanetary medium parameters, *Space Sci. Rev.* 59, 83-165, 1992.

Haerendel, G., Towards an artificial comet, *ESA Journal*, 7 (2), 135-44, 1983.

Hill, T.W., Rotationally induced Birkeland current systems, in *Magnetospheric Currents*, edited by T.A. Potemra, pp. 340-9, Amer. Geophys. U., Washington, DC, 1983.

Hill, T.W., A.J. Dessler, and F.C. Michel, Configuration of the Jovian magnetosphere, *Geophys. Res. Lett.*, 1, 3-6, 1974.

Huddleston, D.E., C.T. Russell, M.G. Kivelson, K.K. Khurana, and L. Bennett, Location and shape of the Jovian magnetopause and bow shock, *J. Geophys. Res.*, 103, 20,075, 1998a.

Huddleston, D.E., R.J. Strangeway, J. Warnecke, C.T. Russell, and M.G. Kivelson, Ion cyclotron waves in the Io torus: Wave dispersion, free energy analysis, and SO2+ source rate estimates, *J. Geophys. Res*, 103, 19,887, 1998b.

Hughes, W.J., The magnetopause, magnetotail, and magnetic reconnection, in *Introduction to Space Physics*, edited by M.G. Kivelson, and C.T. Russell, pp. 227-287, Cambridge University Press, New York, 1995.

Iijima, T., and T. A. Potemra, The amplitude distribution of field aligned currents at northern high latitudes observed by TRIAD, *J. Geophys. Res.*, 81, 2165, 1976.

Ingersoll, A.P., A.R. Vasavada, B. Little, D.D. Anger, S.J. Bolton, C. Alexander, K.P. Klassen, W.K. Tobiska, and the Galileo SSI Team, Imaging Jupiter's aurora at visible wavelengths, *Icarus*, 135, 251-264, 1998.

Kane, M., Hot ions in Jupiter's magnetodisc: A model for Voyager 2 low-energy charged particle measurments, *J. Geophys. Res.*, 100, 19,473-486, 1995.

Kane, M., D.J. Williams, B.H. Mauk, R.W. McEntire, and E.C. Roelof, Galileo Energetic Particles Detector measurements of hot ions in the neutral sheet region of Jupiter's magnetodisk, *Geophys. Res. Lett.*, 265-8, 1999.

Kennel, C.F., and F.V. Coroniti, Is Jupiter's magnetosphere like a pulsar's or Earth's?, in *The Magnetospheres of the Earth and Jupiter*, edited by V. Formisano, pp. 451-78, D. Reidel, Boston, 1975.

Khurana, K.K., Euler potential models of Jupiter's magnetospheric field, *J. Geophys. Res.*, 102, 11,295, 1997.

Khurana, K.K., and M.G. Kivelson, Inference of the angular velocity of plasma in the Jovian magnetosphere from the sweepback of magnetic field, *J. Geophys. Res.*, 98, 67-79, 1993.

Khurana, K.K., M.G. Kivelson, D.J. Stevenson, G. Schubert, C.T. Russell, R.J. Walker, S. Joy, and C. Polanskey, Induced magnetic fields as evidence for subsurface oceans in Europa and Callisto, *Nature*, 395, 777, 1998.

Kivelson, M.G., Pulsations and magnetohydrodynamic waves, in *Introduction to Space Physics*, edited by M.G. Kivelson, and C.T. Russell, pp. 330-55, Cambridge University Press, New York, 1995.

Kivelson, M.G., and F. Bagenal, Planetary magnetospheres, in *The Encyclopedia of the Solar System*, edited by P. Weissman, L.-A. McFadden, and T.V. Johnson, pp. 477, Academic Press, 1999.

Kivelson, M.G., P.J. Coleman, and L. Froidevaux, A time dependent model of the Jovian current sheet, *J. Geophys. Res.*, 83, 4823-29, 1978.

Kivelson, M.G., K.K. Khurana, F.V. Coroniti, S. Joy, C.T. Russell, R.J. Walker, J. Warnecke, L. Bennett, and C. Polanskey, The magnetic field and magnetosphere of Ganymede, *Geophys. Res. Lett*, 24, 2155, 1997.

Kivelson, M.G., J. Warnecke, L. Bennett, S. Joy, K.K. Khurana, J.A. Linker, C.T. Russell, R.J. Walker, and C. Polanskey, Ganymede's magnetosphere: Magnetometer overview, *J. Geophys. Res.*, 103, 19,963, 1998.

Kivelson, M.G., K.K. Khurana, D.J. Stevenson, L. Bennett, S. Joy, C.T. Russell, R.J. Walker, C. Zimmer, and C. Polanskey, Europa and Callisto: Induced or intrinsic fields in a periodically varying plasma environment, *J. Geophys. Res.*, 104, 4609, 1999.

Krupp, N., J. Woch, A. Lagg, B. Wilken, S. Livi, and D.J. Williams, Energetic particle bursts in the predawn Jovian magnetotail, *Geophys. Res. Lett.*, 25, 1249-52, 1998.

Lepping, R.P., L.F. Burlaga, and L.W. Klein, Jupiter's magnetopause, bow shock, and 10 hour modulated magnetosheath : Voyagers 1 and 2, *Geophys. Res. Lett.*, 8, 99-102, 1981.

Louarn, P., A. Roux, S. Perraut, W. Kurth, and D. Gurnett, A study of large scale dynamics of the Jovian magnetosphere using the Galileo Plasma Wave Experiment, *Geophys. Res. Lett.*, 25, 2905-8, 1999.

Mauk, B.H., and S.M. Krimigis, Radial force balance within Jupiter's dayside magnetosphere, *J. Geophys. Res.*, 92, 9931, 1987.

Mauk, B.H., R.W. McEntire, D.J. Williajms, A. Lagg, E.C. Roelof, S.M. Krimigis, T.P. Armstrong, T.A. Fritz, L.J. Lanzerotti, J.G. Roederer, and B. Wilken, Galileo-measured depletion of near-Io hot ring current plasmas since the Voyager epoch, *J. Geophys. Res.*, 103, 4715-22, 1998.

McNutt, R.L., Jr., Force balance in outer planet magnetospheres, in *Physics of Space Plasmas*, edited by T. Chang, MIT Center for Theoretical Geo/Cosmo Plasma Physics, Cambridge, MA, 1982.

McPherron, R.L., Magnetospheric dynamics, in *Introduction to Space Physics*, edited by M.G. Kivelson, and C.T. Russell, pp. 227-287, Cambridge University Press, New York, 1995.

Neubauer, F.M., Non-linear standing Alfven wave current system at Io: Theory, *J. Geophys. Res.*, 85, 1171, 1980.

Newton, R.S., D.J. Southwood, and W.J. Hughes, Damping of geomagnetic pulsations by the ionosphere, *Planet. Space Sci*, 26, 201-9, 1978.

Nishida, A., Reconnection in the Jovian magnetosphere, *Geophys. Res. Lett.*, 10, 6, 451-454, 1983.

Ogino, T., R.J. Walker, and M.G. Kivelson, A global magnetohydrodynamic simulation of the Jovian magnetosphere, *J. Geophys. Res.*, 103, 225, 1998.

Paranicas, C.P., B.H. Mauk, and S.M. Krimigis, Pressure anisotropy and radial stress balance in the Jovian neutral sheet, *J. Geophys. Res.*, 96, 21,135-40, 1991.

Potemra, T.A., *Magnetospheric Currents, Geophys. Mono.*, pp. 357, Amer. Geophys. U., Washington, D.C., 1984.

Russell, C.T., K.K. Khurana, D.E. Huddleston, and M.G. Kivelson, Localized reconnection in the near Jovian magnetotail, *Science*, *280* (15 May), 1061, 1998.

Siscoe, G.L., Towards a comparative theory of magnetospheres, in *Space Plasma Physics: The Study of Solar-System Plasmas*, pp. 557-687, National Academy of Sciences, Washington, DC, 1979.

Spiro, R.W., and R.A. Wolf, Electrodynamics of convection in the inner magnetosphere, in *Magnetospheric Currents*, edited by T.A. Potemra, pp. 247-268, Amer. Geophys. U., Washiington, D.C., 1984.

Smith, E.J., R.W. Fillius, and J.H. Wolfe, Compression of Jupiter's magnetosphere by the solar wind, *J. Geophys. Res.*, *83*, 4733-4742, 1978.

Strobel, D.F., and S.K. Atreya, Ionosphere, in *Physics of the Jovian Magnetosphere*, edited by A.J. Dessler, pp. 51-67, Cambridge Univ. Press, New York, 1983.

Vasyliunas, V.M., Plasma distribution and flow, in *Physics of the Jovian Magnetosphere*, edited by A.J. Dessler, pp. 395-453, Cambridge Univ. Press, New York, 1983.

Vasyliunas, V.M., Role of plasma acceleration time in the dynamics of the Jovian magnetosphere, *Geophys. Res. Lett.*, *21*, 401, 1994.

Volwerk, M., M.G. Kivelson, K.K. Khurana, and R.L. McPherron, Probing Ganymede's magnetosphere with field line resonances, *J. Geophys. Res.*, *104*, 14,729-38, 1999.

Walker, R.J., G. Ballester, K.K. Khurana, M.G. Kivelson, T. Ogino, and M. Sharlow, Studying Jovian aurora by using a global magnetohydrodynamic simulation of the interaction of the solar wind with the magnetosphere, *EOS Trans. AGU*, *79*, F549, 1998.

Warnecke, J., M.G. Kivelson, K.K. Khurana, D.E. Huddleston, and C.T. Russell, Ion cyclotron waves observed at Galileo's Io encounter: Implications for neutral cloud distribution and plasma composition, *Geophys. Res. Lett.*, *24(17)*, 2139, 1997.

Williams, D.J., Energetic particle environment in the Jovian magnetosphere, in *Magnetospheres of the Outer Planets*, pp. 19, Paris, France, 1999.

Williams, D.J., B. Mauk, and R.W. McEntire, Trapped electrons in Ganymede's magnetic field, *Geophys. Res. Lett.*, *24*, 2953-6, 1997.

Woch, J., N. Krupp, K.K. Khurana, M.G. Kivelson, A. Roux, S. Perraut, P. Louarn, A. Lagg, D.J. Williams, S. Livi, and B. Wilken, Plasma sheet dynamics in the Jovian magnetotail: Signatures for substormlike processes?, *Geophys. Res. Lett.*, *26*, 2137-40, 1999.

Wolf, R.A., Magnetospheric configuration, in *Introduction to Space Physics*, edited by M.G. Kivelson, and C.T. Russell, pp. 288-329, Cambridge University Press, New York, 1995.

---

Institute of Geophysics and Planetary Physics and Department of Earth and Space Sciences, University of California, Los Angeles, CA 90095-1567. e-mail: mkivelson@igpp.ucla.edu

# Rotational Current Systems and the Offset Io Plasma Torus

Duane H. Pontius, Jr.

*Birmingham-Southern College, Birmingham, Alabama*

In the Jovian magnetosphere, the Io plasma torus exhibits a distinct and substantial displacement toward the dawn side. Several authors have demonstrated that a large-scale dawn-to-dusk electric field will perturb particle drift paths and produce the required plasma distribution. The source of this imposed electric field is unknown, but it is thought to arise from antisunward plasma flow down the Jovian magnetotail. Such a displacement would presumably divert the centrifugal drift currents at the inner edge of the torus and produce thin sheets of magnetic field-aligned currents. These Birkeland currents are intriguing candidates for powering the intense EUV emissions observed in this region. However, determining the complete current divergence requires including the contributions from all inertial currents, which leads to a surprising result: the net current produces no concentrated Birkeland currents unless there is differential rotation in Jupiter's atmosphere.

## INTRODUCTION

Both optical and ultraviolet emissions from the Io plasma torus exhibit a persistent dawn-dusk (or, equivalently, east-west) asymmetry. Unlike many characteristics of the inner Jovian magnetosphere that rotate with magnetic-field oriented System III coordinates, this phenomenon remains relatively fixed in Sun-Jupiter coordinates. The generally accepted explanation is that interactions with the solar wind drive plasma convection down the magnetotail that somehow imposes a large scale dawn-to-dusk electric field in the inner magnetosphere and perturbs the drift paths of torus plasma. However, this proposed explanation is incomplete because it does not reconcile the perturbing electric field with currents produced upon disturbing the plasma distribution. A similar phenomenon at earth shields the inner magnetosphere from imposed solar wind convection,

and treating this interaction self-consistently is a crucial feature for modeling solar wind-magnetosphere interactions. At Jupiter, the initial hypothesis still awaits a comprehensive theory, even after fifteen years.

The first indication of a dawn-dusk asymmetry came from inbound Voyager UV observations that identified a pronounced brightness enhancement on the west or dusk ansa of the torus [*Broadfoot et al.*, 1979; *Sandel and Broadfoot*, 1982]. Several researchers [*Barbosa and Kivelson*, 1983; *Ip and Goertz*, 1983] proposed that brightening arises when flux tubes are brought closer to Jupiter on its dusk side. The stronger magnetic field there compresses them, thereby increasing electron temperature and the consequent emissions. Barbosa and Kivelson related the offset to the ratio of the dawn-dusk electric field $E_o$ to the corotation field $E_\Omega$ at Io's orbit. The drift path of a particle is offset from the center of Jupiter by the product of $E_o/E_\Omega$ times $r$, the distance to the rotation axis, so the total offset between dawn and dusk is $\Delta r = 2(E_o/E_\Omega)r$.

Apart from the indirect evidence provided by brightness variances, persistent morphological features in the torus are consistently observed closer to Jupiter on the dusk side than on the dawn. Indeed, the torus offset has been directly identified in all observations technically able to distinguish it. *Dessler and Sandel* [1992] fit Voyager 2 Ultra-Violet Spectrometer measurements to model emission profiles and found an average total offset of $\Delta r = 0.38 R_J$ between the two ansa, where $R_J = 7.14 \times 10^7$m is Jupiter's radius. They also reported oscillations in the offset positions that are organized in System III, some of which are probably due to deviations from axisymmetry in Jupiter's magnetic field. *Schneider and Trauger* [1995] found a value of $\Delta r = 0.28 \pm 0.05 R_J$ in their ground-based observations of optical torus emissions from $S^+$. Similar System III modulations were apparent as well, and their paper lists several other observational studies with comparable offsets.

These sets of observations both reveal a prominent torus feature called the "ribbon." The ribbon appears to be a thin and remarkably bright structure extended along the local magnetic field about the centrifugal equator just inside Io's orbit. Although it is difficult to determine its structure in front of Jupiter, on each ansa of the torus projection along the line of sight shows the ribbon as a tall, narrow structure. *Dessler and Sandel* [1993] calculated that this narrow structure is actually the source region for most of the torus emissions, roughly $2 \times 10^{12}$ Watts, mostly in the extreme ultraviolet. *Schneider and Trauger* [1995] studied the shape and motion of the ribbon in detail. Based on their detailed observations, they developed a simplified description of the ribbon as a circle about Jupiter, offset slightly toward the dawn with small variations in System III. However, they stressed that the real problem lies with the origin of the ribbon, a point re-emphasized by *Schneider et al.* [1997] who stated, "There is little agreement on origin of the ribbon."

The original candidate for powering torus emissions was mass injection via ionization of Iogenic neutrals [*Broadfoot et al.*, 1979]. Io injects roughly a ton of mass per second, and the amount of energy required to bring that material from relatively slow Keplerian speeds to corotation is approximately $10^3$kg/s $\times$ $(5.7 \cdot 10^4 \text{m/s})^2/2 \approx 1.6 \cdot 10^{12}$ Watts. Mass loading produces a radial pick-up current when newly ionized ions adjust to the ambient electric and magnetic fields [*Pontius and Hill*, 1982]. The divergence of this pick-up current must be balanced by Birkeland currents that travel into Jupiter's ionosphere, thus providing the causal connection by which energy is transferred. The ionization process also injects thermal energy because a newly produced ion receives a gyration speed equal to its guiding-center speed.

However, there are problems with the hypothesis that mass loading directly powers the torus that have become increasingly apparent. Evidence now suggests that plasma is predominantly added in Io's immediate vicinity [see *Hill and Pontius*, 1998, and references therein], so most of the pick-up currents should occur there as well. For those currents to power torus emissions, which are widely distributed in longitude, the injected energy would have to persist for many hours while flux tubes convect far from Io. However, a correlation analysis of observed UV brightness within the torus implies a cooling time of approximately 2 hours [*Volwork et al.*, 1997], much less than required. Moreover, very close to Io the plasma flow is strongly stagnated, and particles ionized there receive only a small fraction of the thermal energy acquired in undisturbed flow. Therefore, much less energy is available to power emissions.

The ultimate source of energy is undoubtedly Jupiter's rotational energy, but the particular means by which that energy is extracted and made available to the torus remains uncertain. If mass-loading and pick-up currents are not viable for powering the torus, we should consider other currents produced within the torus plasma and examine where and how they necessarily diverge. For example, *Hill* [1979] identified a system of global currents associated with the net outflow of plasma to the outer magnetosphere. *Dessler* [1980] calculated the amount of energy involved at about $2 \times 10^{12}$ Watts. However, the accompanying Birkeland currents are distributed throughout the plasma sheet, not where the torus emissions are located.

Another mechanism for generating Birkeland currents that has not been previously proposed is closely tied to the dawn-dusk offset discussed above. To maintain the inertially fixed offset, plasma must move alternately toward and away from Jupiter as it moves around the planet. Plasma shifts inward on the day side, so simple conservation of angular momentum implies an azimuthal velocity above the local corotation speed. Rotational energy must therefore be removed to maintain corotation. On the antisunward side, plasma moves back outward and rotational energy must be added to overcome the lag implied by angular momentum conservation. Therefore the dawn-dusk offset implies that energy is continually being transferred from plasma moving inward on the day side to plasma moving outward on the night side. The electric current system responsible for transferring this energy will remain fixed in the inertial frame and is an excellent candidate for powering torus emissions.

The average amount of energy involved in shifting the torus back and forth can be calculated straightforwardly, as follows: During one Jovian rotation period, $T_J = 10$ hours, each parcel of plasma in the torus completes a full cycle and moves between its innermost and outermost positions. At any given moment, half the torus is gaining energy, while the other half is losing it, and the rate varies sinusoidally with longitude. The power averaged across, say, the entire dayside is simply the total amount of energy required to shift half the torus inward divided by half the Jovian rotation period. Denoting the entire torus mass as $M = 10^9$ kg and the net inward displacement by $\Delta r = 0.4 R_J$, implies an average power expenditure of

$$\frac{\Delta(\frac{1}{2}(M/2)\Omega_J^2 r^2)}{T_J/2} = \frac{M}{2T_J}(\Omega_J r)^2 \frac{2\Delta r}{r}$$
$$= 10^{13} \text{ Watts}, \quad (1)$$

where $\Omega_J = 1.74 \times 10^{-4} \text{s}^{-1}$ is Jupiter's angular rotation frequency, and the major radius of the torus has been approximated by Io's orbital radius, $r = 6R_J$. This value is substantially larger than the total EUV emissions of $2 \times 10^{12}$ Watts. In steady state, the energy lost on the dayside is exactly matched by energy gained on the nightside with no need for additional energy input from Jupiter. However, the magnetic stresses that continually adjust rotating torus plasma to maintain the dawn-dusk offset are unlikely to be perfectly efficient. The amount of energy involved is so large that a 20% loss into heating torus plasma could power the entire torus emissions.

The rest of this paper focuses on current systems associated with the dawn-dusk offset. I will not address the particulars of how emissions can be excited by Birkeland currents but instead concentrate on determining the spatial distribution of such currents. Consistent with the usual picture of the dawn-dusk offset, I will assume that the perturbation flow field is much smaller than corotation, is directed tailward, and has only negligible variation along plasma streamlines. Surprisingly, under these conditions a simple mathematical calculation shows that the total perpendicular drift current produces no sharply localized Birkeland currents unless differential rotation is included.

BIRKELAND CURRENTS WITHIN AN IDEALIZED MODEL

Consider currents within a rotating magnetosphere that maintains a time-stationary plasma distribution in the nonrotating, inertial frame. The frozen-in flux condition, $\mathbf{E} + \mathbf{V} \times \mathbf{B} = 0$, implies that the amount of plasma mass per unit magnetic flux, or flux tube content $\eta$, is conserved along streamlines of $\mathbf{V}$. Thus for an arbitrary, stationary flow pattern one can simply assign an arbitrary value of $\eta$ to each streamline. However, this kinematic relation does not guarantee that the situation is dynamically stable. To maintain a steady state, the plasma distribution must be consistent with the adopted flow pattern. Typically, currents will diverge from the plasma sheet and flow along field lines into the ionosphere. To close the current via Pedersen currents, the ionosphere responds by producing an electric field that is generally different from the one originally applied. When mapped back to the equator the modified convection field perturbs the plasma distribution and hence the Birkeland currents, and in this way the system evolves. A stable configuration requires that the Birkeland currents produced by plasma in the assumed flow match those produced when the associated convection electric field is mapped into the ionosphere. A trivial example is a perfectly axisymmetric plasma corotating with the planet's ionosphere, which can persist indefinitely because the perpendicular currents do not diverge.

To address this matter, start with a simplified mathematical model of the inner Jovian magnetosphere where the magnetic field is dipolar and aligned with Jupiter's rotation axis. Because the thermal energy of Iogenic plasma is small relative to rotational energy, the plasma is confined near the equatorial plane in a thin sheet [*Vasyliūnas*, 1983]. This thin-disk approximation reduces the analysis to two dimensions, wherein variables are averaged along field lines through the plasma sheet. As long as there is sufficient additional plasma between the plasma sheet and ionosphere, we may assume that magnetic field lines are electric equipotentials and use that to map electric fields between distant regions.

When needed, we will use non-rotating cylindrical polar coordinates $(r, \phi, \theta)$ aligned with Jupiter's spin axis. The momentum equation in the inertial frame is

$$\rho \frac{d\mathbf{V}}{dt} = \mathbf{j} \times \mathbf{B} - \nabla P \quad (2)$$

where $\mathbf{j}$ is the electric current, $\mathbf{B}$ is the magnetic field, $\rho$ is mass density, and $P$ is the plasma pressure. The velocity in the inertial frame $\mathbf{V}$ is related to that in the rotating frame $\mathbf{v}$ by $\mathbf{V} = \mathbf{v} + \mathbf{\Omega} \times \mathbf{r}$, and substitution implies

$$\rho \frac{d\mathbf{v}}{dt} = \mathbf{j} \times \mathbf{B} - \nabla P + \rho \Omega^2 \mathbf{r} + 2\rho \mathbf{v} \times \mathbf{\Omega}. \quad (3)$$

The total current perpendicular to the magnetic field is

$$\mathbf{j}_\perp = \left[-\nabla P + \rho\Omega^2 \mathbf{r} + 2\rho \mathbf{v}\times\mathbf{\Omega} - \rho\frac{d\mathbf{v}}{dt}\right]\times\frac{\mathbf{B}}{B^2} \quad (4)$$

The first term on the right is the pressure gradient current,

$$\mathbf{j}_{\nabla P} = \hat{\mathbf{b}}\times\frac{\nabla P}{B}, \quad (5)$$

while the remainder gives the total inertial current

$$\begin{aligned}\mathbf{j}_{\text{inertial}} &= \frac{\rho}{B}\hat{\mathbf{b}}\times\frac{d\mathbf{V}}{dt} \\ &= \frac{\rho}{B}\left[\Omega^2 r\hat{\phi} + 2\Omega\mathbf{v} + \hat{\mathbf{b}}\times\frac{d\mathbf{v}}{dt}\right],\end{aligned} \quad (6)$$

where $\hat{\mathbf{b}} \equiv \mathbf{B}/B$ and $\hat{\phi}$ is the unit vector along the $\phi$ coordinate. Iogenic plasma is relatively cool, so $\nabla P$ exerts a negligible perpendicular force and will be neglected in the remainder. Because plasma resides in a relatively thin equatorial sheet, we can assume that $r$ is approximately constant along a field lines when integrating quantities along $\mathbf{B}$ through the plasma sheet. Define the height-integrated sheet current $\mathbf{J}_\perp$ and the plasma content $\eta$ by

$$\mathbf{J}_\perp \equiv \int \mathbf{j}_\perp ds \qquad \eta \equiv \int\frac{\rho}{B}ds. \quad (7)$$

Therefore, the inertial drift current in the thin disk approximation is given by

$$\mathbf{J}_{\text{inertial}} = \eta\left[\Omega^2 r\hat{\phi} + 2\Omega\mathbf{v} + \hat{\mathbf{b}}\times\frac{d\mathbf{v}}{dt}\right]. \quad (8)$$

*Hill* [1984] identified the three terms on the right as the centrifugal, Coriolis, and acceleration currents.

If coupling to the ionosphere is fairly effective, so that any departure from corotation is met with a substantial ionospheric response, then the magnitude of such departures will generally be kept small. This suggests a natural expansion parameter

$$\varepsilon \equiv \frac{v}{\Omega r} \ll 1. \quad (9)$$

Inspecting (8) shows that centrifugal current is of order unity, while the Coriolis current is of order $\varepsilon$. For a plasma distribution that remains constant in the rotating frame, the acceleration current is proportional to $\mathbf{v}\cdot\nabla\mathbf{v}$, so it is of order $\varepsilon^2$. Although the distribution under present consideration remains steady in the inertial frame, and a corotating observer would measure temporal variations, let us pursue the expansion as indicated. In this situation, the centrifugal current dominates the Coriolis current, which in turn dominates the acceleration current.

Having identified the dominant current, plasma sheet dynamics can be studied just as they are in the terrestrial magnetosphere: From an initial plasma distribution, one calculates the centrifugal current density, whose divergence must be balanced by Birkeland currents. The perpendicular electric field is determined by requiring ionospheric currents to close the circuit. Finally, the frozen-in flux theorem determines the flow field, which moves the plasma distribution. This sequence is identical to the logic outlined by *Vasyliūnas* [1970], which provides the foundation for the Rice convection model (RCM) and has successfully explained much about plasma dynamics in the Earth's inner magnetosphere. The only difference is that the usual gradient-curvature drift currents have been replaced by the centrifugal drift current. *Chen* [1977] derived the appropriate governing equations originally, while *Hill et al.* [1981] and *Siscoe and Summers* [1981] discussed additional important consequences. The methodology has been used by a number of authors to study particular problems in the Jovian magnetosphere. For instance, *Pontius* [1987] solved for the motion and evolution of an isolated flux tube having a deficit of plasma content (cf. *Pontius and Hill* [1989]). Such structures have recently been identified in Galileo particle [*Thorne et al.*, 1997] and magnetic field [*Kivelson et al.*, 1997] measurements. *Huang and Hill* [1991] treated the problem of centrifugal interchange instability analytically and found very rapid growth. *Yang et al.* [1992, 1994] performed direct simulations of convection in Jupiter's plasma sheet using an appropriately modified version of the Rice convection model, the RCM-J.

The dawn-dusk offset problem is in some ways simpler than these problems because the configuration is approximately time stationary. An important caveat on the present results is that the configuration is modeled as being exactly stationary in the inertial frame. Although the dawn-dusk offset is a persistent feature of the torus, there are additional temporal variations in the position of features within the torus [*Schneider and Trauger*, 1995]. Such additional complications are beyond the scope of the present work.

*Barbosa and Kivelson* [1983] showed that the particle drift paths producing the dawn-dusk offset are consistent with a large-scale dawn-dusk electric field superimposed upon the rotation electric field imposed by

Jupiter itself. They proposed plasma flow down the distant magnetotail as the source of this electric field, but they did not include the effects of any perturbation electric field arising from the displacement. The dominant term in equation (8), the centrifugal current, is strictly azimuthal and will diverge when the plasma content is not symmetric in longitude, which is exactly what the offset produces.

Let us first examine the pattern of Birkeland currents produced at the inner edge of the torus. The divergence of the centrifugal current is

$$\nabla \cdot \mathbf{J}_{\text{cent}} = \Omega^2 \frac{\partial \eta}{\partial \phi} = -j_\parallel, \qquad (10)$$

i.e., it diverges from the plasma sheet only where plasma content varies with azimuth. A dawn-dusk offset of an originally axisymmetric distribution implies longitudinal variations in $\eta$ that should drive Birkeland currents. Because the gradient in $\eta$ is very steep at the inner edge of the torus, centrifugal currents should diverge within a narrow band and supply thin, intense sheets of field-aligned current. Furthermore, because the flow pattern is fixed in the inertial frame, the resulting Birkeland currents will be stationary in the same frame, as is required to power the observed, asymmetric emissions.

Given the observed dawn-dusk offset, the divergence of the centrifugal drift current by itself would produce sharply localize field-aligned currents directed away from Jupiter on the sunward side and toward on the night, as shown in Figure 1. At dawn and dusk, the divergence vanishes and hence the field-aligned current density is zero there as well. However, because of viewing geometry, the observed emissions are most apparent on the dusk and dawn ansae. For Birkeland currents to produce the emissions observed there requires a mechanism that remains intense well after the generating currents have diminished from their maximum values along the Jupiter-Sun meridian.

Before searching for such a mechanism, let us investigate the magnitude of some currents omitted in the above analysis, such as the Coriolis currents. In the absence of differential rotation

$$\begin{aligned}\nabla \cdot \mathbf{J}_{\text{Cor}} &= 2\Omega \nabla \cdot \eta \mathbf{v} \\ &= 2\Omega \left[\mathbf{v} \cdot \nabla \eta + \eta \nabla \cdot \mathbf{v}\right]\end{aligned} \qquad (11)$$

Because the $\eta$ distribution is steady in the inertial frame,

$$\mathbf{V} \cdot \nabla \eta = \Omega \frac{\partial \eta}{\partial \phi} + \mathbf{v} \cdot \nabla \eta = 0, \qquad (12)$$

while for a static dipole magnetic field and frozen-in flow

$$\nabla \cdot \mathbf{v} = \mathbf{v} \cdot (\kappa \hat{\mathbf{r}} - \nabla \ln B) \cong 6 \frac{v_r}{r}, \qquad (13)$$

where $\kappa$ is field line curvature at the equator. The second term in (11) is therefore of order $\varepsilon$ relative to the first and can be neglected. Substitution thus implies

$$\begin{aligned}\nabla \cdot \mathbf{J}_{\text{Cor}} &\cong -2\Omega^2 \frac{\partial \eta}{\partial \phi} \\ &= -2\nabla \cdot \mathbf{J}_{\text{cent}}\end{aligned} \qquad (14)$$

Surprisingly, the Coriolis current contribution has exactly the same spatial distribution as the centrifugal current but with twice the magnitude and the opposite sign. Although the Coriolis current is of order $\epsilon$ relative to the centrifugal current, the divergences are of the same order because the gradient in $\eta$ is mainly radial. Summing these two contributions gives a net Birkeland current exactly opposite to what was found originally. That is, if the centrifugal and Coriolis force were the only important currents, there would be intense sheets of Birkeland currents toward Jupiter on the day side and away on the night.

One might continue this procedure and try to refine the analysis by including the acceleration current. However, the unexpected importance of the Coriolis term makes expansion in $\varepsilon$ suspicious. At the very least, it suggests the possibility of a more direct approach. Let us return to the complete expression for the inertial current in the nonrotating frame:

$$\begin{aligned}\mathbf{J}_{\text{inertial}} &= \eta \hat{\mathbf{b}} \times \left[\frac{\partial \mathbf{V}}{\partial t} + \mathbf{V} \cdot \nabla \mathbf{V}\right] \\ &= \eta \hat{\mathbf{b}} \times \mathbf{V} \cdot \nabla \mathbf{V}\end{aligned} \qquad (15)$$

Expand $\mathbf{V}$ in terms of $\mathbf{v}$ and $\Omega$

$$\begin{aligned}\mathbf{V} \cdot \nabla \mathbf{V} &= \left(\Omega r \hat{\phi} + \mathbf{v}\right) \cdot \nabla \left(\Omega r \hat{\phi} + \mathbf{v}\right) \\ &= \Omega r \hat{\phi} \cdot \nabla \left(\Omega r \hat{\phi}\right) + \mathbf{v} \cdot \nabla \left(\Omega r \hat{\phi}\right) \\ &\quad + \left(\Omega r \hat{\phi} + \mathbf{v}\right) \cdot \nabla \mathbf{v}\end{aligned} \qquad (16)$$

The first term equals $-\Omega^2 \mathbf{r}$, which leads to the centrifugal drift current. The second term is half the ordinary Coriolis term, with an additional contribution that depends on possible differential rotation. The rest of the Coriolis term usually arises from $\Omega r \hat{\phi} \cdot \nabla \mathbf{v}$, which has been grouped into the third term.

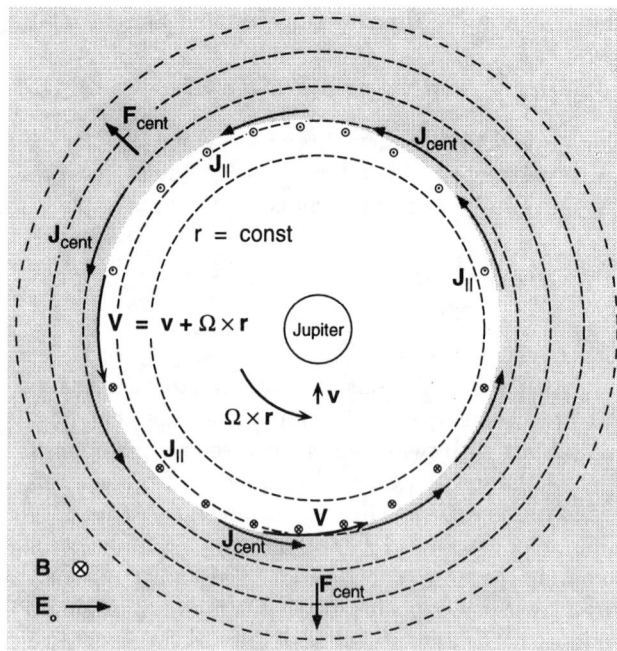

**Figure 1.** Jupiter's equatorial plane. Dashed circles are centered on Jupiter, the sunward direction is toward the bottom, and plasma content $\eta$ is constant throughout the shaded regions. Centrifugal drift currents are indicated, as are the Birkeland currents $J_\parallel$ their divergence feeds (out of the figure plane). None of the other currents included in later calculations are shown here, but the direction of the Coriolis current is along the perturbation flow vector $\mathbf{v}$.

The final term is simply the convective derivative of $\mathbf{v}$ along streamlines of the total velocity $\mathbf{V}$. In the explanation proposed by *Barbosa and Kivelson* [1983] and *Ip and Goertz* [1983], the perturbation velocity is driven by a large-scale uniform electric field $\mathbf{E}_o$. That flow has constant direction tailward and only weakly varying magnitude. As the displacement is very small, the magnetic field along the drift path is nearly constant, so variations in $\mathbf{E}_o \times \mathbf{B}$ along streamlines of $\mathbf{V}$ may be neglected. Unless the total current divergence indicates an additional, substantial perturbation electric field, this means that the vector $\mathbf{v}$ varies very weakly in the inertial frame and that the final term in (17) can be neglected. Again, azimuthal variations that are stationary in an inertial frame produce temporal variations in the rotating frame. Viewing the present situation from a rotating perspective, half the Coriolis current would be cancelled by an acceleration current that includes contributions proportional to $\partial \eta / \partial t$.

Reassembling the full convective derivative and substituting for the inertial current gives the total drift current,

$$\begin{aligned} \mathbf{J}_{\text{inertial}} &= \eta \, \hat{\mathbf{b}} \times \left[ -\Omega^2 \mathbf{r} - \mathbf{v} \times \mathbf{\Omega} + r \hat{\phi} \mathbf{v} \cdot \nabla \Omega \right] \\ &= \eta \Omega^2 r \hat{\phi} + \eta \Omega \mathbf{v} + \eta (\mathbf{v} \cdot \nabla \Omega) \, \mathbf{r} \quad (17) \\ &= \eta \Omega \mathbf{V} + \eta (\mathbf{v} \cdot \nabla \Omega) \, \mathbf{r}. \end{aligned}$$

In the absence of differential rotation, $\nabla \Omega = 0$, and the second term vanishes. Therefore, in a steady state with a uniform rotational frequency $\Omega$, imposing a uniform large-scale electric field in the plasma sheet produces a total inertial current of $\mathbf{J}_{\text{inertial}} = \eta \Omega \mathbf{V}$.

The current divergence is

$$\nabla \cdot \mathbf{J}_{\text{inertial}} = \Omega \mathbf{V} \cdot \nabla \eta + \eta \Omega \nabla \cdot \mathbf{V}. \quad (18)$$

Once again, in the situation under consideration the $\eta$ distribution is stationary, so its convective derivative along streamlines of $\mathbf{V}$ is zero. This leaves only the second term, and $\nabla \cdot \mathbf{V} = \nabla \cdot \mathbf{v}$ implies

$$\begin{aligned} \nabla \cdot \mathbf{J}_{\text{inertial}} &= \eta \Omega \nabla \cdot \mathbf{v} \\ &\cong 6 \eta \Omega \frac{v_r}{r}. \quad (19) \end{aligned}$$

We have arrived at a curious result somewhat at odds with expectations: Upon examining the full inertial current density under the present circumstances, we find that its divergence is of order $\varepsilon$ relative to what the dominant centrifugal current would produce by itself. Moreover, imposing a uniform electric field displaces plasma contours without generating intense Birkeland currents at the inner edge of the plasma sheet. There are Birkeland currents with density proportional to $\nabla \cdot \mathbf{v}$, but these are smoothly distributed within the plasma sheet and not concentrated near the inner edge of the torus. That is, they are proportional to $\eta$ rather than to its gradient as originally expected. The perturbation potential closing such currents through the ionosphere will not exhibit abrupt gradients (which supports our earlier neglect of $\mathbf{V} \cdot \nabla \mathbf{v}$ a posteriori). Such currents are unlikely to be responsible for powering the intense, highly localized emissions from the ribbon.

Before moving on to consider another factor that can produce narrow current sheets, please note that the previous arguments hold only for configurations that are approximately stationary in the inertial frame. It does not apply to structures that evolve slowly in a frame that corotates with Jupiter, such as those related to the interchange instability.

## BIRKELAND CURRENTS FROM DIFFERENTIAL ROTATION

At the first Chapman Conference on magnetospheric currents, *Vasyliūnas* [1984] made the following observation:

> ...one should not forget that [associating the current from the momentum equation with the first-order guiding center description] is no more than a calculational device, and its various terms, even if sometimes dignified by the names "magnetization current," "gradient drift current," etc., have no independent physical existence: no one of them, taken by itself, represents a current that could be measured as such either by a magnetometer or a particle detector.

In other words, what man has put asunder, he should not forget to bring back together. The present situation has considered an analogous situation, where a current description is often divided into distinct parts based on our familiarity with rotating mechanical systems. The analysis in the last section shows how treating one term in isolation can lead to a quite incorrect conclusion.

However, so long as we heed Vasyliūnas' dictum, it can still be instructive to consider how such a decomposition guides our intuition about underlying physical mechanisms. Considering any term in (8) by itself effectively treats the other terms as maintaining balance among themselves. For example, the centrifugal current is simply that required to resist the centrifugal force. However, plasma that falls outward does not completely resist the centrifugal force, but the implied reduction in the total azimuthal current would typically appear in the acceleration current. The Coriolis current is that required to resist the Coriolis force, which tends to make radially moving plasma depart from corotation [*Pontius*, 1997]. Again, that term is smaller if plasma does depart from corotation. Considering the Coriolis current in isolation tacitly assumes that the system enforces corotation to some good approximation.

But corotation with what? If the ionosphere rotates with a single angular frequency, the answer is obvious. However, there is indirect but compelling evidence that mass loading induces differential rotation in Jupiter's ionosphere. Mass loading takes place over a limited range of magnetic latitude, so the torque responsible for accelerating new torus plasma is exerted within narrow circumpolar bands in each hemisphere. The atmosphere must exert a viscous torque to transfer angular momentum from lower altitudes, which requires a substantial vertical shear in horizontal velocity [*Huang and Hill*, 1989]. Therefore, the high-altitude neutral atmosphere within this narrow band of magnetic latitude has a substantial westward velocity (as viewed in rotating System III coordinates), called the Jovian flywheel effect [*Pontius*, 1995]. Its huge moment of inertia implies a sluggish reaction when the electromagnetic torque changes as mass loading in the torus varies. Indeed, an unpublished analysis of rotational velocities in the torus imply a response time of many months [Mike Brown, personal communication]. Therefore, even though magnetospheric plasma is tightly coupled to the Jovian ionosphere, it does not rotate at a single, uniform angular frequency, and we must treat $\Omega$ as a function of magnetic latitude and not simply the canonical System III rotation rate $\Omega_J$.

Although it is not presently feasible to measure the lagging winds directly, the conjecture has additional observable consequences for plasma dynamics within the magnetosphere. Indeed, *Pontius et al.* [1998] presented analytic calculations and RCM-J simulations showing how this phenomenon may explain the persistence of the torus itself against the centrifugal interchange instability. This puzzle has persisted since the discovery of the torus, and earlier proposed resolutions have not been supported by measurements. Differential rotation disrupts the growth of unstable perturbations by shearing them sideways. Therefore, because the flywheel effect imposes a strong shear in angular velocity at the outer edge of the torus, the centrifugal interchange instability is strongly hampered in its ability to transport plasma outward.

With the flywheel effect in mind, we should consider one additional source of Birkeland currents in (17), that is, the differential rotation current:

$$\mathbf{J}_{\nabla\Omega} = \eta\,(\mathbf{v}\cdot\nabla\Omega)\,\mathbf{r} \qquad (20)$$

The role of the ionosphere in enforcing corotation is central to understanding the origin of the differential rotation current. The Coriolis current maintains corotation, but because it depends on $\Omega_J$ and not $\nabla\Omega_J$, its radial component is predicated on maintaining a constant angular velocity for radially moving plasma. The differential rotation current may be viewed as adjusting the radial current when the ionosphere actually tries to enforce an inhomogeneous angular velocity. For example, when plasma moves outward, its angular velocity tends to decrease to conserve angular momentum ($\Omega_{\text{plasma}} \propto 1/r^2$). If there is a region where the ionospheric rotation frequency happens to decline at the same rate ($\Omega \propto 1/r^2$) because of the flywheel effect, the ionosphere would not need to exert any torque on outward moving plasma. The Coriolis current is insensitive to this effect and would be nonzero, but the differential rotation current would contribute an exactly opposite current to eliminate the net radial current.

Although the above analysis is instructive, it is incomplete in neglecting the acceleration current. However, using the entire inertial current avoids that problem. We could calculate the divergence of (17) directly, but the derivative of the differential rotation current is cumbersome, and there is an easier way to estimate whether the gradient in $\Omega$ causes a significant effect. The magnitude of the differential rotation current can be related to the radial component of the first term in (17) (denoted $\tilde{J}_{\text{inertial}}$) by

$$\begin{aligned} J_{\nabla\Omega} &= \eta v_r \frac{d\Omega}{dr} r \\ &\cong \left[ \frac{\Delta\Omega}{\Omega} \frac{r}{\lambda_\Omega} \right] \tilde{J}_{\text{inertial},r} \end{aligned} \quad (21)$$

where the angular frequency changes by $\Delta\Omega$ over a distance $\lambda_\Omega$. *Brown* [1994] provided a radial profile of angular velocity observed within the torus. The gradient is particularly steep at the inner edge near $r = 6R_J$, where the maximum change in $\Omega$ is $\Delta\Omega = 0.02\,\Omega_J$ over a distance of about $\lambda_\Omega = 0.25\,R_J \cong 0.04\,r$. Therefore, within this region the differential rotation current is within a factor of two of canceling the radial component of the remaining inertial current. The true gradient is probably even steeper than the published results because of artifacts from light-of-sight effects [Brown, personal communication, 1996].

Within such a region, the radial current would be substantially reduced, or even eliminated if the velocity shear is large enough. When differential rotation was neglected, we found that the divergences of the radial and azimuthal currents effectively cancelled each other. Therefore, where the radial current alone is reduced, intense, nonvanishing Birkeland currents will be produced by divergences in the azimuthal current, most of which corresponds to the familiar centrifugal current. Observationally, this region corresponds quite well to the radial range of the ribbon, i.e., just inside Io's orbit. Further considerations of the physical connection between concentrated Birkeland currents and the narrow, intense ribbon emissions are consigned to future work.

*Acknowledgments.* I thank T. W. Hill, M. G. Kivelson, and R. A. Wolf for enlightening discussions and the editors for their patience. This research was supported by NASA SR&T grant NAG5-6553, administered by the Bartol Research Institute.

## REFERENCES

Barbosa, D. D., and M. G. Kivelson, Dawn-dusk electric field asymmetry of the Io plasma torus, *Geophys. Res. Lett.*, *10*, 210, 1983.

Broadfoot, A. L., et al., Extreme ultraviolet observations from Voyager 1 encounter with Jupiter, *Science*, *204*, 979, 1979.

Brown, M. E., Observation of mass loading in the Io plasma torus, *Geophys. Res. Lett.*, *21*, 847-850, 1994.

Chen, C.-K., Topics in planetary plasmaspheres, Ph.D. thesis Thesis, University of California Los Angeles, 1977.

Dessler, A. J., Mass-injection rate from Io into the Io plasma torus, *Icarus*, *44*, 291, 1980.

Dessler, A. J., and B. R. Sandel, System III variations in apparent distance of Io plasma torus from Jupiter, *Geophys. Res. Lett.*, *19*, 2099–2103, 1992.

Dessler, A. J., and B. R. Sandel, Reply to comment by D. D. Barbosa, *Geophys. Res. Lett.*, *20*, 2489–2490, 1993.

Hill, T. W., Inertial limit on corotation, *J. Geophys. Res.*, *84*, 6554–6558, 1979.

Hill, T. W., Rotationally-induced Birkeland current systems, in *Magnetospheric Currents*, edited by T. A. Potemra, 340–349, American Geophysical Union, 1984.

Hill, T. W., A. J. Dessler, and L. J. Maher, Corotating magnetospheric convection, *J. Geophys. Res.*, *86*, 9020–9028, 1981.

Hill, T. W., and D. H. Pontius, Jr., Plasma injection near Io, *J. Geophys. Res.–Planets*, *103*, 19,879–19,885, 1998.

Huang, T. S., and T. W. Hill, Corotation lag of the Jovian atmosphere, ionosphere, and magnetosphere, *J. Geophys. Res.*, *94*, 3761–3765, 1989.

Huang, T. S., and T. W. Hill, Drift wave instability in the Io plasma torus, *J. Geophys. Res.*, *96*, 14075–14083, 1991.

Ip, W.-H., and C. K. Goertz, An interpretation of the dawn-dusk asymmetry of UV emission from the Io plasma torus, *Nature*, *302*, 232–233, 1983.

Kivelson, M. G., K. K. Khurana, C. T. Russell, and R. J. Walker, Intermittent short-duration magnetic field anomalies in the Io torus: Evidence for plasma interchange?, *Geophys. Res. Lett.*, *24*, 2127–2130, 1997.

Pontius, D. H., Jr., A Model for Plasma Transport in a Corotation-Dominated Magnetosphere, Ph.D. Thesis, Rice University, 1987.

Pontius, D. H., Jr., Implications of variable mass loading in the Io torus: The Jovian flywheel, *J. Geophys. Res.*, *100*, 19531-19539, 1995.

Pontius, D. H., Jr., The influence of Coriolis forces on the centrifugal interchange instability, *Geophys. Res. Lett.*, *24*, 2961-2964, 1997.

Pontius, D. H., Jr., and T. W. Hill, Departure from corotation of the Io plasma torus: Local plasma production, *Geophys. Res. Lett.*, *9*, 1321–1324, 1982.

Pontius, D. H., Jr., and T. W. Hill, Rotation driven plasma transport: The coupling of macroscopic motion and microdiffusion, *J. Geophys. Res.*, *94*, 15,041–15,053, 1989.

Pontius, D. H., Jr., R. A. Wolf, T. W. Hill, R. W. Spiro, Y. S. Yang, and W. H. Smyth, Velocity shear impoundment of the Io plasma torus, *J. Geophys. Res.-Planets*, *103*, 19,935, 1998.

Sandel, B. R., and A. L. Broadfoot, Io's hot plasma torus-A synoptic view from Voyager, *J. Geophys. Res.*, *87*, 212–218, 1982.

Schneider, N. M., M. H. Taylor, F. J. Crary, and J. T. Trauger, On the nature of the $\lambda_{III}$ brightness asymmetry in the Io torus, *J. Geophys. Res.*, *102*, 19823–19,833, 1997.

Schneider, N. M., and J. T. Trauger, The structure of the Io torus, *Astrophys. J.*, *450*, 450–462, 1995.

Siscoe, G. L., and D. Summers, Centrifugally driven diffusion of Iogenic plasma, *J. Geophys. Res.*, *86*, 8471–8479, 1981.

Thorne, R. M., D. J. Williams, R. W. McEntire, T. P. Armstrong, S. Stone, S. J. Bolton, D. A. Gurnett, and M. G. Kivelson, Galileo evidence for rapid interchange transport in the Io torus, *Geophys. Res. Lett.*, *24*, 2131–2134, 1997.

Vasyliūnas, V. M., Mathematical models of magnetospheric convection and its coupling to the ionosphere, in *Particles and Fields in the Magnetosphere*, edited by B. M. McCormac, p. 60, D. Reidel, Norwell, Mass., 1970.

Vasyliūnas, V. M., Plasma distribution and flow, in *Physics of the Jovian Magnetosphere*, edited by A. J. Dessler, 395–453, Cambridge University Press, New York, 1983.

Vasyliūnas, V. M., Fundamentals of current description, in *Magnetospheric Currents*, edited by T. A. Potemra, 63–66, 28, AGU, 1984.

Volwerk, M., M. E. Brown, A. J. Dessler, and B. R. Sandel, Evidence for short cooling time in the Io plasma torus, *Geophys. Res. Lett.*, *24*, 1147, 1997.

Yang, Y. S., R. A. Wolf, R. W. Spiro, and A. J. Dessler, Numerical simulation of plasma transport driven by the Io torus, *Geophys. Res. Lett.*, *19*, 957–960, 1992.

Yang, Y. S., R. A. Wolf, R. W. Spiro, T. W. Hill, and A. J. Dessler, Numerical simulation of torus-driven plasma transport in the Jovian magnetosphere, *J. Geophys. Res.*, *99*, 8755–8770, 1994.

---

Duane H. Pontius, Jr., Department of Physics, Box 549022, Birmingham-Southern College, Birmingham, Alabama 35254 (dpontius@bsc.edu).

# MHD Simulations of Current Systems in Planetary Magnetospheres: Mercury and Saturn

Tamas I. Gombosi*

*International Space Science Institute, Hallerstr. 6, CH-3012 Bern, Switzerland*

Darren L. DeZeeuw, Clinton P.T. Groth, Kenneth C. Hansen, Konstantin Kabin, Kenneth G. Powell

*The University of Michigan, Ann Arbor, Michigan*

The study of planetary magnetospheres can provide valuable insight into a wide range of phenomena and processes acting in the terrestrial magnetosphere. This paper uses global MHD simulations to investigate the large scale configuration and current systems in two very different planetary magnetospheres. Mercury has a weak intrinsic magnetic moment and it is exposed to a high density, very variable solar wind. It has no ionosphere or plasmasphere and its slow rotation is unimportant. Under typical conditions in the solar wind, Mercury's magnetosphere is very "open", the last closed field-line being at a latitude of 40°–50°. At the other end of the spectrum, the interplay between solar wind and planetary rotation driven processes, combined with strong plasma sources, generates new and interesting configurations and current systems in the magnetosphere of Saturn. In particular, our simulations predict two current systems connecting plasma sources in the rings/icy satellites region and in Titan's torus to the high-latitude regions of Saturn's ionosphere.

## 1. INTRODUCTION

The planets Mercury, Earth, Jupiter, Saturn, Uranus and Neptune possess significant intrinsic magnetic fields. Since the underlying plasma-physical processes are the same in these magnetospheres, qualitative or quantitative analogies with the terrestrial magnetosphere can help us to understand poorly sampled planetary magnetospheres. Conversely, similarities in planetary magnetospheres, in spite of differences in configurations, plasma and energy sources and sinks, can help us to understand better the full range of phenomena and processes in the Earth's magnetosphere. Planetary magnetospheres help us to extend observed magnetospheric phenomena into different parameter regimes. The study of magnetospheric current systems is a very powerful tool in gaining a global perspective of the multiscale coupling between various regions in planetary magnetospheres.

In this paper we investigate the large-scale magnetospheric configuration and associated current systems in two vastly different planetary magnetospheres. Among magnetized planets Mercury has the weakest intrinsic magnetic field (its magnetic moment is about 3,000 times smaller than the terrestrial magnetic moment), while Saturn's magnetic moment is the second largest among all solar system planets (580 times larger than the magnetic moment of Earth). Magnetospheric current systems in these two planetary magnetospheres clearly exhibit many of the most interesting features relevant to Earth and many astrophysical magnetospheres.

This paper uses global MHD simulations to study the large-scale configuration of the Hermean and Kronian magnetospheres. The high performance BATS-R-US simulation code has been developed at the University of Michigan and it solves the equations of ideal magnetohydrodynamics on an adaptively refined grid [*Powell et al.*, 1999]. The code has been successfully used to simulate solar system plasmas ranging from the 3D expansion of solar wind [*Groth et al.*, 1999] to the interaction of the heliosphere with the magnetized interstellar medium [*Linde et al.*, 1999], to the magnetospheres of Earth [*Gombosi et al.*, 1998; 1999], Venus [*Bauske et al.*, 1998], Saturn [*Hansen et al.*, 1998], comets [*Gombosi et al.*, 1996; *Häberli et al.*, 1997], and to the magnetospheric interaction of planetary satellites, such as Io [*Combi et al.*, 1998], Europa [*Kabin et al.*, 1998a] and Titan [*Kabin et al.*, 1998b].

## 2. MODEL

The BATS-R-US (Block Adaptive-Tree Solar-wind Roe-type Upwind Scheme) code solves the governing equations of ideal magnetohydrodynamics. The high-resolution finite volume solution scheme is based on an approximate Riemann solver for magnetohydrodynamics [*Powell*, 1994; *Powell et al.*, 1995]. In this approach, the hydrodynamic and electromagnetic effects are solved for in a fully three-dimensional tightly coupled manner, rather than in separate steps [*Gombosi et al.*, 1994; *Powell et al.*, 1995; 1999].

The code uses a limited reconstruction that ensures second-order accuracy away from discontinuities, while simultaneously providing the stability that ensures nonoscillatory solutions. In addition, the code employs two accurate approximate Riemann solvers: the *Roe* [1981] scheme [*Powell*, 1994] and the *Linde* [1998] solver. The resulting scheme works equally well across a range of several orders of magnitude in plasma $\beta$ ($\beta$ is the ratio of the kinetic and magnetic pressures).

The basic data structure used in the BATS-R-US approach is that of adaptive blocks [*Stout et al.*, 1997; *Powell et al.*, 1999]. Adaptive blocks partition space into regions, each of which is a regular Cartesian grid of cells, called a block. If the region needs to be refined, then the block is replaced by 8 child subblocks (one for each octant of the parent block), each of which is a Cartesian grid of cells containing the same number of cells as the parent block. If coarsening is needed, then the 8 children are replaced by their parent. The blocks in the grid, at their various levels of refinement, are stored in a tree data structure.

BATS-R-US was specially designed to handle objects with strong intrinsic magnetic fields. It achieves improved solution accuracy by solving for $\mathbf{B}_1$, which is the measure of the deviation of the full magnetic field from the intrinsic field, $\mathbf{B}_o$, and is defined as $\mathbf{B}_1 = \mathbf{B} - \mathbf{B}_o$. This approach was first employed by *Ogino and Walker* [1984] and *Ogino* [1986] (and later applied to Godunov-type schemes by *Tanaka* [1995]) and can lead to improved numerical solutions by alleviating the necessity of resolving the often large spatial gradients associated with the intrinsic fields and by ensuring that the divergence of the intrinsic component of the magnetic field is by definition zero. Note that in this formulation $\mathbf{B}_1$ does not have to be small, therefore this decomposition is completely general.

## 3. MERCURY

Mercury's magnetosphere is surprisingly poorly studied, and consequently, poorly understood. It has been visited only by Mariner 10, which encountered Mercury three times between March 1974 and March 1975 (however, one of the three encounters was so far upstream that it completely missed the magnetosphere).

The first estimate of Mercury's magnetic moment was between 284 and 358 nT $R_M^3$ [*Ness*, 1975]. However, the poor spatial coverage during the two magnetospheric flybys of Mariner 10 make the evaluation of the intrinsic moment of Mercury very complicated and the results are highly dependent on the physical assumptions [*Ness*, 1975; *Connerney and Ness*, 1988].

A tenuous atmosphere was observed at Mercury by the Mariner 10 spacecraft [*Broadfoot et al.*, 1976]. This atmosphere is mainly composed of hydrogen, helium and oxygen, with traces of sodium and potassium. The H, He, and O components are thought to be of solar wind origin, while sodium and potassium atoms are probably produced by sputtering [*Lammer and Bauer*, 1997]. The total column density of the neutral atmosphere is estimated to be less than $10^{12}$ cm$^{-2}$, which is essentially negligible from the perspective of magnetospheric interaction. In the Mercury simulations shown in this paper all atmospheric effects are neglected.

In our simulation Mercury was approximated by a solid nonmagnetic sphere with a radius of $R_M = 2,440$ km. The intrinsic magnetic field was approximated by a dipole moment of 330 nT $R_M^3$, with the magnetic moment vector pointing south (the same sense as Earth). The surface conductivity was assumed to be zero, thus neglecting the effects of a photoelectron sheath which may form around Mercury [*Grard*, 1997]. Magnetic flux was allowed to cross the surface of the planet, but no particle flux could penetrate the surface. The boundary conditions on the surface were imposed by utilizing cut cells [*DeZeeuw and Powell*, 1992], which allows second order (piecewise linear) reconstruction of the boundary geometry.

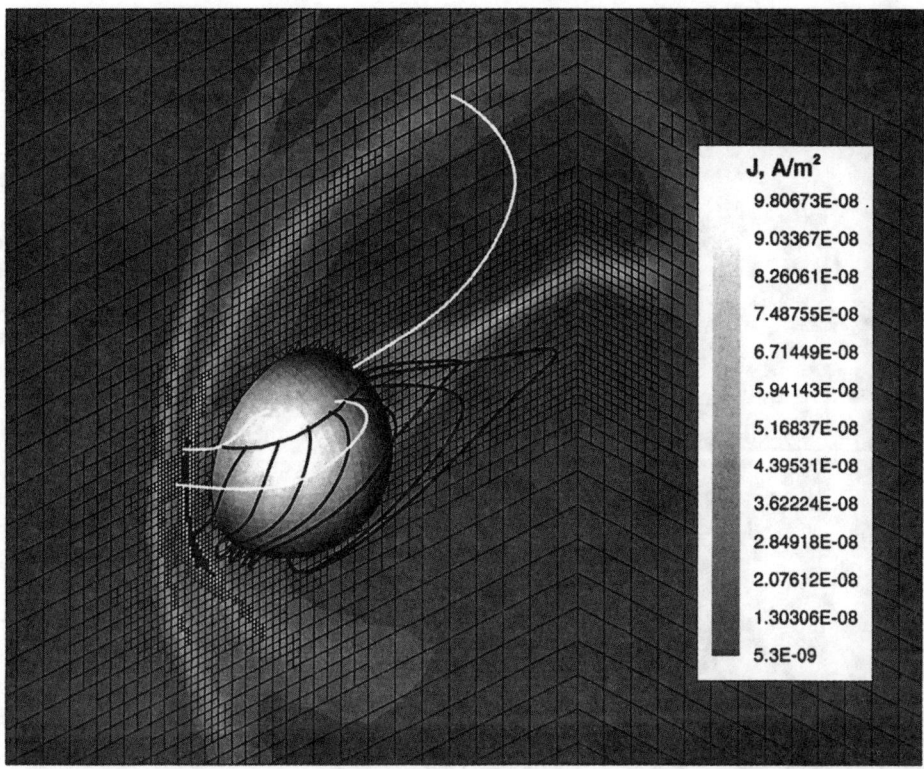

**Figure 1.** Three-dimensional rendering of Mercury's magnetosphere. The electric current density is shown in gray scale in the noon-midnight meridian and in a cross sectional plane in the tail. Closed magnetic field lines are shown by dark solid lines, while open magnetic field lines connected to the northern high latitude region are represented by light solid lines. The open-closed magnetic field boundary is marked by a circle.

The upstream plasma flow conditions around Mercury may vary significantly depending on several factors, including the position of the planet on its trajectory and solar activity. In this paper we present simulations for "typical" solar wind conditions near perihelion: $n = 73$ cm$^{-3}$, $T = T_e + T_i = 14$ eV, $B = 46$ nT, $u = 430$ km/s, ion-acoustic sound speed $a = 74.2$ km/s, Alfvén speed $V_A = 120$ km/s, mean molecular mass $\sim 1$ amu, and specific heat ratio 1.67. The corresponding ion-acoustic Mach number is 5.8 and Alfvénic Mach number is 3.6. At Mercury's orbit the Parker spiral magnetic field forms an angle of 20° with the solar wind direction. However, we must emphasize that these parameters may vary significantly. A parametric study of the effect of various solar wind conditions on the Hermean magnetosphere will be presented in a later paper.

The size of the simulation box was $900 \times 600 \times 600$ $R_M$ and it was divided into $\sim$270,000 computational cells. The smallest cell was 0.02 $R_M$ in each direction, and we used 11 levels of refinement. The inner boundary of the simulation was located at the surface of Mercury. The Mercury simulation was carried out with a version of BATS-R-US which runs on high-end workstations.

Figure 1 shows a three-dimensional rendering of the simulation results. The electric current density is shown in gray scale in the noon-midnight meridian and in a cross sectional plane located at 4 $R_M$ behind Mercury in the magnetic tail. Closed magnetic field lines are shown by dark solid lines, while open magnetic field lines originating from the high latitude region are represented by light solid lines. The open-closed magnetic field boundary is marked by a circle on the surface of the planet. The solar wind is approaching the planet from the left. The figure also shows the adaptively refined computational cells in the two planes.

Inspection of Figure 1 clearly reveals two major current systems in Mercury's magnetosphere: the magnetopause current and the cross tail current. An additional current is also present due to the magnetic field jump across the bow shock. It can be seen that the closed magnetic field lines are

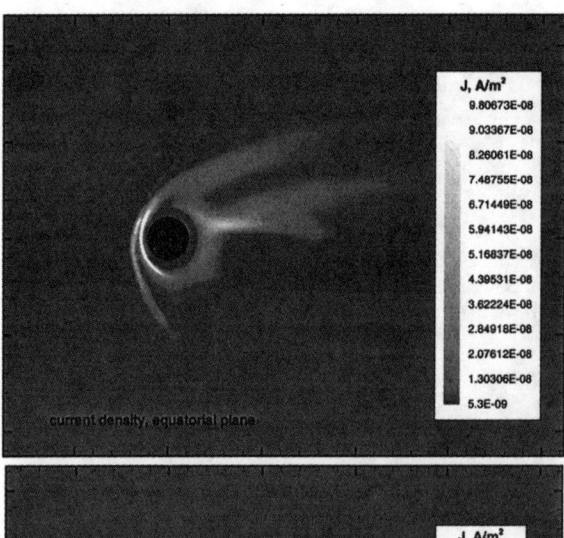

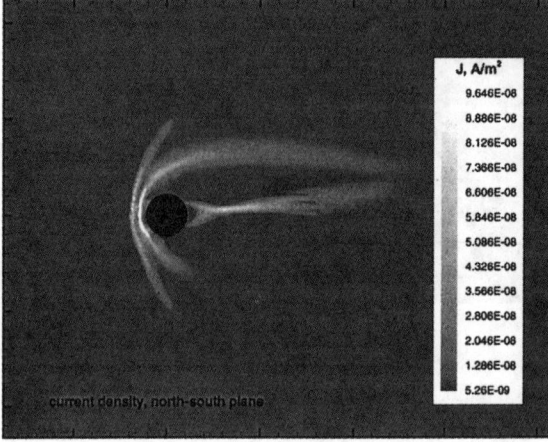

**Figure 2.** Equatorial (top panel) and noon midnight meridian (bottom panel) cuts of the Hermean currenet system. The IMF vector is in the equatorial plane.

rotated and twisted due to the presence of a $B_y$ component in the IMF. This effect has been observed at Earth [*Kaymaz and Siscoe*, 1998] and reproduced by global MHD simulations of the terrestrial magnetosphere [*Gombosi et al.*, 1999].

The interplanetary magnetic field vector is in the equatorial plane ($B_z = 0$) and it points 20 degrees from the radial direction. The $B_y$ component is from dawn to dusk. Due to the presence of a $B_y$ component the closed magnetic field lines are pushed down on the dawn side and pushed up on the dusk side. A similar twist can be also observed for the open magnetic field lines. In addition, the axis of the magnetotail is also shifted duskward. The twist and rotation of the magnetotail can be well observed in the equatorial and non-midnight cuts shown in Figure 2.

Figure 2 shows the equatorial (left panel) and noon-midnight meridian (right panel) cuts of the simulated Hermean current system. The IMF vector is in the equatorial plane.

The cross-tail and magnetopause currents clearly form a unified current system and the cross tail current closes through the magnetopause current. It is interesting to note that on the dawn side the near terminator magnetopause and the quasi-parallel segment of the planetary bow shock are closely pushed together, while they are clearly separated on the dusk side.

In the equatorial plane the current is strongest in a relatively narrow region on the dawn side. This is the region where the field lines strongly bend as they enter the magnetotail. It should be noted that due to the presence of a significant $B_y$ component the cross tail current sheet is twisted into an S shape and rotated. The strong current density on the dawn side is the region where the strong cross tail current is close to the equatorial plane.

The right panel of Figure 2 shows the noon-midnight cross section of the current system. The cross-tail current intersects this plane just below the equatorial plane. This is another manifestation of the twist and rotation of the cross-tail current. It is interesting to note that in this plane the magnetopause current is stronger on the northern side. This is also due to the distortion of the tail configuration.

Overall, our simulation indicates that Mercury has a highly compressed magnetosphere which is dominated by the global magnetopause-tail current system. Peculiar features include very large areas of the open field lines on the surface of Mercury, very small size of the magnetosphere compared to the size of planet itself and a nearly merged quasi-parallel shock and magnetopause on the dawn side.

## 4. SATURN

Saturn's magnetosphere is quite complex due to the influence and interaction of the solar wind, the planetary rotation and the plasma sources, each of which is in some way drastically different from Earth. The solar wind, for example, has quite different characteristics at Saturn than at Earth. The density is much lower, the magnetic field magnitude is smaller, and the nominal Parker spiral gives a magnetic field that is almost completely azimuthal ($\theta = 85°$). All these factors combine to determine the overall size and configuration of the magnetosphere. In addition, Saturn's rotation modifies the structure of the inner magnetosphere. Saturn's large radius and short rotation period produce enough torque on the plasma of the inner magnetosphere to cause it to corotate with the planet. Finally, the Kronian system has several neutral gas sources. These include Saturn, the rings, the icy satellites, Titan's neutral gas torus and Titan itself. The neutrals from these sources are ionized by various processes including charge exchange and photoionization and lead to a significant source of plasma in the inner magnetosphere.

In order to model the important physical processes while at the same time limiting the complexity of our calculation, several simplifications have been made. First, we assume that the planetary magnetic field is a pure dipole aligned with the rotation axis. For Saturn this is a very good approximation. Second, we approximate the plasma sources in the Saturnian system by neglecting the sources and sinks associated with Saturn itself but include sources corresponding to the rings, icy satellites and the neutral gas torus associated with Titan (but not Titan itself). In a later publication the neglected sources will be taken into account. The MHD source terms associated with the mass loading effects are given by *Gombosi et al.* [1996] and *Combi et al.* [1998].

The neutral cloud that results from Titan's presence in the magnetosphere is modeled as an axially symmetric torus centered around the orbit of Titan. The torus is taken to be filled with neutral particles of mass $m_n = 14$ amu. The model takes into account the effects of photoionization, electron impact ionization and charge transfer. Details about the neutral gas distribution in the Titan torus and the associated plasma source terms are given by *Barbosa* [1987] and *Ip* [1992]. The total mass loading rate from the Titan torus is assumed to be $\sim 2.3 \times 10^{26}$ s$^{-1}$ with a peak production rate of $\sim 0.3 \times 10^{-6}$ cm$^{-3}$s$^{-1}$ at Titan's orbit (20.3 $R_s$).

The plasma source associated with the icy satellites and rings were taken from *Ip* [1997] and *Richardson* [1998]. These rates refer to the equatorial plane of Saturn. We assumed that outside the equatorial plane the source decreases as $\exp(-z^2/H^2)$, where the scale height, $H$ is 0.35 $R_s$. The total mass loading rate from the rings and icy satellites is assumed to be $\sim 1.5 \times 10^{27}$ s$^{-1}$ with a peak production rate of $\sim 160 \times 10^{-6}$ cm$^{-3}$s$^{-1}$ at 4.6 $R_s$.

The solar wind parameters used in the simulation are the following: heliocentric distance 9.54 AU, $n = 0.1$ cm$^{-3}$, $u = 400$ km/s, $T = 1.8 \times 10^5$ K, $a = 50$ km s$^{-1}$, and $B = 0.5$ nT. At Saturn's orbit the IMF is assumed to be in the $y$ direction (Parker spiral) pointing from dawn to dusk. These parameters correspond to an ion-acoustic Mach number of 8 and an Alfvénic Mach number of 3.9. The specific heat ratio is 5/3.

The planetary parameters were taken to be the following: radius $R_s = 60,268$ km, angular velocity $\Omega = 1.66 \times 10^{-4}$ s$^{-1}$, equatorial magnetic field $B_e = 0.208$ G, the dipole is non-tilted and oppositely oriented than the terrestrial dipole.

The outer boundaries of the simulation domain are located at 192 $R_s$ upstream and 588 $R_s$ downstream. The other boundaries are located at 192 $R_s$. This large computational domain ensures that the influence of the boundaries on the solution is negligible. The boundary conditions applied at the outer boundaries are those of the free streaming solar wind. The inner boundary associated with Saturn is applied at 3 $R_s$. At this radius the density and pressure are allowed to float and the magnetic field is taken to be only the intrinsic planetary dipole. The velocities are fixed to coincide with the rotation of Saturn. The simulation involved about $10^6$ computational cells. The smallest cell was $\sim 0.19 R_s$ in each direction, and we used 8 levels of refinement.

Figure 3 is a 3D representation of the solution. The grayscale represents the mass density in the equatorial and noon-midnight meridian planes. The Figure also shows the computational grid in these two planes, indicating the refinement near the planet. White lines represent plasma flow lines in the inner magnetosphere.

It can be seen in Figure 3 that a bow shock forms upstream of the planet. The subsolar distance of the shock is about 30 $R_s$. The magnetopause separates the shocked solar wind from the region dominated by the planetary magnetic field. The subsolar point of the magnetopause is located at around 20 $R_s$. Since the Titan torus is centered around 20 $R_s$, a significant plasma source is located in the magnetosheath between the shock and the dayside magnetopause. This plasma source results in additional deceleration and heating of the shocked plasma flow, thus compressing the Kronian magnetosphere.

Inspection of Figure 3 reveals that the plasma density is highest near the equatorial plane. This is due to the combined effect of rapid rotation and the concentration of plasma sources near the equatorial plane. Near the equatorial plane the plasma within Titan's orbit rotates with the planet. Beyond Titan's orbit the corotation breaks down due to mass loading and weakening magnetic field. In this region the plasma exhibits a complicated convection pattern as can be seen by examining the convection streamlines in Figure 3.

The plasma convection pattern in the equatorial plane is shown in the left panel of Figure 4. The panel also shows the plasma mass density distribution (grayscale coded). It can be seen that the plasma density is significantly enhanced in the innermost region (within about 10 $R_s$), where mass loading from the rings and icy satellites plays a major effect. The density has a minimum between about 10 and 15 $R_s$ where there is no major plasma source. This minimum extends to Titan's orbit at around 1400 LT. This density minimum is formed because the magnetopause is near Titan's orbit in this region and the incoming mass loaded plasma is diverted into the magnetosheath (outside the magnetopause) or diverted towards the tail through high latitudes (inside the magnetopause).

At around 2000 LT a density maximum is formed near Titan's orbit. This maximum is a consequence of the interplay between solar wind driven convection and corotation: the two effects nearly cancel and the plasma nearly stagnates. The low plasma velocity naturally leads to high densities.

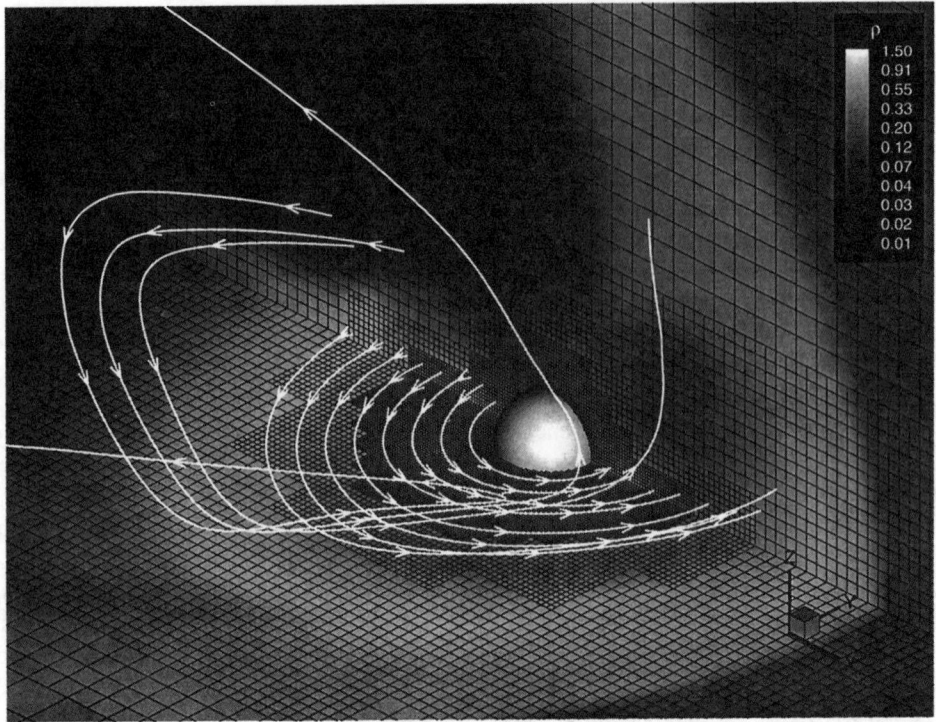

**Figure 3.** 3D rendering of Saturn magnetosphere. The grayscale represent the mass density in the equatorial and noon-midnight meridian planes. The Figure also shows the computational grid in these two planes, indicating the refinement near the planet. White lines represent plasma flow lines in the inner magnetosphere.

Tailward from about $x = -20\,R_s$ the effects of corotation become increasingly negligible and solar wind driven magnetospheric convection starts to dominate. We note that a magnetic X-line is formed at around this distance (not shown in the Figures).

The right panel in Figure 4 shows the grayscale coded electric current density in the noon-midnight meridian. One of the most prominent features seen in this panel is a thin equatorial current ring extending from about $4\,R_s$ to approximately $12\,R_s$. This current ring is formed by the pickup current associated to the inner plasma source due to the rings and icy satellites. The plasma source peaks at around $5\,R_s$. At this distance the current is diverted from the current disc and connects to the high-latitude ionosphere along magnetic field lines.

Another current system seen in Figure 4 is associated with mass loading in the Titan torus. This current system connects to the high-latitude ionosphere poleward from the icy satellite/ring current system. It is interesting to note that on the nightside the Titan torus current system extends towards the magnetotail and connects to the cross-tail current. On the dayside the Titan torus current is split into two parts: one is connected to the source inside the magnetopause, while the second one connects to the torus source outside the magnetopause. The two parts of the current system are separated by a gap. The Titan torus current system is also a pickup generated current.

## 5. SUMMARY

We presented global MHD simulations of the interaction of the solar wind with two very different magnetized planets, Mercury and Saturn. The simulations were carried out with BATS-R-US, a newly developed high performance adaptive MHD code. Similarities in planetary magnetospheres, in spite of differences in ionospheric conductances, configurations, plasma and energy sources and sinks, can help us to understand better the full range of phenomena and processes in the Earth's magnetosphere. Planetary magnetospheres help us to extend observed magnetospheric phenomena into different parameter regimes.

Both simulations considered the interaction of the "nominal" solar wind (with Parker spiral IMF) with the two planets. Mercury was assumed to be a non-conductive body with negligible plasma sources in the Hermean environment. On the other hand Saturn was assumed to have two major

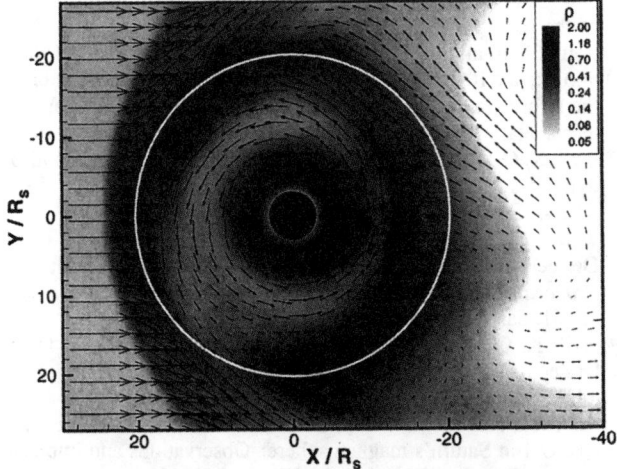

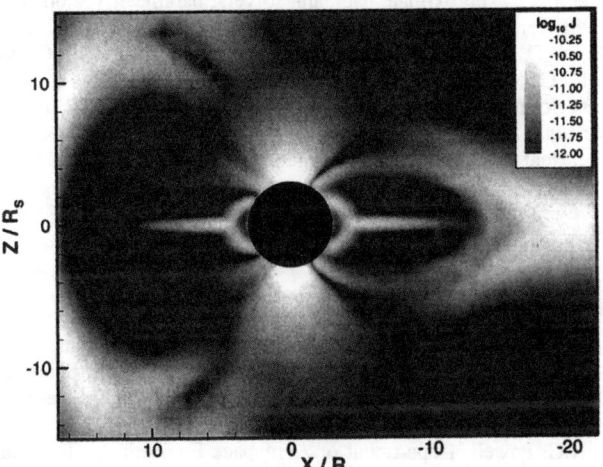

**Figure 4.** Top panel: Grayscale coded mass density distribution in Saturn's inner magnetosphere. Arrows represent plasma velocity vectors. The circle shows Titan's orbit. Bottom panel: Grayscale coded current density distribution in the noon-midnight meridian plane of Saturn's inner magnetosphere.

plasma source regions: one associated with the rings and icy satellites and the other with the neutral torus around Titan's orbit.

In the case of Mercury the intrinsic magnetic field is very weak, so that most of the field lines emanating from the surface of the planet are connected to the IMF either upstream or downstream. Closed field lines appear only at latitudes less than 50°. The whole magnetosphere is twisted and rotated due to the presence of a significant dawn to dusk $B_y$ component of the IMF. The simulation clearly shows the formation of a delicately connected current system consisting of the tail and magnetopause currents.

In the case of Saturn a very interesting and complicated configuration arises from the interplay between mass loading, rapid planetary rotation and solar wind driven convection. In the equatorial plane corotation was obtained within Titan's orbit, while outside Titan's orbit solar wind driven convection dominates the plasma motion. Mass loading generates two current systems which connect the ring/icy satellite current disk and the Titan torus to the high-latitude ionosphere of Saturn.

*Acknowledgments.* This work was supported by the NSF-NASA-AFOSR interagency grant NSF ATM-9318181 and by the NASA HPCC Computational Grand Challenge cooperative agreement award CAN NCCS5-146. One of us (TIG) gratefully acknowledges the support provided by the International Space Science Institute during his sabbatical leave.

## REFERENCES

Barbosa, D., Titan's atomic nitrogen torus: Inferred properties and consequences for the Saturnian magnetosphere, *Icarus*, *72*, 53–61, 1987.

Bauske, R., A. F. Nagy, T. I. Gombosi, D. L. DeZeeuw, K. G. Powell, and J. G. Luhmann, A three-dimensional MHD study of solar wind mass loading processes at Venus: Effects of photoionization, electron impact ionization, and charge exchange, *J. Geophys. Res.*, *103*(A10), 23625–23638, 1998.

Broadfoot, A. L., D. E. Shemansky, and S. Kumar, Mariner 10: Mercury atmosphere, *Geophys. Res. Lett.*, *3*, 577–580, 1976.

Combi, M. R., K. Kabin, T. Gombosi, D. DeZeeuw, and K. Powell, Io's plasma environment during the Galileo flyby: Global three-dimensional MHD modeling with adaptive mesh refinement, *J. Geophys. Res.*, *103*(A5), 9071–9081, 1998.

Connerney, J. E. P., and N. F. Ness, Mercury's magnetic field and its interior, in *Mercury*, edited by F. Vilas, C. R. Chapman, and M. S. Matthews, pp. 494–513, The University of Arizona Press, 1988.

DeZeeuw, D. L., and K. G. Powell, An adaptively-refined Cartesian mesh solver for the Euler equations, *J. Comput. Phys.*, *104*, 55–68, 1992.

Gombosi, T. I., K. G. Powell, and D. L. DeZeeuw, Axisymmetric modeling of cometary mass loading on an adaptively refined grid: MHD results, *J. Geophys. Res.*, *99*, 21525, 1994.

Gombosi, T. I., D. L. DeZeeuw, R. M. Häberli, and K. G. Powell, Three-dimensional multiscale MHD model of cometary plasma environments, *J. Geophys. Res.*, *101*(A7), 15233–15253, 1996.

Gombosi, T. I., D. L. DeZeeuw, C. P. T. Groth, K. G. Powell, and P. Song, The length of the magnetotail for northward IMF: Results of 3D MHD simulations, in *Phys. Space Plasmas (1998)*, edited by T. Chang, and J. R. Jasperse, vol. 15, pp. 121–128, MIT Press, 1998.

Gombosi, T. I., D. L. DeZeeuw, C. P. T. Groth, and K. G. Powell, Magnetospheric configuration for Parker-spiral IMF conditions: Results of a 3D AMR MHD simulation, *Adv. Space Res.*, 1999, in press.

Grard, R., Photoemission on the surface of Mercury and related electrical phenomena, *Planet. Space Sci.*, *45*, 62–72, 1997.

Groth, C. P. T., D. L. DeZeeuw, T. I. Gombosi, and K. G. Powell, Three-dimensional MHD simulation of coronal mass ejections, *Adv. Space Res.*, 1998, submitted.

Häberli, R. M., T. I. Gombosi, D. L. DeZeuuw, M. R. Combi, and K. G. Powell, Modeling of cometary x-rays caused by solar wind minor ions, *Science*, 276, 939–942, 1997.

Hansen, K. C., T. I. Gombosi, D. L. DeZeeuw, C. P. T. Groth, and K. G. Powell, A 3D global MHD simulation of Saturn's magnetosphere, *Adv. Space Res.*, 1999, submitted.

Ip, W., The nitrogen tori of Titan and Triton, *Adv. Space Res.*, 12(8), (8)73–(8)79, 1992.

Ip, W., On the neutral cloud distribution in the Saturnian magnetosphere, *Icarus*, 126, 42–57, 1997.

Kabin, K., M. R. Combi, T. I. Gombosi, A. F. Nagy, D. L. DeZeeuw, and K. G. Powell, On Europa's magnetospheric interaction: An MHD simulation of the E4 flyby, *J. Geophys. Res.*, 104, 1999a, in press.

Kabin, K., T. I. Gombosi, D. L. DeZeeuw, K. G. Powell, and P. L. Israelevich, Interaction of the Saturnian magnetosphere with Titan: Results of a 3D MHD simulation, *J. Geophys. Res.*, 104(A2), 2451–2458, 1999b.

Kaymaz, Z., and G. L. Siscoe, Open geometry of the magnetotail cross section, *J. Geophys. Res.*, 103(A7), 14829–14836, 1998.

Lammer, H., and S. J. Bauer, Mercury's exosphere: origin of surface sputtering and implications, *Planet. Space Sci.*, 45, 73–79, 1997.

Linde, T. J., A three-dimensional adaptive multifluid MHD model of the heliosphere, Ph.D. thesis, University of Michigan, Ann Arbor, Michigan, 1998.

Linde, T. J., T. I. Gombosi, P. L. Roe, K. G. Powell, and D. L. DeZeeuw, The heliosphere in the magnetized local interstellar medium: Results of a 3D MHD simulation, *J. Geophys. Res.*, 103(A2), 1889–1904, 1998.

Ness, N. F., The magnetosphere of Mercury, in *Solar system plasma physics. Vol II*, edited by C. Kennel, L. J. Lanzerotti, and E. N. Parker, pp. 183–206, North Nolland, Amsterdam, 1979.

Ness, N. F., K. W. Behannon, R. P. Lepping, and Y. C. Whang, The magnetic field of Mercury, *J. Geophys. Res.*, 80, 2708, 1975.

Ogino, T., A three-dimensional MHD simulation of the interaction of the solar wind with the Earth's magnetosphere: The generation of field-aligned currents, *J. Geophys. Res.*, 91, 6791, 1986.

Ogino, T., and R. J. Walker, A magnetohydrodynamic simulation of the bifurcation of tail lobes during intervals with a northward interplanetary magnetic field, *Geophys. Res. Lett.*, 11, 1018, 1984.

Powell, K. G., An approximate Riemann solver for magnetohydrodynamics (that works in more than one dimension), Tech. Rep. 94-24, ICASE, Langley, VA, 1994.

Powell, K. G., P. L. Roe, R. S. Myong, T. I. Gombosi, and D. L. DeZeeuw, An upwind scheme for magnetohydrodynamics, in *Proceedings of AIAA 12th Computational Dynamics Conference*, p. 661, San Diego, AIAA-95-1704, 1995.

Powell, K. G., P. L. Roe, T. J. Linde, T. I. Gombosi, and D. L. DeZeeuw, A solution-adaptive upwind scheme for ideal magnetohydrodynamics, *J. Comput. Phys.*, 1999, in press.

Richardson, J. D., A. Eviatar, M. A. McGrath, and V. M. Vasyliunas, OH in Saturn's magnetosphere: Observations and implications, *J. Geophys. Res.*, 103, 20,245–20,255, 1998.

Roe, P. L., Approximate Riemann solvers, parameter vectors, and difference schemes, *J. Comput. Phys.*, 43, 357–372, 1981.

Stout, Q. F., D. L. DeZeeuw, T. I. Gombosi, C. P. T. Groth, H. G. Marshall, and K. G. Powell, Adaptive blocks: A high-performance data structure, in *Proc. Supercomputing'97*, 1997.

Tanaka, T., Generation mechanisms for magnetosphere-ionosphere current systems deduced from a three-dimensional MHD simulation of the solar wind-magnetosphere-ionosphere coupling process, *J. Geophys. Res.*, 100(A7), 12,057–12,074, 1995.

---

T.I. Gombosi, D.L. DeZeeuw, C.P.T. Groth, K.C. Hansen and K. Kabin, Department of Atmospheric and Oceanic Sciences, The University of Michigan, Ann Arbor, MI 48109. (e-mail: tamas@umich.edu)

K.G. Powell, Department of Aerospace Engineering, The University of Michigan, Ann Arbor, MI 48109.

# Currents in Mercury's Magnetosphere

## Karl-Heinz Glassmeier

*Institut für Geophysik und Meteorologie, Technische Universität Braunschweig, Federal Republic of Germany*

Electric currents play an important role in the terrestrial magnetosphere as an agent to couple different plasma regimes and to mediate stresses. Much effort has been put into unravelling the structure of the current system of our planets magnetosphere. Though observations are rather sparse some important aspects on the magnetospheric current systems of planet Mercury are discussed in the present paper, both using theoretical knowledge from Earth as well as satellite measurements near Mercury. Magnetopause and magnetotail currents are found to be of comparable size to their terrestrial counterparts. The Hermean ring current is estimated to be rather small and insignificant. Reconfiguration currents, that is currents associated with waves due to a continuous reshaping of the magnetosphere because of changing solar wind conditions, are speculated to be very important at Mercury. Reconfiguration associated kinetic Alfvén waves are hypothesized as a major process to energize the Hermean multi-component plasma. The Hermean substorm current wedge is found to close via magnetospheric currents at plasma and magnetic field gradients. Finally, the importance of currents induced within the planetary interior is discussed.

## THE MAGNETOSPHERE OF PLANET MERCURY

Since the Mariner 10 spacecraft visited planet Mercury in 1974/1975 we know that this inner planet of our solar system possesses a magnetosphere much as planets Earth, Jupiter, Saturn, Uranus, and Neptun do. However, compared to these other planetary interaction regions with the solar wind the Hermean magnetosphere is rather small due to the rather large solar wind dynamic pressure and the small planetary magnetic field. Solar wind density, flow velocity, and magnetic field magnitude are of the order $50$–$70 \cdot 10^6$ protons/m$^3$, 500–700 km/s, and 50–70 nT, respectively [e.g., *Russell, 1989*; *Glassmeier, 1997*]. The surface magnetic field is of the order 340 nT, in the dipole approximation [*Ness, 1979*].

Concerning plasma measurements only electrons were measured. It was found that the plasmas in the lobes, the plasma sheet, and the polar magnetosphere all appeared similar to those observed at the Earth, but with their densities enhanced by about the same $1/r^2$ factor relative to 1 AU as the external solar wind [*Ogilvie et al., 1977*].

Large solar wind dynamic pressure and low planetary magnetic field cause a magnetopause distance of about one planetary radius $R_p$ above the surface (1 $R_p$ = 2450 km). *Siscoe and Christopher [1975]* pointed out that due to changing solar wind conditions the magnetopause distance may vary between 0.1 and 1.6 $R_p$. Occasionally the magnetopause would even reach the planets surface. However, *Hood and Schubert [1979]* as well as *Suess and Goldstein [1979]* pointed out that magnetopause compression would be strongly opposed by induced currents in the planet's outer mantle. But, as *Slavin and Holzer [1979]* show, dayside reconnection would significantly increase the chances of the solar wind to impact the Hermean surface. Thus, the Hermean magnetosphere may be subject to rather severe solar wind induced variations.

Ground-based spectroscopic observations indicate that the magnetosphere of Mercury is populated by vast amounts of heavy ions. These ions are the result of surface sputtering processes due to the precipitation of energetic

Magnetospheric Current Systems
Geophysical Monograph 118
Copyright 2000 by the American Geophysical Union

magnetospheric ions and electrons. One of the dominant heavy ion population is made out of sodium, which relative number density has been estimated as large as 20% [*Ip, 1986*; *Cheng et al., 1987*; *Othmer et al., 1999*].

Mercury is a slowly rotating planet (rotation period 57 days). This and the small planetary magnetic field are the reason for no plasmasphere existing in this magnetosphere. Any reasonable estimate of a plasmapause position gives a location within the planetary interior [*Brice, 1967*]. However, a strong planetary magnetic field is only the necessary condition for the existence of a plasmasphere. Even if the Hermean field would be stronger no plasmasphere would exist as the sufficient condition, existence of an atmosphere and ionosphere, are not fulfilled. An atmosphere and an ionosphere are required to guarantee angular momentum transport between the rotating planet and the plasmaspere.

As the planet does not have any atmosphere, but is rather surrounded by an exosphere, no ionosphere exists. However, neutral particles sputtered out of the surface, may become ionized and picked up by any magnetospheric plasma, thus forming a thin pickup ion-sphere.

Concerning the dynamics of the Hermean magnetosphere low-frequency pulsations [*Russell, 1989*] as well as substorm-like features have been reported [*Siscoe et al., 1975*; *Baker et al., 1986*; *Christon et al., 1986*].

One may summarize that the interaction region of planet Mercury in many respects is similar to that of the Earth with the solar wind. The major common feature is the existance of a magnetosphere at all. However, major differences, such as outlined above, also exist. This makes the Hermean magnetosphere an interesting object of comparative planetology. In this review I shall concentrate on knowledge and speculations above current systems in the innermost planet's magnetosphere.

## MAGNETOPAUSE AND MAGNETOTAIL CURRENTS

Mariner 10 traversed the magnetopause at Mercury four times during its flybys 1974/75. Fig. 1 displays magnetic field measurements taken at March 10, 1974. The spacecraft enters the magnetosphere at the dawn-side at a planetary distance about 2 $R_p$. After traversing the bowshock at around 20:27 UT, the spacecraft entered the magnetosphere at about 20:37 UT. The interplanetary magnetic field increased from about 20 nT in the upstream region to values of about 60 nT just behind the bowshock. After decaying in the magnetosheath to values of about 38 nT, a clear jump of about 24 nT across the magnetopause has been observed (Fig. 1). This jump occurs on a temporal scale of about $T_{MP} = 12.5$ s. The spacecraft velocity during the magnetopause traversal is about 10 km/s. Assuming that the magnetopause is not in motion gives one a magnetopause thickness $L_{MP} =$ 125 km. This value is a *rough estimate* for the width as it neglects any rapid motion of the magnetopause. Approximating $\nabla \times \vec{B} \simeq \delta \vec{B}/L_{MP}$, a magnetopause current density of $j_{MP} \simeq 1.5 \cdot 10^{-7}$ A/m$^2$ results. However, this value should be used with care as the Mariner 10 observations indicate multiple crossings of the magnetopause, that is rapid motions of the boundary.

Mariner 10 traversed the Hermean magnetotail shortly after its closest approach on March 29, 1974. A clear tail current sheet was observed (Fig. 2). *Whang [1977]* estimated the width of this sheet at about 150 km and the associated magnetic field jump at 89 nT. This gives one a tail current density of the order of $5 \cdot 10^{-7}$ A/m$^2$, a value again comparable to terrestrial ones. Stresses imposed by the solar wind onto the magnetospheric plasma are controlling this tail current sheet density.

## THE HERMEAN RING CURRENT

No observations of the Hermean ring current strength are available. This would require detailed measurements of the plasma pressure tensor [e.g., *Williams, 1986*]. However, the Dessler-Parker-Sckopke theorem [e.g., *Sckopke, 1966*] may be used to estimate the ring current magnetic field at the planetary surface as well as the ring current density. The Dessler-Parker-Sckopke theorem relates the ring current induced magnetic field perturbation at the Earth equator, $\delta H$, to the total kinetic energy of the ring current particles, $E_{Par}$, via

$$\left.\frac{\delta H}{H_0}\right|_{Eq} = \frac{2}{3} \cdot \frac{E_{Par}}{E_{Mag}} \qquad (1).$$

and is valid for a non-conducting planet [e.g. *Wolf, 1995*]. Here, $H_0$ is the surface magnetic field and $E_{Mag}$ the total energy content of the planetary magnetic field.

To estimate $E_{Par}$ I shall assume a particle density of $3 \cdot 10^6$ particles/m$^3$ [e.g., *Russell, 1989*] and, somewhat arbitrarily, a particle energy of 1 keV. Furthermore, the ring current region is approximated as a torus of diameter 1 $R_p$, located at 0.5 $R_p$ above the planetary surface. With these assumptions an estimate $E_{Par} \approx 1.7 \cdot 10^{11}$ J results. With $H_0 = 340$ nT $E_{Mag} = 10^{13}$ J, and the ring current surface field is estimated as $\delta H \simeq 3.8$ nT, that is about 1/100 of the planetary field at the surface. The terrestrial ratio is about 1/1000.

## FIELD-ALIGNED CURRENT OBSERVATIONS

The concept of field-aligned currents plays an important role in our understanding of magnetospheric physics ever since Kristian Birkeland first proposed it in 1909, and field-aligned currents have been a matter of much debate [*Akasofu, 1983*]. Field-aligned currents serve to couple different plasma regimes, both electrodynamically

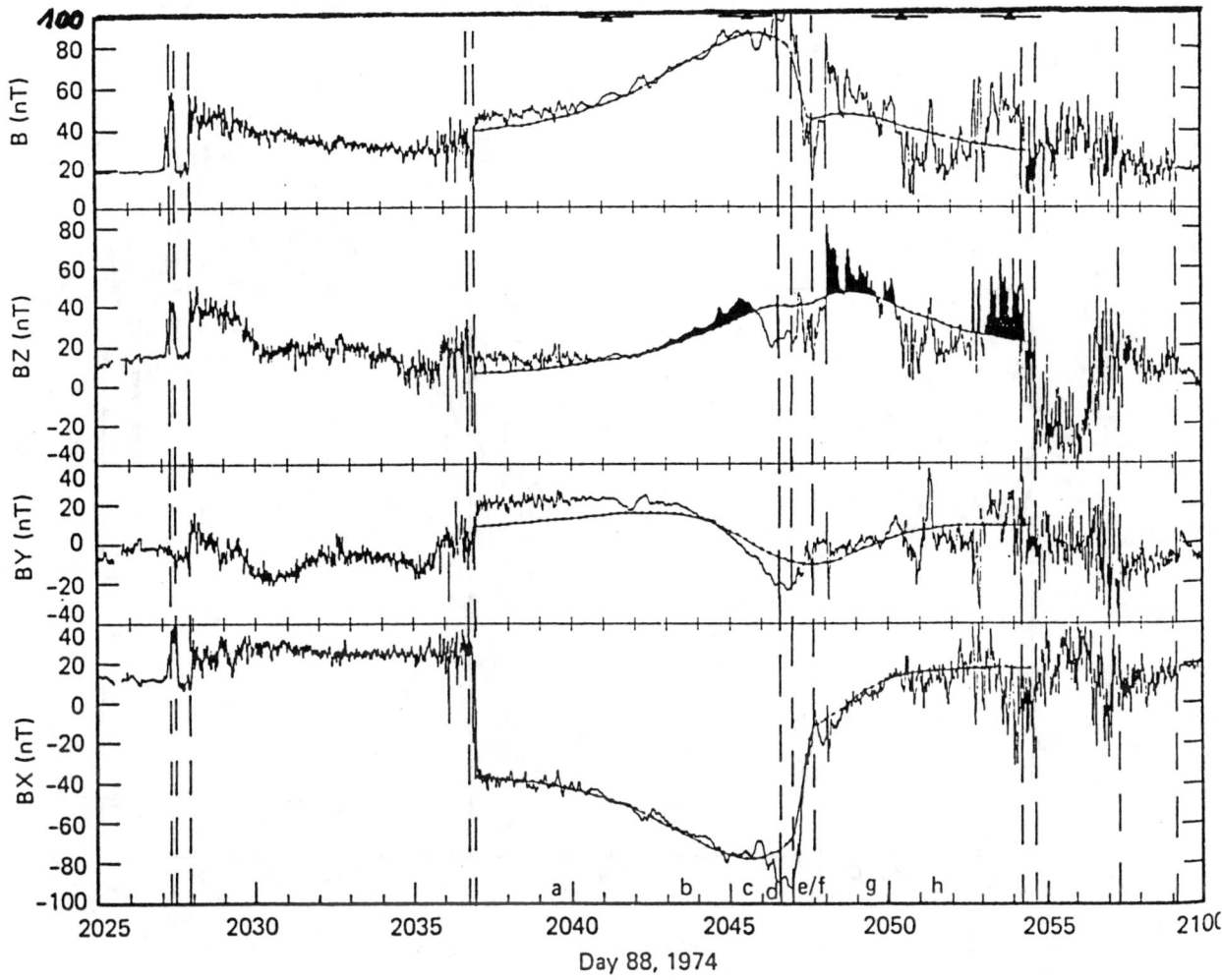

**Figure 1.** Mariner 10 magnetic field measurements taken at March 14, 1974 [after Ness, 1979].

as well as mechanically. Magnetosphere-ionosphere coupling is a very important example of coupling mediated by field-aligned currents.

*Slavin et al. [1997]* have recently reanalyzed the Mariner 10 data searching for field-aligned current signatures in Mercury's magnetosphere. Such parallel currents manifest themselves as transverse magnetic field variations in spacecraft magnetic field observations. The most prominent transverse field variation observed during the Mariner 10 Mercury flyby is displayed in Fig. 3. At around 20:51 UT the spacecraft traversed a region of changing $B_y$ component which is reminiscent of field-aligned current flow. Assuming that the observed perturbations are caused by an infinite current sheet, a sheet current density of about 50 mA m$^{-1}$ results. If this current sheet were quasi-aligned with a constant L-shell, and given the Mariner trajectory, the average estimated field-aligned current density is $7 \cdot 10^{-7}$ A/m$^2$ [*Slavin et al., 1997*]. Again, this value is comparable to terrestrial values [e.g., *Cummings and Dessler, 1967; Glassmeier, 1987*]. This is a somewhat surprising result as no ionosphere exists where this current can be closed and/or dissipated. Closure within the magnetospheric plasma is required. The associated magnetic field variations of $\delta B \simeq 60$ nT are comparable to the ambient magnetic field itself. Nonlinear effects are certainly involved in generating this large field-aligned current. Reconfigurations of the magnetospheric system due to changing solar wind conditions may be involved.

## RECONFIGURATION CURRENTS

As mentioned already above the Hermean magnetosphere suffers from major reconfigurations due to changing solar wind conditions [*Siscoe and Christopher, 1975;*

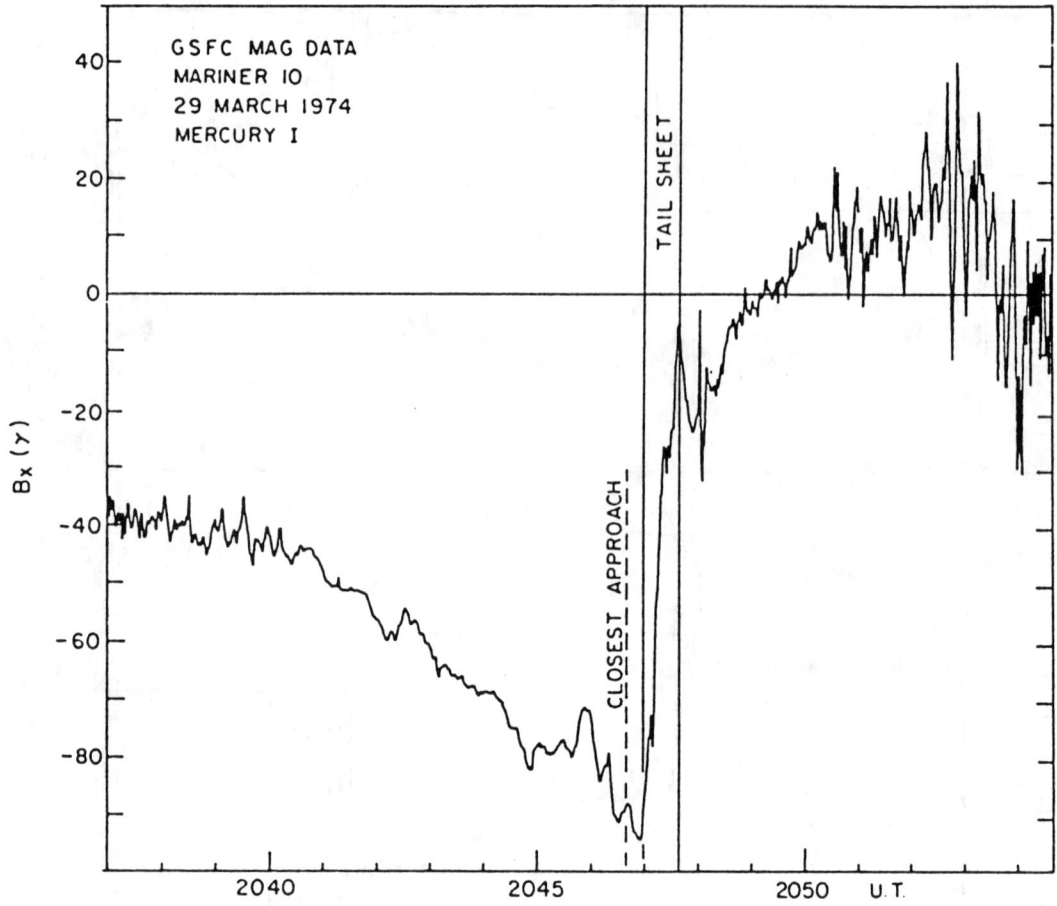

**Figure 2.** Mariner 10 magnetic field measurements during the traversal of the Hermean magnetotail [after Whang, 1977].

*Luhmann et al., 1998*]. Fig. 4 exhibits possible magnetopause Positions during Mercury's perihelion passages. Positions of the magnetopause between 0.1 and 1.6 $R_p$ are rather likely. Corresponding changes of the magnetotail are also expected. Such strong responses to solar wind dynamics are due to the small planetary magnetic field. Solar wind dynamic pressure induced dynamics thus plays a major role in the Hermean magnetosphere.

Reconfigurations of a magnetospheric system requires low-frequency waves to communicate changes of e.g. the Chapman-Ferraro current system or the tail current system. A prominent example of such reconfiguration establishing waves are Pi2 pulsations of the terrestrial magnetosphere associated with magnetospheric substorms [e.g., *Baumjohann and Glassmeier, 1984*]. The low-frequency waves at hand are magnetohydrodynamic waves. In a normal two-component plasma, consisting of electrons and protons, field-aligned current changes are carried by electrons, while transverse polarization currents are carried by the much heavier protons [e.g., *Fejer and Kan, 1969*]. This still holds in a multicomponent plasma such as in the magnetosphere of Mercury, where heavy ions such as sodium are a major part of the ion population [*e.g. Othmer et al., 1999*]. The transverse polarization current $\vec{j}_P$ in a multicomponent plasma is given by [e.g., *Marsch, 1990*]

$$\vec{j}_P(\omega \ll \Omega_i) \simeq \sum_i \frac{n_i m_i}{B_0^2} \cdot \frac{d\vec{E}_\perp}{dt} \qquad (2)$$

where $\Omega_i$ is the ion gyrofrequency, $n_i$ and $m_i$ are the ion density and mass, respectively. The background magnetic field and the wave electric field are denoted by $\vec{B}_0$ and $\vec{E}_\perp$, respectively.

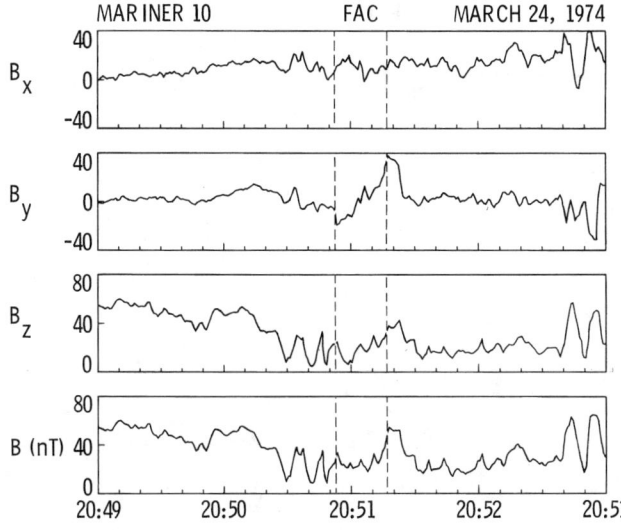

**Figure 3.** Mariner 10 magnetic field measurements indicating the presence of field-aligned currents in the Hermean magnetosphere [after Slavin et al., 1997].

Taking, as an example, sodium ions, and assuming a 20 of the magnetospheric ion population, one finds that $n_{Na} m_{Na} \approx 2 n_P m_P$ in the Hermean magnetospheric plasma. Thus, transverse currents associated with magnetospheric reconfigurations, communicated by low-frequency waves, are carried mainly by heavy ions rather than protons.

This has important consequences. Assuming proton temperatures corresponding to 1 keV, a value as used to estimate to ring current strength, and assuming temperate equilibrium between protons and the heavier plasma ions, allows to estimate a sodium thermal ion gyroradius to be of the order of 150 km. The scale of the magnetospheric system, however, is of the order of $R_p = 2450$ km. Thus, the ratio of system scale to heavy ion gyroradius is about 16 in the Hermean magnetosphere. The corresponding terrestrial value is of the order of 600.

A ratio as small as 16 suggests that reconfiguration associated waves are propagating in the kinetic Alfvén wave mode, not via MHD Alfvén wave modes. Kinetic Alfvén waves are strongly dispersive due to electron inertia and finite gyroradius effects. Their dispersion relation is given by [*Hasegawa and Uberoi, 1982; Lysak, 1990*]

$$\omega^2 \approx v_A^2 k_\parallel^2 \cdot \left(1 + k_\perp^2 r_{ion}^2\right) \quad (3)$$

where $v_A^2 = B_0^2/\mu_0 \rho$ is the classical Alfvén speed, $r_{ion}$ the ion gyroradius, and $k_\perp$ and $k_\parallel$ are the wave vector components transverse and parallel to the ambient magnetic field $\vec{B}_0$.

Finite electron inertia and ion gyroradius cause kinetic Alfvén waves to carry a significant parallel electric field $E_\parallel$. For a low $\beta$ plasma this parallel component is related to the transverse field $E_\perp$ via [e.g., *Goertz and Boswell, 1979; Lysak, 1990*]

$$E_\parallel \approx r_{GIA}^2 \cdot \frac{\partial^2 E_\perp}{\partial s_\perp \partial s_\parallel} \quad (4)$$

where $r_{GIA} = m_i v_e / e B_0$ is the ion acoustic gyroradius, and $\partial s_\perp$ and $\partial s_\parallel$ denote derivates along the transverse and longitudinal directions with respect to $\vec{B}_0$. Approximating $\frac{\partial^2 E_\perp}{\partial s_\perp \partial s_\parallel} \simeq \frac{\delta E_\perp}{L_\perp L_\parallel}$, assuming $L_\perp = L_\parallel = R_p$, and with $r_{GIA} \simeq 0.2\, R_p$, one has

$$\delta E_\parallel \simeq 0.04\, \delta E_\perp \quad (5).$$

The transverse electric field may be estimated via

$$\frac{\delta E_\perp}{\delta B} \simeq v_A \quad (6).$$

With $v_A = 1000$ km/s [*Russell, 1989*] and $\delta B = 5$ nT (see Fig. 3) a value of $\delta E_\perp \simeq 5$ mV/m is estimated. Thus, parallel electric fields of up to 0.2 mV/m and potential drops of about 500 eV may easily be associated with solar wind induced reconfigurations of the Hermean magnetosphere.

Particle acceleration due to such kinetic Alfvén waves will help to damp and dissipate the reconfiguration assciated low-frequency waves. In the terrestrial magnetosphere reconfiguration associated waves are dissipated mainly by Joule heating in the ionosphere. As Mercury does not possess any significant ionosphere electron acceleration by kinetic Alfvén waves replaces ionospheric dissipation. Magnetospheric reconfigurations thus lead to a direct energy transfer from the solar wind to the magnetospheric plasma.

## SUBSTORMS AT MERCURY

Dipolarization events of the Hermean magnetotail as well as energetic electron events have been conjectured as clear indications of substorm activity in Mercury's magnetosphere by a number of authors such as *Siscoe et al.* [1975], *Baker et al.* [1986], and *Christon et al.* [1986]. In contrast to this *Luhmann et al.* [1998] suggested that

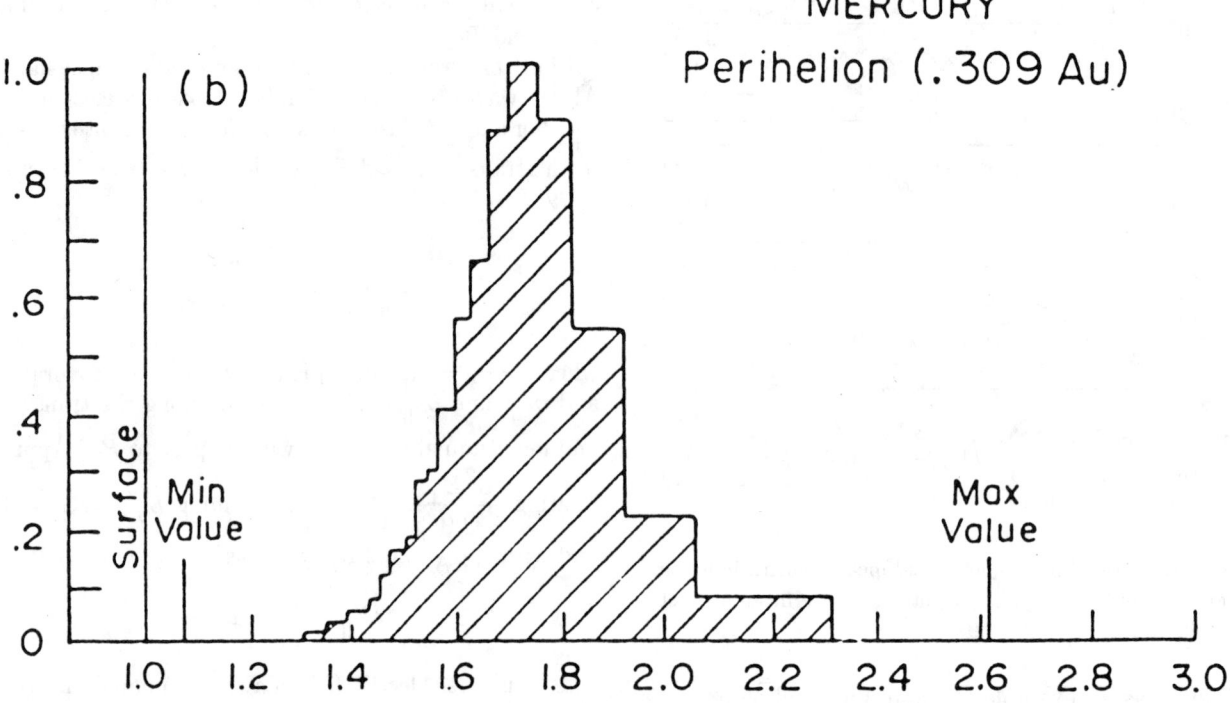

**Figure 4.** Possible locations of the Hermean magnetopause due to changing solar wind conditions [after Siscoe and Christopher, 1975].

the putative substorm events in the Hermean magnetosphere are driven events, due to IMF changes. *Glassmeier [1996, 1997]* has posed the question how the associated substorm current wedge, if it exists, can be closed if no ionosphere exists. As closure within the planets crust and mantle is unlikely *Glassmeier [1996, 1997]* suggested that magnetospheric reconnection in the Hermean tail may be difficult.

Tail reconnection and substorm formation is associated with major topological changes of the tail magnetic field, driving a localized, reconnection associated current $\vec{j}_R$ opposing the tail current. Once switched on this current needs to be closed by field-aligned currents, these being closed via transverse polarization currents $\vec{j}_P$. This Alfvén wave pulse travels towards the ionosphere where it is reflected. The reflection coefficient $R$ at the polar ionosphere is given by [*Scholer, 1970; Mallinckrodt and Carlson, 1978; Glassmeier, 1984*]

$$R = \frac{\Sigma_A - \Sigma_P}{\Sigma_A + \Sigma_P} \qquad (7)$$

where $\Sigma_P$ is the height-integrated Pedersen conductivity and $\Sigma_A = 1/\mu_0 v_A$ is the magnetosphere wave conductance. The field-aligned current of the Alfvén pulse reflecting back and forth between the ionosphere and an assumed voltage generator in the tail is given by [*Lysak, 1990; Glassmeier, 1997*]

$$j_\parallel^\infty = j_{\parallel,i} \cdot \frac{1-R}{1+R}$$

Here, the index "∞" indicates that the field-aligned current after many bounces is discussed. The field-aligned current carried by the initial pulse is given by $j_{\parallel,i}$. Two different cases may be discussed, that of a poorly conducting ionosphere with $R \simeq 1$, and the one with $R \simeq -1$, that is a highly conducting ionosphere. For $R \simeq -1$, $j_\parallel^\infty$ tends toward a finite value, that is the Alfvén wave pulse serves to build up the substorm current system. For $R \simeq 1$, $j_\parallel^\infty$ becomes zero: a non conducting boundary does not allow to build up the current wedge. Alfvén waves reflected off such an ionosphere oppose the processes generating the pulse in the tail.

However, the problem has a different possible solution. The equation of motion for a fluid may be written as

$$\rho \frac{d\vec{v}}{dt} = -\nabla p + \eta \vec{v} + \vec{j} \times \vec{B} \qquad (9)$$

where $\rho$, $\vec{v}$, $p$ denote plasma mass density, flow velocity, and pressure, respectively. A viscous term with $\eta \sim \tau_C$, where $\tau_C$ is a collision frequency between e.g. ions and neutrals, has been introduced to allow for ionospheric effects. Crossing equation (9) from the right with $\vec{B}$ gives one a transverse current

$$\vec{j}_\perp \simeq \sigma_P \vec{E}_\perp + \frac{1}{\mu_0 v_A^2} \frac{d\vec{E}_\perp}{dt} - \frac{1}{B^2} \nabla p \times \vec{B} \quad (10)$$

Here $\sigma_P = \eta/B^2$ and the approximation $\vec{E}_\perp = -\vec{v} \times \vec{B}$ has been used. Now, applying the transverse divergence operator onto equation (10) results into

$$-\frac{\partial j_\parallel}{\partial s_\parallel} = \nabla_\perp \cdot \vec{j}_\perp \simeq \sigma_P \nabla_\perp \cdot \vec{E}_\perp +$$

$$\frac{1}{\rho v_A^2} \nabla p \cdot \vec{j}_B - \nabla_\perp \left(\frac{1}{B^2}\right) \cdot \nabla p \times \vec{B} +$$

$$\frac{1}{\mu_0 v_A^2} \frac{d}{dt} \nabla_\perp \cdot \vec{E}_\perp + \nabla_\perp \left(\frac{1}{\mu_0 v_A^2}\right) \cdot \frac{d\vec{E}_\perp}{dt} \quad (11)$$

This approximative expression for the horizontal divergence of the transverse currents, that is for the variation of the field-aligned current along the ambient magnetic field helps to clarify the substorm problem at Mercury. If $\sigma_P$ is large, such as in the terrestrial ionosphere, any divergence of the transverse polarization currents may be balanced via divergences of the Pedersen currents $\sigma_P \vec{E}_\perp$. However, if the background plasma carries currents, that is $\vec{j}_B = \frac{1}{\mu_0} \nabla \times \vec{B} \neq 0$, or if the plasma is inhomogenous, that is $\nabla_\perp \left(\frac{1}{B^2}\right) \neq 0$, than divergences of the polarization currents can be balanced via divergences of the diamagnetic current $\vec{j}_{Dia} = \frac{1}{B^2} \nabla p \times \vec{B}$.

Thus, an ionosphere is not required to close the substorm current wedge, provided a proper plasma distribution and magnetic field topology exists or can be build up. Closure via diamagnetic currents as suggested here has already been observed in first simulations of magnetic reconnection in the Hermean magnetosphere [M. Hesse, pers. communication; Barker et al., 1997].

## INDUCED CURRENTS IN THE PLANETARY INTERIOR

Induced currents in the Earth crust and mantle due to rapidly changing externally generated magnetic fields were and are subject of enumerous scientific studies [e.g., Chapman and Bartels, 1940; Rikitake, 1966; Jacobs, 1987]. Inducing magnetic field variations are ULF pulsaltions, geomagnetic variations such as magnetic substorms, or Sq variations. As a rule of thumb one may conclude that induced magnetic fields are of minor importance for the structure of the terrestrial magnetosphere, but serve as an important tool to infer the electrical conductivity of the Earth crust and mantle.

The importance of induced magnetic fields for the structure of the magnetospheres of e.g. Jupiters moons Europa and Ganymed has recently been highlighted by the flyby of the Galileo spacecraft at these moons [e.g., Neubauer, 1998; Kivelson et al., 1999]. Whether induction is of importance for the structure and dynamics of the Hermean magnetosphere can be discussed as follows. For slightly different treatments reference is made to Hood and Schubert [1979] and Suess and Goldstein [1979].

Solar wind variations cause significant temporal changes of the magnetopause position and require major reconfigurations of the magnetosphere [Siscoe and Christopher, 1975; see also Fig. 4]. These variations will be an important source of inducing magnetospheric fields. As an example changing Chapman-Ferraro currents shall be discussed. For this purpose the magnetopause currents will be approximated as a ring current at the magnetopause location. The strength $I_{CF}$ of this Chapman-Ferraro ring current may be estimated as

$$I_{CF} \simeq \pi \cdot j_{MP} \cdot L_{MP} \cdot R_P \quad (12)$$

where $L_{MP}$ is the thickness of the magnetopause current layer and the factor $\pi R_P$ approximates current integration in meridional direction of the magnetopause surface. With values of $j_{MP} = 1.5 \cdot 10^{-7}$ A/m$^2$ and $L_{MP} = 125$ km, an equivalent ring current strength of $I_{CF} \simeq 1.5 \cdot 10^5$ A results. To first order such a ring current causes a magnetic field of about 70 nT at the surface of Mercury, that is the magnetopause currents significantly contribute to the surface magnetic field. Temporal variations of the Chapman-Ferraro currents may thus constitute a significant cause for induction effects in Mercury's conducting interior.

Electromagnetic induction in a spherical body has at length been discussed by Lahiri and Price [1939] (see also Chapman and Bartels, [1940] or Rikitake [1966]). Here I shall shortly review major aspects in determining the induced fields. To describe these fields in the planetary interior a vector potential $\vec{A}$ with $\vec{B} = \nabla \times \vec{A}$ is introduced. Using a Coulomb gauge the induction equation

$$\frac{\partial \vec{A}}{\partial t} = \frac{1}{\mu_0 \sigma} \Delta \vec{A} \quad (13)$$

results, where $\sigma(\vec{r})$ denotes the electrical conductivity structure of the planetary body. For a spherically symmetric conductivity distribution $\sigma(\vec{r}) = \sigma(r)$ only the toroidal component of the vector potential becomes important, as its poloidal part gives rise to an induced toroidal magnetic field entirely confined to the planet. With $\vec{A} = \vec{r} \times \nabla u$, the induction equation thus reads

$$\frac{\partial u}{\partial t} = \frac{1}{\mu_0 \sigma(r)} \Delta u \qquad (14)$$

A convenient ansatz for its solution is [e.g., *Rikitake, 1959*]

$$u(\vec{r}, t) = e^{pt} \cdot \sum_{n=1}^{\infty} R_n(r,p) \cdot Y_n(\varphi, \Theta) \qquad (15)$$

where $R_n$ describes the radial variation of $u(\vec{r}, t)$ and $Y_n$ are spherical harmonic functions. The parameter p denotes a complex frequency.

Outside the planet the magnetic induction may be described by a magnetic potential $W$, provided that this region is current free:

$$\Delta W = 0 . \qquad (16)$$

An appropriate ansatz for this potential is $W = \sum_n W_n$ with

$$W_n = \left( g_{e,n} r^n + g_{i,n} r^{-n-1} \right) Y_n \qquad (17)$$

where the first term in brackets denotes the external field due to e.g. magnetopause currents. The second term describes the contribution of the internally induced currents.

At the planetary boundary appropriate matching conditions are continuity of the tangential magnetic field and of the normal component of the magnetic induction. With these boundary conditions $u$ and $W$ may be compared provided the radial function $R(r,p)$ is known. For the special case $\sigma(r) = \sigma_0 =$ const., $R(p,r)$ may be expressed by modified spherical Bessel functions giving one a ratio at the planetary surface of [*Rikitake, 1966*]

$$\frac{g_{i,n}}{g_{e,n}} = \frac{n}{n+1} \cdot \left( 1 - \left( \frac{k\, r_P}{2n+1} \frac{I_{n-1/2}}{I_{n+1/2}} \right)^{-1} \right) . \qquad (18)$$

Here $k^2 = \mu_0 \sigma_0 p$ is an induction parameter. Thus, if $g_{e,n}$ is known, the induced field may be determined, if $\sigma_0$ and $p$ are also known.

The inducing magnetic field of the assumed magnetopause ring current is approximately a uniform field with

$$\vec{H} = \left( -\frac{2\pi I_{CF}}{r_P} \cos\Theta, 0, \frac{2\pi I_{CF}}{r_P} \sin\Theta \right) . \qquad (19)$$

Thus, at the planetary surface

$$g_{e,1} = \frac{2\pi \mu_0 I_{CF}}{r_P} . \qquad (20)$$

To simplify matters all other expansion coefficients are assumed negligible. Therefore, one has [*Rikitake, 1966*]

$$\frac{g_{i,1}}{g_{e,1}} = \frac{1}{2} \cdot \left( 1 - \left( \frac{k\, r_P}{3} \frac{I_{1/2}}{I_{3/2}} \right)^{-1} \right) \qquad (21)$$

Quantifying this induction ratio requires to determine $k$, that is $\sigma_0$ and $p$. For a periodically varying magnetopause position, one assumes $p = i\omega = i\frac{2\pi}{T}$. Typical periods of the order of one day should be assumed. Estimates of the electrical conductivity are more difficult due to lacking knowledge of Mercury's conducting interior. Conductivity of common Earth materials such as olivine and magnetite have conductivities of about $10^{-4}$ Sm$^{-1}$ and $10^3$ Sm$^{-1}$ at 300 K, respectively [*Parkinson and Hutton, 1989*]. Assuming that the regolith at Mercury is similar in its electric properties to that of the Moon, lunar values are of interest. *Dyal et al.* [*1974*] have estimated an upper conductivity limit of about $10^{-9}$ Sm$^{-1}$, while *Hobbs et al.* [*1984*] report about values of $10^{-4}$ Sm$^{-1}$. For the lunar mantle *Hobbs et al.* [*1984*] have estimated values of about $10^{-2}$ Sm$^{-1}$. Fig. 5 displays the magnitude of the induction ratio for a variety of conductivity and period values. As a most likely set of parameters we assume $\sigma_0 = 10^{-2}$ Sm$^{-1}$ and $T = 1$ day. The corresponding induction ratio is $g_{i,1}/g_{e,1} \approx 0.62$. Thus, diurnal variations of the location of the magnetopause and associated magnetic field variations at the Hermean surface of about 70 nT (see above) may well give rise to induced magnetic fields of the order of 43 nT. This implies that about one third of the observed field might be due to magnetospheric induction and direct contributions from magnetopause currents.

## SUMMARY AND CONCLUSIONS

This paper has been prepared on account of the 25th anniversary of the first encounter of a human spacecraft with planet Mercury. A concise review of known properties of the current systems in the Hermean magnetosphere and conjectures about their characteristics and importance have been presented. Comparing with the terrestrial magnetosphere it is found that magnetopause and magnetotail current densities are similar to their terrestrial counterparts.

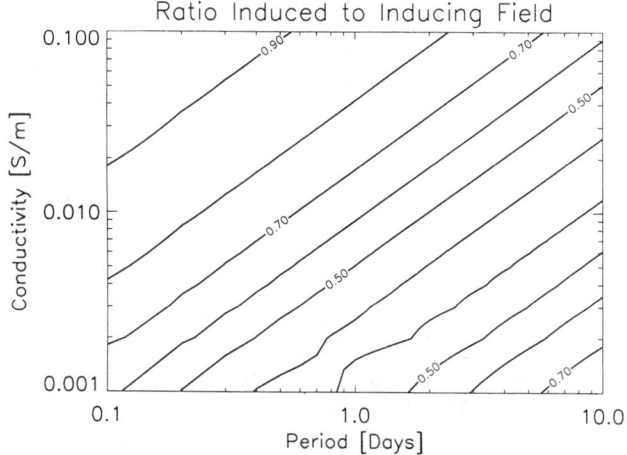

**Figure 5.** The magnitude of the induced to the inducing field for a spherical body with uniform electrical conductivity as a function of period T and conductivity.

Substorm current wedges may form as well. However, closure of the current system is not via any conducting magnetospheric boundary such as an ionosphere but within the magnetosphere, where substorm associated changes of the plasma density and the magnetic field topology and strength cause diamagnetic closure currents. The magnetosphere acts as a self-adaptive system.

The small size, due to a weak planetary magnetic field, causes the magnetosphere to react rather volatile to solar wind changes. As heavy ions such as sodium constitute a substantial part of the ion population in the Hermean magnetosphere [e.g. Cheng et al., 1987] kinetic Alfvén waves communicate these magnetospheric reconfigurations and may cause significant particle acceleration.

A first estimate of induction effects in the planet due to changing magnetopause locations indicates that about one third of the planets magnetic field at its surface might be due to induction and direct contributions from the magnetopause currents. Even if this conjecture seems to be too optimistic it demonstrates that induced currents play a much more important role than at Earth.

In any case it indicates how important a dual spacecraft mission to planet Mercury is, with the one satellite observing the solar wind field and the other measuring the induced field. Such a dual spacecraft mission would furthermore allow to infer the electrical conductivity structure of the planet. Thus, magnetospheric physics and the study of Hermean current systems will be able to contribute significantly to our understanding of the Hermean interior proper.

*Acknowledgements.* I am especially grateful to Andre Balogh who stimulated my interest for planet Mercury. Special thanks are to Shin Ohtani and Ryo Fuji, who organized a wonderful Chapman Conference on Big Island. This work was financially supported by the Deutsches Zentrum für Luft- und Raumfahrt (DLR) and the Deutsche Forschungsgemeinschaft (DFG).

## REFERENCES

Akasofu, S.-I., Evolution of ideas in solar-terrestrial physics, Geophys. J. R. astr. Soc., 74, 257-299, 1983.

Baker, D. N., J. A. Simpson, and J. H. Eraker, A model of impulsive acceleration and transport of energetic particles in Mercury's magnetosphere, J. Geophys. Res. 91, 8742-8748, 1986.

Barker, P., J. Birn, J. Borovsky, J. Bonnell, R. Elphic, S. Maurice, H. Funsten, and M. Hesse, Modelling substorm effects in Mercury's magnetosphere, EOS, Trans. AGU, 78, 1163, 1997.

Baumjohann, W., and K. H. Glassmeier, The transient response mechanism and Pi2 pulsations at substorm onset, Planet. Space Sci., 32, 1361-1370, 1984.

Brice, N. M., Bulk motion of the magnetosphere, J. Geophys. Res. 72, 5193, 1967.

Chapman, S. and J. Bartels, Geomagnetism, Oxford University Press, 1940.

Cheng, A. F., R. E. Johnson, S. M. Krimigis, and L. J. Lanzerotti, Magnetosphere, Exosphere, and Surface of Mercury, Icarus, 71, 430-440, 1987.

Christon, S. P., A comparison of the Mercury and Earth magnetospheres: electron measurements and substorm time scales, Icarus, 71, 448, 1987.

Cummings, W. D. and A. J. Dessler, Field-aligned currents in the magnetosphere, J. Geophys. Res. 72, 1007-1013, 1967.

Dyal, P., C. W. Parkin, and W. D. Daily, Magnetism and the interior of the Moon, Rev. Geophys. Space Phys., 12, 568-591, 1974.

Fejer, J. A. and J. R. Kan, A guiding centre Vlasov equation and its application to Alfven waves, J. Plasma Phys., 3, 331-351, 1969.

Glassmeier, K. H., On the influence of ionospheres with nonuniform conductivity distribution on hydromagnetic waves, J. Geophys., 54, 125-138, 1984.

Glassmeier, K. H., Ground-based observations of field-aligned currents in the auroral zone: Methods and results, Ann. Geophys. 5, 115-126, 1987.

Glassmeier, K. H., Concerning substorms in the Hermean magnetosphere, Proc. of the Third International Conference on Substorms (ICS-3), Versailles, ESA SP-389, 707-712, 1996.

Glassmeier, K. H., The Hermean magnetosphere and its ionosphere-magnetosphere coupling, Planet. Space Sci., 45, 119-125, 1997.

Hasegawa, A. and Ch. Uberoi, *The Alfvén wave*, Tech. Inf. Center, U.S. Depart. of Int., Oak Ridge, 1982.

Hobbs, B. A., L. L. Hood, F. Herbert, and C.P. Sonett, Low-frequency electromagnetic induction in the Moon: Linearized inverse theory and Lunar core calculations, *Geophys. J. R. astr. Soc., 79,* 691-696, 1984.

Hood, L. L. and G. Schubert, Inhibition of solar wind impingement on Mercury by planetary induction currents, J. Geophys. Res. *84,* 2641, 1979.

Ip, W.-H. (1986) The sodium exosphere and magnetosphere of Mercury, Geophys. Res. Lett. *13,* 423-426, 1986.

Jacobs, J. A. (ed.), *Geomagnetism,* Academic Press, London, 1987.

Kivelson, M. G. et al., Europa and Callisto: Induced or intrinsic fields in a periodically varying plasma environment, J. Geophys. Res. *104,* 4609-4617, 1999.

Lahiri, B. N. and A. T. Price, Electromagnetic induction in non-uniform conductors and the determination of the conductivity of the earth from terrestrial magnetic variations, *Phil. Trans. Roy. Soc. London, Ser. A, 237,* 509-540, 1939.

Lysak, R. L., Electrodynamic coupling of the magnetosphere and ionosphere, *Space Sci. Rev., 52,* 33, 1990.

Luhmann, J. G., C. T. Russell, and N. A. Tsyganenko, Disturbances in Mercurys magnetosphere: Are the Mariner 10 substormssimply driven?, J. Geophys. Res. *103,* 10391-10397, 1998.

Mallinckrodt, A. J. and C. W. Carlson, Relations between transverse electric fields and field-aligned currents, J. Geophys. Res. *83,* 1426, 1978.

Marsch, E., Plasmawellen und Welle-Teilchen-Wechselwirkungen in K. H. Glassmeier and M. Scholer (eds.), *Plasmaphysik im Sonnensystem,* pp. 248-283, Mannheim, 1991.

Ness, N. F., The magnetosphere of Mercury in C. F. Kennel, L. J. Lanzerotti, and E. N. Parker (eds.), *Solar System Plasma Physics,* Vol. II, pp. 185-206, North-Holland Publishing Company, 1979.

Neubauer, F. M., The sub-Alfvénic interaction of the Galilean satellites with the Jovian magnetosphere, J. Geophys. Res. *103,* 19843-19866, 1998.

Othmer, C., K. H. Glassmeier, and R. Cramm, Concerning field line resonances in Mercury's magnetosphere, J. Geophys. Res. *104,* 10369-10378, 1999.

Parkinson, W. D., and V. R. S. Hutton, The electrical conductivity of the Earth, in *Geomagnetism,* Vol. III, edited by J.A. Jacobs, Academic, London, 1989.

Rikitake, T., *Electromagnetism and the Earths Interior,* Elsevier Publishing Company, Amsterdam, 1966.

Russell, C. T., ULF waves in the Mercury Magnetosphere, Geophys. Res. Lett. *16,* 1253-1256, 1989.

Scholer, M., On the Motion of artificial ion clouds in the magnetosphere, *Planet. Space Sci., 18,* 977-1004, 1970.

Sckopke, N., A general relation between the energy of trapped particles and the disturbance field near the Earth, J. Geophys. Res. *71,* 3225-3230, 1966.

Siscoe, G. and L. Christopher, Variations in the solar wind stand-off distance at Mercury, Geophys. Res. Lett. *2,* 158-160, 1975.

Siscoe, G. L., N. F. Ness, and C. M. Yeates, Substorms on Mercury? J. Geophys. Res. *80,* 4359, 1975.

Slavin, J. A. and R. E. Holzer, The effect of erosion on the solar wind stand-off distance at Mercury, J. Geophys. Res. *84,* 2076, 1979.

Slavin, J. A., J. C. J. Owen, J. E. P. Connerney, and S. P. Christon, Mariner 10 observations of field-aligned currents at Mercury, *Planet. Space Sci., 45,* 133-141, 1997.

Suess, S. T. and B. E. Goldstein, Compression of the Hermean magnetosphere by the solar wind, J. Geophys. Res. *84,* 3306-3312, 1979.

Williams, D. J., The Earth's ring current: Present situation and future thrusts, *Physica Scripta, T18,* 140-151, 1986.

Whang, Y. C., Magnetospheric Magnetic Field of Mercury, J. Geophys. Res. *82,* 1024-1030, 1977.

Wolf, R. A., Magnetospheric configuration, in: M. G. Kivelson, C. T. Russell (eds.), *Introduction to Space Physics,* Cambridge University Press, Cambridge, 1995.

---

K.-H. Glassmeier, Institut für Geophysik und Meteorologie, Technische Universität Braunschweig, Mendelssohnstr. 3, D-38106 Braunschweig, Federal Republic of Germany; email: kh.glassmeier@tu-bs.de

# A new Technique for the Mapping of Ionospheric Field-Aligned Currents From Satellite Magnetometer Data

Daniel R. Weimer

*Mission Research Corporation, Nashua, New Hampshire*

A new, state of the art technique is being developed for mapping the large-scale, field-aligned currents above the ionosphere. The method uses satellite measurements of the magnetic field, but unlike previous techniques, which have usually relied upon the "infinite current sheet assumption," the new method makes no assumption about the current geometry. The new technique is an adaptation of one used with much success to model the electric potential distribution over the polar caps. Previous electric potential maps were derived from measurements of the potential along multiple satellite passes, by using a least-error fit of spherical harmonic coefficients. The models of the currents are derived from similar maps of "magnetic Euler potential." These Euler potentials are obtained by a path integral of the cross product of a radial vector with the measured magnetic field perturbations. Surface potential maps are derived by a least-error-fit of several paths, and the field-aligned current density is then calculated from the horizontal surface Laplacian of the potential functions. Maps that are derived using this technique show unprecedented details. It is planned that a comprehensive set of field-aligned current maps will be developed for various orientations of the interplanetary magnetic field (IMF).

## INTRODUCTION

In order to understand the interregional processes and driving mechanisms of magnetospheric current systems, it is necessary to first have detailed maps of the currents. For example, a figure by *Iijima and Potemra* [1976] shows a map of the locations of the Region 1 and Region 2 field-aligned currents (FAC) at geomagnetically quiet times. This figure has become one of the most often reproduced illustrations in space physics, and the terminology "Region 1" and "Region 2" originated with the *Iijima and Potemra* [1976] and associated publications.

Of course, much additional work on the subject of current patterns has been done since then. Most of the more recent discussions have been concerned mainly with the morphology of the so-called "Region 0" or "cusp currents" on the dayside. A few examples are by *Cowley et al.* [1991], *de la Beaujardiere et al.* [1993], *Saunders* [1992], *Erlandson et al.* [1988], *Ohtani et al.* [1995a], *Yamauchi et al.* [1993], and *Ohtani et al.* [1995b], to name just a few. Given the differences of opinion evident in these papers, it is evident that the morphology of the cusp currents is a controversial subject. It is clear is that the $Y$ component of the Interplanetary Magnetic Field (IMF) does have a strong influence on the topology of the field-aligned currents near noon. Surprisingly, there have been very few discussions about the topology of the current systems on the night side as a function of the IMF $B_Y$, particularly since it is well known that the electric potential patterns have significant distortions near midnight which are also influenced a great deal by IMF $B_Y$ [*Heppner and Maynard*, 1987; *Weimer*, 1995].

Magnetospheric Current Systems
Geophysical Monograph 118
Copyright 2000 by the American Geophysical Union

When the IMF $B_Z$ component is strongly positive the electric potentials change to a four-cell configuration, where there is a pair of "reversed" convection cells near the polar, surrounded by a pair of "normal" cells at lower latitudes. Likewise, *Iijima et al.* [1984] and *Zanetti et al.* [1984] have found that for northward $B_Z$, there are a pair of reversed-polarity field-aligned currents poleward of the Region 1 currents. These currents are associated with reversed convection cells, and the configuration is referred to as the "NBZ" Birkeland current system [*Iijima et al.*, 1984; *Iijima and Shibaji*, 1987].

The relationship between the FAC systems and the electric potential patterns is another issue. *Cowley et al.* [1991] show a superposition of the current systems on the electric potential contours for two signs of $B_Y$. In their illustration there are three current systems only on one side of noon. In contract, *Ohtani et al.* [1995a] show how three current systems on both sides of noon might exist with a symmetrical convection pattern. Perhaps the most quantitative comparisons between the FACs and electric potentials have been derived from inversions of ground-based magnetograms. Just of couple of many examples are by *Friis-Christensen et al.* [1985] and *Lu et al.* [1995]. The major disadvantage of the inversion technique is that it must rely on estimates of the ionospheric conductivity, which can have a significant influence on both the derived electric fields and FACs.

To summarize, the current state of knowledge about the high-latitude field-aligned current systems consists primarily of qualitative sketches of the configurations for a few generalized IMF orientations. The models that do exist are contradictory in some cases. There do not exist any quantitative mappings of the FACs for a complete set of IMF orientations or dipole tilt angle, as there does for the high-latitude electric potentials [*Weimer*, 1995]. The purpose of this paper is to introduce a new technique for the analysis of satellite magnetometer data in order to create detailed maps of the field-aligned currents for various IMF, tilt angle, and solar wind conditions.

## METHOD

The new technique uses many of the same methods of that have been used in the past with much success for the analysis of electric potential patterns, involving a determination of the patterns by a least-error-fit of spherical harmonic coefficients [*Weimer*, 1995, 1996, 1999]. In the past, nearly all calculations of field-aligned current density from satellite magnetometer data have relied upon the "infinite current sheet approximation"; the new technique requires no such assumption about the geometry of the currents.

The current patterns are obtained from a scalar potential field, which is determined over a two-dimensional spherical surface, using multiple satellite measurements along different tracks. It is shown by *Stern* [1970] that any three-dimensional magnetic field can be represented as

$$\mathbf{B} = \nabla \alpha \times \nabla \beta \qquad (1)$$

where $\alpha$ and $\beta$ are referred to as "Euler potentials." In spherical coordinates a pure toroidal field may be represented by one Euler potential [*Stern*, 1970]:

$$\mathbf{B} = \nabla \psi \times \nabla(\tfrac{1}{2}r^2) = \nabla \psi \times \mathbf{r} \qquad (2)$$

*Backus* [1986], hereafter referred to as B86, considers the modeling of magnetic fields and currents in spherical shells, and also describes a theory of surface operators. Using the same notation as B86, the total magnetic field in a spherical shell with a non-zero current can be separated into two parts:

$$\mathbf{B} = \mathbf{P} + \mathbf{Q} \qquad (3)$$

$$\mathbf{P} = \nabla \times (\mathbf{r} \times \nabla P) \qquad (4)$$

$$\mathbf{Q} = \mathbf{r} \times \nabla Q \qquad (5)$$

"The fields $\mathbf{P}$ and $\mathbf{Q}$ are called poloidal and toroidal vector fields, respectively, and $P$ and $Q$ are called poloidal and toroidal scalars for $\mathbf{B}$. ... By definition, the curl of a toroidal field is poloidal. It is a useful fact about spheres that the curl of a poloidal field is toroidal" [B86]. The toroidal scalar defined by Backus in (5) is identical to Stern's Euler potential in (2), with a sign change.

The current $\mathbf{J}$ in the shell also can be represented by poloidal and toroidal vector fields, and "the poloidal magnetic field $\mathbf{P}$ comes entirely from the toroidal current and the toroidal magnetic field $\mathbf{Q}$ comes entirely from the poloidal current. ... The poloidal current is also the radial component of the total current, since the toroidal current has no radial component" [B86]. In his Equation (71) Backus shows that $J_r$, the radial component of the current crossing the shell, can be found from

$$\mu_o r J_r = \nabla_1^2 Q \qquad (6)$$

where $\nabla_1^2$ is the "dimensionless surface Laplacian", which in spherical coordinates is

$$\nabla_1^2 Q = \frac{1}{\sin\theta} \frac{\partial}{\partial \theta}\left(\sin\theta \frac{\partial Q}{\partial \theta}\right) + \frac{1}{\sin^2\theta} \frac{\partial^2 Q}{\partial \varphi^2} \qquad (7)$$

The scalar $Q$, as used in B86, has the same units as the magnetic field, such as Tesla. Equation (6) does not depend on an assumption that the radial currents extend to infinity. For an example of a geophysical application of the formulas in B86, see *Olsen* [1997].

Equation (5) can be rearranged to the form

$$\hat{r} \times \mathbf{Q} = -r\nabla_S Q \qquad (8)$$

where $\hat{r}$ is a unit vector, $\nabla_S$ is the horizontal surface gradient, and $r$ is constant on a spherical shell. For the application being developed here it has been found that the combined quantity $rQ$, which has units of Tesla-meters, is more convenient to use. This quantity will be referred to as the Euler potential $\psi$. If $\psi$ is used instead of $Q$ then Equation (6) translates to:

$$\mu_o J_r = \nabla_s^2 \psi \qquad (9)$$

where

$$\nabla_s^2 = r^{-2} \nabla_1^2 \qquad (10)$$

The vector $\hat{r} \times \mathbf{Q}$ is equal to the gradient of $\psi$. Just the same as where an electric field that is measured on a satellite can be integrated along the satellite path to obtain the electric potential [*Weimer*, 1995], the magnetic Euler potential $\psi$ along a satellite path can be obtained by integrating the measured magnetic field perturbations $\mathbf{Q}$ that are due to the field-aligned currents, after a cross-product operation with $\hat{r}$.

The theory in B86 was presented for generic curvilinear coordinate systems, with examples given for spherical coordinates. The above equations should also be valid for curvilinear dipole coordinates. It seems intuitive that for the case of magnetic field-aligned currents a dipole geometry would be a more appropriate choice, where $\hat{r}$ is always in the direction of the geomagnetic field and current. In this case the poloidal current is the total current, not just the radial component of the total current. Where field-aligned current is conserved the quantity $\psi$ is constant along a given magnetic field line. Although the current density does change with altitude, the corresponding change in the flux tube dimensions exactly cancels the density variations in the evaluation of the Euler potential $\psi$.

To demonstrate how the magnetic Euler potential can be used to derive field-aligned currents from satellite data, we start with one example case of magnetometer measurements [*Farthing et al.*, 1981] on the DE-2 satellite [*Hoffman et al.*, 1981]. The data are from measurements of the full vector magnetic field from 18:19 UT to 19:04 UT on January 16, 1982. Figure 1 shows the measured magnetic field after subtracting the IGRF magnetic field model that is calculated at the satellite position. The coordinate system is spacecraft centered, where $X$ is in the direction of the velocity vector, $Y$ is either up or down to the center of the earth, and $Z$ is in the perpendicular direction and also contains the majority of the perturbations due to the field aligned currents.

If the model field were perfect and if the satellite attitude had precise pointing accuracy, then the delta-B components would be nearly flat outside of the auroral current zone. But

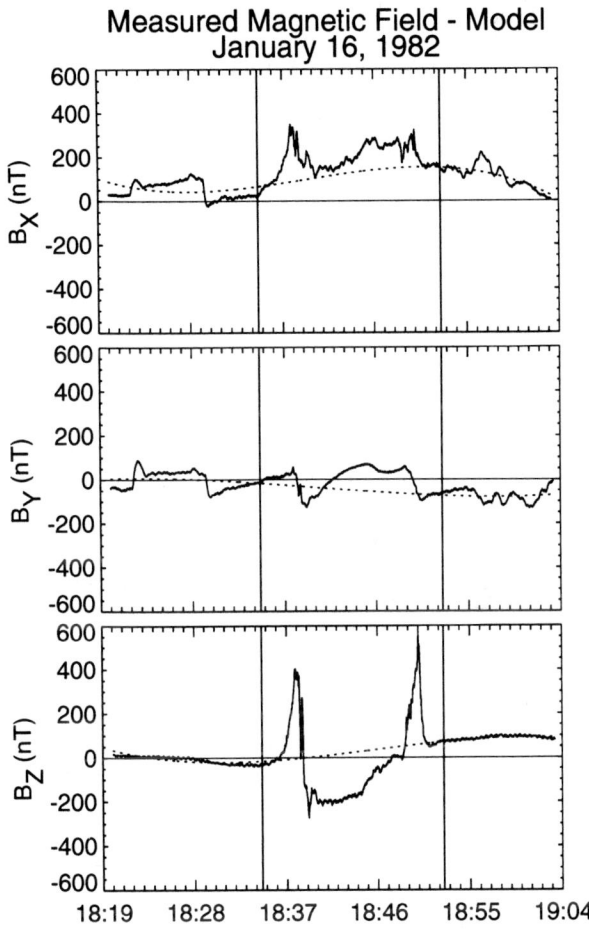

**Figure 1.** Example of magnetic field measured across the polar cap, by the Dynamics Explorer 2 satellite on January 16, 1982. All three vector components are shown, in a spacecraft-centered coordinate system., with the IGRF model field subtracted from the measured field. The vertical lines mark the boundaries of the auroral zone. The superimposed dotted lines show the results of fitting these data with polynomials, using only the portions outside of the vertical lines.

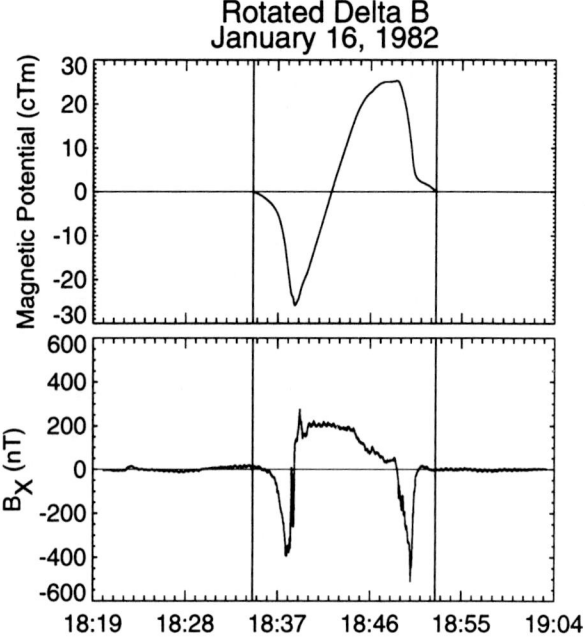

**Figure 2.** The bottom plot shows the $X$ component, in the direction of the satellite's velocity, of the vector obtained by first subtracting the polynomial lines from the data shown in Figure 1, and then taking a cross product with a unit vector in the direction of the main geomagnetic field. The upper plot shows the magnetic Euler potential result from integrating this component along the satellite's path within the auroral zone.

since the attitude knowledge is imperfect then there are offsets, which must be corrected before further processing can be done. These offsets are the bane of magnetometer data analysis. Polynomial equations are derived by a least-error fit to each component of delta-B, as shown in the dotted lines in Figure 1 (others have used spline fits). So that the polynomial fits do not attempt to match the perturbations due to the field aligned currents, which is the information which is desired to be retained, then the fitting must use only the portion of the satellite orbit that is below the latitude of the currents. The region of interest is marked with the vertical lines.

The next step is to subtract these polynomial fits to obtain a corrected delta-B vector, which corresponds to **Q** in the B86 notation. Delta-B is then used in a cross product with a unit vector in the direction of the local geomagnetic field, which does a rotation. The result of this calculation is shown in the bottom part of Figure 2, where just the $X$ component is shown, as this component is in the direction of spacecraft motion. The majority of the contribution to the resulting $X$ component is from the original $Z$ component, but all three components of delta-B were actually used. The other components may contribute where the geomagnetic field is not

exactly along $Y$. As mentioned previously, the Euler potential is now obtained by integrating the vector cross product along the spacecraft track. The top of Figure 2 shows the result of this integration, using only the data between the lines. When $\psi$ is converted to units of cTm the results are comparable to the electric potentials in units of kV.

When the similar procedure is used to obtain satellite measurements of electric potentials, it is desired to have the potentials constant along and outside the low-latitude boundary of the auroral zone, since the electric fields there are presumed to be zero in a co-rotating frame of reference. A small offset constant may be used in the integration to force both ends of each pass to match, to correct for temporal variations. In the case of the magnetic potentials it is also desired to have no potential gradients along or outside the low-latitude boundary, which is presumed to be outside the influence of the auroral currents. The boundary also circles a region which is presumed to have a total net current of zero passing through it. In the example just shown a small offset of 35 nT was subtracted from the rotated delta-$B_X$ in order to make both end points equal to zero, correcting for the time variations during the satellite's passage.

In order to obtain a two-dimensional map of the magnetic Euler potential field it is necessary to combine data from multiple satellite passes during times of similar IMF conditions, hopefully sampling all local times. Next these randomly positioned samples are analyzed by the technique of fitting a series of spherical harmonic coefficients to the data by the method of least squares. The result is an expression for the magnetic Euler potential on a two-dimensional spherical surface in the form:

$$\psi(\theta,\varphi) = \sum_{l=0}^{4} \sum_{m=0}^{\mathrm{Min}(l,2)} (A_{lm}\cos m\varphi + B_{lm}\sin m\varphi)P_l^m(\cos\theta) \quad (11)$$

where $\varphi$ is the MLT angle, $\theta$ is a function of the co-latitude, and $P_l^m$ is the "associated Legendre function." The $A_{lm}$ and $B_{lm}$ are the spherical harmonic coefficients which contain the description of the two-dimensional function.

The outcome of this process with one group of data is shown in the contour map in the right column of the middle row of Plate 1. This particular group of data were obtained when the average IMF during the preceding 40 minutes was oriented within ±22.5° of the $+Y_{GSM}$ direction and had a total $Y$-$Z$ magnitude in the range of 5.2 to 7.25 nT, to facilitate a comparison with earlier electric potential results shown in *Weimer* [1995]. The interval between contour levels is 4 cTm, as indicated in the upper-right corner of the box, starting at ±2 cTm. The minimum and maximum values of the potential are shown in the lower left and right corners respectively.

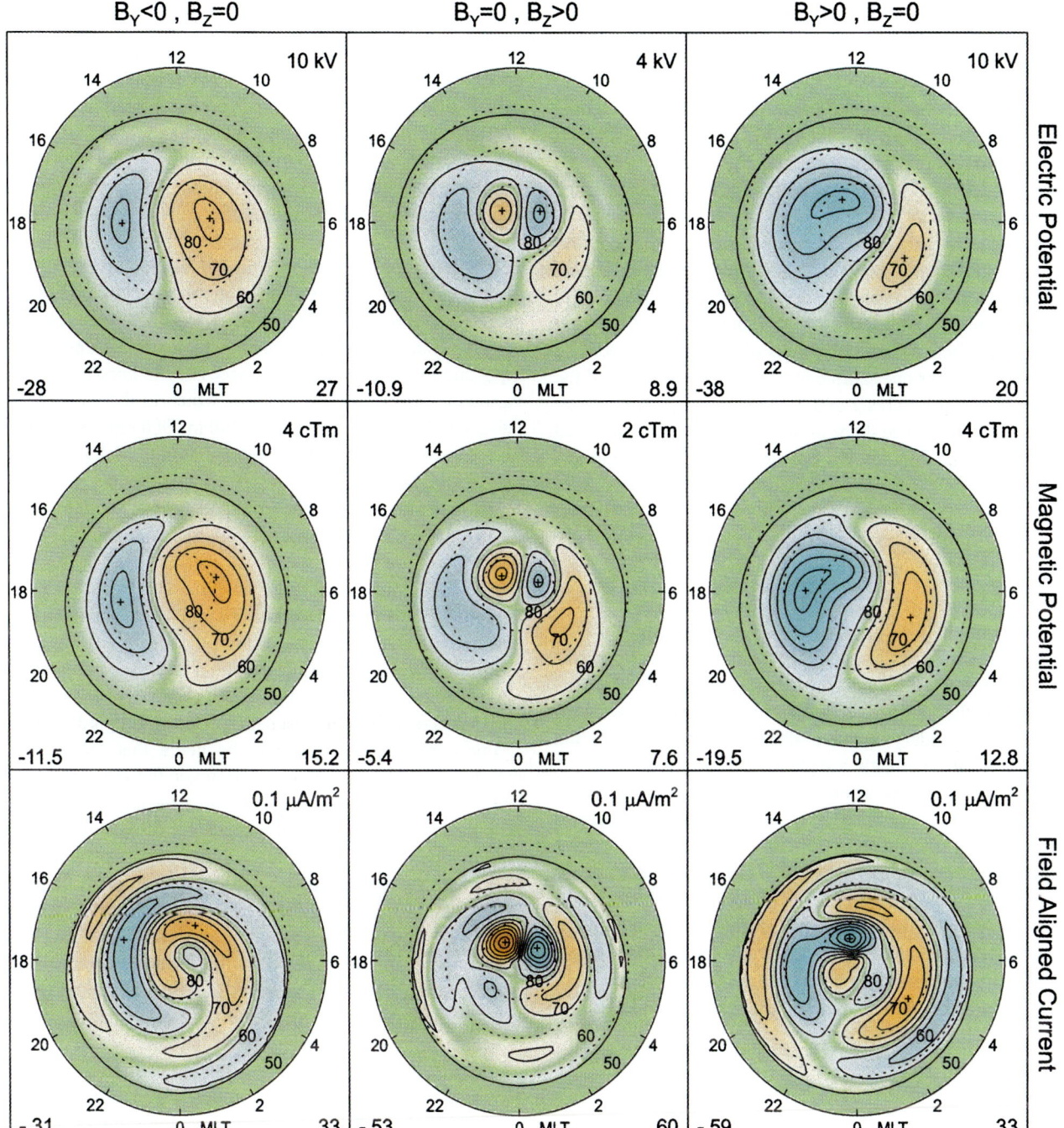

**Plate 1.** The middle row shows the magnetic Euler potentials that result from a least-error-fit of data from multiple satellite passes. The left column shows the results for data associated with IMF in the -Y direction, the middle column shows the results for +Z IMF, and the right column is for +Y IMF. The upper row shows the electric potentials for the same cases, and the bottom row shows the maps of the field-aligned currents that are obtained from the magnetic Euler potentials.

For a comparison, the electric potentials derived from electric field data on the same group of satellite passes (give or take a few passes where both instruments were not operating at the same time) is shown in the upper right corner of Plate 1, with a contour interval of 10 kV. There are many similarities between the magnetic Euler and electric potential contour patterns, and in fact if the ionospheric Pedersen conductivity were perfectly uniform then these two patterns would be identical in appearance, having magnitudes related by the conductivity.

Both the magnetic Euler and electric potential contour patterns in Plate 1 are surrounded by slightly elliptical boundaries that are offset from the center. These boundaries show the low-latitude limit of the input data, and they are determined from the ending points of the orbit tracks. The end points for our one sample pass are at the locations of the vertical lines in Figures 1 and 2, which were determined by visual inspection. The shape of these boundaries are specified by a co-latitude $\theta_{Max}$ that is a function of the MLT angle $\varphi$:

$$\theta_{Max}(\varphi) = \sum_{n=0}^{2} (C_n \cos n\varphi + D_n \sin n\varphi) + O \qquad (12)$$

where again the $C_n$ and $D_n$ coefficients are determined by a least-error fit using the end point locations. The boundary that is determined by this fit goes through the median co-latitude of the end points; a constant offset $O$ is added to extend the boundary so that it also encompasses points that are within approximately a standard deviation away.

The purpose of this boundary is to improve the spherical harmonic fits. Spherical harmonics normally represent functions that span an entire spherical surface having a polar angle that extends from 0° to 180°. However, the polar cap area that we are interested in is all within approximately 35° from the pole. Fitting this limited data to a function that covers the entire sphere has disadvantages in that a larger number of coefficients (the maximum $l$ value) are required, and the end result tends to have undesired harmonic oscillations, which are particularly noticeable at latitudes below where there were any input data. To eliminate these difficulties the polar angle $\theta$ in (12) is specified as

$$\theta = \frac{\lambda}{\theta_{Max}(\varphi)} \pi \qquad (13)$$

where $\lambda$ is the actual geomagnetic colatitude. In essence this procedure does a "conformal mapping" or stretching of the data space to span a sphere. With this being done, then the order of the spherical harmonic coefficients needs to be only four, and the overall result is smoother.

The next step is to calculate the field-aligned currents from the magnetic Euler potential, using Equation (9). The results are shown in the bottom row of Plate 1, scaled to an ionospheric altitude of 150 km. The contour interval is 0.1 µA-m$^{-2}$, and the minimum and maximum values, also in µA-m$^{-2}$, are indicated in the lower left and right corners. Positive, downward currents are indicated with the yellow shading and negative, upward currents are indicated with the cyan shading.

It is important to note that the calculation of the current density is the only place where the radius or altitude really needs to be considered. If one were to use satellite measurements to make a statistical map of the magnetic field perturbations due to currents, then it is necessary to account for the variations in the satellite altitude, which in the case of the DE-2 satellite ranges from about 300 to 1000 km. In order to determine how the different components of the delta-B vector vary with altitude it is also necessary to assume a geometry for the currents, such as the infinite sheet model. However, the magnetic Euler potential mapping does not depend on the geometry of the current.

The two dimensional Euler potential maps contain all of the information about the vector magnetic field perturbations that one satellite pass alone cannot determine. In essence, sampling the magnetic Euler potential along separated paths yields information about the variation in the magnetic field and the currents between the paths, in the unsampled regions.

For completeness it should be noted that the delta-B magnetic field that is integrated to obtain the Euler potential or toroidal scalar $\psi$ may also contain a component due to a poloidal field from external currents. In this situation the poloidal field would be due to ionospheric Hall currents. However, it has been determined that if a poloidal field, as described in Equation (4), is used to derive the scalar $\psi$ then the current that is computed with Equation (9) is zero. Even if this were not the case, then the influence of the Hall currents as a source of error in the field-aligned current measurements would be no greater than with techniques that use the infinite current sheet approximation.

One other aspect of this technique that should be mentioned is that it is not sensitive to the small attitude oscillations that occasionally contaminate satellite magnetic field data, particularly in the case of DE 2. These oscillations, which have a period of a few seconds, would make it very hard to measure currents by using the derivatives of the measured magnetic field, unless the oscillations are first filtered out. It has been found that the integration procedure automatically eliminates the influence of the oscillations on a scale greater than one period.

## RESULTS

The same procedure just described has also been applied to two other test sets of data. In one set the IMF conditions were nearly the same as in the first case, except the sign of the $Y$ component is changed from positive to negative. The results for $-Y$ IMF are shown in the left column of Plate 1. The scales are the same as in the right column. The other set used data where the IMF had a very strong $+Z$ component and a small $Y$ component. The magnitudes were over 7.25 nT, with an orientation within 22.5° of purely northward. The results for $+Z$ IMF are shown in the middle column. In this case the electric and magnetic Euler potentials are smaller, so the contour intervals have been reduced, as indicated in the upper right corners. However, the resulting field-aligned current intensities are just as strong, and even higher, as in the other two cases.

The maps of the field aligned currents in Plate 1 are much more detailed than previously published figures, which for the most part have consisted of rough sketches showing simply regions of up and down, rather than magnitudes. These newer maps are also in a numerical form which could easily be imported into other types of calculations, such as numerical simulations or "space weather" forecasting. Models which exist only as drawings are not as useful.

The maps in Plate 1 are particularly helpful in showing the relationships between the Region 1 and Region 2 currents in the $+Y$ and $-Y$ IMF orientations. In particular, for the $-Y$ orientation (left column in Plate 1) the Region 1 current on the dawn side wraps around through noon to become part of the "Region 0" current on the post-noon side. There appears to be an isolated "Region 0" of opposite polarity on the pre-noon side. A mirror-like situation exists for the $+Y$ orientation (right column), where now the dusk side Region 1 current passes through noon to become part of Region 0 on the pre-noon side, and again there is an isolated, opposite polarity Region 0 current on the post-noon side. Differences in magnitudes may perhaps be due to statistical variations in the input data in the IMF magnitudes or perhaps the (seasonal) dipole tilt angle, which appears to have a strong influence. These differences will be resolved after more data can be processed.

There is also an interesting relationship between the currents near midnight. For the $-Y$ orientation there is a continuous band of downward current which links the dawn's Region 1 with the dusk's Region 2, whereas for the $+Y$ orientation the upward current has more continuity. This aspect of the current geometry has not generally been appreciated up until now.

The northward IMF case (center column) is even more interesting. The magnetic Euler potentials have a four-cell pattern like the electric potentials, and in fact the four cells are even more distinct in the magnetometer data. The currents that are associated with this pattern consist of a pair of intense, cylindrically shaped regions, having opposite polarity, nearly centered around noon. At lower latitudes there are weaker currents with the conventional Region 1 and 2 shapes and polarity. This is the NBZ system described by *Iijima et al.* [1984] and *Zanetti et al.* [1984], only now the picture is much clearer. The geometry of all current patterns in the bottom row of Plate 1 also have excellent agreement in comparison to the corresponding patterns that are shown by *Friis-Christensen et al.* [1985].

Again it is emphasized that the currents that are derived from the magnetic Euler potentials do not depend on the geometry of the currents, and this is particularly important in the NBZ case. If a satellite passes through the pair of cylindrical current regions in the dusk-dawn direction then the cross-track magnetometer component will have a W-shaped pattern. Using the traditional infinite current sheet approximation, this pattern could be interpreted as four alternating current belts instead of two cylindrical regions.

What the magnetometers will see for a given current pattern is readily apparent in the Euler potential contours. The Euler potentials have the property that "since $\nabla\alpha$ and $\nabla\beta$ are perpendicular to **B**, surfaces of constant $\alpha$ and $\beta$ are tangential to the field at all points, and the same holds for the lines along which surfaces intersect. Such lines are therefore magnetic field lines" [*Stern*, 1970]. In other words, the Euler potential contour lines show the direction of the magnetic field perturbation vectors that are due to the field aligned currents. They also indicate the direction of the **J**×**B** force or "magnetic tangential stress" that is transmitted from the magnetosphere to the ionosphere by the currents, as shown by *Iijima* [manuscript submitted to this monograph, 1999].

## SUMMARY

A new technique for using satellite measurements of the magnetic field to derive maps of the field-aligned currents above the ionosphere has been shown. Unlike the traditional approach which relies upon the "infinite current sheet assumption", this new method makes no assumption about the current geometry. The models of the currents are derived from maps of magnetic Euler potential, which is obtained from a cross product of the measured magnetic field perturbations with the geomagnetic field unit vector, and then integrating the result along the satellite path. Many of the same spherical harmonic fitting techniques that were previously developed for modeling electric potential patterns are used in calculating the magnetic Euler potential, and the Euler potentials actually have many similarities to the electric potentials. The field aligned current density is then calculated from the surface Laplacian of the Euler potential, which is equivalent to the curl of the two-dimensional perturbation vector field.

Maps that are derived with this technique show unprecedented detail. Much can be learned about the configurations of the currents for various orientations of the IMF, and these results could be useful for other modeling and prediction efforts as well. Presently the technique has only been in the developmental stage, and only a few groups of test data have been analyzed. Since the data from each and every pass needs to be manually processed to correct for attitude errors, processing of a sufficiently large data base is a time consuming process. The next step is to process all of the DE-2 magnetometer data having simultaneous IMF available. Eventually measurements from the DMSP satellites and MAGSAT could also be included, and perhaps that from other current or future missions. (It is known that the accuracy of the MAGSAT data is much better than on DE-2, but it could not be used for these test maps because the MAGSAT orbit tracks only went from dawn to dusk and did not adequately sample all MLT.) It is planned that a comprehensive set of field-aligned current maps will be developed for various orientations of the interplanetary magnetic field (IMF). These maps may be pertinent to numerical modeling and space weather forecasting.

*Acknowledgments.* This research was supported by NSF grant ATM-9701868 to Mission Research Corporation. The data from the DE 2 satellite were provided by the National Space Science Data Center on NSSDC Data Set ID 81-070B-01E/-02H.

## REFERENCES

Backus, G., Poloidal and toroidal fields in geomagnetic field modeling, *Rev. Geophys., 24,* 75, 1986.

Cowley, S. W. H., J. P. Morelli, and M. Lockwood, Dependence of convective flows and particle precipitation in the high-latitude dayside ionosphere on the X and Y components of the interplanetary magnetic field, *J. Geophys. Res., 96,* 5557, 1991.

de la Beaujardiere, O., J. Watermann, P. Newell, and F. Rich, Relationships between Birkeland current regions, particle precipitation, and electric fields, *J. Geophys. Res., 98,* 7711, 1993.

Erlandson, R. E., L. J. Zanetti, T. A. Potemra, P. F. Bythrow, and R. Lundin, IMF $B_Y$ dependence of region 1 Birkeland currents near noon, , *J. Geophys. Res., 93,* 9804, 1988.

Farthing, W. H., M. Sugiura, B. G. Ledley, and L. J. Cahill, Jr., Magnetic field observations on DE-A and -B, *Space Sci. Instrum., 5,* 551-560, 1981.

Friis-Christensen, E., Y. Kamide, A. D. Richmond, and S. Matsushita, Interplanetary magnetic field control of high-latitude electric fields and currents determined from Greenland magnetometer data, *J. Geophys. Res., 90,* 1325, 1985.

Heppner, J. P. and N. C. Maynard, Empirical high-latitude electric field models, *J. Geophys. Res., 92,* 4467-4489, 1987.

Hoffman, R. A., and E. R. Schmerling, Dynamics Explorer program: an overview, *Space Sci. Instrum., 5,* 345, 1981.

Iijima, T. and T. A. Potemra, Large-scale characteristics of field-aligned currents associated with substorms, *J. Geophys. Res., 81,* 3999, 1976.

Iijima, T., T. A. Potemra, L. J. Zanetti, and P. F. Bythrow, Large-scale Birkeland currents in the dayside polar region during strongly northward IMF: A new Birkeland current system, *J. Geophys. Res., 89,* 7441, 1984.

Iijima, T., and T. Shibaji, Global characteristics of northward IMF-associated (NBZ) field-aligned currents, *J. Geophys. Res., 92,* 2408, 1987.

Lu, G., L. R. Lyons, P. H. Reiff, W. F. Denig, O. de la Beaujardiere, H. W. Kroehl, P. T. Newell, F. J. Rich, H. Opgenoorth, M. A. L. Persson, J. M. Ruohoniemi, E. Friis-Christensen, L. Tomlinson, R. Morris, G. Burns, and A. McEwin, Characteristics of ionospheric convection and field-aligned current in the dayside cusp region, *J. Geophys. Res., 100,* 11,845, 1995.

Ohtani, S., T. A. Potemra, P. T. Newell, L. J. Zanetti, T. Iijima, M. Watanabe, M. Yamauchi, R. D. Elphinstone, O. de la Beujrdiere, and L. G. Blomberg, Simultaneous pre-noon and post-noon observations of three field-aligned current systems from Viking and DMSP-F7, *J. Geophys. Res., 100,* 119, 1995a.

Ohtani, S., T. A. Potemra, P. T. Newell, L. J. Zanetti, T. Iijima, M. Watanabe, L. G. Blomberg, R. D. Elphinstone, J. S. Murphree, M. Yamauchi, and J. G. Woch, Four large-scale field-aligned current systems in the dayside high-latitude region, *J. Geophys. Res., 100,* 137, 1995b.

Olsen, N., Ionospheric *F* region currents at middle and low latitudes estimated from Magsat data, *J. Geophys. Res., 102,* 4563, 1997.

Saunders, M. A., The morphology of dayside Birkeland currents , *J. Atoms. Terr. Phys., 54,* 457, 1992.

Stern, D. P., Euler potentials, *Am. J. Phys., 38,* 494, 1970.

Weimer, D. R., Models of high-latitude electric potentials derived with a least error fit of spherical harmonic coefficients, *J. Geophys. Res., 100,* 19,595-19,607, 1995.

Weimer, D. R., A flexible IMF dependent model of high-latitude electric potentials having "space weather" applications, *Geophys. Res. Lett., 23,* 2549-2553, 1996.

Weimer, D. R., Substorm influence on the ionospheric electric potentials and currents, *J. Geophys. Res., 104,* 185, 1999.

Yamauchi, M., R. Lundin, and J. Woch, The IMF $B_Y$ effects on large-scale field-aligned currents near local noon: Contributions from cusp part and noncusp part, *J. Geophys. Res., 98,* 5761, 1993.

Zanetti, L. J., T. A. Potemra, T. Iijima, W. Baumjohann, and P. F. Bythrow, Ionospheric and Birkeland current distributions for northward interplanetary magnetic field: Inferred polar convection, *J. Geophys. Res., 89,* 7453, 1984.

---

Daniel R. Weimer, Mission Research Corporation, 1 Tara Blvd., Nashua, NH 03062. (e-mail: dweimer@mrcnh.com)

# Automatic Identification of Large-Scale Field-Aligned Current Structures and its Application to Night-Side Current Systems

T. Higuchi

*The Institute of Statistical Mathematics, Tokyo, Japan*

S. Ohtani

*The Johns Hopkins University Applied Physics Laboratory, Laurel, Maryland, U.S.A.*

The present paper reports a newly developed procedure to automatically identify a spatial structure of large-scale field-aligned currents (FACs) from satellite magnetic field measurements. The procedure is based on the concept of the first-order B-spline fitting with variable (i.e., movable) node positions, which may be envisioned as fitting line segments to a plot of a magnetic field component. The number of node points is optimized for each current sheet crossing so that a certain information criterion is minimized. The properties of a FAC system such as its location and intensity can be easily calculated from the information of node points. The procedure is applied to the entire set of nightside magnetic field measurements (1339 days) made by the DMSP-F7 satellite and for demonstration, the intensity of nightside region-1 FACs is examined as a function of magnetic latitude. The results not only confirm the general expectation that intense region 1 currents tend to be observed at lower latitudes but also quantify the relationship between the current intensity and latitude. The capability of such quantification should be useful for future studies including practical applications such as space weather forecast.

## 1. INTRODUCTION

The importance of FACs in the electromagnetic coupling between the magnetosphere and ionosphere cannot be overstated. The characteristics of FACs have been examined for more than three decades since FACs were initially detected by a low-altitude satellite at the very early stage of satellite observation [e.g., *Zmuda et al.*, 1966; *Cummings and Dessler*, 1967]. Large-scale FACs are classified into three systems, that is, region 2 (R2), region 1 (R1), and region 0 (R0) systems from equatorward to poleward [*Iijima and Potemra*, 1976]. The R0 system is distributed in the midday sector, whereas the R2 and R1 currents encircle a magnetic pole but are off-centered toward midnight. The R1 FAC tends to flow toward and away from the ionosphere in the morning and evening sectors, respectively, and at a given local time the polarity of a R2 current is the opposite to that of a R1 current.

Although such average properties of large-scale FACs are well accepted, it is also known that the spatial structure and intensity of these current systems are far from stationary and depend on both external and internal conditions. However, since the analysis of FACs has been

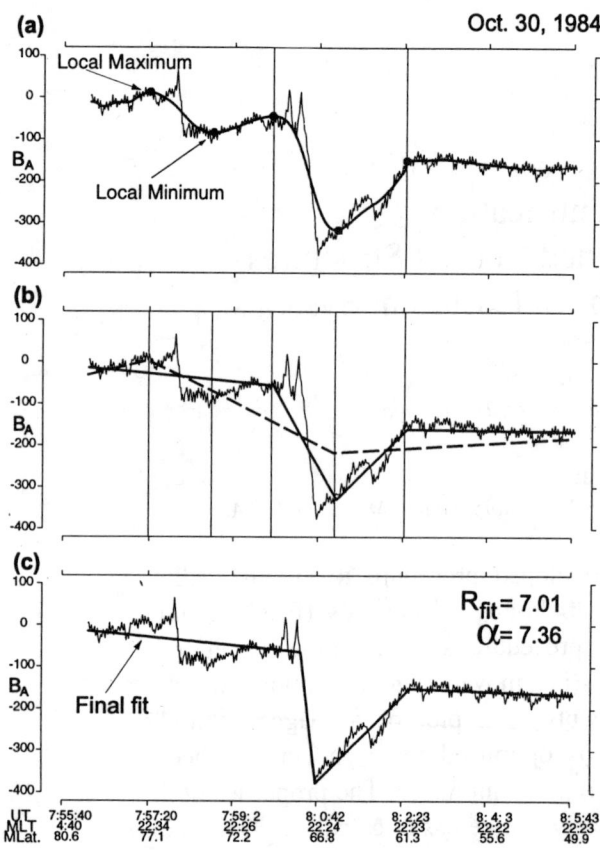

**Figure 1.** Application of the procedure to the DMSP-F7 data of October 30, 1984. (a) Determination of node-point candidates (local minima and maxima) based on the smoothed data. (b) Search for the optimum combination of node points. (c) The final adjustment based on the original data.

strongly depending on visual inspection of satellite magnetic field data, a large statistical study not only requires significant effort but also tends to lack quantification. Considering that the archive of satellite data keeps growing, we believe that a computational procedure to identify large-scale FACs is most needed for performing a data analysis of the next level, which eventually will provide better understanding of the electrodynamic processes of the magnetosphere. In the following we explain a procedure we developed for this purpose and present the results of an initial application to nightside current systems.

## 2. AUTOMATIC IDENTIFICATION OF LARGE-SCALE FAC

In this section we briefly explain the scheme we developed to automatically identify the spatial structure of large-scale FACs from satellite magnetic field data. Measurements of two magnetic field components transverse to the background magnetic field are used for this procedure. We assume that a data file to be processed includes a single crossing of a FAC region. The procedure consists of three tasks, that is, the determination of the interval of a FAC crossing, the rotation of magnetic field vectors to the direction of the maximum variance, and fitting of line segments to an (imaginary) plot of that component [*Hiragi et al.*, 1985; *Kitagawa and Higuchi*, 1998]. Each task consists of a few steps, which are explained in the following:

*Step 1*: An interval of a FAC crossing is preparatorily determined from parameterized characteristics such as variances of magnetic field variations. Only the east–west magnetic component is used at this step.

*Step 2*: The principal component analysis (PCA) (also known as the minimum variance analysis) is applied to magnetic field measurements during a FAC crossing. The maximum-to-minimum ratio of two eigenvalues, $\alpha$, represents the extent to which the spatial structure of FACs can be approximated by an extending sheet. Measured magnetic vectors are rotated, and only the component in the maximum variance direction, which is usually close to the azimuthal direction, is used in the subsequent procedure. We refer to this component as $B_A$.

(Note: Step 1 is necessary before Step 2 because if the PCA is applied to the whole crossing, a gradual background variation affects the result of Step 2. Such a variation is attributed partly to differences between actual magnetic field and model magnetic field subtracted and partly to a smearing effect of FACs.)

*Step 3*: The interval of a FAC crossing is redefined based on the maximum-variance component.

*Step 4*: A smoothing filter [*Higuchi*, 1991] is constructed by examining the autocorrelation of magnetic variations for each of the three intervals, that is, intervals equatorward, inside, and poleward of the FAC crossing. After smoothing the data with the constructed filters, all local maximum and minimum points plus two points at both edges of the FAC crossing are registered as candidates for the boundaries of FAC sheets; these points are called node points. Figure 1a shows a plot of $B_A$ (thin line), as an example, for a DMSP-F7 nightside pass of October 30, 1984. The two vertical lines indicate the poleward and equatorward edges of the FAC crossing, which were determined at Step 3. The thick line plots the smoothed data. The solid circles indicate the node points.

*Step 5*: For an assumed number of FAC sheets (the optimum number of FAC sheets will be determined at the next step), all possible combinations of node points are selected. For a two-sheet structure, for example, three node points are required to fit segments to the plot: two points at the beginning and end of the FAC region and another point at the interface between the two systems. If there are five node points, we have 10 (= $_5C_3$) such combinations. The goodness of each combination is evaluated by the residual sum of squares (RSS) between measurements and segments connecting the selected node points. The combination with the minimum RSS is registered as the optimum one for each number of FAC sheets. For the example shown in Figure 1b, the fit by the thick line segments approximates the actual data much better than the fit by the broken line segments, giving a smaller RSS; the former actually provides the best fit for this number of FAC sheets and the best fit overall.

*Step 6*: An information criterion called Akaike Information Criterion (AIC) [*Akaike*, 1974; *Sakamoto et al.*, 1986] is introduced to determine the number of FAC sheets, $J$, which is expressed as

$$AIC(J) = N \log\left(\frac{RSS(J)}{N}\right) + 4J + Const.,$$

where $N$ is the number of data points.

This parameter evaluates the efficiency of fitting in terms of the number of fitting parameters, $J$ in the present case. RSS is a decreasing function with $J$; in other words, the measurements can be fitted better with more node points. Thus the value of RSS alone does not determine $J$. The number of FAC sheets is determined to be the one that minimizes AIC, and accordingly the optimum combination of node points is selected.

*Step 7*: The values of the smoothed data at local maxima or minima do not always agree with those of the original data. The procedure adopts the quasi-Newton method to make the fitted segments trace more closely to the original data. The thick line in Figure 1c shows the result of this refinement for the given example.

*Step 8*: The properties of each FAC sheet such as its location and intensity are calculated from the locations of node points and are registered. The goodness of the optimum fit, $R_{fit}$, is defined as the standard deviation for the optimum fit divided by the intensity of the most intense FAC and is also recorded. For a magnetic slope to be identified as a FAC sheet, (1) the magnitude of the associated magnetic change must be larger than both 50 [nT] and 30% of the magnitude of the magnetic change associated with the most intense FAC for each pass, and (2) the slope must be steeper than 1.2 [nT/s], which corresponds to a current density of 0.128 [μA/m$^2$], if the trajectory is perpendicular to a current sheet.

There are a few other techniques for calculating the spatial distribution of FACs. One popular example is the assimilative mapping of ionospheric electrodynamics (AMIE) technique [*Richmond and Kamide*, 1988]. Although the AMIE is most useful for examining the time development of the global FAC distribution, it requires magnetic field data from numerous ground stations, and it has to assume ionospheric conductance. A new technique reported by Weimer at the conference [*Weimer*, 1999] uses spherical harmonic functions for presenting the magnetic potential and does not require any assumption about the geometry of FACs. This technique, however, requires data assembled from many satellite passes to calculate the global distribution of FACs. In contrast with these techniques, the method we developed is basically the automation of the way we visually examine a plot of satellite magnetic field data. Because the method can be applied even to a single satellite pass, it may be easily implemented in a real-time processing of data. It can also be applied to an entire set of satellite data, allowing us to perform a large statistical study. The trade-off of our method is that it assumes the sheet structure of FACs, although the extent to which this assumption is justified is quantified by the principal component analysis. Thus, each technique has both pros and cons, and it is most important to choose the right one based on the purpose of a study.

## 3. APPLICATIONS

We have applied the procedure to the entire set of 1-s magnetometer data acquired by the DMSP-F7 satellite from December 1983 to January 1988 (1339 days). DMSP-F7 is a Sun-synchronous satellite with a nearly circular polar pass at about 835 km in altitude, with its ascending and descending nodes at 1030 and 2230 local time (LT), respectively. The orbital period is about 101 min. The satellite crossed the nightside auroral oval more than 35,000 times during its entire mission. After checking data gap and screening of crossings by the principal component analysis and the segment fitting (we use only crossings with $\alpha > 2$ and $R_{fit} < 8$), the procedure identified the structure of FACs for 5,590 and 4,386 crossings in the northern and southern hemispheres, respectively. This task is doable with an ordinary PC machine, as demonstrated with a notebook PC at the conference. With a 260-MHz Pentium-II processor it takes less than 5 s to process one crossing or a minute to transact a 1-day data file.

Table 1 lists the number of crossings for each number of FAC sheets. Note that for each number of FAC sheets there are two distinct types in terms of the polarities (upward or downward) of FACs. Figure 2 shows an example of each

**Table 1.** Number of Passes for Different FAC Structures Crossed by DMSP-F7 on the Nightside

| | \multicolumn{6}{c|}{Number of FAC sheets} | |
|---|---|---|---|---|---|---|---|
| | 1 | 2 | 3 | 4 | 5 | 6 | Total |
| North | 53 | 3,381 | 1,561 | 483 | 110 | 2 | 5,590 |
| South | 338 | 2,673 | 956 | 318 | 101 | 0 | 4,386 |

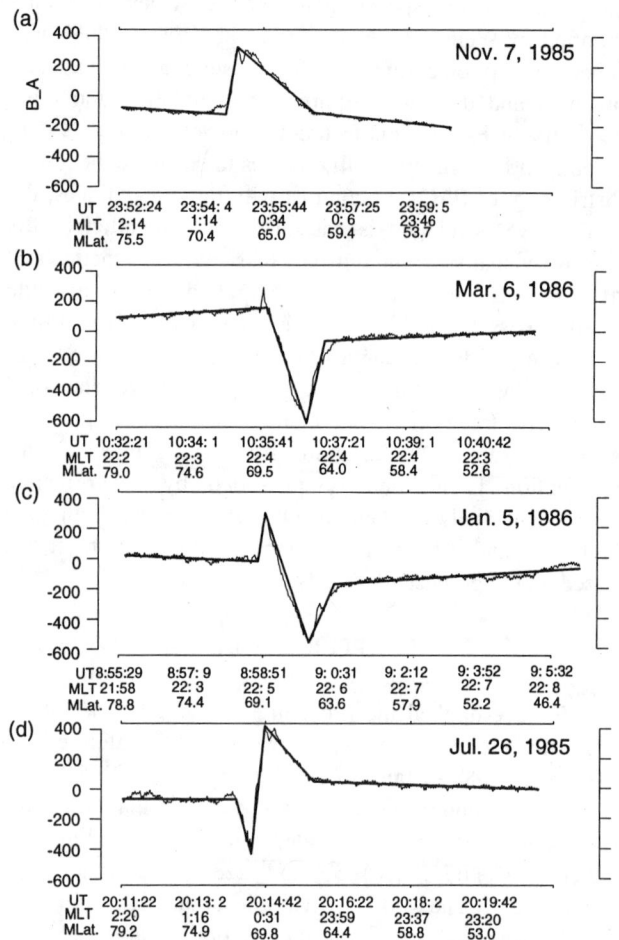

**Figure 2.** Examples of (a and b) two and (c and d) three field-aligned current sheet crossings. The result of the automatic procedure (thick line) is superposed on the plot of the original data (thin line).

type for two- and three-sheet structures. The result of the automatic procedure (thick line) is superposed on the actual measurement (thin line).

To demonstrate the potential of the developed procedure for studying FACs, we examine the relationship between the latitude and the intensity of nightside R1 FACs. We used data only from the northern hemisphere (the northern passes of DMSP-F7 cover both premidnight and postmidnight sectors, but the southern passes do not) and focused on downward-flowing R1 FACs observed as a part of a two- or three-sheet structure. Examples of such events are shown in Figures 2a and 2d. A pair of downward-flowing R1 and upward-flowing R2 currents should be accompanied by a westward-flowing electrojet in the ionosphere, which is an important element of the magnetospheric substorm. Here we note that the term "region 1" or "region 2" in a general sense without distinction in terms of their association with the substorm process.

Figure 3 plots the location of the center of region 1 FAC sheets in the frame of magnetic latitude (MLat) vs. magnetic local time (MLT) for a total of 2061 events. The solid circles represent two-sheet structure events with $\Delta B < 200$ nT ($\Delta B$ is the magnitude of a change in $B_A$ throughout the R1 FAC sheet crossing), whereas the crosses and dots represent three-sheet structure events and other two-sheet structure events, respectively. The solid curve represents the optimum fit of a quadratic function for the solid-circled events, which is expressed as

$$\text{MLat}(\Delta\text{MLT}) = 68.94 - 0.1133 \times (\Delta\text{MLT}) + 1.276 \times (\Delta\text{MLT})^2$$

$$\Delta\text{MLT} = \text{MLT (for MLT} > 0) / \text{MLT} - 24 \text{ (for MLT} < 24)$$

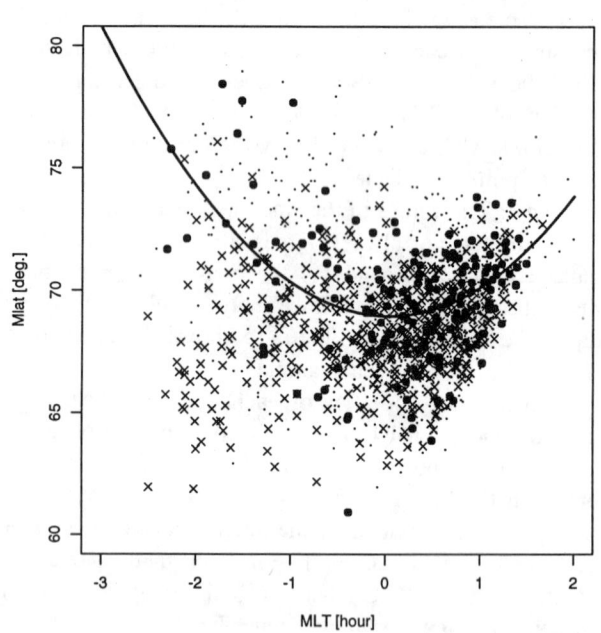

**Figure 3.** Positions of the center of R1 FAC sheet crossings in the MLat vs. MLT format for two-sheet structure events with $\Delta B < 200$ nT (solid circles), other two-sheet structure events (dots) and three-sheet structure events (crosses). The solid line represents the result of the quadratic fit for the solid circles.

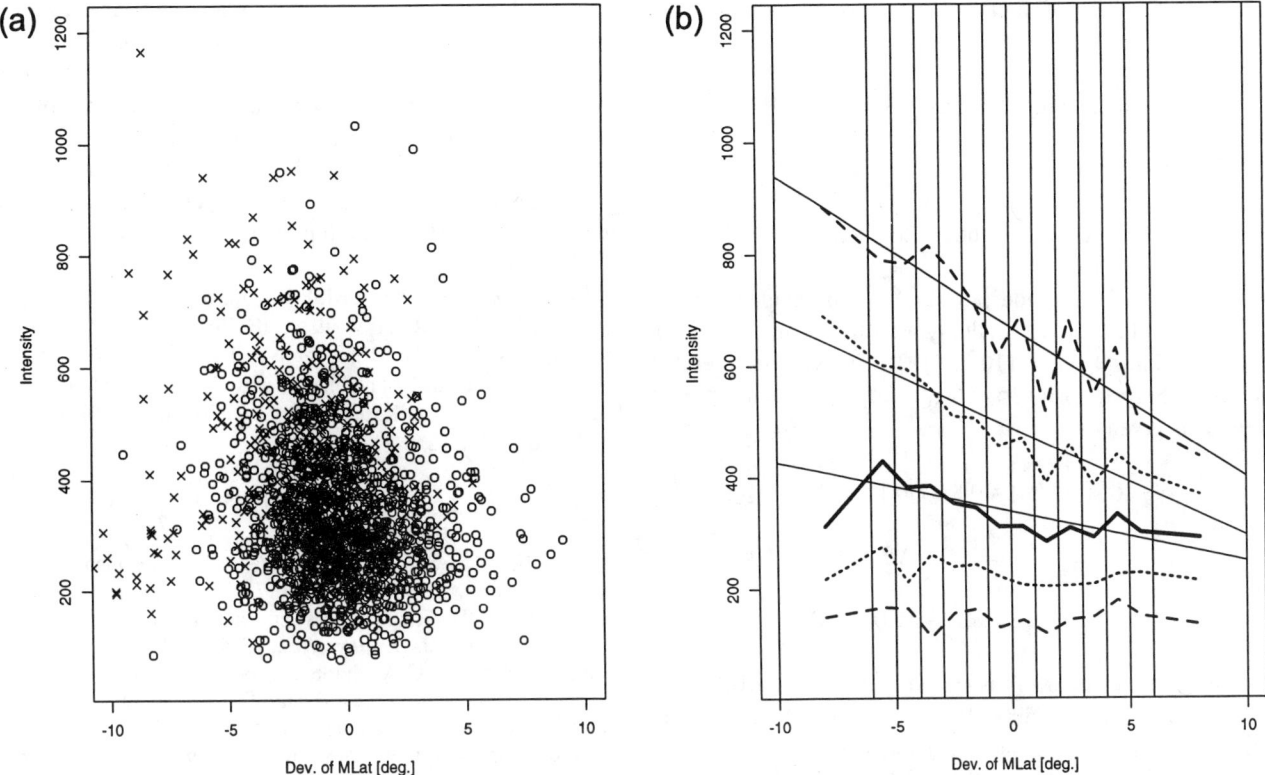

**Figure 4.** (a) The magnitude of magnetic perturbations ($\Delta B$) associated with R1 currents plotted against $\Delta$MLT for two-sheet (open circles) and three-sheet (crosses) structure events. (b) The values of $\Delta B$ at the medium, $\pm\sigma$, and $\pm 2\sigma$, along with the result of the linear fit for the medium, $+\sigma$, and $+2\sigma$ values.

We did the same fitting to events with $\Delta B < 300$ nT but found that the result is not very different. We use MLat($\Delta$MLT) as a reference and evaluate the latitude of a R1 FAC as a deviation from MLat($\Delta$MLT), which we denote as $\Delta$MLat.

Figure 4a plots $\Delta B$ against $\Delta$MLat. Again the crosses indicate events with three-sheet structure. Events with small values of $\Delta B$ are absent because we required $\Delta B > 50$ nT at Step 8 of our procedure. The data points scatter significantly. Nevertheless, it is clear that the lower bound of $\Delta B$ does not depend on $\Delta$MLat, whereas the upper bound increases as $\Delta$MLat decreases. This can be interpreted in the following way in terms of the substorm-associated dynamics. As the substorm growth phase proceeds, the auroral oval shifts equatorward [*Feldstein and Starkov*, 1967] and so does the R1 current. Although the intensity of such a R1 current may be as small as that found at higher latitude during quiet periods, the intensity of a R1 current after a substorm onset, even if located at the same latitude, can be significantly larger. The extent of the intensification is inferred to depend on the preonset energy storage, which is positively correlated with the size of the polar cap and therefore with the extent of the equatorward shift of the R1 current. The correlation between the intensity and latitude of the westward electrojet reported previously [*Kamide and Akasofu*, 1974] is consistent with this idea. It is also found that for larger negative values of $\Delta$MLat, all intense R1 currents are a part of the three-sheet structure. By combining the fact that three-sheet structure events tend to take place in the premidnight sector (Figure 3), we infer that such intense R1 currents are related to the development of the Harang discontinuity.

The large number of events of our data set allows us to quantify the relationship of $\Delta B$ vs. $\Delta$MLat. In Figure 4b $\Delta$MLat is binned with a 1° increment from –6° to +6° leaving two bins outside of this range. For each bin, the values of $\Delta B$ at the medium, $\pm\sigma$, and $\pm 2\sigma$ are calculated based on the deduced likelihood and are connected by different lines. In contrast with the value at $-2\sigma$, which is almost constant irrespective of $\Delta$MLat, the value at $+2\sigma$ increases with decreasing $\Delta$MLat. The solid lines superposed on the plot for the medium, $+\sigma$, and $+2\sigma$

represent the result of the least-square fit and are expressed as

$$\Delta B(\text{medium}) = 338 - 8.9 \times \Delta \text{Mlat},$$

$$\Delta B(+\sigma) = 488 - 19.4 \times \Delta \text{Mlat},$$

$$\Delta B(+2\sigma) = 669 - 26.9 \times \Delta \text{MLat}.$$

$\Delta B$ at the DMSP altitude (835 km) can be mapped to the ionosphere (110 km) by multiplying a factor of 1.17 (= [(1 $R_E$ + 835 km)/(1 $R_E$ + 110 km)]$^{1.5}$). The current intensity corresponding to $\Delta B(+2\sigma)$, $J_{\text{FAC}}(+2\sigma)$ at the ionospheric altitude is given by

$$J_{\text{FAC}}(+2\sigma) \text{ [mA/m]} = 0.8 \times \Delta B \text{ [nT; at 110 km]}$$

$$= 0.93 \times \Delta B \text{ [nT; at 865 km]}$$

$$= 623 - 25.0 \times \Delta \text{MLat}.$$

Provided that this current is closed with an upward-flowing FAC through a latitudinal Pedersen current and that the Hall conductivity is twice the Pedersen conductivity, the intensity of the westward electrojet, which we assume is mostly a Hall current, can be estimated by

$$J_{\text{Hall}}(+2\sigma) \text{ [mA/m]} = 1246 - 50 \times \Delta \text{MLat}.$$

The above formula provides an estimate of the intensity of the most intense westward electrojet expected as a function of $\Delta$MLat. The above formula, however, may overestimate the electrojet intensity, since, in the three-sheet structure, a part of the region 1 current is closed with a current poleward of the region 1 current.

## 4. SUMMARY AND FUTURE PROSPECTS

In this study we developed an automatic procedure to identify the structure of large-scale FAC systems from low-altitude satellite magnetic field measurements. We successfully applied the procedure to the DMSP-F7 magnetometer data and examined the relationship between the latitude and the intensity of the nightside R1 current. It is demonstrated that this procedure allows us to quantitatively examine a subject that has been addressed qualitatively.

There are two important areas in which we expect this procedure to prove useful. One is an extremely large statistical study of FACs, as we demonstrated for the nightside R1 current. A statistical comparison of FAC signatures with other measurements such as particle precipitation is one of the most reasonable applications. The other area is the real-time monitor of the space environment. The procedure works at a reasonable speed even on an ordinary PC platform. Thus, it is possible to determine and announce the location and intensity of FACs as soon as data are transmitted to the ground. An automatic procedure such as reported in this paper could change the way we study FACs in the future.

*Acknowledgments.* The DMSP-F7 magnetometer data were provided by F. J. Rich. Work at JHU/APL was supported by NASA, NSF, and the Office of Naval Research. Work at the Institute of Statistical Mathematics was in part carried out under the ISM cooperative Research Program (H9-ISM.CRP-B7 and H10-ISM.CRP-B12).

## REFERENCES

Akaike, H., A new look at the statistical model identification, *IEEE Trans. Autom. Control, AC 19*, 716, 1974.

Cummings, W. D., and A. J. Dessler, Field-aligned currents in the magnetosphere, *J. Geophys. Res.*, 72, 1007, 1967.

Feldstein, Y. I., and G. V. Starkov, Dynamics of auroral and polar geomagnetic disturbances, *Planet. Space. Sci.*, 15, 209, 1967.

Higuchi, T., Frequency domain characteristics of linear operator to decompose a time series into multi-components, *Ann. Inst. Stat. Math.*, 43, 469, 1991.

Hiragi, Y., Urakawa, H., and Tanabe, K., Statistical procedure for deconvoluting experimental data, *J. Appl. Phys.*, 58, 5, 1985.

Iijima, T., and T. A. Potemra, Field-aligned currents in the dayside cusp observed by Triad, *J. Geophys. Res.*, 81, 5971, 1976.

Kamide, Y., and S.-I. Akasofu, Latitudinal cross section of the auroral electrojet and its relation to the Interplanetary magnetic field polarity, *J. Geophys. Res.*, 79, 3755, 1974.

Kitagawa, G., and T. Higuchi, Automatic transaction of signal via statistical modeling, *Proc. First Int. Conf. Discovery Sci.*, Springer-Verlag Lecture Notes in Artificial Intelligence Series, *1532*, 375, 1998.

Richmond, A. D., and Y. Kamide, Mapping electrodynamic features of the high-latitude ionosphere from localized observations, Technique, *J. Geophys. Res.*, 93, 5741–5759, 1988.

Sakamoto, Y., M. Ishiguro, and G. Kitagawa, *Akaike Information Criterion Statistics: Mathematics and its Application*, Boston, D. Reidel, 1986.

Weimer, D. R., A new technique for the mapping of ionospheric field-aligned currents from satellite magnetic field data, in this issue, 1999.

Zmuda, A. J., J. H. Martin, and F. T. Heuring, Transverse magnetic disturbances at 1100 km in the auroral region, *J. Geophys. Res.*, 71, 5033, 1966.

---

T. Higuchi, The Institute of Statistical Mathematics, Tokyo 106, Japan.

S. Ohtani, The Johns Hopkins University Applied Physics Laboratory, 11100 Johns Hopkins Road, Laurel, MD, 20723-6099, U.S.A.

# Symmetry Breaking and Nonlinear Wave-Wave Interaction in Current Disruption: Possible Evidence for a Phase Transition

Giuseppe Consolini

*Istituto di Fisica dello Spazio Interplanetario, CNR, Rome, Italy*

Anthony T. Y. Lui

*The Johns Hopkins University Applied Physics Laboratory, Laurel, Maryland*

Magnetospheric substorms are generally accompanied by sporadic and intermittent large magnetic fluctuations in the magnetotail. These fluctuations are related to the disruption of the cross-tail current. Here, we have applied different techniques to a case study of this phenomenon. In detail, we have investigated the change on the scaling properties of the magnetic field fluctuations and the occurrence of short-lived nonlinear coupling in the magnetic field fluctuations as current disruption progresses. Clear evidence is found for a change of the scaling properties during the current disruption and the existence of short-lived phase-coupling structures, supporting the hypothesis of the occurrence of a reorganization process. These characteristics about current disruption are discussed in the framework of self-organized criticality and phase transitions out of equilibrium.

## 1. INTRODUCTION

One of the major displays of the Earth's magnetospheric dynamics is the magnetic substorm, an aggregate of phenomena spreading over a vast volume in the near-Earth space [*Lui*, 1991]. Several studies of substorm-associated phenomena suggested that many processes are often transient, localized, and intermittent. For example, large magnetic fluctuations and rapid particle energization are noted to occur briefly at a given location in the tail current sheet during the substorm expansion phase [*Takahashi et al.*, 1987; *Lui et al.*, 1988].

Recent analysis of "burst size" statistics for the auroral electrojet index (AE) has revealed a power law distribution extending over many orders of magnitude, meaning that no characteristic scale exists in the magnetospheric substorm dynamics [*Consolini*, 1997, 1999]. As suggested by *Chang* [1998a, b and references therein], this result could be connected with the possible occurrence of self-organized criticality (SOC) [see as a review *Bak*, 1997; *Jensen*, 1998] during magnetospheric substorms [*Consolini*, 1997, 1999, *Chapman et al.* 1998, *Uritsky & Pudovkin*, 1998].

Without any doubt, the growth of a current wedge plays a key role in association with substorm onset. This current wedge, linked to ionospheric region 2 current system and responsible for the *magnetosphere-ionosphere coupling* during magnetic substorms, is generally viewed as the diversion or disruption of the near-Earth cross-tail current system. The magnetospheric substorm onset is, indeed, generally associated with the change of the tail magnetic field from a stressed to a more dipolar configuration. This change is interpreted in term of a decrease in the tail current intensity due to the development of a new current wedge [*Atkinson*, 1967; *Akasofu*, 1972].

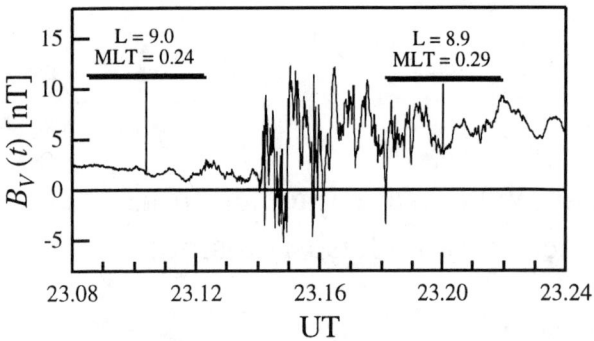

**Figure 1.** The $B_V$ magnetic field component as observed by AMPTE/CCE spacecraft during the magnetospheric tail current disruption on June 1, 1985.

The *current disruption* (CD) phenomenon is characterized by large-amplitude and turbulent magnetic field fluctuations and has been widely studied [e.g.; *Lui et al.*, 1992; *Ohtani et al.*, 1995]. These studies showed that a *broadband spectrum* of waves is excited during CD, with frequencies ranging from below to above the ion cyclotron frequency $f_{ci}$, and that the characteristic frequency component is several times below $f_{ci}$. Moreover, *Lui and Najmi* [1997], using a wavelet time-frequency decomposition of the magnetic field fluctuations, showed that the high-frequency fluctuations seem to be linked with a broadband excitation at the beginning of the event, cascading down to lower frequencies with time, and that the highest frequency components show an intermittent behavior.

In this paper, we investigate the change on the scaling properties of the magnetic field fluctuations prior to and after a current disruption, using the 1$^{st}$ order structure function, and we also examine the occurrence of short-lived nonlinear coupling in the magnetic field fluctuations. The observed features are discussed in the framework of self-organized criticality and order-disorder phase transition. This goal has not been addressed in previous works on current disruption, especially in terms of its analogy to a phase transition.

## 2. DATA DESCRIPTION AND ANALYSIS

Data used in this work refer to $B_V$ component of the magnetic field in a tail current disruption event of June, 1, 1985, as observed by the AMPTE/CCE spacecraft. During this CD event the CCE was near its apogee $\approx 8.8$ $R_E$ in the near-Earth neutral sheet region (MLT $\approx 24$; field latitude angle $\approx 90°$) for an extended period, allowing us to examine the temporal evolution of the magnetic field fluctuations near the onset of current disruption. Here, the choice of VDH coordinate system is made because of the proximity to the Earth at the point of measurement which is greatly influenced by the Earth's dipole field. Anyway, the $B_V$ component - V points radially outward and is parallel to the magnetic equator - is very similar to the $B_X$ component in GSM, and is directly related to the intensity of the cross-tail current. Previous studies have clearly demonstrated that CD is a multiscale phenomenon [*Ohtani et al.*, 1995; *Lui & Najmi*, 1997].

Figure 1 shows the actual data for the magnetic field component $B_V$ during the current disruption event considered in this work. The magnetic field evolution suggests that during CD the cross-tail current is drastically reduced in coincidence of the observed magnetic field fluctuations. As a matter of fact, before the CD-onset, the field magnitude was very small ($\approx 8$ nT) if compared with the contribution of the Earth's dipole field ($\approx 45$ nT). This evidence is the signature of an intense cross-tail current sheet. Conversely, after the CD, the field magnitude recovered to $\approx 35$ nT, which is very close to the dipolar field, meaning that the intense cross-tail current is disrupted. Consequently, during CD the topological features of the magnetic field change. This change could be read as a sort of phase transition, during which a new order appears to relax the stored energy. As a consequence of this transition, we might expect a change in the symmetry properties of the magnetic field.

In order to check this point, we have investigated the scaling properties of the magnetic field fluctuations prior to and after the CD by means of the 1$^{st}$ order structure function $S_1(t)$ [*Kolmogorov*, 1941; *Bunde & Havlin*, 1994]:

$$S_1(\tau) = \left\langle \left| \phi(t+\tau) - \phi(t) \right| \right\rangle \quad (1)$$

where the <...> denotes the time average. Moreover, for a self-affine signal we obtain that:

$$S_1(\tau) \approx \tau^h \quad (2)$$

where the scaling exponent $h$ is named *Holder exponent*.

Figure 2 shows the 1$^{st}$ order structure function $S_1(t)$ in the case of the magnetic field fluctuations for two selected intervals before (from 23:08 UT to 23:12 UT) and after (from 23:20 UT to 23:24 UT) current disruption. A scaling region, covering the range from $\approx 0.4$s to $\approx 20$s, is recovered in both intervals. This scaling region extends below and above the characteristic cyclotron frequency $f_{ci}$ meaning that there is no characteristic time scale in the magnetic field fluctuations before and after current disruption. In other words, the statistical properties of the geomagnetic tail fluctuations are scale invariant at least over a scale range, even if the *Holder* exponent $h$ changes, being $h=[0.48\pm0.02]$ and $h=[0.70\pm0.02]$ before and after CD, respectively. Since it was realized that a scale/time free behavior in statistical systems is due to *criticality* (*Kadanoff et al.*, 1967), we may infer that the scale invariance of the magnetic fluctuations could be related to a near-critical configuration for the geomagnetic near-Earth tail. Moreover, the change of the scaling exponent $h$ indicates that the scaling features of the magnetic fluctuations change during CD. As a matter of fact, while the *Holder* exponent before CD indicates that the magnetic fluctuations are statistically equivalent to a stochastic signal ($h=0.5$ for Brownian motions), the higher value of $h$ after CD indicates that the statistical features of the geomagnetic fluctuations are equivalent to a long persistent signal (here persistency means that to a positive/negative fluctuation follows another positive/negative fluctuation with a probability higher than the case of Brownian motions [*Bunde & Havlin*, 1994]). In other words, after CD fluctuations are longer correlated. This evidence should be taken as an indication for the occurrence of a reorganization phenomenon during CD, which is the consequence of the appearance of a new ordered dynamical phase. Anyway, because the scale-invariance is generally due to a symmetry property of the equations describing the physical system under a scale-transformation, the change of the *Holder* exponent means that the new ordered phase possesses different symmetry properties. Therefore, current disruption could be regarded as a phenomenon during which a symmetry-breaking occurs (see also *Consolini & Lui*, 1999). As clearly stated by *Chang* [1992a, b], symmetry-breaking can generally occur in stochastic systems, perturbed from a near-critical configuration. Therefore, the existence of scaling features for the magnetic field fluctuations and the change of the scaling exponent $h$ during CD seem to support the hypothesis that the near-tail region exists in a critical configuration (perhaps self-organized).

As a consequence of the above concepts, if CD can be taken as a sort of order-disorder transition [*Careri*, 1987, and references therein] in a system out of the equilibrium, we might expect that an overall fluctuation (usually named *slow mode* in statistical mechanics), characterized by a long decaying time, would appear during the critical transition as

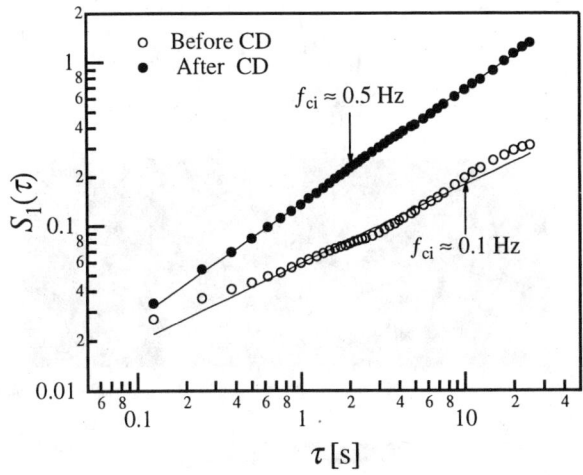

**Figure 2.** The 1$^{st}$ order structure function $S_1(\tau)$ for two selected intervals of 4 minutes before (from 23:08 UT to 23.12 UT) and after (from 23:20 UT to 23.24 UT) current disruption. Solid lines are power-law best fits. Scaling exponents $h$ are 0.48±0.02 and 0.70±0.02 before and after current disruption, respectively.

a consequence of the decaying of the local instabilities. This means that, if the aforementioned framework is plausible, during CD magnetic field fluctuations must strongly interact to produce an overall fluctuation which is responsible for the appearance of an order parameter that characterizes the new order phase.

In order to check the existence of this overall fluctuation, characterized by a long decaying time and emerging in the transition, we have performed the *wavelet transform* $W_f(a,t)$ [*Fargé*, 1992; *Kumar & Foufoula-Georgiou*, 1997] that decomposes a signal into wavelet components that depend on both scale $a$ and time $t$:

$$W_f(a, t) = \frac{1}{a^n} \int f(t')\Psi_a(t' - t)dt', \quad a > 0, \; t' \in \Re \quad (3)$$

where the factor $1/a^n$ may be chosen to best reveal the structure of the signal under consideration [*Arneodo et al.*, 1988]. Here, the factor $1/a^n$ has been chosen so that the *wavelet transform amplitude* $|W_f(a,t)|$ is equal to the Fourier one.

Wavelet analysis is a powerful tool to investigate turbulent signals and short lived structures [*Argoul et al*, 1989]. In order to do this, we choose the *Morlett wavelet*:

$$\Psi_a(t) = \frac{1}{\sqrt{a}} \exp\left[i\frac{2\pi t}{a} - \frac{1}{2}\left(\frac{t}{a}\right)^2\right] \quad (4)$$

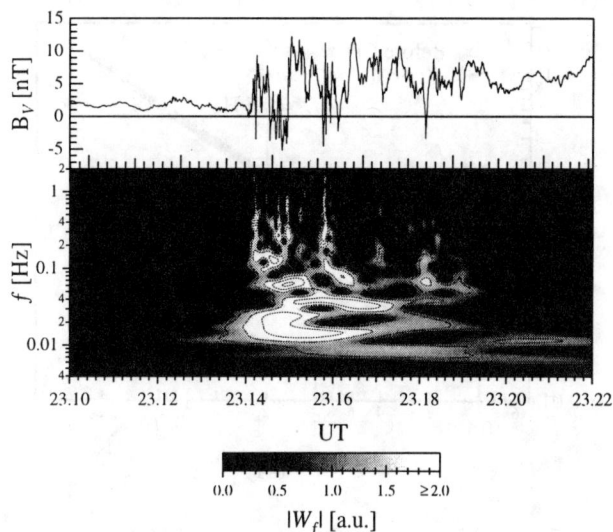

**Figure 3.** The *scalogram* (lower panel) of the $B_v$ magnetic field evaluated on the basis of the *wavelet transform* [eq. 3] using the *Morlett wavelet*, in comparison with the actual measurements of the $B_v$ magnetic field component (upper panel) [see also *Lui and Najimi*, 1997].

where $a$ is the time scale. This wavelet shows a single peak at the frequency $\omega = 2\pi/a$ with a FWHM of $\Delta\omega = \omega/4$ in the Fourier space. In other words, a wavelet of characteristic scale $a$ may be, consequently, read as representing a frequency $\omega = 2\pi/a \pm \Delta\omega/2$.

Figure 3 shows the *scalogram* - the amplitude of the wavelet transform - of the CD in the interval from 23.10 UT to 23.22 UT (a similar analysis was performed by *Lui & Najimi*, 1997, using a slight different mother wavelet on a more limited time interval). This figure clearly shows that CD is a multiscale and intermittent phenomenon, and that low-frequency excitations ($f \leq 0.1$ Hz) persist after CD. As a matter of fact, a rich frequency structure of excitations are active during CD. As already stated by *Lui & Najimi* [1997], the high-frequency components seem to be linked to a broadband excitation, and cascading down to lower frequencies with time (high-frequency fluctuations appears in time before than low-frequency ones). Therefore, the low-frequency excitations could be the result of a coalescence process of high-frequency seed excitations, perhaps generated from some local plasma instability. In this framework we might expect that during CD *waves* must strongly interact to give rise to these low-frequency fluctuations.

To investigate the occurrence of this wave-coupling, we have evaluated the time evolution of the *spectral bicoherence* using a wavelet approach [see *van Milligen at al.*, 1995], that is able to resolve short-lived events, and pulses in the phase coupling. The *bicoherence*, generally investigated by Fourier transform, is indeed a measure of the phase coupling, and of the *coupling constant* in some quadratic wave-interaction model [*Ritz et al.*, 1989; *Hidalgo et al.*, 1993; *Mattor & Terry*, 1992]. Unfortunately, the application of Fourier analysis to highly-intermittent and turbulent data is unable to detect short-lived phase coupling, because it presupposes *waves* in the physical system that endure in time. On the contrary *wavelet bicoherence technique* is able to solve this problem.

Here, according to *van Milligen et al.* [1995], we define the *wavelet bicoherence* as

$$[b^w(a_1, a_2)]^2 = \frac{|B^w(a_1, a_2)|^2}{\int |W_f(a_1,\tau)W_f(a_2,\tau)|^2 d\tau \int |W_f(a,\tau)|^2 d\tau} \quad (5)$$

with $B^w(a_1, a_2)$ being the *normalized squared wavelet bispectrum*:

$$B^w(a_1, a_2) = \int W_f^*(a,\tau) W_f(a_1,\tau) W_f(a_2,\tau) d\tau \quad (6)$$

where $W_f(a,t)$ is the *wavelet transform*, and integration is performed over a finite time interval $T$: $t_0 \leq t \leq t_1$, and satisfying the following sum rule:

$$1/a = 1/a_1 + 1/a_2 \quad (7)$$

that corresponds to a frequency sum rule, $\omega = \omega_1 + \omega_2$, since the time scale length can be read as an inverse frequency $\omega = 2\pi/a$.

Figure 4 shows the time development of the *total wavelet bicoherence*, defined according to:

$$(b^w)^2 = \sum_{a_1,a_2} [b^w(a_1, a_2)]^2 \quad (8)$$

and evaluated using a moving time window of $T = 20$s (161 points), with a frequency resolution of $\approx 0.02$ Hz.

Although numerical values of *total wavelet bicoherence* are not fundamental, this quantity is useful to summarize the information and can be used to compare cases analyzed in the same numerical condition. Note how the coherency is generally higher during CD. The occurrence of bursts of coherency is indicative of intermittency in phase-coupling, implying a strong interaction between the *waves* during CD [*van Milligen et al.*, 1995].

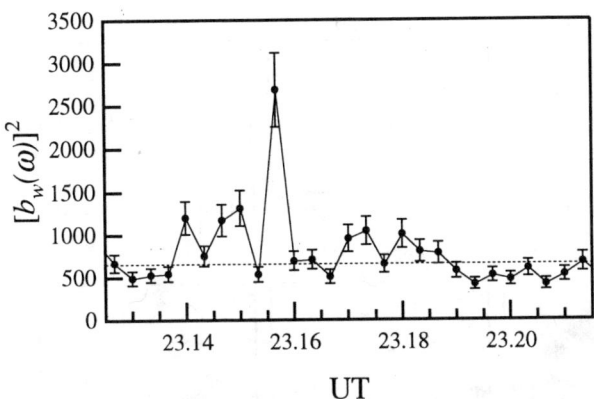

**Figure 4.** The total wavelet bicoherence, evaluated on the basis of equation (8), vs. universal time (UT). Horizontal dashed line refers to the noise level evaluated according to equation (7) in *van Milligen et al.* [1995]. Error bars were computed using equation (6) in *van Milligen et al.* [1995].

Figure 5 shows five sections of the *bicoherence* in the time interval from 23.14 UT to 23:17 UT to illustrate the time development of this *phase-coupling* in more detail. The upper plots illustrate the *summed bicoherence*, defined as:

$$\left[b^w(a)\right]^2 = \sum_{a_1, a_2} \left[b^w(a_1, a_2)\right]^2 \delta_{1/a=1/a_1+1/a_2} \quad (9)$$

in the five selected sections, while the lower plates show the *bichoerence* in the plane ($\omega_1$, $\omega_2$) limited by Nyquist frequency and symmetry considerations. A large amount of details are present in the *bicoherence plates*. Anyway the time evolution seen in these plates indicates the existence of *short-lived structures* that are responsible for the intermittent phase-coupling. As the horizontal and diagonal ridges related to very-low frequencies, the strong values of the bicoherence near these ridges are indicative of coupling of single frequencies to a broadband spectrum (see the 23:15:40 UT plate). In addition, the short-lived diffuse structures at high-frequencies are mainly responsible for the intermittency in the phase-coupling. In other words, these short-lived diffuse structures are due to three wave coupling which occurs locally in time and allows the energy transfer [*van Milligen et al.*, 1995]. The existence of these structures supports the previous results by *Lui & Najmi* [1997] that inferred the existence of an inverse cascading process. Furthermore, if we suppose that the microscopic mechanism of the current disruption is the decay of a local instability, then our results support the hypothesis that during CD this decaying moves from the small scales towards the big scales like a sort of inverse cascade, and avalanche phenomenon.

## 3. SUMMARY AND CONCLUSIONS

The results of our analyses have clearly shown that a change on the statistical features of the magnetic field fluctuations occurs during current disruption and that the magnetic field fluctuations show a strong phase coupling. Moreover, this nonlinear interaction is time intermittent, as evidenced by the short-lived structures in the bicoherence plane, and after CD only low-frequency fluctuations seem to survive. In other words, during current disruption a *slow fluctuation* grows through a sort of inverse cascading process that leads to the formation of large-scale features from small-scale ones. In this framework this slow-fluctuation could account for the topological changes and the appearance of a new dynamical phase.

From the point of view of the statistical mechanics of far-from equilibrium systems these results agree with the possible occurrence of a phase transition (perhaps disorder-order) in a system out of equilibrium near a critical point, and confirm the preliminary results by *Consolini & Lui* [1998, 1999] on the possible occurrence of a symmetry breaking and higher-order phase transition. If current disruption is consistent with a transition during which a new order phase appears to dissipate the excess of energy, we might expect that the symmetry features of the new phase change. Roughly speaking, a similar behavior can be found in the fluid dynamic convection, where Bénard convection cells appear as the system moves from a subcritical configuration to a critical (or super-critical) one [*Nicolis and Prigogine*, 1987].

We believe that our results support *Chang's* hypothesis [1992] of a near-critical configuration (perhaps self-organized) of the plasma confined in the tail regions. As a matter of fact, the high values of the *Holder* exponent and the presence of *slow fluctuations* after current disruption may be indicative of a cooperative response of the whole system as in SOC systems, when long-range correlations are expected.

Finally, we want to emphasize that this preliminary work represents an informative approach to the time-evolution of CD and substorm onset from a macroscopic point of view based on the new concepts developed for complex system dynamics, which is clearly not in contradiction to the more classical microscopic point of view of

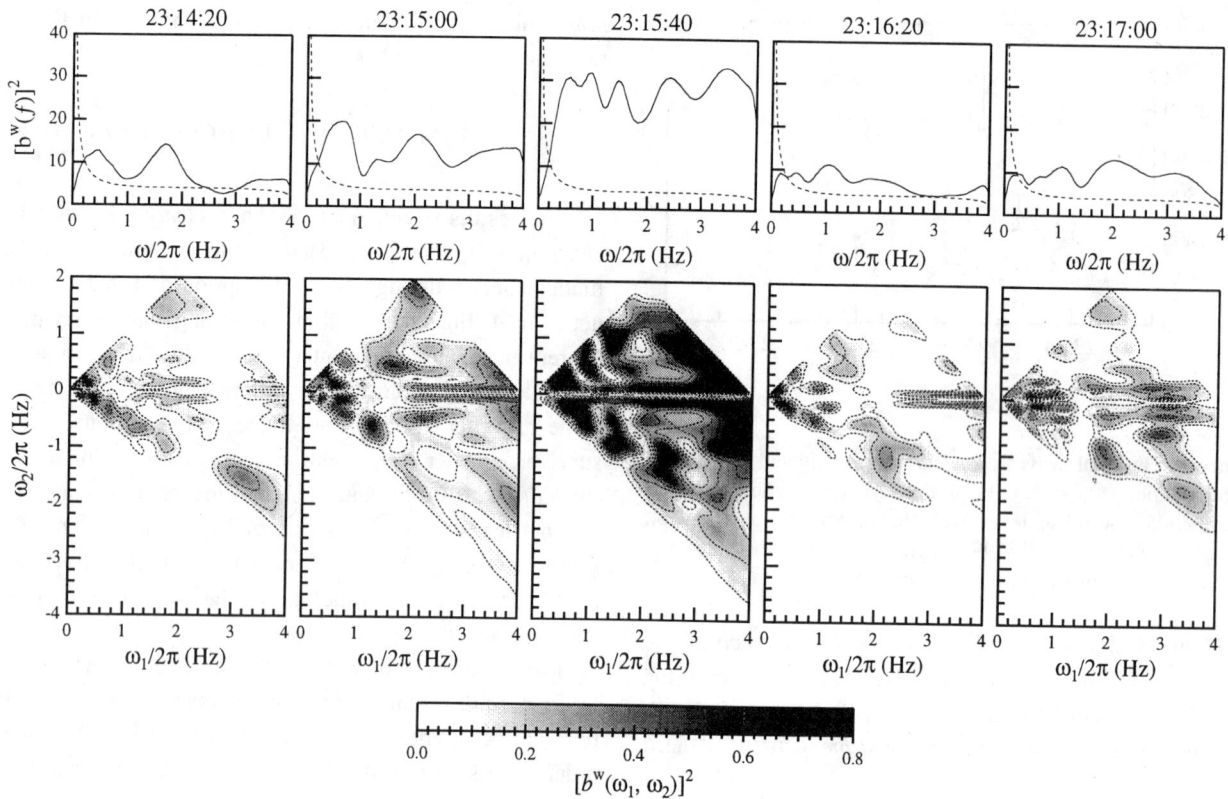

**Figure 5.** Example of the time development of the intermittent phase-coupling structures revealed by the wavelet bicoherence analysis. Upper plots refer to the summed bicoherence at five selected times, evaluated on the basis of equation (9). Dashed line refers to the noise level, according to equation (7) in *van Milligen et al.* [1995]. Lower plates show the bicoherence plots at the same five selected times. The plotting area was limited by the symmetries in the bicoherence definition and by Nyquist frequency. Note the strong coupling at high frequencies, responsible of the intermittency, and its time evolution.

plasma instabilities, that could be responsible of the energy source.

*Aknowledgements.* G. Consolini thanks B. Bavassano of the IFSI/CNR (Italy) for the useful discussions and comments. A.T.Y. Lui thanks the Atmospheric Sciences Section of the National Science Foundation (grant ATM-9622080) and National Aeronautics and Space Administration (grant NAG5-7797) for their financial supports.

## REFERENCES

Akasofu, S.-I., Magnetospheric substorm, a model, *Solar Terrestrial Physics,* Part I, D. Dryer ed., p.131, D. Reidel, Norwell, Mass. 1972.

Argoul, F., et al., Wavelet analysis of turbulence reveals the multifractal nature of the Richardson cascade, *Nature,* 388, 51, 1989.

Arneodo, A., G. Grasseau, and M. Holschneider, Wavelet transform of multifractals, *Phys.Rev.Lett.,* 61, 2281, 1988.

Atkinson, G., An approximate flow equation for geomagnetic flux tubes and its application to polar substorms, *J. Geophys. Res.,* 72, 5373, 1967.

Bak, P., *How Nature Works: The science of Self-Organised Criticality.* Oxford University Press, Oxford, England, 1997.

Bunde, A., & S. Havlin (eds.), *Fractals in Science,* Springer-Verlag, 1994

Careri, G., *Ordine e disordine nella materia,* Gius. Laterza & Figli Spa, Roma-Bari, 1982.

Chang, T.S., Low-dimensional behavior and symmetry breaking of stochastic systems near criticality - can these effects be observed in space and in laboratory ?, *IEEE Trans. Plasma Sci.,* 20, 691, 1992a.

Chang, T.S., Path Integral Approach to Stochastic Systems

Near Self-Organized Criticality, in *Research Trends in Physics: Nonlinear Space Plasma Physics*, edited by H. Alfvén, K. Quest, and R.Z. Sagdeev, La Jolla International School of Physics, AIP, 1992b.

Chang, T.S., Intermittent Turbulence in the Magnetotail, AGU 1998 Spring Meeting, *EOS Trans., 79 suppl.,* S328, 1998a.

Chang, T.S., Sporadic Reconnection and Multiscale Intermittent Turbulence in the Magnetotail, in *Encounter between Global Observations and Models in the ISTP Era*, edited by J.L. Horwitz, D.L. Gallagher, and W.K. Peterson, AGU Monograph, Washington, D.C. American Geophysical Union, 1998b.

Chapman S.C., et al., A simple avalanche model as an analogue for magnetospheric activity, *Geophys. Res. Lett., 25,* 2397, 1998.

Consolini G., Sandpile cellular automata and magnetospheric dynamics, in *Proc. vol. 58, "Cosmic Physics in the Year 2000"*, S. Aiello et al. (eds.), SIF, Bologna, Italy, 1997.

Consolini G., and A.T.Y. Lui, Sign-singularity analysis of current disruption and possible evidence that substorm is a Self-Organized Criticality phenomenon, , AGU 1998 Fall Meeting, *EOS Trans., 79 suppl.,* F787, 1998.

Consolini G., and A.T.Y. Lui, Sign-Singularity Analysis of Current Disruption, *Geophys. Res. Lett., 26,* 1673, 1999.

Consolini G., Avalanche, scaling and $1/f$ noise in magnetospheric dynamics, submitted to *Phys. Rev. Lett.,* 1999.

Fargé, M., Wavelet transforms and their applications to turbulence, *Ann. Rev. Fluid Mech., 24,* 395, 1992.

Hidalgo C. et al., Experimental evidence of three-wave coupling on plasma turbulence, *Phys. Rev. Lett., 71,* 3127, 1993

Jensen, H.J.K., *Self-Organised Criticality: Emergent Complex Behaviour in Physical and Biological Systems*. Cambridge University Press, Cambridge, England, 1998.

Kadanoff, L.P. et al., Static phenomena near critical points: theory and experiments, *Rev. Mod. Phys., 39,* 395, 1967.

Kolmogorov, A.N., The local structure of turbulence in incompressible viscous fluid for very large Reynolds number, *Dokl. Akad. Nauk SSSR, 30,* 9, 1941.

Kumar, P., and E. Foufoula-Georgiou, Wavelet analysis for geophysical applications, *Rev. Geophys., 35,* 385, 1997.

Lui A.T.Y. et al., A case study of magnetotail current sheet disruption and diversion, *Geophys. Res. Lett., 15,* 721, 1988.

Lui A.T.Y., A synthesis of magnetospheric substorm models, *J. Geophys. Res., 96,* 1849, 1991.

Lui A.T.Y. et al., Current disruptions in the near-Earth neutral sheet region, *J. Geophys. Res., 97,* 1461, 1992.

Lui A.T.Y. and A.-H. Najmi, Time-frequency decomposition of signals in a current disruption event, *Geophys. Res. Lett., 24,* 3157, 1997.

Mattor, N., and P.W. Terry, Frequency spectrum in drift wave turbulence, *Phys. Fluids B, 4,* 1126, 1992.

Nicolis, N. and I. Prigogine, Exploring complexity. An Introduction, Monaco: R.Piper GmbH and Co. KG, 1987.

Ohtani, S., T. Higuchi, A.T.Y. Lui, and K. Takahashi, "Magnetic fluctuations associated with tail current disruption: Fractal analysis, *J. Geophys. Res., 100,* 19135, 1995.

Ritz, Ch.P, E.J. Powers, and R.D. Bengtson, Experimental measurement of three-wave coupling and energy cascading, *Phys. Fluids B, 1,* 153, 1989.

Takahashi, K., L.J. Zanetti, R.E. Lopez, R.W. McEntire, T.A. Potemra, and K. Yumoto, Disruption of the magnetotail current sheet observed by AMPTE CCE, *Geophys. Res. Lett., 14,* 1019, 1987.

Uritsky, V.M., and M.I. Pudovkin, Low frequency $1/f$-like fluctuations of AE-index as a possible manifestation of self-organized criticality in the magnetosphere, *Ann. Geophys., 16,* 1580, 1998.

van Milligen, B.Ph., C. Hidalgo, and E. Sanchez, Nonlinear Phenomena and Intermittency in Plasma Turbulence, *Phys. Rev. Lett., 74,* 395, 1995.

---

G. Consolini, Istituto di Fisica dello Spazio Interplanetario - CNR, Via Fosso del Cavaliere 100, 00133 Roma, Italy. (e-mail: consolini@ifsi.rm.cnr.it).

A.T.Y. Lui, The Johns Hopkins University Applied Physics Laboratory, Laurel, MD, 20723 (e-mail:Anthony.Lui@jhuapl.edu).